# QUARK CONFINEMENT AND THE HADRON SPECTRUM VII

To learn more about the AIP Conference Proceedings, including the
Conference Proceedings Series, please visit the webpage
**http://proceedings.aip.org/proceedings**

# QUARK CONFINEMENT AND THE HADRON SPECTRUM VII

7th Conference on Quark Confinement
and the Hadron Spectrum

QCHS7

Ponta Delgada, Açores, Portugal    2 – 7 September 2006

*EDITOR*
José Emílio F. T. Ribeiro
*Center for Physics of Fundamental Interactions*
*Lisbon, Portugal*

**SPONSORING ORGANIZATIONS**
Foundation for Science and Technology (FCT)
Regional Secretary of Economy of Azores
Gulbenkian Foundation
Italian Institute of Culture of Lisbon

AMERICAN INSTITUTE OF PHYSICS

Melville, New York, 2007
AIP CONFERENCE PROCEEDINGS ■ VOLUME 892

# Editor

José Emílio F. T. Ribeiro
CFIF-IST, Edifício Ciência
Av. Rovisco Pais P-1049-001
Lisbon, Portugal
E-mail: EmilioRibeiro@netcabo.pt

L.C. Catalog Card No. 2007921351
ISBN 978-0-7354-0396-3
ISSN 0094-243X

Printed in the United States of America

# CONTENTS

## PLENARY SESSION PRESENTATIONS

## SUMMARIES

## PARALLEL SESSION CONTRIBUTIONS

### SESSION A: THE VACUUM STRUCTURE OF QCD AND THE MECHANISM OF CONFINEMENT

## SESSION B: LIGHT QUARKS (AND GLUONIA)

### SESSION C: HEAVY QUARKS (AND GLUONIA)

## SESSION D: DECONFINEMENT

## SESSION E: QCD, NEW PHYSICS AND EXPERIMENTS

## SESSION F: QCD AND NUCLEAR PHYSICS, QCD AND ASTROPHYSICS

### POSTER SESSION
(Organizer: M. Creutz)

# PREFACE

The seventh edition of the conference series "Quark confinement and the Hadron Spectrum" was held in the Ponta Delgada Campus of the University of the Azores. Starting in 1994, in the beautiful location of Como, Italy, this series of Conferences aims to bring together, every two years, people working on the subject of strong interactions. The present edition was sponsored by the Autonomous Government of the Azores, the Town Hall of Ponta Delgada, the gracious host of the Gala Dinner at the Coliseum of Ponta Delgada, and the national Portuguese funding agency FCT. In particular, I must acknowledge the excellent support given to this Conference by the University of the Azores. This edition followed the usual pattern of invited talks, contributed talks and posters. The contribution papers were presented in parallel sessions. In addition to the usual sessions, this time we had a new section devoted to the connections between Nuclear Physics, Astrophysics and QCD. They were:

Section A: The vacuum structure of QCD and the mechanism of confinement
Conveners: M. Faber (Wien), J. Greensite (San Francisco), M. Polikarpov (ITEP).

Section B: Light Quarks (and Gluonia)
Conveners: J. Goity (JLAB), H. Sazdjian (Orsay), M. Testa (Roma).

Section C: Heavy Quarks (and Gluonia)
Conveners: G. Bodwin (ANL), J. Soto (Barcelona), A. Vairo (Milano).

Section D: Deconfinement
Conveners: Y. Foka (GSI), C. Pajares (Santiago Compostela), J. Rafelski (Arizona).

Section E: QCD, New Physics and Experiments
Conveners: F. Harris (Hawaai), G. Nardulli (Bari), A. Sczcepaniak (Indiana).

Section F: QCD and nuclear physics, QCD and astrophysics
Conveners: M. Alford (Saint Louis), T. Cohen (Maryland).

The Poster session was organized, in the usual efficient manner, by M. Creutz. In this edition, the after-dinner talk - a superb talk - was also delivered by M Creutz. Finally, M. Shifman gave us a very interesting lecture on the life of the late Yuval Neeman, a distinguished physicist and a dear participant of past editions of this series of conferences. In the tradition of the previous conferences, the last day was devoted to the summary talks of the various sections given by J. Greensite, D. Melikhov, A. Vairo, C. Pajares, G. Nardulli and T. Cohen.

Last but probably not least, the wine tasting, from the vineyards of the World, the reception cocktail, a generous offer from the Secretaria Regional de Economia, and the trip to the Terra Nostra Park, with the volcano-cooked meal, provided as many different opportunities, as it has been the hallmark of this series of conferences, for contacts among the participants.

And now, as it is customary, some words about the Azores.

The Carta Régia, dated from 2nd July 1439, was the first known official document reporting the settlement of the Azores. Sta. Maria and S. Miguel were the first islands to be settled with families coming from Estremadura, Alto Alentejo, Algarve and later from Madeira. The settlement of Terceira may have started in 1450, under the command of the Flemish Jácome de Bruges. The island Graciosa was settled under Pedro Correia and Vasco Gil Sodré. As for the islands Faial and Pico, they were donated, just before 1466, to the Flemish Josse Van Huertere (Joz de Utra). Through the centuries, the Azores saw the coming of Italians, Spanish, French, English and, since 1776, Americans. In fact, George Washington appointed John Street as the first U.S. Consul in 1795 when Thomas Jefferson was the Secretary of State. Therefore, the U.S. Consulate in Ponta Delgada is the oldest continuously operating U.S. Consulate in the world.

The Azores have played their part in History. Between the end of the 15th century and a great part of the following century, the Archipelago of Azores was a key actor in voyages of discovery towards the West. From the Azores sailed people like the Corte Real brothers, João Fernandes Labrador and others.

The 15th century was fertile in magical tales of discovery: the legendary islands of S. Brandão, the existence of the Seven Cities, of Antília and many other geographical chimeras. However, this alchemist impetus was to lay the foundations for a more scientific endeavor. As physicists know only too well, to discover something presupposes to charter the method to tell others the steps needed to repeat the experience. For instance, we know today-through the existence of archeological evidence- that the Vikings arrived in Newfoundland during the 10th and 11th centuries. However, it is unclear what happened to those colonies. In the 15th century, this western route was all but lost, with the only remaining vestiges being the legends of lost lands sometimes associated with the lost paradise. With the settlement of the Azores, a solid logistic basis for the systematic research of the Western pathways was, for the first time, established. It should therefore come as no surprise that in the return of his voyage, in 1492, Colon should have sojourned for quite some time on the Island of Santa Maria-Azores, before setting course to Spain.

In those years, the popular theory of everything was to get to India sailing west. With respect to this, a not too well known João Fernandes Lavrador, of the island of Terceira in the Azores, deserves to be mentioned. It is known that he had long standing business connections with the English port of Bristol, and that he made one or more voyages to the New World. He chartered the southern coasts of Greenland and it is quite possible that before 1500 he had reached what we now know as the Coast of Lavrador in Canada. It is also quite probable that he had made an acquaintance with the Italian Giovanni Cabotto, who in 1497, in the same town of Bristol, had a small ship equipped to sail west to look for the Eastern Island of Zipango, believed then to be the source of all spices. Following Colon, Cabotto looked for a Passage to India (or China). Probably this was not the objective of the Corte Real brothers: they knew that Vasco da Gama had solved the problem of finding a sea path to India. This family was connected to the captaincy of Angra do Heroísmo, in the island of Terceira, with interests in the Algarve and the North of Africa. Some of the most notable voyages of discovery made in the icy Norwest Atlantic seas were made in 1500, 1501 and 1502, by two of the Corte Real brothers. The father, João Vaz Corte Real, himself a navigator, may have touched what is now Canada in 1472, although the documentation is scarce. The experimental existence of these new lands was published, for the first time, in the "Cantino" Portuguese Planisphere of 1502. They were "pushed" towards the east, in order to be within the Portuguese portion of the Tordesilhas Treaty. Throughout the centuries up to our days, the Azores kept playing an important role in Portuguese history, which cannot be summarized in this preface.

You were on the island of S. Miguel and I am certain that you now look at the North Atlantic with different eyes. It is no longer a barren, storm prone, large mass of water to be flown over by planes. It contains the archipelago of the Azores and, hence, some of the roots of what has become the History of the West.

I would like to thank the co-organizers of this edition, K. Maung, G. Prosperi, N. Bramblla and A. Vairo for their support and to express my thanks to all the participants.

José Emílio F.T. Ribeiro

Chairman of the Conference

**International Advisory Board**

K. Baker (JLAB/Hampton)
M. Baker (Seattle)
G. Bodwin (ANL)
M. Creutz (Brookhaven)
J. Dias de Deus (Lisboa)
G. Ecker (Wien)
E. Eichten (FNAL)
M. Faber (Wien)
H. Georgi (Harvard)
D. Gromes (Heidelberg)
G. Krein (São Paulo)
H. Leutwyler (Bern)
W. Lucha (Wien)

M. Lüscher (CERN)
A. Manohar (San Diego)
G.Martinelli (Roma)
Y. Ne'eman (Tel Aviv)
M. Neubert (Cornell)
E. Predazzi (Torino)
F. Schöberl (Wien)
M. Shifman (Minnesota)
J. Soto (Barcelona)
M. Testa (Roma)
H. Toki (RCNP Osaka)
N. A. Törnqvist (Helsinki)
F. J. Ynduráin (Madrid)

## Organizing Committee

N. Brambilla (Milano)[Scientific Secretary]
G. Prosperi (Milano)
A. Vairo (Milano)
K. M. Maung (Hattiesburg)
J.E. Ribeiro (Lisboa) [Chair]

## Conference Supported by

FCT-Foundation for Science and Technology
Secretaria Regional de Economia dos Açores
Gulbenkian Foundation
Instituto Italiano de Cultura de Lisboa

**FIGURE 1.** Lattices abound in the real world. Here is lattice of daisies.

# So you want to be a lattice theorist?

## Michael Creutz

*Physics Department, Brookhaven National Laboratory, Upton, NY 11973, USA*

**Abstract.** For this after dinner talk I intersperse images of real lattices with a discussion of the motivations for lattice gauge theory and some current unresolved issues.

**Keywords:** lattice gauge theory
**PACS:** 11.15.Ha

Although lattices are frequently seen in the real world, as in Figure 1, to the particle theorist they are nothing but a mathematical trick. We constrain quarks so that rather than following arbitrary world lines, they only move in discrete hops between lattice sites. As they hop they get spun around in group space by the gauge fields, which are restricted to the lattice bonds. It is a nice framework for exploring confinement, which is related to this spinning; quarks act like kangaroos, strongly preferring to hop together in mobs.

Since the vacuum is not a crystal, this seems at first sight a rather strange thing to do. However, the lattice has several advantages, primarily in allowing calculations in situations where other methods fail. In particular, one can go far beyond the realms of perturbation theory or semi-classical methods. Furthermore, the predictions can have

CP892, *Quark Confinement and the Hadron Spectrum VII*
edited by J. E. F. T. Ribeiro
© 2007 American Institute of Physics 978-0-7354-0396-3/07/$23.00

**FIGURE 2.**  Lattices can have good flavors. But beware of lurking tastes.

crucial experimental implications. These extend to many areas of particle and nuclear physics, from extracting weak matrix elements in processes involving large hadronic corrections, to understanding the behavior of matter under the extreme conditions of heavy ion collisions, and to detailed studies of hadronic structure.

And of course we get to have fun playing with big computers. Indeed, these themselves are large lattices of processors, such as the six dimensional torus that makes up the QCDOC supercomputer dedicated to lattice gauge theory. There are also more abstract reasons to study lattice gauge theory. As shown in Figure 2, lattices can have good flavors. However one should be careful of any harmful lurking tastes. Lattices are frequently seen in cities, such as the lattice of trees seen in Figure 3.

One of the fun things about lattice gauge theory is the addictive power it gives over the system. Entire lattice configurations are stored in the computer memory, and you are free to measure anything you want. In the process uncertainties can arise, and the theorist is in the unusual situation of having error bars. First of all, since we are using Monte Carlo methods, there will be statistical errors. These can be reduced by massive applications of computer time. There are also several sources of systematic error, some of which we have control over. These include finite volume and finite lattice spacing corrections, which can also be reduced by increased computer time. In practice using quarks with physical masses is quite computer intensive; so, we usually simulate with heavier than normal quarks and then do an extrapolation.

There are also some sources of error that are basically uncontrolled. One is the so-called "valence" or "quenched" approximation, wherein the feedback of internal quark loops is ignored. This is a tempting approximation since it saves a couple of orders of magnitude of computer time. But fortunately the continuing growth in computer power is now alleviating the need for this inexact approach.

Another uncontrolled source of error comes from extrapolations in the number of quark flavors. Again to save computer time, it is popular, mainly in the US, to start with a fermion formulation that has some of the naive doubling issues remaining and then do

**FIGURE 3.**   A lattice of trees surrounded by other lattices in New York City.

an extrapolation down to the desired number of quark species. This is done by replacing the fermion determinant by a non integer power. Since the starting determinant is not a power, this procedure has not been theoretically justified. Indeed, it explicitly gives incorrect behavior in the chiral limit of small masses. I will return to this issue later.

Sometimes the lattice can reveal rather subtle issues. In particular, for many years the way chiral symmetry worked on the lattice was puzzling. We know chiral symmetry is important to the lightness of the pion, which is theoretically tied to the lightness of the up and down quarks. The lattice removes all infinities, and thus issues such as anomalies coming from divergences can be tricky. Ignoring these anomalies forces the theory to cancel them with extra species, known as doublers. But recent years have seen the development of elegant approaches that solved these problems. One tack considers our four dimensional world as an interface in five dimensions [1, 2]. An alternative extracts the essence of this interface into the slightly non-local overlap operator [3]. This satisfies an elegant modification of naive chiral symmetry. So, as indicated in Figure 4, the lattice and chiral symmetry now get along nicely.

Despite these advances, there remain some subtle unsolved problems in lattice gauge theory. One of these involves the standard model, where the weak gauge fields are coupled in a parity violating manner. Neutrinos are experimentally known to spin only to the left, but all known lattice formulations also bring in right handed partners. For example, with domain wall fermions there is naturally present an anti-wall which couples with equal strength to the gauge fields. Ad hoc Higgs fields can give the mirror particles a different mass, but they are always there. To the extend that the lattice is a technique to define a field theory, this raises worries that the usual standard model might be incomplete or even not well defined.

The other major unsolved problem involves the properties of matter at high baryon density. Here there are no practical known algorithms for simulations. Monte Carlo methods fail because there is no positive measure for the path integral. All existing attempts to circumvent this issue require computer time growing exponentially with the

**FIGURE 4.** After recent advances, the lattice now embraces chiral symmetry.

system size. This is particularly frustrating in light of the rich phase diagram expected at high density, filled with exotic phenomena such as color superconductivity.

There are some lattice topics which are highly controversial. I will illustrate the issue starting from a conventional continuum discussion of how chiral symmetry works in three flavor QCD. Here a longstanding tool comes from effective chiral Lagrangians. The physics of the light pseudoscalars is nicely modeled in terms of an effective field $\Sigma$ which lies in the group $SU(3)$. Incorporating quark masses into this picture involves a potential of the form $V(\Sigma) = -\mathrm{Tr}\, M\Sigma$, where the mass matrix is

$$M = \begin{pmatrix} m_u & 0 & 0 \\ 0 & m_d & 0 \\ 0 & 0 & m_s \end{pmatrix} \qquad (1)$$

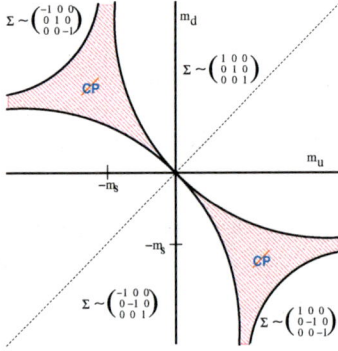

**FIGURE 5.** The phase structure expected for three flavor QCD as the up and down quark masses are varied at fixed strange quark mass. Spontaneous CP violation occurs in regions where the up and down quark masses differ in sign. No structure appears when just a single quark mass vanishes.

xxii

**FIGURE 6.** Controversial ideas came to the front at Lattice 2006.

As we vary the quark masses, minimizing this potential predicts a rich phase structure [4], sketched in Figure 5. Indeed, I discussed this structure at length during the previous meeting in this series [5]. Striking features are the regions of spontaneous CP violation where the minima of the potential are doubly degenerate at complex values of $\Sigma$.

An important feature of this diagram is the absence of any special features when only a single quark mass vanishes. The presence of the other quark masses is sufficient to stabilize the vacuum value for $\langle\Sigma\rangle$, which is real and not accompanied by any exact massless modes. This is a consequence of the anomaly at work; massless Goldstone particles require more than one quark mass to vanish at the same time.

The controversy concerns a numerical algorithm that is incapable of seeing this structure. The feelings here are rather strong, as shown in Figure 6 from Lattice 2006. The "staggered cabal" promotes using a technique known as "rooted staggered quarks." This is the procedure mentioned above of starting with extra particles and taking a fractional power of the fermion determinant. The issue that arises is that the starting staggered formulation has an exact chiral symmetry when any single quark mass vanishes. This symmetry survives the rooting process, and demands the existence of a massless Goldstone mode where the simple effective chiral Lagrangian says there is none. Indeed, this is in direct contradiction with known anomalies [6].

The condoners of this algorithm [7] suggest, without proof, that these evils will drop away in the continuum limit as long as one avoids the zero quark mass axes in Figure 5. They argue that there is actually a plethora of extra particles, one of which is this unwanted Goldstone mode, but their total contribution cancels as the continuum limit is taken. For three flavors using independent rooted staggered quarks, there are 144 pseudoscalar bosons, out of which only the usual 9 should survive the continuum limit. This requires a loss of unitarity so that the total cross sections to produce some of these extra particles can be negative. Also the extra massless particle induces long range forces that make the algorithm non-local. And all of these unproven conjectures are being made just to save some computer time over other algorithms, such as Wilson, domain wall,

**FIGURE 7.** A lattice of palapas at the 2004 meeting in this series, held in Villasimius, Sardinia.

or overlap fermions, that do not so severely mutilate the qualitative chiral behavior, I conclude that rooting can be unhealthy, although the extreme contortions being tried to rescue the approach might be amusing enough to warrant a movie.

I conclude with one final reason one might want to be a lattice theorist. We often meet in very nice places to search out new lattices, such as the marble/basalt arrays here in the Azores or the environment shown in Figure 7 from the 2004 meeting in this series. And of course, as you will see tomorrow night, this meeting has a strong tradition of taking poster sessions seriously!

## ACKNOWLEDGMENTS

This manuscript has been authored under contract number DE-AC02-98CH10886 with the U.S. Department of Energy. Accordingly, the U.S. Government retains a non-exclusive, royalty-free license to publish or reproduce the published form of this contribution, or allow others to do so, for U.S. Government purposes.

## REFERENCES

1. D. B. Kaplan, Phys. Lett. B **288**, 342 (1992) [arXiv:hep-lat/9206013].
2. V. Furman and Y. Shamir, Nucl. Phys. B **439**, 54 (1995) [arXiv:hep-lat/9405004].
3. H. Neuberger, Phys. Lett. B **417** (1998) 141 [arXiv:hep-lat/9707022].
4. M. Creutz, Phys. Rev. Lett. **92**, 201601 (2004) [arXiv:hep-lat/0312018].
5. M. Creutz, AIP Conf. Proc. **756**, 143 (2005) [arXiv:hep-lat/0410043].
6. M. Creutz, arXiv:hep-lat/0603020; M. Creutz, arXiv:hep-lat/0608020.
7. C. Bernard, M. Golterman, Y. Shamir and S. R. Sharpe, arXiv:hep-lat/0603027.

# PLENARY SESSION PRESENTATIONS

## SUMMARIES

Session A: A Partial Summary of Session A (Jeff Greensite)
Session D: Deconfinement (Carlos Pajares)
Session E: QCD, New Physics and Experiment (Giuseppe Nardulli)
Session F: Summary of Section F: QCD in Nuclear Physics and Astrophysics (Thomas D. Cohen)

## PARALLEL SESSION CONTRIBUTIONS

### Session A: The vacuum structure of QCD and the mechanism of confinement

(I) Vacuum Configurations (vortices, monopoles, calorons,...) and other lower-dimensional structures in the QCD vacuum; Eigenmode Spectrum of covariant (Dirac, Laplacian) operators, connection to confinement and topology; Ghost/Gluon propagators and confinement criteria; New analytic approaches to confinement; Numerical studies of the QCD string.

(II) The interface between perturbative and nonperturbative QCD; renormalons and power corrections.

**Conveners**: M. Faber (Wien), J. Greensite (San Francisco), M. Polikarpov (ITEP).

### Session B: Light Quarks (and Gluonia)

Chiral effective theories; sum rules; lattice; Schwinger-Dyson equations; masses of light quarks; light-quark loops; phenomenology of light-hadron form factors, spectra and decays; exotics and glueballs; experiments.

**Conveners**: J. Goity (JLAB), H. Sazdjian (Orsay), M. Testa (Roma).

### Session C: Heavy Quarks (and Gluonia)

Effective theories for heavy quarks (HQET, NRQCD, pNRQCD, vNRQCD); sum rules; lattice; heavy quark masses and renormalons; phenomenology of spectra and decays; glueballs; experiments.

**Conveners:** G. Bodwin (ANL), J. Soto (Barcelona), A. Vairo (Milano).

### Session D: Deconfinement

QCD at finite temperature; quark-gluon plasma; QCD phases; lattice, imaginary chemical potentials; experiments at RHIC and CERN.

**Conveners**: Y. Foka (GSI), C. Pajares (Santiago Compostela), J. Rafelski (Arizona).

### Session E: QCD, New Physics and Experiments

Hints on the confinement mechanism from supersymmetric and string theories. Precision calculations in QCD with respect to experiments and possible new physics. Applications of QCD nonperturbative methods into different fields.

**Conveners**: F. Harris (Hawaai), G. Nardulli (Bari), A. Sczcepaniak (Indiana).

## Session F: QCD and nuclear physics, QCD and astrophysics

Color superconducting quark matter; neutron and compact stars; effective field theories for Nuclear Physics; Nucleon-Nucleon interaction; nuclear forces.

**Conveners**: M. Alford (Saint Louis), T. Cohen (Maryland).

# Lattice QCD simulations with light dynamical quarks

## Sinya AOKI

*Graduate School of Pure and Applied Sciences, University of Tsukuba, Tsukuba, Ibaraki 305-8571, Japan*
*Riken BNL Research Center, Physics 510A, BNL, Upton, NY11973, USA*

**Abstract.** I report recent results from full QCD simulations by CP-PACS and JLQCD collaborations.

**Keywords:** Lattice QCD, numerical simulations, dynamical quarks
**PACS:** 11.15.Ha, 12.38.Gc, 11.30.Rd

## INTRODUCTION

Lattice QCD is a powerful tool to understand the strong interaction of hadrons from the first principles of QCD for quarks and gluons with the aid of numerical simulations. Physical quantities calculated with the method range from the spectrum of light hadrons to electroweak matrix elements. Systematic errors such as finite lattice volume and spacing, and the use of the quenched approximation are gradually being reduced thanks to development of computer power as well as simulation algorithms. Among these systematics, a current main concern is the effect of light dynamical quarks.

As a member of CP-PACS and JLQCD collaborations in Japan, I have been working on large scale lattice QCD simulations for many years. In this talk I report recent results of our collaborations in lattice QCD simulations with light dynamical quarks.

Let me first explain the necessity of dynamical quark effects, by presenting the quenched light hadron spectrum in the left panel of Fig. 1. These results have been obtained by the CP-PACS collaboration after taking the continuum limit[1]. In this calculation the experimental $\rho$ and $\pi$ meson masses are used to fix the lattice spacing $a$ and the light quark mass $m_l$, where up and down quark masses are assumed to be equal ($m_u = m_d = m_l$). For the strange quark mass, two choices are compared, one employing the $K$ meson mass (filled symbols ; $K$-input) and other with the $\phi$ meson mass (open symbols; $\phi$-input). Experimental values are given by horizontal lines. This figure shows an overall agreement of the light hadron spectrum in the quenched lattice QCD at a 5–10% level. However, it is also clear that there are systematic deviations between the quenched spectrum and experiments beyond statistical errors of 2–3%. In particular, the hyperfine splitting between the $\phi$, $K^*$ meson masses and the $K$ meson mass is smaller than the experimental one. This indicates that full QCD simulations are indeed necessary for more accurate results.

We then performed a large scale 2 flavor full QCD simulations[2]. The right panel of Fig. 1 shows the $\phi$ meson mass from the $K$ input as a function of the lattice spacing $a$.

CP892, *Quark Confinement and the Hadron Spectrum VII*
edited by J. E. F. T. Ribeiro
© 2007 American Institute of Physics 978-0-7354-0396-3/07/$23.00

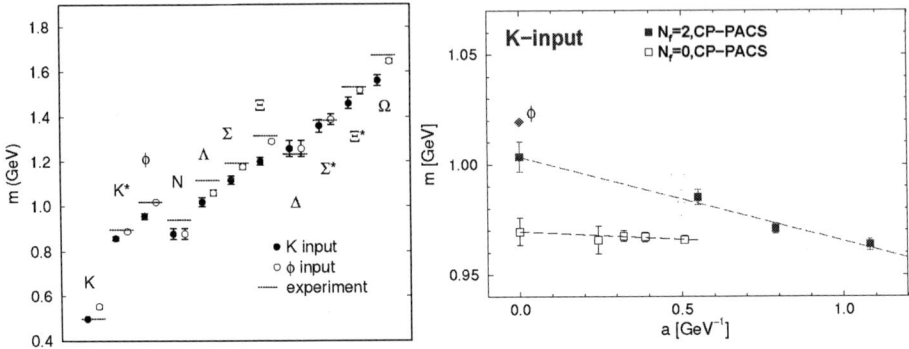

**FIGURE 1.** Left: Light hadron spectra in quenched QCD. Right: $\phi$ meson masses from $K$ input as a function of $a$ in $N_f = 2$ full QCD (solid), together with the quenched results (open).

As can been seen from the figure, the deviation from an experimental value in quenched QCD is much reduced in the $N_f = 2$ full QCD after the continuum extrapolation. The effect of dynamical sea quarks is really important for reproducing the correct spectrum.

# $N_f = 2 + 1$ FULL QCD SIMULATIONS

The success of 2 flavor full QCD simulations motivates us to perform more accurate calculations, $N_f = 2 + 1$ flavor full QCD simulations, in order to remove the systematic error associated with the absence of the dynamical strange quark.

# $N_f = 2 + 1$ full QCD project

We have started 2+1 full QCD simulations as a joint project of CP-PACS and JLQCD collaborations [3, 4, 5], employing the RG improved gauge action and the Wilson-type quark action. In order to reduce the effect of the explicit chiral symmetry violation in the Wilson quark, we introduce the non-perturbative $O(a)$ improvement. A necessary parameter $c_{SW}$ has already been determined by our collaborations[6], prior to large scale simulations. We employ the standard Hybrid Monte-Carlo (HMC) algorithm to simulate degenerate up and down quarks, while polynomial HMC algorithm for the dynamical strange quark. The latter algorithm has been developed by us to simulate odd number of dynamical quarks[7].

## Simulations and analyses

We take 3 values of lattice spacing, $a \simeq 0.07$, 0.10, 0.12 fm, equally spaced in $a^2$, in order to perform the continuum extrapolation, with the $(2 \text{ fm})^3$ spatial volume. We accumulate more than 5,000 HMC trajectories at each lattice spacing. We take 5 values

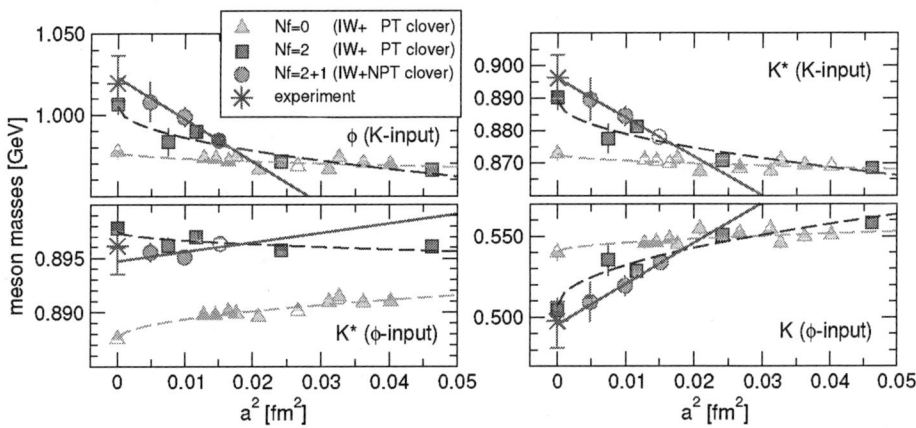

**FIGURE 2.** Left-Top: $\phi$ meson mass as a function of $a^2$ with $K$ input. Left-Bottom: $K^*$ meson mass with $K$ input. Right-Top: $K$ meson mass with $\phi$ input. Right-Bottom: $K^*$ with $\phi$ input.

of the degenerate up and down quark mass ranging between $m_{PS}/m_V \simeq 0.6$ and $0.78$, where $m_{PS}$ and $m_V$ are the pseudo-scalar meson mass and the vector meson mass, respectively. For the strange quark mass, we take 2 values around $m_{PS}/m_V \simeq 0.7$. Note that our light quark mass is much heavier than the experimental value, $m_{PS}/m_V = 0.18$, while the strange quark mass is close to the value, $m_{PS}/m_V \simeq 0.68$, estimated by the 1-loop chiral perturbation theory.

For the chiral extrapolation of meson masses, we have used polynomial functions in quark masses, including up to quadratic terms with an interchange symmetry among 3 sea quarks and that among 2 valence quarks. Chiral fits are made for light-light(LL), light-strange(LS) and strange-strange(SS) mesons simultaneously. Polynomial functions describe quark mass dependences of data very well[5].

In order to estimate the systematics of the polynomial chiral extrapolation, we also employ another fit function, obtained by the Wilson chiral perturbation theory (WChPT)[9, 10, 11], which contains both chiral loop and finite lattice spacing effects. No difference between the WChPT fit and the polynomial fit is observed for PS meson masses, while a slight difference is detected in the small quark mass region for V meson masses[5]. This analysis suggests that the effect of chiral log is small in the region of the light quark mass employed in our simulations. Note however that our light quark mass may be too heavy to apply the NLO formula. Further analysis including data with lighter quark mass will be required for the definite conclusion on the effect of the chiral log to meson masses.

## Continuum extrapolation

Now let me consider the continuum extrapolation of some quantities.

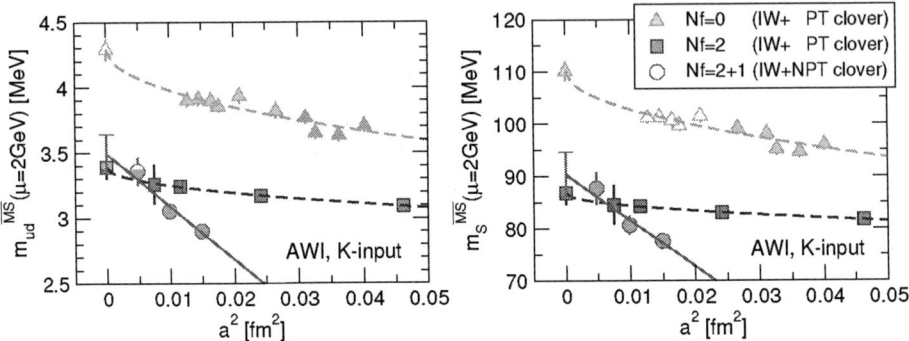

**FIGURE 3.** Left: The light quark mass $m_{ud}$ as a function of $a^2$. Right: The strange quark mass $m_s$.

## Light meson spectra

Left two panels of Fig.2 show vector ($\phi$ and $K^*$) meson masses as the function of $a^2$ with the strange quark mass from the $K$ input, while right panels are $K^*$ and $K$ meson masses with the strange quark mass from the $\phi$ input. Circles represent 2+1 flavor full QCD results, while 2 flavor and quenched results are given for comparison by squares and triangles, respectively. 2+1 flavor results are consistent with experimental values within 2% statistical errors after the continuum extrapolation. This agreement is encouraging. It is difficult, however, to pin down the effect of the dynamical strange quark on meson spectra, since 2% errors are much larger than those of 2 flavor results.

## Quark mass

Quark masses are determined for the $\overline{\text{MS}}$ scheme at the scale $\mu = 2$ GeV. Lattice results are translated to the $\overline{\text{MS}}$ scheme at $\mu = a^{-1}$ using tadpole-improved one-loop matching factor [8], and then evolved to $\mu = 2$ GeV using the four-loop RG-equation. Quark mass results are shown in Fig. 3. As already observed in $N_f = 2$ QCD [2], values of the strange quark mass determined for either the $K$- or the $\phi$-inputs, while different at finite lattice spacings, extrapolate to a common value in the continuum limit. Therefore the quark masses in the continuum limit is estimated from a combined fit to data with the $K$- and the $\phi$-inputs. We finally obtain[5]

$$m_{ud}^{\overline{\text{MS}}}(\mu = 2 \text{ GeV}) = 3.50(14)\binom{+26}{-15} \text{ MeV}, \quad m_s^{\overline{\text{MS}}}(\mu = 2 \text{ GeV}) = 91.8(3.9)\binom{+6.8}{-4.1} \text{ MeV}.$$
$$(1)$$

Dynamical up and down quarks reduce significantly the quark masses. The effect of strange quark is less dramatic, and we do not see deviations from the $N_f = 2$ results, $m_{ud} = 3.44^{+0.14}_{-0.22}$ Mev, $m_s = 88^{+6}_{-6}$ Mev ($K$ input)[2] beyond statistical errors.

4

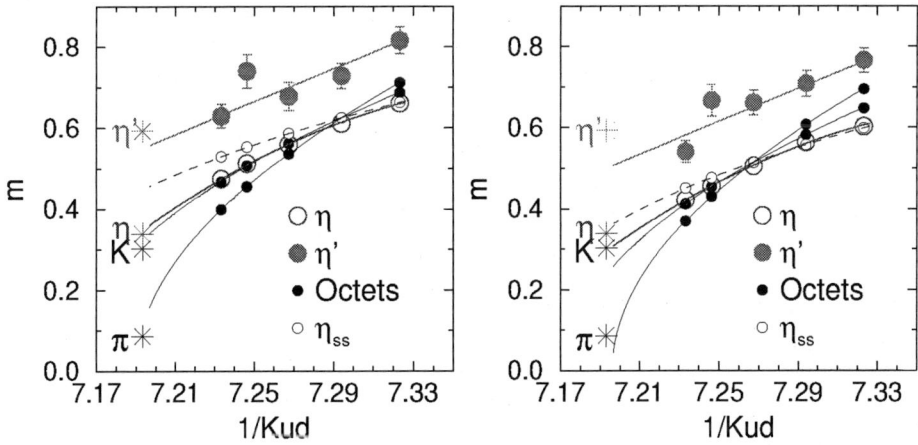

**FIGURE 4.** PS meson masses including $\eta'$ and $\eta$ as a function of $1/K_{ud}$ at $K_s = 0.13710$(Left) and 0.13760(Right).

## Pseudo-scalar meson decay constant

PS meson decay constants are estimated using matching factor determined by tadpole-improved one-loop perturbation theory. The results with the $K$-input are

$$f_\pi = 140.7(9.3) \text{ MeV}, \quad f_K = 160.9(9.1) \text{ MeV}, \quad f_K/f_\pi = 1.142(17). \qquad (2)$$

We recall that in our $N_f = 2$ QCD calculation, the magnitude of scaling violation was so large that we were not able to estimate values in the continuum limit [2]. The situation is much better in the present case and $f_\pi$ and $f_K$ turn out to be consistent with experiment. The errors are large, however. Furthermore, the ratio $f_K/f_\pi$ differs significantly from the experimental value, 1.223(12). A long chiral extrapolation is a possible cause of the discrepancy.

## Flavor singlet mesons

In this subsection, I briefly present a preliminary result on the flavor singlet meson mass[12]. In Fig. 4 we present PS meson masses including $\eta'$ and $\eta$ as a function of the light quark mass ($1/K_{ud}$ where $K$ is the hopping parameter) for 2 values of the strange quark mass ($1/K_s$) at $a \simeq 0.12$ fm. Small solid circles denote LL and LS meson results while small open circles correspond to SS meson results without disconnected diagrams. Once we correctly include contributions from disconnected diagrams, the SS state is mixed with the flavor singlet state, so that the mass of the SS state becomes a little lighter, as shown by large open circles in the figure. The singlet $\eta'$, denoted by large solid circles in the figure, appears much heavier than other PS mesons.

By the polynomial chiral extrapolation to the physical point, we obtain $m_\eta = 545(16)$ MeV, consistent with the experimental value (550 MeV), while $m_{\eta'} = 871(46)$ MeV, which is much larger than octet PS meson masses and is smaller than the experimental value (960 MeV) only by 100 MeV (2 $\sigma$). The U(1) problem seems to be solved. More studies at two other lattice spacings, however, will be required for the final conclusion.

# SUMMARY AND OUTLOOK

## Summary

CP-PACS and JLQCD collaborations has performed the 2+1 full QCD project, using the RG improved gauge action and non-perturbatively $O(a)$ improved clover quark action. Configuration generations have already been completed and the analyses are now being finalized. Light meson masses agree with experimental values after the continuum extrapolation assuming that the $a^2$ contribution dominates the scaling violation. Values of the up-down quark mass and the strange quark mass are determined in the continuum limit. We observe that the dynamical strange quark effect is much small than that of the up-down quarks.

Currently there are several on-going analyses, which include the non-perturbative determination of renormalization factors to remove an ambiguity of 1-loop estimates, the flavor singlet meson mass as presented, and heavy quark quantities using a relativistic heavy quark action.

## PACS-CS project

We have just started a new project, PACS-CS project, which uses a new cluster PACS-CS. The PACS-CS starts operating this July with the peak speed of 14.3 Tflops[13]. In order to remove the most serious ambiguity due to the chiral extrapolation, the PACS-CS collaboration wishes to go down to lighter up-down quark masses with the clover fermion, employing the domain decomposed HMC algorithm proposed by Lüsher[14]. Our preliminary test study indicates that we can go down to as small as 15 MeV quark mass[15].

## Nucleon force

Last but not least, I briefly introduce an interesting application of lattice QCD technique to the nucleon force (the potential between two nucleons). Recently we try to extract this $NN$ potential on the lattice from the Bethe-Salpeter wave function $\phi$ and the effective Schrödinger equation as

$$V(r) = E + \frac{1}{m_N} \frac{\nabla^2 \phi(r)}{\phi(r)}.$$

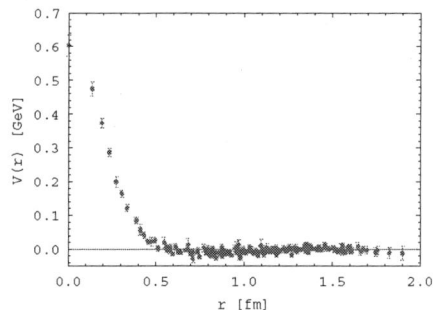

**FIGURE 5.** The $NN$ potential.

Fig. 5 gives the $NN$ potential for the $(J^P, I) = (0^+, 1)$ channel, obtained in quenched QCD at $a \simeq 0.14$ fm and $m_\pi \simeq 880$ MeV[16]. We clearly observe the strong repulsive force at the short distances. Although errors are still too large to see an expected attractive force at the intermediate distance, this method seems promising. Currently we investigate systematics of this method.

## ACKNOWLEDGMENTS

I am grateful to all members of CP-PACS and JLQCD collaborations. In particular I would like to thank Drs. N. Ishii, T. Ishikawa, Y. Kuramashi and T. Yoshié for providing me data and figures used in this talk. This work is supported in part by Grant-in-Aid of the Ministry of Education (Nos. 13135204,15540251).

## REFERENCES

1.  CP-PACS Collaboration: S. Aoki *et al.*, *Phys. Rev. Lett.* **84**, 238-241 (2000); *Phys. Rev.* **D67**, 034503 (2002).
2.  CP-PACS Collaboration: A. Ali Khan *et al.*, *Phys. Rev. Lett.* **85**, 4674 (2000); *Eratum-ibid.* **90**, 029902 (2003); *Phys. Rev.* **D65**, 054505 (2002); *Eratum-ibid.* **D67**, 059901 (2003).
3.  CP-PACS and JLQCD Collaborations: T. Ishikawa *et al.*, *Nucl. Phys.* **B**(Proc. Suppl.)**140**, 225 (2005).
4.  CP-PACS and JLQCD Collaborations: T. Ishikawa *et al.*, *PoS* **LAT2005**, 057 (2005).
5.  CP-PACS and JLQCD Collaborations: T. Ishikawa *et al.*, *PoS* **LAT2006**, 181 (2006).
6.  CP-PACS and JLQCD Collaborations: S. Aoki *et al.*, *Phys. Rev.* **D73**, 034501 (2006).
7.  JLQCD Collaboration: S. Aoki *et al.*, *Phys. Rev.* **D65**, 094507 (2002).
8.  S. Aoki *et al.*, *Phys. Rev.* **D58**, 074505 (1998).
9.  S. Aoki, *Phys. Rev.* **D68**, 054508 (2003).
10. S. Aoki, O. Bär, T.~Ishikawa and S. Takeda, *Phys. Rev.* **D73**, 014511 (2006) .
11. S. Aoki, O. Bär and S. Takeda, *Phys. Rev.* **D73**, 094501 (2006) .
12. S. Aoki *et al.*, *PoS* **LAT2006** (2006) (hep-lat/0610021).
13. PACS-CS Collaboration: A. Ukawa *et al.*, *PoS* **LAT2006**, 039 (2006).
14. *JHEP* **05**, 052 (2003); *Comput. Phys. Commun.* **156**, 209 (2004); *ibid* **165**, 199 (2005).
15. PACS-CS Collaboration: Y. Kuramashi *et al. PoS* **LAT2006**, 029 (2006).
16. N. Ishii, S. Aoki and T. Hatsuda, *PoS* **LAT2006**, (2006) (hep-lat/061002).

# Neutron stars and quark matter

Gordon Baym

*Department of Physics, University of Illinois at Urbana-Champaign*
*1110 W. Green Street, Urbana, IL, 61801 USA*

**Abstract.**
Recent observations of neutron star masses close to the maximum predicted by nucleonic equations of state begin to challenge our understanding of dense matter in neutron stars, and constrain the possible presence of quark matter in their deep interiors.

**Keywords:** neutron stars, quark matter
**PACS:** 97.60.Jd, 26.60.+c, 12.38.Mh

## Introduction

Neutron stars – highly compact stellar objects with masses $\sim$ 1-2 $M_\odot$ (solar masses), radii of order 10-12 km, and temperatures well below one MeV – are natural laboratories to study cold ultradense matter [1]. Indeed, the inner cores of neutron stars are the only known sites where one could expect degenerate quark matter in nature. Figure 1 shows the cross section of a neutron star interior. The mass density, $\rho$, increases with increasing depth in the star. The crust is typically $\sim$1 km thick, and consists, except in the molten outer tens of meters, of a lattice of bare nuclei immersed in a sea of degenerate electrons, as in a normal metal. The matter becomes more neutron rich with increasing density, a result of the increasing electron Fermi energy favoring electron capture on protons, $e^- + p \rightarrow n + \nu_e$. Beyond the *neutron drip* point, $\rho_{drip} \sim 10^{11}$g/cm$^3(= 2 \times 10^{-4}$ fm$^{-3})$, the matter becomes so neutron rich that the continuum neutron states begin to be filled, and the still solid matter becomes permeated by a sea of free neutrons in addition to the electron sea. At a density of order half nuclear matter density, $n_0 \simeq 0.16$fm$^{-3}$, the matter dissolves into a uniform liquid composed primarily of neutrons, plus $\sim$5% protons and electrons, and a sprinkle of muons.

The nature of the extremely dense matter in the cores of neutron stars, while determining the gross structure of neutron stars, e.g., density profiles $\rho(r)$, radii $R$, moments of inertia, and the maximum neutron star mass, $M_{max}$, remains uncertain. Scenarios, from nuclear and hadronic matter, to exotic states involving pionic [2] or kaonic [3] Bose-Einstein condensation, to bulk quark matter and quark matter in droplets, including superconducting states, as well as strange quark matter, have been proposed. Ultra-relativistic heavy ion collision experiments at RHIC, and soon at ALICE and CMS at the LHC, probe hot dense matter, from which one can gain hints of the properties of cold matter. The uncertainies in the properties of matter at densities much greater than $n_0$ are reflected in uncertainties in $M_{max}$, important in distinguishing possible black holes from a neutron stars by measurement of their masses, and in inferring whether an independent family of denser quark stars, composed essentially of quark matter, can exist.

CP892, *Quark Confinement and the Hadron Spectrum VII*
edited by J. E. F. T. Ribeiro
© 2007 American Institute of Physics 978-0-7354-0396-3/07/$23.00

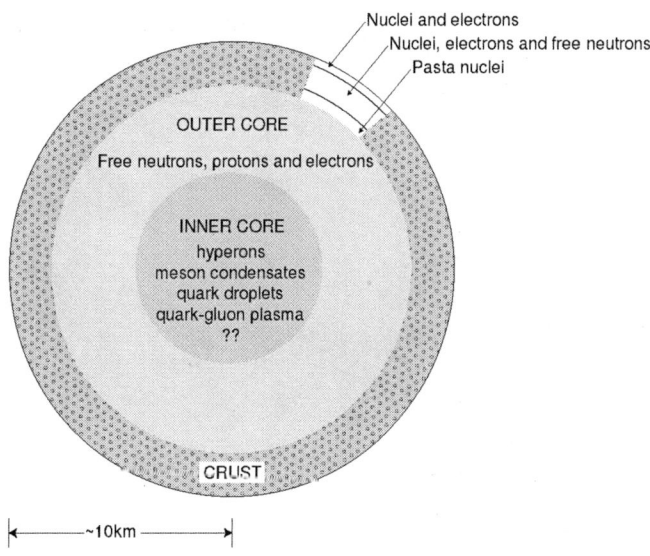

Nuclei and electrons
Nuclei, electrons and free neutrons
Pasta nuclei

OUTER CORE
Free neutrons, protons and electrons

INNER CORE
hyperons
meson condensates
quark droplets
quark-gluon plasma
??

CRUST

~10km

**FIGURE 1.** Schematic cross section of a neutron star.

## Nuclear matter in the interior

The properties of the liquid near $n_0$ can be readily determined by extrapolation from laboratory nuclear physics. The most reliable equations of state of nuclear matter in neutron stars are based on extracting nucleon-nucleon interactions from pp and pn scattering experiments at energies below $\sim 300$ MeV, constrained by fitting the properties of the deuteron, and solving the many-body Schrödinger equation numerically via variational techniques to find the energy density as a function of baryon number, e.g., [4, 5]. The most complete two-body potential is the Argonne A18 (with 18 different components, such as central, spin-orbit, etc., of the interactions).

Two-body potentials predict a reasonable binding energy of nuclear matter; however the calculated equilibrium density is too high. Similarly, two-body potentials fail to produce sufficient binding of light nuclei [6]. The binding problems indicate that one must take into account intrinsic three-body forces acting between nucleons, such as the process in which two of the nucleons scatter becoming internally excited to an intermediate isobar state ($\Delta$) while the third nucleon scatters from one of the isobars. The three-body forces must increase the binding in the neighborhood of $n_0$, but, to avoid overbinding nuclear matter, they must become repulsive at higher densities. This repulsion leads to a stiffening of the equation of state of neutron star matter at higher densities over that computed from two-body forces alone.

Figure 2 shows the energy per baryon of neutron matter as a function of baryon density [5] with the A18 two-body potential, and Urbana UIX three-body potential, together with relativistic boost corrections ($\delta v$), accurate to order $(v/c)^2$. This equation of state, taking into account all two-nucleon data, and data from light nuclei, is currently

9

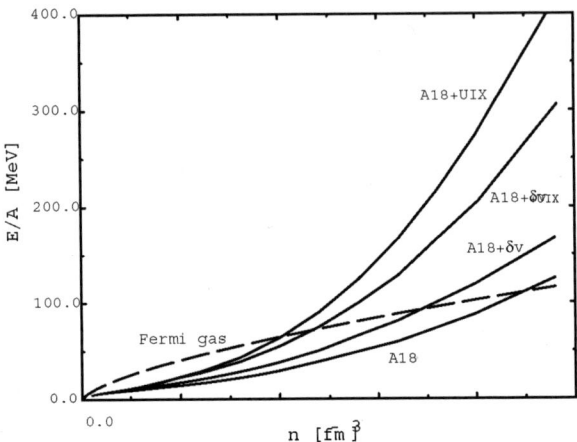

**FIGURE 2.** Energy per baryon of pure neutron matter as a function of baryon density, $n$, calculated with the A18 two-body potential with and without the Urbana IX (UIX) three-body potential, and lowest order relativistic corrections, $\delta v$. From [5].

the best available for $n \gtrsim n_0$. (Nuclear equations of state based on the more accurate "Illinois" three-body potentials [7] will be reported shortly [8].) One sees here the stiffening of the equation of state from inclusion of three-body forces, slightly mitigated by relativistic effects. Figure 3a shows the gravitational mass vs. central density for families of stars calculated by integrating the Tolman-Oppenheimer-Volkoff equation for the same equation of state as in Fig. 2, with beta equilibrium of the nucleons included. The maximum mass for the nucleonic equation of state, A18+$\delta v$+UIX, is $\simeq$ 2.2 $M_\odot$, marginally consistent with observed neutron star masses. By contrast, without three-body forces, the maximum mass is $\sim 1.6 M_\odot$, below some observed masses. The corresponding mass vs. radius of the families of models is shown in Fig. 3b; the radii of these models vary little with mass, and are in the range 10-12 km, except at the extremes.

An equation of state based on nucleon interactions alone, while accurately describing neutron star matter in the neighborhood of $n_0$, has several fundamental limitations. One should not expect beyond a few times $n_0$ that the forces between particles can be described in terms of static few-body potentials. Since the characteristic range of the nuclear forces is $\sim 1/2m_\pi$, the parameter measuring the relative importance of three and higher body forces is of order $n/(2m_\pi)^3 \sim 0.4n/fm^3$, so that at densities well above $n_0$ a well defined expansion in terms of two-, three-, four-, ..., body forces no longer exists. The nucleonic equation of state furthermore does not take into account the rich variety of hadronic ($\Delta$, hyperonic, mesonic, etc.) and quark degrees of freedom in the nuclear system which become important with increasing density. Nor can one continue to assume at higher densities that the system can even be described in terms of well-defined "asymptotic" laboratory particles. As one sees in Fig. 3a, the density in the central cores rises well above $n_0$; equations of state and neutron star models based on consideration of nuclear matter alone should not be regarded as definitive.

10

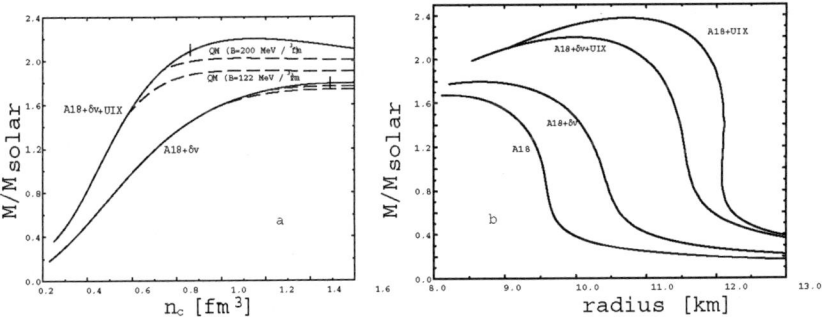

**FIGURE 3.** a) Neutron star mass vs. central baryon density for the equations of state shown in Fig. 2, including beta equilibrium. The curves labelled QM show the effect of allowing for a transition to quark matter described in the simple MIT bag model, with bag constants $B$ = 122 and 200 MeV/fm$^3$. b) Mass vs. radius of neutron stars for the same models.

## Quark matter and quark droplets

Nuclear matter is expected to turn into a quark-gluon plasma at sufficiently high baryon density. Figure 3a shows effects of including quark matter cores, naively calculated in the simple MIT bag model, with bag parameter $B$ =122 and 200 MeV/fm$^3$. Because of the well-known technical problems in implementing lattice gauge theory calculations at non-zero baryon density, we do not to date have a reliable estimate of the transition density at zero temperature or even a compelling answer as to whether there is a sharp phase transition or a crossover. Lattice approaches have included a canonical framework [9], which suggests a phase transition at $n \sim 3n_0$ in the hadronic phase to $\sim 10n_0$ in the quark phase; and the density of states method [10], which yields a transition at baryon chemical potential $\mu_b \sim 750$MeV, as well as giving a triple point in the phase diagram at finite temperature. See also [11]. Representative field-theoretic calculations have been carried out in effective NJL theories [12]; in strong coupling qcd [13], and in terms of instanton overlap [14]. Although estimates of the density range of the transition, $\sim 5-10n_0$, are possibly above the central density found in neutron stars models based on nuclear equations of state, the question of whether the dominant degrees of freedom of the matter in the deep cores of neutron stars are quark-like remains open. In the absence of information about the equation of state at very high densities, the issue of whether a distinct family of quark stars with higher central densities than neutron stars can exist also remains open.

If pure neutron matter and quark matter are distinct phases with a first order transition between them, the transition occurs at nucleonic density $n_q$ where the energy per baryon of quark matter crosses below that in neutron matter. However, the transition in neutron matter with a small admixture of protons and electrons in beta equilibrium must proceed through a mixed phase [15] starting at density below $n_q$. The mixed phase should consist of large droplets of quark matter immersed in a sea of hadronic matter [16, 17]. Formation of droplets is favored because the presence of s and d quarks allows reduction

11

of the electron density, and hence electron Fermi energy, and because it consequently permits an increase in the proton concentration in the hadronic phase. The onset of the droplet phase could, for favorable model parameters of the quark phase, be at a density as low as $\sim 2n_0$. A typical droplet is estimated to have a radius of $\sim 5$ fm, and contain $\sim 100$ u, and $\sim 300$ d as well as s quarks, and thus having a net negative charge $\sim 150$, but the results are very model dependent.

## Neutron star masses

Observations of neutron star masses constrain the equation of state in the cores of neutron stars. The general rule obeyed by families of neutron stars generated at various central densities from a given equation of state is that the stiffer the equation of state, the higher is the maximum mass that a neutron star can have, but the lower is the central mass density, $\rho_c$ at the maximum mass. Lower central density means that there is less room for exotic matter in the interior including $\pi$ and K meson condensates, as well as quark matter. Observations of millisecond binary radio pulsars, consisting of two orbiting neutron stars, have permitted accurate determinations of their neutron star masses, as well as confirmed the existence of gravitational radiation; the masses lie in a relatively narrow interval, $\sim 1.35 \pm 0.04 M_\odot$ [18], a mass reminiscent of the Chandrasekhar core mass of the pre-supernova star. Were the maximum neutron star mass of order $1.4 M_\odot$, the central densities could be sufficiently large to allow substantial exotica in the interior.

Even though the range of measured masses of neutron stars in binary neutron-star systems is tightly constricted, not all neutron stars must have such small masses. The constriction reflects the narrow evolutionary track that allows the two neutron stars to remain bound after their predecessors undergo supernova explosions [19]. A number of determinations of late of neutron star masses in compact binary x-ray sources call into question whether the maximum mass is indeed of order $1.4 M_\odot$. The first is that of the neutron star in the x-ray binary, Vela X-1, with mass deduced to lie in the range $1.86 \pm 0.33 M_\odot$ [20, 21]. (Also [22] which infers $1.75 M_\odot < M < 2.44 M_\odot$.) The uncertainties in the measurement arise from uncertainties in the dynamic behavior of the atmosphere of the B-supergiant companion star, HD 77581; while the reported mass, if confirmed, would rule out very soft equations of state, e.g., those based on kaon condensation, the uncertainties in the determination do not allow one to make a definitive conclusion. A second measurement is that of the neutron star mass in the low mass x-ray binary, Cyg X-2, $1.78 \pm 0.23 M_\odot$ [23]. Such higher masses in x-ray binaries would allow for some exotic matter to be present in neutron stars.

More recently Nice et al. [24] have reported mass determination of the neutron star in the 3.4ms pulsar PSR J0751+1807 in a close circular (6h) binary orbit about a helium white dwarf. The measured neutron star mass is $2.1 \pm 0.2 M_\odot$, almost at the limit of compatibility with the nucleonic equation of state! The companion mass is $0.191 \pm 0.015 M_\odot$. While white dwarfs do not give rise to the uncertainties present in the Vela X-1 companion, the thermal structure of the white dwarf irradiated by the pulsar is incompletely understood [25].

Another very promising approach to measuring neutron star masses is through understanding the origins of the KHz quasiperiodic oscillations (QPO's) observed in low mass x-ray binary neutron star systems [26]. The power spectra of these sources are characterized by pairs of peaks with a nearly constant frequency difference. The QPO's arise from gas accreted from the companion star working its way through a disk down to orbits very close to the neutron star surface, where general relativistic effects are crucial. If, as is strongly suggested by detailed models [27], the upper QPO frequency $v_2$ is that of orbital motion of the accreted matter in the innermost circular stable orbit about the neutron star, at radius $R_{\text{ISCO}} = 6MG/c^2 = 3R_{\text{Schwarzschild}}$, then one directly infers $M = c^3/12\sqrt{6}\pi G v_2$. For the QPO system 4U1820-30, with $v_2 \simeq 1070$ Hz [28], one would deduce a mass $\simeq 2.0M_\odot$.

Finally, Özel, analyzing the neutron star in the low mass x-ray binary EXO0748-676, a thermonuclear burst source, finds a mass $\simeq 2.1 \pm 0.28 M_\odot$, and radius $R \simeq 13.8 \pm 1.8$km – on the outer edge of the radii predicted by the Akmal et al. equation of state. These results suggest that the equation of state may be even stiffer at lower densities, $\lesssim 4n_0$.

## Conclusions

Observations of masses close to the maximum, $\simeq 2.2M_\odot$, predicted by the nucleonic equation of state begin to challenge our knowledge of the physics of neutron star interiors. The existence of high mass neutron stars immediately indicates that the equation of state must be very stiff, whether produced by interacting nucleons or other physics. Central densities are unlikely to be well above $\sim 7n_0$. If there is a sharp deconfinement phase transition below this density, then neutron stars could have quark matter cores as long as the quark matter is itself very stiff. Could $M_{max}$ be larger? Given the equation of state up to a mass density $\rho_f$, the maximum possible mass is produced if the equation of state above $\rho_f$ has sound velocity $c_s = \sqrt{\partial P/\partial \rho}$ equal to the speed of light [30]. The $c_s$ predicted by the nucleonic equation of state [5] reaches $c$ at $n \sim 7n_0$; modifying the physical input, e.g., by including further degrees of freedom such as hyperons, mesons, or quarks at densities in this neighborhood would lower the energy per baryon and tend to decrease the stiffness. Larger $M_{max}$ would require larger $c_s$ at lower densities; e.g., if one maintains the nuclear equation of state only up to $\rho_f = 2\rho_0$, then the maximum mass can be as large as $2.9\, M_\odot$ [31]. Such stiffening would lead to larger radii, perhaps more consistent with that reported for the neutron star in EXO0748-676 [29].

Nonetheless, quarks degrees of freedom – not accounted for by interacting nucleons interacting via static potential – are expected play a role in neutron stars. As nucleons begin to overlap, quark degrees of freedom should become more important, as stressed by Horowitz [32]. Indeed, once nucleons overlap considerably the matter should percolate, opening the possibility of their quark constituents propagating throughout the system (although near the onset of percolation valence quarks may prefer to remain bound in triplets, mimicking nucleons, and leaving the matter a color insulator) [33, 34]. Furthermore, the transition from hadronic to quark matter at low temperature is likely a crossover from BCS-paired superfluid hadronic matter to superfluid quark matter [35, 36]. A firm assessment of the role of quarks in neutron stars must await a better understanding of mechanisms of quark deconfinement with increasing baryon density.

# ACKNOWLEDGMENTS

I would like to dedicate this talk to the memory of my dear friend and colleague, Vijay Pandharipande, who contributed so much to my understanding of neutron star interiors. This work was supported in part by NSF Grant No. PHY03-55014.

# REFERENCES

1. G. Baym, and C. Pethick, *Ann. Rev. Nucl. Sci.* **25** (1975) 27, *Ann. Rev. Astron. Astrophys.* **17** (1979) 415; C. J. Pethick and D. G. Ravenhall, *Ann. Rev. Nucl. Part. Sci.* **45**, 429 (1995).
2. A.B. Migdal, *Rev. Mod. Phys* **50** (1978) 107; G.E. Brown and W. Weise, *Phys. Rept.* **27** (1976) 1; G. Baym and D. K. Campbell, in *Mesons in Nuclei*, v. 3, edited by M. Rho and D. Wilkinson, North-Holland Publ. Co., Amsterdam, 1979, p. 1031.
3. D. B. Kaplan and A. E. Nelson, *Phys. Letters* **B175** (1986) 57; G. E. Brown, K. Kubodera, M. Rho, and V. Thorsson, *Phys. Letters* **B291** (1992) 355.
4. R. B. Wiringa, V. Fiks, and A. Fabrocini, *Phys. Rev. C* **38** (1988) 1010.
5. A. Akmal, V.R. Pandharipande, and D.G. Ravenhall, *Phys. Rev. C* **58**, 1804 (1998).
6. S.C. Pieper, R.B. Wiringa, and J. Carlson, *Phys. Rev. C* **70**, 054325 (2004).
7. S. C. Pieper, V. R. Pandharipande, R. W. Wiringa and J. Carlson, *Phys. Rev. C* **64**, 014001 (2001).
8. J. Morales and D.G. Ravenhall, to be published.
9. Ph. de Forcrand and S. Kratochvila, hep-lat/0602024; and these proceedings.
10. C. Schmidt, Z. Fodor, and S. Katz, hep-lat/0510087,0512032; C. Schmidt, hep-lat/0610116.
11. S. Ejiri, F. Karsch, E. Laermann, and C. Schmidt, *Phys. Rev. D* **73**, 054506 (2006).
12. S.B. Rüster, V. Werth, M. Buballa, I. A. Shovkovy, and D. H. Rischke, nucl-th/0602018.
13. N. Kawamoto, K. Miura, A. Ohnishi, and T. Ohnuma, hep-lat/0512023.
14. A.R. Zhitnitsky, hep-ph/0601057.
15. N. Glendenning, *Phys. Rev. D* **46** (1992) 1274.
16. H. Heiselberg, C. J. Pethick, and E. F. Staubo, *Phys. Rev. Letters* **70** (1993) 1355.
17. V. R. Pandharipande and E. F. Staubo, in *Int. Conf. on Astrophys.*, edited by B. Sinha, World Scientific, Singapore, 1993.
18. S.E. Thorsett and D. Chakrabarty, *Ap. J.* **512**, 288 (1999).
19. B. Willems and V. Kalogera, *Ap. J. Letters* **603**, L101 (2004).
20. O. Barziv, L. Kaper, M. H. van Kerkwijk, J. H. Telting, and J. Van Paradijs, *Astron. Astrophys.* **377** 925 (2001).
21. M. H. van Kerkwijk, in *Compact Stars: Quest for New States of Dense Matter*, edited by D. K. Hong et al., World Scientific, Singapore, 2004, p. 116; astro-ph/0403489.
22. H. Quaintrell et al., *Astron. Astrophys.* **401**, 313 (2003).
23. J.A. Orosz, and E. Kuulkers, *MNRAS* **305**, 132 (1999).
24. D.J. Nice, E.M. Splaver, I.H. Stairs, O. Löhmer, A. Jessner, M. Kramer and J.M. Cordes, *Ap. J.* **634**, 1242 (2005).
25. C. G. Bassa, M. H. van Kerkwijk, S. Kulkarni, astro-ph/0601205.
26. M. van der Klis, *Ann. Rev. Astron. Astrophys.* **38**, 717 (2000).
27. M.C. Miller, F.K. Lamb, and D. Psaltis, *Astrophys. J.* **508**, 791 (1998); *Nucl. Phys.* **B69**, 123 (1999).
28. W. Zhang, A.P. Smale,, T.E. Strohmayer, and J.H. Swank, *Astrophys. J.* **500**, L171 (1998).
29. F. Özel, astro-ph/0605106.
30. C.E. Rhoades, Jr. and R. Ruffini, *Phys. Rev. Letters* **32**, 324 (1974).
31. V. Kalogera and G. Baym, *Ap. J. Letters* **469** (1996) L61.
32. C. Horowitz, these proceedings.
33. G. Baym, *Physica* **96A**, 131 (1979).
34. H. Satz, *Rept. Prog. Phys.* **63**, 1511 (2000).
35. K. Rajagopal, *Nucl. Phys. A* **661** 150c, (1999); M. Alford, K. Rajagopal, and F. Wilczek, *Nucl. Phys. B* **537**, 443 (1999).
36. T. Hatsuda, M. Tachibana, N. Yamamoto, and G. Baym, *Phys. Rev. Letters* **97**, 122001 (2006).

# Experimental Review on Light Meson Physics

## C.Bini

*Università "La Sapienza" and INFN Roma*

**Abstract.** Some recent experimental results concerning the properties of the lowest mass mesons are reviewed and discussed. Some prospects in this field are also presented.

**Keywords:** Mesons, quark
**PACS:** 14.40.-n

## INTRODUCTION

In the mass region below 1 GeV there are two well defined meson nonets: the pseudoscalar nonet ($\pi$, K, $\eta$ and $\eta'$) and the vector nonet ($\rho$, K$^*$, $\omega$ and $\phi$). Both nonets are well interpreted as quark-antiquark states. In the case of the pseudoscalar mesons, the quark and the antiquark are in a $L = 0$ and $S = 0$ state giving the quantum numbers $J^{PC} = 0^{-+}$. In the case of the vector nonet, the quark and the antiquark are in a $L = 0$ but $S = 1$ state giving the vector quantum numbers $J^{PC} = 1^{--}$.

Other mesons below 1 GeV are the scalar mesons. In this sector the situation is much less clear. Infact two of these states, namely the $f_0(600)$ or $\sigma$ and the $K_0^*(800)$ or $\kappa$ are still controversial. Moreover the nature of the only two well estabilished scalar mesons below 1 GeV, namely the $a_0(980)$ and the $f_0(980)$ is also matter of discussion. Since a long time it was proposed [1] a diquark-antidiquark structure for the whole lowest mass scalar nonet, or alternatively, due to the closeness of the $a_0$ and $f_0$ mass to twice the kaon mass, it was suggested that these two states could be two different levels of a kaonic molecule [2].

Above 1 GeV the scenario is complicated by the presence of several states whose existence and quantum number assignment is still controversial.

The experimental investigation of the low mass pseudoscalar and vector mesons has been in the recent years mostly devoted to precision mesurements of masses and mixing angles. These measurements are reviewed in sections 2 and 3. In the scalar sector many measurements have been done recenlty mostly devoted to two main questions: the assessment of the existence of the $\sigma$ and $\kappa$ and the nature of the $a_0$ and $f_0$. This is the subject of section 4. Few recent observations concerning mesons above 1 GeV are reported in section 5.

## THE PSEUDOSCALAR NONET

A major parameter in the pseudoscalar mesons sector is the $\eta$-$\eta'$ mixing angle [3]. It can be defined in two different ways: as the angle $\varphi_P$ in the quark flavour basis or as

CP892, *Quark Confinement and the Hadron Spectrum VII*
edited by J. E. F. T. Ribeiro
© 2007 American Institute of Physics 978-0-7354-0396-3/07/$23.00

the angle $\theta_P$ in the octet-singlet basis. Recently two precise measurements have been reported, pushing the uncertainty on this quantity below $1°$.

The first is due to the KLOE experiment at Frascati that measures the ratio:

$$R = \frac{B.R.(\phi \to \eta'\gamma)}{B.R.(\phi \to \eta\gamma)} \tag{1}$$

that can be simply related to $\varphi_P$ [4]. In fact the $\phi$ meson is essentially an $s\bar{s}$ state so that, due to the OZI rule, the ratio between the $\phi\eta'$ and $\phi\eta$ couplings measures the amount of $|s\bar{s} >$ in the $\eta'$ and $\eta$ wave-functions that, in turn, is related to the angle $\varphi_P$. KLOE has published a first result based on a small statistic sample in 2002 [5] and has recently presented the updated result [6]:

$$\varphi_P = (41.4 \pm 0.3_{stat} \pm 0.7_{syst} \pm 0.6_{th})° \tag{2}$$

where the dominant contribution to the uncertainty is due to the knowledge of other quantities affecting the angle calculation.

The second result has been obtained by the BES experiment at Bejing with the measurement of the ratio

$$R = \frac{B.R.(J/\psi \to \eta'\gamma)}{B.R.(J/\psi \to \eta\gamma)} \tag{3}$$

from which the angle $\theta_P$ is extracted according to the formulation of Ref.[7]. BES obtains the value [8]:

$$\theta_P = (-15.9 \pm 1.2)° \tag{4}$$

The KLOE result can be translated in the octet-singlet basis using the relation $\theta_P = \varphi_P - \text{arctg}\sqrt{2}$, obtaining $\theta_P = (-13.3 \pm 1.0)°$ 1.7 standard deviations away from the BES result. The average of the two values, that is by now the best estimate of $\theta_P$ is:

$$< \theta_P >= (-14.6 \pm 0.8)°. \tag{5}$$

Related to this issue is the question of the presence of gluonium content in the $\eta'$ wave-function. This is done by KLOE that has combined several experimental results as shown in Fig.1. In the figure, $X_{\eta'}$ and $Y_{\eta'}$ are the weights of the $1/\sqrt{2}|u\bar{u} + d\bar{d} >$ and $|s\bar{s} >$ components in the $\eta'$ wave-function and each band shown corresponds to a different experimental observable. If the $X_{\eta'}^2 + Y_{\eta'}^2 = 1$ circumference crosses the intersection of the bands, no gluonium content is needed. In case the intersection lies within the circle, an extra contribution in the $\eta'$ is required. KLOE reports as a preliminary result of this analysis, $X_{\eta'}^2 + Y_{\eta'}^2 = 0.93 \pm 0.06$ that constrains the gluonium content of the $\eta'$, if any, to be below the few percent level. Notice that some experimental informations used in this test can be improved in future experiments so that the test can become stricter (in particular $\Gamma(\eta')$, $B.R.(\eta' \to \omega\gamma)$ and $\Gamma(\pi^0 \to \gamma\gamma)$ are all known at 8% or worse).

In the last 4 years the measurement of the $\eta$ mass has seen a strong improvement, from the few per mil level to well below the per mil level. However the two most recent and

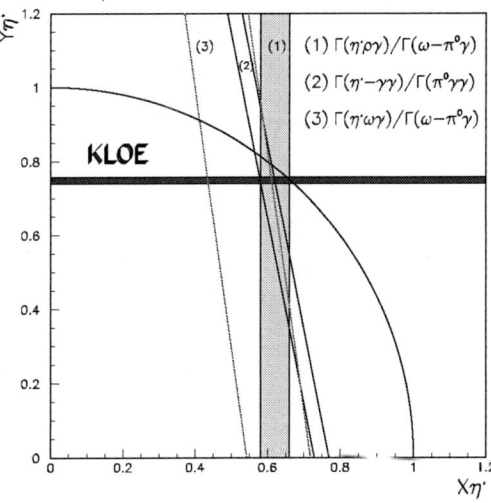

**FIGURE 1.** Bounds in the $Y_{\eta'}$-$X_{\eta'}$ plane coming from KLOE (horizontal band) and from the other experimental quantities indicated in the insert.

most accurate measurements are in disagreement with each other:

$$m_\eta = (547.843 \pm 0.030 \pm 0.041)\,\text{MeV} \quad (\text{NA48}) \tag{6}$$
$$m_\eta = (547.311 \pm 0.028 \pm 0.032)\,\text{MeV} \quad (\text{GEM}) \tag{7}$$

The NA48 experiment at CERN has looked at the decay $\eta \to \pi^0\pi^0\pi^0$ from a high energy (about 110 GeV) $\eta$ sample produced by protons on a beryllium target [9]. The GEM experiment has measured the missing mass distribution of the reaction $p + d \to {}^3He + \eta$ [10]. Very recently KLOE has also presented a precision measurement of the $\eta$ mass based on the decay chain $\phi \to \eta\gamma$ with $\eta \to \gamma\gamma$ [6]. The KLOE result is

$$m_\eta = (547.822 \pm 0.005 \pm 0.069)\,\text{MeV} \quad (\text{KLOE}) \tag{8}$$

that confirms the NA48 value and is in disagreement with the GEM value. At the moment there is no explanation for this inconsistency between different kinds of measurements of the same quantity.

A wide experimental program on $\eta$ and $\eta'$ precision physics is expected in the next years. At the MAMI laboratories in Mainz the joint Crystal Ball / TAPS experiment has started to study some rare $\eta$ and $\eta'$ decays that allow tests of fundamental invariances and of the chiral perturbation theory; the WASA detector at COSY, Julich, will be mostly concentrated on the isospin violation in the $\eta'$ and $\eta$ decays in 3 pions. Very high fluxes of $\eta$ and $\eta'$ in the range of $10^6 \div 10^7$ mesons per day are expected in both experiments. KLOE has collected about $9 \times 10^7$ $\eta$s and $5 \times 10^5$ $\eta'$s so that further refinements of its

17

results are expected. A program to improve the accelerator DAFNE is also under scrutiny [11] and a proposal concerning also the $\eta$ and $\eta'$ physics has been presented [12].

## THE VECTOR NONET

New precision measurements of the $\rho^0$, $\omega$ and $\phi$ masses and widths have been recently published by the Novosibirsk experiments CMD-2 and SND [13] based on the fits of the $\sqrt{s}$ dependencies of the $e^+e^- \to \pi^+\pi^-$ and $e^+e^- \to \pi^+\pi^-\pi^0$ cross-sections in the $\sqrt{s}$ range 300÷1400 MeV. KLOE has also published a measurement of the masses and widths of the three states of charge of the $\rho$ [14] by a fit of the Dalitz plot of the $\phi \to \pi^+\pi^-\pi^0$ decay. In general the relative precisions in the mass measurements range between the $10^{-3}$ level for the $\rho$ and the $10^{-5}$ level for the $\phi$, the latter is obtained by CMD-2 using the beam depolarization method [13]. On the other hand the widths are known not better than at the 1% level.

The vector mesons are also used as probes in the study of the behaviour of dense nuclear matter [15]. In fact the observation of a modification in the vector meson line-shapes (mass shift and broadening) when the vector mesons are produced in dense nuclear medium can be related to the properties of the nuclear matter. A positive evidence has been reported by the TAPS experiment at Bonn [16] as a modification of the $\omega$ line-shape in the reaction $\gamma + A \to \omega + X$ with $\omega \to \pi^0\gamma$ and by the KEK PS-E325 [17] experiment in Japan that has measured the vector meson line-shapes using the invariant masses of $e^+e^-$ pairs produced in proton collisions on different heavy targets. Preliminary results of the JLab g4 experiment are presented at this conference [18].

## THE SCALARS

The idea of a non standard quark structure for the lowest mass scalar mesons is based on the analysis of the mass spectrum that appears to be "inverted" if compared to the pseudoscalar and vector ones. Fig.2 shows the scalar meson mass spectrum and compares it to the pseudoscalar and vector spectrum. An "inverted spectrum" can be easily explained assuming a 4-quark structure or, more precisely, a diquark-antidiquark structure [1]. It is well known that the attractive interaction between color triplet diquarks and anti-diquarks gives rise to color singlet 4-quark mesons. The building rule of these mesons implies an "inverted spectrum" [1]. This picture requires first of all that the $\sigma$ and the $\kappa$ are firmly established as meson states. Moreover it requires the $f_0(980)$ and $a_0(980)$ to have a large s-quark content in their wave-functions. The most recent experimental investigations on the scalar mesons are indeed related to the assessment of the $\sigma$ and $\kappa$, and the measurement of the $f_0(980)$ and $a_0(980)$ coupling to the s-quark.

The BES experiment has published two analyses of the Dalitz plots of the decays $J/\psi \to \omega\pi^+\pi^-$ [19] and $J/\psi \to K^*K^+\pi^-$ [20] respectively. They claim the existence of the $\sigma$ to describe the $\omega\pi^+\pi^-$ decay and of the $\kappa$ to describe the $K^*K^+\pi^-$ decay. The pole positions found are respectively: $(541 \pm 39) - i(252 \pm 42)$ MeV and $(841 \pm$

18

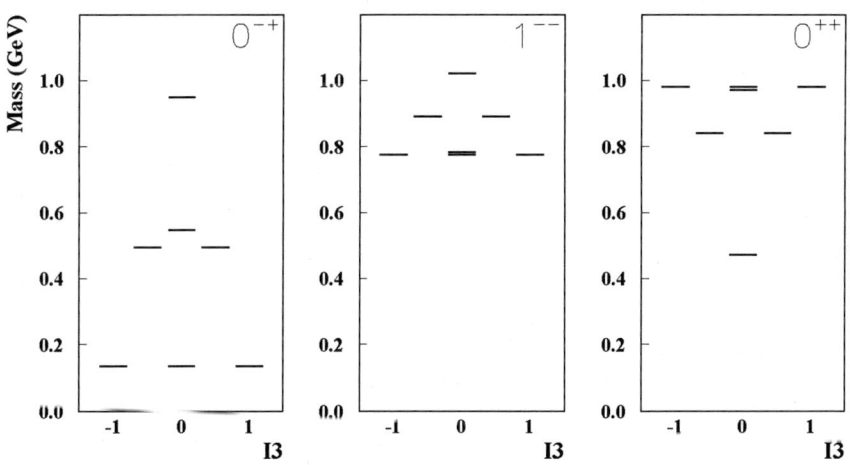

**FIGURE 2.** From left to right the mass spectra of the three lowest mass meson nonets: the pseudoscalar, the vector and the scalar. I3 is the third isospin component.

$30^{+81}_{-73}) - i(309 \pm 45^{+48}_{-72})$ MeV. A recent more sophisticated analysis of the BES $\omega\pi^+\pi^-$ data [21] that includes also the KLOE data on $\phi$ decays [22], lowers the $\sigma$ mass value down to $472 \pm 35$ MeV in agreement with a previous evidence reported by E791 [23]. Recently a theoretical evaluation of the $\sigma$ has been presented in Ref.[24]. Using a procedure based on the Roy equations, the authors claim the presence of a pole in the $\pi\pi$ scattering amplitude, and give a very precise estimate of the pole position: $(441^{+16}_{-8}) - i(272 \pm 45^{+9}_{-12})$ MeV. The value of the mass is in substantial agreement with the recent experimental observations quoted above so that there is a convergence on the existence of a scalar isoscalar resonance in with a mass around 450 MeV. On the other hand, for what concern the $\kappa$ the evidence is still weak.

The properties of the $f_0(980)$ are studied by KLOE in $\phi$ radiative decays [25, 26] and by BES in $J/\psi$ decays [27]. KLOE has observed the $f_0(980)$ looking at the Dalitz plot of the $\phi \to \pi^0\pi^0\gamma$ decays and at the $M_{\pi^+\pi^-}$ spectrum of the $\pi^+\pi^-\gamma$ final states. The fits done [28, 29] indicate the following:

- the $f_0(980)$ is strongly coupled to the kaons, the ratio $g_{f_0K^+K^-}/g_{f_0\pi^+\pi^-}$ being larger than one, in agreement with previous experiments [30];
- the $f_0(980)$ is also strongly coupled to the $\phi$, the coupling $g_{\phi f_0\gamma}$ is in the region $1 \div 3$ GeV$^{-1}$, one order of magnitude larger than the same coupling for the $\pi^0$ $g_{\phi\pi^0\gamma} \sim 0.12$ GeV$^{-1}$.

Similar conclusions are obtained by BES that compares the spectra of the decays

19

$J/\psi \to \phi\pi^+\pi^-$ and $J/\psi \to \omega\pi^+\pi^-$. The $f_0$ peak is well evident in the first case and is suppressed in the second case (where on the contrary the $\sigma$ is rather well evident). All these results are in agreement with a significant "affinity" of the $f_0(980)$ with the $\phi$, hence with a large s-quark content for the $f_0(980)$.

Among the prospects in this field I list the following.

- $\gamma\gamma$ physics: the $\gamma\gamma$ widths of the $f_0$ and $a_0$ are poorly known, and the $\gamma\gamma \to \sigma$ has never been clearly observed [31, 32]. All these informations are very significant in the assessment of the nature of these particles (see the discussion in Ref.[33] and more recently Ref.[34]).
- Quark counting by measuring the asymptotic behaviour of the $e^+e^- \to \phi f_0$ cross-section compared to the $e^+e^- \to \phi\eta$ cross-section [35].
- Quark counting by measuring the high $p_T$ behaviour of the elliptic flow in heavy ions experiments [36].

BABAR and BELLE data can help in the first two points using radiative return and $\gamma\gamma$ tagging. Moreover one of the priorities in the physics program of the upgrade of DAFNE is the study of the $\gamma\gamma$ physics in the lowest energy region.[11, 12]. The possibility to observe $f_0$ in heavy ion experiments is still to be discussed (see the discussion in Ref.[36]).

## THE 1÷2 GEV REGION

In the mass region between 1 and 2 GeV the following scalar states are present: the I=1 $a_0(1450)$ and I=1/2 $K_0^*(1430)$ states, both well established, and the three I=0 states $f_0(1370)$, $f_0(1500)$ and $f_0(1710)$. Trying to arrange these states in a multiplet structure, one finds that again the spectrum is "inverted". Moreover there is one extra I=0 state. The possibility that this extra-state could be a scalar glueball has been investigated by many authors (recent reviews on this subject are Refs.[37, 38]). A recent analysis of this spectrum [39] suggests a 4-quark structure even for this multiplet together with the presence of a scalar glueball mixed with the two 4-quark I=0 states. Finally notice that BES has recently claimed a further scalar state $f_0(1790)$ (see Ref.[40]).

The mass region around 1.9 GeV, that is twice the mass of the nucleon, is of particular interest because is the place where to look for nucleon-antinucleon states like the long searched baryonium. Recently several interesting results have been reported in this mass range.

From the analysis of $J/\psi$ decays to different final states like $p\bar{p}$, $\pi^+\pi^-\eta'$ and $\phi\omega$, BES claims a resonance with a mass between 1820 and 1840 MeV [41, 42] and a width below 100 MeV with properties similar to those of the $\eta'$. The quantum numbers are most probably $J^{PC} = 0^{-+}$. At the same time BABAR [43] using the technique of radiative return, claims a resonance that gives rise to a clear dip in the $e^+e^- \to 6\pi$ cross-section and a small peak in the $e^+e^- \to 4\pi$ at a mass approximately between 1850 and 1900, a width around 150 MeV and vector quantum numbers. A similar effect had been previously observed by FENICE at Frascati [44] and by E687 at Fermilab

[45]. The two resonances are slightly different in mass and width and have different quantum numbers. However, both lies across the nucleon - antinucleon threshold and are significantly coupled to the $p\bar{p}$ channel (see the discussion in Ref.[44]).

# REFERENCES

1. R. L. Jaffe, *Phys. Rev.* **D15** 15 (1977).
2. J. Weinstein and N. Isgur, *Phys. Rev. Lett.* **48** 10 (1982).
3. T. Feldmann, hep-ph/9907491.
4. A. Bramon et al., *Eur. Phys. J.* **C7** 24 (1999).
5. A. Aloisio et al. (KLOE collaboration), *Phys. Lett.* **B541**, 45 (2002).
6. C. Bini et al., (KLOE collaboration) talk given at the Quark and Nuclear Physics 06 Conference, Madrid, June 2006.
7. D. Gross, S. Treiman and F. Wilczek, *Phys. Rev.* **D19** 2188 (1979).
8. M. Ablikim et al. (BES collaboration), *Phys. Rev.* **D73** 052008 (2006).
9. A. Lai et al. (NA48 collaboration), *Phys. Lett.* **B533** 196 (2002).
10. M. Abdel-Bary et al. (GEM collaboration), *Phys. Lett.* **B619** 281 (2005).
11. F. Ambrosino et al., submitted to *Eur. Phys. J. C.*, hep-ex/0603056.
12. See the web site: *http://www.lnf.infn.it/lnfadmin/direzione /roadmap.*
13. M. N. Achasov, talk given at the International Workshop $e^+e^-$ Collisions from $\phi$ to $\psi$, Novosibirsk, March 2006, hep-ex/0604051.
14. A. Aloisio et al., (KLOE collaboration)*Phys. Lett.* **B561** 55 (2003).
15. V. Bernard and U-G. Meissner, *Nucl. Phys.* **A489** 647 (1988).
16. D. Trnka et al., *Phys. Rev. Lett.* **94** 192303 (2005).
17. R. Muto et al., *J.Phys.* **G30** S1023 (2004).
18. See the talk of C.Djalali at this conference.
19. M. Ablikim et al., (BES collaboration)*Phys. Lett.* **B598** 149 (2004).
20. M. Ablikim et al., (BES collaboration)*Phys. Lett.* **B633** 681 (2006).
21. D.Bugg, hep-ph/0608081.
22. A. Aloisio et al., (KLOE collaboration)*Phys. Lett.* **B537** 21 (2002).
23. E. M. Aitala et al., (E791 collaboration)*Phys. Rev. Lett.* **86** 770 (2001).
24. I. Caprini, G. Colangelo and H. Leutwyler, *Phys. Rev. Lett.* **96** 132001 (2006).
25. F. Ambrosino et al. (KLOE collaboration), *Phys. Lett.* **B634**, 148 (2006).
26. F. Ambrosino et al. (KLOE collaboration), submitted to *Eur. Phys. J. C*, hep-ex/0609009.
27. M. Ablikim et al., *Phys. Lett.* **B607** 243 (2005).
28. N. N. Achasov and V. N. Ivanchenko, *Nucl. Phys.* **B315** 465 (1989).
29. G. Isidori et al., *JHEP* **0605** 049(2006).
30. D. Barberis et al., (WA102 collaboration)*Phys. Lett.* **B462** 462 (1999).
31. H.Marsiskie et al., *Phys. Rev.* **D41** 3324 (1990).
32. F. Nguyen, F. Piccinini and A. D. Polosa, *Eur. Phys. J* **C47** 65 (2006).
33. M. Boglione and M. R. Pennington, *Eur. Phys. J.* **C9** 11 (1999).
34. M. R. Pennington, *Phys. Rev. Lett.* **97** 0011601 (2006).
35. S. Pacetti talk given at the Quark and Nuclear Physics 06 Conference, Madrid, June 2006.
36. L. Maiani et al., hep-ph/0606217.
37. C. Amsler and N. .A. Tornqvist, *Phys. Rep.* **389** 61 (2004).
38. D. V. Bugg, *Phys. Rep.* **397** 257 (2004).
39. L. Maiani et al., hep-ph/0604018.
40. H. Liao et al., (BES collaboration) talk given at the Quark and Nuclear Physics 06 Conference, Madrid, June 2006.
41. M. Ablikim et al., (BES collaboration)*Phys. Rev. Lett.* **95** 262001 (2005).
42. M. Ablikim et al., (BES collaboration)*Phys. Rev. Lett.* **96** 162002 (2006).
43. B. Aubert et al., (BABAR collaboration)*Phys. Rev.* **D73** 052003 (2006).
44. A. Antonelli et al., (FENICE collaboration)*Phys. Lett.* **B365** 427 (1996).
45. P. L. Frabetti et al., (E687 collaboration)*Phys. Lett.* **B514** 240 (2001).

# Lattice Results in Coulomb Gauge

Attilio Cucchieri

*Instituto de Física de São Carlos, Universidade de São Paulo, Caixa Postal 369,*
*13560-970 São Carlos, SP, Brazil*

**Abstract.** We discuss recent numerical results obtained for gluon and ghost propagators in lattice Coulomb gauge and the status of the so-called Gribov-Zwanziger confinement scenario in this gauge. Particular emphasis will be given to the eigenvalue spectrum of the Faddeev-Popov matrix.

**Keywords:** Yang-Mills theory; confinement; Green's functions; Gribov copies; Coulomb gauge
**PACS:** 11.15.Ha 12.38.Aw

## INTRODUCTION

In recent years several groups have studied confinement of quarks and gluons using lattice simulations in Coulomb gauge [1]–[11]. This gauge has several advantages, even though it breaks the Lorentz symmetry explicitly. Indeed, it is a physical gauge [12], i.e. there are no unphysical degrees of freedom. Also, a nice confinement scenario [13]–[19] is available for this gauge, based on Gribov's classical work [13]. Finally, Coulomb gauge is well-suited for the Hamiltonian approach and the study of hadron physics by variational methods [20].

Here we review lattice numerical studies in Coulomb gauge. We divide these studies in two periods. In the first period — called here the *classical era* — the studies focused on the infrared (IR) behavior of propagators (gluon and ghost) [1]–[4] and on the long-distance behavior of the color-Coulomb potential [3]–[8]. On the other hand, in the second period — the *modern era* — the eigenvalue spectrum of the Faddeev-Popov (FP) operator became the main subject of investigation [9, 10].

## COULOMB GAUGE AT THE CLASSICAL LEVEL

In a classical Yang-Mills theory [12], Gauss's law is written as $(D_i E_i)^a (\vec{x},t) = \rho_{qu}^a (\vec{x},t)$, where $D_i$ is the gauge-covariant derivative, $E_i^a(\vec{x},t)$ is the color-electric field and $\rho_{qu}^a(\vec{x},t)$ is the quark color-charge density. (Here the sum over repeated indices is always understood.) In Coulomb gauge, i.e. $(\partial_i A_i)^a(\vec{x},t) = 0$, the color-electric field can be decomposed into its transverse and longitudinal parts using $E_i^a(\vec{x},t) \equiv (E^{tr})_i^a(\vec{x},t) - \partial_i \phi^a(\vec{x},t)$, where $\phi^a(\vec{x},t)$ is the so-called *color-Coulomb potential* and $(\partial_i E_i^{tr})^a(\vec{x},t) = 0$. Then, Gauss's law becomes $(\mathcal{M} \phi^a)(\vec{x},t) = \rho^a(\vec{x},t)$, where $\mathcal{M} \equiv -D_i \partial_i$ is the 3-dimensional FP operator and $\rho^a(\vec{x},t) = \rho_q^a(\vec{x},t) - f^{abc} A_i^b(\vec{x},t) (E_i^{tr})^c(\vec{x},t)$ is the total color-charge density. It follows that the color-Coulomb potential $\phi^a(\vec{x},t)$ can be expressed by means of the instantaneous and non-local operator $(\mathcal{M}^{-1})^{ab}(\vec{x},\vec{y};t)$, namely $\phi^a(\vec{x},t) = (\mathcal{M}^{-1}\rho)^a(\vec{x},t) = \int d^3y \, (\mathcal{M}^{-1})^{ab}(\vec{x},\vec{y};t) \rho^b(\vec{y},t)$. At the same time, the

CP892, *Quark Confinement and the Hadron Spectrum VII*
edited by J. E. F. T. Ribeiro
© 2007 American Institute of Physics 978-0-7354-0396-3/07/$23.00

classical Hamiltonian $\mathcal{H} = \int d^3x \, (E_i^2 + B_i^2)/2$ can be written as $\mathcal{H} = \mathcal{H}_{Coul} + \int d^3x \, [(E_i^{tr})^2 + B_i^2]/2$. Here

$$\mathcal{H}_{Coul} = \frac{1}{2} \int d^3x \, (\partial_i \phi)^2 = \frac{1}{2} \int d^3x \int d^3y \, \rho^a(\vec{x}) \, \mathcal{V}^{ab}(\vec{x}, \vec{y}) \, \rho^b(\vec{y}) \,, \tag{1}$$

$\mathcal{V}^{ab}(\vec{x}, \vec{y}) = \left[ \mathcal{M}^{-1}(-\Delta) \mathcal{M}^{-1} \right]^{ab}(\vec{x}, \vec{y})$ is the color-Coulomb-potential energy functional and we indicate with $\Delta$ the usual Laplacian.

Clearly, the static color-Coulomb potential $\phi^a(\vec{x}, t)$ is closely related to the 3d FP (equal-time) ghost propagator $G(\vec{x} - \vec{y}, t)\, \delta^{a,b} \equiv \langle (\mathcal{M}^{-1})^{ab}(\vec{x}, \vec{y}; t) \rangle$. In particular, if we consider the Fourier transform $\widetilde{G}(\vec{k}, t)$ we obtain

$$\phi^a(\vec{x}, t) \approx \frac{1}{(2\pi)^3} \int d^3k \int d^3y \, \widetilde{G}(\vec{k}, t) \, \exp\left[ ik_j(x_j - y_j) \right] \rho^a(\vec{y}, t) \,. \tag{2}$$

Thus, if the ghost propagator has a $k^{-4} = |\vec{k}|^{-4}$ singularity at small momenta we get, in the limit of large separation $x = |\vec{x}|$, a linearly rising potential, i.e. $\phi^a(\vec{x}, t) \sim x$.

## THE GRIBOV-ZWANZIGER CONFINEMENT SCENARIO

A non-perturbative investigation of QCD in the IR limit is necessary in order to get an understanding of color confinement. Of course, in developing non-perturbative techniques, one has to deal with the redundant gauge degrees of freedom of the theory. The gauge-fixing technique developed by Faddeev and Popov assumed that one could find a gauge-fixing condition that uniquely determines a gauge field on each gauge orbit. However, in Ref. [13] Gribov showed that the Coulomb and the Landau gauge conditions do not fix the gauge fields uniquely, namely there exist gauge-related field configurations that satisfy the gauge condition (*Gribov copies*) [13, 19].

In order to get rid of the problem of spurious gauge copies, Gribov proposed the use of additional gauge conditions. In particular, for Coulomb gauge, he proposed the restriction of the physical configuration space (on each time-slice $t$) to the region $\Omega_t \equiv \{ A : (\partial_i A_i)^a(\vec{x}, t) = 0, \mathcal{M}^{ab}(\vec{x}, \vec{y}; t) \geq 0 \}$. Thus, inside the region $\Omega_t$, the FP operator has no negative eigenvalues. This region is delimited by the so-called *first Gribov horizon* $\partial\Omega_t$, where the smallest non-trivial eigenvalue of the FP operator $\mathcal{M}^{ab}(\vec{x}, \vec{y}; t)$ is zero. On the lattice, given a thermalized lattice configuration $\{U(x)\}$, a configuration belonging to the region $\Omega_t$ can be obtained by finding a gauge transformation $\{g(x)\}$ that brings the functional[1] $\mathcal{E}_{\mathrm{hor}, U}[g] = -\sum_{i=1}^{3} \sum_{\vec{x}, t} \mathrm{Tr}\left[ g(\vec{x}, t) U_i(\vec{x}, t) g^\dagger(\vec{x} + a e_i, t) \right]$ to a local minimum. Recall that, in the $SU(N_c)$ case, both the link variables $U_\mu(x)$ and the gauge transformation matrices $g(x)$ are elements of the $SU(N_c)$ group (in the fundamental $N_c \times N_c$ representation).

---

[1] In this review we do not discuss results related to the (possible) spontaneous symmetry breaking of the residual gauge freedom $g(t)$ [6, 7, 11] and to the so-called $\lambda$ gauge [21], which interpolates between the Landau gauge ($\lambda = 1$) and a Coulomb-gauge like condition ($\lambda \to 0$).

The additional gauge condition added by Gribov is not significant for the high-frequency vacuum fluctuations, i.e. for the perturbative regime, but it suppresses the low-frequency fluctuations, modifying the (non-perturbative) IR regime [13, 14]. In particular, one can show that, when the functional integration is restricted to the region $\Omega_t$, then (on each time slice) the ghost propagator $G(\vec{k},t)$ is IR enhanced. On the other hand, the transverse gluon propagator $D^{tr}(\vec{k},t)$ may go to zero in the IR limit, implying a maximal violation of reflection positivity. The latter result may be viewed as an indication of gluon confinement [22]. Analytic results for the IR behavior of propagators and vertices using Dyson-Schwinger equations have been presented in Ref. [23].

At the same time, the 44-component of the gluon propagator can be written [15] as $D_{44}(\vec{x}-\vec{y},t) = V_{Coul}(\vec{x}-\vec{y})\,\delta(t) + P(\vec{x}-\vec{y},t)$, where $V_{Coul}(\vec{x}-\vec{y})\,\delta^{ab} = \langle \mathscr{V}^{ab}(\vec{x},\vec{y}) \rangle$ is anti-screening and should yield a linearly rising potential, while $P(\vec{x}-\vec{y},t)$ is the vacuum-polarization term, i.e. it is responsible for screening and for the breaking of the string between color sources. One can show that these three quantities [e.g. $D_{44}(\vec{x}-\vec{y},t)$, $V_{Coul}(\vec{x}-\vec{y})$ and $P(\vec{x}-\vec{y},t)$] are renormalization-group invariant [15, 16]. One can also define the running coupling constant

$$g_{Coul}^2(\vec{k}) = \frac{11N_c - 2N_f}{12N_c}\, k^2\, V_{Coul}(\vec{k})\,. \tag{3}$$

Clearly, if the color-Coulomb potential $V_{Coul}(x)$ is linearly rising at large separation $x$, then in the IR limit we find $V_{Coul}(\vec{k}) \sim 1/k^4$ and $g_{Coul}^2(\vec{k}) \sim 1/k^2$. Also, it has been shown [17] that the Coulomb energy of static sources is an upper bound for the static inter-quark potential $V(\vec{x})$, i.e. if at large $x$ one has $V_{Coul}(\vec{x}) = \sigma_{Coul}\,x$ and $V(\vec{x}) = \sigma\,x$ then we find $(N_c^2 - 1)\sigma_{Coul}/(2N_c) \geq \sigma$. Analytic results for the long distance behavior of $V_{Coul}(\vec{x})$ have been presented in Ref. [18].

Summarizing [24], in the Gribov-Zwanziger confinement scenario (in Coulomb gauge), the long-range force, responsible for color confinement, is carried by an instantaneous static color-Coulomb field. In particular, the linearly rising potential is related to the IR divergence of the ghost-propagator (at equal time). At the same time, the propagator of three-dimensionally transverse (would-be physical) gluons is IR suppressed and the gluons are absent from the spectrum.

## THE CLASSICAL ERA: RESULTS

The analytic predictions described above for the gluon propagators $D^{tr}(\vec{k})$ and $D_{44}(\vec{k})$ have been verified for the $SU(2)$ group in Refs. [1]–[4]. In particular, from Fig. 1 of Ref. [1] it is evident that, in the IR limit, the transverse propagator is suppressed, while $D_{44}(\vec{k})$ blows up. Moreover, in the infinite-volume limit, it has been found [1, 2] that $D^{tr}(\vec{k})$ is well described a Gribov-like propagator characterized by a pair of purely imaginary poles $m^2 = \pm iy$. Numerically, at $\beta = 2.2$ and in the infinite-volume limit, one finds $y = 0.33 \pm 0.14\ \text{GeV}^2$. As for the ghost propagator $G(\vec{k},t)$, it has been studied up to now only in Ref. [4]. There, it is shown that $G(\vec{k},t)$ has indeed an IR divergence stronger than $1/k^2$. At the same time, the running coupling $g_{Coul}^2(\vec{k})$, defined in Eq. (3) above, seems to be consistent [3, 4] with an IR behavior of the type $1/k^2$. These analyses have also obtained $\sigma_{Coul} \approx \sigma$.

In Ref. [5], the color-Coulomb potential $V_{Coul}(\vec{x})$ has been evaluated [for the $SU(2)$ group][2] as a function of the separation $x$, using correlators of two time-like Wilson lines of length 1 (in lattice units). It was found that $V_{Coul}(\vec{x})$ increases linearly with $x$, in agreement with the $1/k^2$ behavior for $g^2_{Coul}(\vec{k})$ obtained in Refs. [3, 4]. However, in this case the estimate for the Coulomb string tension was $\sigma_{Coul} \approx 2 - 3\,\sigma$. Moreover, if one removes the so-called center vortices [5], then the color-Coulomb potential $V_{Coul}(\vec{x})$ goes to a constant at large $x$ and $\sigma_{Coul} = 0$. This suggests a strong relation between these center vortices and the (Coulomb) confinement mechanism. Note that similar effects have been observed in the gluon and in the ghost propagators in Landau gauge [25] after removing the center vortices.

It is also interesting that, when the temperature is turned on [7, 8], the color-Coulomb potential $V_{Coul}(\vec{x})$ is not screened and it is still a linearly rising function of $x$. Moreover, the Coulomb string tension $\sigma_{Coul}$ shows a magnetic-like behavior [8], i.e. $\sigma^{1/2}_{Coul} \sim g^2(T)T$. This implies that the Coulomb string tension cannot be used as an order parameter for confinement. This conclusion can be understood by observing that the temperature is defined by compactifying the time direction and that the Coulomb gauge is defined on the subspace orthogonal to the time direction. Thus, there is no reason for the system in Coulomb gauge to be sensitive to the deconfining transition.

## THE MODERN ERA

In Ref. [9] the authors evaluate the gauge-field excitation energy $\mathscr{E}$ (above the ground state energy) of a single (point-like) static color charge in Coulomb gauge. Considering that long-range effects should be related to the non-local-interaction term $\mathscr{H}_{Coul}$, one finds

$$\mathscr{E} \propto \mathscr{V}^{aa}(x,x) = \left[\mathscr{M}^{-1}(-\Delta)\mathscr{M}^{-1}\right]^{aa}(x,x) . \qquad (4)$$

A necessary condition for confinement is that $\mathscr{E}$ should diverge in the infinite-volume limit, due to IR effects. (Ultraviolet divergences are regulated by the lattice cut-off.) For the (Coulomb) FP matrix $\mathscr{M}^{ab} = -\delta^{ab}\Delta - f^{acb}A^c_\mu \partial_\mu$ one can consider (inside the Gribov region $\Omega_t$) the eigenvalues $\lambda > 0$ and the corresponding eigenfunctions $\Phi^a_{\lambda,x}$. Then, Eq. (4) can be written as $\mathscr{E} \propto \langle \sum_\lambda F_\lambda/\lambda^2 \rangle$ with $F_\lambda = V_s^{-1} \sum_{xy}(\Phi^a_{\lambda,x})^*(-\Delta)_{x,y}(\Phi^a_{\lambda,y})$. (Here, $V_s$ is the 3d spatial volume of the lattice.) For a sufficiently large volume, the sums can be approximated by integrals and $\mathscr{E} \propto \langle \int_{\lambda_{min}}^{\lambda_{max}} d\lambda\, \rho(\lambda)F(\lambda)/\lambda^2 \rangle$, with $\int d\lambda\, \rho(\lambda) = 1$. In the infinite-volume limit, the volume of the Gribov region gets concentrated near the Gribov horizon [9, 17, 26], i.e. $\lambda_{min} \to 0$. In the same limit, the gauge-field excitation energy $\mathscr{E}$ blows up if

$$\lim_{\lambda \to 0} \frac{\rho(\lambda)F(\lambda)}{\lambda} > 0 . \qquad (5)$$

---

[2] Similar results were obtained for the $SU(3)$ group in Refs. [7, 8].

25

Thus, a necessary condition for confinement is the enhancement of $\rho(\lambda)F(\lambda)$ at small momenta.

In Appendix A of Ref. [9], an interesting analysis based on a random-matrix model shows that, for small eigenvalues, one should have $\rho(\lambda) = c\lambda^{\alpha}$ if the eigenvalues $\lambda$ scale as $V_s^{-1/(1+\alpha)}$ if the volume is increased. Numerically they find $\alpha = 0.25(5)$, implying $\lambda_{min} \sim 1/L^{2.4}$. At the same time, they obtain $F(\lambda_{min}) \sim 1/L \sim \lambda_{min}^{0.38}$ and the confinement criterion (5) is clearly fulfilled [9, 10]. A similar result is obtained when considering the so-called "vortex-only" configurations [9]. On the other hand, after removing the center vortices, one recovers a Laplacian-like eigenvalue spectrum for the FP operator $\mathcal{M}_{xy}^{ab}$ with $\lim_{\lambda \to 0} \rho(\lambda)F(\lambda)/\lambda = 0$. Thus, in agreement with the findings reported in the previous Section, the enhancement of $\rho(\lambda)F(\lambda)$ at small eigenvalues $\lambda$ and the confinement mechanism in Coulomb gauge seem to be strictly related to the properties of the center-vortex configurations. One can also show [9] that center-vortex configurations are (infinitely many) distinguished points on the Gribov horizon. The relation between these configurations and the Gribov-Zwanziger scenario in Coulomb gauge is then clear if, in the infinite-volume limit, the center-vortex configurations are sufficiently dense on the Gribov horizon.

## AN OPEN QUESTION FOR THE FUTURE

In order to understand fully the Gribov-Zwanziger confinement scenario one should consider a generic gauge condition $\mathscr{F}[A] = 0$, imposed by minimizing a functional $E[U]$. Then, from the second variation of $E[U]$, we can always define the FP matrix $\mathcal{M}_{xy}^{ab}$. Clearly, when we are at a (local) minimum of $E[U]$, the (non-trivial) eigenvalues of $\mathcal{M}_{xy}^{ab}$ are positive and we can define a Gribov region $\Omega$ and the first Gribov horizon $\partial\Omega$. Moreover, since the configuration space has a very large dimensionality, entropy should favor (in the limit of large volumes) configurations near the Gribov horizon [9, 17, 26], i.e. $\lambda_{min}$ should go to zero in the same limit. This is indeed the case in 3d [27] and 4d Landau gauge [28], in 4d Coulomb gauge [9] and in 4d Maximally Abelian gauge (MAG) [29]. Since the FP matrix develops a null eigenvalue at the Gribov horizon $\partial\Omega$, we should also expect the corresponding ghost propagator $G(k)$ to blow up at small momenta in the infinite-volume limit. This result should in turn introduce a long-range effect in the theory, being probably related to the color-confinement mechanism. Indeed, we know from several numerical studies that the ghost propagator $G(k)$ is IR enhanced in 3d [27] and 4d Landau gauge [25, 30, 31] and in 4d Coulomb gauge [4]. On the other hand, recent numerical results in MAG [29] suggest an IR finite $G(k)$. Thus, the line of thinking reported above cannot be completely correct. Of course, one does not expect the ghost propagator to be particularly important for confinement in MAG, since in this case the accepted scenario is that confinement is related to Abelian dominance and (therefore) to the IR behavior of the diagonal gluon propagator [29, 32]. In any case, we should try to answer the following question: *what makes the ghost propagator IR enhanced in Coulomb and in Landau gauge but IR finite in MAG?*

A possible solution comes from the observation that in Landau [27, 28] and in Coulomb gauge [9] $\lambda_{min} \sim 1/L^{2+\alpha}$ with $\alpha > 0$, i.e. it goes to zero faster than in the case of the Laplacian. On the contrary, in MAG [29] one has $\lambda_{min} \sim 1/L^{2-\alpha}$ (with $\alpha > 0$), i.e. it goes to zero more slowly than for the Laplacian. This (unproven) hypothesis seems to be supported by the following observation. Using the same notation introduced in the previous section and in the limit of a large volume, we can write the ghost propagator $G(k)$ as

$$G(k) = \int_{\lambda_{min}}^{\lambda_{max}} d\lambda \, \frac{\rho(\lambda)f_\lambda(k)}{\lambda} \, , \qquad f_\lambda(k) = \frac{1}{N_c^2 - 1} \sum_a |\Phi_\lambda^a(k)|^2 \, . \qquad (6)$$

If we consider a FP matrix of the type $\mathcal{M}^{ab} = -\delta^{ab}\Delta - K^{ab}$ (this is the case in Landau, Coulomb and MAG) then we have

$$G(k) = \int_{\lambda_{min}}^{\lambda_{max}} \frac{d\lambda}{\lambda} \rho(\lambda) \frac{1}{N_c^2 - 1} \sum_a |\Phi_\lambda^a(k)|^2 \, , \quad \Phi_\lambda^a(k) = \frac{1}{k^2 - \lambda} \sum_{x,y} e^{-ikx} K_{xy}^{ab} \Phi_{\lambda,y}^b \, . \quad (7)$$

In a numerical simulation we look at $G(k_{min})$ when the volume increases (and $\lambda_{min}$ decreases). Thus, the IR behavior of $G(k_{min})$ depends on the quantity

$$\Phi_{\lambda_{min}}^a(k_{min}) = \frac{1}{k_{min}^2 - \lambda_{min}} \sum_{x,y} e^{-ik_{min}x} K_{xy}^{ab} \Phi_{\lambda_{min},y}^b \qquad (8)$$

and there is a clear competition between the smallest eigenvalue of the Laplacian $k_{min}^2 \sim L^{-2}$ and the smallest eigenvalue of the FP operator $\lambda_{min}$.

It is interesting to notice that using Eq. (6) we can easily explain why finite-size effects are small[3] when the ghost propagator $G(k)$ is evaluated numerically. Indeed, it is sufficient to have $\rho(\lambda)f_\lambda(k)/\lambda \sim \lambda^\beta$ with $\beta > -1$ in the limit of small eigenvalues $\lambda$. In Ref. [28] it has been obtained (for 4d Landau gauge) that the quantity $R(\overline{\lambda}) = \int_{\lambda_{min}}^{\overline{\lambda}} d\lambda \, \rho(\lambda)f_\lambda(k)\lambda^{-1}/G(k)$ behaves as $\overline{\lambda}^\nu$, with $\nu > 0$, for small $\overline{\lambda}$ considering the two smallest nonzero momenta $k$. This implies $\beta = \nu - 1 > -1$.

## CONCLUSIONS

We believe that the study of the spectral properties of the FP operator in different gauges can help us understand the general features of the Gribov-Zwanziger confinement scenario. In particular, it would be important to clarify for which gauge conditions the confinement mechanism can be related to an enhancement of the ghost propagator in the IR limit.

---

[3] Note that in Ref. [31] there are, actually, strong finite-size effects for the ghost propagator $G(k)$. However, in that case the effects are probably due to the use of strong asymmetric lattices, with different ratios of the spatial over the temporal extension of the lattice.

# ACKNOWLEDGMENTS

The author thanks the organizers for the invitation to present this review at QCHS7 and A. Maas, T. Mendes and D. Zwanziger for helpful discussions. This work was partially supported by FAPESP (under grants # 00/05047-5 and 06/57316-6) and by CNPq.

# REFERENCES

1. A. Cucchieri and D. Zwanziger, Phys. Rev. D **65**, 014001 (2002).
2. A. Cucchieri and D. Zwanziger, Phys. Lett. B **524**, 123 (2002); A. Cucchieri, T. Mendes and D. Zwanziger, Nucl. Phys. Proc. Suppl. **106**, 697 (2002).
3. A. Cucchieri and D. Zwanziger, Nucl. Phys. Proc. Suppl. **119**, 727 (2003).
4. K. Langfeld and L. Moyaerts, Phys. Rev. D **70**, 074507 (2004).
5. J. Greensite and S. Olejnik, Phys. Rev. D **67**, 094503 (2003).
6. J. Greensite, S. Olejnik and D. Zwanziger, Phys. Rev. D **69**, 074506 (2004).
7. A. Nakamura and T. Saito, Prog. Theor. Phys. **115**, 189 (2006).
8. Y. Nakagawa et al., Phys. Rev. D **73**, 094504 (2006).
9. J. Greensite, S. Olejnik and D. Zwanziger, JHEP **0505**, 070 (2005).
10. Y. Nakagawa et al., hep-lat/0610128.
11. M. Grady, hep-lat/0607013; M. Grady, hep-lat/0610042.
12. See for example *Particle Physics and Introduction to Field Theory*, T. D. Lee (Harwood Academic Publishers, New York, 1981).
13. V. N. Gribov, Nucl. Phys. B **139**, 1 (1978).
14. D. Zwanziger, Phys. Lett. B **257**, 168 (1991); D. Zwanziger, Nucl. Phys. B **364** (1991) 127.
15. D. Zwanziger, Nucl. Phys. B **518** (1998) 237; A. Cucchieri and D. Zwanziger, Phys. Rev. D **65**, 014002 (2002).
16. A. Cucchieri and D. Zwanziger, Nucl. Phys. Proc. Suppl. **106**, 694 (2002).
17. D. Zwanziger, Phys. Rev. Lett. **90**, 102001 (2003).
18. D. Zwanziger, Phys. Rev. D **70**, 094034 (2004).
19. Y. L. Dokshitzer and D. E. Kharzeev, Ann. Rev. Nucl. Part. Sci. **54**, 487 (2004); R. F. Sobreiro and S. P. Sorella, hep-th/0504095.
20. D. Zwanziger, Nucl. Phys. B **485**, 185 (1997); D. Zwanziger, hep-th/9710157; A. Cucchieri and D. Zwanziger, Nucl. Phys. Proc. Suppl. **53**, 815 (1997); A. Cucchieri and D. Zwanziger, Phys. Rev. Lett. **78**, 3814 (1997); A. P. Szczepaniak and E. S. Swanson, Phys. Rev. D **65**, 025012 (2002); A. P. Szczepaniak, Phys. Rev. D **69**, 074031 (2004); C. Feuchter and H. Reinhardt, Phys. Rev. D **70**, 105021 (2004); S. M. Antunes et al., Braz. J. Phys. **35** (2005) 877; A. P. Szczepaniak and P. Krupinski, Phys. Rev. D **73**, 034022 (2006); A. P. Szczepaniak and P. Krupinski, Phys. Rev. D **73**, 116002 (2006).
21. L. Baulieu and D. Zwanziger, Nucl. Phys. B **548**, 527 (1999); A. Cucchieri and T. Mendes, hep-lat/9902024; C. S. Fischer and D. Zwanziger, Phys. Rev. D **72**, 054005 (2005); A. Maas, A. Cucchieri and T. Mendes, arXiv:hep-lat/0610123.
22. R. Alkofer and L. von Smekal, Phys. Rept. **353**, 281 (2001).
23. W. Schleifenbaum, M. Leder and H. Reinhardt, Phys. Rev. D **73**, 125019 (2006).
24. See for example D. Zwanziger, hep-ph/0610021.
25. J. Gattnar, K. Langfeld and H. Reinhardt, Phys. Rev. Lett. **93**, 061601 (2004).
26. D. Zwanziger, Nucl. Phys. B **412**, 657 (1994); A. Cucchieri, Nucl. Phys. B **521**, 365 (1998).
27. A. Cucchieri, A. Maas and T. Mendes, Phys. Rev. D **74**, 014503 (2006).
28. A. Sternbeck, E. M. Ilgenfritz and M. Muller-Preussker, Phys. Rev. D **73**, 014502 (2006).
29. T. Mendes, A. Cucchieri and A. Mihara, hep-lat/0611002.
30. A. Cucchieri, Nucl. Phys. B **508**, 353 (1997); S. Furui and H. Nakajima, Phys. Rev. D **69**, 074505 (2004); J. C. R. Bloch et al., Nucl. Phys. B **687**, 76 (2004); A. Sternbeck et al., Phys. Rev. D **72**, 014507 (2005); A. Cucchieri and T. Mendes, Phys. Rev. D **73**, 071502 (2006).
31. O. Oliveira and P. J. Silva, arXiv:hep-lat/0609036.
32. Z. F. Ezawa and A. Iwazaki, Phys. Rev. D **25**, 2681 (1982); K. Amemiya and H. Suganuma, Phys. Rev. D **60**, 114509 (1999); V. G. Bornyakov et al., Phys. Lett. B **559**, 214 (2003).

# Localization properties of fermions and bosons

Philippe de Forcrand

*Institut für Theoretische Physik, ETH Zürich, CH-8093 Zürich, Switzerland*
*CERN, Physics Department, TH Unit, CH-1211 Geneva 23, Switzerland*

**Abstract.** The topological structure of the QCD vacuum can be probed by monitoring the spatial localization of the low-lying Dirac eigenmodes. This approach can be pursued on the lattice, and unlike the traditional one requires no smoothing of the gauge field. I review recent lattice studies, attempting to extract a consistent description. What emerges is a picture of the vacuum as a "topological sandwich" of alternating, infinitely thin $3d$ layers of opposite topological charge.

**Keywords:** Lattice QCD, vacuum structure, numerical simulations
**PACS:** 11.15.Ha, 12.38.Gc, 12.38.Aw, 11.30.Rd

"Understanding" confinement, by identifying the relevant infrared degrees of freedom of the gauge field, has been a long-standing theoretical goal. It is natural to associate this non-perturbative phenomenon with non-perturbative, topological excitations. The standard list of potentially relevant excitations consists of instantons, Abelian monopoles and center vortices, with co-dimension 4, 3, and 2 respectively. They each have received a fluctuating degree of attention over the years. One may hope that a proper lattice study may unambiguously identify the right excitation. However, in the past it has been necessary to filter out UV fluctuations of the gauge field in order to reveal the large-scale structure. This is accomplished by a smoothing/cooling/smearing procedure which reduces the action, and inevitably drives the gauge field towards an action minimum, i.e. a classical instanton solution. Recently, as outlined below, a different strategy has been followed, which avoids such bias. The localization properties of low-lying Dirac eigenmodes presumably tell us about the underlying gauge field excitations, responsible for chiral symmetry breaking and confinement.

## LOCALIZATION: ANDERSON AND DIAKONOV-PETROV

Anderson [1] considered the Hamiltonian $H = \Delta + V$, where $\Delta$ is a nearest-neighbor hopping operator (a discretized Laplacian) and $V$ a random potential. This Hamiltonian mimics that of a crystal doped with random impurities. An eigenmode $\psi(\vec{r})$ can be *localized*, meaning that $|\psi(\vec{r})|$ decays exponentially for large $|\vec{r}|$: the electron cannot hop to infinity, and this mode does not contribute to the electric conductivity of the material. Otherwise, the eigenmode is *extended*. Anderson showed that eigenmodes were always localized, if the disorder in $V$ was large enough, or the energy low enough. This is intuitively clear: at very high energy, the random potential plays no role and eigenmodes are extended, plane-wave-like; but at low energy, sufficient randomness in the potential may forbid hopping to any of the neighboring sites. Thus, the spectrum looks generically as in Fig. 1, with a *mobility edge* $\lambda_c$ separating the localized from the extended regime.

CP892, *Quark Confinement and the Hadron Spectrum VII*
edited by J. E. F. T. Ribeiro

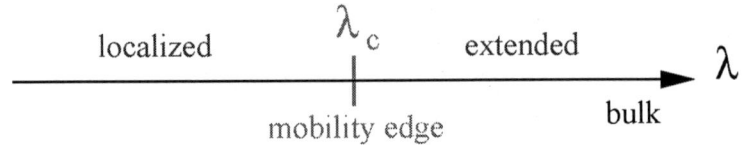

FIGURE 1.    Localization properties of boson eigenstates in a random medium

$\lambda_c$ may lie above or below the ground-state energy depending on the disorder, allowing for a so-called quantum transition at zero temperature.

Diakonov and Petrov [2] proposed an analogous explanation for the QCD chiral transition at temperature $T_c$. Since $\langle \bar{\psi}\psi \rangle = -\pi \rho(0)$ [3], the spectral properties of the Dirac operator, described by the density $\rho(\lambda)$, must change with temperature. Recalling that an instanton supports a chiral zero mode $\psi^I$ (for each fermion flavor), one considers the Dirac spectrum for a linear superposition of Instantons and Antiinstantons. The exact zero modes are now displaced. The infrared spectrum results from diagonalizing the effective Dirac matrix $\begin{pmatrix} 0 & T_{IA} \\ T_{IA}^\dagger & 0 \end{pmatrix}$ of overlap elements $T_{ij} = \langle \psi_i^I | \psi_j^A \rangle$ between individual Instantons and Antiinstantons zeromodes. At $T = 0$, this overlap decreases as $\frac{1}{|r_{ij}|^3}$, and gives rise to extended modes having support on all instantons and antiinstantons, and an essentially uniform spectrum near zero. Such extended modes can be observed on the lattice (see Fig. 2). As the temperature increases, the density of instantons decreases (their action $8\pi^2/g^2(T)$ increases) and $T_{ij}$ now decays exponentially $\sim \exp(-\pi r_{ij}T)$ in spatial directions. Both factors may trigger a transition to localization, which suppresses near-zero eigenvalues and restores chiral symmetry.

As in Anderson's bosonic case, this is a disorder-driven transition. But Dirac eigenvalues come in pure imaginary pairs $\pm i\lambda$, and the focus here is on the center of the spectrum $\lambda = 0$. Correspondingly, the transition can be modeled by *chiral* random matrices [5].

It is important to note that the Diakonov-Petrov scenario does not require instantons, but only chiral zero modes. It turns out that other topological defects - domain-walls, monopoles, vortices - also support chiral zero modes. Under the working assumption that extended modes have support on the union of topological defects, this opens the possibility to determine the topological vacuum structure from the spatial distribution of low-lying eigenmodes. This approach is gauge-invariant and requires no smoothing/cooling of the gauge field.

## LATTICE STUDIES

To characterize how localized or extended a mode $\psi(x)$ is, one uses the *inverse participation ratio* $IPR \equiv V \frac{\sum_x |\psi(x)|^4}{(\sum_x |\psi(x)|^2)^2}$. This ratio of moments is equal to 1 if $\psi(x) = \delta(x_0)$ is completely localized, and to $V$ if $\psi(x) = const. \; \forall x$ is completely delocalized. If $\psi(x) = 1$ on a fraction $f$ of the sites, 0 elsewhere, then $IPR = 1/f$, which justifies its name.

Of course one should consider the continuum limit $a \to 0$ of the lattice study. Note

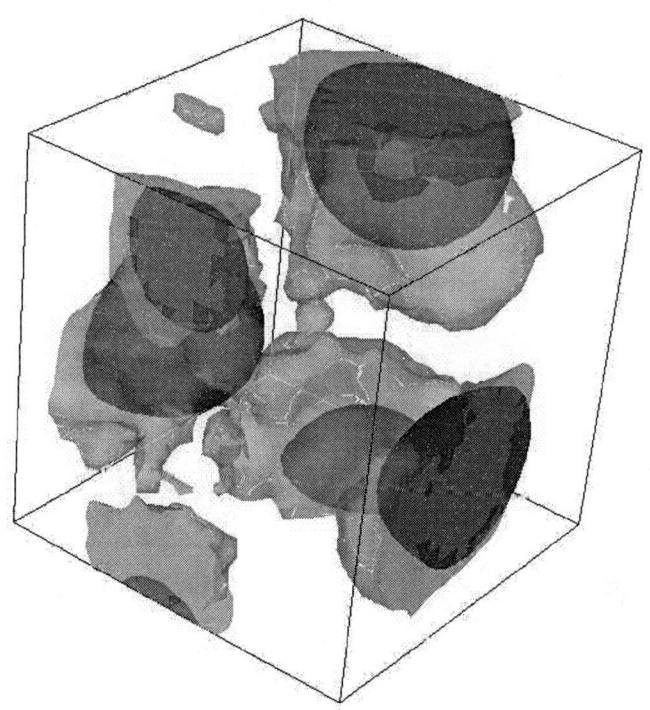

**FIGURE 2.** Magnitude of lowest-lying Dirac eigenmode in the presence of an Instanton-Antiinstanton pair, from [4]. The gauge field was cooled to identify the $I - A$ pair, but *not* to obtain the Dirac eigenmode.

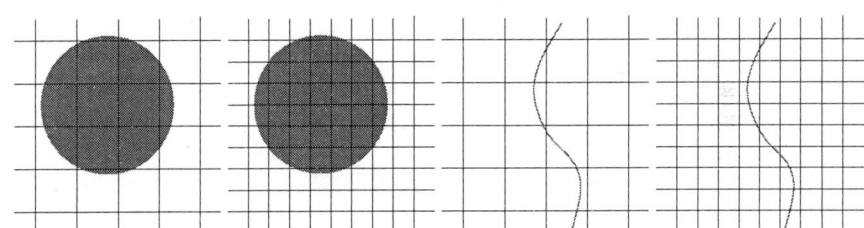

**FIGURE 3.** The fraction $f$ of occupied lattice sites, hence the *IPR*, remains unchanged as $a \rightarrow 0$ in the presence of *any* thick object (*left*). It goes to zero as $a^{d-4}$ for a thin object of dimension $d$ (*right*).

that the fraction $f$ of occupied sites, hence the IPR, remains constant as $a \rightarrow 0$ for any type of macroscopic object. Conversely, if the object is "thin" and lives on a submanifold of dimension $d$ and volume $\mathcal{V}_d$, then $f = \frac{\mathcal{V}_d/a^d}{V/a^4} \sim a^{4-d}$, and $IPR \sim a^{d-4}$, which diverges as $a \rightarrow 0$ (see Fig. 3). $d = 0, 1, 2$ then characterizes "thin" instantons (point-like), monopoles (line-like) and vortices (surface-like).

• The first study, by the MILC collaboration [6] for $SU(3)$, indicated $d \sim 3$: the vacuum, it seems, is made of infinitely thin domain-walls! (see Fig. 4, left).

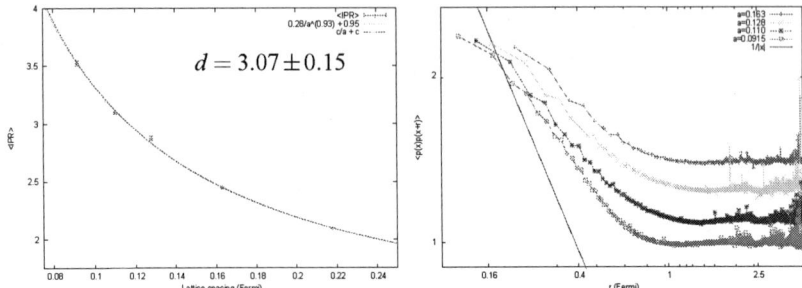

**FIGURE 4.** (*left*) Most recent *IPR* data from MILC Collaboration [6] versus lattice spacing, for *SU*(3) with Asqtad Dirac operator and Symanzik gauge action. The dimension of the supporting manifold is about 3. (*right*) Correlator $\langle |\psi(0)| |\psi(\vec{x})| \rangle$ of magnitude of low-lying Dirac eigenmode, for different lattice spacings. The straight line shows what would happen if $|\Psi| = 1$ on a $3d$ fractal, 0 elsewhere.

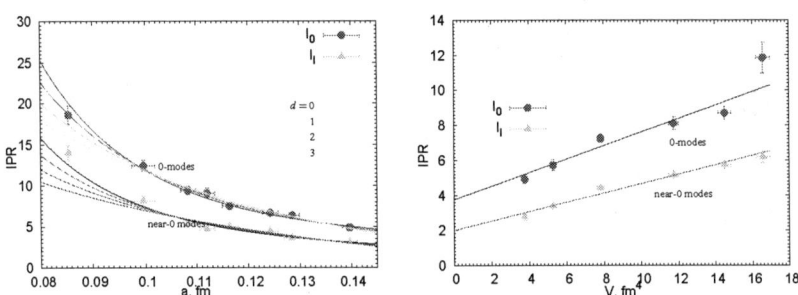

**FIGURE 5.** *IPR* versus lattice spacing (*left*) and volume (*right*), for *SU*(2) with overlap Dirac operator and Wilson gauge action [8].

- This surprising result was quickly checked in [8]. There, the IPR diverges even faster, favoring point-like instantons (see Fig. 5). That study used the overlap discretization of the Dirac operator, with exact chiral properties, but used the Wilson action for the gauge group $SU(2)$. Unfortunately, it is known since [7] that with this choice of gauge action, the action of size $a$ defects ("dislocations") is insufficient to compensate for their entropy $\sim \log a^{-4}$, and they become dense as $a \to 0$. The observations of [8] may then be entirely caused by lattice artifacts.

- A third study, also with the overlap Dirac operator, but for $SU(3)$ and Lüscher-Weisz gauge action, shows results (Fig. 6, left) roughly consistent with [6]. The IPR for the low-lying modes grows more or less as $1/a$, consistent again with thin domain-wall structures (Fig. 6, right).

It should not have come as a surprise that the IPR diverges. Even if classical lumps of size $1/\Lambda_{QCD}$ are relevant to the QCD vacuum structure, these lumps do not look smooth at small distance due to quantum fluctuations (see Fig. 7). Besides, this result can be related to a curious property of the topological charge density operator $\langle q(0)q(\vec{x}) \rangle$ [10]. On one hand, this correlator is negative for any $\vec{x} \neq 0$. This can be seen from reflection-positivity, or simply by realizing that $q(\vec{x}) \sim \vec{E} \cdot \vec{B}$ acquires an extra factor $i$ in Euclidean

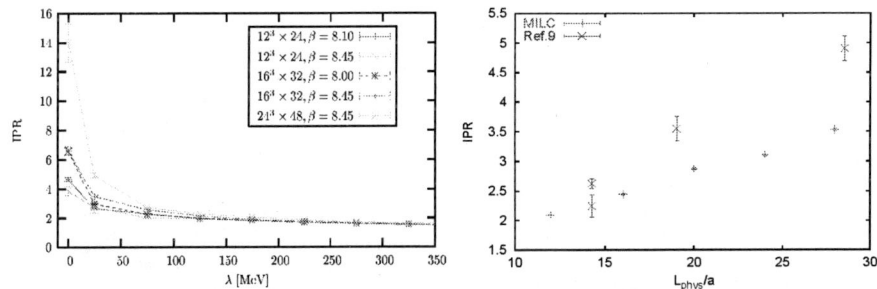

**FIGURE 6.** (*left*) *IPR* versus eigenvalue and lattice spacing, for $SU(3)$ with overlap Dirac operator and Lüscher-Weisz gauge action [9]. (*right*) For the lowest non-zero modes, the *IPR* grows $\propto 1/a$ as for MILC.

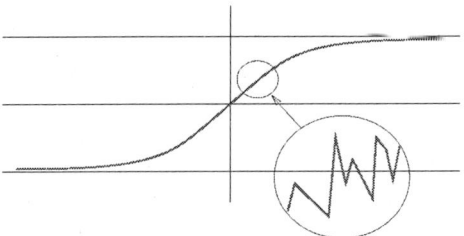

**FIGURE 7.** A classical kink is smooth. The actual kink looks like a random walk at short distances.

space. On the other hand, $\langle \int d^4x\, q(0)q(\vec{x}) \rangle = \chi_{\text{top}}$ is finite and about $(190\text{MeV})^4$. Hence, a contact term $\delta(\vec{x})$ must compensate the negative $\vec{x} \neq 0$ integral. Moreover, the canonical dimension of $q(\vec{x})$ is 4, so that one expects $\langle \int_{\vec{x} \neq 0} d^4x\, q(0)q(\vec{x}) \rangle$ to diverge. This divergence is exactly compensated by the contact term to leave the finite piece $\chi_{\text{top}}$.

This behaviour has recently been observed on the lattice: compare Fig. 8 left and right.

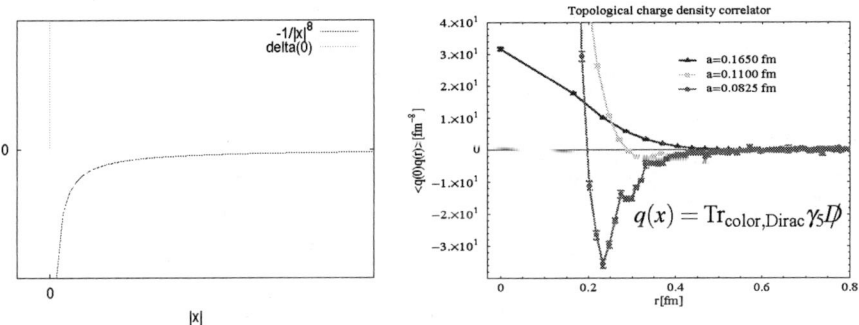

**FIGURE 8.** Correlator of the topological charge density: (*left*) as expected in the continuum and (*right*) as measured on the lattice [11]. The negative part and divergent contact term both become visible as $a \to 0$.

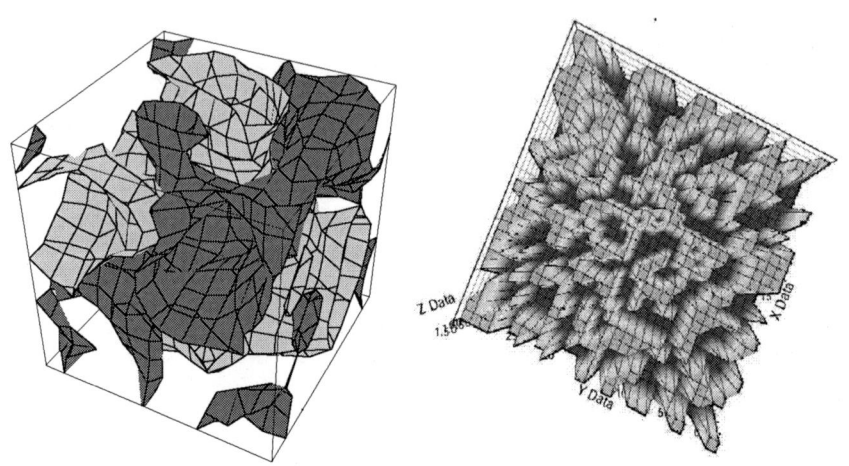

**FIGURE 9.** Topological charge density: *(left)* in QCD [9] and *(right)* in the $(1+1)dCP^3$ model [12].

One has then to understand how the vacuum becomes "topologically antiferromagnetic", with $\langle q(0)q(\vec{x})\rangle$ going to $-\infty$ as $|\vec{x}| \rightarrow 0$. The Kentucky group [11] claims to see space-filling $3d$ structures of transverse size $\mathcal{O}(a)$ and of opposite topological charges. Analogous structures are seen in Ref. [9], and also in the $CP^3$ model [12] (see Fig. 9).

As a final check, the MILC collaboration has measured the correlator $\langle |\psi(0)||\psi(\vec{x})|\rangle$, where $\psi$ is a low-lying Dirac eigenmode. If $|\psi(\vec{x})| = 1$ on a $3d$ fractal and 0 elsewhere, this correlator would decay as $1/|\vec{x}|$. Measurements [6] are not too different (Fig. 4, right), until $|\vec{x}|$ reaches distances $\mathcal{O}(1/\Lambda_{QCD})$.

• Most recently, this kind of lattice study has been extended to bosonic fields and eigenmodes of the covariant Laplacian in various representations [13], with yet more surprises. In the adjoint representation, the mobility edge is found to rise unexpectedly as $1/a$ (see Fig. 10). If true, all finite-energy modes in the continuum would be localized, contradicting perturbation theory. One wonders if this strange result also can be blamed on the dense dislocations caused by the choice of Wilson action for gauge group $SU(2)$.

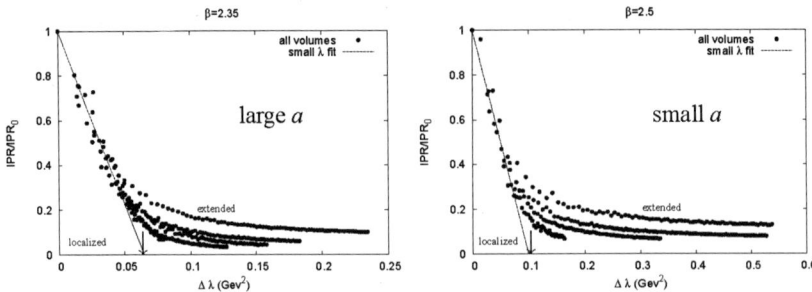

**FIGURE 10.** The mobility edge (arrow) for an adjoint Higgs appears to go to $\infty$ as $a \rightarrow 0$ [13], for $SU(2)$ with Wilson gauge action.

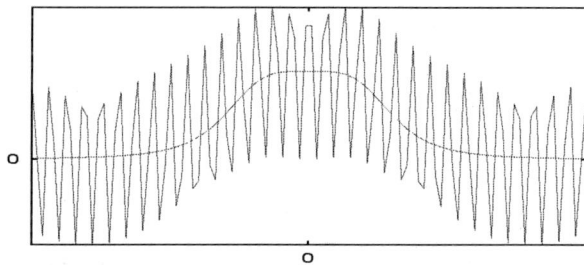

**FIGURE 11.** The "topological sandwich" structure observed at short distance (blue) is not inconsistent with another structure (red) at scale $1/\Lambda_{QCD}$.

## CONCLUSION

The picture which emerges (Figs. 4, 6, 8, 9) is that of a "topological sandwich", with alternating, infinitely thin $3d$ layers of opposite topological charge density. While bizarre, this picture is not forbidden, and is supported by some theoretical arguments [10]. But it is far from the usual picture of a dilute gas of classical excitations. One should stress that these two descriptions are not mutually exclusive, as sketched Fig. 11: one applies at UV scales, the other may apply at scales $\mathscr{O}(1/\Lambda_{QCD})$. The structure of the QCD vacuum depends on the scale considered.

## REFERENCES

1.  P. W. Anderson, Phys. Rev. **109** (1958) 1492.
2.  D. Diakonov and V. Y. Petrov, Nucl. Phys. B **272** (1986) 457; D. Diakonov, arXiv:hep-ph/9602375.
3.  T. Banks and A. Casher, Nucl. Phys. B **169** (1980) 103.
4.  P. de Forcrand, M. Garcia Perez, J. E. Hetrick, E. Laermann, J. F. Lagae and I. O. Stamatescu, Nucl. Phys. Proc. Suppl. **73** (1999) 578 [arXiv:hep-lat/9810033].
5.  A. M. Garcia-Garcia and J. C. Osborn, Phys. Rev. Lett. **93** (2004) 132002 [arXiv:hep-th/0312146].
6.  C. Aubin *et al.* [MILC Collaboration], Nucl. Phys. Proc. Suppl. **140** (2005) 626 [arXiv:hep-lat/0410024]; C. Bernard *et al.*, PoS **LAT2005** (2006) 299 [arXiv:hep-lat/0510025].
7.  D. J. R. Pugh and M. Teper, Phys. Lett. B **224** (1989) 159.
8.  F. V. Gubarev, S. M. Morozov, M. I. Polikarpov and V. I. Zakharov, arXiv:hep-lat/0505016; M. I. Polikarpov, F. V. Gubarev, S. M. Morozov and V. I. Zakharov, PoS **LAT2005** (2006) 143 [arXiv:hep-lat/0510098].
9.  Y. Koma, E. M. Ilgenfritz, K. Koller, G. Schierholz, T. Streuer and V. Weinberg, PoS **LAT2005** (2006) 300 [arXiv:hep-lat/0509164]; V. Weinberg, E. M. Ilgenfritz, K. Koller, Y. Koma, G. Schierholz and T. Streuer, arXiv:hep-lat/0610087.
10. E. Seiler and I. O. Stamatescu, MPI-PAE/PTh 10/87; E. Vicari, Nucl. Phys. B **554** (1999) 301 [arXiv:hep-lat/9901008]; E. Seiler, Phys. Lett. B **525** (2002) 355 [arXiv:hep-th/0111125].
11. I. Horvath *et al.*, Phys. Lett. B **617** (2005) 49 [arXiv:hep-lat/0504005].
12. S. Ahmad, J. T. Lenaghan and H. B. Thacker, Phys. Rev. D **72** (2005) 114511 [arXiv:hep-lat/0509066].
13. J. Greensite, S. Olejnik, M. Polikarpov, S. Syritsyn and V. Zakharov, Phys. Rev. D **71** (2005) 114507 [arXiv:hep-lat/0504008]; J. Greensite, A. V. Kovalenko, S. Olejnik, M. I. Polikarpov, S. N. Syritsyn and V. I. Zakharov, Phys. Rev. D **74** (2006) 094507 [arXiv:hep-lat/0606008].

# Restoration of QCD classical symmetries in excited hadrons

## L. Ya. GLOZMAN

*Institute for physics/Theoretical physics, University of Graz,*
*Universitätsplatz 5, A-8010 Graz, Austria*

**Abstract.** Restoration of chiral and $U(1)_A$ symmetries in excited hadrons is reviewed. A connection of these restorations with the semiclassical regime in highly excited hadron is discussed. A solvable confining field-theoretical toy model that exhibits chiral restoration is presented. Implications of the string description of the highly excited hadrons that suggests an additional dynamical symmetry of the spectra on the top of $U(2)_L$ $U(2)_R$ are presented.

**Keywords:** Chiral symmetry, Hadrons, QCD, Strings
**PACS:** 11.30.Rd, 12.38.Aw, 14.40.-n

## PHENOMENOLOGICAL FACTS, CLASSIFICATION AND DEFINITIONS

The experimental spectrum of excited hadrons, both baryons [1, 2, 3] and mesons [4, 5] suggests that the highly excited hadrons in the $u, d$ sector fall into approximate multiplets of $SU(2)_L \times SU(2)_R$ and $U(1)_A$ groups that are compatible with the Poincaré invariance, for a short overview see ref. [6]. If confirmed by discovery of still missing states, this phenomenon is referred to as effective chiral symmetry restoration or chiral symmetry restoration of the second kind.

It is important to precisely characterize what is implied under effective restoration, because sometimes it was (is) erroneously interpreted in the sense that the highly-excited hadrons are in the Wigner-Weyl mode. This confusion was a source of a controversy [7, 8], which has been overcome, however [9]. The mode of symmetry is defined only by the properties of the vacuum. If symmetry is spontaneously broken in the vacuum, then it is the Nambu-Goldstone mode and the *whole* spectrum of excitations on the top of the vacuum is in the Nambu-Goldstone mode. However, it may happen that the role of the chiral symmetry breaking condensates of the vacuum becomes progressively irrelevant in excited states. This means that the chiral symmetry breaking effects (dynamics) become less and less important in the highly excited hadrons and asymptotically the states approach the regime where their properties are determined by the underlying unbroken chiral symmetry (i.e. by the symmetry in the Wigner-Weyl mode). One of the particular consequences of the chiral symmetry restoration in excited hadrons is that they should gradually decouple from the Goldstone bosons [3, 8, 9, 10, 11, 12]. A hint for such a decoupling is indeed observed phenomenologically since the coupling constant for $h^* \to h + \pi$ decreases very fast higher in the spectrum (because a decay width increases with the mass of the resonance much slower than the phase space factor).

By definition this effective chiral symmetry restoration means the following. All hadrons that are created by the given interpolator, $J_\alpha$, appear as intermediate states in the two-point correlator,

$$\Pi = \imath \int d^4x \, e^{\imath qx} \langle 0|T\{J_\alpha(x)J_\alpha^\dagger(0)\}|0\rangle. \tag{1}$$

Consider two interpolating fields $J_1(x)$ and $J_2(x)$ which are connected by a chiral transformation (or by a $U(1)_A$ transformation), $J_1(x) = U J_2(x) U^\dagger$. Then, in the Wigner-Weyl mode, $U|0\rangle = |0\rangle$, it follows from (1) that the spectra created by the operators $J_1(x)$ and $J_2(x)$ would be identical. We know that in QCD one finds $U|0\rangle \neq |0\rangle$. As a consequence the spectra of the two operators must be in general different. However, it happens that the noninvariance of the vacuum becomes unimportant (irrelevant) high in the spectrum. Then the spectra of both operators become close at large masses and asymptotically identical. One could say, that the valence quarks in high-lying hadrons *decouple* from the quark condensate of the vacuum.

More precisely the effective symmetry restoration is defined to occur if two conditions are satisfied: (i) the states fall into approximate multiplets of $SU(2)_L \times SU(2)_R$ (and of $U(1)_A$) and the splittings within the multiplets ($\Delta M = M_+ \quad M_-$) vanish at $n \to \infty$ and/or $J \to \infty$ ; (ii) the splitting within the multiplet is much smaller than between the two subsequent multiplets [4, 5, 6]. This definition is very restrictive, because the structure of the chiral multiplets is nontrivial and is different for mesons with different spins. The latter is a consequence of the requirement that the chiral multiplets must satisfy the Poincaré invariance [5]. In particular, given the set of the standard quantum numbers $I, J^{PC}$ the meson multiplets of $SU(2)_L \times SU(2)_R$ are

**J = 0**

$$
\begin{aligned}
(1/2,1/2)_a &\; : \; 1,0^{-+} \longleftrightarrow 0,0^{++} \\
(1/2,1/2)_b &\; : \; 1,0^{++} \longleftrightarrow 0,0^{-+},
\end{aligned} \tag{2}
$$

**J = 2k,  k=1,2,...**

$$
\begin{aligned}
(0,0) &\; : \; 0,J^{--} \longleftrightarrow 0,J^{++} \\
(1/2,1/2)_a &\; : \; 1,J^{-+} \longleftrightarrow 0,J^{++} \\
(1/2,1/2)_b &\; : \; 1,J^{++} \longleftrightarrow 0,J^{-+} \\
(0,1) \oplus (1,0) &\; : \; 1,J^{++} \longleftrightarrow 1,J^{--}
\end{aligned} \tag{3}
$$

**J = 2k-1,  k=1,2,...**

$$\begin{aligned}
(0,0) &: \quad 0,J^{++} \longleftrightarrow 0,J^{--} \\
(1/2,1/2)_a &: \quad 1,J^{+-} \longleftrightarrow 0,J^{--} \\
(1/2,1/2)_b &: \quad 1,J^{--} \longleftrightarrow 0,J^{+-} \\
(0,1) \oplus (1,0) &: \quad 1,J^{--} \longleftrightarrow 1,J^{++}
\end{aligned} \tag{4}$$

The $U(1)_A$ symmetry connects the opposite parity states with the same isospin from the distinct $(1/2,1/2)_a$ and $(1/2,1/2)_b$ multiplets of $SU(2)_L \times SU(2)_R$.

The recent data on highly excited mesons from the $\bar{p}p$ annihilation at LEAR [13, 14] do support the $SU(2)_L \times SU(2)_R$ and $U(1)_A$ restorations as can be seen from the high-lying $\bar{n}n$ mesons with $J = 2$.

**(0,0)**

| $\omega_2(0,2^{--})$ | $f_2(0,2^{++})$ |
|---|---|
| $1975 \pm 20$ | $1934 \pm 20$ |
| $2195 \pm 30$ | $2240 \pm 15$ |

$(1/2,1/2)_a$

| $\pi_2(1,2^{-+})$ | $f_2(0,2^{++})$ |
|---|---|
| $2005 \pm 15$ | $2001 \pm 10$ |
| $2245 \pm 60$ | $2293 \pm 13$ |

$(1/2,1/2)_b$

| $a_2(1,2^{++})$ | $\eta_2(0,2^{-+})$ |
|---|---|
| $2030 \pm 20$ | $2030 \pm ?$ |
| $2255 \pm 20$ | $2267 \pm 14$ |

**(0,1)+(1,0)**

| $a_2(1,2^{++})$ | $\rho_2(1,2^{--})$ |
|---|---|
| $1950^{+30}_{-70}$ | $1940 \pm 40$ |
| $2175 \pm 40$ | $2225 \pm 35$ |

Note, that the chiral symmetry requires a doubling of some of the radial and angular Regge trajectories for $J > 0$. This is a highly nontrivial prediction of chiral symmetry. For example, asymptotically some of the $\rho$-mesons lie on the trajectory that is characterized by the chiral index $(0,1) \oplus (1,0)$ and have as their chiral partners the $a_1$ mesons, while the other $\rho$-mesons have $h_1$ mesons as their chiral partners and lie on the other independent trajectory with the chiral index $(1/2,1/2)_b$.

If we look carefully at the data one notices that all possible different chiral multiplets with the same $J$ are approximately degenerate [5]. Then it means that all these states fall into a reducible representation

$$(0,1/2) \oplus (1/2,0)] \times [(0,1/2) \oplus (1/2,0) \tag{5}$$

which combines all possible representations for the system of massless quark and anti-quark. Such a degeneracy is consistent with the view of the excited hadron as a string with massless quarks with definite chirality at the end points of the string [3].

There are still some missing states in the multiplets with $J = 0, 1, 3, 4$ [4, 5] and it is a very important experimental task to find them. This can be done in particular with the polarized target formation experiment in $\bar{p}p$ at the NESR low-energy antiproton ring at GSI, which will have similar or better characteristics than LEAR.

## ORIGINS OF CHIRAL AND $U(1)_A$ RESTORATIONS

An important question is a physical origin of chiral and $U(1)_A$ restorations. If the spectrum is strictly continuous and the function $R$ approaches a constant value at large $s$, then the asymptotic freedom at large space-like momenta together with a dispersion relation do allow us to claim that the chiral symmetry is manifest in such a spectral function, as it is observed e.g. in $e^+e^- \rightarrow jets$. However, it is a trivial case and not what we actually need. We have to consider a (quasi)discrete spectrum where the given hadron state is isolated. The conjecture of ref. [2] was that may be the chiral restoration is true in the regime where the spectrum is quasidiscrete and saturated mainly by resonances.

One would expect that the Operator Product Expansion (OPE) could help us to find the correct spectrum of the high-lying hadrons. This is not so, however. This is because the OPE is only an asymptotic expansion. While such a kind of expansion is very useful in the space-like region, it does not define any analytical solution which could be continued to the time-like region. This means that while the real (correct) spectrum of QCD must be consistent with the OPE, there is an infinite amount of incorrect spectra that can also be consistent with the OPE. Then, if one wants to get some information about the spectrum, one needs to assume something else on the top of the OPE. Clearly a result then is crucially dependent on these additional assumptions, for the recent activity in this direction see refs. [15, 16, 17]. This implies that in order to really understand chiral symmetry restoration one needs a microscopic insight and theory that would incorporate *at the same time* chiral symmetry breaking and confinement.

A fundamental insight into phenomenon can be obtained from the semiclassical expansion of the functional integral directly in the time-like region [6]. We know that the axial anomaly as well as the spontaneous breaking of chiral symmetry in QCD is an effect of quantum fluctuations of the quark field. The latter can generally be seen from the definition of the quark condensate, which is a closed quark loop. This closed quark loop explicitly contains a factor $\hbar$. The chiral symmetry breaking, which is necessarily a nonperturbative effect, is actually a (nonlocal) coupling of a quark line with the closed quark loop, which is a graphical representation of the Schwinger-Dyson (gap) equation.

At large $n$ (radial quantum number) or at large angular momentum $J$ we know that in quantum systems the *semiclassical* approximation *must* work. Physically this approximation applies in these cases because the de Broglie wavelength of particles in the system is small in comparison with the scale that characterizes the given problem. In such a system as a hadron the scale is given by the hadron size while the wavelength of valence quarks is given by their momenta. Once we go high in the spectrum the size of hadrons increases as well as the typical momentum of valence quarks.

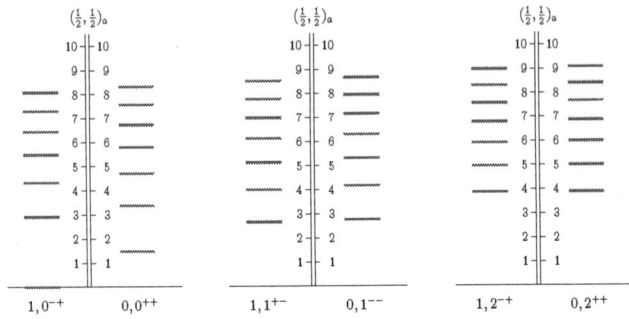

**FIGURE 1.** J=0,1 and 2 spectra for mesons in $(1/2, 1/2)_a$ representations.

The semiclassical approximation applies when the action in the system $S \gg \hbar$. In this case the whole amplitude (path integral) is dominated by the classical path (stationary point) and those paths that are infinitesimally close to the classical path. In other words, in the semiclassical case the quantum fluctuations effects are strongly suppressed and vanish asymptotically. Then the correlation function can be expanded in powers of $\hbar/S$. The leading contribution is a tree-level contribution to the path integral and keeps chiral symmetries of the classical Lagrangian. It contains no quantum fluctuations of the valence quark lines. Its contribution is of the order $(\hbar/S)^0$. The quantum fluctuations of the quark lines as well as the vacuum fermion loops contribute at the subleading orders in $(\hbar/S)$ and hence are suppressed in hadrons with large intrinsic action $S$. Then it follows that in a hadron with large enough radial quantum number $n$ or $J$, where action is large, the loop contributions must be relatively suppressed and vanish asymptotically. Hence in such systems both the chiral and $U(1)_A$ symmetries should be approximately restored. This is precisely what we see phenomenologically. Note that the semiclassical expansion is not an expansion in the coupling constant, which is large in the nonperturbative regime.

## A SOLVABLE TOY MODEL

While the argument presented above is general and solid enough, a detailed microscopical picture is missing. Then to see how all this works one needs a solvable field-theoretical model. The model must be chirally symmetric and confining. Such a model is known [22, 23, 24]. This model can be considered as a generalization of the large $N_c$ 't Hooft model (QCD in 1+1 dimensions) [25] to 3+1 dimensions. In both models the only interaction between quarks is the instantaneous infinitely raising Lorentz-vector linear potential. Then chiral symmetry breaking is described by the standard summation of the valence quarks self-interaction loops (the Schwinger-Dyson or gap equations), while mesons are obtained from the Bethe-Salpeter equation for the quark-antiquark bound states. Restoration of chiral symmetry in excited heavy-light mesons has been previously studied with the quadratic confining potential [19].

40

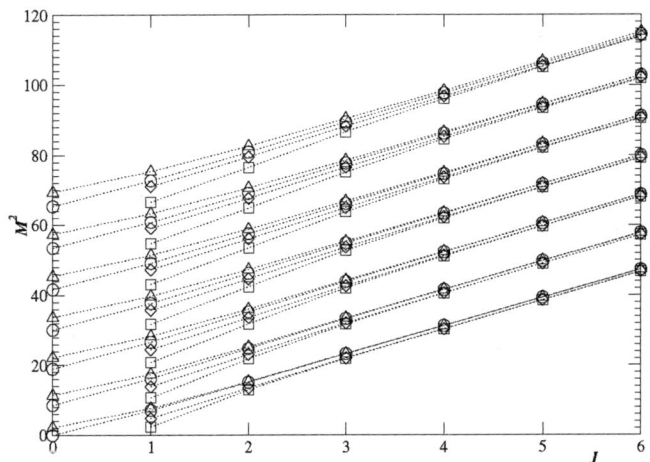

**FIGURE 2.** Angular Regge trajectories for isovector mesons with $M^2$ in units of $\sigma$. Mesons of the chiral multiplet $(1/2, 1/2)_a$ are indicated by circles, of $(1/2, 1/2)_b$ by triangles, and of $(0, 1) \oplus (1, 0)$ by squares ($J^{++}$ and $J^{--}$ for even and odd $J$, respectively) and diamonds ($J^{--}$ and $J^{++}$ for even and odd $J$, respectively).

The results for excited light-light mesons with the linear potential are reported in ref. [26] and presented in Fig. 1. The excited states fall into approximate chiral multiplets and a very fast restoration of both $SU(2)_L \times SU(2)_R$ and $U(1)_A$ symmetries with increasing of $J$ and essentially more slow restoration with increasing of $n$ is seen. In Fig. 2 the angular Regge trajectories are shown. They exhibit deviations from the linear behavior. This fact is obviously related to the chiral symmetry breaking effects for lower mesons. In the limit $n \to \infty$ and/or $J \to \infty$ one observes a complete degeneracy of all multiplets, which means that the states fall into representation (5). This means that in this limit the quark loop effects disappear completely [6, 18].

A few comments about physics which is behind these results. The chiral symmetry breaking Lorentz-scalar dynamical mass of quarks arises via loop dressing of quarks and represents effects of quantum fluctuations of the quark field [18]. A key feature of this dynamical mass is that it is strongly momentum dependent and vanishes at large quark momenta. When one increases excitation energy of a hadron, one also increases a typical momentum of valence quarks. Consequently, the chiral symmetry violating Lorentz-scalar dynamical mass of quarks becomes small and asymptotically vanishes in highly excited hadrons. Hence, chiral and $U(1)_A$ symmetries get approximately restored [1, 18, 10].

Exactly the same reason implies a decoupling of these hadrons from the Goldstone bosons [3, 10]. The coupling of the Goldstone bosons to the valence quarks is regulated via the axial current conservation by the Goldberger-Treiman relation. Then the coupling constant must be proportional to the Lorentz-scalar dynamical mass of valence quarks and vanishes at larger momenta. This represents a microscopical mechanism of decoupling which is required by the general considerations of chiral symmetry in the

Nambu-Goldstone mode [8, 11, 12].

## CHIRAL MULTIPLETS AND STRING

There are certain phenomenological evidences that the chiral multiplets of excited baryons and mesons with *different* spins cluster at some energies [3, 27, 12, 28]. This implies that one observes higher symmetry that includes chiral $U(2)_L \times U(2)_R$ as a subgroup. Presumably this additional degeneracy reflects a dynamical symmetry of the string. Indeed, like the Veneziano amplitude, the spectrum of the open bosonic string is degenerate with respect to the orbital momentum of the string [29]. A mathematical description of a hadron as a string with quarks at the ends that have definite chirality [3] is an open question.

This work was supported by the Austrian Science Fund (projects P16823-N08 and P19168-N16).

## REFERENCES

1. L. Ya. Glozman, Phys. Lett. B **475**, 329 (2000).
2. T. D. Cohen and L. Ya. Glozman, Phys. Rev. D **65**, 016006 (2002); Int. J. Mod. Phys. A **17**, 1327 (2002).
3. L. Ya. Glozman, Phys. Lett. B **541**, 115 (2002).
4. L. Ya. Glozman, Phys. Lett. B **539**, 257 (2002).
5. L. Ya. Glozman, Phys. Lett. B **587**, 69 (2004).
6. L. Ya. Glozman, Int. J. Mod. Phys. A. **21**, 475 (2006).
7. R. L. Jaffe, D. Pirjol, A. Scardicchio, Phys. Rev. Lett. **96**, 121601 (2006).
8. T. D. Cohen and L. Ya. Glozman, Mod. Phys. Lett. **A 21**, 1939 (2006).
9. R. L. Jaffe, D. Pirjol, A. Scardicchio, Phys. Rev. **D74**, 057901 (2006).
10. L. Ya. Glozman, A. V. Nefediev, Phys. Rev. D **73**, 074018 (2006).
11. R. L. Jaffe, D. Pirjol, A. Scardicchio, hep-ph/0602010.
12. M. Shifman, A. Vainstein, to be published.
13. A. V. Anisovich et al, Phys. Lett. **B491**, 47 (2000); **B517**, 261 (2001); **B542**, 8 (2002); **B542**, 19 (2002); **B513**, 281 (2001).
14. D. V. Bugg, Phys. Rep. **397**, 257 (2004).
15. S.S. Afonin et al, J. High. Energy. Phys. **04**, 039 (2004).
16. M. Shifman, hep-ph/0507246.
17. O. Cata, M. Golterman, S. Peris, hep-ph/0602194.
18. L. Ya. Glozman, A. V. Nefediev, J.E.F.T. Ribeiro, Phys. Rev. D **72**, 094002 (2005).
19. Yu. S. Kalashnikova, A. V. Nefediev, J.E.F.T. Ribeiro, Phys. Rev. D **72**, 034020 (2005).
20. T. DeGrand, Phys. Rev. D **64**, 074024 (2004).
21. T. D. Cohen, hep-ph/0605206.
22. A. Le Yaouanc, L. Oliver, O. Pene, and J. C.Raynal, Phys. Rev. D **29**, 1233 (1984); **31**, 137 (1985).
23. P. Bicudo and J. E. Ribeiro, Phys. Rev. D **42**, 1611 (1990); **42**, 1625 (1990).
24. F. J. Llanes-Estrada and S. R. Cotanch, Phys. Rev. Lett., **84**, 1102 (2000)
25. G. t't Hooft, Nucl. Phys. B **75**, 461 (1974).
26. R. F. Wagenbrunn, L. Ya. Glozman, hep-ph/0605247.
27. S. S. Afonin, hep-ph/0606310
28. A. Zaitsev, Plenary talk at ICHEP' 06 (Moscow); V. A. Rubakov, Conclusions and Outlook at ICHEP' 06 (Moscow)
29. J. Polchinski, String Theory, Cambridge University Press, 1998

# The Color Glass Condensate

Edmond Iancu

*Service de Physique Théorique, CE Saclay, F-91191 Gif-sur-Yvette, France*
*E-mail: iancu@spht.saclay.cea.fr*

**Abstract.** I give a brief overview of recent theoretical developments within perturbative QCD concerning the high–energy dynamics in the vicinity of the unitarity limit.

## MOTIVATION: GLUONS AT HERA

One of the main motivations for the recent developments in QCD at high energy comes from the experimental results at HERA, which suggest that, when increasing the energy, QCD evolves towards a regime of *high parton density* and thus of *weak coupling*. As visible, e.g., on the H1 data shown in the l.h.s. of Fig. 1, the gluon distribution $xG(x, Q^2)$ rises very fast when decreasing $x$ at fixed $Q^2$ (roughly, as a power of $1/x$), and also when increasing $Q^2$ at fixed $x$. (Recall that $x \simeq Q^2/s$, with $Q^2$ the virtuality of the photon exchanged in the lepton–proton collision, and $s$ the invariant energy squared.) Physically, $xG(x, Q^2)$ is the number of gluons in the proton wavefunction which are localized within an area $\Delta x_\perp \sim 1/Q^2$ in the transverse plane and carry a fraction $x = k_z/P_z$ of the proton longitudinal momentum. Thus, the results at HERA imply the physical picture illustrated in the right plot in Fig. 1: The number of partons increases both with $Q^2$ and with $1/x$, but whereas in the first case (increasing $Q^2$) the transverse area $\sim 1/Q^2$ occupied by every parton decreases very fast and more than compensates for the increase in their number — so, the proton is driven towards a regime which is more and more dilute —, in the second case (decreasing $x$) all the partons produced by the evolution have roughly the same transverse area, hence their density is necessarily increasing.

By asymptotic freedom, a high–density phase of QCD is characterized by *weak coupling*, and thus it can be studied from first principles. This observation gave a new impulse to the theoretical efforts, initiated several decades ago [1, 2], which aim at understanding the high–energy dynamics within perturbative QCD. This remains a formidable problem though, since high–density means also *strong collective phenomena*: at sufficiently high energy, the smallness of the coupling can be compensated by the high density of the quanta participating in the interaction, and then the dynamics becomes *fully non–linear*. This complexity appeared as a challenge for the theorists and stimulated vigorous studies over the recent years, with important results. New formalisms have been developed, which encompass the non–linear dynamics at high–energy to lowest order in $\alpha_s$ and allow for a unified picture of various high–energy phenomena ranging from DIS to heavy–ion, or proton–proton, collisions, and to cosmic rays. Here, I would like to provide a brief introduction to such new ideas, with emphasis on the physical picture and its consequences for deep inelastic scattering (DIS) at high energy.

CP892, *Quark Confinement and the Hadron Spectrum VII*
edited by J. E. F. T. Ribeiro
© 2007 American Institute of Physics 978-0-7354-0396-3/07/$23.00

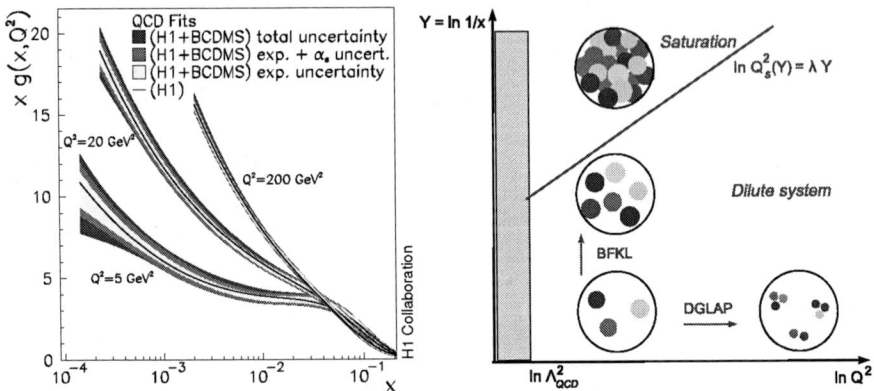

**FIGURE 1.** *Left: Gluon distribution extracted at HERA (here, data from H1), as a function of x in three bins of $Q^2$. Right: The 'phase–diagram' for QCD evolution suggested by the HERA data; each colored blob represents a parton with transverse area $\Delta x_\perp \sim 1/Q^2$ and longitudinal momentum $k_z = xP_z$.*

## DIS IN THE DIPOLE PICTURE

In DIS at small $x$, the struck quark is typically a 'sea' quark produced at the very end of a gluon cascade. It is then convenient to work in the 'dipole frame' in which the struck quark appears as an excitation of the virtual photon, rather than of the proton (see Ref. [3] for more details). In this frame, the proton still carries most of the total energy, while the virtual photon has just enough energy to dissociate long before the scattering into a 'color dipole' (a $q\bar{q}$ pair in a color singlet state), which then scatters off the gluon fields in the proton. The non–trivial, hadronic, part of DIS is then fully encoded in the cross–section for dipole–proton scattering, which at high energy can be computed as

$$\sigma_{\mathrm{dipole}}(x,r) \;=\; 2\int \mathrm{d}^2 b \;\; T(r,b,Y). \tag{1}$$

Here, $T(r,b,Y)$ is the *forward scattering amplitude* for a dipole with size $r$ and impact parameter $b$. The *unitarity* of the $S$–matrix requires $T \le 1$, with the upper limit $T = 1$ corresponding to total absorbtion, or 'black disk limit'.

But the unitarity constraint can be easily violated by an incomplete calculation, as we show now. To lowest order, $T$ involves the exchange of two gluons between the dipole and the target. Each exchanged gluon brings a contribution $gt^a \boldsymbol{r} \cdot \boldsymbol{E}_a$, where $\boldsymbol{E}_a$ is the color electric field in the target. Thus, $T \sim g^2 r^2 \langle \boldsymbol{E}_a \cdot \boldsymbol{E}_a \rangle_x$, where the expectation value is recognized as the number of gluons per unit transverse area in the proton wavefunction:

$$T(x,r,b) \;\sim\; \alpha_s r^2 \, \frac{xG(x,1/r^2)}{\pi R^2} \;\equiv\; \alpha_s n(x, Q^2 \sim 1/r^2). \tag{2}$$

We have also introduced here the *gluon occupation number*: $n(x,Q^2) = $ [number of gluons $xG(x,Q^2)$] times [the area $1/Q^2$ occupied by each gluon] divided by [the proton transverse area $\pi R^2$]. Eq. (2) applies so long as $T \ll 1$ and shows that weak scattering corresponds to low gluon occupancy $n \ll 1/\alpha_s$. But if naively extrapolated to very small values of $x$, this formula leads to *unitarity violations* : $T$ could become larger than one ! Before this happens, however, new physical phenomena are expected to come into play and restore unitarity. As we shall see, these are *non–linear* phenomena, and are of two types: (i) *multiple scattering*, i.e., the exchange of more than two gluons between the dipole and the target, and (ii) *gluon saturation*, i.e., non–linear effects in the proton wavefunction which tame the rise of the gluon distribution at small $x$.

Eq. (2) also provides a criterion for the onset of unitarity corrections: These become important when $T(x,r) \sim 1$ or $n(x,Q^2) \sim 1/\alpha_s$. This condition can be solved for the *saturation momentum*, which is the value of the transverse momentum below which saturation effects are expected to be important in the gluon distribution. One thus finds[2]

$$Q_s^2(x) \simeq \alpha_s \frac{xG(x,Q_s^2)}{\pi R^2} \sim \frac{1}{x^\lambda},$$

(3)

which grows with the energy as a power of $1/x$, since so does the gluon distribution before reaching saturation. In logarithmic units, the *saturation line* $\ln Q_s^2(Y) = \lambda Y$ is therefore a *straight* line, as illustrated in the right hand side of Fig. 1.

## BFKL EVOLUTION

Within perturbative QCD, the emission of small–$x$ gluons is amplified by the infrared sensitivity of the bremsstrahlung process, whose iteration leads to the BFKL evolution. Consider the emission of a gluon which carries a small fraction $x \ll 1$ of the longitudinal momentum of its parent quark. The differential probability for this emission reads

$$dP_{\text{Brem}} \simeq \frac{\alpha_s C_F}{2\pi^2} \frac{d^2k_\perp}{k_\perp^2} \frac{dx}{x},$$

(4)

which is singular as $x \to 0$. Introducing the rapidity $Y \equiv \ln(1/x)$, and hence $dY = dx/x$, Eq. (4) shows that there is a probability of $\mathcal{O}(\alpha_s)$ to emit one gluon per unit rapidity. The same would hold for the emission of a soft photon from an electron in QED. However, unlike the photon, the child gluon is itself charged with 'colour', so it can further emit an even softer gluon, with longitudinal fraction $x' \ll x$. When the rapidity is large, $\alpha_s Y \gg 1$, such successive emissions lead to the formation of gluon cascades, in which the gluons are ordered in rapidity and which dominate the small–$x$ part of the hadron wavefunction (see Fig. 2). So long as the density is not too high, these gluons don't 'see' each other and the evolution remains *linear* : when increasing the rapidity in one step $(Y \to Y + dY)$, the gluons created in the previous steps *incoherently* act as *color sources* for the emission of a new gluon. This leads to the following evolution equation

$$\frac{\partial n}{\partial Y} \simeq \omega \alpha_s n \quad \Longrightarrow \quad n(Y) \propto e^{\omega \alpha_s Y},$$

(5)

which predicts the exponential rise of $n$ with $Y$. This is a schematic version of the BFKL equation [1] which captures the main feature of this evolution: the unstable growth of

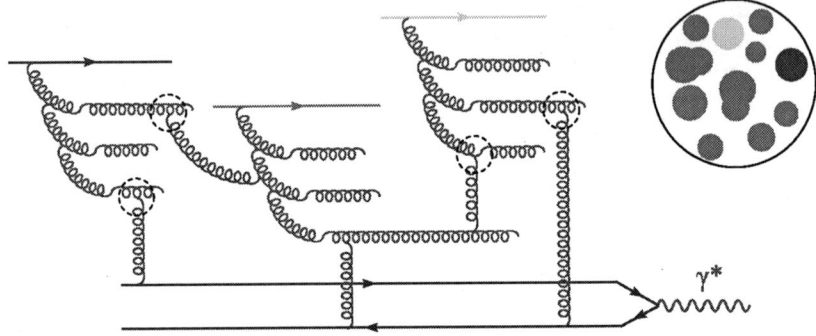

**FIGURE 2.** *DIS in the presence of BFKL evolution, saturation and multiple scattering.*

the gluon distribution. One knows by now that this growth is considerably tempered by NLO effects [4], but the basic fact that the gluon density increases exponentially with $Y$ is expected to remain true to all orders in $\alpha_s$ so long as one neglects the *non–linear* effects, or 'gluon saturation', in the evolution.

## JIMWLK EQUATION AND THE CGC

Non–linear effects appear because gluons carry colour charge, so they can interact with each other by exchanging gluons in the $t$–channel, as illustrated in Fig. 2. These interactions are amplified by the gluon density and thus they should become more and more important when increasing the energy. Back in 1983, L. Gribov, Levin and Ryskin [2] suggested that gluon saturation should proceed via $2 \to 1$ 'gluon recombination', which is a process of order $\alpha_s^2 n^2$ (cf. Fig. 2). To take this into account, they proposed the following, *non–linear*, generalization of Eq. (5) :

$$\frac{\partial n}{\partial Y} \simeq \alpha_s n - \alpha_s^2 n^2 = 0 \quad \text{when} \quad n = \frac{1}{\alpha_s} \gg 1 \tag{6}$$

which has a fixed point $n_{\text{sat}} = 1/\alpha_s$ at high energy, as indicated above. That is, when $n$ is as high as $1/\alpha_s$, the emission processes (responsible for the BFKL growth) are precisely compensated by the recombination ones, and then the gluon occupation factor saturates at a fixed value.

Twenty years later, we know that the actual mechanism for gluon saturation in QCD is more subtle than just gluon recombination and that its mathematical description is considerably more involved than suggested by Eq. (6). This mechanism, as encoded in the effective theory for the Color Glass Condensate [3] and its central evolution equation, the JIMWLK equation [5, 6], is the *saturation of the gluon emission rate* : At high density, the gluons are not independent color sources, rather they are strongly correlated with each other in such a way to ensure *color neutrality* over a distance $\Delta x_\perp \sim 1/Q_s$. Accordingly, the soft gluons with $k_\perp \lesssim Q_s$ are *coherently* emitted from a quasi–neutral gluon distribution, and then the emission rate $\partial n/\partial Y$ saturates at a constant value, implying that the gluon occupation factor keeps growing, but only *linearly* in $Y$.

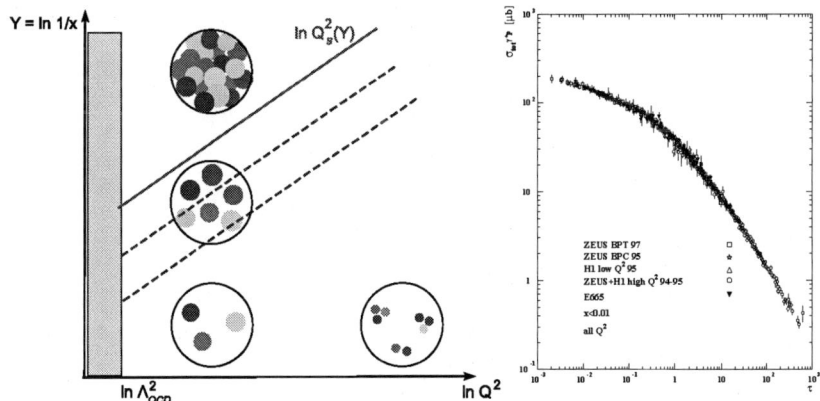

**FIGURE 3.** *Left: Line of constant gluon occupancy at high $Q^2$. Right: Geometric scaling in the HERA data for $\sigma_{\gamma^* p}$ at $x \leq 0.01$ [10]; $\tau$ is the scaling variable, $\tau \equiv Q^2/Q_s^2(Y)$.*

## UNITARITY & GEOMETRIC SCALING

We now discuss the consequences of this non–linear evolution for the dipole scattering, and thus for DIS. The first observation is that, when the energy is so high that saturation effects become important on the dipole resolution scale (this requires $r \gtrsim 1/Q_s(Y)$, cf. Eq. (2)), then *multiple scattering* becomes important as well (cf. Fig. 2) : e.g., the double–scattering $T^{(2)} \sim (\alpha_s n)^2$ is of $\mathcal{O}(1)$ in this regime, so like the single–scattering $T^{(1)} \sim \alpha_s n$. Together, multiple scattering and the saturation of the gluon distribution in the target limit the rise of the scattering amplitude with the energy and thus restore unitarity. Within the CGC formalism, the high–energy evolution of the average dipole amplitude in the presence of unitarity corrections is described by a non–linear equation with the following schematic structure :

$$\partial_Y \langle T \rangle = \alpha_s \langle T \rangle - \alpha_s \langle T^2 \rangle . \tag{7}$$

A priori, this is not a closed equation — the amplitude $\langle T \rangle$ for one dipole is related to the amplitude $\langle T^2 \rangle$ for two dipoles — but only the first equation in an infinite hierarchy, originally obtained by Balitsky [7]. A closed equation, known as the Balitsky–Kovchegov (BK) equation [8], can be obtained by assuming factorization: $\langle T^2 \rangle \approx \langle T \rangle \langle T \rangle$.

Eq. (7) has the fixed point $\langle T \rangle = 1$ at high energy and thus it respects unitarity, as announced. Remarkably, the growth of $\langle T \rangle_Y$ with $Y$ before saturation is entirely determined by the linearized version of the BK equation, i.e., the BFKL equation. This is important since, unlike the BK equation, the BFKL equation is presently known to NLO accuracy [4]. By using the latter together with the condition $\langle T(r) \rangle_Y = \mathcal{O}(1)$ when $r \gtrsim 1/Q_s(Y)$, Triantafyllopoulos has computed [9] the *saturation exponent* $\lambda$ (i.e., the slope of the saturation line; cf. Fig. 1) to NLO accuracy and thus found a value $\lambda \simeq 0.3$.

Another crucial consequence of the non–linear evolution towards saturation is the property known as *geometric scaling* : Physics should be invariant along trajectories which run parallel to the saturation line (cf. left plot in Fig. 3) because these are

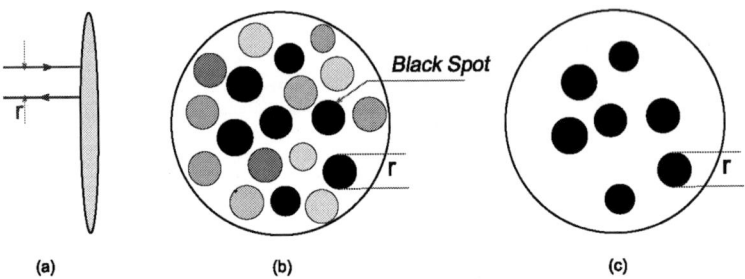

**FIGURE 4.** *Dipole–hadron scattering in the fluctuation–dominated regime at very high energy. (a) a view along the collision axis; (b) a transverse view of the hadron, as 'seen' by a small dipole impinging at different impact parameters; (c) the simplified, black&white, picture of the hadron which is relevant for the average dipole amplitude.*

lines of *constant gluon occupancy*. This implies that, up to relatively large momenta $Q^2 \gg Q_s^2(Y)$, the observables should depend only upon the difference $\ln Q^2 - \ln Q_s^2(Y)$ from the saturation line, i.e., they should *scale* upon the ratio $\tau \equiv Q^2/Q_s^2(Y)$, rather than separately depend upon $Q^2$ and $Y$. The outstanding feature of this scaling is the fact that this is a consequence of saturation which manifests itself up to relatively large transverse momenta, well above $Q_s$ [11]. This is consistent with the HERA data, which show approximate scaling for $x < 0.01$ and $Q^2 \leq 450$ GeV$^2$, whereas the saturation scale estimated from these data is relatively low: $Q_s \simeq 1$ GeV for $x \sim 10^{-4}$ [10] (cf. right plot in Fig. 3). Remarkably, the value for the saturation exponent coming out from such scaling fits to HERA is in agreement with its theoretical estimate [9] $\lambda \simeq 0.3$.

## FLUCTUATIONS & POMERON LOOPS

Although the experimental results at HERA appear to be consistent, as we have seen, with theoretical expectations based on the BK or JIMWLK equations, the latter are nevertheless incomplete (even to lowest order in $\alpha_s$) and thus cannot describe the actual dynamics in QCD at very high energies. Indeed, as recognized in [12, 13, 14], these equations miss the effects of *gluon–number fluctuations* in the dilute regime, which are however important for the evolution with increasing energy. Once this failure has been recognized, new equations have been proposed (the 'Pomeron loop' equations) [14, 15], which encompass both saturation and fluctuations in the limit where $N_c$ is large.

This strong sensitivity to fluctuations can be understood as follows: Non–linear phenomena like gluon saturation and multiple scattering involve the simultaneous exchange of several gluons in the $t$–channel (cf. Fig. 2), and thus probe *correlations* in the gluon distribution. At high energy, the most important such correlations are those generated via *gluon splitting* in the dilute regime: the 'child' gluons produced after a splitting are correlated with each other because they 'remember' about their common parent. These correlations manifest themselves in the difference $\langle nn \rangle - \langle n \rangle \langle n \rangle$ between the average pair density $\langle nn \rangle$ and its mean–field piece $\langle n \rangle \langle n \rangle$.

Fig. 4 illustrates a striking consequence of the evolution with pomeron loops, as probed in DIS at very high energy. The small blobs, grey or black, represent regions of the target disk which are explored by the dipole with size $r$ at various impact parameters. A light grey spot denotes weak scattering, $T(r,b) \ll 1$, and hence a region with low gluon density, a white region means almost no gluons at all, and a black spot represents a region where the gluon density is so high (on the resolution scale $Q^2 = 1/r^2$ of the incoming dipole) that the black disk limit is reached: $T(r,b) \approx 1$. Thus, when probed on a fixed resolution scale $Q^2$, a hadron with very high energy may look extremely inhomogeneous. This inhomogeneity is the result of gluon–number fluctuations in the high–energy evolution. What is most remarkable about this picture is that, for sufficiently high energy, the *average* amplitude $\langle T(r) \rangle_Y$ (and thus the DIS cross–section) is completely dominated by black spots up to very large values of $Q^2$, well above the *average* saturation momentum $\langle Q_s^2 \rangle_Y$. That is, although the target looks dilute *on the average* — meaning that $\langle T(r) \rangle_Y \ll 1$ —, its scattering is in fact controlled by *rare fluctuations* with unusually large density, for which $T \sim 1$. For the incoming dipole, the hadron disk looks either black ($T \simeq 1$) or white ($T \simeq 0$), as illustrated in Fig. 4.c.

This physical picture has interesting consequences for the phenomenology of DIS in the high energy limit. For instance, it predicts that, at sufficiently high energy, geometric scaling should be washed out by fluctuations [12] and replaced by a new type of scaling [13, 14], known as *diffusive scaling* [16] : instead of being a function of the ratio $Q^2/Q_s^2(Y)$, the DIS cross–section at high–energy should rather scale as a function of $\ln[Q^2/\langle Q_s^2 \rangle_Y]/\sqrt{DY}$, up to very large values $Q^2 \gg \langle Q_s^2 \rangle_Y$. ($D$ is a diffusion coefficient which measures the dispersion in the gluon distribution due to fluctuations [13].) A similar scaling law holds for the diffractive DIS cross–section [16] and also for the cross–section for gluon production in proton–proton scattering at forward rapidity [17], a process to be studied at the LHC.

# REFERENCES

1. L.N. Lipatov, *Sov. J. Nucl. Phys.* **23** (1976) 338; E.A. Kuraev, L.N. Lipatov and V.S. Fadin, *Zh. Eksp. Teor. Fiz* **72**, 3 (1977); Ya.Ya. Balitsky and L.N. Lipatov, *Sov. J. Nucl. Phys.* **28** (1978) 822.
2. L.V. Gribov, E.M. Levin, and M.G. Ryskin, *Phys. Rept.* **100** (1983) 1.
3. E. Iancu and R. Venugopalan, hep-ph/0303204.
4. V.S. Fadin and L.N. Lipatov, *Phys. Lett.* **B429** (1998) 127; G. Camici and M. Ciafaloni, *Phys. Lett.* **B430** (1998) 349.
5. J. Jalilian-Marian, A. Kovner, A. Leonidov and H. Weigert, *Nucl. Phys.* **B504** (1997) 415; *Phys. Rev.* **D59** (1999) 014014; H. Weigert, *Nucl. Phys.* **A703** (2002) 823.
6. E. Iancu, A. Leonidov and L. McLerran, *Nucl. Phys.* **A692** (2001) 583; *Phys. Lett.* **B510** (2001) 133; E. Ferreiro et al, *Nucl. Phys.* **A703** (2002) 489.
7. I. Balitsky, *Nucl. Phys.* **B463** (1996) 99; *Phys. Lett.* **B518** (2001) 235.
8. Yu.V. Kovchegov, *Phys. Rev.* **D60** (1999) 034008; *ibid.* **D61** (1999) 074018.
9. D.N. Triantafyllopoulos, *Nucl. Phys.* **B648** (2003) 293.
10. A.M. Stasto, K. Golec-Biernat, J. Kwiecinski, *Phys.Rev.Lett.* **86** (2001) 596.
11. E. Iancu, K. Itakura, and L. McLerran, *Nucl. Phys.* **A708** (2002) 327.
12. A.H. Mueller and A.I. Shoshi, *Nucl. Phys.* **B692** (2004) 175.
13. E. Iancu, A.H. Mueller and S. Munier, *Phys. Lett.* **B606** (2005) 342.
14. E. Iancu and D.N. Triantafyllopoulos, *Nucl. Phys.* **A756** (2005) 419; *Phys. Lett.* **B610** (2005) 253.
15. A.H. Mueller, A.I. Shoshi, and S.M.H. Wong, *Nucl. Phys.* **B715** (2005) 440.
16. Y. Hatta et al, *Nucl. Phys.* **A773** (2006) 95.
17. E. Iancu, C. Marquet, and G. Soyez, arXiv:hep-ph/0605174.

# Effective field theory as the bridge between lattice QCD and nuclear physics

## David B. Kaplan

*Institute for Nuclear Theory, Seattle, WA, 98195-1550, USA*

**Abstract.** A confluence of theoretical and technological developments are beginning to make possible contributions to nuclear physics from lattice QCD. Effective field theory plays a critical role in these advances. I give several examples.

**Keywords:** nuclear physics, lattice QCD, effective field theory, chiral perturbation theory
**PACS:** 12.38.-t,12.39.Fe,12.38.Gc

## INTRODUCTION

While it is unrealistic to expect to see a solution of the structure of a uranium nucleus from QCD within our lifetimes, it is not unreasonable to predict that lattice QCD will make significant contributions to nuclear physics over the next couple of decades. Simultaneous progress in computer technology, computational algorithms, and advances in theory have made it feasible to begin such a program in lattice nuclear physics now.

The limitation one faces is the computational cost of a realistic simulation. L. Giusti presented the following formula at Lattice '06 for the cost (in Tflops-yrs) for generating gauge field configurations with dynamical Wilson fermions (http://www.physics.arizona.edu/lattice06/):

$$\text{Cost} \sim 0.15 \cdot \left[\frac{\#\text{configs}}{1000}\right] \cdot \left[\frac{m_q}{20\,\text{MeV}}\right]^{-1} \cdot \left[\frac{V}{32\,\text{fm}^4}\right]^{\frac{5}{4}} \cdot \left[\frac{a}{0.08\,\text{fm}}\right]^{-6} \qquad (1)$$

Here $a$ is the lattice spacing, $m_q$ is the light quark mass, $V$ is the lattice volume.

Significant advances in algorithms have occurred in recent years, with the discovery in the 1990's of how to simulate chiral fermions [1, 2], and the improvement of methods for including light dynamical fermions (for example, the power of the mass dependence in the above formula has dropped from $m_q^{-6}$ to $m_q^{-1}$ since the development of algorithms in refs. [3, 4]). Technological advances continue unabated, and machines currently exist operating in the $10^2$ Tflops range, and Pflops computing will exist before long.

Nevertheless, technology plus algorithms do not by themselves add up to advances in nuclear theory in the near future because of the daunting computation costs of a realistic simulation. To avoid the disadvantages of non-chiral lattice fermions, such as the Wilson formulation, one should use domain wall or overlap fermions, incurring in the cost another factor of $\sim 100\times$; the correct light quark masses are $m_u \simeq 2.5$ MeV, and $m_d \simeq 5$ MeV, not 20 MeV; the box size should be ample enough to accommodate the hadrons of interest (the Compton wavelength of the pion is about 1.4 fm, while the scattering length for the deuteron is about 5 fm); and the lattice spacing of the real world

CP892, *Quark Confinement and the Hadron Spectrum VII*
edited by J. E. F. T. Ribeiro
© 2007 American Institute of Physics 978-0-7354-0396-3/07/$23.00

is, of course, zero. Finally, the above cost estimate only covers generation of lattice configurations; one must also account for the cost of generating quark propagators, the number of which grows factorially with the number of quarks involved — an unfortunate fact highly relevant to the study of even the smallest nuclei! Lattice QCD studies of a helium nucleus, for example, require $6!^2 = 518,400$ quark propagator contractions. It is easy to see that a brute force approach on a Pflops machine will not provide useful information about the $\alpha$ particle at realistic quark masses.

Effective field theory is the tool that will allow us to extract useful information from available technology, giving us the ability to simulate real systems at unrealistic lattice parameters.

In particular, we will have the opportunity to learn about fundamental properties of matter which are not directly obtainable from experiment, and which are necessary inputs for reliable nuclear structure or equation-of-state calculations. These include an improved understanding of three-body forces, such as in the experimentally inaccessible $I - 3/2$ channel, and the interactions between hyperons and nucleons. The thesis of this talk is that progress in these directions will need an intense effort by theorists in order to extract physically relevant quantities from feasible lattice calculations, and that the basic tool for this effort will be effective field theory.

Effective field theory (EFT) in all its forms is basically a perturbative expansion in the ratio of two length scales. As such, its validity requires that there be small ratios to exploit. Chiral perturbation theory has been the basic EFT exploited in continuum QCD, making use of the mass gap between the pion and the heavier hadrons. In addition, an effective field theory for nuclear physics has been in the making over the past 15 years, which incorporates an additional small ratio, the QCD length divided by the *NN* scattering length.

What is new when working with *lattice* QCD is that there are a host of additional dimensionful scales which do not exist in the real world, but which can be profitably exploited. These include the lattice spacing, the lattice size, and independently varied masses for valence and sea quarks. EFT allows one to

- extrapolate to smaller quark mass than is feasible to simulate
- parametrize and correct for finite lattice spacing errors
- parametrize and correct for finite volume errors
- extract physics from "cheaper" fermions
- determine $S$-matrix elements from Euclidean simulations by measuring volume dependence of the spectrum
- extract useful physical quantities from complicated multi-hadron systems

## USES OF CHIRAL PERTURBATION THEORY

### Quark mass extrapolation

Chiral perturbation theory is an expansion of the Lagrangian for low energy QCD about the chiral limit, $m_q = 0$. As such, it is obviously useful to extrapolate from lattice simulations at somewhat heavy quark masses, down to realistic quark masses. For this to work, the lattice quark mass has to be light enough so that the chiral expansion still converges. The chiral expansion parameter for mesonic processes is $m_\pi^2/\Lambda^2$, where $\Lambda \sim m_\rho$ is not far from 1 GeV. A light quark mass of 20 MeV, for example, corresponding to $m_\pi \sim 325$ MeV, should be within the range of validity of chiral perturbation theory. I will not dwell on this conventional and important application of chiral perturbation theory which is widely familiar (see lectures by S. Sharpe [5] for a comprehensive introduction to lattice applications of chiral perturbation theory).

## Lattice spacing extrapolation

Another application of chiral perturbation theory is to account for finite lattice spacing errors. One first matches the lattice action to the "Symanzik action" – a continuum theory with all operators allowed by the lattice symmetries, suppressed by powers of the lattice spacing $a$ appropriate to the dimension of the operator. For example, with Wilson fermions (which do not possess a chiral symmetry), the leading operators in the Symanzik action not present in continuum QCD include

- dimension-3 chiral symmetry violation: $a^{-1}\bar{q}q$
- dimension-5 chiral symmetry violation: $a\bar{q}\sigma_{\mu\nu}G_{\mu\nu}q$
- dimension-6 Lorentz violation: $a^2\bar{q}D_\mu^3\gamma^\mu q$

In order to determine the effects of finite lattice spacing on low energy hadronic physics, on can then match the Symanzik action onto a generalized chiral Lagrangian, which includes the effects of these finite lattice spacing operators [6, 7, 8]. The coefficients of these operators may be determined by making measurements at several lattice spacings, and then the extrapolation to $a = 0$ may be improved.

This program is versatile and can be applied to different lattice fermion formulations. For Wilson fermion the chiral symmetry violating operators give rise to the leading $O(a)$ corrections, even when the dimension-3 operator is fine-tuned away. For staggered fermions, corrections begin at $O(a^2)$, but the effective theory is complicated by the presence of additional "tastes", with an approximate $SU(4)$ taste symmetry, broken by finite lattice spacing operators. The analysis of the chiral Lagrangian is simplest for chiral lattice fermions, such as domain wall or overlap fermions, which automatically avoid the $O(a)$ operators without incurring spurious fermion tastes. The computational price of dynamical chiral fermions is about a factor of 100, which is severe.

## Partially quenched chiral perturbation theory

Quark masses appear in two distinct ways in the calculation of a correlation function in lattice QCD: either in the fermion determinant, which controls the gauge field configuration one generates; or in the fermion propagators on sews together in the gauge field background to compute the desired Green function. The former is called the "sea quark mass", the latter the "valence quark mass". In the real world they are the same, but in a lattice calculation they can be different. By making the valence quark mass light

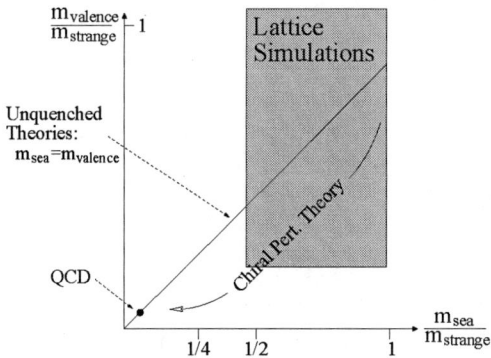

**FIGURE 1.** Schematic representation of parameter space in partially quenched theories, from [11].

while keeping the sea quark mass heavier, one obtains the benefits of chiral symmetry to leading order in the gauge coupling, without paying the light fermion price in generating the gauge field configurations.

One can think of this unphysical partition function arising from unphysically heavy $u$ and $d$ "sea" quarks with mass $m_S$, plus two flavors of light "valence" quarks with mass $m_V$, plus two flavors of ghosts with mass $m_V$ to cancel the valence quark contribution to the fermion determinant. Physical hadrons are then those made of $VV$ quarks (as well as the strange quark), in the limit $m_V = m_S$. The theory has additional unphysical mixed states of $VS$ and $SS$ content. A chiral Lagrangian can be constructed for this system, which contains new operators, but also all the operators of real QCD [9, 10]. Provided that $m_S$ is light enough for chiral perturbation theory to apply, by varying $m_S$ and $m_V$ independently, and by making use of generalized chiral perturbation theory for the various types of correlation functions one can compute, it is possible to isolate and measure quantities of interest in QCD, such as the Gasser-Leutwyler coefficients for *NLO* chiral perturbation theory [11]. See fig. 1 for a picture of the expanded parameter space. For an example of how to use the partially quenched method to determine the up quark mass, see [12]; there it was shown that by computing meson masses in the combination $(M_{VV}^2 + M_{SS}^2 - 2M_{SV}^2)$ one can extract the combination of Gasser-Leutwyler coefficients $(2L_8 - L_5)$ for QCD (where $S$ and $V$ label the two propagators used).

# Mixed action

There has been much work done recently with staggered fermions, employing "the fourth root trick" to reduce the number of tastes from four to one per physical quark flavor. This results in a nonlocal theory at finite lattice spacing, and there has been a controversy about whether or not the resulting lattice theory is in the same universality class as QCD and is capable of delivering an approximation to continuum QCD [13, 14]. In addition, the EFT for staggered fermions is extremely complex and difficult to work with for baryons, due to the multiplicity of tastes and taste symmetry violating operators.

53

The benefits of staggered fermions are their computational cheapness, and so at least for now, they are widely employed for mesons (see, for example, [15]).

It is possible to improve the utility of staggered fermions by working with mixed actions, where the sea quarks are staggered, while the valence quarks are domain wall fermions. This approach benefits from combining the speed of staggered fermions and the chiral symmetry of domain wall fermions, and has been used in recent calculations of $g_A$ at the $\sim 10\%$ level [16], the ratio $f_\pi/f_K$ to $\sim 1\%$ [17], $\pi - \pi$ scattering [18], $K - \pi$ scattering [19], and two-nucleon properties [20]. For mixed action computations one can use the partially quenched chiral perturbation machinery described above to extract physical results from the lattice calculations [21].

## Volume dependence

An additional handle on QCD provided by the lattice is the ability to manipulate the volume. Chiral perturbation theory is useful for understanding volume dependence of physical quantities, because it is the lightest modes that are most sensitive to finite volume. For a sufficiently small lattice, the zero-momentum pion modes become collective coordinates corresponding to global rotations of the chiral condensate. This occurs when $m\langle \bar{q}q \rangle V < 1$, defining the $\varepsilon$-regime [22]. This regime was recently cleverly exploited to extract information about QCD in the infinite volume continuum for the $\Delta I = 1/2$ rule [23], and for nucleon properties [24]. Both references make extensive use of chiral perturbation theory.

## NUCLEAR EFFECTIVE FIELD THEORY

EFT will play a critical role in computing nuclear physics properties on the lattice. Just as for meson interactions, one would like to find strategies to measure on the lattice the coefficients of the most relevant operators which control the interactions of nucleons, and then use that effective theory to compute nuclear properties. This program requires (i) that there be a sensible EFT for the interactions of nucleons, and (ii) that one can relate lattice measurements in Euclidean space to experimentally measurable quantities, such as scattering lengths.

Nuclear effective theory was pioneered by Weinberg [25], first explored in ref. [26], and then developed by many subsequent authors (for a review, see [27]). For low energy nucleon interactions, the pion may be considered as heavy, and the pion-less EFT developed in [28, 29, 30, 31] consists of contact interactions $\sim C_0(N^\dagger N)^2 + C_2(N^\dagger N)(N^\dagger \nabla^2 N) + O(p^4) \ldots$ The momentum expansion treats $C_{2n} = O(p^{n-1})$ when renormalized at a scale $\mu = O(p)$, and $C_0$ is summed to all orders as the leading contribution, while the higher $C_{2n}$ are inserted perturbatively. For $p \gtrsim m_\pi/2$, the pion has to be included explicitly. Unfortunately, unlike the case for chiral perturbation theory, the power counting scheme for the effective theory for nucleons interacting via pions is somewhat controversial. The reason is in part because $NN$ scattering is nonperturbative, and so the actual scaling of an operator does not match its naive dimension, making it difficult to construct a consistent power counting scheme. The original Weinberg scheme

54

suffers inconsistencies, where counterterms appear at higher orders than the divergences they are supposed to cancel. One example of this was given in [32] where it was shown that a quark mass-dependent counterterm for the non-derivative $NN$ vertex was required at leading order; a more recent numerical analysis of $NN$ scattering by the tensor force demonstrated that counterterms are needed at leading order in all partial waves where the interaction is attractive [33], even though such counterterms would be subleading in Weinberg's expansion. By working at a fixed and not too large cutoff, this problem can be swept under the rug, but this procedure in effect corresponds to constructing a model for short distance physics rather than performing a *bona fide* EFT calculation. This point of view is not universally accepted, and for recent contributions on various sides of the controversy see refs. [34, 35, 36]. The KSW expansion [29, 30] was offered as an alternative to Weinberg's expansion, but was found not to converge well for two nucleons in the $s = 1$ channel [37]. I believe that the most consistent expansion currently available is that of ref. [38], with generalizations to account for three nucleon forces [39].

There have been many notable successes of the nuclear effective theory, most remarkably at very low energy. A nice example is the isolation of the EFT coupling $L_{1A}$, the analogue of $g_A$ for the axial isovector two-nucleon current. By fitting it to data one can compute the neutrino-deuteron breakup cross section, thereby reducing a major source of systematic error in the analysis of data from the Sudbury Neutrino Observatory [40].

In order to study multi-nucleon states on the lattice, one approach is to compare EFT and numerical results in Euclidean space, and determine the EFT couplings. Another approach is to compare lattice results for S-matrix elements with the predictions of the EFT. An important contribution to the problem of extracting S-matrix elements from Euclidean lattice theory was devised by Lüscher, who showed how the volume dependence of the energy for a 2-particle state in a box yields the scattering lengths [41]. A nonrelativistic formulation found in ref. [42, 20] goes as follows: The Feynman scattering amplitude for two nucleons has the form $\mathscr{A} = (4\pi/M)/(p\cot\delta(p) - ip)$, where $p = \sqrt{ME}$, $E$ being the energy in the center of mass. When formulated in a box, the energy eigenvalues correspond to zeros of $Re[(\mathscr{A})^{-1}]$. So the energy eigenvalues solve

$$0 = Re[(\mathscr{A})^{-1}]_{\text{box}} = Re[(\mathscr{A})^{-1}]_{L=\infty} + \left(Re[(\mathscr{A})^{-1}]_{\text{box}} - Re[(\mathscr{A})^{-1}]_{L=\infty}\right) . \quad (2)$$

The first term on the right is just proportional to $p\cot\delta$ in an infinite box. The second term is the difference between the bubble diagram for two nucleons scattering off each other computed in an infinite box versus computed in a finite box. This is a finite and computable function of the box size $L$ and the eigenenergy $E_n$, and one arrives at the formula

$$p_n\cot\delta(p_n) = \frac{1}{\pi L}S(\eta_n) , \quad p_n = \sqrt{E_n M} , \quad \eta_n = (p_n L/2\pi)^2 , \quad (3)$$

$$S(\eta) = \lim_{\Lambda\to\infty}\left[\sum_{|\vec{j}|<\Lambda}\frac{1}{|\vec{j}|^2 - \eta} - 4\pi\Lambda\right] , \quad (4)$$

where the $\vec{j}$ are integer triplets. By measuring the energy eigenvalues for two nucleons in a box, one can then in principle solve the above equation for $\delta(p_n)$. A pioneering measurement of two nucleon scattering lengths using this method was performed in ref. [20], where they concluded that one would need a box of size $5 - 15$ fm in order to study properties of the deuteron. This is a large box, but much smaller than one would conclude directly for Lüscher's work, which would seem to indicate $L \gg a$.

It is important to push two nucleon studies much further, technically and theoretically. Using EFT techniques, one should learn how to measure matrix elements of currents in the two-nucleon state, study hyperon-nucleon interactions, and prepare the groundwork for the study of three-nucleon states on Pflops machines.

## THE FRONTIER

Lattice QCD can make substantial and useful contributions by predicting properties of baryons which are not experimentally accessible. I believe that in the foreseeable future, useful prediction are feasible in four areas:

- The masses and couplings of QCD resonances and hybrids;
- Strangeness physics, such as $\bar{K}N$, $YY$ and $YN$ interactions, where $\bar{K}$ is the anti-kaon, $Y$ is a hyperon, and $N$ is a nucleon;
- Determination of 3-body interactions, such as in the $I = 3/2$ channel;
- Quark mass dependence of nuclear properties.

The first category could make an important contribution to the JLab experimental program; information the second category could answer basic questions about dense matter in neutron stars, such as which hadronic channel is favored when strange quarks first appear, the $K^-$, $\Lambda$, or $\Sigma^-$? Accomplishing the third task would be an important milestone, whereby lattice QCD could inform the so-called *ab initio* nuclear structure calculations and make them significantly more *ab initio*. The fourth category could be of interest in understanding how fine-tuned is our world, and could be important in certain cosmological theories where the quark masses are dynamically determined quantities [43].

All of these projects could be very rewarding, and will require an intensive theoretical effort that further develops the 3-nucleon EFT, extends Lüscher's work to three particles in a box (where inelastic thresholds could cause problems [44]), and improves the available computational algorithms. Lattice QCD will clearly play an important role in the future progress of nuclear theory, and EFT will be a vital component of the program.

## ACKNOWLEDGMENTS

I wish to thank G. Martinelli, C. Sachrajda, S. Sharpe and M. Savage for useful conversations, and to the organizers of QCHSVII for their hospitality. This work was supported by the US Department of Energy grant DE-FG02-00ER41132.

## REFERENCES

1.  D. B. Kaplan, *Phys. Lett.* **B288**, 342–347 (1992), hep-lat/9206013.

56

2. H. Neuberger, *Phys. Lett.* **B417**, 141–144 (1998), hep-lat/9707022.
3. M. Luscher, *JHEP* **05**, 052 (2003), hep-lat/0304007.
4. M. Luscher, *Comput. Phys. Commun.* **165**, 199–220 (2005), hep-lat/0409106.
5. S. R. Sharpe (2006), hep-lat/0607016.
6. S. R. Sharpe, and J. Singleton, Robert L., *Phys. Rev.* **D58**, 074501 (1998), hep-lat/9804028.
7. G. Rupak, and N. Shoresh, *Phys. Rev.* **D66**, 054503 (2002), hep-lat/0201019.
8. O. Bar, G. Rupak, and N. Shoresh, *Phys. Rev.* **D70**, 034508 (2004), hep-lat/0306021.
9. C. W. Bernard, and M. F. L. Golterman, *Phys. Rev.* **D49**, 486–494 (1994), hep-lat/9306005.
10. S. R. Sharpe, and R. S. Van de Water, *Phys. Rev.* **D69**, 054027 (2004), hep-lat/0310012.
11. S. R. Sharpe, and N. Shoresh, *Nucl. Phys. Proc. Suppl.* **83**, 968–970 (2000), hep-lat/9909090.
12. A. G. Cohen, D. B. Kaplan, and A. E. Nelson, *JHEP* **11**, 027 (1999), hep-lat/9909091.
13. M. Creutz (2006), hep-lat/0608020.
14. S. R. Sharpe (2006), hep-lat/0610094.
15. A. S. Kronfeld, et al., *PoS* **LAT2005**, 206 (2006), hep-lat/0509169.
16. R. G. Edwards, et al., *Phys. Rev. Lett.* **96**, 052001 (2006), hep-lat/0510062.
17. S. R. Beane, P. F. Bedaque, K. Orginos, and M. J. Savage (2006), hep-lat/0606023.
18. S. R. Beane, P. F. Bedaque, K. Orginos, and M. J. Savage, *Phys. Rev.* **D73**, 054503 (2006), hep lat/0506013.
19. S. R. Beane, et al. (2006), hep-lat/0607036.
20. S. R. Beane, P. F. Bedaque, K. Orginos, and M. J. Savage, *Phys. Rev. Lett.* **97**, 012001 (2006), hep-lat/0602010.
21. O. Bar, C. Bernard, G. Rupak, and N. Shoresh, *Phys. Rev.* **D72**, 054502 (2005), hep-lat/0503009.
22. J. Gasser, and H. Leutwyler, *Nucl. Phys.* **B307**, 763 (1988).
23. P. Hernandez, and M. Laine (2006), hep-lat/0607027.
24. W. Detmold, and M. J. Savage, *Phys. Lett.* **B599**, 32–42 (2004), hep-lat/0407008.
25. S. Weinberg, *Phys. Lett.* **B251**, 288–292 (1990).
26. C. Ordonez, L. Ray, and U. van Kolck, *Phys. Rev.* **C53**, 2086–2105 (1996), hep-ph/9511380.
27. P. F. Bedaque, and U. van Kolck, *Ann. Rev. Nucl. Part. Sci.* **52**, 339–396 (2002), nucl-th/0203055.
28. U. van Kolck (1997), hep-ph/9711222.
29. D. B. Kaplan, M. J. Savage, and M. B. Wise, *Phys. Lett.* **B424**, 390–396 (1998), nucl-th/9801034.
30. D. B. Kaplan, M. J. Savage, and M. B. Wise, *Nucl. Phys.* **B534**, 329–355 (1998), nucl-th/9802075.
31. J.-W. Chen, G. Rupak, and M. J. Savage, *Nucl. Phys.* **A653**, 386–412 (1999), nucl-th/9902056.
32. D. B. Kaplan, M. J. Savage, and M. B. Wise, *Nucl. Phys.* **B478**, 629–659 (1996), nucl-th/9605002.
33. A. Nogga, R. G. E. Timmermans, and U. van Kolck, *Phys. Rev.* **C72**, 054006 (2005), nucl-th/0506005.
34. E. Epelbaum, and U. G. Meissner (2006), nucl-th/0609037.
35. M. Rho (2006), nucl-th/0610003.
36. E. Ruiz Arriola, and M. Pavon Valderrama (2006), nucl-th/0609080.
37. S. Fleming, T. Mehen, and I. W. Stewart, *Nucl. Phys.* **A677**, 313–366 (2000), nucl-th/9911001.
38. S. R. Beane, P. F. Bedaque, M. J. Savage, and U. van Kolck, *Nucl. Phys.* **A700**, 377–402 (2002), nucl-th/0104030.
39. P. F. Bedaque, G. Rupak, H. W. Griesshammer, and H.-W. Hammer, *Nucl. Phys.* **A714**, 589–610 (2003), nucl-th/0207034.
40. J.-W. Chen, K. M. Heeger, and R. G. H. Robertson, *Phys. Rev.* **C67**, 025801 (2003), nucl-th/0210073.
41. M. Luscher, *Nucl. Phys.* **B354**, 531–578 (1991).
42. S. R. Beane, P. F. Bedaque, A. Parreno, and M. J. Savage, *Phys. Lett.* **B585**, 106–114 (2004), hep-lat/0312004.
43. S. R. Beane, and M. J. Savage, *Nucl. Phys.* **A717**, 91–103 (2003), nucl-th/0208021.
44. C. J. D. Lin, G. Martinelli, C. T. Sachrajda, and M. Testa, *Nucl. Phys.* **B619**, 467–498 (2001), hep-lat/0104006.

# On the dispersion theory of $\pi\pi$ scattering

## H. Leutwyler

*Institute for Theoretical Physics, University of Bern, Sidlerstr. 5, CH-3012 Switzerland*

**Abstract.** Recent developments in low energy pion physics are reviewed, emphasizing the strength of dispersion theory in this context. As an illustration of the method, I discuss some consequences of the forward dispersion relation obeyed by the isoscalar component of the scattering amplitude.

Pions play a crucial role whenever the strong interaction is involved at low energies – the Standard Model prediction for the muon magnetic moment provides a good illustration. My talk dealt with the remarkable theoretical progress made in low energy pion physics in recent years.

In the first part, I discussed the current theoretical and experimental knowledge of the S-wave $\pi\pi$ scattering lengths $(a_0^0, a_0^2)$ in some detail, because these play a central role. Simulations of QCD on a lattice now reach sufficiently small quark masses for a meaningful extrapolation to the values of physical interest to become possible [1, 2]. Chiral perturbation theory not only describes the dependence on the quark masses, but also allows one to calculate the leading finite volume effects [3]. The values obtained for the coupling constants $\ell_3$ and $\ell_4$ are consistent with the estimates given in [4, 5]. Since these control the leading corrections to the low energy theorems for $a_0^0$ and $a_0^2$, the lattice results at the same time provide a rough check of the remarkably precise predictions for these quantities obtained in [5] (in the following, this paper is referred to as CGL). The exotic scattering length $a_0^2$ can also be extracted directly from the volume dependence of the energy levels occurring on the lattice. The result obtained in [6] is in good agreement with the prediction as well. The current state of our knowledge concerning $a_0^0$ and $a_0^2$ is briefly summarized in [7].

The second part of the talk covered recent results established with dispersive methods. The upshot of this development is that, in the threshold region, the $\pi\pi$ scattering amplitude is now known to an amazing degree of accuracy [5]. In particular, we know how to calculate mass and width of the lowest resonance of QCD [8]. The actual uncertainty in the pole position is smaller than the estimate given in the 2006 edition of the Review of Particle Physics [9], by more than an order of magnitude. The progress made in this field heavily relies on the fact that the dispersion theory of $\pi\pi$ scattering is particularly simple: the s-, t- and u-channels represent the same physical process. As a consequence, the scattering amplitude can be represented as a dispersion integral over the imaginary part and the integral exclusively extends over the physical region [10]. The projection of the amplitude on the partial waves leads to a dispersive representation for these, the Roy equations. For a detailed discussion, I refer to [11].

CP892, *Quark Confinement and the Hadron Spectrum VII*
edited by J. E. F. T. Ribeiro
© 2007 American Institute of Physics 978-0-7354-0396-3/07/$23.00

In the present article, I wish to explain the essence of the dispersive approach, avoiding technical machinery as much as possible. Although the Roy equations represent an optimal and comprehensive framework for the low energy analysis of the $\pi\pi$ scattering amplitude, the main points can be seen in a simpler context: forward dispersion relations [13]. More specifically, I consider the component of the scattering amplitude with $s$-channel isospin $I = 0$, which I denote by $T^0(s,t)$. It satisfies a twice subtracted fixed-$t$ dispersion relation in the variable $s$. In the forward direction, $t = 0$, this relation reads

$$
\operatorname{Re} T^0(s,0) = c_0 + c_1 s + \frac{s(s-4M_\pi^2)}{\pi} P \int_{4M_\pi^2}^{\infty} \frac{dx \operatorname{Im} T^0(x,0)}{x(x-4M_\pi^2)(x-s)} + \tag{1}
$$
$$
+ \frac{s(s-4M_\pi^2)}{\pi} \int_{4M_\pi^2}^{\infty} \frac{dx \{\operatorname{Im} T^0(x,0) - 3 \operatorname{Im} T^1(x,0) + 5 \operatorname{Im} T^2(x,0)\}}{3x(x-4M_\pi^2)(x+s-4M_\pi^2)}.
$$

The symbol $P$ indicates that the principal value must be taken. The first integral accounts for the discontinuity across the right hand cut, while the second represents the analogous contribution from the left hand cut, where the components of the scattering amplitude with $I = 1,2$ also show up. According to the optical theorem, the imaginary part of the forward scattering amplitude represents the total cross section: in the normalization of [11], we have $\operatorname{Im} T^I(s,0) = \sqrt{s(s-4M_\pi^2)}\, \sigma_{tot}^I(s)$. The subtraction constants are also determined by physical quantities – the $S$-wave scattering lengths:

$$
c_0 + c_1 s = 32\pi \left\{ a_0^0 + (2a_0^0 - 5a_0^2)\frac{s-4M_\pi^2}{12M_\pi^2} \right\}. \tag{2}
$$

A dispersion relation of the above type also holds for other processes. What is special about $\pi\pi$ is that the contribution from the crossed channels can be expressed in terms of observable quantities – total cross sections in the case of forward scattering.

The right hand side of equation (1) can be evaluated with the available representations of the scattering amplitude. The contribution from the left hand cut is dominated by the $\rho$-meson, which generates a pronounced peak in the total cross section with $I = 1$. This contribution is known very accurately from the process $e^+e^- \to \pi^+\pi^-$. Since the channel with $I = 2$ is exotic, it does not contain any resonances and – at low energies – only generates a minor correction. In the physical region, $s > 4M_\pi^2$, the entire contribution from the crossed channels is a smooth function that varies only slowly with the energy. Note, however, that this contribution is by no means small.

The angular momentum barrier suppresses the higher partial waves: at low energies, the first term in the partial wave decomposition

$$
\operatorname{Re} T^0(s,0)/(32\pi) = \operatorname{Re} t_0^0(s) + 5 \operatorname{Re} t_2^0(s) + \dots, \tag{3}
$$

represents the most important contribution. In the vicinity of the threshold, where the contribution from the $D$-wave is small, the dispersion relation (1) thus essentially determines the real part of the isoscalar $S$-wave. For brevity, I refer to this wave as $S^0$.

Figure 1a is based on the representation of the scattering amplitude in CGL, where the low energy behaviour of the $S$- and $P$-wave phase shifts was determined by solving the Roy equations below 800 MeV. That calculation required input for (a) the imaginary

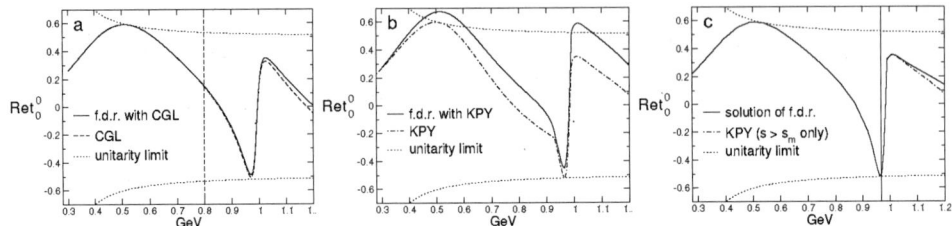

**FIGURE 1.** Real part of the isoscalar $S$-wave from forward dispersion relation.

parts of the higher partial waves, (b) the imaginary parts of the $S$- and $P$-waves above 800 MeV and (c) the $S$-wave scattering lengths. For (a) and (b), we relied on the literature [12], while for (c), we used the low energy theorems of chiral perturbation theory. The Roy equations then yield an approximate representation for the real parts of all partial waves, throughout their region of validity. Together with the input used for the imaginary parts, this also fixes the phase shifts and elasticities in that region. The dashed lines in figures 1a, 2 and 3b are calculated in this way. I emphasize that, above 800 MeV, these curves amount to an extrapolation. The uncertainties in the representation of the scattering amplitude are discussed in detail in CGL, but are not shown in the figures, which are calculated with the central values. In the threshold region, the uncertainties are tiny, but they grow with the energy. In particular, Figure 3b shows that the extrapolation overestimates the inelasticity in the region between 800 MeV and $2M_K$ – in reality, a significant amount of inelasticity only arises when the $K\bar{K}$ channel opens.

In order to demonstrate that the representation of the scattering amplitude in CGL is consistent with equation (1), I evaluate the function $\mathrm{Re}\,T^0(s,0)$ with it and remove the $D^0$-wave, setting $\mathrm{Re}\,t_0^0(s)|_{f.d.r.} = \mathrm{Re}\,T^0(s,0)/(32\pi) - 5\,\mathrm{Re}\,t_2^0(s)$ and thereby ignoring partial waves with $\ell \geq 4$. Figure 1a shows that, below 800 MeV, the f.d.r. is indeed obeyed very well. No wonder: if the partial waves satisfy the Roy equations, then the sum over <u>all</u> of these automatically obeys the f.d.r. The difference between the two curves arises from the neglected higher partial waves, which become more important if the energy is increased.

Figure 1b shows the result obtained if the representation in CGL is replaced by the one in [13] (in the following, this paper is referred to as KPY). Visibly, the forward dispersion relation is not obeyed well, but this is to be expected: that parametrization does not rely on dispersion theory, but represents a phenomenological fit to data sets that are subject to large errors. Note also that the figure does not show the uncertainties of the partial wave analysis in KPY, which are considerable: the difference between the two curves is in the noise of that analysis.

As discussed in [13, 14], the forward dispersion relations can be used to improve the partial wave representation. In the following, I apply the method of [11], which is very suitable for the purpose: it suffices to replace the Roy equation for $\mathrm{Re}\,t_0^0$ by the f.d.r. for $\mathrm{Re}\,T^0$. If the subtraction constants and all partial waves except $t_0^0$ are treated as known, the problem may be given the following mathematical form:

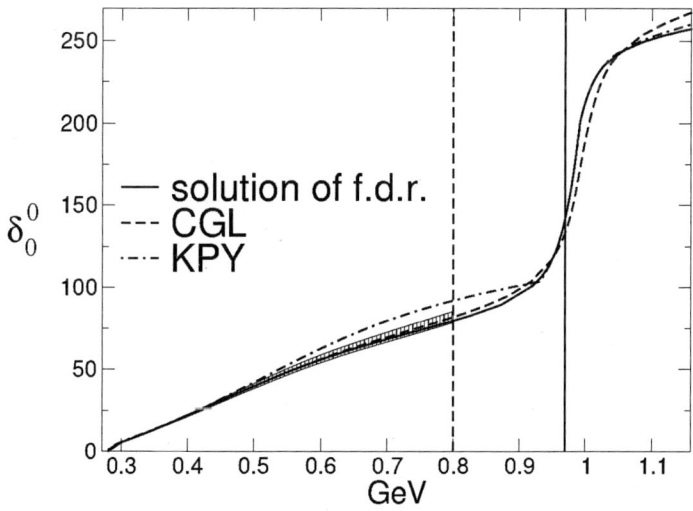

**FIGURE 2.** Phase of the isoscalar S-wave.

*Choose a "matching point" $s_m$ and prescribe the function $\mathrm{Im}\,t_0^0(s)$ above $s_m$ as well as the elasticity $\eta_0^0(s)$ below $s_m$. Find solutions of equation (1) for $s < s_m$ that respect the unitarity relation between $\mathrm{Re}\,t_0^0$, $\mathrm{Im}\,t_0^0$ and $\eta_0^0$.*

In the framework of the Roy equations, this problem was discussed in detail in [11]. If the matching point is taken below the energy where the phase goes through 90°, then there is exactly one solution. This is the situation considered in CGL, where the matching point and the central value of the phase at that point were set equal to 800 MeV and 82.3°, respectively. The dashed lines in figures 1a, 2 and 3b represent the resulting unique solution.

Next, I observe that, near 970 MeV, the representations given for $S^0$ in KPY and CGL are very similar – in either case, the real part reaches the lower unitarity limit in the immediate vicinity of that energy. For this reason, I now discuss the situation for the case where the matching point is taken at $\sqrt{s_m} = 970$ MeV (full vertical line) and where the input for the imaginary parts is taken from KPY: Above 1.42 GeV, I use the Regge parametrization of the forward scattering amplitudes given in that reference. At lower energies, the partial wave decomposition is used. With the exception of $S^0$ below 970 MeV, all of the partial waves are taken from KPY (central values of the parameters, throughout). Finally, below 970 MeV, the elasticity of $S^0$ is set equal to 1, while the phase is left open – the f.d.r. is used to determine it. In order to be able to compare the result with the one obtained with CGL, I keep the subtraction constants fixed at the central values in CGL. The range used for $a_0^0, a_0^2$ in KPY is consistent with that.

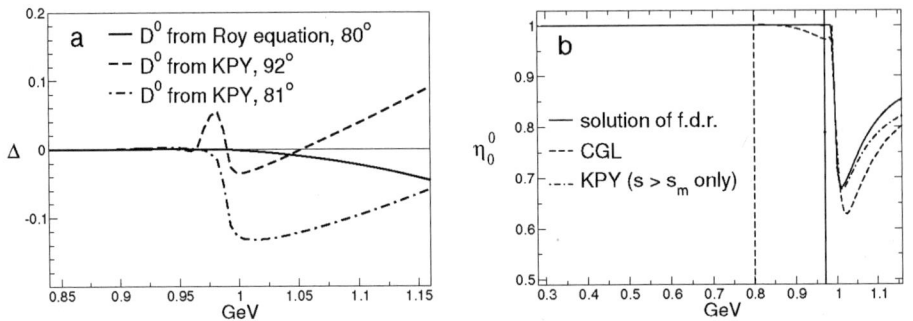

**FIGURE 3.** Violation of f.d.r. for $T^0$ and elasticity of the isoscalar $S$-wave.

With this input, the value of the phase at the matching point is $142°$. As discussed in detail in [11], the mathematical problem posed above leads to a curiosity if the phase at the matching point exceeds $90°$: the solution of the forward dispersion relation then fails to be unique – there is an entire family of solutions. In the present case, there is a one-parameter family, which may be labeled with the value of the phase somewhere below the matching point, at 800 MeV, for example. In particular, we may select the solution for which the phase at 800 MeV agrees with the central parametrization in KPY, $\delta_0^0(800\,\text{MeV}) = 92°$.

This particular solution is physically unacceptable, however, because it contains a strong cusp at the matching point. An enlargement of the region around this point is shown in figure 3a, where the difference $\Delta$ between the left and right hand sides of equation (1) is plotted as a function of the energy. By construction, the difference vanishes below the matching point – within the accuracy to which the solutions are worked out. The cusp manifests itself as a spike in the vicinity of the matching point. The occurrence of a cusp is a generic feature of the manifold of solutions to the mathematical problem specified above. The amplitude of the cusp decreases if the phase at 800 MeV is lowered. There is a unique solution for which a cusp does not occur, in the sense that the first derivative is continuous at the matching point. This solution is reached if the phase at 800 MeV is lowered to about $81°$. Figure 3a shows that the violation of the f.d.r. is then postponed to $K\bar{K}$ threshold, but it persists: with the input specified above, equation (1) does not have a physically acceptable solution.

The problem originates in $D^0$, the isoscalar $D$-wave – indeed, in KPY, the parametrization used for the elasticity of this wave is mentioned as a potential culprit. Around $K\bar{K}$ threshold, the real part of $D^0$ is by no means negligible and it is essential that the representation used for it is consistent with analyticity. In this regard, there is a difference between the mathematical problem specified above and the Roy equations, where the input of the calculation exclusively involves imaginary parts. As was noted already in the pioneering work of Basdevant, Froggatt and Petersen [15], dispersion theory imposes very strong constraints on the low energy behaviour of the higher partial waves.

The representation for $D^0$ in KPY is not consistent with these constraints. In particular, phase space strongly suppresses the inelasticity generated by the $K\bar{K}$ states: $1 - \eta_2^0$ only grows with the fifth power of the kaon momentum.

The problem encountered with the behaviour of the solutions around $K\bar{K}$ threshold disappears if the representation for $D^0$ is taken from the solution of the Roy equation for $\mathrm{Re}\,t_2^0$, retaining the parametrization of KPY only for the other waves. Alternatively, the parametrization of $D^0$ given in [16] can be used – this is less accurate, but barely makes any difference around $K\bar{K}$ threshold. The full lines in figures 3a and 3b show that equation (1) does now admit physically acceptable solutions. As before, there is a one-parameter family of solutions, which differ in the strength of the cusp at the matching point. The particular solution shown is obtained by minimizing the difference between the right and left hand sides of equation (1) on the entire interval from $2M_\pi$ to $2M_K$. For this solution, the cusp is too small to be visible in the figures.

As was to be expected, the solution represents a compromise between the two curves in figure 1b. Below $2M_K$, the difference between the left and right hand sides of equation (1) is less than $10^{-3}$, but above 1050 MeV, the real part of the solution departs from the real part of the parametrization in KPY, which is indicated as a dash-dotted line (shown only above 970 MeV, where the imaginary part of that parametrization is used as an input). The real part and the phase of the solution are displayed as full lines in figures 1c and 2. At low energies, the solution of the forward dispersion relation runs within the shaded region, which represents the uncertainty band in CGL. The value of the phase at the upper end is $\delta_0^0(800\,\mathrm{MeV}) = 80°$. This confirms one of the main results in CGL: below 800 MeV, the solution is not sensitive to the behaviour at high energies. On the other hand, above the matching point, the phase very closely follows the parametrization of the $S^0$-wave used in the input. As discussed in detail in [11], in the context of the Roy equations, this happens whenever the input stems from an analytic parametrization of the partial wave in question. As can be seen in figure 3b, shifting the matching point up has the effect of removing the unitarity violation in the elastic region – the output for $\eta_0^0$ now differs appreciably from KPY only above 1050 MeV.

I conclude that – in the specific framework used here, where all partial waves except the isoscalar $S$-wave are treated as known – the constraint imposed on the scattering amplitude by the forward dispersion relation for $T^0(s,t)$ is essentially equivalent to the Roy equation for $\mathrm{Re}\,t_0^0(s)$. Likewise, the Roy equations for $\mathrm{Re}\,t_1^1(s)$ and $\mathrm{Re}\,t_0^2(s)$ can be replaced by the forward dispersion relations for $T^1(s,t)$ and $T^2(s,t)$. In all three cases, the subtraction constants are determined by the $S$-wave scattering lengths. The forward dispersion relations do not provide a handle on the higher partial waves, but the Roy equations do. As we have pushed the matching point up, the dispersive framework now correlates the value of the phase at 800 MeV with the behaviour of the scattering amplitude above $K\bar{K}$ threshold. In particular, if the representation in KPY represents a good approximation above $2M_K$ and if the theoretical predictions for $a_0^0$ and $a_0^2$ are correct, then the phase at 800 MeV must be in the vicinity of $80°$.

In the above, I did not discuss the experimental information at all. Instead, I merely showed that – in view of the sharp theoretical predictions for the subtraction constants – the properties of the amplitude above $K\bar{K}$ threshold very strongly constrain the behaviour at lower energies. A more systematic investigation is under way, which aims at extending the work described in CGL to somewhat higher energies. The matching point can be pushed up all the way to the limit of validity of the Roy equations. The price to pay is that the contributions from the high energy region then become more important. We are using a Regge representation for the asymptotic domain and invoke experimental information as well as sum rules to pin down the residue functions. A short account of this work is given in [17] and more should soon be ready for publication, in particular also the application to the electromagnetic form factor of the pion, for which an accurate representation is needed in connection with the Standard Model prediction for the magnetic moment of the muon.

It is a pleasure to thank the organizers of the meeting for their kind hospitality during a very pleasant stay on the Azores islands. Also, I acknowledge Irinel Caprini and Gilberto Colangelo for a most enjoyable and fruitful collaboration and thank Hanqing Zheng for many informative discussions.

## REFERENCES

1.  C. Aubin *et al.*, [MILC Collaboration], Phys. Rev. D **70** (2004) 114501.
2.  L. Del Debbio, L. Giusti, M. Lüscher, R. Petronzio and N. Tantalo, hep-lat/0610059.
3.  G. Colangelo, S. Dürr and C. Haefeli, Nucl. Phys. B **721**, 136 (2005).
4.  J. Gasser and H. Leutwyler, Annals Phys. **158** (1984) 142.
5.  G. Colangelo, J. Gasser and H. Leutwyler, Nucl. Phys. B **603** (2001) 125, referred to as CGL.
6.  S. R. Beane, P. F. Bedaque, K. Orginos and M. J. Savage [NPLQCD Collaboration], Phys. Rev. D **73** (2006) 054503.
7.  H. Leutwyler, in Proc. MESON 2006, Cracow, Poland, to be published, hep-ph/0608218.
8.  I. Caprini, G. Colangelo and H. Leutwyler, Phys. Rev. Lett. **96** (2006) 132001.
9.  W.-M. Yao et al. [Particle Data Group], Journal of Physics G **33** (2006) 1.
10. S. M. Roy, Phys. Lett. B **36** (1971) 353.
11. B. Ananthanarayan et al., Phys. Rept. **353** (2001) 207.
12. In particular, the asymptotics was borrowed from M. R. Pennington, Annals Phys. **92** (1975) 164.
13. R. Kaminski, J. R. Pelaez and F. J. Yndurain, Phys. Rev. D **74** (2006) 014001 [Erratum-ibid. D **74** (2006) 079903], referred to as KPY.
14. R. Kaminski, J. R. Pelaez and F. J. Yndurain, arXiv:hep-ph/0610315.
15. J. L. Basdevant, C. D. Froggatt and J. L. Petersen, Phys. Lett. B **41** (1972) 173; *ibid.* B **41** 178; Nucl. Phys. B **72** (1974) 413.
16. B. Hyams *et al.*, Nucl. Phys. B **64** (1973) 134.
17. I. Caprini, G. Colangelo and H. Leutwyler, Int. J. Mod. Phys. A **21** (2006) 954.

# Hadron Physics and the Dyson–Schwinger Equations of QCD

Pieter Maris

*Dept. of Physics and Astronomy, University of Pittsburgh, Pittsburgh, PA 15260*

**Abstract.** We use the Bethe–Salpeter equation in rainbow-ladder truncation to calculate the ground state mesons from the chiral limit to bottomonium, with an effective interaction that was previously fitted to the chiral condensate and pion decay constant. Our results are in reasonable agreement with the data, as are the vector and pseudoscalar decay constants. The meson mass differences tend to become constant in the heavy-quark limit. We also present calculations for the pion and rho electromagnetic form factors, and for the single-quark form factors of the $\eta_c$ and $J/\psi$.

**Keywords:** Bethe–Salpeter equation, meson, quark propagator, electromagnetic form factor
**PACS:** 11.10.St, 12.38.Lg, 13.40.Gp, 14.40.-n

## INTRODUCTION

Hadrons are color-singlet bound states of quarks, antiquarks, and gluons. Bound states appear as poles in the $n$-point functions of a quantum field theory. Thus a study of the poles in the $n$-point functions of QCD will tell us something about hadrons.

In the ultraviolet region, these $n$-point functions can be calculated using perturbation theory. For hadronic observables however, we need to understand the nonperturbative, infrared behavior of the $n$-point functions of QCD. The Dyson–Schwinger equations [DSEs], which are the equations of motion of a quantum field theory, provide us with a tool to study the $n$-point functions nonperturbatively. For reviews on the DSEs and their use in hadron physics, see [1–5].

## MESON PHYSICS

Mesons can be described by solutions of the homogeneous Bethe–Salpeter equation

$$\Gamma(p_{\text{out}}, p_{\text{in}}; P) = \int \frac{d^4 k}{(2\pi)^4} K(p_{\text{out}}, p_{\text{in}}; k_{\text{out}}, k_{\text{in}}) \chi(k_{\text{out}}, k_{\text{in}}; P), \qquad (1)$$

with $p_{\text{in}}$, $p_{\text{out}}$ the 4-momenta of the quark and antiquark, subject to momentum conservation: $p_{\text{in}} - p_{\text{out}} = P$, $\Gamma$ the Bethe–Salpeter amplitude [BSA], and $\chi(k_{\text{out}}, k_{\text{in}}; P) = S(k_{\text{out}}) \Gamma(k_{\text{out}}, k_{\text{in}}; P) S(k_{\text{in}})$; the kernel $K$ is the $q\bar{q}$ scattering kernel. This integral equation has solutions $\Gamma$ at discrete values of $P^2 = -M^2$ (in Euclidean metric) of the total meson 4-momentum $P$. Different types of mesons, such as pseudoscalar or vector mesons, are characterized by different Dirac structures. The properly normalized BSA $\Gamma(p_{\text{out}}, p_{\text{in}}; P)$ completely describes the meson as a $q\bar{q}$ bound state.

CP892, *Quark Confinement and the Hadron Spectrum VII*
edited by J. E. F. T. Ribeiro
© 2007 American Institute of Physics 978-0-7354-0396-3/07/$23.00

Since Eq. (1) has solutions at discrete values of $P^2 = -M_i^2$, one does not obtain the "complete" spectrum, including the excited states, by solving a matrix equation once; instead, one has to repeatedly solve Eq. (1) at different values of $P^2$ in order to find the mass spectrum. The ground state in any particular spin-flavor channel corresponds to the solution with the lowest mass, $M_0$. Excited states can be found by looking for solutions of Eq. (1) with a larger mass $M_i > M_0$, and this can indeed be done [6, 7].

## Rainbow-ladder truncation

A viable truncation of the infinite set of DSEs has to respect relevant (global) symmetries of QCD such as chiral symmetry, Lorentz invariance, and renormalization group invariance. Here we use the so-called rainbow-ladder truncation, in which the $q\bar{q}$ scattering kernel is replaced by an effective one-gluon exchange

$$K(p_{\text{out}}, p_{\text{in}}; k_{\text{out}}, k_{\text{in}}) \quad \rightarrow \quad -4\pi\,\alpha(q^2)\,D_{\mu\nu}^{\text{free}}(q)\tfrac{\lambda^i}{2}\gamma_\mu \otimes \tfrac{\lambda^i}{2}\gamma_\nu, \tag{2}$$

where $q = p_{\text{out}} - k_{\text{out}} = p_{\text{in}} - k_{\text{in}}$, and $\alpha(q^2)$ is an effective running coupling. The corresponding truncation of the quark DSE is

$$S(p)^{-1} \;=\; i\,\slashed{p}\,Z_2 + m_q(\mu)\,Z_4 + \tfrac{4}{3}\int \frac{d^4k}{(2\pi)^4}\,4\pi\alpha(q^2)\,D_{\mu\nu}^{\text{free}}(q)\,\gamma_\mu\,S(k)\,\gamma_\nu, \tag{3}$$

where $S(p) = Z(p^2)/[i\,\slashed{p} + M(p^2)]$ and $q = k - p$. This truncation is the first term in a systematic expansion [8] of the quark-antiquark scattering kernel $K$; asymptotically, it reduces to leading-order perturbation theory. Furthermore, these two truncations are mutually consistent in the sense that the combination produces vector and axial-vector vertices satisfying their respective Ward identities.

For the effective interaction we use the 2-parameter model of Ref. [10]

$$\frac{4\pi\alpha(q^2)}{k^2} = \frac{4\pi^2 D k^2}{\omega^6}\,e^{-k^2/\omega^2} + \frac{4\pi^2 \gamma_m\,\mathscr{F}(k^2)}{\tfrac{1}{2}\ln\left[e^2 - 1 + \left(1 + k^2/\Lambda_{\text{QCD}}^2\right)^2\right]}, \tag{4}$$

with $\mathscr{F}(s) = (1 - e^{-s})/s$, $\gamma_m = 12/(33 - 2N_f)$, and fixed parameters $N_f = 4$ and $\Lambda_{\text{QCD}} = 0.234\,\text{GeV}$, $\omega = 0.4$ GeV and $D = 0.93\,\text{GeV}^2$, were fitted in [10] to reproduce a chiral condensate of $(240\,\text{MeV})^3$ and $f_\pi = 131$ MeV. The first term in Eq. (4) models the infrared enhancement of the effective $q\bar{q}$ scattering kernel necessary to generate the experimentally observed amount of dynamical chiral symmetry breaking [11]. It was introduced in [10] as a finite-width representation of a $\delta$-function [12], which can be interpreted as a regularized $1/p^4$ singularity in $K$ [13, 14]. The second term ensures the correct perturbative behavior in the ultraviolet region.

# Meson spectroscopy

In Table 1 we give our results for the equal-mass ground states in each spin channel. The masses of the light quarks where fitted in [10] to the pion mass (using equal $u$ and $d$ quark masses) and to the kaon mass. The light vector and pseudoscalar mesons are described very well by this model: not only their masses, but also a wide range of other observables agree with experiments, without adjusting any of the parameters, see [4] and references therein. Here we apply this model to heavy quarks as well, and use the vector mesons $J/\psi$ and $\Upsilon$ to fix the $c$ and $b$ masses.

The mass splitting between the pseudoscalar and vector mesons is too large for the heavy quarkonium states, but the decay constants are in reasonable agreement with available data. On the other hand, the mass splitting between the vector and the scalar mesons is too small; and the scalar-pseudoscalar mass difference is reasonable. Also the axialvector masses are too small, but the mass difference between the scalar and the $1^{++}$ states is in agreement with data, both for the light and for the $c$ and $b$ quarks. Similar results for the light quark sector and for the charmonium states were found in Ref. [17] with a slightly different model interaction. Presumably corrections beyond ladder truncation are necessary for the scalar and axialvector masses: there are significant cancellations between these corrections in the pseudoscalar and vector channels [8], but not necessarily in the scalar and axialvector channels.

Over the entire mass range from the chiral limit up to the bottomonium states, the pseudoscalar, vector, and scalar masses can be fitted by

$$M_{meson}^2 = C_0 + C_1\, m_q + C_2\, m_q^2, \tag{5}$$

where $m_q$ is the current quark mass at our renormalization point $\mu = 19$ GeV. The fit parameters are $C_0 = 0$ and $C_1 = 6.94$ for the pseudoscalars, $C_0 = 0.51$ and $C_1 = 7.27$ for the vectors, and $C_0 = 0.38$ and $C_1 = 8.65$ for the scalar mesons, with a common parameter $C_2 \approx 4.6$. The fact that the trajectories can all be fitted with (approximately) the same value for $C_2$ means that for large masses, the meson mass differences become constant: in the limit $m_q \to \infty$ the above fit suggests $\Delta M \to \frac{1}{2}\Delta C_1/\sqrt{C_2}$. Thus this global fit indicates that the mass difference $M_V - M_{PS}$ approaches 0.07 GeV, whereas $M_S - M_{PS}$ approaches 0.4 GeV for heavy quarks; our numerical results however do not exclude that

TABLE 1. Masses and leptonic decay constants for equal-mass ground state $J^{PC}$ mesons. Experimental data are from Ref. [15], with the exception of $f_{\eta_c}$ [16].

| quark flavor | $M_{PS}$ | $f_{PS}$ | $M_V$ | $f_V$ | $M(0^{++})$ | $M(1^{+-})$ | $M(1^{++})$ |
|---|---|---|---|---|---|---|---|
| up/down | 0.1385 | 0.131 | 0.743 | 0.207 | 0.672 | 0.83 | 0.91 |
| expt. | 0.135,0.140 | 0.131 | 0.775 | 0.221 | 0.985 | 1.23 | 1.23 |
| strange | 0.697 | 0.183 | 1.076 | 0.260 | 1.081 | 1.17 | 1.25 |
| expt. | — | — | 1.020 | 0.229 | — | — | — |
| charm | 2.908 | 0.381 | 3.098 | 0.421 | 3.250 | 3.26 | 3.33 |
| expt. | 2.980 | $0.335 \pm 0.075$ | 3.097 | 0.416 | 3.415 | | 3.51 |
| bottom | 9.38 | 0.66 | 9.46 | 0.62 | 9.72 | 9.73 | 9.75 |
| expt. | 9.30 | | 9.46 | 0.715 | 9.86 | | 9.89 |

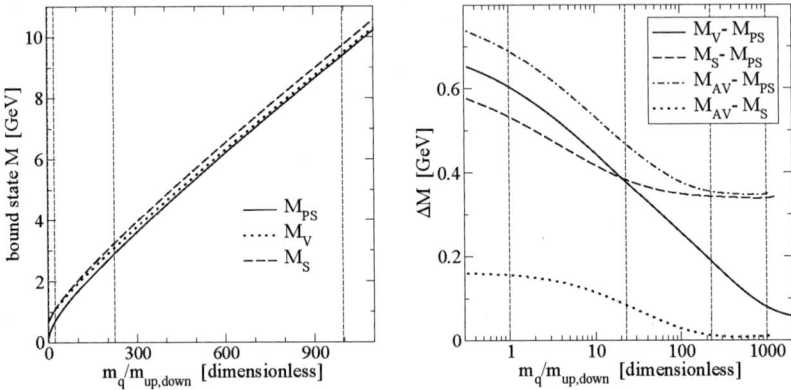

**FIGURE 1.** Meson masses (left) and mass differences (right) as function of current quark mass, normalized to the up and down quark masses. The vertical dashed lines indicate physical quark masses.

the coefficients $C_1$ are identical for the pseudoscalar and vector mesons, in which case this mass difference vanishes in the heavy quark limit.

This is indeed consistent if we look at the actual mass differences we find, see the right panel of Fig. 1: the mass difference $M_V - M_{PS}$ decreases with increasing quark mass, it is about $\Delta_M \approx 0.06$ GeV for at $2m_b$, and still decreasing. Similarly, the mass difference $M_{AV} - M_S$ appears to vanish in the heavy quark limit, but the differences $M_S - M_{PS}$ and $M_{AV} - M_{PS}$ clearly remain nonzero and appear to go to a constant of about $\Delta M \approx 0.35$ GeV. However, one should keep in mind that the model was fitted to the pion decay constant and the chiral condensate; implicitly we may have incorporated corrections beyond the ladder kernel in our model for the effective $q\bar{q}$ scattering kernel. Higher-order corrections affect light quarks differently than heavy quarks [9].

The corresponding quark mass functions are shown in Fig. 2, and summarized in Table 2. Our current quark masses are in good agreement with conventional values [15] of both the light and the heavy quark masses. For the light quarks, the nonperturbative mass function $M_q(p^2)$ is significantly larger than the perturbative quark mass $m_q(\mu)$ at $p = 2 = \mu$, indicating that chiral symmetry breaking sets in well above this scale. The momentum dependence of $M_{c,b}(p^2)$ is much less dramatic. Nevertheless, there is a

**TABLE 2.** Current quark masses $m_q(\mu)$ at $\mu = 19$ GeV, scaled down to $\mu = 2$ GeV and to $\mu = m_q$ using one-loop pQCD, together with the dynamical mass function $M(p^2)$ at several values of $p^2$.

| $m_q(19)$ | $m_q(2)$ | $m_q(m_q)$ | $M_q(p^2 = M_q(p^2)^2)$ | $M_q(p^2 = 4)$ | $M_q(p^2 = 0)$ | $M_q(p^2 = -\frac{1}{4}M_V^2)$ |
|---|---|---|---|---|---|---|
| chiral limit | | | 0.392 | 0.010 | 0.477 | 0.594 |
| 0.0037 | 0.005 | | 0.401 | 0.017 | 0.499 | 0.610 |
| 0.0838 | 0.118 | | 0.556 | 0.168 | 0.689 | 0.845 |
| 0.827 | 1.17 | 1.30 | 1.42 | 1.31 | 1.61 | 2.00 |
| 3.68 | 5.65 | 4.46 | 4.30 | 4.46 | 4.52 | 5.33 |

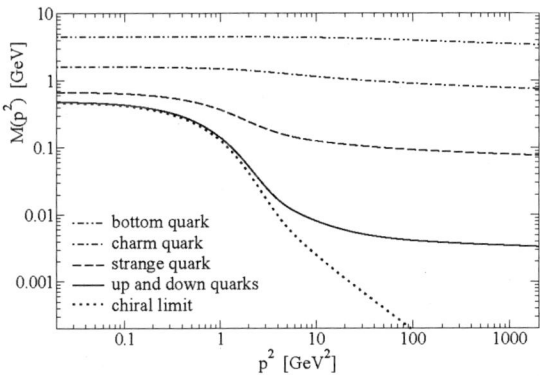

**FIGURE 2.** Dynamical quark mass function $M_q(p^2)$ for $u = d$, $s$, $c$, $b$ and chiral quarks.

significant difference between the dynamical mass in the region relevant for $q\bar{q}$ bound states, namely $p^2 \sim -\frac{1}{4}M_{\text{meson}}^2$ in the timelike region, and $M_{c,b}(p^2)$ in the spacelike region, even for $b$ quarks. For $0 < p^2 < -\frac{1}{4}M_{\text{meson}}^2$, the mass function of the heavy quarks is in fact quite close to the typical pole masses used in non-relativistic calculations of charmonium, $m_c^{\text{pole}} \approx 1.47$ to $1.83$ GeV and bottomonium, $m_b^{\text{pole}} \approx 4.7$ to $5.0$ GeV [15].

## Electromagnetic form factors

The $q\bar{q}\gamma$ vertex is the solution of the renormalized inhomogeneous Bethe–Salpeter equation with the same kernel $K$ as Eq. (1). Thus for photon momentum $Q$, we have

$$\Gamma_\mu(p_{\text{out}}, p_{\text{in}}) = Z_2\,\gamma_\mu + \int \frac{d^4k}{(2\pi)^4} K(p_{\text{out}}, p_{\text{in}}; k_{\text{out}}, k_{\text{in}})\, S(k_{\text{out}})\, \Gamma_\mu(k_{\text{out}}, k_{\text{in}})\, S(k_{\text{in}})\,, \quad (6)$$

with $p_{\text{out}}$ and $p_{\text{in}}$ the outgoing and incoming quark momenta, respectively, and similarly for $k_{\text{out}}$ and $k_{\text{in}}$, with $p_{\text{out}} - p_{\text{in}} = k_{\text{out}} - k_{\text{in}} = Q$. The ladder truncation for Eq. (6), in combination with the rainbow truncation for the quark propagators and impulse approximation for electromagnetic form factors, satisfies the vector Ward–Takahashi identity and electromagnetic current conservation is guaranteed.

Also note that solutions of the *homogeneous* version of Eq. (6) define vector meson bound states with masses $M_V^2 = -Q^2$ at discrete timelike momenta $Q^2$. It follows that $\Gamma_\mu$ has poles at those locations. Thus the effects of intermediate vector meson states on electromagnetic processes can be unambiguously incorporated by using the properly dressed $q\bar{q}\gamma$ vertex rather than the bare vertex $\gamma_\mu$ [18].

Consider for example the 3-point function describing the coupling of a photon with momentum $Q$ to the quark $a$ of a meson $a\bar{b}$, with initial and final momenta $P \pm \frac{1}{2}Q$

$$\Lambda_\mu^a(P, Q) = iN_c \int \frac{d^4k}{(2\pi)^4} \text{Tr}\big[\Gamma_\mu^a(q_-, q_+)\chi^{a\bar{b}}(q_+, q)\, S^b(q)^{-1}\bar{\chi}^{\bar{b}a}(q, q_-)\big]\,, \quad (7)$$

**TABLE 3.** Static electromagnetic properties of pseudoscalar and vector $ud$ mesons ($\pi$ and $\rho$) and $c\bar{c}$ mesons (fictitious).

|  | $r_{PS}^2$ | $r_{V,E}^2$ | $\mu$ | $r_{V,M}^2$ | $\mathcal{Q}$ |
|---|---|---|---|---|---|
| up/down | 0.44 | 0.54 | 2.01 | 0.49 | $-0.41$ |
| charm | 0.048 | 0.052 | 2.13 | 0.047 | $-0.28$ |
| lattice [23] | 0.063(1) | 0.066(2) | 2.10(3) |  | -0.23(2) |

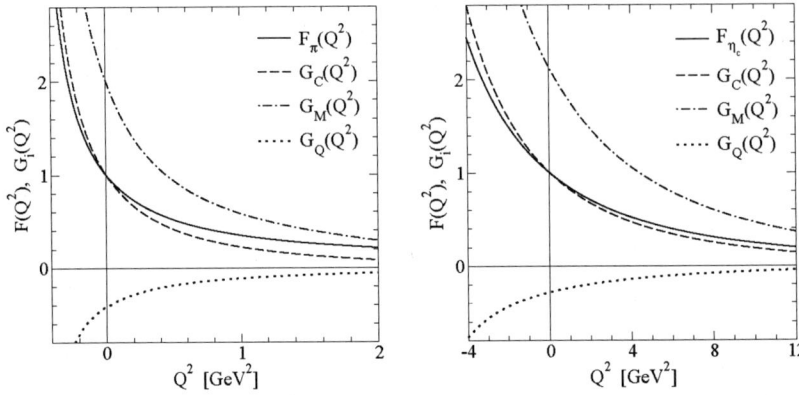

**FIGURE 3.** Single-quark form factors: $\pi$ and $\rho$ (left) and $c\bar{c}$ pseudoscalar and vector mesons (right).

with $q = k - \frac{1}{2}P$ and $q_\pm = k + \frac{1}{2}P \pm \frac{1}{2}Q$. The corresponding single-quark elastic form factor $F^a$ of a pseudoscalar meson is defined by

$$2 P_\mu F^a(Q^2) = \Lambda_\mu^a(P,Q). \tag{8}$$

Vector mesons have three elastic form factors, commonly referred to as the electric, magnetic, and quadrupole form factors $G_E(Q^2)$, $G_M(Q^2)$, and $G_Q(Q^2)$. The electric monopole moment (i.e. the electric charge), magnetic dipole moment and the electric quadrupole moment follow from the values of these form factors in the limit $Q^2 \to 0$: $G_E(0) = 1$ (constrained by current conservation), $G_M(0) = \mu$, and $G_Q(0) = \mathcal{Q}$.

Our results for the pion form factor [18, 19] are in good agreement with the data, both in the spacelike region [20] and in the timelike region; the charge radius agrees very well with the experimental value $\langle r_\pi^2 \rangle = 0.44 \pm 0.01$ fm$^2$ [21], see Table 3. The vector charge radius [22] is slightly larger than the pseudoscalar radius, both for light quarks and for charm quarks. This suggests that the vector states are broader than the corresponding pseudoscalar states, assuming that the charge distribution is indicative of the physical size of the bound state. This agrees with the naive intuition that a more tightly bound state is more compact than a heavier state with the same constituents. For charm quarks this difference is significantly smaller than for up and down quarks, in agreement with recent lattice calculations [23]. The magnetic moment appears to be surprisingly independent of the quark mass; the quadrupole moment decreases with increasing quark mass [22]. Recent lattice simulations [23] agree quite well with our results for the moments of the single-quark form factors of the $J/\psi$.

In Fig. 3 we see that both the pseudoscalar $F_\pi$ and all three vector form factors $G_i^\rho$ diverge in the timelike region as $Q^2 \to -0.55$ GeV$^2$, corresponding to the vector-meson poles in the dressed quark photon vertex. Similarly, the single-quark form factors of the $\eta_c$ and $J/\psi$ diverge as $Q^2 \to -9.5$ GeV$^2$. However, it is only the pion form factor that can be described by a vector meson dominance [VMD] curve, $F_\pi \approx M_\rho^2 / [Q^2 + M_\rho^2]$, over the entire $Q^2$-region shown. The $\rho$ form factors $G_i^\rho$ drop significantly faster [22] than a VMD curve, as do the $c\bar{c}$ form factors, both for pseudoscalar and vector states.

## ACKNOWLEDGMENTS

This work was supported by the US Department of Energy, contract No. DE-FG02-00ER41135, and benefited from the facilities of the NSF Terascale Computing System at the Pittsburgh Supercomputing Center.

## REFERENCES

1.  C.D. Roberts and A.G. Williams, Prog. Part. Nucl. Phys. **33**, 477 (1994) [arXiv:hep-ph/9403224].
2.  C.D. Roberts and S.M. Schmidt, Prog. Part. Nucl. Phys. **45S1**, 1 (2000) [arXiv:nucl-th/0005064].
3.  R. Alkofer and L. von Smekal, Phys. Rept. **353**, 281 (2001) [arXiv:hep-ph/0007355].
4.  P. Maris and C.D. Roberts, Int. J. Mod. Phys. E **12**, 297 (2003) [arXiv:nucl-th/0301049].
5.  C.S. Fischer, J. Phys. G **32**, R253 (2006) [arXiv:hep-ph/0605173].
6.  A. Holl, A. Krassnigg and C.D. Roberts, Phys. Rev. C **70**, 042203 (2004) [arXiv:nucl-th/0406030].
7.  A. Holl *et al.*, Phys. Rev. C **71**, 065204 (2005) [arXiv:nucl-th/0503043].
8.  A. Bender, C.D. Roberts and L. Von Smekal, Phys. Lett. B **380**, 7 (1996) [arXiv:nucl-th/9602012]; A. Bender *et al.*, Phys. Rev. C **65**, 065203 (2002) [arXiv:nucl-th/0202082].
9.  M.S. Bhagwat, A. Holl, A. Krassnigg, C.D. Roberts and P.C. Tandy, Phys. Rev. C **70**, 035205 (2004) [arXiv:nucl-th/0403012].
10. P. Maris and P.C. Tandy, Phys. Rev. C **60**, 055214 (1999) [arXiv:nucl-th/9905056].
11. F.T. Hawes, P. Maris and C.D. Roberts, Phys. Lett. B **440**, 353 (1998) [arXiv:nucl-th/9807056].
12. P. Maris and C.D. Roberts, Phys. Rev. C **56**, 3369 (1997) [arXiv:nucl-th/9708029].
13. D.W. McKay and H.J. Munczek, Phys. Rev. D **55**, 2455 (1997) [arXiv:hep-th/9607075].
14. R. Alkofer, C.S. Fischer and F.J. Llanes-Estrada, arXiv:hep-ph/0607293.
15. W. M. Yao *et al.* [Particle Data Group], J. Phys. G **33**, 1 (2006).
16. K.W. Edwards *et al.* [CLEO Collaboration], Phys. Rev. Lett. **86**, 30 (2001) [arXiv:hep-ex/0007012].
17. R. Alkofer, P. Watson and H. Weigel, Phys. Rev. D **65**, 094026 (2002) [arXiv:hep-ph/0202053].
18. P. Maris and P.C. Tandy, Phys. Rev. C **61**, 045202 (2000) [arXiv:nucl-th/9910033].
19. P. Maris and P.C. Tandy, Phys. Rev. C **62**, 055204 (2000) [arXiv:nucl-th/0005015].
20. V. Tadevosyan *et al.* [Fpi-1 Collaboration], arXiv:nucl-ex/0607007; T. Horn *et al.* [Fpi2 Collaboration], arXiv:nucl-ex/0607005.
21. S. R. Amendolia *et al.* [NA7 Collaboration], Nucl. Phys. B **277**, 168 (1986).
22. M.S. Bhagwat and P. Maris, in preparation.
23. J.J. Dudek, R.G. Edwards and D.G. Richards, Phys. Rev. D **73**, 074507 (2006) [arXiv:hep-ph/0601137].

# Dispersive chiral approach to Meson-meson dynamics: Spectroscopy results for light scalars and precision studies

J. R. Peláez

*Departamento de Física Teórica II, Universidad Complutense, 28040 Madrid, SPAIN*

**Abstract.** Dispersive approaches provide model independent description of meson-meson scattering. We first review here the use of dispersion relations to obtain a model independent unitarization of Chiral Perturbation Theory amplitudes, that establish the existence of light scalar mesons and whose leading $1/N_c$ behavior suggest they ahve a non $\bar{q}q$ dominant component. We also review the forward dispersion relation checks on conflicting experimental data and the resulting very precise $\pi\pi$ scattering scattering amplitudes.

**Keywords:** Chiral Perturbation Theory, Dispersive approach, Unitarization, ligt scalar mesons
**PACS:** 11.30.Rd,11.55.Fv,12.39.Fe,13.75.Lb,14.40.-n

Over the last years there has been a renewed interest in meson-meson scattering. The reasons are that its low energy description is relevant for understanding the QCD vacuum, and that the description below 1.2 GeV requires the existence of some poles in the amplitudes, related to light scalar mesons. Such resonances are the subject of a strong debate, first on their existence, that seems to be settling, but also on their nature.

Data on meson-meson scattering is obtained indirectly from other processes involving nucleons or the decays of other heavy mesons. Thus, very often the existing data on meson-meson amplitudes are extracted with strong model dependent assumptions (one-pion exchange, absorption models, rescattering, pole extrapolations to define the initial meson-meson state, etc...), and are therefore plagued with systematic uncertainties much larger than statistical errors. There are also $K \to \pi\pi l\nu$ experiments, known as $K_{l4}$ decays, where the pion phase shifts are extracted in a particular combination of isospin 0 and 1, free of the previous systematic uncertainties and yield very precise and reliable results, but limited to invariant masses smaller than the kaon mass.

Concerning the theoretical description, we have two model independent approaches. On the one hand, we have Chiral Perturbation Theory (ChPT), which is the effective Lagrangian of QCD, written as an expansion in masses and derivatives of pions, kaons and etas, which are the Goldstone bosons of the spontaneous chiral symmetry breaking of QCD. The only caveat to this systematic expansion is that it can only be applied at low energies. On the other hand, it is possible to use the usual S-matrix constraints of causality, analyticity, unitarity, crossing, etc... to write dispersion relations for the different meson-meson channels. We review here how we have recently applied this approach to check the consistency of different data sets and to obtain a precise pion-pion scattering amplitudes in the whole energy range. Of course, both approaches can also be combined to obtain a model independent description of data that also incorporates the low energy chiral symmetry constraints (see. H. Leutwyler's talk on Roy equations

CP892, *Quark Confinement and the Hadron Spectrum VII*
edited by J. E. F. T. Ribeiro

in this conference). We review here how this has been done by means of the Inverse Amplitude Method (IAM), which has the advantage that only uses ChPT input in the dispersive integrals and therefore allows for a relation between the resulting fits and QCD. Since the IAM generates the light resonances that appear in scattering, it is then possible to study their nature in terms of QCD parameters, like the number of colors.

## THE INVERSE AMPLITUDE METHOD FROM DISPERSION THEORY

We will review here how the *one-channel* Inverse Amplitude Method (IAM) [1, 2, 3] for pion-pion scattering is obtained just by using ChPT up to a given order inside a dispersion relation. There are no further assumptions and therefore the approach is model independent and provides an elastic amplitude that satisfies unitarity and has the correct ChPT expansion up to that given order.

To fix ideas, let us consider the ChPT series for a pion-pion scattering partial wave amplitude of definite isospin $I$ and angular momentum $J$, namely, $t_{IJ} = t_{IJ}^{(2)} + t_{IJ}^{(4)} + ...$ where $t_2 = O(p^2)$, $t_4 = O(p^4)$ and $p$ stands for the pion mass or momentum. For the complete partial wave $t_{IJ}(s)$, it is possible to write a dispersion relation

$$t_{IJ}(s) = C_0 + C_1 s + C_2 s^2 + \frac{s^3}{\pi} \int_{s_{th}}^{\infty} \frac{\text{Im}\, t_{IJ}(s')ds'}{s'^3(s'-s-i\varepsilon)} + LC(t_{IJ}), \qquad (1)$$

that, for convenience, we have subtracted three times. Note we have explicitly written the integral over the right hand cut (or physical cut, extending from threshold, $s_{th}$ to infinity) but we have abbreviated by $LC$ the equivalent expression for the left cut (from 0 to $-\infty$). We could do similarly with other cuts, if present, as in the $\pi K$ case.

We can also write dispersion relations for $t^{(2)}$ and $t^{(4)}$, but remembering that $t^{(2)}$ is a pure tree level amplitude and it does not have imaginary part nor cuts:

$$t_{IJ}^{(2)} = a_0 + a_1 s, \qquad t_{IJ}^{(4)} = b_0 + b_1 s + b_2 s^2 + \frac{s^3}{\pi} \int_{s_{th}}^{\infty} \frac{\text{Im}\, t_{IJ}^{(4)}(s')ds'}{s'^3(s'-s-i\varepsilon)} + LC(t_{IJ}^{(4)}). \qquad (2)$$

We now recall that unitarity, for physical values of $s$ in the elastic region implies:

$$\text{Im}\, t_{IJ} = \sigma |t_{IJ}|^2 \quad \Rightarrow \quad \text{Im}\, \frac{1}{t_{IJ}} = -\sigma \quad \Rightarrow \quad t_{IJ} = \frac{1}{\text{Re}\, t_{IJ}^{-1} - i\sigma}, \qquad (3)$$

where $\sigma = 2p/\sqrt{s}$. Therefore, the imaginary part of the *inverse amplitude* is *exactly* known in the elastic regime. We can then write a dispersion relation like that in (1) but now for the auxiliary function $G = (t_{IJ}^{(2)})^2/t_{IJ}$, i.e.,

$$G(s) = G_0 + G_1 s + G_2 s^2 + \frac{s^3}{\pi} \int_{s_{th}}^{\infty} \frac{\text{Im}\, G(s')ds'}{s'^3(s'-s-i\varepsilon)} + LC(G) + PC, \qquad (4)$$

where now $PC$ stands for possible pole contributions in $G$ coming from zeros in $t_{IJ}$. It is now straightforward to expand the subtraction constants and use that $\text{Im}\, t_{IJ}^{(2)} = 0$

and $\mathrm{Im}\, t_{IJ}^{(4)} = \sigma |t_{IJ}^{(2)}|^2$, so that $\mathrm{Im}\, G = -\mathrm{Im}\, t_{IJ}^{(4)}$. In addition, up to the given order, $LC(G) \simeq -LC(t_{IJ}^{(4)})$, whereas $PC$ is of higher order and can be neglected. Thus

$$\frac{t_{IJ}^{(2)2}}{t_{IJ}} \simeq a_0 + a_1 s - b_0 - b_1 s - b_2 s^2 - \frac{s^3}{\pi} \int_{s_{th}}^{\infty} \frac{\mathrm{Im}\, t_{IJ}^{(4)}(s')ds'}{s'^3(s'-s-i\varepsilon)} - LC(t_{IJ}^{(4)}) \simeq t_{IJ}^{(2)} - t_{IJ}^{(4)}. \quad (5)$$

We have thus arrived to the so-called Inverse Amplitude Method (IAM):

$$t_{IJ} \simeq t_{IJ}^{(2)2} / (t_{IJ}^{(2)} - t_{IJ}^{(4)}), \quad (6)$$

that provides an elastic amplitude that satisfies unitarity and has the correct low energy expansion of ChPT up to the order we have used. It is straightforward to extend it to other elastic channels or to higher orders [3]. Note also that, by looking at (3), it seems that it can also be derived by replacing $\mathrm{Re}\, t_{IJ}^{(-1)}$ by its $O(p^4)$ ChPT expansion. But, strictly speaking, (3) is only valid in the real axis, whereas our derivation allows us to consider the amplitude in the complex plane, and, in particular, look for poles of the associated resonances. Actually, already ten years ago [3], with the single channel IAM we were able to generate poles for the $\rho(770)$, $K^*(892)$ and most interestingly, the controversial $\sigma$ (also called $f_0(600)$), *without any model dependent assumptions*.

Note that **the above one-channel IAM derivation is model independent**, and that contrary to a wide belief in the community **contains a left cut** and *respects crossing symmetry up to*, of course, the order in the ChPT expansion that has been used.

The confusion may come from the fact that the IAM has also been applied in a coupled channel formalism, for which *there is still no dispersive derivation*, and sometimes with furtehr approximations. Indeed one can arrive to (6) in a matrix form, ensuring coupled channel unitarity, just by expanding the real part of the inverse T matrix. *For the coupled channel case* different approximations to $ReT^{-1}$ have been used:

• The fully renormalized one-loop ChPT calculation of $ReT^{-1}$ provides the correct ChPT expansion in all channels, also with left cuts approximated to $O(p^4)$ [4, 5]. Indeed, using ChPT parameters consistent with previous determinations within standard ChPT, it was possible [5, 6] to describe below 1.2 GeV all the scattering channels of two body states made of pions, kaons or etas. Simultaneously, this approach [6] generates poles associated to the $\rho(770)$, $K^*(892)$ vector mesons, and the $f_0(980)$, $a_0(980)$, $\sigma$ and $\kappa$ (also called $K_0(800)$) scalar resonances.

• Originally [7], the coupled channel IAM was used neglecting the crossed loops and tadpoles. This approach is considerably simpler, and although it is true that the left cut is absent, its numerical influence was shown to be rather small, since the meson-meson data are nicely described with very reasonable chiral parameters and generates all the poles enumerated above. Let us remark that this approximation keeps the s-channel loops but also the tree level up to $O(p^4)$, and that this tree level encodes the effect of heavier resonances, like the rho. Thus, contrary to some common belief, this approach still incorporates, for instance, the low energy effects of t-channel rho exchange.

• Finally, if one is interested in describing just the scalar meson-meson channels, it is possible to use just one cutoff (or even a dimensional regularization scale) that numerically mimics the combination of chiral parameters that appear in those scalar channels. This has become very popular, even beyond the meson-meson interaction realm, due to its great simplicity but remarkable success [8].

# NATURE OF LIGHT SCALARS FROM UNITARIZED CHPT

One of the big advantages of the unitarization approaches described in the previous sections is that when they use the fully renormalized ChPT amplitudes, they therefore have the correct chiral and flavor symmetry structure, including both the spontaneous and explicit symmetry breaking. Furthermore we also have the correct dependence on QCD parameters like the number of colors, which is of particular interest, since there are sharp predictions on how the mass and width of $\bar{q}q$ resonances should behave in a large $N_c$ QCD expansion. In particular, $M \simeq O(1)$ whereas $\Gamma \simeq O(1/N_c)$.

The $1/N_c$ leading behavior of all ChPT parameters is known and model independent, so that they can be varied accordingly to study [9] the $N_c$ dependence of the amplitudes. In particular, we can study the $N_c$ behavior of all the poles generated in the IAM. It has been shown that both the mass and width of the vector mesons generated with the IAM follow remarkably well the expected $\bar{q}q$ behavior. However, *light scalars do not follow a dominant $\bar{q}q$ $N_c$ behavior* [9], at least for $N_c$ not too far from real life, $N_c = 3$. In figure 1 we illustrate these two different behaviors calculated at with the $O(p^4)$ IAM, for the $\rho$ $K^*(892)$, and for the $\sigma$, and $\kappa$, whose poles can be obtained with the one-channel IAM, and therefore in a model independent way. Similar plots can be found for the other scalars in [9].

At this point it is worth noting that two-meson loop diagrams are subdominant at large $N_c$. Indeed, the above results imply that for vectors the meson loops play a small role in the cancellations that lead to poles in the amplitude, whereas for scalars such loop diagrams are very important at $N_c = 3$. Since these diagrams become smaller and smaller one could wonder about the influence of higher order effects in scalars. Thus recently, we have performed [10] the full two-loop IAM analysis of the $\sigma$ channel confirming that, as it happened for the $O(p^4)$ amplitude, close to $N_c = 3$ the sigma behaves rather differently than expected for a $\bar{q}q$ state, but that as the loop diagrams are suppressed, a subdominant $\bar{q}q$ behavior is recovered at larger $N_c$. Remarkably, this $\bar{q}q$ behavior arises slightly above 1 GeV, as it can be seen in Figure 1 (right plot). This seems to support the suggestion [11] that there is a non-$\bar{q}q$ scalar nonet below roughly 1 GeV, and another $\bar{q}q$ nonet above, but using a mode independent framework based on dispersive integrals, ChPT and the large $N_c$ QCD behavior.

# PRECISE AMPLITUDES FROM DISPERSION RELATIONS AND ROY EQUATIONS

In the previous approach we used ChPT inside the integrals of partial wave dispersion relations. This has the advantage that we can control all the parameters of our input, and allow for their variation in order to understand the nature of poles or the dependence on certain QCD parameters. However ChPT does not necessarily give a good description of the integrands at high energies. Thus, even though we performed several subtractions to suppress the high energy regime we cannot claim to have very precise amplitudes, but a qualitative and most likely a semi-quantitative description.

There are however other approaches where one can use directly the data inside the integrals. Of course, now it is much harder, if possible at all, to interpret changes in

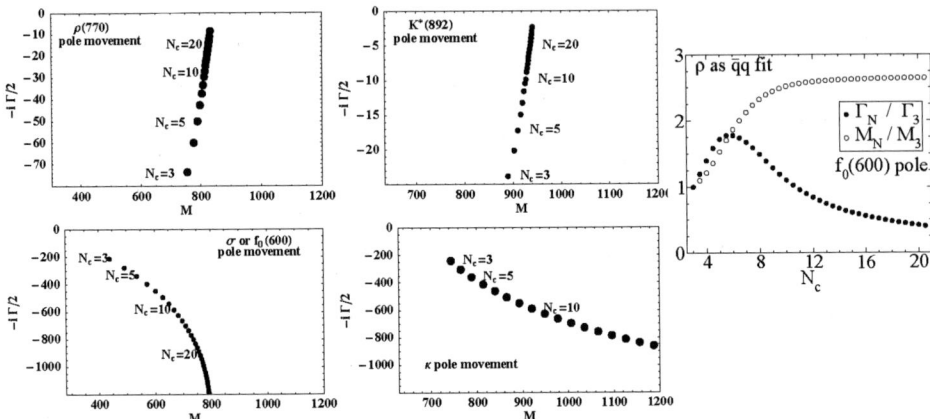

**FIGURE 1.** Leading $N_c$ behavior of resonance poles generated IAM. Left and center columns: with one-loop $O(p^4)$ ChPT Both the $\rho$ and $K^*$ poles in the imaginary plane behave as expected for $\bar{q}q$ states, namely $M \simeq O(1)$ and $\Gamma \simeq O(1/N_c)$, respectively. In contrast, the light scalars $\sigma$ and $\kappa$ do not behave predominantly as $\bar{q}q$. Right: with two-loop $O(p^6)$ ChPT. We plot the mass $M$ and width $\Gamma$ evolution with $N_c$ calculated from the pole position, normalized to the $N_c = 3$ case. Once more the dominant behavior is not that of a $\bar{q}q$ but a subdominant $\bar{q}q$ behavior emerges at larger $N_c$ around 1 GeV in mass.

terms of parameters in terms of QCD, but we can get extremely precise results for the amplitude and other observables like scattering lengths, poles, etc... Recently, ChPT constraints and data have been included in single channel dispersion relations for $\pi\pi$ and $\pi K$ *partial waves* [12], confirming the existence of poles for $\sigma$ and $\kappa$ resonances, introducing some cutoffs on the dispersive integrals. For partial waves, the left cut is always a very delicate issue, because the large $t$ behavior is not well known. Two possible model independent approaches to this problem have been given in the literature: One is to rewrite the left cut integrals in terms of a coupled set of integral equations relating different partial waves, known as Roy equations for $\pi\pi$ scattering and Roy-Steininger equations for $\pi K$ scattering, which have been recently used to obtain precise determinations of the $\sigma$ [13] and $\kappa$ [14] poles, respectively.

Here I will comment on the other approach, namely, to use *Forward* Dispersion Relations (FDR), that is, to avoid using partial waves and use full amplitudes setting $t = 0$. In a recent analysis we have used the following set of dispersion relations that form a complete isospin set: by choosing either $F = F_{00}$ or $F = F_{0+}$ in

$$F(s) - F(4M_\pi^2) = \frac{s(s - 4M_\pi^2)}{\pi} \text{P.P.} \int_{4M_\pi^2}^{\infty} ds' \frac{(2s' - 4M_\pi^2)\text{Im}F(s')}{s'(s' - s)(s' - 4M_\pi^2)(s' + s - 4M_\pi^2)}. \quad (7)$$

we have two dispersion relations which are *twice subtracted* . Thus, the weight of the high energy part is quite suppressed, indeed as much as in Roy equations, but with the advantage of having always *positive* contributions to the integrand, a fact that makes them much more precise. In addition, by setting $s = 2M_\pi^2$, and $F = F_{00}$, we find two sum rules important to fix the Adler zeros. Finally, for the t-channel exchange of isospin 1,

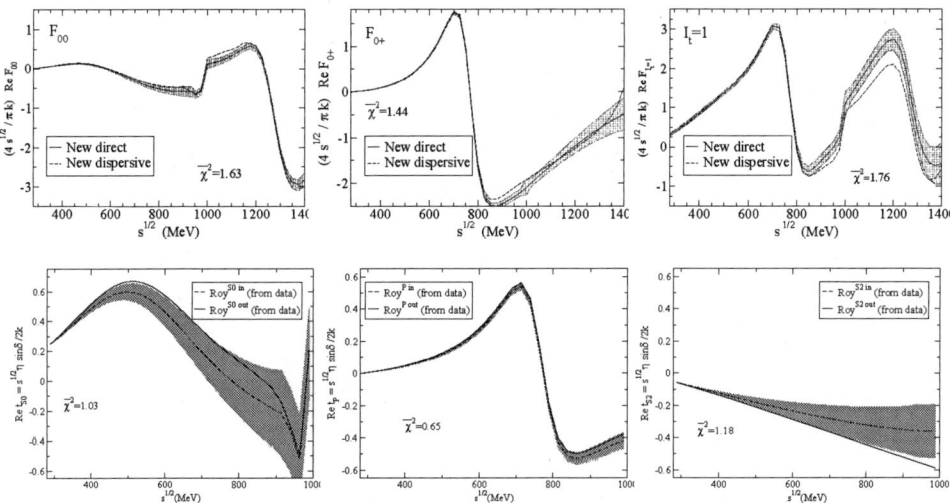

**FIGURE 2.** The data fits obtained in [16, 15] satisfy remarkably well the complete set of FDR and Roy Equations below $KK$ threshold and fairly well above. In this plots we compare the result of the direct parametrization (direct or "in" curves) versus the integral representation (dispersive or "out"). We emphasize these curves come from fits to data, without FDR or Roy equations as constraints, so that the agreement within uncertainties between continuous and dashed lines is even more remarkable.

which does not require subtractions, we use,

$$F^{(I_t=1)}(s,0) = \frac{2s - 4M_\pi^2}{\pi} \text{P.P.} \int_{4M_\pi^2}^{\infty} ds' \frac{\text{Im} F^{(I_t=1)}(s',0)}{(s'-s)(s'+s-4M_\pi^2)}. \tag{8}$$

At threshold this is known as the Olsson sum rule.

First of all, we have found [15] that only a few sets of data available in the literature, satisfy reasonably well the above forward dispersion relations within errors. Some of the most widely used experimental phase shift determinations indeed fail, on average for each FDR, by more than 2 standard standard deviations. The sets that satisfy FDR better are those closer to the fit using just the low energy data on $K_{l4}$ decays commented in the introduction. All these sets have a marked "hunchback" in the 400 to 900 MeV region that undoubtedly is caused by the $\sigma$ pole. We show in Figure 2 (upper row) the comparison between the FDR left hand sides ("direct" calculation) and FDR right hand sides ("dispersive" calculation). Note that the agreement below $KK$ threshold is astonishing given the tiny uncertainties and that only data has been fitted, not the FDR themselves. We have recently checked [18] that our direct fits to data also satisfy remarkably well Roy equations, as shown in Figure 2 (lower row).

In [15] we went further than this and improved the $\pi\pi$ amplitudes by constraining the fits to different partial waves to satisfy the FDR below 950 MeV. In this way we obtained an even more precise representation that lies not too far from the best fits to data only, precisely those showing the $\sigma$ "hunchback" in the S0 wave.

Compared with the solution from Roy equations in [17], the phase shift solution that satisfy better the FDR representation in [15] agrees within errors in the low energy region for the S and P waves, but the S0 phase shifts above 400 MeV lie about 2 standard deviations higher than in [17]. We also have a discrepancy of about two standard deviations in the D wave scattering lengths and the Regge residue of the $\rho$.

At present, and using our very recent improved fits above $KK$ threshold, we are working on an even more precise $\pi\pi$ amplitude by constraining our fits to satisfy simultaneously FDR and Roy equations, together with Froissart-Gribov and other sum rules. This parametrization will be used to obtain a precise determination of the $\sigma$ pole as well as of threshold parameters to determine the low energy ChPT constants.

In summary, I believe that, dispersive approaches combined with ChPT are the most powerful technique at our disposal to describe precisely and understand in terms of QCD the meson-meson interactions and light scalar spectroscopy.

## ACKNOWLEDGMENTS

I thank the organization staff, and particularly E. Ribeiro, for creating such a nice working environment and offering me the chance to participate in this wonderful Conference.

## REFERENCES

1. T. N. Truong, Phys. Rev. Lett. **61** (1988) 2526. A. Dobado, M. J. Herrero and T. N. Truong, Phys. Lett. B **235**, 134 (1990).
2. A. Dobado and J. R. Pelaez, Phys. Rev. D **47**, 4883 (1993)
3. A. Dobado and J. R. Pelaez, Phys. Rev. D **56**, 3057 (1997)
4. F. Guerrero and J. A. Oller, Nucl. Phys. B **537**, 459 (1999) [Erratum-ibid. B **602**, 641 (2001)]
5. A. Gomez Nicola and J. R. Pelaez, Phys. Rev. D **65**, 054009 (2002)
6. J. R. Pelaez, Mod. Phys. Lett. A **19**, 2879 (2004)
7. J. A. Oller, E. Oset and J. R. Pelaez, Phys. Rev. Lett. **80**, 3452 (1998). Phys. Rev. D **59**, 074001 (1999) [Erratum-ibid. D **60**, 099906 (1999)]
8. J. A. Oller and E. Oset, Nucl. Phys. A **620**, 438 (1997) [Erratum-ibid. A **652**, 407 (1999)]
9. J. R. Pelaez, Phys. Rev. Lett. **92**, 102001 (2004)
10. J. R. Pelaez and G. Rios, arXiv:hep-ph/0610397. To appear in Phys. Rev. Lett.
11. E. Van Beveren, *et al.* Z. Phys. C **30**, 615 (1986) and hep-ph/0606022. E. van Beveren and G. Rupp, Eur. Phys. J. C **22** (2001) 493, J. A. Oller and E. Oset, Phys. Rev. D **60** (1999) 074023. F. E. Close and N. A. Tornqvist, J. Phys. G **28**, R249 (2002) (See. N.A. Tornqvist talk in this workshop)
12. Z. Y. Zhou and H. Q. Zheng, Nucl. Phys. A **775**, 212 (2006). Z. Y. Zhou, *et al.*,JHEP **0502**, 043 (2005)
13. I. Caprini, G. Colangelo and H. Leutwyler, Phys. Rev. Lett. **96**, 132001 (2006) (See. H. Leutwyler's talk in this conference)
14. S. Descotes-Genon and B. Moussallam, arXiv:hep-ph/0607133.
15. J. R. Pelaez and F. J. Yndurain, Phys. Rev. D **71**, 074016 (2005)
16. R. Kaminski, J. R. Pelaez and F. J. Yndurain, Phys. Rev. D **74**, 014001 (2006) [Erratum-ibid. D **74**, 079903 (2006)]
17. G. Colangelo, J. Gasser and H. Leutwyler, Nucl. Phys. B **603**, 125 (2001)
18. R. Kaminski, J. R. Pelaez and F. J. Yndurain, in preparation.

# Hybrid potentials versus gluelumps

Antonio Pineda

Grup de Física Teòrica and IFAE, Universitat Autònoma de Barcelona, E-08193 Bellaterra,
Barcelona, Spain

**Abstract.** A potential model description of heavy quarkonium can be rigorously deduced from
QCD under some circumstances. The potentials can be unambiguously related with Wilson loops
with gluonic insertions, the spectral decomposition of which is a function of the spectrum and
matrix elements solution of the static limit of NRQCD. This spectrum is nothing but the static
singlet potential and the hybrid potentials (which correspond to the gluonic excitations). We will
quantitatively show that the latter unambiguously relate to the gluelumps at short distances using
effective field theories.

**Keywords:** Potentials, heavy quarkonium, effective theories
**PACS:** 12.39.Pn, 12.39.Hg, 12.38.Gc

## INTRODUCTION

Potentials appear in a natural way in the description of systems with slow moving particles like in heavy quarkonium. It is now possible [1, 2] to obtain a potential model-like description of the heavy quarkonium dynamics from first principles using effective field theories under some circumstances (for a review see [3]). The study of these potentials (the static [4] and the relativistic [5, 6, 7] corrections) becomes then crucial as they dictate the dynamic of the heavy quarkonium[1]. They can be obtained from the solution (spectrum and matrix elements) of NRQCD in the static limit. The spectrum of NRQCD directly provides the static singlet potential and the hybrid potentials. Therefore it is interesting to study them on their own. In particular, the study of the hybrid potentials may become important in order to discern whether the recently found resonances near threshold can be understood as hybrid states (see [8] for a review on the different possible interpretation of these new states).

In this review we closely follow [9]. We will focus on the short distance limit of the hybrid potentials, specially on the aspects that can be fixed on symmetry arguments only, and we will quantitatively relate them with the physics of gluelumps.

---

[1] One may wonder why bother to compute these potentials in the lattice, as one may try to do a direct computation of the heavy quarkonium properties in the lattice. Leaving aside the technical difficulties of these direct computations (the computation of the potentials is, technically, much easier), in the best of the worlds they only provide a very limited information on the properties of the heavy quarkonium. Mainly a few masses (for the ground state) and, maybe, some inclusive decays. On the other hand, with the potentials it is possible to obtain a detailed information of the shape of the heavy quarkonium, one can then compute the complete spectrum, all the inclusive decays and also opens the possibility to consider differential decay rates, which are far beyond the possibility of present direct lattice simulations.

CP892, *Quark Confinement and the Hadron Spectrum VII*
edited by J. E. F. T. Ribeiro
© 2007 American Institute of Physics 978-0-7354-0396-3/07/$23.00

# STATIC LIMIT OF PNRQCD

The static limit of NRQCD at short distances can be studied with the static version of pNRQCD. In this limit new symmetries arise.

The pNRQCD Lagrangian at leading order in $1/m$ and in the multipole expansion reads [10, 11],

$$L_{\text{pNRQCD}} = \int d^3r\, d^3\mathbf{R}\, \text{Tr}\left[ S^\dagger \left(i\partial_0 - V_s\right)S + O^\dagger \left(iD_0 - V_o\right)O \right] - \int d^3\mathbf{R}\, \frac{1}{4}F^a_{\mu\nu}F^{\mu\nu a} + O(r). \tag{1}$$

All the gauge fields in Eq. (1) are evaluated in $\mathbf{R}$ and $t$, in particular $F^{\mu\nu a} \equiv F^{\mu\nu a}(\mathbf{R},t)$ and $iD_0O \equiv i\partial_0 O - g[A_0(\mathbf{R},t),O]$. The singlet and octet potentials $V_i$, $i = s,o$ are to be regarded as matching coefficients, which depend on the scale $v_{us}$ separating soft gluons from ultrasoft ones. In the static limit "soft" energies are of $O(1/r)$ and "ultrasoft" energies are of $O(\alpha_s/r)$. Notice that the hard scale, $m$, plays no rôle in this limit. The only assumption made so far concerns the size of $r$, i.e. $1/r \gg \Lambda_{\text{QCD}}$, such that the potentials can be computed in perturbation theory. Also note that throughout this paper we will adopt a Minkowski space-time notation.

The spectrum of the singlet state reads,

$$E_s(r) = 2m_{\text{OS}} + V_s(r) + O(r^2), \tag{2}$$

where $m_{\text{OS}}$ denotes an on-shell (OS) mass. One would normally apply pNRQCD to quarkonia and in this case $m_{\text{OS}}$ represents the heavy quark pole mass. For the static hybrids, the spectrum reads

$$E_H(r) = 2m_{\text{OS}} + V_o(r) + \Lambda_H^{\text{OS}} + O(r^2), \tag{3}$$

where

$$\Lambda_H^{\text{OS}} \equiv \lim_{T \to \infty} i\frac{\partial}{\partial T} \ln\langle H^a(T/2)\phi(T/2,-T/2)H^b(-T/2)\rangle. \tag{4}$$

$$\begin{aligned}\phi(T/2,-T/2) &\equiv \phi(T/2,\mathbf{R},-T/2,\mathbf{R}) \\ &= \text{P}\exp\left\{-ig\int_{-T/2}^{T/2} dt\, A_0(\mathbf{R},t)\right\},\end{aligned} \tag{5}$$

denotes the Schwinger line in the adjoint representation and $H$ represents some gluonic field, for examples see Table 1.

Eq. (3) allows us to relate the energies of the static hybrids $E_H$ to the energies of the gluelumps,

$$\Lambda_H^{\text{OS}} = [E_H(r) - E_s(r)] - [V_o(r) - V_s(r)] + O(r^2). \tag{6}$$

This equation encapsulates one of the central ideas of this paper. The combination $E_H - E_s$ is renormalon-free in perturbation theory [up to possible $O(r^2)$ effects], and can be calculated unambiguously non-perturbatively: the ultraviolet (UV) renormalons related to the infrared (IR) renormalons of twice the pole mass cancel each other. However, $\Lambda_H$ contains an UV renormalon that corresponds to the leading IR renormalon of $V_o$.

# Symmetries of hybrid potentials and gluelumps

The spectrum of open QCD string states can be completely classified by the quantum numbers associated with the underlying symmetry group, up to radial excitations. In this case, these are the distance between the endpoints, the gauge group representation under which these endpoints transform (in what follows we consider the fundamental representation), and the symmetry group of cylindrical rotations with reflections $D_{\infty h}$. The irreducible representations of the latter group are conventionally labelled by the spin along the axis, $\Lambda$, where $\Sigma, \Pi, \Delta$ refer to $\Lambda = 0, 1, 2$, respectively, with a subscript $\eta = g$ for gerade (even) $PC = +$ or $\eta = u$ for ungerade (odd) $PC = -$ transformation properties. All $\Lambda \geq 1$ representations are two-dimensional. The one-dimensional $\Sigma$ representations have, in addition to the $\eta$ quantum number, a $\sigma_v$ parity with respect to reflections on a plane that includes the two endpoints. This is reflected in an additional $\pm$ superscript. The state associated with the static singlet potential transforms according to the representation $\Sigma_g^+$ while the two lowest lying hybrid potentials are within the $\Pi_u$ and $\Sigma_u^-$ representations, respectively.

In contrast, point-like QCD states are characterised by the $J^{PC}$ of the usual $O(3) \otimes \mathscr{C}$ rotation group as well as by the gauge group representation of the source. In the pure gauge sector, gauge invariance requires this representation to have vanishing triality, such that the source can be screened to a singlet by the glue. States created by operators in the singlet representation are known as glueballs, octet states as gluelumps. In contrast to gluelump states, where the octet source propagates through the gluonic background, the normalization of glueball states with respect to the vacuum energy is unambiguous.

Since $D_{\infty h} \subset O(3) \otimes \mathscr{C}$, in the limit $r \to 0$ certain hybrid levels must become degenerate. For instance, in this limit, the $\Sigma_u^-$ state corresponds to a $J^{PC} = 1^{+-}$ state with $J_z = 0$ while the $\Pi_u$ doublet corresponds to its $J_z = \pm 1$ partners. The gauge transformation property of the hybrid potential creation operator will also change in this limit, $3 \otimes 3^* = 1 \oplus 8$, such that hybrids will either approach gluelumps [cf. Eq. (3)] or glueballs, in an appropriate normalization. In the case of glueballs the correct normalization can be obtained by considering the difference $E_H(r) - E_s(r)$ from which the pole mass cancels. We will discuss the situation with respect to gluelumps in detail below.

In perturbation theory, the ground state potential corresponds to the singlet potential while hybrid potentials will have the perturbative expansion of the octet potential, which have recently been computed to two loops [15].

# Hybrid and gluelump mass splittings

We would like to establish if lattice data on hybrid potentials reproduces the degeneracies expected from the above discussion in the short distance region. In the limit $r \to 0$, any given $\Lambda \geq 1$ hybrid potential can be subduced from any $J^{PC}$ state with $J \geq \Lambda$ and $PC = +$ for $\eta = g$ or $PC = -$ for $\eta = u$ representations. For instance the $\Pi_u$ is embedded in $1^{+-}, 1^{-+}, 2^{+-}, 2^{-+}, \cdots$. The situation is somewhat different for $\Lambda = 0$ states, which have the additional $\sigma_v$ parity: the $\Sigma_g^+$ representation can be obtained from $0^{++}, 1^{--}, 2^{++}, \cdots$, $\Sigma_g^-$ from $0^{--}, 1^{++}, \cdots$, $\Sigma_u^+$ from $0^{+-}, 1^{-+}, \cdots$ and $\Sigma_u^-$

**TABLE 1.** Expected degeneracies of hybrid potentials at short distance, based on the level ordering of the gluelump spectrum. Note that if the $3^{+-}$ gluelump turned out to be lighter than the $2^{+-}$ then the $\Sigma_u^{-'}, \Pi_u', \Delta_u, \Phi_u$ potentials would approach the $3^{+-}$ state while the $\Sigma_u^+, \Pi_u'', \Delta_u'$ potentials would approach the $2^{+-}$ instead.

| point particle $J^{PC}$ | $H$ | $\Lambda_H^{RS} r_0$ | $\Lambda_H^{RS}$/GeV | open string $\Lambda_\eta^{\sigma_v}$ |
|---|---|---|---|---|
| $1^{+-}$ | $B_i$ | 2.25(39) | 0.87(15) | $\Sigma_u^-, \Pi_u$ |
| $1^{--}$ | $E_i$ | 3.18(41) | 1.25(16) | $\Sigma_g^{+'}, \Pi_g$ |
| $2^{--}$ | $D_{\{i}B_{j\}}$ | 3.69(42) | 1.45(17) | $\Sigma_g^-, \Pi_g', \Delta_g$ |
| $2^{+-}$ | $D_{\{i}E_{j\}}$ | 4.72(48) | 1.86(19) | $\Sigma_u^+, \Pi_u', \Delta_u$ |
| $3^{+-}$ | $D_{\{i}D_jB_{k\}}$ | 4.72(45) | 1.86(18) | $\Sigma_u^{-'}, \Pi_u'', \Delta_u', \Phi_u$ |
| $0^{++}$ | $B^2$ | 5.02(46) | 1.98(18) | $\Sigma_g^{+''}$ |
| $4^{--}$ | $D_{\{i}D_jD_kB_{l\}}$ | 5.41(46) | 2.13(18) | $\Sigma_g^{-'}, \Pi_g'', \Delta_g', \Phi_g, \Gamma_g$ |
| $1^{-+}$ | $(B\wedge E)_i$ | 5.45(51) | 2.15(20) | $\Sigma_u^{+'}, \Pi_u'''$ |

from $0^{-+}, 1^{+-}, \cdots$. We list all combinations of interest to us in Table 1. The ordering of low lying gluelumps has been established in Ref. [12] and reads with increasing mass: $1^{+-}, 1^{--}, 2^{--}, 2^{+-}, 3^{+-}, 0^{++}, 4^{--}, 1^{-+}$, with a $3^{--}$ state in the region of the $4^{--}$ and $1^{-+}$. The $2^{+-}$ and $3^{+-}$ as well as the $4^{--}$ and $1^{-+}$ states are degenerate within present statistical uncertainties[2]. The continuum limit gluelump masses are displayed as circles at the left of Fig. 1, where we have added the (arbitrary) overall constant $2.26/r_0$ to the gluelump splittings to match the hybrid potentials.

Juge, Kuti and Morningstar [13, 14] have comprehensively determined the spectrum of hybrid potentials. We convert their data, computed at their smallest lattice spacing $a_\sigma \approx 0.2$ fm, into units of $r_0 \approx 0.5$ fm. Since the results have been obtained with an improved action and on anisotropic lattices with $a_\tau \approx a_\sigma/4$, one might expect lattice artifacts to be small, at least for the lower lying potentials. Hence we compare these data, normalized to $E_{\Sigma_g^+}(r_0)$, with the continuum expectations of the gluelumps [12]. The full lines are cubic splines to guide the eye while the dashed lines indicate the gluelumps towards which we would expect the respective potentials to converge.

The first 7 hybrid potentials are compatible with the degeneracies suggested by Table 1. The next state is trickier since it is not clear whether $2^{+-}$ or $3^{+-}$ is lighter. In the figure we depict the case for a light $2^{+-}$. This would mean that $(\Sigma_u^+, \Pi_u', \Delta_u)$ approach the $2^{+-}$ while $(\Sigma_u^{-'}, \Pi_u'', \Delta_u', \Phi_u)$ approach the $3^{+-}$. Note that of the latter four potentials only data for $\Pi_u''$ and $\Phi_u$ are available. Also note that the continuum states $\Pi_u', \Pi_u''$ and $\Phi_u$ are all obtained from the same $E_u$ lattice representation. For the purpose of the figure we make an arbitrary choice to distribute the former three states among the $E_u', E_u''$ and $E_u'''$ lattice potentials. To firmly establish their ordering one would have to investigate radial excitations in additional lattice hybrid channels and/or clarify the gluelump spectrum in more detail. Should the $2^{+-}$ and $3^{+-}$ hybrid levels be inverted then $(\Sigma_u^{-'}, \Pi_u', \Delta_u, \Phi_u)$ will converge to the $3^{+-}$ while $(\Sigma_u^+, \Pi_u'', \Delta_u')$ will approach the $2^{+-}$. We note that the

---

[2] The splittings between all states with respect to the $1^{+-}$ ground state have been extrapolated to the continuum limit in Ref. [12] and we add our own extrapolations for the $4^{--}$ and $1^{-+}$ states to these, based on the tables of this reference.

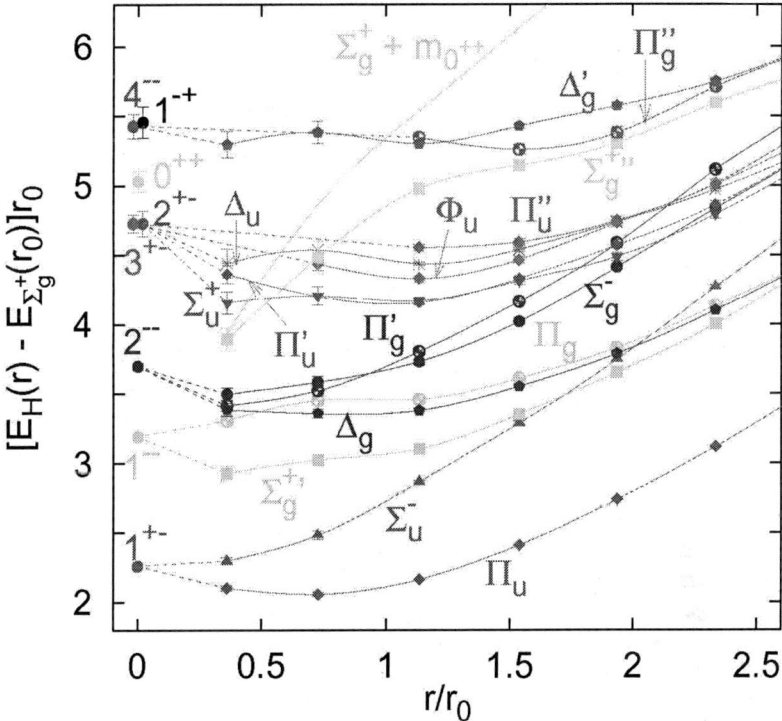

**FIGURE 1.** Different hybrid potentials [14] at a lattice spacing $a_\sigma \approx 0.2$ fm $\approx 0.4 r_0$, where $r_0 \approx 0.5$ fm, in comparison with the gluelump spectrum, extrapolated to the continuum limit [12] (circles, left-most data points). The gluelump spectrum has been shifted by an arbitrary constant to adjust the $1^{+-}$ state with the $\Pi_u$ and $\Sigma_u^-$ potentials at short distance. In addition, we include the sum of the ground state ($\Sigma_g^+$) potential and the scalar glueball mass $m_{0^{++}}$. The lines are drawn to guide the eye.

ordering of the hybrid potentials, with a low $\Sigma_u^+$, makes the first interpretation more suggestive.

Finally the $\Sigma_g^{+\prime\prime}$ potential seems to head towards the $0^{++}$ gluelump but suddenly turns downward, approaching the (lighter) sum of ground state potential and scalar glueball instead. The latter type of decay will eventually happen for all lattice potentials but only at extremely short distances. We also remark that all potentials will diverge as $r \to 0$. This does not affect our comparison with the gluelump results, since we have normalized them to the $\Pi_u/\Sigma_u^-$ potentials at the shortest distance available. (The gluelump values are plotted at $r = 0$ to simplify the figure.)

On a qualitative level the short-distance data are very consistent with the expected degeneracies. From the figure we see that at $r \approx 2 r_0 \approx 1$ fm the spectrum of hybrid potentials displays the equi-distant band structure one would qualitatively expect from a string picture. Clearly this region, as well as the cross-over region to the short-distance behaviour $r_0 < r < 2 r_0$, cannot be expected to be within the perturbative domain: at best one can possibly imagine perturbation theory to be valid for the left-most 2 data points.

With the exception of the $\Pi_u$, $\Pi'_u$ and $\Phi_u$ potentials there are also no clear signs for the onset of the short distance $1/r$ behaviour with a positive coefficient as expected from perturbation theory. Furthermore, most of the gaps within multiplets of hybrid potentials, that are to leading order indicative of the size of the non-perturbative $r^2$ term, are still quite significant, even at $r = 0.4\,r_0$; for instance the difference between the $\Sigma_u^-$ and $\Pi_u$ potentials at this smallest distance is about $0.28\,r_0^{-1} \approx 110$ MeV.

## The difference between the $\Pi_u$ and $\Sigma_u^-$ hybrids

From the above considerations it is clear that for a more quantitative study we need lattice data at shorter distances, which have been obtained for the lowest two gluonic excitations, $\Pi_u$ and $\Sigma_u^-$ in Ref. [9]. We display their differences in the continuum limit in Fig. 2. We see how these approach zero at small $r$, as expected from the short distance expansion. pNRQCD predicts that the next effects should be of $O(r^2)$ (and renormalon-free). In fact, we can fit the lattice data rather well with a $E_{\Pi_u} - E_{\Sigma_u^-} = A_{\Pi_u - \Sigma_u^-} r^2$ ansatz for short distances, with slope (see Fig. 2),

$$A_{\Pi_u - \Sigma_u^-} = 0.92^{+0.53}_{-0.52}\, r_0^{-3}\,, \tag{7}$$

where the error is purely statistical (lattice). This fit has been done using points $r \lesssim 0.5\,r_0$. By increasing the fit range to $r \lesssim 0.8\,r_0$ the following result is obtained,

$$A_{\Pi_u - \Sigma_u^-} = (0.83 \pm 0.29)\, r_0^{-3}\,, \tag{8}$$

indicating stability of the result Eq. (7) above.

In order to estimate systematic errors one can add a quartic term: $b r^4$ (only even powers of $r$ appear in the multipole expansion of this quantity). If the result is stable, our determination of $A_{\Pi_u - \Sigma_u^-}$ should not change much. Actually this is what happens. If we fit up to $r \lesssim 0.5\,r_0$, we obtain the central value $A_{\Pi_u - \Sigma_u^-} r_0^3 = 0.93$ with a very small quartic coefficient, $b r_0^5 = -0.05$. If we increase the range to $r \lesssim 0.8\,r_0$, we obtain the same central value, $A_{\Pi_u - \Sigma_u^-} r_0^3 = 0.93$, but with a slightly bigger quartic term, $b r_0^5 = -0.18$. Introducing the quartic term enhances the stability of $A_{\Pi_u - \Sigma_u^-}$ under variations of the fit range. From this discussion we conclude that the systematic error is negligible, in comparison to the error displayed in our result Eq. (7).

We remark that within the framework of static pNRQCD and to second order in the multipole expansion, one can relate the slope $A_{\Pi_u - \Sigma_u^-}$ to gluonic correlators of QCD.

## CONCLUSIONS

In conclusion, we have shown that the short distance behavior of the hybrid potentials is consistent with perturbation theory, the operator product expansion (effective field theories), and with the description of the leading non-perturbative effects in this regime in terms of the gluelump masses. For more details see [9].

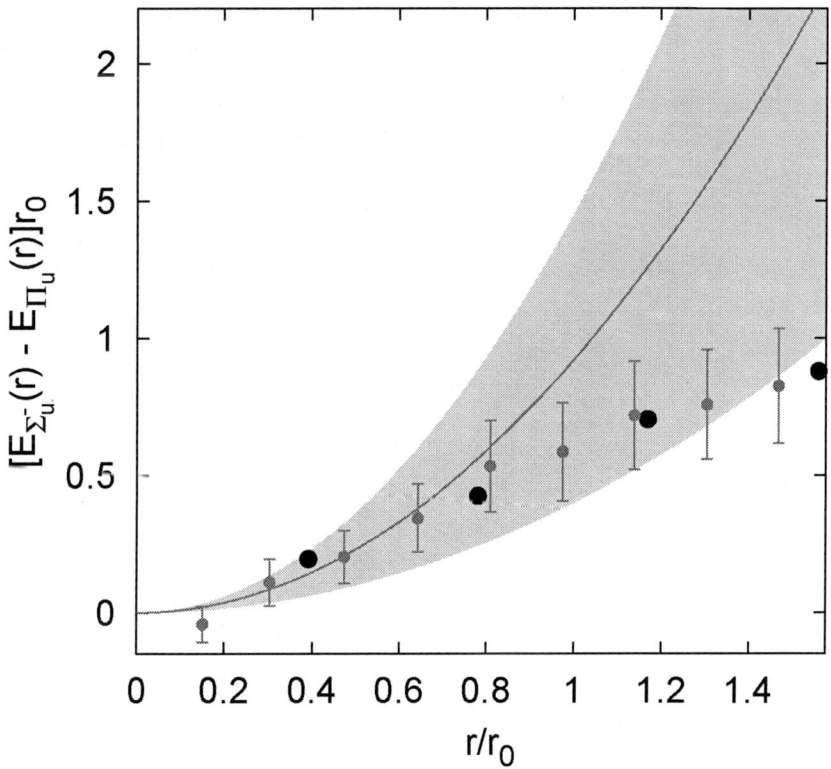

**FIGURE 2.** Splitting between the $\Sigma_u^-$ and the $\Pi_u$ potentials, extrapolated to the continuum limit, and the comparison with a quadratic fit to the $r \lesssim 0.5\,r_0$ data points ($r_0^{-1} \approx 0.4$ GeV). The big circles correspond to the data of Juge et al. [14], obtained at finite lattice spacing $a_\sigma \approx 0.39\,r_0$. The errors in this case are smaller than the symbols.

## REFERENCES

1. N. Brambilla, A. Pineda, J. Soto and A. Vairo, Phys. Rev. D **63**, 014023 (2001).
2. A. Pineda and A. Vairo, Phys. Rev. D **63**, 054007 (2001) [Erratum-ibid. D **64**, 039902 (2001)].
3. N. Brambilla, A. Pineda, J. Soto and A. Vairo, Rev. Mod. Phys. **77**, 1423 (2005).
4. S. Necco and R. Sommer, Nucl. Phys. B **622**, 328 (2002).
5. G. S. Bali, K. Schilling and A. Wachter, Phys. Rev. D **56**, 2566 (1997).
6. Y. Koma, M. Koma and H. Wittig, arXiv:hep-lat/0607009.
7. Y. Koma and M. Koma, arXiv:hep-lat/0609078.
8. E. Eichten, these proceedings.
9. G. S. Bali and A. Pineda, Phys. Rev. D **69**, 094001 (2004).
10. A. Pineda and J. Soto, Nucl. Phys. Proc. Suppl. **64**, 428 (1998).
11. N. Brambilla, A. Pineda, J. Soto and A. Vairo, Nucl. Phys. B **566**, 275 (2000).
12. M. Foster and C. Michael [UKQCD Collaboration], Phys. Rev. D **59**, 094509 (1999).
13. K. J. Juge, J. Kuti and C. J. Morningstar, Nucl. Phys. Proc. Suppl. **63**, 326 (1998).
14. K. J. Juge, J. Kuti and C. Morningstar, Phys. Rev. Lett. **90**, 161601 (2003).
15. B.A. Kniehl, A.A. Penin, Y. Schroder, V.A. Smirnov and M. Steinhauser, Phys. Lett. B **607**, 96 (2005).

# QCD Factorization for heavy quarkonium production at collider energies

## Jian-Wei Qiu

*Department of Physics and Astronomy, Iowa State University*
*Ames, Iowa 50011, U.S.A.*
*Physics Department, Brookhaven National Laboratory*
*Upton, New York 11973-5000, U.S.A.*

**Abstract.** In this talk, I briefly review several models of the heavy quarkonium production at collider energies, and discuss the status of QCD factorization for these production models.

**Keywords:** Heavy quarkonium, factorization
**PACS:** 12.38.Bx, 12.39.St, 13.87.Fh, 14.40Gx

## INTRODUCTION

The production of bound states of heavy quark pairs has been the subject of a vast theoretical literature and of intensive experimental study, and offers unique perspectives into the formation of QCD bound states [1]. The first step in quarkonium production, the inclusive creation of a pair of heavy quarks, is an essentially perturbative process, and takes place at a distance scale much smaller than the physical size of a quarkonium. Therefore, the transition from the produced heavy quark pairs to bound mesons is unlikely to be instantaneous. The "long" lifetime of the produced pairs allows the dynamics of QCD confinement to evolve, and provides us with a window of opportunities to probe the formation of QCD bound states via the interaction between the pairs and the medium where they were produced.

Much of the predictive content of QCD perturbation theory is contained in factorization theorems [2]. They allow us to separate long-distance physics from short-distance interactions in hadronic cross sections, and to provide physical content for the uncalculable long-distance quantities, so that they can be measured independently or calculated numerically. In this talk, I first briefly review several models of heavy quarkonium production, and then, discuss the status of QCD factorization for these production models, and finally, give a brief summary and outlook.

## PRODUCTION MODELS

In order to produce a heavy quarkonium in hadronic collisions, the energy exchange in the collisions has to be larger than the invariant mass of the produced quark pair ($\geq 2m_Q$ with heavy quark mass $m_Q$). The pairs should be produced at a distance scale $\Delta r \leq 1/2m_Q \leq 0.1$ fm for charmonia or 0.025 fm for bottomonia. Since the binding energy of a heavy quarkonium of mass $M$ is much less than heavy quark mass, ($M^2 -$

CP892, *Quark Confinement and the Hadron Spectrum VII*
edited by J. E. F. T. Ribeiro
© 2007 American Institute of Physics 978-0-7354-0396-3/07/$23.00

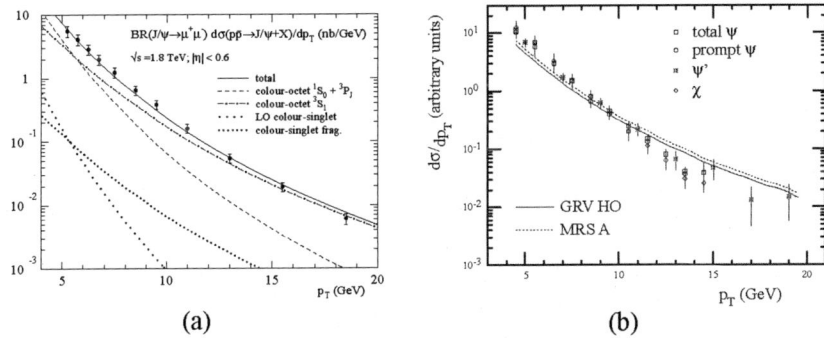

**FIGURE 1.** Charmonium cross section as a function of $p_T$ along with the CDF data points [7] and the theory curves from NRQCD model from Ref. [6] (a), and CEM from Ref. [9] (b).

$4m_Q^2)/4m_Q^2 \ll 1$, the transition from the pair to a meson is sensitive to soft physics. The quantum interference between the production of the heavy quark pairs and the transition process is powerly suppressed by the heavy quark mass, and the production rate for a heavy quarkonium state, $H$, up to corrections in powers of $1/m_Q$, can be factorized as,

$$\sigma_{A+B\to H+X} \approx \sum_n \int d\Gamma_{Q\bar{Q}} \, \sigma_{A+B\to Q\bar{Q}[n]+X}(\Gamma_{Q\bar{Q}}, m_Q) \, F_{Q\bar{Q}[n]\to H}(\Gamma_{Q\bar{Q}}) \qquad (1)$$

with a sum over possible $Q\bar{Q}[n]$ states and an integration over available $Q\bar{Q}$ phase space $d\Gamma_{Q\bar{Q}}$. Neglecting the "high twist" interaction between incoming hadrons and quarkonium formation, the nonperturbative transition probability $F$ for a pair of off-shell heavy quark ($\psi$) and antiquark ($\chi$) to a quarkonium state $H$ is proportional to the Fourier transform of following matrix elements

$$\sum_N \langle 0|\chi^\dagger(y_1) \, \mathcal{K}_n \, \psi(y_2)|H+N\rangle \langle H+N|\psi^\dagger(\tilde{y}_2) \, \mathcal{K}_n' \, \chi(\tilde{y}_1)|0\rangle \,, \qquad (2)$$

where $y_i(\tilde{y}_i)$ are coordinates, and $\mathcal{K}_n$ and $\mathcal{K}_n'$ are local combinations of color and spin matrices for the $Q\bar{Q}$ state $n$. A proper insertion of Wilson lines to make the operators in Eq. (2) gauge invariant is implicit [3]. In Eq. (2), $\sigma_{A+B\to Q\bar{Q}[n]+X}$ represents the production of a pair of on-shell heavy quarks and is calculable in perturbative QCD [4]. The debate on the production mechanism has been focusing on the transition from the pair to the meson.

*Color-singlet model* The color-singlet model (CSM) assumes that only a color singlet heavy quark pair with the right quantum number can become a quarkonium of the same quantum number and the transition from the pair to a meson is given by the quarkonium wave function [5]. By neglecting the dependence on the pair's relative momentum in $\sigma_{A+B\to Q\bar{Q}[n]+X}$, $\int d\Gamma_{Q\bar{Q}}F_{Q\bar{Q}\to H}$ in Eq. (1) is set to equal the matrix element in Eq. (2) evaluated at $y_i(\tilde{y}_i) = 0$, which is proportional to the square of coordinate-space quarkonium wave function at the origin, $|R_H(0)|^2$ [1], and therefore,

$$\sigma_{A+B\to H+X}^{\text{CSM}} \propto \sigma_{A+B\to Q\bar{Q}[H]+X}(m_Q) \, |R_H(0)|^2 \,. \qquad (3)$$

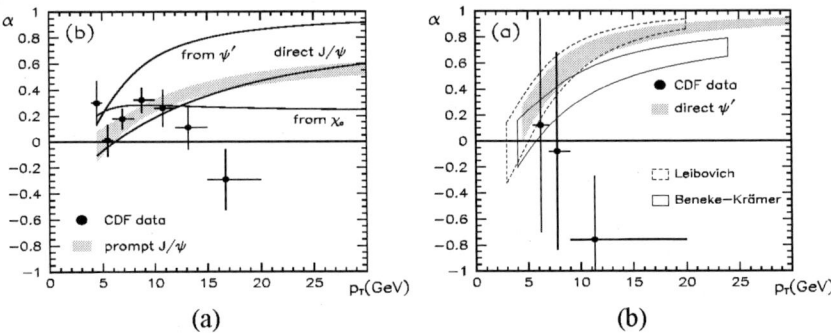

**FIGURE 2.** From Ref. [6], NRQCD predictions of charmonium polarizations are compared with the CDF data [11].

The same wave function appears in both production and decay, and the model provides absolutely normalized predictions. It works well for $J/\psi$ production in deep inelastic scattering, photon production, and some low energy experiments [1], but fails to predict the CDF data, as demonstrated by the dotted lines in Fig. 1(a) [6].

*Color evaporation model*  The color evaporation model (CEM) takes a very different approach to handle the nonperturbative transition. It assumes that all $Q\bar{Q}$ pairs with invariant mass less than the threshold of producing a pair of open-flavor heavy mesons, regardless their color, spin, and invariant mass, have the same probability to become a quarkonium [8]. That is, the $F_{Q\bar{Q}[n]\rightarrow H}$ in Eq. (1) is a constant for a given quarkonium state, $H$, and therefore,

$$\sigma_{A+B\rightarrow H+X}^{\text{CEM}} \approx f_H \int_{2m_Q}^{2M_Q} dm_{Q\bar{Q}} \, \sigma_{A+B\rightarrow Q\bar{Q}+X}(m_{Q\bar{Q}}) \tag{4}$$

with open-flavor heavy meson mass $M_Q$ and a constant $f_H$ [8]. With a proper choice of $m_Q$ and $f_H$, the model gives a reasonable description of almost all data including the CDF data, as shown in Fig. 1(b).

*Nonrelativistic QCD model*  The Nonrelativistic QCD (NRQCD) model is based on the fact that the typical heavy quark rest-frame kinetic energy and binding energy, $m_Q v^2$, in a heavy quarkonium is much smaller than the heavy quark mass. The model separates the physics at scales of order $m_Q$ and higher from the dynamics of the binding by using NRQCD, an effective field theory [10]. It provides a systematic prescription to calculate the physics at $m_Q$ order-by-order in powers of $\alpha_s$, and expands the nonperturbative dynamics in terms of local matrix elements ordered in power series of heavy quark velocity $v$ [6, 10],

$$\sigma_{A+B\rightarrow H+X}^{\text{NRQCD}} = \sum_n \hat{\sigma}_{A+B\rightarrow Q\bar{Q}[n]+X} \left\langle \mathcal{O}_n^H \right\rangle, \tag{5}$$

where the $\mathcal{O}_n^H$ are NRQCD operators for the state $H$ [10],

$$\mathcal{O}_n^H(0) = \chi^\dagger(0)\, \mathcal{K}_n \psi(0) \left(a_H^\dagger a_H\right) \psi^\dagger(0)\, \mathcal{K}_n' \chi(0), \tag{6}$$

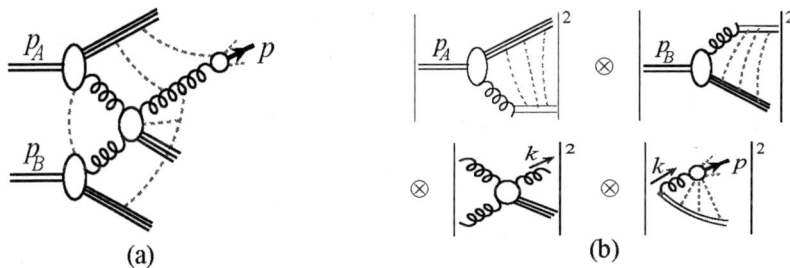

**FIGURE 3.** Sketch for process $A + B \rightarrow H(p) + X$: (a) sample scattering amplitude with possible factorization breaking soft interactions indicated by dashed lines, and (b) factorization with Wilson lines indicated by thin double lines.

where $a_H^{\dagger}$ is the creation operator for $H$, $\chi$ ($\psi$) are two component Dirac spinors, and $\mathcal{K}_n$ and $\mathcal{K}_n'$ are defined in Eq. (2) and can also involve covariant derivatives. At higher orders in $v$, the operator $\mathcal{O}_n^H$ can have additional dependence on field strength as well as more fermion fields. The factorization in Eq. (5) could be understood from Eq. (1) by expanding the $\sigma_{A+B \rightarrow Q\bar{Q}[n]+X}$ at heavy quark relative momentum, $q = (p_Q - p_{\bar{Q}})/2 = 0$. The moments, $\int d\Gamma_{Q\bar{Q}} \, q^N F_{Q\bar{Q} \rightarrow H}$, lead to local matrix elements with high powers of $v$.

The NRQCD model allows every $Q\bar{Q}[n]$ state to become a bound quarkonium, while the probability is determined by corresponding nonperturbative matrix elements $\langle \mathcal{O}_n^H \rangle$. It in principle includes the physics of CSM. Its octet contribution is the most important one for high $p_T$ quarkonium production at collider energies [6]. The NRQCD model has been most successful in interpreting data [1, 6], as shown in Fig. 1(a).

*Quarkonium polarization and other models* The key difference between the NRQCD model and the CEM is the prediction on quarkonium polarization. Once the matrix elements $\langle \mathcal{O}_n^H \rangle$ are determined, the NRQCD model can systematically calculate the polarization of produced heavy quarkonia. On the other hand, CEM does not provide a systematic prescription to calculate the polarization because of the uncontrolled radiations from the quark pair. The polarization of quarkonia at large $p_T$ was considered a definite test of the NRQCD model [6]. But, the model has failed the test, as seen in Fig. 2, if the CDF data hold [11]. Several improved and new models have been proposed to address the issues of polarization [12].

# FACTORIZATION

The heavy quarkonium production in hadronic collisions involves both perturbative and nonperturbative scales. The nonperturbative physics appears not only in the transition from the heavy quark pair to a bound state but also in incoming hadron wave functions. A typical scattering amplitude for quarkonium production, as sketched in Fig. 3(a), can have soft and nonperturbative interactions between incoming hadrons as well as between the spectators and the formation process. These soft interactions may introduce process dependence to the nonperturbative matrix elements, and consequently, spoil the predictive powers of Eqs. (4) and (5).

A proof of the factorization needs to: 1) show that the square of the scattering amplitude in Fig. 3(a), after summing over all amplitudes with the same initial and final states, can be expressed as a convolution of the probabilities, as sketched in Fig. 3(b); each probability represents a square of sub-amplitudes and is evaluated at its own momentum scale(s); 2) show that the piece evaluated at perturbative scale(s) is infrared safe and those evaluated at nonperturbative scales are universal.

As argued in Ref. [4], cross sections for producing on-shell heavy quark pairs can be computed in terms of QCD factorization. Therefore, the right-hand-side of the CEM formalism in Eq. (4) can be factorized into a convolution of a calculable partonic cross section of producing the heavy quark pair and two universal parton distributions from respective hadrons. However, the factorization statement here does not provide justification that the $F_{Q\bar{Q}[n] \to H}$ in Eq. (1) is independent of the pair's invariant mass $m_{Q\bar{Q}}$, spin, and other quantum numbers. For quarkonium production at a large $p_T$, the $f_H$ should be universal *within the model* because soft interactions in Fig. 3(a) are suppressed by powers of $1/p_T$. When $p_T \ll m_Q$, the $f_H$ may not be universal.

Fully convincing arguments have not yet been given for NRQCD factorization formalism in Eq. (5) [6, 10]. Since the interaction between the beam jet and the jet of heavy quark pair should be suppressed by powers of $1/p_T$, one might expect the NRQCD factorization formalism to work at large $p_T$. When $p_T \gg m_Q$, the heavy quarkonium production is similar to the single light hadron production, and is dominated by parton fragmentation. The cross section is proportional to the universal parton-to-hadron fragmentation functions [3],

$$\sigma_{A+B \to H+X}(p_T) = \sum_i \hat{\sigma}_{A+B \to i+X}(p_T/z, \mu) \otimes D_{H/i}(z, m_Q, \mu) + \mathcal{O}(m_H^2/p_T^2). \quad (7)$$

Here, $\otimes$ represents a convolution in the momentum fraction $z$. The cross section $\hat{\sigma}_{A+B \to i+X}$ includes all information on the incoming state, including convolutions with parton distributions of hadrons $A$ and $B$ at factorization scale $\mu$, as sketched in Fig. 3(b). As a necessary condition for NRQCD factorization in Eq. (5), the following factorization relation,

$$D_{H/i}(z, m_Q, \mu) = \sum_n d_{i \to Q\bar{Q}[n]}(z, \mu, m_Q) \langle \mathcal{O}_n^H \rangle, \quad (8)$$

is required to be valid to all orders in $\alpha_s$ and all powers in $v$-expansion for all parton-to-quarkonium fragmentation functions [3]. In Eq. (8), $d_{i \to Q\bar{Q}[n]}$ describes the evolution of an off-shell parton into a heavy quark pair in state $[n]$, including logarithms of $\mu/m_Q$, and should be infrared safe [3].

The factorization relation in Eq. (8) was tested up to next-to-next-to-leading order (NNLO) in $\alpha_s$ at $v^2$ order in Ref. [3], as well as at finite $v$ in Ref. [13]. Consider representative NNLO contributions to the fragmentation process of transforming a color octet heavy quark pair to a singlet, as sketched in Fig. 4. The individual classes of diagrams in Fig. 4(I) and (II), for which two gluons are exchanged between the quarks and the Wilson line, satisfy the infrared cancellation conjecture of Ref. [10], by summing over the possible cuts and connections to quark and antiquark lines, as do diagrams that have three gluon-eikonal vertices on the quark pair and one on the Wilson line [3]. For Fig. 4(III) type of diagrams, however, with a three-gluon interaction, this cancellation

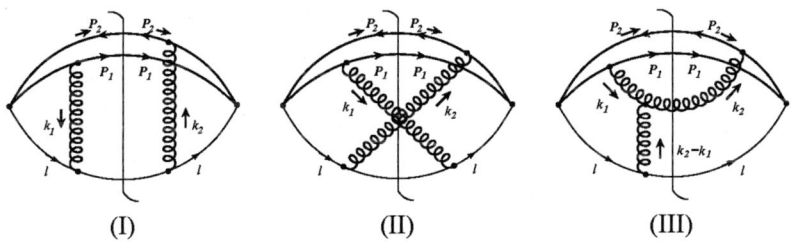

**FIGURE 4.** Representative NNLO contributions to $g \to Q\bar{Q}$ fragmentation in eikonal approximation, see Ref. [3] for the details.

fails. By summing over all contributions to second order in the relative momentum $q$, it was found that the fragmentation function has a noncanceling real pole [3]

$$\mathscr{S}_2^{(8 \to 1)}(v) - -\alpha_s^2 \frac{1}{3\varepsilon} v^2. \tag{9}$$

From Fig. 4(III), this infrared divergence is not topologically factorizable and can not be absorbed into the matrix elements $\langle \mathscr{O}_n^H \rangle$. As demonstrated in Ref. [3], as a necessary condition for restoring the NRQCD factorization at NNLO, the conventional $\mathscr{O}(v^2)$ octet NRQCD production matrix elements $\langle \mathscr{O}_n^H \rangle$ must be modified by incorporating Wilson lines that make them manifestly gauge invariant, so that the infrared divergence in Eq. (9) can be absorbed into the gauge-completed matrix elements.

For quarkonium production, all heavy quark pairs with invariant mass less than a pair of open-flavor heavy mesons could become a bound quarkonium. Therefore, the effective velocity of heavy quark pair in the production could be much larger than that in decay. For charmonium production, charm quark velocity, $v_c \sim |\vec{q}_c|/m_c \le \sqrt{(4M_D^2 - 4m_c^2)/(4m_c^2)} \sim 0.88$, is not small, and therefore, the velocity expansion for charmonium production may not be a good approximation, unless one can identify and resum large contributions to all order in $v$ or have a factorized formalism at finite $v$. It was found in Ref. [13] that with the gauge-completed matrix elements, infrared singularities in the fragmentation function for a color octet pair to a singlet at NNLO are consistent with NRQCD factorization to all orders in $v$ or to a finite $v$,

$$\mathscr{S}^{(8 \to 1)}(v) = \frac{\alpha_s^2}{4\varepsilon} \left[ 1 - \frac{1}{2f(v)} \ln \left[ \frac{1 + f(v)}{1 - f(v)} \right] \right], \tag{10}$$

where $f(v) = 2v/(1 + v^2)$. The result in finite $v$ is remarkably compact and intriguing, and should encourage further work on the factorization theorem.

## SUMMARY AND OUTLOOK

After more than 30 years since the discovery of J/$\psi$, we still have not been able to fully understand the production mechanism of heavy quarkonium in high energy collisions, in particular, the transition from the heavy quark pair to a bound quarkonium. Although there are good reasons for each production model, none of the factorized production formalism, including that of the NRQCD model, has been proved theoretically. Further work on the factorization theorem is critical.

The heavy quarkonium production has a "long" transition time from the produced pairs to bound mesons, and can be a good probe for the properties of newly formed hot and dense matter of quarks and gluons, or the quark-gluon plasma, in ultrarelativistic heavy ion collisions [14, 15], if we can calibrate the production. Since the medium interaction is sensitive to the quantum state of the proporgating heavy quark pair, nuclear matter could be an effective filter to distinguish the production mechanism [16]. Detailed studies of nuclear dependence of heavy quarkonium production in hadron-nucleus collisions should provide invaluable information on the formation of heavy quarkonia in hadronic collisions.

## ACKNOWLEDGMENTS

The author thanks G. T. Bodwin, G. Nayak, and G. Sterman for discussions, and nuclear theory group at Brookhaven National Laboratory for its hospitality during the writing of this contribution. This work is supported in part by the U.S. Department of Energy, Grant No. DE-FG02-87ER40371, and Contract No. DE-AC02-98CH10886.

## REFERENCES

1. N. Brambilla et al. [Quarkonium Working Group], arXiv:hep-ph/0412158.
2. J. C. Collins, D. E. Soper and G. Sterman, Adv. Ser. Direct. High Energy Phys. **5**, 1 (1988).
3. G. C. Nayak, J. W. Qiu and G. Sterman, Phys. Lett. B **613**, 45 (2005); Phys. Rev. D **72**, 114012 (2005).
4. J. C. Collins, D. E. Soper and G. Sterman, Nucl. Phys. B **263**, 37 (1986).
5. C. H. Chang, Nucl. Phys. B **172**, 425 (1980); R. Baier and R. Ruckl, Phys. Lett. B **102**, 364 (1981); E. L. Berger and D. L. Jones, Phys. Rev. D **23**, 1521 (1981).
6. E. Braaten, S. Fleming and T. C. Yuan, Ann. Rev. Nucl. Part. Sci. **46**, 197 (1996); M. Kramer, Prog. Part. Nucl. Phys. **47**, 141 (2001); G. T. Bodwin, Int. J. Mod. Phys. A **21**, 785 (2006).
7. F. Abe et al. [CDF Collaboration], *Phys. Rev. Lett.* **79**, 572 (1997).
8. H. Fritzsch, "Producing Heavy Quark Flavors In Hadronic Collisions: A Test Of Quantum Phys. Lett. B **67**, 217 (1977); F. Halzen, Phys. Lett. B **69**, 105 (1977); M. Gluck, J. F. Owens and E. Reya, Phys. Rev. D **17**, 2324 (1978).
9. J. F. Amundson, O. J. P. Eboli, E. M. Gregores and F. Halzen, Phys. Lett. B **390**, 323 (1997).
10. G. T. Bodwin, E. Braaten and G. P. Lepage, Phys. Rev. D **51**, 1125 (1995) [Erratum-ibid. D **55**, 5853 (1997)].
11. A. A. Affolder et al. [CDF Collaboration], Phys. Rev. Lett. **85**, 2886 (2000); CDF Collaboration, Notes 8212 and 8424 (06-06-22).
12. J. P. Lansberg, Int. J. Mod. Phys. A **21**, 3857 (2006); G. T. Bodwin, in this proceedings.
13. G. C. Nayak, J. W. Qiu and G. Sterman, Phys. Rev. D **74**, 074007 (2006).
14. I. Arsene et al. [BRAHMS Collaboration], Nucl. Phys. A **757**, 1 (2005) [arXiv:nucl-ex/0410020]; B. B. Back et al., Nucl. Phys. A **757**, 28 (2005) [arXiv:nucl-ex/0410022]; J. Adams et al. [STAR Collaboration], Nucl. Phys. A **757**, 102 (2005) [arXiv:nucl-ex/0501009]; K. Adcox et al. [PHENIX Collaboration], Nucl. Phys. A **757**, 184 (2005) [arXiv:nucl-ex/0410003].
15. T. Matsui and H. Satz, Phys. Lett. B **178**, 416 (1986); M. Nardi, Nucl. Phys. A **774**, 353 (2006); and references therein.
16. for example, see J. W. Qiu, J. P. Vary and X. f. Zhang, Phys. Rev. Lett. **88**, 232301 (2002); D. Kharzeev and K. Tuchin, Nucl. Phys. A **770**, 40 (2006); and references therein.

# Hamiltonian approach to Yang-Mills theory in Coulomb gauge[1]

H. Reinhardt, D. Epple, W. Schleifenbaum

*Universität Tübingen, Institut für Theoretische Physik, Auf der Morgenstelle 14, 72076 Tübingen, Germany*

**Abstract.** Recent results obtained within the Hamiltonian approach to continuum Yang-Mills theory in Coulomb gauge are reviewed.

**Keywords:** Yang-Mills theory, Coulomb gauge
**PACS:** 11.10Ef, 12.38Aw, 12.38Lg

## INTRODUCTION

There have been many attempts in the past to solve the Yang-Mills Schrödinger equation using gauge invariant wave functionals. Unfortunately, all these approaches have not been blessed with much success[2]. The reason is that it is extremely difficult to work with gauge invariant wave functionals. A much more economic way is to explicitly resolve Gauss' law (which ensures gauge invariance of the wave functional) by fixing the gauge. For this purpose, the Coulomb gauge $\vec{\partial} \cdot \vec{A} = 0$ is particularly convenient. In my talk, I would like to present a variational solution of the Yang-Mills Schrödinger equation in Coulomb gauge [2]. Let me start by briefly summarizing the essential ingredients of the quantization of Yang-Mills theory in Coulomb gauge.

In Coulomb gauge the space of (transversal) gauge orbits has a non-trivial metric, which is given by the Faddeev-Popov determinant $J(A) = Det(-\hat{D}_i \partial_i)$, where $\hat{D}_i^{ab} = \delta^{ab} \partial_i + \hat{A}_i^{ab}$, $\hat{A}_i^{ab} = f^{acb} A_i^c$ denotes the covariant derivative in the adjoint representation of the gauge group ($f^{acb}$ is the structure constant). In Coulomb gauge the Yang-Mills Hamiltonian is given by [3]

$$H = \frac{1}{2} \int J^{-1} \Pi J \Pi + \frac{1}{2} \int B^2 + \frac{g^2}{2} \int J^{-1} \rho (-\hat{D}\partial)^{-1} (-\partial^2)(-\hat{D}\partial)^{-1} J \rho \,, \tag{1}$$

where $\Pi_i^a(x) = \delta / i \delta A_i^a(x)$ denotes the momentum operator, representing the color electric field, $B$ is the color magnetic field and $\rho^a(x) = -\hat{A}_i^{ab}(x)\Pi_i^b(x)$ is the non-Abelian color charge of the gauge field. The first term in the Hamiltonian is the Laplacian in curved space, the second term represents the potential and the last term arises from the longitudinal momentum part of the kinetic energy after resolving Gauss' law. This term

---

[1] Invited plenary talk given by H. Reinhardt at the conference "Quark Confinement and the hadron spectrum VII", Ponta del Gada, Portugal, 2.-7.9.2006.
[2] Recent progress has been made, however, in $D = 2 + 1$, see ref. [1].

CP892, *Quark Confinement and the Hadron Spectrum VII*
edited by J. E. F. T. Ribeiro

is usually referred to as the Coulomb term, since in the Abelian case it reduces to the ordinary Coulomb potential.

## VARIATIONAL SOLUTION OF THE YANG-MILLS SCHRÖDINGER EQUATION

We have performed a variational solution of the Yang-Mills Schrödinger equation for the vacuum using the following ansatz for the wave functional [2]

$$\Psi[A] = \mathcal{N} J[A]^{-\alpha} \exp\left(-\frac{1}{2}\int A\,\omega A\right) \tag{2}$$

with[3] $\alpha = \frac{1}{2}$, where $\omega(|x-x'|)$ is the variational kernel determined by minimizing the energy

$$\langle\Psi|H|\Psi\rangle = \int \mathcal{D}A\,J[A]\Psi^*[A]\,H\,\Psi[A]\,. \tag{3}$$

This gives rise to a set of coupled Dyson-Schwinger equations for the ghost propagator

$$G^{ab}(x,x') = \langle\Psi|\langle x,a|(-\hat{D}\partial)^{-1}|x',b\rangle|\Psi\rangle = \frac{1}{g}(-\partial^2)^{-1}d(x,x') \tag{4}$$

with $d$ being the ghost form factor, and the gluon propagator

$$D^{ab}_{ij}(x,x') = \langle\Psi|A^a_i(x)A^b_j(x)|\Psi\rangle = \frac{1}{2}\delta^{ab}t_{ij}(x)\omega^{-1}(x,x')\,,\ \ t_{ij}(x) = \delta_{ij} - \frac{\partial^x_i \partial^x_j}{\partial^2_x}\,. \tag{5}$$

The gluon Dyson-Schwinger equation has the form of the dispersion relation of a relativistic particle

$$\omega^2(k) = k^2 + \Sigma^2(k)\,, \tag{6}$$

where the gluon self-energy $\Sigma$ is dominated by the ghost loop

$$\chi^{ab}_{ij}(x,x') = -\frac{1}{2}\langle\Psi|\frac{\delta^2 \ln J[A]}{\delta A^a_i(x)\delta A^b_j(y)}|\Psi\rangle\,, \tag{7}$$

which is a measure for the curvature of the space of transversal gauge orbits and which in the following will be referred to as curvature. Due to the ansatz for the ghost form factor $d$ the coupling constant $g$ drops out from the Dyson-Schwinger equations which is a specific feature of the one-loop approximation.

---

[3] This ansatz is motivated by the form of the wave function of a point particle in a spherical symmetric potential in a zero angular momentum (s-)state $\Psi(r) = \frac{u(r)}{r}$, where $J(r) = r^2$. In ref. [4] the ansatz (2) with $\alpha = 0$ was used.

In ref. [5] an infrared analysis of the Dyson-Schwinger equations was performed without resorting to the angular approximation. Using the infrared ansätze

$$D(p) = \frac{1}{p^{2+\alpha}} , \quad G(p) = \frac{B}{p^{2+\beta}} \tag{8}$$

and implementing the horizon condition $d^{-1}(0) = 0$, first given in ref. [6], one finds from the ghost Dyson-Schwinger equation the sum rule

$$\alpha + 2\beta = d - 4 , \tag{9}$$

where $d$ is the number of spatial dimensions, see also ref. [7]. The sum rule (9) is due to the non-renormalization of the ghost-gluon vertex which is a feature of both the Landau gauge [8] and the Coulomb gauge [9, 5]. Inserting this relation into the gluon Dyson-Schwinger equation, one obtains for the infrared exponent a unique solution $\beta = 0.4$ in $d = 2$ and two solutions in $d = 3$,

$$\beta_1 = 0.796 , \quad \beta_2 = 1.0 . \tag{10}$$

Only $\beta_2$ was previously found in the angular approximation [2], while $\beta_1$ corresponds to the numerical solutions reported in ref. [2], where $\beta = 0.85$ was found. Recently the full (numerical) solution corresponding to the infrared exponent $\beta_2$ was also found [10] and is shown in fig. 1.

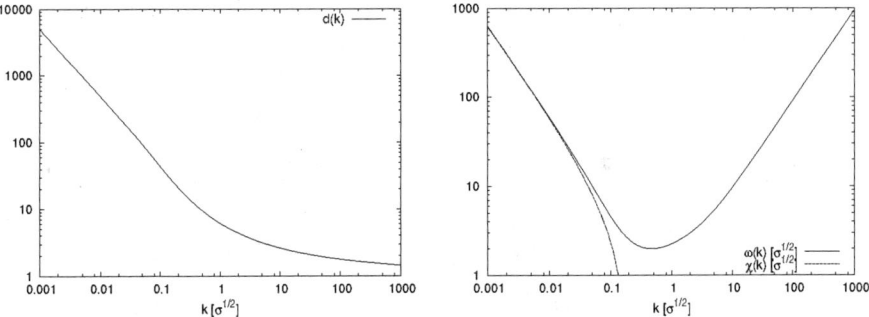

**FIGURE 1.** The ghost form factor $d(k)$ (left panel) and the gluon energy $\omega(k)$ and the curvature $\chi(k)$ (right panel)

The infrared exponent $\beta = 1$ together with the sum rule (9) implies an infrared linearly diverging gluon energy, see fig. 1, which is a manifestation of confinement, and also produces a linear rising static color potential, shown in fig. 2. Let me stress that it is absolutely crucial to keep fully the Faddeev-Popov determinant $J[A]$ in both the Hamiltonian and the integration measure. The Faddeev-Popov determinant converts the flat integration measure $\int \mathcal{D}A$ into the Haar measure of the gauge group [11]. When the Faddeev-Popov determinant is neglected [4] the gluon energy becomes infrared finite and the linear rise in the static color potential is lost. On the other hand, the obtained infrared behavior is quite robust with respect to changes of the variational wave functional. Extending the variational ansatz, eq. (2), to arbitrary $\alpha$ (where $\alpha = \frac{1}{2}$ and

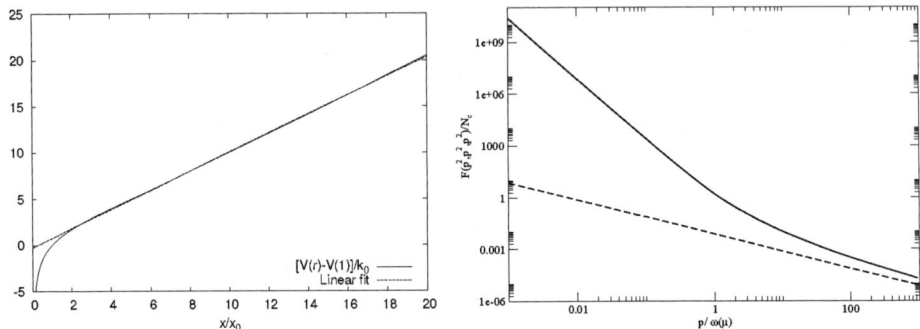

**FIGURE 2.** Left panel: The static Coulomb potential. Right panel: The three-gluon vertex at the symmetric point, full line with non-perturbative propagators, dashed line with perturbative propagators.

$\alpha = 0$ correspond to the wave functional used in refs. [2] and [4], respectively), it can be shown that up to two loops in the energy, the variational solution and in particular the ghost and gluon propagators are actually independent of $\alpha$, see ref. [12]. Other quantities like the three-gluon vertex are, however, more sensitive than the energy and do depend on $\alpha$.

The leading contribution to the three-gluon vertex is given by the triangle diagram with an internal ghost loop. In this order the three-gluon vertex has the same color structure as the bare one and can be expressed by a single form factor

$$\Gamma_{ijk}^{abc}(p_1, p_2, p_3) = f^{abc}\left(i(p_1)_j \delta_{ik} F(p_1, p_2, p_3) + \text{permutations}\right) . \tag{11}$$

The form factor $F$ is shown in fig. 2 for the symmetric point $p_1 = p_2 = p_3$. The infrared analysis performed in ref. [5] shows $F(p^2) \sim (p^2)^{\frac{d}{2}-2-\frac{3}{2}\beta}$. Inserting here $d = 3$ and $\beta = 0.85$ from ref. [2], one finds $F(p^2) \sim (p^2)^{-1.775}$, which is in perfect agreement with the numerical result shown in fig. 2, from which one extracts $F(p^2) \sim (p^2)^{-1.77}$. We have also investigated the ghost-gluon vertex and found that, like in Landau gauge, it becomes bare when one of the external ghost momenta vanishes, so that its dressing can be ignored, at least in the infrared.

We can define the RG-invariant running coupling constant,

$$\alpha(k) = \frac{16}{3}\frac{g_R^2}{4\pi} p^5 G^2(p) D(p) , \tag{12}$$

from the ghost-gluon vertex [9], where $G(p)$ and $D(p)$ are the ghost and gluon propagators, respectively. Due to the sum rule (9) for the infrared exponents this coupling has an infrared fixed point which for the DSE solutions with $\beta = 1.0$ is given by $\alpha(0) = 16\pi/(3N_c) \approx 16.76/N_c$, see ref. [5]. This value obtained in the infrared analysis is in excellent agreement with the value of $\alpha(0) = 16.94/N_c$, see [10], which is obtained from the numerical solution of the DSE presented in fig. 1. Let us also stress that this result disagrees with the value for the infrared fixed point obtained in the Coulomb gauge limit of the interpolating gauges [9].

96

# THE 'T HOOFT LOOP

The class (2) of wave functionals yield (up to two loops in the energy) the same Dyson-Schwinger equations independent of $\alpha$. To test our wave functional and to get more detailed information on the structure of the Yang-Mills vacuum, observables more sensitive than the energy density should be calculated. Moreover, the confining Coulomb potential obtained above does not necessarily guarantee that our wave functional describes indeed a confining vacuum state. The reason is that the Coulomb potential is only an upper-bound to the true static quark potential [13]. Thus a confining Coulomb potential is a necessary but not a sufficient condition for a confining vacuum.

In pure Yang-Mills theory the order parameter of confinement is the temporal Wilson loop, which obeys an area law in the confined phase and a perimeter law in the deconfined phase. Unfortunately, this quantity is difficult to calculate in the continuum theory due to the path ordering. Fortunately, there exists another (dis-)order parameter of Yang-Mills theory which is easier to calculate in the continuum theory. This is the spatial 't Hooft loop which is dual to the temporal Wilson loop in the sense that it obeys a perimeter law in the confined phase and an area law in the deconfined phase [14].

The operator $V(C)$ of 't Hooft's disorder parameter is defined for a spatial loop $C$ by the following relation [14]:

$$V(C_1)W(C_2) = Z^{L(C_1,C_2)}W(C_2)V(C_1) \quad , \quad W(C) = \frac{1}{N}tr P\exp(i \oint_c A) \tag{13}$$

where $W(C)$ is the spatial Wilson loop, $Z$ is a (non-trivial) center element of the gauge group ($Z = -1$ for SU(2)) and $L(C_1,C_2)$ denotes the Gaussian linking number between the two spatial loops $C_1$ and $C_2$. Unfortunately 't Hooft did not give an explicit realization of the operator $V(C)$ in the continuum theory, but defined the operator implicitly as having the effect of a singular (on the loop $C$) gauge transformation.

The physical meaning of the 't Hooft loop defined by equation (13) is recognized by noticing that a time independent (spatial) center vortex $\mathscr{A}(C_1)$ whose magnetic flux is localized on the closed loop $C_1$, produces the Wilson loop $W[\mathscr{A}(C_1)](C_2) = Z^{L(C_1,C_2)}$. It follows that the 't Hooft loop operator $V(C)$ can be interpreted as a center vortex creation operator. The operator $V(C)$ which generates a center vortex gauge field $\mathscr{A}(C)$ i.e., $V(C)\Psi(A) = \Psi(A + \mathscr{A}(C))$, is given by

$$V(C = \partial\Sigma) = \exp\left(i \int d^3x \mathscr{A}_i^a(\Sigma,x)\Pi_i^a(x)\right). \tag{14}$$

In reference [15] it was proven that the operator (14) with the vortex field

$$\mathscr{A}_i^a(\Sigma,x) = \zeta^a \int_\Sigma d^2\tilde{\sigma}_i \delta^3(x - \bar{x}(\sigma)) \tag{15}$$

with $\zeta = \zeta^a T_a$ being a co-weight vector defined by $\exp(i\zeta) = Z$ actually satisfies the defining relation (13). Since $\Pi_i^a(x)$ is the operator of the electric field, the 't Hooft loop measures the electric flux (projected on the co-weight $\zeta = \zeta^a T_a$) through the closed loop $C$, while the (spatial) Wilson loop $W(C)$ measures the magnetic flux through $C$. The operator $V(C)$ is not manifestly gauge invariant, but produces a gauge invariant result, when acting on gauge invariant states $\Psi(A^U) = \Psi(A)$.

The wave functional in Coulomb gauge satisfies Gauss' law (which ensures gauge invariance) and has, hence, to be considered as the restriction of the gauge invariant wave functional to transverse gauge fields. Therefore by implementing the Coulomb gauge by the standard Faddeev-Popov method, we obtain for the 't Hooft loop

$$\langle V(C) \rangle = \int \mathcal{D}A J(A) \Psi^*(A) \Psi(A + \mathcal{A}^\perp(C)) , \qquad (16)$$

where $\mathcal{A}^\perp(C = \partial\Sigma)$ is the transverse part of the center vortex field $\mathcal{A}(\Sigma)$ (15) given by [16]

$$\mathcal{A}^\perp(\partial\Sigma, x) = -\zeta \oint_{\partial\Sigma} d\tilde{\sigma}_{ik} \partial_k^{\bar{x}} \frac{1}{4\pi|x - \bar{x}|} \quad , \quad d\tilde{\sigma}_{ik} = \Sigma_{ik\ell} d\bar{x}_\ell \qquad (17)$$

which manifestly depends only on the loop $C = \partial\Sigma$ but not on the enclosed surface $\Sigma$ as the field (15) does. It is straightforward to calculate from eq. (17) 't Hooft's disorder parameter inserting for $\Psi(A)$ the wave functional determined previously [2]. After straightforward calculations, one finds

$$\langle V(C) \rangle = \exp(-S) , \; S(C) = \int_0^\infty dq K(q) h(C, q) , \qquad (18)$$

where

$$K(q) = \frac{1}{2}\omega(q) \left(1 - \frac{\chi^2(q)}{\omega^2(q)}\right) \qquad (19)$$

contains the whole information about Yang-Mills vacuum while the quantity $h(C, q)$ is exclusively determined by the geometry of the considered loop. To simplify the calculations of this quantity, we consider a planar circular loop with radius $R$. A detailed analysis shows that the $R$ dependence of $S(R)$ is determined by the infrared behavior of $K(q)$. Given that $\omega(q)$ and $\chi(q)$ are both infrared divergent and differ only by a finite constant, one finds

$$K(q \to 0) = \left(\omega(q) - \chi(q)\right)_{q \to 0} = c \qquad (20)$$

where $c$ is a finite renormalization constant which can be chosen arbitrarily. Choosing $c = 0$, which implies that the infrared wave functional reduces to the strong coupling limit $\Psi[A] = 1$, one finds indeed a perimeter law $S(R) = \check{\kappa}R$, while for $c \neq 0$ one obtains $SR \sim R \ln(R/R_0)$. Finally, if one ignores the curvature ($\chi = 0$) in eq. (19), one finds an area law $S(R) = \tilde{\rho}R^2$. Thus again we find the curvature to be crucial for the confinement properties of the theory.

# SUMMARY AND CONCLUSIONS

In my talk I have discussed recent results obtained in the Hamilton approach to Yang-Mills theory in Coulomb gauge. Using ghost dominance, an infrared analysis of the Schwinger-Dyson equations has revealed both quark and gluon confinement. The gluon energy is infrared diverging implying the absence of asymptotic gluon states from the physical spectrum, and the static Coulomb potential is linearly rising. These results are confirmed by a full numerical solution of the Schwinger-Dyson equations. While the ghost-gluon vertex remains basically unrenormalized and in particular becomes the bare one when the ghost momentum vanishes, the three-gluon vertex is strongly infrared divergent. As a first non-trivial test of the variationally determined vacuum wave functional, we have calculated 't Hooft's disorder parameter choosing the renormalization constants of the gap equation in such a way that the wave functional reproduces in the infrared the strong coupling limit. Then the 't Hooft loop shows indeed a perimeter law, indicating a confining vacuum.

## ACKNOWLEDGMENTS

Discussions with M. Quandt and P. Watson are gratefully acknowledged. This work was supported by DFG-Re856/6-1,2.

## REFERENCES

1. R. G. Leigh, D. Minic and A. Yelnikov, Phys. Rev. Lett. **96** (2006) 222001 [arXiv:hep-th/0512111].
2. C. Feuchter and H. Reinhardt, Phys. Rev. D **70**, 105021 (2004) [arXiv:hep-th/0408236].
3. N. H. Christ and T. D. Lee, Phys. Rev. D **22**, 939 (1980) [Phys. Scripta **23**, 970 (1981)].
4. A. P. Szczepaniak and E. S. Swanson, Phys. Rev. D **65**, 025012 (2002) [arXiv:hep-ph/0107078].
5. W. Schleifenbaum, M. Leder and H. Reinhardt, Phys. Rev. D **73**, 125019 (2006) [arXiv:hep-th/0605115].
6. D. Zwanziger, Nucl. Phys. B **485** (1997) 185 [arXiv:hep-th/9603203].
7. D. Zwanziger, Phys. Rev. D **65** (2002) 094039 [arXiv:hep-th/0109224], C. Lerche and L. von Smekal, Phys. Rev. D **65** (2002) 125006 [arXiv:hep-ph/0202194].
8. J. C. Taylor, Nucl. Phys. B **33** (1971) 436, W. Schleifenbaum, A. Maas, J. Wambach and R. Alkofer, Phys. Rev. D **72** (2005) 014017 [arXiv:hep-ph/0411052].
9. C. S. Fischer and D. Zwanziger, Phys. Rev. D **72**, 054005 (2005) [arXiv:hep-ph/0504244].
10. D. Epple, H. Reinhardt and W. Schleifenbaum, to be published.
11. H. Reinhardt, Mod. Phys. Lett. A **11**, 2451 (1996) [arXiv:hep-th/9602047].
12. H. Reinhardt and C. Feuchter, Phys. Rev. D **71**, 105002 (2005) [arXiv:hep-th/0408237]
13. D. Zwanziger, Phys. Rev. Lett. **90** (2003) 102001 [arXiv:hep-lat/0209105], J. Greensite, S. Olejnik and D. Zwanziger, Phys. Rev. D **69** (2004) 074506 [arXiv:hep-lat/0401003].
14. G. 't Hooft, Nucl. Phys. B **138**, 1 (1978).
15. H. Reinhardt, Phys. Lett. B **557**, 317 (2003) [arXiv:hep-th/0212264].
16. H. Reinhardt, Nucl. Phys. B **628**, 133 (2002) [arXiv:hep-th/0112215], M. Engelhardt and H. Reinhardt, Nucl. Phys. B **567** (2000) 249 [arXiv:hep-th/9907139].

# Selected Topics in Lattice Phenomenology

*School of Physics and Astronomy, University of Southampton, Southampton SO17 1BJ, UK*

**Abstract.** I discuss three topics in lattice phenomenology: the use of *twisted* boundary conditions to improve the momentum resolution; the evaluation of the amplitude for $K_{\ell 3}$ semileptonic decays and the theory of finite-volume corrections in the computation of $K \to \pi\pi$ decay amplitudes.

**Keywords:** Quantum Chromodynamics, Lattice Simulations, Flavourdynamics, CKM Matrix
**PACS:** 11.15.Ha, 12.15.Ff, 12.15,Hh, 12.38.-t, 12.38.Gc, 13.20.-v, 13.20.Eb, 13.20.He

## INTRODUCTION

One of the most important approaches to testing the Standard Model of Particle Physics and searching for signatures of new physics is *flavourdynamics*; the study of a large number of weak decays, CP-asymmetries and mixing amplitudes to obtain information about the unitarity triangle and to check its consistency. The precision with which this check can be accomplished is limited by non-perturbative QCD effects and lattice QCD provides the opportunity to quantify these effects without model assumptions. In this talk I briefly describe three topics in lattice phenomenology: the use of *twisted* boundary conditions to improve the momentum resolution; the evaluation of $K_{\ell 3}$ decay amplitudes from which the CKM matrix element $V_{us}$ is obtained and the finite-volume effects in $K \to \pi\pi$ decays.

## IMPROVING THE MOMENTUM RESOLUTION ON THE LATTICE

Numerical simulations of lattice QCD are necessarily performed on a finite spatial volume, $V = L^3$. Providing that $V$ is sufficiently large, we are free to choose any consistent boundary conditions for the fields $\phi(\vec{x},t)$, and it is conventional to use periodic boundary conditions, $\phi(x_i + L) = \phi(x_i)$ ($i = 1, 2$ or 3). This implies that components of momenta are quantized to take integer values of $2\pi/L$. Taking a typical example of a lattice with 24 points in each spatial direction, $L = 24a$, with a lattice spacing $a = 0.1$ fm so that $a^{-1} \simeq 2$ GeV, we have $2\pi/L = .52$ GeV. The available momenta for phenomenological studies (e.g. in the evaluation of form-factors) are therefore very limited, with the allowed values of each component $p_i$ separated by about 1/2 GeV. The momentum resolution in such simulations is therefore very poor.

Bedaque [1] has advocated the use of *twisted* boundary conditions for the quark fields $q(\vec{x})$ e.g.

$$ q(x_i + L) = e^{i\theta_i} q(x_i) \quad \text{with momentum spectrum} \quad p_i = n_i \frac{2\pi}{L} + \frac{\theta_i}{L}, \qquad (1) $$

CP892, *Quark Confinement and the Hadron Spectrum VII*
edited by J. E. F. T. Ribeiro

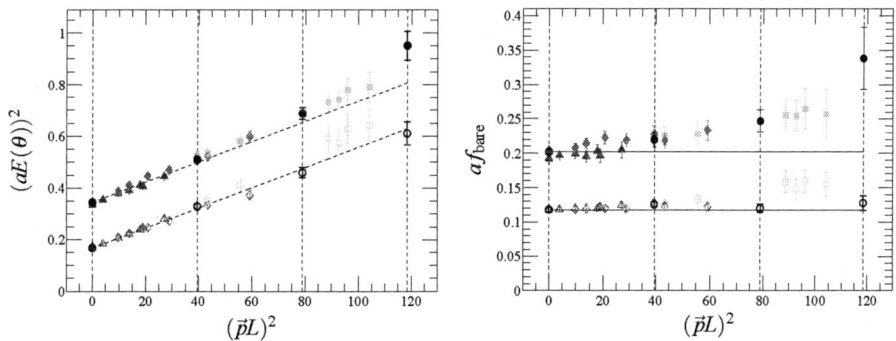

**FIGURE 1.** Plots of the Dispersion Relation (left) and Decay Constants (right) as a function of the momentum $\vec{p}$ of the mesons. In both cases the top (bottom) plot corresponds to the $\rho$-meson ($\pi$-meson).

with integer $n_i$. Modifying the boundary conditions changes the finite-volume effects, however for quantities which do not involve *Final State Interactions* (e.g. hadronic masses, decay constants, form-factors) these errors remain exponentially small also with twisted boundary conditions [2]. Since we usually neglect such errors when using periodic boundary conditions, we can use twisted boundary conditions with the same precision. Moreover the finite-volume errors are also exponentially small for *partially twisted boundary conditions* in which the sea quarks satisfy periodic boundary conditions but the valence quarks satisfy twisted boundary conditions [2, 3]. This is of significant practical importance, since it implies that we do not need to generate new gluon configurations for every choice of twisting angle $\{\theta_i\}$.

The use of partially twisted boundary conditions opens up many interesting phenomenological applications, solving the problem of poor momentum resolution. To illustrate its effectiveness in simulations, consider the plots in fig. 1, obtained using an unquenched (2 flavours of sea quarks) UKQCD simulation on a $16^3 \times 32$ lattice, with a spacing of about 0.1 fm. The plots correspond to a value for the light-quark masses for which $m_\pi/m_\rho = 0.7$ [4]. The lower (upper) left-hand plot shows the energy of the $\pi$ ($\rho$) as a function of the momentum of the meson, and the right-hand plot shows the bare values of the leptonic decay constants $f_\pi$ and $f_\rho$. The x-axis denotes $(|\vec{p}|L)^2$. The results are beautifully consistent with expectations (particularly for $pL \leq 2\pi$ where lattice artifacts are small); the predicted dispersion relation is satisfied and the extracted decay constants are independent of the momenta. Using periodic boundary conditions only the results at values of $\vec{p}$ indicated by the dashed lines are accessible. With partially twisted boundary conditions all momenta are reachable. The use of partially twisted boundary conditions therefore solves the problem of poor momentum resolution on the lattice and can be applied to the evaluation of a large number of phenomenologically important hadronic matrix elements.

# $K_{\ell 3}$ DECAYS

A relatively new area of investigation for lattice simulations is the evaluation of non-perturbative QCD effects in $K \to \pi \ell \nu_\ell$ decays, from which the CKM matrix element $V_{us}$ can be determined. The QCD contribution to the amplitude is contained in two invariant form-factors $f^0(q^2)$ and $f^+(q^2)$ defined by

$$\langle \pi(p_\pi)|\bar{s}\gamma_\mu u|K(p_K)\rangle = f^0(q^2)\frac{M_K^2 - M_\pi^2}{q^2}q_\mu + f^+(q^2)\left[(p_\pi + p_K)_\mu - \frac{M_K^2 - M_\pi^2}{q^2}q_\mu\right],$$

where $q = p_K - p_\pi$. (Parity Invariance implies that only the vector current from the $V - A$ charged current contributes to the decay.) A useful reference value for $f^+(0)$ comes from the 20-year old prediction of Leutwyler and Roos [5], $f^+(0) = 1 + f_2 + f_4 + \cdots = 0.961(8)$ where $f_n = O(M_{K,\pi,\eta}^2)$. $f_2 = -0.023$ is well determined, whereas the higher order terms in the chiral expansion require model assumptions. A more recent alternative calculation (described by G.Ecker at this conference), based on the evaluation of the scalar-pseudoscalar-pseudoscalar correlation function using the $1/N_c$ expansion and truncating the hadron spectrum to the lowest lying resonances gave a larger central value, $f^+(0) = 0.984(12)$ [6].

To be useful in extracting $V_{us}$ from experimental measurements we need to be able to evaluate $f^0(0) = f^+(0)$ to better than about 1% precision. This would appear to be impossible for lattice calculations until one notes that it is possible to compute $1 - f^+(0)$, so that an error of 1% on $f^+(0)$ is actually an error of O(25%) on $1 - f^+(0)$. The calculation follows a similar strategy to that proposed in ref. [7] for the form-factors of $B \to D$ semileptonic decays (which in the heavy quark limit are also close to 1), starting with a computation of double ratios such as

$$\frac{\langle \pi|\bar{s}\gamma_0 l|K\rangle\langle K|\bar{l}\gamma_0 s|\pi\rangle}{\langle \pi|\bar{l}\gamma_0 l|\pi\rangle\langle K|\bar{s}\gamma_0 s|K\rangle} = [f^0(q_{\max}^2)]^2\frac{(m_K + m_\pi)^2}{4m_K m_\pi}, \tag{2}$$

where all the mesons are at rest and $q_{\max}^2 = (M_K - M_\pi)^2$. The ratio on the right-hand side of eq.(2), and hence $f^0(q_{\max}^2)$, can be evaluated with remarkable precision for the quark masses used in current simulations. In order to obtain the physical values of the form factor at $q^2 = 0$, extrapolations in quark masses and $q^2$ need to be performed.

The first lattice calculation of this quantity was performed in 2004 by the SPQR collaboration in the quenched approximation [8], with the result $f^+(0) = 0.960(9)$. The techniques developed in [8] have since been applied to unquenched calculations, with the following quoted results

|  |  |
|---|---|
| FNAL/MILC/HPQCD(2004 Preliminary) [9] | $f^+(0) = 0.962\,(6)\,(9)$ |
| JLQCD($N_f = 2$, 2005) [10] | $f^+(0) = 0.952\,(6)$ |
| RBC($N_f = 2$, 2006) [11] | $f^+(0) = 0.968(9)(6)$ |
| RBC/UKQCD($N_f{=}2{+}1$, 2006 Preliminary) [12] | $f^+(0) = 0.9680(16)$. |

The lattice results are in remarkable agreement with that of Leutwyler and Roos [5].

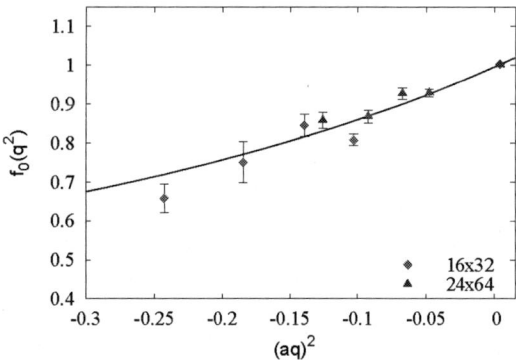

**FIGURE 2.** Illustration of the $q^2$ behaviour of $f^0(q^2)$ for an intermediate value of the quark mass from the simulations of ref. [12]. The red diamonds and blue triangles correspond to $16^3 \times 32$ and $24^3 \times 64$ lattices respectively.

For the remainder of this section I briefly discuss the extrapolation from $q^2 = q^2_{max}$ to $q^2 = 0$. A typical example of the behaviour of $f^0(q^2)$ with $q^2$ is shown in fig. 2 (from ref. [12]). The right-most point is at $q^2 = q^2_{max}$ where the pion and kaon are at rest and the form factor is determined very precisely, whereas the remaining points have one of the meson's with a non-zero momentum and the determined form factors have larger errors as shown. A notable feature is that $q^2_{max}$ is very close to 0; for the masses corresponding to the figure $q^2_{max}a^2 \simeq 4 \times 10^{-3}$ and the inverse lattice spacing is given by $a^{-1} \simeq 1.6$ GeV. Thus here the extrapolation to $q^2 = 0$ is a short one. As the light quark mass ($m_{u,d}$) decreases $q^2_{max}$ grows (for physical values of the masses $q^2_{max}a^2 \simeq 0.05$) and the extrapolation to $q^2 = 0$ is longer.

The standard procedure for performing the extrapolation to $q^2 = 0$ is to make an ansatz for the $q^2$ behaviour of the form factor, obtain the parameters by fitting to the lattice data and then take the value at $q^2 = 0$. Such a procedure has been employed in obtaining the results quoted above. By using partially twisted boundary conditions however, it is possible to determine the form factor directly at $q^2 = 0$ and hence avoid the uncertainties due to the choice of ansatz [13]. (The behaviour of the form-factor at small $q^2$ has been investigated in the quenched approximation using twisted boundary conditions in ref.[14].) Setting $\vec{p}_\pi = 0$ ($\vec{p}_K = 0$) one can readily determine the values of $\vec{p}_K$ ($\vec{p}_\pi$) which give $q^2 = 0$ and choose the twisting angles accordingly. Defining

$$R^\mu(p_\pi, p_k) \equiv \frac{\langle \pi(p_\pi)|V^\mu|K(p_K)\rangle \langle K(p_K)|V^\mu|\pi(p_\pi)\rangle}{\langle \pi(p_\pi)|V^\mu|\pi(p_\pi)\rangle \langle K(p_K)|V^\mu|K(p_K)\rangle}, \qquad (3)$$

for the momenta which give $q^2 = 0$ we have

$$R^4(\vec{p}_K, 0) = \frac{[f^+(0)(E_K + m_\pi) + f^-(0)(E_K - m_\pi)]^2}{4m_\pi E_K}$$

$$R^4(0, \vec{p}_\pi) = \frac{[f^+(0)(m_K + E_\pi) + f^-(0)(m_K - E_\pi)]^2}{4E_\pi m_K}.$$

In this way we obtain the form factors directly at $q^2 = 0$ and avoid the uncertainties due to the choice of ansatz. We are currently investigating the precision with which this procedure can be used [13].

## Finite Volume Effects in $K \to \pi\pi$ Decays

A quantitative understanding of non-perturbative effects in $K \to \pi\pi$ decays will be an important future milestone for lattice QCD. Two particularly interesting challenges are:
i) an understanding of the empirical $\Delta I = 1/2$ rule, which states that the amplitude for decays in which the two-pion final state has isospin I=0 is larger by a factor of about 22 than that in which the final state has $I = 2$;
ii) a calculation of $\varepsilon'/\varepsilon$, whose experimental measurement with a non-zero value, $(17.2 \pm 1.8) \times 10^{-4}$, was the first observation of direct CP-violation.
The two challenges require the computation of the matrix elements of the $\Delta S = 1$ operators which appear in the effective Weak Hamiltonian.

For most of the quantities being calculated in lattice simulations (such as $K_{\ell 3}$ decays described above) there is at most a single hadron in the initial and final state. In $K \to \pi\pi$ decays there are two-pions and the interactions between the two pions induce finite-volume corrections which do not vanish exponentially. An indirect approach which avoids this complication is to calculate $K \to \pi$ and $K \to$ vacuum matrix elements and to estimate the $K \to \pi\pi$ decay amplitudes using lowest order chiral perturbation theory [15, 16]. For the direct evaluation of the $K \to \pi\pi$ amplitudes, the theory of finite-volume effects for two-hadron states in the elastic regime is now fully understood, both in the centre-of-mass and moving frames. In this section I describe the status of our understanding based on the pioneering work of Lüscher [17] and subsequent papers [18] – [22]. For a discussion of two-hadron states in nucleon-nucleon systems see ref. [23] and references therein.

Consider the two-hadron correlation function represented by the diagram

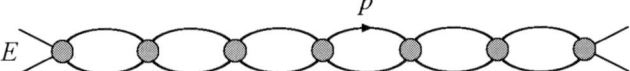

where the shaded circles represent two-particle irreducible contributions in the s-channel. For simplicity let us take the two-hadron system to be in the centre-of mass frame and assume that only the s-wave phase-shift is significant (the discussion can be extended to include higher partial waves). Consider the loop integration/summation over $p$ (see the figure). Taking the time extent of the lattice to be infinite, we can perform the $p_0$ integration by contours and obtain a summation over the spatial momenta of the form:

$$\frac{1}{L^3} \sum_{\vec{p}} \frac{f(p^2)}{p^2 - k^2} \qquad (4)$$

where $E$ and $k$ are the relative momentum and total energy, $E^2 = 4(m^2 + k^2)$, the function $f(p^2)$ is non-singular and (for periodic boundary conditions) the summation is over momenta $\vec{p} = (2\pi/L)\vec{n}$ where $\vec{n}$ is a vector of integers. In infinite volume the summation in eq. (4) is replaced by an integral and it is the difference between the summation and integration which gives the finite-volume corrections. The relation between finite-volume sums and infinite-volume integrals is the *Poisson Summation Formula*, which (in 1-dimension) is:

$$\frac{1}{L}\sum_p g(p) = \sum_{l=-\infty}^{\infty} \int \frac{dp}{2\pi} e^{ilLp} g(p). \tag{5}$$

If the function $g(p)$ is non-singular, the oscillating factors on the right-hand side ensures that only the term with $l = 0$ contributes, up to terms which vanish exponentially with $L$. The summand in eq. (4) on the other hand is singular (there is a pole at $p^2 = k^2$) and this is the reason why the finite-volume corrections only decrease as powers of $L$. The full derivation of the formulae for the finite-volume corrections can be found in refs. [17] – [22] and is beyond the scope of this talk. Here I just sketch the key ingredients. Following ref. [21], it is convenient to start by rewriting the expression eq. (4) in a form without singularities so that, up to exponential precision in the volume,

$$\frac{1}{L^3}\sum_{\vec{p}} \frac{f(p^2) - f(k^2)e^{\alpha(k^2 - p^2)}}{p^2 - k^2} = \int \frac{d^3 p}{(2\pi)^3} \frac{f(p^2) - f(k^2)e^{\alpha(k^2 - p^2)}}{p^2 - k^2}. \tag{6}$$

The exponential factors in eq.(6) are introduced to ensure ultra-violet convergence ($\alpha$ is the cut-off). Eq.(6) can then readily be rewritten in the form

$$\frac{1}{L^3}\sum_{\vec{p}} \frac{f(p^2)}{p^2 - k^2} = \int \frac{d^3 p}{(2\pi)^3} \frac{f(p^2)}{p^2 - k^2 - i\varepsilon}$$
$$- \frac{ik}{4\pi} f(k^2) + f(k^2) \left\{ \frac{1}{L^3}\sum_{\vec{p}} \frac{e^{\alpha(k^2 - p^2)}}{p^2 - k^2} - \mathscr{P}\int \frac{d^3 p}{(2\pi)^3} \frac{e^{\alpha(k^2 - p^2)}}{p^2 - k^2} \right\}, \tag{7}$$

where $\mathscr{P}$ represents the principal value. The finite-volume correction exhibited above appears in every loop in diagrams such as that in the figure, and upon resummation give a geometric series. In infinite volume there is a two-particle cut with a branch point at the two-pion threshold. In finite volume the cut is replaced by a series of poles and the positions of these poles correspond to the allowed energy levels (i.e. the Lüscher Quantization Condition). Note that the finite-volume corrections in eq.(7) depend on the function $f$ evaluated at the external energy corresponding to $k^2$, which allows us to express the positions of the poles in terms of the physical amplitude (or phase-shift $\delta(k^2)$) and kinematic factors; specifically the poles occur at values of $k$ satisfying:

$$\tan(\delta(k^2)) = -\frac{k}{4\pi} \left\{ \frac{1}{L^3}\sum_{\vec{p}} \frac{e^{\alpha(k^2 - p^2)}}{k^2 - p^2} - \mathscr{P}\int \frac{d^3 p}{(2\pi)^3} \frac{e^{\alpha(k^2 - p^2)}}{k^2 - p^2} \right\}^{-1}. \tag{8}$$

The finite-volume corrections to the matrix elements have also been obtained both in the centre-of-mass frame [19] and in moving frames [21, 22].

# SUMMARY AND CONCLUSIONS

In this talk I discussed three topics in lattice phenomenology with the following conclusions:

1. The use of (partially) twisted boundary conditions enables an improved momentum resolution for lattice phenomenology, with many potential applications.
2. The evaluation of $f^+(0)$ for $K_{\ell 3}$ decays is possible with about the required accuracy and much work is being done to improve the precision.
3. Finite volume effects for the two-pion spectrum and $K \to \pi\pi$ decay amplitudes are understood in both the rest and moving frames. For $I = 2$ final states, we have the ingredients necessary to calculate the matrix elements reliably. For $I = 0$ $\pi\pi$ final states we still need to learn how to calculate the disconnected diagrams.

# REFERENCES

1. P. F. Bedaque, *Phys. Lett. B* **593** 82 (2004) [arXiv:nucl-th/0402051].
2. C. T. Sachrajda and G. Villadoro, *Phys. Lett. B* **609** 73 (2005) [arXiv:hep-lat/0411033].
3. P. F. Bedaque and J. W. Chen, *Phys. Lett. B* **616** 208 (2005) [arXiv:hep-lat/0412023].
4. J. M. Flynn, A. Jüttner and C. T. Sachrajda [UKQCD Collaboration], *Phys. Lett. B* **632** 313 (2006) [arXiv:hep-lat/0506016]; J. Flynn, A. Jüttner, C. Sachrajda and G. Villadoro, PoS **LAT2005** (2005) 352 [arXiv:hep-lat/0509093].
5. H. Leutwyler and M. Roos, *Z. Phys. C* **25** 91 (1984).
6. V. Cirigliano, G. Ecker, M. Eidemuller, R. Kaiser, A. Pich and J. Portoles, *JHEP* **0504** 006 (2005) [arXiv:hep-ph/0503108]; G.Ecker, these proceedings.
7. S. Hashimoto, A. X. El-Khadra, A. S. Kronfeld, P. B. Mackenzie, S. M. Ryan and J. N. Simone, *Phys. Rev. D* **61** 014502 (2000) [arXiv:hep-ph/9906376].
8. D. Becirevic *et al.*, *Nucl. Phys. B* **705** 339 (2005) [arXiv:hep-ph/0403217].
9. M. Okamoto [Fermilab Lattice Collaboration], arXiv:hep-lat/0412044.
10. N. Tsutsui *et al.* [JLQCD Collaboration], PoS **LAT2005** (2005) 357 [arXiv:hep-lat/0510068].
11. C. Dawson, T. Izubuchi, T. Kaneko, S. Sasaki and A. Soni, arXiv:hep-ph/0607162.
12. UKQCD/RBC Collaboration, D. J. Antonio *et al.*, arXiv:hep-lat/0610080.
13. P.A. Boyle, J.M. Flynn, A. Jüttner, C.T.Sachrajda and J. Zanotti (in preparation)
14. D. Guadagnoli, F. Mescia and S. Simula, Phys. Rev. D **73**, 114504 (2006) [arXiv:hep-lat/0512020].
15. T. Blum *et al.* [RBC Collaboration], *Phys. Rev. D* **68** 114506 (2003) [arXiv:hep-lat/0110075].
16. J. I. Noaki *et al.* [CP-PACS Collaboration], *Phys. Rev. D* **68** 014501 (2003) [arXiv:hep-lat/0108013].
17. M. Luscher, *Commun. Math. Phys.* **104** 177 (1986); *Commun. Math. Phys.* **105** 153 (1986); *Nucl. Phys. B* **354** 531 (1991); *Nucl. Phys. B* **364** 237 (1991).
18. K. Rummukainen and S. A. Gottlieb, *Nucl. Phys. B* **450** 397 (1995) [arXiv:hep-lat/9503028].
19. L. Lellouch and M. Luscher, *Commun. Math. Phys.* **219** 31 (2001) [arXiv:hep-lat/0003023].
20. C. J. D. Lin, G. Martinelli, C. T. Sachrajda and M. Testa, *Nucl. Phys. B* **619** 467 (2001) [arXiv:hep-lat/0104006].
21. C. h. Kim, C. T. Sachrajda and S. R. Sharpe, *Nucl. Phys. B* **727** (2005) 218 [arXiv:hep-lat/0507006]; PoS **LAT2005** (2005) 359 [arXiv:hep-lat/0510022].
22. N. H. Christ, C. Kim and T. Yamazaki, *Phys. Rev. D* **72** 114506 (2005) [arXiv:hep-lat/0507009].
23. D. Kaplan, these proceedings.

# Status of Flavor Physics

## Ian Shipsey

*Department of Physics, Purdue University,*
*West Lafayette, IN 47907, U.S.A.*

**Abstract.** The role of charm in testing the Standard Model description of quark mixing and CP violation through measurements of lifetimes, decay constants and semileptonic form factors is reviewed. Together with Lattice QCD, charm has the potential this decade to maximize the sensitivity of the entire flavor physics program to new physics. and pave the way for understanding physics beyond the Standard Model at the LHC in the coming decade. The status of indirect searches for physics beyond the Standard Model through charm mixing, CP-violation and rare decays is also reported, as are recent discoveries in charm spectroscopy.

**Keywords:** beauty quark; charm quark; weak interaction; strong interaction.
**PACS:** 11.25.Hf, 123.1K

## INTRODUCTION

ICHEP2006 devoted 180 minutes to experimental plenary summaries of "flavor physics" in addition there were theory talks. As I have 35 minutes I will be selective. As Guido Martinelli focused on "beauty from experiments and the lattice" in his plenary presentation, this talk will be mostly about charm. See the talk of Estia Eichten for theoretical aspects of spectroscopy. See also the parallel session talks at this conference by Kalashnikova, Kwon, Lasiak, Pakhlova, Seth, Shuxian, Stelzer, Swanson, Torres, and Verde-Valasco.

## BIG QUESTIONS IN FLAVOR PHYSICS

The big questions in quark flavor physics are: (1) "What is the dynamics of flavor?" The gauge forces of the standard model (SM) do not distinguish between fermions in different generations. The electron, muon and tau all have the same electric charge, quarks of different generations have the same color charge. Why generations? Why three? (2) "What is the origin of baryogenesis?" Sakharov gave three criteria, one is CP-violation [1]. There are only three known examples of CP-violation: the Universe, and the beauty and kaon sectors. However, SM CP-violation is too small, by many orders of magnitude, to give rise to the baryon asymmetry of the Universe. Additional sources of CP-violation are needed. (3) "What is the connection between flavor physics and electroweak symmetry breaking?" Extensions of the SM, for example supersymmetry, contain flavor and CP-violating couplings that should show up at some level in flavor physics but precision measurements and precision theory are required to detect the new physics.

CP892, *Quark Confinement and the Hadron Spectrum VII*
edited by J. E. F. T. Ribeiro
© 2007 American Institute of Physics 978-0-7354-0396-3/07/$23.00

# SPECTROSCOPY

A primary experimental tool for understanding QCD is heavy quark hadron spectroscopy, recall atomic spectroscopy and its role in understanding QED. This year there has been a number of new charm meson, charm baryon and states above $D\overline{D}$ threshold observed. Due to limited space I will simply list them here, see the conference transparencies on the web for greater experimental detail and interpretation. Two new mesons at 2690 MeV and 2860 MeV were observed in the process $e^+e^- \rightarrow DKX$ by BABAR [2]. A new meson was seen in the process $B \rightarrow \overline{D^0}D^0K^+$ from Belle with $J^P = 1^-$ at a mass of 2715 MeV which is consistent with the BABAR observation at 2690 MeV so it is presumably the same particle [3]. I will call this particle the $D_{sJ}(2700)$. These two mesons are presumably radial excitations of the $D_{sJ}$. For the $D_{sJ}(2700)$ the possibilities are the $c\overline{s} \, 2^3S_1(2S)$ which is predicted to have a mass of 2711 MeV [4], or the chiral doubler state to the $1^+$ state $D_{sJ}(2356)$ which is predicted to have a mass of $2721 \pm 10$ MeV [5]. It is important to determine the $J^P$ of the 2860 state. What we do know is that as both a $D$ and a $K$ are pseudoscalars the allowed $J^P$ values are $0^+, 1^-, 2^+, 3^-$. The modified quark model predicts the $^3P_0(2P)$ state to have a mass of 2817 MeV [4], other identifications include the $c\overline{s}$ scalar [6], and a $c\overline{s}$ with $J^P = 3^-$ [7].

BABAR has seen a new charm baryon in the process $\Lambda_c(2940) \rightarrow D^0p$ [8]. The absence of an isospin partner decaying to $D^+p$ indicates this particle is a $\Lambda_c$. Belle has also seen the state in the process $\Lambda_c(2940) \rightarrow \Sigma_c(2455)\pi$. It would be valuable to know the ratio of these two decay modes. Belle has seen new charm baryons $\Xi_{cx}(3077)^+$ and $\Xi_{cx}(2980)^+$ in the $\Lambda_cK^-\pi^+$ final state [9]. The isopartner of the former has also been seem by Belle in the $\Lambda_cK_S\pi^-$ final state. BABAR confirms these states [10]. The mass and width are consistent between the two experiment for the $\Xi_{cx}(3077)$. There is some spread in mass for the $\Xi_{cx}(2980)$ between BABAR and Belle, and this is presumably due to the differing functional forms used in the fits, and the proximity of this state to threshold. BABAR has made the first observation of the $\Omega_c^*$. They find $m(\Omega_c^*) - m(\Omega_c^0) = (70.8 \pm 1.0 \pm 1, 1)$ MeV which is in reasonable agreement with theoretical predictions from HQET, LQCD and non-relativistic quark models. The SELEX doubly charmed baryons have not been confirmed.

In 2002 it had been more than 25 years since the discovery of a new charmonium state. Since then the $\eta_c(2S)$ and the $h_c$ have finally been discovered below $D\overline{D}$ threshold. There have also been six new states observed above $D\overline{D}$ threshold. These states are not understood. The states which have been seen by more than one experiment and that I view as well-established are (1) the $X(3872)$, $J^{PC} = 1^{++}$ or $2^{-+}$ [11] [12]. This particle is very close indeed to the $DD^*$ threshold which suggests it may be a $\overline{D^0}D^{*0}$ molecule. However a critical test of this idea is to check the sign of the binding energy, for which it is vital to improve the precision of the $X(3872)$ mass, until such a measurement is made it is fair to say there is no entirely satisfactory explanation of the nature of the $X(3872)$. (2) The $Y(4260)$, $J^{PC} = 1^{--}$ was discovered by BABAR in ISR in $e^+e^-$ annihilation at 10 GeV in the decay mode $Y(4260) \rightarrow J/\psi\pi\pi$ [13]. The $Y(4260)$ has now been seen by CLEO and Belle in ISR as well. Like the $X(3872)$ the $Y(4260)$ is near a two-particle threshold, this time the $\overline{DD_1}$. There is no obvious mass assignment in the conventional $c\overline{c}$ spectrum and this state may be hybrid or tetraquark. The other four states are the $Y(3940)$ which may be a hybrid, the $Z(3940)$ which may be the $\chi_{c2}$, the $X(3940)$ which may be the $\eta_c(3S)$ and the $Y(4350)$ which has $J^{PC} = 1^{--}$. With the discovery os a plethora of new states, some have argued that we witnessing the birth of a new spectroscopy. I believe it is too early to say. It is very challenging to establish a new spectroscopy as the region has many expected but unidentified charmonium states. More data, refinements to theory, and time will bring resolution

## CHARM IN CKM PHYSICS

This is the decade of precision flavor physics. The goal is to over-constrain the CKM matrix with a range of measurements in the quark flavor changing sector of the SM at the per cent level. If inconsistencies are found between, for example, measurements of the sides and angles of the $B_d$ unitarity triangle, it will be evidence for new physics. Many experiments will contribute including BaBar and Belle, CDF, D0 at Fermilab, ATLAS, CMS, and LHC-b at the LHC, BESIII, CLEO-c, and experiments studying rare kaon decays.

However, the study of weak interaction phenomena, and the extraction of quark mixing matrix parameters remain limited by our capacity to deal with non-perturbative strong interaction dynamics. Current constraints on the CKM matrix are shown in Fig. 1(a). The widths of the constraints, except that of $\sin 2\beta$, are dominated by the error bars on the calculation of hadronic matrix elements. Recent advances in LQCD have produced calculations of non-perturbative quantities such as $f_\pi$, $f_K$, and heavy quarkonia mass splittings that agree with experiment [14]. Several per cent precision in charm and beauty decay constants and form factors is hoped for, but the path to higher precision is hampered by the absence of accurate charm data against which to test lattice techniques. This is beginning to change with the BES II run at the $\psi(3770)$, and the start of data taking at the charm and QCD factory CESR-c/CLEO-c [15]. Later in the decade BES III at the new double ring accelerator BEPC-II will also turn on [16]. CLEO-c is in the process of obtaining charm data samples one to two orders of magnitude larger than any previous experiment, and the BES III data set is expected to be $\sim \times 20$ larger than CLEO-c. These data sets have the potential to provide unique and crucial tests of LQCD, and other QCD technologies such as QCD sum rules and chiral theory, with accuracies of 1-2%.

If LQCD passes the charm factory tests, we will have much greater confidence in lattice calculations of decay constants and semileptonic form factors in $B$ physics. When these calculations are combined with 500 fb$^{-1}$ of $B$ factory data, and improvement in the direct measurement of $|V_{tb}|$ at the Tevatron [17], they will allow a significant reduction in the size of the errors on $|V_{ub}|, |V_{cb}|, |V_{td}|$ and $|V_{ts}|$, quantitatively and qualitatively transforming knowledge of the $B_d$ unitarity triangle, see Fig. 1(b), and thereby maximizing the sensitivity of heavy quark physics to new physics.

Equally important, LQCD combined with charm data allows a significant advance in understanding and control over strongly-coupled, non-perturbative quantum field theories in general. Field theory is generic, weak coupling is not. Two of the three known interactions are strongly coupled: QCD and gravity (string theory). Understanding strongly coupled theories may be a crucial to interpret new phenomena at the high

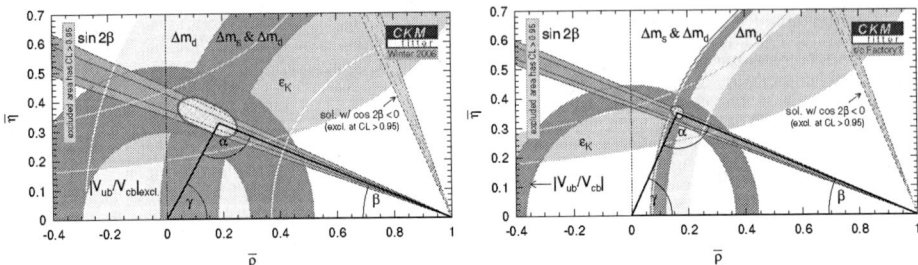

**FIGURE 1.** Lattice impact on the $B_d$ unitarity triangle from $B_d$ and $B_s$ mixing, $|V_{ub}|/|V_{cb}|$, $\varepsilon_K$, and $\sin 2\beta$. (a) Winter 2006 status of the constraints including the recent observation of $B_s$ mixing. (b) Prospects under the assumption that LQCD calculations of $B$ system decay constants and semileptonic form factors achieve the the precision projected in Table 4.

energy frontier.

## Decay Constants

The $B_d$ ($B_s$) meson mixing probability can be used to determine $|V_{td}|$ ($|V_{ts}|$).

$$\Delta m_d \propto |V_{tb}V_{td}|^2 f_{B_d}^2 B_{B_d} \tag{1}$$

The $B_d$ mixing rate is measured with exquisite precision (1%) [18] but the decay constant is calculated with a precision of about 10-15%. If theoretical precision could be improved to 3%, the error on $|V_{td}|$ would be about 5%.

Since LQCD hopes to predict $f_B/f_{D^+}$ with a small error, measuring $f_{D^+}$ would allow a precision prediction for $f_B$. Hence a precision extraction of $|V_{td}|$ from the $B_d$ mixing rate becomes possible. Similar considerations apply to $B_s$ mixing now it has been observed i.e. a precise determination of $f_{D_s^+}$ would allow a precision prediction for $f_{B_s}$ and consequently a precision measurement of $|V_{ts}|$. Finally the ratio of the two neutral $B$ meson mixing rates determines $|V_{td}|/|V_{ts}|$, but $|V_{ts}| = |V_{cb}|$ by unitarity and $|V_{cb}|$ is known to a few per cent, and so the ratio again determines $V_{td}$. Which method of determining $|V_{td}|$ will have the greater utility depends on which combination of hadronic matrix elements have the smallest error.

Charm leptonic decays measure the charm decay constants $f_{D_s^+}$ and $f_{D^+}$ because $|V_{cs}|$ and $|V_{cd}|$ are known from unitarity to 0.1% and 1% respectively.

$$\frac{\mathscr{B}(D^+ \to \mu\nu_\mu)}{\tau_{D^+}} = (\text{const.})f_{D^+}^2 |V_{cd}|^2 \tag{2}$$

(Charge conjugation is implied throughout this paper.) The measurements are also a precision test of the LQCD. At the start of 2004 $f_{D^+}$ was experimentally undetermined and $f_{D_s^+}$ was known to 33%.

# Semileptonic form factors

$|V_{ub}|$ is determined from beauty semileptonic decay

$$\frac{d\Gamma(B \to \pi e^- \bar{v}_e)}{dq^2} = (\text{const.})|V_{ub}|^2 f_+^{B\pi}(q^2)^2 \tag{3}$$

The differential rate depends on a form factor, $f_+(q^2)$ that parameterizes the strong interaction non-perturbative effects. A representative value of $|V_{ub}|$ determined from $B \to \pi \ell^- \bar{v}_e$ is [19]:

$$|V_{ub}| = (3.76 \pm 0.16^{+0.87}_{-0.51}) \times 10^{-3} \tag{4}$$

where the uncertainties are experimental statistical and systematic, and from the LQCD calculation of the form factor, respectively. The experimental errors are expected to be reduced to 5% with $B$ factory data samples of 500 fb$^{-1}$ each, and the theory error will dominate.

Again, because the charm CKM matrix elements are known from unitarity, the differential charm semileptonic rate

$$\frac{d\Gamma(D \to \pi e^+ v_e)}{dq^2} = (\text{const.})|V_{cd}|^2 f_+^{D\pi}(q^2)^2 \tag{5}$$

tests calculations of charm semileptonic form factors. Thus, a precision measurement tests the LQCD calculation of the $D \to \pi$ form factor. As the form factors governing $B \to \pi e^- \bar{v}_e$ and $D \to \pi e^+ v_e$ are related by heavy quark symmetry, the charm test gives confidence in the accuracy of the $B \to \pi$ calculation. The $B$ factories can then use a tested LQCD prediction of the $B \to \pi$ form factor to extract a precise value of $|V_{ub}|$. At the start of 2004, $\mathscr{B}(D \to \pi e^+ v_e)$ had been determined to 45% [18, 20], and the absolute value of the $D \to \pi$ form factor had not been measured.

## ABSOLUTE CHARM BRANCHING RATIOS

We reviewed above the importance of absolute charm leptonic and semileptonic branching ratios. The absolute hadronic branching ratios $\mathscr{B}(D^+ \to K^- \pi^+ \pi^+)$, $\mathscr{B}(D^0 \to K^- \pi^+)$, and $\mathscr{B}(D_s^+ \to \phi \pi^+)$ are also important as, currently, all other $D^+$, $D^0$ and $D_s^+$ branching ratios are determined from ratios to one or the other of these branching fractions [18]. In consequence, nearly all branching fractions in the $B$ and $D$ sectors depend on these reference modes. While charm lifetimes are precisely measured, absolute charm branching ratios are poorly known.

## BES II AND CLEO-C AT THE $\psi(3770)$

In 2003 the venerable BES II detector accumulated an integrated luminosity of 33 pb$^{-1}$ at and around the $\psi(3770)$. CLEO-c has accumulated 281 pb$^{-1}$ at the $\psi(3770)$ ($1.8 \times 10^6 D\bar{D}$ pairs) and about 300 pb$^{-1}$ at $\sqrt{s} \sim 4170$ MeV for $D_s$ physics. These $\psi(3770)$

datasets exceed those of the BESII (Mark III) experiments by factors of 30 (15). CLEO-c expects to approximately triple each data set by 2008. The BEPCII Project will be commissioned soon. Luminosity is expected to be $10^{33}$cm$^{-2}$s$^{-1}$ at 1.89 GeV $6 \times 10^{32}$cm$^{-2}$s$^{-1}$ at 1.55 GeV and $6 \times 10^{32}$cm$^{-2}$s$^{-1}$ at 2.1 GeV. At 5/fb/yr or 15/fb/3yrs, there will be $90 \times 10^6 D\bar{D}$ pairs or a factor 20 greater than the full CLEO-c data sample.

In the longer term proposed Super B Factories at KEK or SuperB or a dedicated charm factory would produce an abundance of charm. For example the SuperB machine at $10^{36}$cm$^{-2}$s$^{-1}$ will produce $10^{10}e^+e^- \to c\bar{c}$ pairs/$10^7$s. Due to the "Linear Collider design" there is an option to lower the energy to 4 GeV with a modest luminosity penalty of a factor 10. In this mode of operation the super B Factory becomes a super flavour factory.

## Analysis Technique

There are decisive advantages to running at charm threshold. As $\psi \to D\bar{D}$, the technique is to fully reconstruct one $D$ meson in a hadronic final state, the tag, and then to analyze the decay of the second $D$ meson in the event to extract inclusive or exclusive properties.

As $E_{\text{beam}} = E_D$, the candidate is required to have energy close to the beam energy, and the beam-constrained candidate mass, $M_D = \sqrt{E_{\text{beam}}^2 - p_{\text{cand}}^2}$, is computed. Charm mesons have many large branching ratios to low multiplicity final states, and so the tagging efficiency is very high, about 25%, compared to much less than 1% for $B$ tagging at a $B$ factory.

Tagging creates a single $D$ meson beam of known momentum. The beam constrained mass for events in which the second $D$ meson is also reconstructed are shown in Fig. 2. These double tag events, which are key to making absolute branching fraction measurements, are pristine. The absolute branching fraction is given by:

$$\mathscr{B}(D^+ \to K^-\pi^+\pi^+) = \frac{N(K^-\pi^+\pi^+)}{\varepsilon(K^-\pi^+\pi^+) \times N(D^-)} \qquad (6)$$

where $N(K^-\pi^+\pi^+)$ is the number of $D^+ \to K^-\pi^+\pi^+$ observed in tagged events, $\varepsilon(K^-\pi^+\pi^+)$ is the reconstruction efficiency and $N(D^-)$ is the number of tagged events.

In a method similar to that pioneered by Mark III [23, 24], CLEO fits to the observed single tag and double tag yields for six $D^+$ and three $D^0$ modes [25]. I will only consider the two most important branching fractions here. For $D^0 \to K^-\pi^+$ the total errors are comparable to previous measurements, see Table 1. This is the most precise measurement of $\mathscr{B}(D^+ \to K^-\pi^+\pi^+)$ to date, see Table 2.

BES II has performed a similar analysis. These recent measurements are in remarkably good agreement with the PDG averages, indicating that the charm, and hence beauty, decay scales, are approximately correct and are now, finally, on a solid foundation.

In consequence the number of events in 100pb$^{-1}$ with two $D$ mesons reconstructed is about the same as the number of events at 10 GeV with 500fb$^{-1}$ with two $B$ mesons reconstructed.

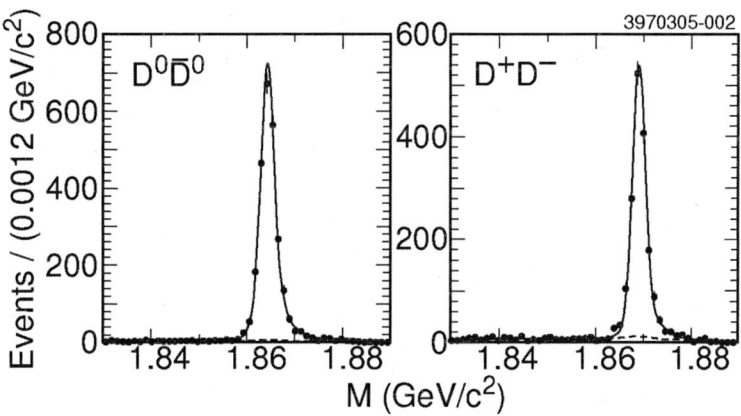

**FIGURE 2.** Beam constrained mass of $D$ mesons in CLEO-c events in which both $D$ mesons have been fully reconstructed.

**TABLE 1.** The $D^0 \rightarrow K^-\pi^+$ absolute charm branching ratio.

| $\mathscr{B}(\%)$ | Error (Source) |
|---|---|
| $3.82 \pm 0.07 \pm 0.12$ | 3.6% (CLEO [27]) |
| $3.90 \pm 0.09 \pm 0.12$ | 3.8% (ALEPH [26]) |
| $3.80 \pm 0.09$ | 2.4% (PDG) |
| $3.91 \pm 0.08 \pm 0.09$ | 3.1 % (CLEO-c [25]) |

## Charm Decay Constant

The measurement of the leptonic decay $D^+ \rightarrow \mu^+\nu_\mu$ benefits from the fully tagged $D^-$ at the $\psi(3770)$. One observes a single charged track recoiling against the tag that is consistent with a muon of the correct sign. Energetic electromagnetic showers unassociated with the tag are not allowed. The missing mass $MM^2 = m_\nu^2$ is computed; it peaks at zero for a decay where only a neutrino is unobserved. Fig. 3 shows the $MM^2$ distribution from CLEO-c [30].

**TABLE 2.** The $D^+ \rightarrow K^-\pi^+\pi^+$ absolute charm branching ratio.

| $\mathscr{B}(\%)$ | Error (Source) |
|---|---|
| $9.3 \pm 0.6 \pm 0.8$ | 10.8% (CLEO [28]) |
| $9.1 \pm 1.3 \pm 0.4$ | 14.9% (MKIII [29]) |
| $9.1 \pm 0.7$ | 7.7% (PDG) |
| $9.52 \pm 0.25 \pm 0.27$ | 3.9 % (CLEO-c [25]) |

113

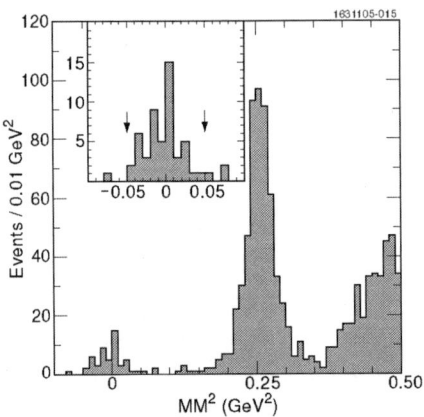

**FIGURE 3.** The $MM^2$ distribution in events with $D^-$ tag, a single charged track of the correct sign, and no additional (energetic) showers. The insert shows the signal region for $D^+ \rightarrow \mu \nu_\mu$. A $\pm 2\sigma$ range is indicated by the arrows.

There are 50 candidate signal events, and $2.81 \pm 0.3^{+0.84}_{-0.22}$ background events. After correcting for efficiency, CLEO-c finds

$$\mathscr{B}(D^+ \rightarrow \mu^+ \nu_\mu) = (4.40 \pm 0.66^{+0.09}_{-0.12}) \times 10^{-4}, \qquad (7)$$

where the uncertainties are statistical and systematic, respectively. Under the assumption of three generation unitarity, and using the precisely known $D^+$ lifetime, CLEO-c obtains

$$f_{D^+} = (222.6 \pm 16.7^{+2.8}_{-3.4}) \text{ MeV}. \qquad (8)$$

This is the most precise measurement of $f_{D^+}$ [30]. The result appeared at Lepton-Photon 2005 just two days after the first unquenched lattice QCD calculation [31] had predicted:

$$f_{D^+} = (201 \pm 3 \pm 17) \text{ MeV}. \qquad (9)$$

The combined experimental error is 8% while the LQCD error is also 8% [31]. The results are in good agreement but errors are still large. The only other positive observation of this decay is by BES II who found three candidate events with a background of 0.25 events in their 33pb$^{-1}$ data sample. They find a branching ratio of $(0.122^{+0.111}_{-0.053} \pm 0.010)\%$ corresponding to $f_{D^+} = (371^{+129}_{-119} \pm 25)$ MeV [32]. The CLEO value is considerably smaller and in better agreement with expectations from the lattice and other theoretical approaches. With 0.75 fb$^{-1}$ a 4.5% error for $f_{D^+}$ is expected. Similar precision is expected for $f_{D_s^+}$ at $\sqrt{s} = 4160$ MeV. BES III will make even more precise measurements achieving a precision of several per cent for both $f_{D^+}$ and $f_{D_s^+}$ which is well matched to the ultimate precision of the LQCD calculations.

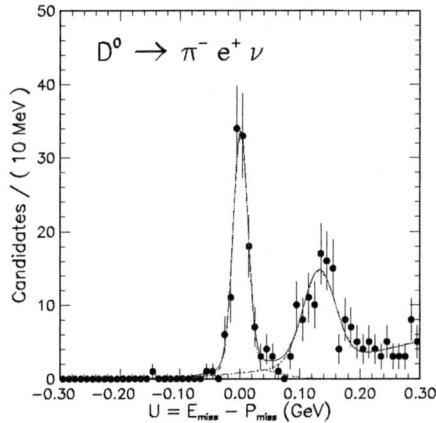

**FIGURE 4.** The $U = E_{miss} - P_{miss}$ distribution in events with a $\bar{D}^0$ tag, a positron, and a single charged track of the correct sign. The peaks at zero and 0.13 GeV correspond to $D^0 \to \pi^- e^+ \nu_e$ and $D^0 \to K^- e^+ \nu_e$ (preliminary.)

**TABLE 3.** Selected CLEO-c charm semileptonic branching ratio measurements in % and a comparison to the PDG.

| Mode | PDG | CLEO-c |
|---|---|---|
| $D^0 \to \pi^- e^+ \nu_e$ | $0.36 \pm 0.06$ | $0.26 \pm 0.03 \pm 0.1$ |
| $D^0 \to K^- e^+ \nu_e$ | $3.58 \pm 0.18$ | $3.44 \pm 0.10 \pm 0.1$ |
| $D^+ \to \pi^0 e^+ \nu_e$ | $0.31 \pm 0.05$ | $0.44 \pm 0.06 \pm 0.01$ |
| $D^+ \to \bar{K}^0 e^+ \nu_e$ | $6.7 \pm 0.9$ | $8.71 \pm 0.38 \pm 0.37$ |

## Measurement of the Charm Semileptonic Form Factors

The measurement of semileptonic decay absolute branching ratios and absolute form factors is also based on the use of tagged events. The analysis procedure, using $D^0 \to \pi^- e^+ \nu_e$ as an example is as follows. A positron and a hadronic track are identified recoiling against the tag. The quantity $U = E_{miss} - P_{miss}$ is calculated, where $E_{miss}$ and $P_{miss}$ are the missing energy and missing momentum in the event. For a tagged event with a semileptonic decay $E_{miss}$ and $P_{miss}$ are the components of the four-momentum of the neutrino. $U$ peaks at zero if only a neutrino is missing. The $U$ distribution in 56 pb$^{-1}$ of CLEO-c data is shown in Fig. 4 where a clean signal of about 100 events is observed for $D \to \pi e^+ \nu_e$ with $S/N$ 20/1 [33]. In previous analyses at $B$ Factories and fixed target experiments the background was usually larger than the signal see for example [34]. BES II have performed similar analyses [35] [36] and results are in good agreement with CLEO-c. Selected CLEO-c absolute semileptonic branching ratio measurements are compared to PDG values in Table 3.

Recently there have been several beautiful measurements of the form factor shape in $D \to K\ell^+ \nu_\ell$ and $D \to \pi \ell^+ \nu_\ell$ by CLEO, FOCUS, Belle, and BABAR. By reconstructing

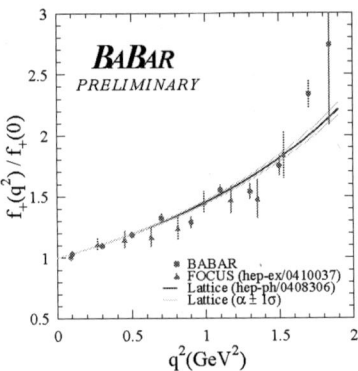

**FIGURE 5.** The differential rate, normalized to the rate at $q^2 = 0$ for the decay $D^0 \rightarrow K^-\ell\nu_\ell$ after removal of phase space factors, compared to the LQCD prediction.

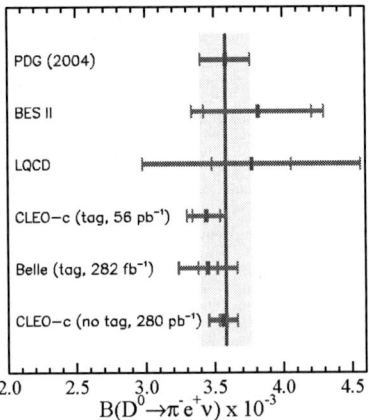

**FIGURE 6.** Measurements of the absolute branching fraction for $D \rightarrow \pi e^+\nu_e$ and comparison to LQCD. A preliminary result from an untagged measurement from CLEO-c has also been included.

two $D$ mesons in $e^+e^- \rightarrow c\bar{c}$ events at 10 GeV Belle are able to make an absolute measurement and so a determination of the form factor magnitude as well. CLEO-c results were presented at ICHEP.

The FOCUS, BABAR, Belle and CLEO-c measurements check the shape of the form factor, for a tabulation see [38]. The best way to measure the normalization it to fit to the differential rate to obtain $f^+(0)V_{cx}$ or from the absolute branching fraction, in both cases using unitarity and the $D$ meson lifetime. A comparison of absolute branching fraction measurements and the LQCD prediction is shown in Figure 6. Agreement is reasonable, although the theory errors are in urgent need of being reduced.

If LQCD passes the experimental tests outlined above it will be possible to use the

LQCD calculation of the $B \to \pi$ form factor with increased confidence at the $B$ factories to extract a precision $V_{ub}$ from $B \to \pi e^- \bar{v}_e$. BaBar and Belle will also be able to compare the LQCD prediction of the shape of the $B \to \pi$ form factor to data as an additional cross check.

Successfully passing the experimental tests allows the charm factories to use LQCD calculations of the charm semileptonic form factors to directly measure $|V_{cd}|$ and $|V_{cs}|$. Using the isospin averaged semileptonic widths $\Gamma(D \to K e^+ v_e)$ and $\Gamma(D \to \pi e^+ v_e)$ [33] and the LQCD prediction of the semileptonic partial width [37] I obtain

$$V_{cs} = 0.957 \pm 0.017 \pm 0.093$$
$$V_{cd} = 0.213 \pm 0.008 \pm 0.029 \tag{10}$$

where the uncertainties are experimental statistical, experimental systematic and from LQCD. The results are consistent with the unitarity values

$$V_{cs} = 0.9745 \pm 0.0008$$
$$V_{cd} = 0.2238 \pm 0.012 \tag{11}$$

$V_{cd}$ has previously been determined from neutrino production of di-muons off of nucleons, and $V_{cs}$ has been determined from $W \to cs$ transitions at LEP to be [18]

$$|V_{cs}| = 0.976 \pm 0.014$$
$$|V_{cd}| = 0.224 \pm 0.012 \tag{12}$$

Due to the large theoretical uncertainties in the CLEO-c numbers the extracted values of $V_{cs}$ and $V_{cd}$ should be considered as tests of LQCD. Nonetheless, they are the single most precise determinations of $V_{cs}$ and $V_{cd}$ to date. With $0.75\text{fb}^{-1}$ of data the CLEO-c precision is expected to be respectively:

$$|V_{cs}| = \sqrt{0.8\% \oplus \delta\Gamma/2\Gamma}$$
$$|V_{cd}| = \sqrt{1.6\% \oplus \delta\Gamma/2\Gamma} \tag{13}$$

Where $\delta\Gamma/\Gamma$ is the uncertainty in the partial rate from theory. This in turn allows new unitarity tests of the CKM matrix. For example, the second row of the CKM matrix can be tested at the few % level. With the current measurements I find:

$$1 - (|V_{cs}|^2 + |V_{cd}|^2 + |V_{cb}|^2) = 0.037 \pm 0.181 \tag{14}$$

which is consistent with unitarity, with an uncertainty dominated by the LQCD charm semileptonic form factor magnitude The measurements also allow the first column of the CKM matrix to be tested with similar precision to the first row (which is currently the most stringent test of CKM unitarity); finally, the ratio of the long sides of the $uc$ unitarity triangle will be tested to a few percent.

Table 4 provides a summary of projections for the precision with which the CKM matrix elements will be determined if LQCD passes the charm factory tests in the $D$ system. In the tabulation the current precision of the CKM matrix elements is obtained

**TABLE 4.** LQCD impact (in per cent) on the precision of CKM matrix elements. A charm factory data set of 3/fb and a B factory data set of 500/fb is assumed.

|      | $V_{cd}$ | $V_{cs}$ | $V_{cb}$ | $V_{ub}$ | $V_{td}$ | $V_{ts}$ |
|------|------|------|------|------|------|------|
| 2004 | 7    | 11   | 4    | 15   | 36   | 39   |
| LQCD | 2    | 2    | 3    | 5    | 5    | 5    |

by considering methods applicable to LQCD, for example the determination of $|V_{cb}|$ and $|V_{ub}|$ from inclusive decays and OPE is not included. The projections are made assuming $B$ factory data samples of 500 fb$^{-1}$ and improvement in the direct measurement of $|V_{tb}|$ expected from the Tevatron experiments [17].

## The bottom line

How can we be sure that if LQCD works for $D$ mesons it will work for $B$ mesons? Or, equivalently, is charm factory data sufficient to demonstrate that lattice systematic errors are under control? There are a number of reasons to answer this question in the affirmative. (1) There are two independent effective field theories: NRQCD and the Fermilab method. (2) The CLEO-c, and later BESIII, data provide many independent tests in the $D$ system; leptonic decay rates, and semileptonic modes with rate and shape information. (3) The $B$ factory data provide additional independent cross checks such as $d\Gamma(B \rightarrow \pi \ell \nu)/dp_\pi$. (4) Unlike models, methods used for the $D/B$ system can be tested in heavy onia with measurements of masses, and mass splittings, $\Gamma_{ee}$ and electromagnetic transitions. (5) The main systematic errors limiting accuracy in the $D/B$ systems are: chiral extrapolations in $m$light, perturbation theory, and finite lattice spacing. These are similar for charm and beauty quarks. In my opinion a combination of CLEO-c and BES III data in the $D$ systems and onia, plus information on the light quark hadron spectrum, can clearly establish whether or not lattice systematic errors are under control.

While this picture is encouraging, experimentalists also have concerns. The lattice technique is all encompassing but LQCD practitioners are very conservative about what can be calculated. For example when there was a hint that $\sin 2\beta(\psi K_S^0) \neq \sin 2\beta(\phi K_S^0)$, and when CP violation was observed in $B \rightarrow K\pi$ [41] the lattice was not able to contribute. There is a pressing need to move beyond the limited set of easy to calculate quantities in the next few years: for example resonances such as $\rho$, $\phi$ and $K^*$ may be difficult to treat on the lattice, but they feature in many important $D$ semileptonic decays which will be well measured by the charm factories. There is also a need to be able to calculate for states near threshold such as $\psi(2S)$ and $D_s(0)^+$, and hadronic weak decays in the $B$ and $D$ systems as well.

# NEW PHYSICS SEARCHES WITH CHARM

In the early part of the 20th Century table top nuclear $\beta$ decay experiments conducted at the MeV mass scale probed the $W$ at the 100 GeV mass scale. In an analogous way can we find violations of the Standard Model by studying low energy processes? The existence of multiple fermion generations appears to originate at very high mass scales and so can only be studied indirectly. Mixing, CP violation, and rare decays may investigate the new physics at these scales through intermediate particles entering loops. Why is charm a good place to look? In the charm sector, the SM contributions to these effects are small, in other words, a background free search for new physics is possible (see caveats below). Typically $D^0 - \overline{D^0}$ mixing $\mathcal{O}(< 10^{-2})$, CP asymmetry $\mathcal{O}(< 10^{-3})$ and rare decays $\mathcal{O}(< 10^{-6})$. In addition, charm is a unique probe of the up-type quark sector (down quarks in the loop). The sensitivity of searches for new physics in charm depends on high statistics rather than high energy. New physics searches in the charm sector involving mixing, $CP$−violation and rare decays have become considerably more sensitive in the past several years, however, all results are null. For a recent review see see [38],[39],[40].

## SUMMARY

Charm spectroscopy is witnessing a renaissance. In 2002 it had been more than 25 years since the discovery of a new charmonium state. Since then the $\eta_c(2S)$ and the $h_c$ have finally been discovered below $D\overline{D}$ threshold. There have also been six new states observed above $D\overline{D}$ threshold. These states are not understood.

In charm's role as a natural testing ground for QCD techniques, there has been solid progress. Data at the $\psi(3770)$ from BESII and CLEO-c, and later BESIII, is finally producing a new era of precision absolute charm branching ratios. This is well-matched to developments in theory, especially the lattice, which has a goal to calculate to a few percent precision in the $D, B, \Upsilon$, and $\psi$ systems. CLEO-c, and later BES III, will provide few per cent precision tests of lattice calculations in the $D$ system and in heavy onia, which will quantify the accuracy for the application of LQCD to the $B$ system. If all goes to plan, BABAR, Belle, CDF, D0, CMS, ATLAS, and LHC-b data, in combination with LQCD will produce a few per cent determinations of $|V_{ub}|, |V_{cb}|, |V_{td}|$, and $|V_{ts}|$ thereby maximizing the sensitivity of the flavor physics program to new physics beyond the SM this decade and aid understanding beyond the SM physics at the LHC in the coming decade.

## ACKNOWLEDGEMENTS

I thank my colleagues on BABAR, Belle, BES II and BES III, CDF, CLEO, D0, FOCUS and LHC-b for many valuable discussions. Bo Xin is thanked for technical assistance. I am particularly grateful to Nora Brambilla, Fred Harris and Emilio Ribeiro for organizing a superb meeting in the hauntingly beautiful setting of the Azores Islands, and for their patience while I completed this manuscript.

# REFERENCES

1. A.D. Sakharov, *JETP Lett.* **6** 24 (1967).
2. The BABAR Collaboration hep-ex/0607082.
3. The Belle Collaboration Belle-CONF-0543 (2006).
4. J. Swanson hep-ph/0608139
5. Novak, Rho and Zahed, Acta Phys. Polon. B 35, 2377 (2004).
6. Beverin and Rupp hep-ph/060611.
7. Colangelo, De Fazio and Nicotri hep-ph/0607245.
8. The BABAR Collaboration hep-ex/0603052
9. The Belle Collaboration hep-ex/0606051
10. The BABAR Collaboration hep-ex/0607042
11. The Belle Collaboration, Phys. Rev. Lett. 91, 262001 (2003).
12. The CDF Collaboration, Phys. Rev. Lett. 93, 072001 (2004).
13. The BABAR Collaboration, Phys. Rev. Lett. 95, 142001 (2005).
14. C.T.H. Davies *et al. Phys. Rev. Lett.* **92** 022001 (2004).
15. CLEO-c/CESR-c Taskforces and CLEO Collaboration, *Cornell LEPP Preprint CLNS* 01/1742 (2001).
16. A decription of the BES III physics program and detector, and the design of the BEPCII accelerator may be found at the *http://bes.ihep.ac.cn/conference/wksp04*
17. J. Swain and I. Taylor, *Phys. Rev. D* **58** 093006 (1998).
18. S. Eidelman *et al. Phys Lett. B* **593** (2004) 1.
19. The Heavy Flavours Averaging Group (HFAG) summer 2005.
20. The PDG reports $\mathscr{B}(D \to \pi e^+ v_e)$ with an error of 20%, but the sole measurement, by the Mark III experiment, reports an error of 45%.
21. G. Bellini, I.I. Bigi and P.J. Dornan, *Phys. Rep.* **289** 1 (1997).
22. BaBar Collaboration, B. Aubert *et al. Phys. Rev. D* **71** 091104 (2005).
23. MARK III Collaboration, R.M. Baltrusaitis *et al., Phys. Rev. Lett.* **56** 2140 (1986).
24. MARK III Collaboration, J. Adler *et al., Phys. Rev. Lett.* **60** 89 (1988).
25. CLEO Collaboration, Q. He *et al., Phys. Rev. Lett.* **95** 121801 (2005).
26. ALEPH Collaboration R. Barate *et al. Phys. Lett. B* **405** 191 (1997).
27. The CLEO measurement of $\mathscr{B}(D^0 \to K^- \pi^+)$ is obtained by combining CLEO Collaboration, D.S. Akerib *et al., Phys. Rev. Lett.* 71 3070 (1993) and CLEO Collaboartion M. Artuso *et al., Phys. Rev. Lett.* **80** 3193 (1998).
28. The CLEO measurement is obtined using [27] and CLEO Collaboration R. Balest *et al., Phys. Rev. Lett.* **72** 2328 (1994).
29. MARK III Collaboratiion, J. Adler *et al., Phys. Rev. Lett.* **60** 89 (1988).
30. CLEO Collaboration, M. Artuso *et al., arXiv:hep-ex/*050857.
31. (MILC Collaboration) C. Aubin *et al. hep-lat/*0506030 (2005).
32. The BES II Collaboration M. Abilikim *et al. Phys. Lett. B* 610, 183 (2005).
33. CLEO Collaboration, G.S. Huang *et al., Phys. Rev. Lett.* **95** 1818101 (2005). CLEO Collaboration T.E. Coan *et al., Phys. Rev. Lett.* **95** 181802 (2005).
34. CLEO Collaboration, G.S. Huang *et al., Phys. Rev. Lett.* **94** 011802 (2005).
35. BES Collaboration, M. Ablikim *et al. hep-ex/*0410030 submitted to *Phys. Lett.* (2004).
36. BES Collaboration, M. Ablikim *et al. Phys. Lett. B* **597** 39 (2004).
37. C.Aubin *et al., Phys. Rev. Lett.* **94** 011601 (2005).
38. I. Shipsey in Proceedings of Charm 2006, to appear in Rev. Mod. Phys. A. (2006) and hep-ex/0607070.
39. G. Burdman and I. Shipsey, *Ann. Rev. Nucl. Part. Sci.,* (2003). *hep-ph/*0310076.
40. D. Asner $D^0 \overline{D^0}$ Mixing Review to appear in *PDG 2006*.
41. BABAR Collaboration, B. Aubert *et al., Phys. Rev. Lett.* **93** 131801 (2004).

# On the Equation of State of the Gluon Plasma

Daniel Zwanziger

*New York University, New York, NY 10003, USA*

**Abstract.** We consider a local, renormalizable, BRST-invariant action for QCD in Coulomb gauge that contains auxiliary bose and fermi ghost fields. It possess a non-perturbative vacuum that spontaneously breaks BRST-invariance. The vacuum condition leads to a gap equation that introduces a mass scale. Calculations are done to one-loop order in a perturbative expansion about this vacuum. They are free of the finite-$T$ infrared divergences found by Lindé and which occur in the order $g^6$ corrections to the Stefan-Boltzmann equation of state. We obtain a finite result for these corrections.

**Keywords:** QCD, Gluon plasma, Spontaneous breaking of BRST
**PACS:** 12.38.Aw, 12.38.Mh, 12.38.-t, 11.10.Wx, 11.15.-q, 11.10.Gh

## INTRODUCTION

In 1980 Lindé [1] showed that standard finite-temperature perturbation theory suffers from infrared divergences. Since then no solution has been found, although the infrared divergences may be avoided by introducing a magnetic mass for the gluon in an ad hoc manner. Some time ago, it was suggested that these divergences arise because the suppression of infrared gluons by the proximity of the Gribov horizon in infrared directions is neglected in standard perturbation theory [2]. We present an approach in which is based on a local action that possesses a non-perturbative vacuum. It appears that the infrared divergences found by Lindé do not arise in this approach because the free propagator, in this vacuum, of 3-dimensional transverse, would-be physical gluons is strongly suppressed in the infrared.

Starting from the local action, we shall derive a gap equation that determines the non-perturbative vacuum. We shall solve the gap equation at high temperature to obtain the leading non-perturbative correction to the Stefan-Boltzmann equation of state of the gluon plasma. Additional results in the present approach, including propagators at the one-loop level, relation to the cut-off at the Gribov horizon, issues of principle concerning unitarity and Lorentz invariance, relation to other approaches, accord of the present approach with numerical studies, and references, may be found in [3].

## LOCAL BRST-INVARIANT ACTION

We shall be interested in pure SU(N) gauge theory at temperature $T$. Finite $T$ is described by a Euclidean action which, for pure SU(N) gauge theory, is of the form

$$S = \int d^D x \, \mathscr{L}; \qquad \mathscr{L} = \mathscr{L}_{YM} + s\xi. \tag{1}$$

CP892, *Quark Confinement and the Hadron Spectrum VII*
edited by J. E. F. T. Ribeiro
© 2007 American Institute of Physics 978-0-7354-0396-3/07/$23.00

where $\mathcal{L}_{YM} = \frac{1}{4} F_{\mu\nu}^2$, and $F_{\mu\nu} = \partial_\mu A_\nu - \partial_\nu A_\mu + gA_\mu \times A_\nu$. Here $g$ is the coupling constant, and we use the notation for the Lie bracket $(A \times B)^a \equiv f^{abc}A^bB^c$, where $f^{abc}$ are the fully anti-symmetric structure constants of the SU(N) group. The color index is taken in the adjoint representation, $a = 1, ..., N^2 - 1$. We shall generally suppress the color index, and leave summation over it implicit. Configurations are periodic in $x_0$, $A_\mu(x_i, x_0) = A_\mu(x_i, x_0 + \beta)$, with period $\beta = 1/T$, where $T$ is the temperature. The integral over $x_0$ always extends over one cycle, $\int dx_0 \equiv \int_0^\beta dx_0$. We are in $D$ Euclidean dimensions. Lower case Latin indices take values, $i = 1, 2, ..., D - 1$, while lower case Greek indices take values $\mu = 0, 1, ..., D - 1$.

In the BRST formulation there are, in addition to $A_\mu$, a pair of Faddeev-Popov ghost fields $c$ and $\bar{c}$ and a Lagrange multiplier field, $b$, on which the BRST operator acts according to

$$sA_\mu = D_\mu c; \qquad sc = -(g/2)c \times c; \qquad s\bar{c} = ib; \qquad sb = 0. \qquad (2)$$

It is nil-potent, $s^2 = 0$. Here $D_\mu$ is the gauge-covariant derivative in the adjoint representation, $D_\mu c \equiv \partial_\mu c + gA_\mu \times c$.

The choice of the gauge-fixing density $\xi$ in (1) is the choice of gauge. Physics is independent of $\xi$, provided that it results in a well-defined calculational scheme. For finite $T$, this is a crucial proviso, because the standard gauge choice leads to infrared divergences [1]. The standard Coulomb gauge is defined by $\xi = \xi_{coul} = \partial_i \bar{c} A_i$, with $s\xi = i\partial_i b A_i - \partial_i \bar{c} D_i c$. The Lagrange-multiplier field $b$ imposes the Coulomb gauge condition $\partial_i A_i = 0$. To avoid the infrared divergences of the standard gauge choice we introduce an additional $s$-exact term $s\xi_{aux}$ that involves a quartet of auxiliary ghost fields on which $s$ acts trivially

$$s\phi_\mu^{ab} = \omega_\mu^{ab}; \qquad s\omega_\mu^{ab} = 0; \qquad s\bar{\omega}_\mu^{ab} = \bar{\phi}_\mu^{ab}; \qquad s\bar{\phi}_\mu^{ab} = 0. \qquad (3)$$

The fields $\phi_\mu^{ab}$ and $\bar{\phi}_\mu^{ab}$ are a pair of bose ghosts, while $\omega_\mu^{ab}$ and $\bar{\omega}_\mu^{ab}$ are fermi ghost and anti-ghost. The indices $a$ and $b$ label components in the adjoint representation of the global gauge group, $a, b = 1, ..., N^2 - 1$, and $\mu$ is a Lorentz index. We take for the gauge-fixing term

$$\xi = \xi_{coul} + \xi_{aux} = \partial_i \bar{c}^a A_i^a + \partial_i \bar{\omega}_\mu^{ab}(D_i \phi_\mu)^{ab}, \qquad (4)$$

where we stipulate that the gauge covariant derivative acts on the *first* color index only, $(D_i \phi_\mu)^{ab} = \partial_i \phi_\mu^{ab} + gf^{acd}A_i^c \phi_\mu^{db}$ etc.

# NON-PERTURBATIVE VACUUM

## Maggiore-Schaden shift

We make a change of variable whereby the bose ghosts are translated by a $c$-number term linear in the spatial coordinate $x_\mu$ [4],

$$\phi_\mu^{ab}(x) = \varphi_\mu^{ab}(x) - \gamma^{1/2} \delta^{ab} x_\mu; \qquad \bar{\phi}_\mu^{ab}(x) = \bar{\varphi}_\mu^{ab}(x) + \gamma^{1/2} \delta^{ab} x_\mu. \qquad (5)$$

Note that $\phi$ and $\varphi$ designate different fields. Here $\gamma$ is a parameter with dimensions of (mass)$^4$ that will be determined by the condition $\frac{\partial W}{\partial \gamma} = 0$, where $W$ is the free energy. We also translate $b$ and $\bar{c}$ by compensating terms,

$$\bar{c}^d = \bar{c}^{\star d} + \gamma^{1/2} g f^{adb} x_\mu \bar{\omega}_\mu^{ab}; \qquad b^c = b^{\star c} - i\gamma^{1/2} g f^{acb} x_\mu \bar{\varphi}_\mu^{ab}, \qquad (6)$$

which are chosen to cancel explicit $x$-dependence in the new action. The BRST operator $s$ acts on the new variables according to

$$s\varphi_\mu^{ab} = \omega_\mu^{ab}; \qquad s\bar{\omega}_\mu^{ab} = \bar{\varphi}_\mu^{ab} + \gamma^{1/2}\, \delta^{ab}\, x_\mu; \qquad s\bar{c}^{\star d} = ib^{\star d}; \qquad sb^{\star d} = 0. \qquad (7)$$

Despite the $x$-dependent shift, remarkably, neither the gauge-density (4) nor the Lagrangian density acquire any explicit $x$-dependence when expressed in terms of the new variables,

$$\xi = \partial_i \bar{c}^{\star a} A_i^a + \partial_i \bar{\omega}_\mu^{ab} (D_i \varphi_\mu)^{ab} - \gamma^{1/2}\, (D_i \bar{\omega}_i)^{aa}. \qquad (8)$$

As before, it is understood that the gauge-covariant derivative acts on the first index only, $(D_i \bar{\omega}_\mu)^{ab} = \partial_i \bar{\omega}_\mu^{ab} + g f^{acd} A_i^c \bar{\omega}_\mu^{db}$. After the shift, the Lagrangian density is given by $\mathcal{L} = \mathcal{L}_{YM} + s\xi$, where

$$
\begin{aligned}
s\xi &= i\partial_i b^{\star a} A_i^a - \partial_i \bar{c}^{\star a} (D_i c)^a + \partial_i \bar{\varphi}_\mu^{ab} (D_i \varphi_\mu)^{ab} - \partial_i \bar{\omega}_\mu^{ab} [\, (D_i \omega_\mu)^{ab} + (gD_i c \times \varphi_\mu)^{ab}\, ] \\
&\quad + \gamma^{1/2}\, (D_i \varphi_i)^{aa} - \gamma^{1/2}\, [\, (D_i \bar{\varphi}_i)^{aa} + (gD_i c \times \bar{\omega}_i)^{aa}\, ] - \gamma\, (N^2 - 1)(D-1), \qquad (9)
\end{aligned}
$$

and $(D_i c \times \varphi_\mu)^{ab} \equiv f^{acd} (D_i c)^c \varphi^{db}$ acts on the first color index, etc. For purposes of expansion in powers of $g$ we shall change indepdendent parameter from $\gamma$ to $m$ according to $\gamma^{1/2} \equiv \frac{m^2}{(2N)^{1/2}g}$, where $m$ has dimensions of mass, and $m$ is taken to be of order $g^0$.

## Gap equation

Henceforth we shall be concerned with the action $S$, regarded as a function of the new fields. The partition function is given by $Z = \int d\Phi \exp(-S)$, where $\Phi \equiv (A_\mu, c, \bar{c}, b, \varphi, \bar{\varphi}, \omega, \bar{\omega})$ is the set of all (new) fields, and $d\Phi$ represents integration over them. For simplicity we have written $\bar{c}$ and $b$ instead of $\bar{c}^\star$ and $b^\star$. The field $\varphi_\mu^{ab}$ is real while $\bar{\varphi}_\mu^{ab}$ is pure imaginary. The classical vacuum occurs where all these fields vanish $\Phi \equiv (A_\mu, c, \bar{c}, b, \varphi, \bar{\varphi}, \omega, \bar{\omega}) = 0$. Finally the value of $\gamma$ is determined by the condition that the free energy $W = \ln Z$ be stationary, $\frac{\partial W}{\partial \gamma^{1/2}} = 0$, or $\left\langle \frac{\partial S}{\partial \gamma^{1/2}} \right\rangle = 0$. There is a non-perturbative vacuum if this equation has a solution with $\gamma \neq 0$. We do not require that $W$ be a maximum because there are unphysical fields present. The last equation reads

$$\langle D_i(\varphi_i - \bar{\varphi}_i)^{aa} - (gD_i c \times \bar{\omega}_i)^{aa} \rangle = 2\gamma^{1/2}(N^2 - 1)(D-1). \qquad (10)$$

The second term vanishes, $\langle (gD_i c \times \bar{\omega}_i)^{aa} \rangle = 0$, because there is no $\bar{c}\omega$ term in the action. Moreover the new Lagrangian density (9) is invariant under space-time translation of the new fields, $\Phi(x) \to \Phi(x+a)$, and the vacuum just found, at $\Phi = 0$, is also. Translation invariance implies that the terms $\partial_i \varphi$ and $\partial_i \bar{\varphi}$ do not contribute to (10), and we obtain

$$\frac{1}{(2N)^{1/2}} \langle f^{abc} A_i^b (\varphi - \bar{\varphi})_i^{ca} \rangle = \frac{m^2}{Ng^2} (D-1)(N^2-1). \tag{11}$$

This gap equation determines $m = m(g, T)$. Invariance under scale transformation is spontaneously broken for $m \neq 0$.

## Spontaneous breaking of BRST symmetry

The quantity $m^2$ is the analog of the vacuum expectation-value $v$ of the Higgs field $\Phi$ that appears in spontaneous symmetry breaking of global gauge invariance. For in the Higgs mechanism one makes the translation $\Phi^a = \Phi^{\star a} + v\delta_3^a$, and the vacuum expectation-value $v$ is determined by the condition that the free-energy be stationary with respect to $v$, $\frac{\partial W}{\partial v} = 0$. Similarly, $m^2$ is determined by the condition $\frac{\partial W}{\partial m} = 0$. In the present case, BRST invariance is spontaneously broken rather than global gauge invariance, because the expectation-value of $s$-exact quantities is non-zero, for example,

$$\langle s\bar{\omega}_\mu^{ab} \rangle = \langle \bar{\varphi}_\mu^{ab} + \gamma^{1/2}\delta^{ab} x_\mu \rangle = \gamma^{1/2}\delta^{ab} x_\mu \neq 0. \tag{12}$$

As in the Higgs case, the spontaneously broken theory inherits renormalizability from the unbroken theory. But (12) shows that we cannot identify observables with equivalence classes of $s$-invariant objects, modulo $s$-exact quantities, as in the standard BRST approach. However, as shown in [3], the present method is formally equivalent to the canonical formulation of Coulomb gauge, with a cut-off at the Gribov horizon. This allows us to identify observables, such as the energy-momentum tensor $T_{\mu\nu}$, with the corresponding quantities in the canonical formulation so, for example, $T_{\mu\nu} = F_{\mu\lambda}F_\nu^\lambda - \frac{1}{4}g_{\mu\nu}F_{\kappa\lambda}F^{\kappa\lambda}$.

## FREE PROPAGATORS

We now develop a perturbative expansion about the new, non-perturbative, vacuum. For this purpose we treat $m$ as an independent parameter of order $g^0$, and calculate perturbatively all one-particle irreducible graphs, including the gap equation, to a given order in $g$. Then $m = m(g, T)$ is determined by solving the the gap equation (11) non-perturbatively.

The first step is to expand the action in powers of $g$, $S = S_{-2} + S_0 + \dots$. The leading term is of order $g^{-2}$,

$$S_{-2} \equiv -\frac{m^4}{2Ng^2} (N^2-1)(D-1)L^3\beta, \tag{13}$$

where the spatial quantization volume is $V = L^3$, and the time extent $\beta = T^{-1}$. Although $S_{-2}$ is independent of the fields, and does not contribute to the propagators, it should

124

not be ignored because, when the gap equation is solved for $m = m(T)$, it gives a $T$-dependent contribution to the free energy. The terms in the action of order $g^0$ are all quadratic in the fields

$$S_0 = \int d^D x \; \left[ \tfrac{1}{4} \partial_\mu A_\nu^a - \partial_\nu A_\mu^a)^2 + i\partial_i b^a A_i^a - \partial_i \bar{c}^a \partial_i c^a \right.$$

$$\left. + \partial_i \bar{\varphi}_\mu^{ab} \partial_i \varphi_\mu^{ab} - \partial_i \bar{\omega}_\mu^{ab} \partial_i \omega_\mu^{ab} + \frac{m^2}{(2N)^{1/2}} f^{abc} A_i^b (\varphi_i - \bar{\varphi}_i)^{ca} \right], \quad (14)$$

and determine the "free" propagators. The term with coefficient $m^2$ causes a mixing of the zero-order transverse gluon and bose-ghost propagators.

To calculate the free propagators, we define the field, $\psi_j^b \equiv \frac{i}{(2N)^{1/2}} f^{abc}(\varphi_j^{ca} - \bar{\varphi}_j^{ca})$, that mixes with $A_i^b$. The orthogonal component $\frac{1}{(2N)^{1/2}} f^{abc}(\varphi_j^{ca} + \bar{\varphi}_j^{ca})$ and other components of $\varphi$ and $\bar{\varphi}$ do not mix with $A_i$. The free propagators are given by

$$D_{A_i A_j}(\mathbf{k}, k_0) = \frac{P_{ij} \, \mathbf{k}^2}{\Delta}; \quad D_{A_i \psi_j}(\mathbf{k}, k_0) = \frac{P_{ij} \, i m^2}{\Delta}; \quad D_{\psi_i \psi_j}(\mathbf{k}, k_0) = \frac{P_{ij} \, (\mathbf{k}^2 + k_0^2)}{\Delta}. \quad (15)$$

where $\Delta \equiv (\mathbf{k}^2 + k_0^2)\mathbf{k}^2 + m^4$, and $P_{ij} \equiv \delta_{ij} - \hat{k}_i \hat{k}_j$ is the transverse projector. Here $k_0 = 2\pi n/\beta$ are the Matsubara frequencies, where $n$ is any integer, and we have suppressed the trivial color factor $\delta^{bc}$. In terms of the variable $\psi_j^a$, the gap equation (11) reads

$$-i \, \langle A_j^c(0) \psi_j^c(0) \rangle = \frac{m^2}{Ng^2} (D-1)(N^2-1). \quad (16)$$

## GAP EQUATION IN ONE-LOOP APPROXIMATION

When the left-hand side of the gap equation is evaluated to zeroth order in $g$, using the mixed propagator (15), it reads

$$\int \frac{d^{D-1}k}{(2\pi)^{D-1}} \, T \sum_{k_0} \frac{D-2}{(\mathbf{k}^2 + k_0^2)\mathbf{k}^2 + m^4} = \frac{D-1}{Ng^2}, \quad (17)$$

where we used $P_{ii}(\mathbf{k}) = D - 2$. The sum over Matsubara frequencies yields

$$\int \frac{d^{D-1}k}{(2\pi)^{D-1}} \frac{(D-2)}{2\mathbf{k}^2 E} \left( 1 + \frac{2}{\exp(\beta E) - 1} \right) = \frac{D-1}{Ng^2}, \quad (18)$$

which holds in $D$ Euclidean space-time dimensions. The first term in parentheses gives the result at $T = 0$, and the second term is a Planck-type finite-temperature correction. For $D < 4$ the integral is convergent. We take the limit $D \to 4$. The first term in parentheses has the limiting form

$$\int \frac{d^{D-1}k}{(2\pi)^{D-1}} \frac{(D-2)}{2\mathbf{k}^2 E} \to \frac{1}{4\pi^2} \left[ \frac{1}{\varepsilon} + \ln\left( \frac{\mu^2}{m^2} \right) \right] \quad (19)$$

where $\varepsilon \equiv (4 - D)/2$, and $\mu = \mu(T)$ is, in general, a temperature-dependent renormalization mass. There is a pole at $D = 4$. We subtract the pole term, and the gap equation reads

$$\frac{1}{2}\ln\left(\frac{\mu}{m}\right) + \int_0^\infty \frac{dx}{u} \frac{1}{\exp(m\beta u) - 1} = \frac{3\pi^2}{Ng^2(\mu)}, \tag{20}$$

where $u \equiv (x^2 + \frac{1}{x^2})^{1/2}$.

We now specialize to high temperature $T$, and take the renormalization mass to be $\mu = T$. This simplifies the high-T case because of asymptotic freedom. The running coupling $g(T)$ is small, and is given approximately by $\frac{1}{g^2(T)} = \frac{11}{24}\frac{N}{\pi^2}\ln\left(\frac{T}{\Lambda_{QCD}}\right)$, where $\Lambda_{QCD}$ is a physical QCD mass scale. At high temperature, the first term in (20) may be neglected, and the second term has the limit

$$\int_0^\infty \frac{dx}{u} \frac{1}{\exp(\beta mu) - 1} \rightarrow \frac{1}{\beta m}\int_0^\infty dx \frac{x^2}{x^4 + 1} = \frac{\pi T}{2^{3/2} m}, \tag{21}$$

so the gap equation at high $T$ simplifies to $\frac{\pi T}{2^{3/2} m} = \frac{3\pi^2}{Ng^2(T)}$, with solution,

$$m(T,g) = \frac{N}{2^{3/2}\,3\,\pi} g^2(T)\,T \qquad T \rightarrow \infty. \tag{22}$$

Thus, in the high-temperature limit, $m(T)$ is proportional to the magnetic mass $g^2(T)T$.

## FREE ENERGY

To order $g^0$, the free energy $W = \ln Z$ is given by

$$\exp W = \int d\Phi \, \exp(-S_{-2} - S_0), \tag{23}$$

where $S_{-2}$, given in (13), is field-independent and of order $g^{-2}$, while $S_0$, given in (14), is quadratic in the fields. We obtain $W = W_{-2} + W_0$ where, for $D - 1 = 3$, $W_{-2} = -S_{-2} = \frac{3m^4}{2Ng^2}(N^2 - 1)L^3\beta$, and $\exp W_0 = \int d\Phi \exp(-S_0)$. The evaluation of the free energy $W_0$ for the quadratic action $S_0$ is straightforward, keeping in mind the $A$-$\varphi$ and $A$-$\bar{\varphi}$ mixing, with the result

$$W_0 = \frac{(N^2 - 1)V\beta}{3\pi^2}\int_0^\infty dk \frac{(k^4 - m^4)}{E\,[\exp(E\beta) - 1]}, \tag{24}$$

where $E = E(k) = (k^2 + \frac{m^4}{k^2})^{1/2}$. We add the term $W_{-2}$ and obtain to order $g^0$ the free energy per unit volume, $w = W/V$,

$$w = (N^2 - 1)\,\beta\left(\frac{3\,m^4}{2Ng^2} + \frac{1}{3\pi^2}\int_0^\infty dk \frac{(k^4 - m^4)}{E\,[\exp(E\beta) - 1]}\right). \tag{25}$$

126

# EQUATION OF STATE AT HIGH TEMPERATURE

We now evaluate $w$ in the high-temperature limit, where $g(T)$ is small, and our expansion should be reliable. We insert the solution, $m = \frac{N \, g^2(T) \, T}{2^{3/2} \, 3 \, \pi}$, of the gap equation and obtain

$$w = (N^2 - 1) \left( \frac{N^3 \, g^6(T)}{2^7 \, 3^3 \, \pi^4} + \frac{1}{3\pi^2} K(\eta) \right) T^3, \tag{26}$$

where $K(\eta) \equiv \int_0^\infty \frac{dy \, (y^4 - \eta)}{u \, (\exp u - 1)}$; $u \equiv (y^2 + \frac{\eta}{y^2})^{1/2}$, and $\eta \equiv \left( \frac{m}{T} \right)^4 = \left( \frac{N \, g^2(T)}{2^{3/2} \, 3 \, \pi} \right)^4$ is a small parameter. With neglect of higher order terms, one obtains $K(\eta) = \frac{\pi^4}{15} - \frac{\pi \, \eta^{3/4}}{2^{1/2}}$, which gives

$$w = (N^2 - 1) \left( \frac{\pi^2}{45} - \frac{N^3}{10{,}368 \, \pi^4} \, g^6(T) \right) T^3. \tag{27}$$

This is the leading correction to the Stefan-Boltzmann limit from the non-perturbative vacuum. The equation of state of the gluon plasma follows from the thermodynamic formulas for the energy per unit volume and pressure, $e = -\frac{\partial w}{\partial \beta}$; $p = \frac{w}{\beta}$, and entropy per unit volume, $s = \frac{e+p}{T}$. To calculate the energy density, we use $-\beta \frac{\partial g}{\partial \beta} = T \frac{\partial g}{\partial T} = \beta$-function $= O(g^3)$, which is of higher order. We thus obtain for the energy density and pressure at high temperature, $e = 3p = w(T)T$, and $s = \frac{4}{3} w(T)$.

We have obtained the leading contribution that comes from the non-perturbative vacuum. Numerically it is a small correction, whereas the correction of order $g^6$ is divergent when calculated with the perturbative vacuum [1]. To this must be added the perturbative contributions, including resumations, that have been calculated at $m = 0$, and that are of lower order in $g$ [5]. Since standard, resummed perturbation theory diverges at order $g^6$, which is precisely the order of the correction we have found, the result obtained here is consistent with standard perturbative calculations.

## ACKNOWLEDGMENTS

I recall with pleasure stimulating conversations about this work with Reinhard Alkofer, Laurent Baulieu, David Dudal, Andrei Gruzinov, Atsushi Nakamura, Alexander Ruten- burg, Martin Schaden, and Silvio Sorella.

## REFERENCES

1. A. D. Lindé, Phys. Lett. **96B**, 289 (1980).
2. Ismail Zahed and Daniel Zwanziger Phys. Rev. D **61**, 037501 (2000) [arXiv:hep-th/9905109].
3. Daniel Zwanziger, arXiv:hep-ph/0610021.
4. Nicola Maggiore and Martin Schaden, Phys. Rev. D **50** 6616 (1994).
5. Joseph I. Kapusta, Finite-Temperature Field Theory, Cambridge U. Press (1989).

# A Partial Summary of Session A

Jeff Greensite

*Physics and Astronomy Dept., San Francisco State University, San Francisco CA 94132 USA*

**Abstract.** I summarize a selection of talks delivered in parallel session A of the QCHS7 meeting.

**Keywords:** confinement, QCD vacuum, non-perturbative methods
**PACS:** 11.15.-q,11.15.Ha,12.38.Aw,12.38.Gc

## 1. INTRODUCTION

Parallel session A at this conference was devoted to the topic of confinement mechanisms and vacuum structure. There were some twenty-nine talks presented in this session, and to summarize all of them in the allotted time is an impossible task, at least for me. So instead I will just briefly describe a selection of the talks we heard, that I happened to find particularly interesting. The selection is, of course, only my personal choice. There were many good talks in session A; far too many to discuss here.

Before going on, its worth asking: why session A? That is, why don't we all agree by now on QCD vacuum structure and the quark confinement mechanism? I think there are two reasons. The first is obvious: QCD has not been solved analytically, at least not yet, and this means that all proposals for large-scale vacuum structure and the confinement mechanism are unproven, and open to dispute. Nevertheless, there exist a number of good ideas about quark confinement, and this brings me to the second point: Nowadays, the bar is set pretty high. Its not enough anymore to just come up with an explanation for the linearity of the static quark potential; any proposal should also try to explain (or at least not contradict)

1. Casimir scaling: At intermediate distance scales, the string tension of static quarks is roughly proportional to the quadratic Casimir of the color charge representation of the quarks.
2. N-ality dependence: Asymptotically, the string tension depends only on the N-ality of the quark color charges.
3. The absence of long-range hadronic van der Waals forces. There should also be no long-range color dipole field associated with a static quark-antiquark pair.
4. Stringy properties: The color electric field between widely separated quarks is collimated into a flux tube with its own degrees of freedom and string-like spectrum of excitations. The Lüscher term, a correction to the linear potential proportional to $-1/R$, and roughening, which is a logarithmic broadening of the flux tube with increasing quark separation, are two well-known aspects of this string-like nature of the QCD flux tube.

CP892, *Quark Confinement and the Hadron Spectrum VII*
edited by J. E. F. T. Ribeiro
© 2007 American Institute of Physics 978-0-7354-0396-3/07/$23.00

This is a tall order, and a good reason for the existence of session A. The talks I will discuss can be grouped in a few broad categories: (i) strings; (ii) field configurations (vortices and calorons) and (iii) Hamiltonian approaches in 2+1 and 3+1 dimensions.

## 2. QCD AND STRINGS

*Bosonic and baryonic string theory in QCD.* As already mentioned, the linearity of the static quark potential is only part of the confinement story. There is also a flux tube, with string-like properties and a string-like spectrum. **Julius Kuti**, in his talk, has shown us that lattice Monte Carlo calculations have attained a level of accuracy which allows us to study both small deviations from linearity in the static quark potential, and the low-lying excitations of the flux tube, and to compare these with the predictions of string theory. In particular, the static quark potential according to string theory goes as

$$V(R) = \sigma R - \frac{\pi(D-2)}{24R} - \frac{1}{2}\left(\frac{\pi(D-2)}{24}\right)^2 \frac{1}{\sigma R^3} + \text{New Physics} \tag{1}$$

The first three terms are derived from the Nambu action, and are expected to be universal in any bosonic string theory. Of course, the Nambu action is pathological in $D = 4$ dimensions, so the correct effective string theory must contain some correction terms. These would result in model-dependent higher powers of $1/R$, labeled "New Physics", and measurements capable of distinguishing such terms would allow us to test proposals for the QCD string. Remarkably, the first three terms appear to fit existing data [1] for the QCD string in D=4 dimensions rather well; there is somewhat more deviation in D=3 dimensions.

The most accurate numerical results for the static quark potential, and for the flux tube excitation spectrum, have been obtained for gauge theories based on the discrete $Z_2$ and $Z_3$ gauge groups. A very clear picture of the "Y" configuration for baryonic flux tubes has been obtained in $Z_3$ lattice gauge theory, and there is now striking evidence for the existence of, e.g., an excited flux tube state with the excitation concentrated at the junction of the Y. No doubt these results will be extended to SU(3) in due course.

*String/Field Duality.* It is an old dream to formulate non-abelian gauge theory in terms of a confining string. The irony is that such a formulation has in fact been achieved, but so far only for a *non*-confining gauge theory: $\mathcal{N} = 4$ supersymmetric Yang-Mills theory. This is, of course, the AdS/CFT correspondance. The static quark potential in $D = 4$ dimensions is calculable at strong couplings, in this formulation, from the minimal area of a worldsheet, bounded by a Wilson loop, and living in a 10-dimensional $AdS_5 \times S_5$ spacetime. The result is that $V(R) \propto -\sqrt{\lambda}/R$, where $\lambda$ is the 't Hooft coupling.

The string worldsheet in $AdS_5 \times S_5$ has an excitation spectrum, and according to AdS/CFT duality this spectrum corresponds to excited states in $\mathcal{N} = 4$ susy Yang-Mills theory with a pair of static quark-antiquark charges. On the AdS side there are an infinite number of discrete energy levels (at $N_{colors} = \infty$) lying between the ground state at $E \propto -\sqrt{\lambda}/R$, up to the start of the continuous spectrum. Can this discrete string excitation spectrum be seen in $\mathcal{N} = 4$ super Yang-Mills?

**Charles Thorn** and collaborators [2] have calculated the excitation spectrum of a static quark-antiquark pair on the super Yang-Mills side, by summing up ladder diagram exchanges. This approximation technique allows them to probe weak 't Hooft couplings, not so far reachable by the AdS/CFT approach. What Charles et al. find is quite interesting: For 't Hooft coupling $\lambda < \frac{1}{4}$, there is a finite gap between the ground state and the continuum spectrum, with no discrete excitations in between. The existence of a gap at weak coupling is already interesting; there is no such gap in QED, for example. At $\lambda > \frac{1}{4}$ an infinite set of discrete states suddenly appears between the ground state and the continuum, in qualitative agreement with the AdS/CFT calculation. This looks as though a gluonic bound state, with a characteristic bound state spectrum, suddenly appears at $\lambda > \frac{1}{4}$. As Charles puts it, the color electric flux in the super Yang-Mills theory appears to be just on the edge of becoming a genuine flux tube.

*D-Branes in QCD.* Monte Carlo calculations carried out by Horvath and collaborators [3] have revealed a very peculiar and surprising short-distance structure in lattice configurations, in both QCD in $D = 4$ dimensions, and in $CP^{N-1}$ models for $N > 4$ and $D = 2$ dimensions. In these models, topological charge is concentrated on membranes of co-dimension = 1, with 3-volumes (QCD) or lines ($CP^{N-1}$) juxtaposed in dipole layers. These membranes are only one or two lattice spacings thick, so regions of positive and negative topological charge are everywhere close; this seems to be an exception to the usual lore about structure on such small scales being a lattice artifact, and appears to be required by the fact that the correlator $\langle q(x)q(y) \rangle$ of topological charge densities is negative. **Hank Thacker**, in his talk, provides an intriguing interpretation of these membranes in QCD in terms of the AdS/CFT correspondence: in this correspondence, topological charge in QCD is AdS/CFT dual to Ramond-Ramond charge, and the peculiar 3-volumes of topological charge found in numerical simulations of the gauge theory are dual to D6 branes in the AdS theory.

# 3. VORTICES AND CALORONS

*Center Vortices in Various Gauge Theories.* There is an effective model for large-scale QCD vacuum structure, based on on center vortices, which has been remarkably successful, both qualitatively and quantitatively, in explaining various non-perturbative features of SU(2) and SU(3) gauge theories. The model, due to Engelhardt and Reinhardt, is really very simple: just a lattice random-surface model with a single coupling (in an extrinsic curvature term) and a short-distance cutoff. This model gets a lot of things right, such as the order of the deconfinement transitions in SU(2) and SU(3), the topological susceptibility, and the chiral condensate [4].

In his talk in session A, **Michael Engelhardt** discussed the extension of this model to the SU(4) and Sp(2) gauge groups (center vortices in SU(4) were also discussed in session A by **Mina Deldar**). The choice of SU(4) is motivated by the fact that this is the smallest group with two types of vortices. It is possible that center monopoles become important as $N_{colors}$ increases, and SU(4) is a good place to begin to look for this effect.

The group Sp(2) is of interest because the gauge group center is $Z_2$, the same as SU(2), yet unlike SU(2) the deconfinement transition is first order, rather than second order. Since the center vortex content of the two groups is the same, does this fact contradict the vortex confinement scenario?

The answer is no. Vortex dynamics depend on both the vortex content *and* the vortex dynamics. In the case of SU(4) and Sp(2), the simple extrinsic curvature action is not sufficient to capture all the relevant features at finite temperatures; some additional terms are required, which favor branchings (center monopoles) and vortex attraction. These random surface actions, requiring more couplings, are of course less predictive than the SU(2) and SU(3) models. It would be good to someday be able to calculate the couplings of these effective models from first principles. In the meantime, I think it is gratifying to see that the effective vortex models are at least consistent with known results.

*Calorons.*   **Pierre van Baal** gave us a nice review of caloron physics. Calorons are instantons at finite temperature, and the Kraan-van Baal-Lee-Lu (KvBLL) calorons are solutions with non-trivial holonomy at distances far from the caloron center. In, e.g., SU(2) gauge theory, this means that the trace of Polyakov lines is zero far away from the caloron. The calorons have magnetic monopole constituents. It has been suggested that calorons might dominate the vacuum state at low temperatures, and account for the area law falloff of Wilson loops necessary to account for quark confinement. The KvBLL calorons have in fact been observed in Monte Carlo simulations at low temperatures on cooled lattices (after the loss of the string tension on the finite lattice).

**Michael Mueller-Preussker** described, in his talk in session A, the numerical results for static quark potentials and spacelike Wilson loops in a caloron-gas model, at temperatures a little below the deconfinement temperature. The static quark potential obtained from Polyakov line correlators looks good in this model: Casimir scaling is obtained at short distances, while adjoint quark color charges are screened at large distances. This is perhaps a consequence of the trace-Polykov line = 0 holonomy, which is a feature of the KvBLL calorons. More challenging, in my opinion, is the string tension of the spacelike Wilson loops. There seems to be no obvious reason for N-ality dependence of the string tension for spacelike loops; is it observed nonetheless? van Baal and Mueller-Preussker showed us some data for the logarithm of spacelike adjoint-representation Wilson loops, plotted versus area, which was very suggestive of color-screening; the simplest interpretation is that the adjoint spacelike string tension is falling to zero at large areas. On the other hand, Creutz ratios plotted versus area, for the spacelike Wilson loops, appear to show some curvature for both the adjoint *and* fundamental representations. These Creutz ratios *seem* to be falling at large areas, although this could well be an illusion, since the errorbars are quite large. The caloron model is promising, but I think we will have to wait for more results at larger loop sizes, to really know if an area-law falloff with the correct N-ality dependence is obtained asymptotically.

## 4. CANONICAL APPROACHES

*New Variables, 2+1 dimensions.*   **Rob Leigh** outlined a very ambitious approach to SU(N) gauge theory in D=2+1, in which the ground state wavefunctional $\Psi_0$ and the glueball mass spectrum are calculated analytically. This approach is based on the gauge-invariant variables in D=2+1 introduced by Karabali and Nair in 1996. In terms of these variables, the Hamiltonian operator of SU(N) gauge theory looks like

$$\mathcal{H}_{KN}[J] = m \left( \int_x J^a(x) \frac{\delta}{\delta J^a(x)} + \int_{z,w} \Omega^{ab}(z,w) \frac{\delta}{\delta J^a(z)} \frac{\delta}{\delta J^b(w)} \right) + \frac{\pi}{mc_A} \int_x \bar{\partial} J^a \bar{\partial} J^a \quad (2)$$

where $m \propto g_{YM}^2$. Leigh and co-workers make the ansatz for the vacuum state

$$\Psi_0 = \exp\left( -\frac{\pi}{2c_A m^2} \int \bar{\partial} J^a \, K\left(\frac{\partial \bar{\partial}}{m^2}\right) \bar{\partial} J^a + \dots \right). \quad (3)$$

where the kernel $K$ is $J$-dependent. Leigh et al. compute $K$ explicitly and non-perturbatively, show that it has a finite range, and go on to compute a glueball spectrum analytically, from correlators of $\partial J$. This spectrum seems to be in impressive agreement with lattice data.

These remarkable results do entail certain assumptions and approximations. For one thing, the eigenstate equation $H\Psi_0 = E_0 \Psi_0$ is not satisfied exactly by the ansatz; some terms that arise on the lhs are dropped, and it is an open question whether those terms are in fact negligible. Also there is a certain operator relationship which is required, but which has not yet been fully derived. Nevertheless, in view of the close agreement with lattice results, this new application of the Karabali-Nair variables by Leigh and his co-workers deserves serious consideration.

*Old Variables, 2+1 Dimensions.* Suppose we take the known ground state solution of $H\Psi_0 = E_0 \Psi_0$ at $g = 0$ in temporal gauge, and make the minimal modification, at $g \neq 0$, such that the physical state (Gauss' Law) condition is satisfied exactly. In D=d+1 dimensions, the state is

$$\Psi_0[A] = \exp\left[ -\frac{1}{2} \int d^d x d^d y \, F_{ij}^a(x) \left( \frac{1}{\sqrt{-\mathcal{D}^2}} \right)_{xy}^{ab} F_{ij}^b(y) \right] \quad (4)$$

where $\mathcal{D}_k$ is the covariant derivative. **Stefan Olejnik and I** claim that this state is confining, in $D = 2+1$ dimensions. The essential point is that in $2+1$ dimensions the kernel $1/\sqrt{-\mathcal{D}^2}$ has a finite range in almost any stochastic background gauge field, and this in turn means that the effective ground state has the form

$$\Psi_0 \sim \exp\left[ -\mu \int d^2 x \, F_{12}^a(x) F_{12}^a(x) \right] \quad (5)$$

which is known to be confining.

*Asymmetric Lattices, 2+1 Dimensions.* **Peter Orland**'s approach is to treat the lattice in $D = 2+1$ dimensions as stack of 1+1 dimensional lattices, with different weak-but-unequal couplings in different directions. He finds Sine-Law scaling of the potential in one direction, and Casimir scaling of the string tension in the other directions.

*Coulomb Gauge, 3+1 Dimensions.* **Adam Szczepaniak**, in previous work with Eric Swanson, derived a linear Coulomb potential from a "mean field" approach, using a trial vacuum state $\Psi_0[A]$ which is gaussian in the $A$-field. This is good but not sufficient, since it is known from numerical studies in Coulomb gauge that the long-range color Coulomb potential is linear, but with a string tension which is at least three times larger than the asymptotic string tension. This means that the physical state in Coulomb gauge, containing just a quark, an antiquark, and the Coulomb field between them, is not the minimal energy state.

It is reasonable, in Coulomb gauge, that the minimal energy quark-antiquark state consists not only of the quark and antiquark, but, for widely separated quarks, also some constituent gluons. The constituent-gluon approach can be seen as a step on the way to a full-fledged "gluon-chain" model of the QCD flux tube. In his talk, Adam reported on some new results along these lines, which allow for a number of constituent gluons in the physical state, in addition to the quark-antiquark sources. The results are not bad: the energy of the state still grows linearly with quark separation, but the string tension is reduced substantially, as was hoped. In addition, the inclusion of constituent gluons provides some new insight into the ordering of excited states that are seen in lattice Monte Carlo simulations.

## 5. CONCLUSIONS

There were many interesting results presented in session A, far more than I have space to discuss here. There are also many basic issues still to be resolved. We look forward to more QCHS meetings, at beautiful locations like this one, to help us resolve them!

## ACKNOWLEDGMENTS

I would like to thank Nora Brambilla, Jose Emilio Ribeiro, and all the other organizers for bringing us together in this enjoyable and productive meeting. I would also like to thank my fellow convenors of session A, Manfried Faber and Mikhail Polikarpov, for a pleasant collaboration. My own work is supported in part by the U.S. Department of Energy under Grant No. DE-FG03-92ER4071.

## REFERENCES

1. M. Luscher and P. Weisz, JHEP **0207**, 049 (2002) [arXiv:hep-lat/0207003].
2. I. R. Klebanov, J. M. Maldacena and C. B. Thorn, JHEP **0604**, 024 (2006) [arXiv:hep-th/0602255].
3. I. Horvath *et al.*, Phys. Rev. D **68**, 114505 (2003) [arXiv:hep-lat/0302009].
4. M. Quandt, H. Reinhardt and M. Engelhardt, PoS **LAT2005**, 320 (2006) [arXiv:hep-lat/0509114].

# Deconfinement[1]

## C. Pajares

*Instituto Galego de Física de Altas Energías, Departamento de Física de Partículas.*
*Universidade de Santiago de Compostela. 15782 Santiago de Compostela, Spain.*

**Abstract.**
   This is an attempt to summarize the talks given at the session on Deconfinement in the Conference "Quark Confinement and hadron spectrum". This talk covers the following topics: Elliptic flow and evidence of nearly perfect fluid of the created matter; High Transverse momentum production, propagation of jets and energy loss; Heavy quarqonia in dense QCD matter; Phase transition, Multiplicity fluctuations and long range correlations; Multiparticle production and thermalization.

**Keywords:** <Enter Keywords here>
**PACS:** <Replace this text with PACS numbers; choose from this list:

In the last few years a large progress has been done in the knowledge of the deconfined phase of QCD. The SPS data already displayed several facts that hinted at the onset of Quark Gluon Plasma (QGP) formation. The RHIC data have conclusively discovered a striking set of new phenomena. Most of these data have been extensively discussed in the different talks of the session of Deconfinement.

   The space limits prevent me from describing all the reported exciting developments, so I will concentrate on some of them.

## ELLIPTIC FLOW

The flow pattern of thousand of particles produced in a heavy ion reaction is the main observable used to look for collective behaviour and its properties. These properties test the conditions necessary for the obtention of QGP. One is the degree of thermalization. The evolution of the matter from the initial conditions can be computed by means of relativistic hydrodynamics if local equilibrium is maintained. These equations can be further approximated by perfect fluid equation when the viscosity correction can be neglected. The second condition is the validity of the equation of state, numerically determined from QCD. The data on elliptic flow show evidence that a fast thermalization is reached at RHIC energy, compatible with a soft equation of state and a low viscosity. The matter created at RHIC behaves as a perfect fluid [1].

   The elliptic flow, $v_2 = < p_x^2 - p_y^2/p_t^2 >$, results from pressure gradients developed in the initial almond-shaped collision zone. That is, the initial transverse coordinate

---

[1] Summary of the Deconfinement Session of the Quark Confinement and Hadron Spectrum VII Conference. Ponta Delgada, Azores(Portugal)

CP892, *Quark Confinement and the Hadron Spectrum VII*
edited by J. E. F. T. Ribeiro

space anisotropy of the collision zone or eccentricity $\varepsilon = <(y^2 - x^2)/(y^2 + x^2)>$ is converted, via hadronic or partonic interactions into an azimuthal momentum anisotropy. Elliptic flow self-quenches due to expansion of the collision zone, therefore in order to achieve relatively large $v_2$ a fast thermalization is required (see Lisa and Bai-Yuting talks [2] and [3]). Therefore it is expected that the elliptic flow scaled by the eccentricity should be proportional to the density of scatterings, $\frac{dN}{dy}\frac{1}{S}$, being $S$ the overlapping collision area. This is well satisfied as fig. 1 shows. It is also shown the hydrodynamics result for a perfect fluid [4], which only is reached for central Au-Au collisions. As a for LHC, for central Au-Au collisions $\frac{1}{s}\frac{dN}{dy} \simeq 80$ it is expected a change on the shape of the curve becoming flat. In fig. 2 we show the agreement of the observed hadron mass dependence of $v_2$ with the hydrodynamics predictions below $p_t = 1$ GeV/c. This result shows that there is a common collective flow velocity. In fig. 3 it is shown the scaling law of $v_2/n$ versus $p_T/n$, $n$ being the number of quarks of the respective hadrons. This scaling law was predicted by coalescence models suggesting that the collective flow is at the partonic stage. The coalescence models, also explain naturally the differences between the inclusive cross sections for baryons and mesons at intermediate transverse momentum as it was explained in the Hippolyte talk [5]. The hydrodynamics perfect fluid prediction [6] for the higher azimuthal momentum $v_4$ is $v_4 = \frac{1}{2}v_2^2$, but the data for $\pi^\pm$ and $p$ and $\bar{p}$ in minimum bias Au-Au collisions gives a factor $3/2$ instead $1/2$. However in a perfect fluid model [7] it is obtained the scaling law $v_4 = v_2^2/2 + k_4 y_T^4$, which is in agreement with data. $k_4$ is a constant depending on the mass of the particle and $y_T = \frac{1}{2}log(m_T + p_T)/(m_T - p_T)$.

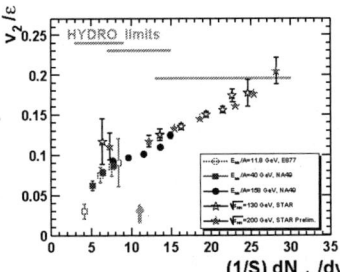

**FIGURE 1.** $\frac{v}{\varepsilon}$ versus $\frac{1}{S}\frac{dN}{dy}$ for different energies and centralities.

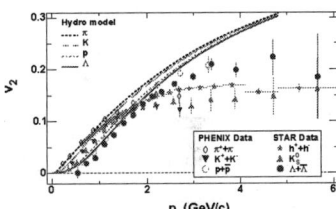

**FIGURE 2.** The elliptic flow $v_2$ versus $p_T$ for different particles together with the hydrodynamics model predictions

## TRANSVERSE MOMENTUM SUPRESSION

One of the exciting results of the RHIC data is the strong suppression of $p_T$ in central heavy ion collisions, consistent with the predicted energy loss [8] [9] of the parent light quarks and gluons when transverse the dense colored medium due to the induced gluon radiation. In fig. 4 is shown the nuclear attenuation factor $R_{AA}(p_T)$. For $\pi^0$ and $\eta$ the data show a suppression factor 5 for $p_T > 4$ GeV/c, compared to the superposition of $NN$ collision, see N. Borghini talk [10]. On the contrary, $R_{AA} = 1$ for direct photons in agreement with perturbative QCD [11] [12].

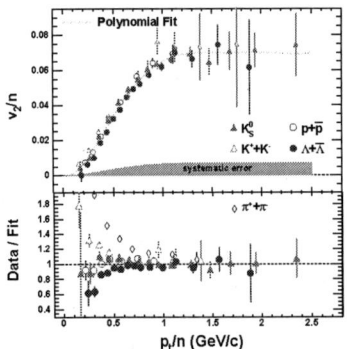

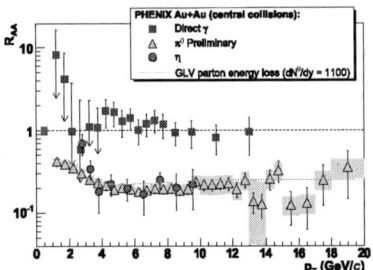

**FIGURE 3.** $\frac{v_2}{n}$ versus $\frac{p_T}{n}$ being n the number of quark consituents for different mesons and baryons

**FIGURE 4.** PHENIX data for central Au-Au collisions of the modified nuclear factor $R_{AA}$ as a function of transverse momentum for photons, $\pi$ and $\eta$

However, the suppression factor for high $p_T$ electrons from semi-leptonic D and B decays is as suppressed as the light hadrons in central Au-Au collision [13],see fig. 5, in conflict with the prediction of radiative energy loss models. This discrepancy may point out to a elastic energy loss for heavy quarks. The study of b and c jets at LHC can be very valuable to clarify this point.

A second exciting phenomena observed at RHIC was the suppression of the back to back jet-like correlation. Jet-like correlations are measured by selecting the highest $p_T$ trigger hadron of the event and measuring the azimuthal $\Delta\Phi = \Phi - \Phi_{trig}$ and rapidity $\Delta\eta = \eta - \eta_{trig}$ distributions of associated hadrons. In $pp$ collisions a dijet signal appears as two back to back Gaussian peaks at $\Delta\Phi \simeq 0$ (near-side) and $\Delta\Phi \simeq \pi$ (away-side). On the contrary, the away-side dihadron azimuthal correlation in central Au-Au collision is clearly suppressed, showing a dip and a double peak structure [14] at $\Delta\Phi \approx \pi \pm 1.1$, for associated hadron in the range $1 \leq p_T \leq 2.5 GeV/c$ (see fig. 6). This double peak structure has been pointed out as due to the emission of energy from the quenched parton at a finite angle respect to the jet axis. Such conical configuration can appear if a fast jet moving in a fluid medium generates a wake of shock wave of Mach Type [15] or Cerenkov Type [16]. In the case of Mach wave, the characteristic angle $\theta$ of the emitted secondaries determines the speed of sound, $c_s = cos\theta$. However the double-peak structure of the away-side correlation is consistent not only with conical emission but also with other scenarios. In order to distinguish between the different mechanisms, 3-particle azimuthal correlations are needed. The three particle results reported at this conference [17] as central Au-Au collisions are consistent with conical emission, but additional studies on the $p_T$ dependence are needed to emission distinguish between Mach cone shock waves and Cerenkov emission.

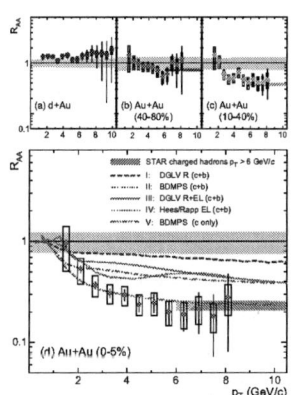

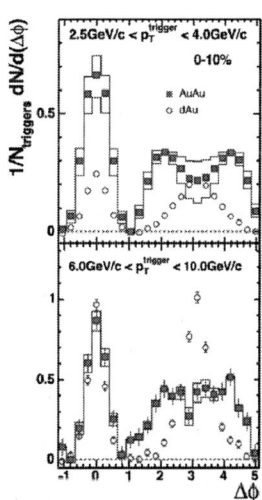

**FIGURE 5.** $R_{AA}$ versus $p_T$ for nonphotonic electrons for d-Au (a), Au-Au 40-80 (b), Au-Au 10-40 (c)and Au-Au 0-5(d) compared with the STAR data for charged hadrons

**FIGURE 6.** STAR data on azimutal distributions of semihard hadrons (associated $p_T = 1 - 2.5$ GeV/c) in central Au-Au and d-Au collisions with respect to a trigger hadron measured of 2.5 GeV/c $< p_T < 4.0$ GeV/c (top) and 6.0 GeV/c $< p_T < 10.0$ GeV/c.

## CHARMONIUM SUPPRESSION

Early predictions were that the two heavy quarks that would form the bound state would be screened from each other in the high-density deconfined medium [18]. These states would melt at different energy densities depending on their size and binding energies. However, recently, lattice QCD has suggested that the $J/\psi$ would not be screened up to $T \geq 1.5 - 2T_c$. On the contrary other charmonium states would be screened around $1.1T_c$ [19].

The $J/\psi$ suppression at RHIC was predicted to be larger than the observed at SPS by most of the models. Contrary to this expectation, additional suppression was not found, fig. 7. In this figure it is also shown the suppression due to normal absorption using for the absorption cross section the values 1, 3 and 4 mb respectively. The usual used value of 4.2 mb is higher than the required by $d - Au$ data which is in the range 1-3 mb. A better understanding of the absorption cross section and in general of cold nuclear matter effects would be welcome.

We are left with two possible theoretical explanations of the RHIC data, namely regeneration [22] and sequential dissociation [23] models. In regeneration models a strong dissociation of the charm pairs due to screening is compensated by the regeneration of bound charm pairs in the later states of the expansion due to the large production of charmed quarks at high density. At LHC a large enhancement is predicted.

In the sequential screening model, the $J/\psi$ is not melt at SPS and RHIC energies as suggested by lattice calculations and the observed suppression comes only from screening of the higher mass resonances $\psi'$ and $\chi_c$ that though feed down normally provide about 40% of the $J/\psi$ production. This picture provides a simple explanation for the similar suppression observed at SPS and RHIC. As at LHC it would be reached temperatures high enough to dissociate $J/\psi$, it is expected stronger suppression at LHC, contrary to the regeneration model expectation.

On the other hand, the sequential dissociation model relates the observed $J/\psi$ production to the higher resonance production once the absorption has been substracted. In fact, denoting by $S_{J/\psi}$ and $S_{\psi'}$, the survival probabilities for the observed total suppression and for the higher resonances respectively, once the absorption has been substracted

$$S_{J/\psi} = 0.6 + 0.4 S_{\psi'} \tag{1}$$

Using the absorption cross sections $\sigma_{\psi'} = 7.1 \pm 1.6 mb$ and $\sigma_{J/\psi} = 4.3 \pm 0.3 mb$ we can test the equation (1) from the data on $J/\psi$ and $\psi'$ production, obtaining a good ag

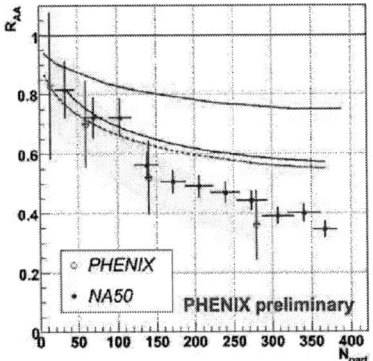

**FIGURE 7.** NA50 data together with PHENIX data of the modified nuclear factor $R_{AA}$ for $J/\psi$ production as a function of the number of participants.

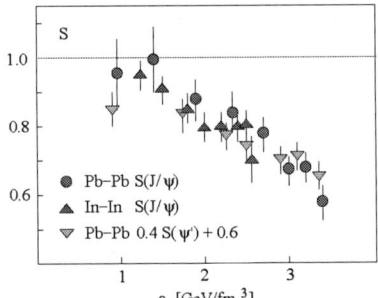

**FIGURE 8.** The survival probability of $J/\psi$ for Pb-Pb and In-In collisions together with expression (1).

## PHASE TRANSITION

Lattice studies indicates that the deconfinement phase transition at $\mu = 0$ is a cross over [24], as at $\mu \neq 0$ it is expected to be of first order, in some point there will be a critical end point. At this conference, were reported studies both experimental and theoretical on this critical end point. From one side, it was explained the proposal RHICII [25] which will explore regions of lower energy, looking for signatures associated to the critical point. In this way, it will join FAIR in the search of the critical point. On the other hand were reported interesting theoretical results. B. Kaempfer [26] in a quasi-particle model on of QCD matter study the critical end point and the equation of state, obtaining with agreement for relevant observables of AA collisions as the elliptical flow for different particles or the rapidity multiplicity distribution. Szabo reported the recent results of lattice at $\mu$ and $\mu \neq 0$, in particular the cross-over at $\mu = 0$. Antonov analytically studied [27] thermodynamics of a heavy quark-antiquark pair in SU(3) QCD, both below and above the deconfinement critical temperature. He derived the effective temperature-dependent string tension, which enabled him to calculate internal energy and entropy of two heavy-light mesons for two flavours. The anomalously large peaks of these two observables around $T_c$, observed recently on the lattice are well described.

Other interesting aspects of the phase transition (dynamics, flux-tube) were discussed by G. Krein [28] and G. Kozlov [29].

## CORRELATIONS

In the early stage of heavy ion collision an extended region with large energy density is produced where quark and gluons degrees of freedom leading to a new partonic phase of matter. In the subsequent evolution, the system dilutes and cools down, hadronizes and finally decays into observed hadrons. These hadrons carry only indirect information about the early stage of the collision. Results as the elliptic flow discussed before suggest that a deconfined phase starts in the early stage of the reaction. The study of correlation and fluctuations can provide additional information on the reaction mechanism. For these reasons correlations between oppositely charge particles and multiplicity and transverse momentum fluctuations have been measured in the last years.

The balance function [30] measured the correlation of the oppositively charged particles. The width of the balance function $< \Delta y >$ is sensitive to the hadronization time.

If the system produced in a heavy-ion collision has undergone a partonic phase, the hadronization will occur at later time and therefore the temperature will be lower and the diffusive interaction with other particles will be lesser than those in the direct hadronization [31]. A delayed hadronization implies stronger correlation in rapidity for the charged particles and therefore a narrower balance function. Indeed a narrowing of the balance function is observed with increasing size of the colliding nuclei by the NA49 and STAR Collaboration [32]. The Hijing model as well as shuffled events retaining only correlations from global charge conservation do not show any decrease of the width. However, other models without delayed hadronization can describe the data [33]. Notice, that the integral of the balance function is related to the event by event charge fluctuations, which are expected to be suppressed in a QGP [34].

The multiplicity correlations have been studied by the NA49 Correlations [35]. In fig. 9, the scaled variance of negative particles is shown as a function of the number of projectile participants together with the results of three different string models. It has been pointed out that for peripheral collisions a significant contribution comes from fluctuations in the number of target participants at fixed projectile participant number, what means that the projectile and target hemisphere are connected. The strings models of fig. 9 are based on Fritiof model where the strings are stretched between partons of the same excited hadron and therefore there is not connection between hemispheres. This does not happen in string models like, dual parton model, quark gluon string model and Venus where there is color exchange and the strings are stretched between partons of the projectile and the target.

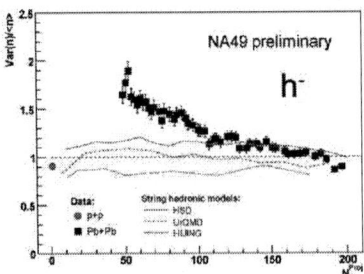

**FIGURE 9.** NA49 experimental data on the scaled variance of negative multiplicity of Pb-Pb collisions as a function of the number of projectile participants together the predictions of different non color exchange string models.

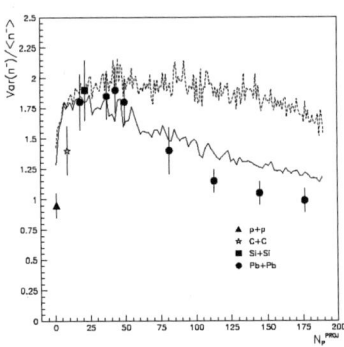

**FIGURE 10.** NA49 preliminary data on scaled variance of negative multiplicity as a function of the number of projectile participants compared with the result of percolation of strings.

In fig. 10 it is shown the result of a percolation color sources model [37] together with previous NA49 data. It is seen that contrary to the above string models in this case reproduced the general trend of data. The percolation framework is able to describe the dependence on the centrality of the transverse momentum fluctuations [38]. In this approach the strings stretched between partons of the projectile and target. In the transverse space these color strings are seen as small circles, with $r_0 \simeq 0.2 - 0.3\, fm$. With growing energy and/or atomic number, the number of strings grows, starting to overlap forming clusters. Each cluster decay into particles with a mean multiplicity and mean transverse momentum which depends on the number of strings of the cluster and the total area of the cluster. These dependence are essentially determined by the Schwinger mechanism and the color field of each cluster. At a certain critical density, a macroscopic cluster appear which marks the percolation phase transition [39]. The fluctuations are understood as follows: At low density, most of the particles are produced by individual strings with the same mean multiplicity and mean $p_T$, therefore small fluctuations. At large density above the critical point essentially there will be only are cluster and therefore the fluctuations are small. The maximum of fluctuations corresponds to the largest number of clusters with different size and number of strings.

The percolation of strings predicts that the long range rapidity correlation measured by $D_{FB}^2 = < n_B n_F > - < n_F > < n_B >$ (between forward F and backward B the rapidity gap should be larger than 1-1.5 to eliminate short range correlations) increases with centrality, being much less than what is expected from superposition models [40]. This is in agreement with STAR preliminary data [41].

In Color Glass Condensate(CGC)[42] we expect also that $D_{FB}^2$ grows with the centrality [43]. In fact, the main contribution is given by

$$< \frac{dN}{dy_1} \frac{dN}{dy_2} > \simeq < \left( \frac{dN}{dy} \right)^2 > = \sim \frac{1}{\alpha_s^2} \pi R^2 Q_S^2 \qquad (2)$$

As centrality increases, $\alpha_s$ decreases and (2) grows.

## THERMALIZATION

It is clear that a fast thermalization of the partonic is required. We learned at beginning of the conference [44] that this can be achieved naturally in the CGC approach. In this approach, the collision of two heavy ion, which are gluon saturated in the initial state, develop strong longitudinal chromoelectric fields [45], which via the Schwinger mechanism produce particles with a thermal spectrum due to the fluctuations of the color field. The temperature T of this spectrum is related to the saturation momentum $Q_s$, $T \simeq Q_s/2\pi$. The thermalization time is $\tau \sim 1/Q_s$.

As it has pointed out [44], a similar picture arises in percolation [46]. In this case, the critical density $\eta_c$ for the non-thermal percolation phase can be related to the critical temperature, $T_c = \frac{<p_T>_1}{\sqrt{2F(\eta_c)}}$, where $< p_T >_1$ is the mean transverse momentum of particles produced in one single string (essentially the string tension) and $F(\eta_c) = \sqrt{\frac{1-e^{\eta_c}}{\eta_c}}$ has geometrical origin and has to do with the fraction of total avalaible area occupied by the cluster $(1 - e^{-\eta_c})$. The shear viscosity is $< p_T >_1 F(\eta)L$ is also determined by the same factor . L is the longitudinal extension, $L \simeq 1 fm$. For reasonable values of $< p_T >_1 \simeq 200 MeV$ and $\eta_c \simeq 1.2 - 1.5$, it is obtained $T_c = 170 - 180 MeV$ and a very low shear viscosity.

In conclusion, theoretical and experimental progress have been achieved in the understanding of deconfinement as it has been reported in the different talks of this session. The future experiments of LHC, FAIR and we expect that also RHIC II will let us to go on this understanding, and clarify some of the questions of the field.

## ACKNOWLEDGMENTS

We thank the organizers for such a nice meeting and Y. Foka, J. Rafelski and M. Leith for helping in the configuration of this talk. The work has done under contract FPA2005-01963 of SPAIN and the support of Xunta de Galicia.

## REFERENCES

1. M. Gyulassy and L. McLerran, *Nucl. Phys. A* **750**, 30 (2005).
2. M. Lisa, this conference.
3. Bai Yuting, this conference.
4. D. Teaney, J. Lauren and E. V. Shuryak, nucl-th/0110037.
5. B. Hippolyte, this conference.
6. N. Borghini and J. Y. Ollitrant, *Phys. Lett. B*, to appear.
7. R. Lacey, *Nucl. Phys. A* **774**, 199 (2006).

8. J. D. Bjorken, Fermilab-Pub 82-059-THY; M. Gyulassy, M. Plumer, M. Thoma and X. N. Wang, *Nucl. Phys. A* **538**, 37c (1992).

9. R. Baier, Y. L. Dokshitzer, A. H. Mueller and D. Schiff, *JHEP* **109**, 033 (2001).

10. N. Borghini, this conference.

11. K. Reygers, this conference.

12. H. Torii, this conference.

13. S. S. Adler et al., PHENIX nucl-ex/010047; J. Bielcik et al., STAR nucl-ex/051105.

14. S. S. Adler et al., PHENIX nucl-ex/0507004.

15. H. Stocker, *Nucl. Phys. A* **750**, 121 (2005); J. Casalderrey, E. Shunyak, D. Teancy, *J. Phys. Conf. Ser.* **27** 22 (2005).

16. V. Koch, A. Majunder and X. N. Wang, *Phys. Rev. Lett.* **96**, 172302 (2006); I. M. Dremin, *Nucl. Phys. A* **767**, 233 (2006).

17. F. Wong, STAR Collaboration. This conference and nucl-ex/0610027.

18. T. Matsui and H. Satz, *Phys. Lett. B* **178**, 416 (1986).

19. F. Karsch, *Eur. Phys. J. C* **43**, 35 (2005).

20. M. Leith, this conference, nucl-ex/0610031.

21. P. Martins, NA60 Collaboration, this conference.

22. L. Grandchamp, R. Rapp and G. E. Brown, *Phys. Rev. Lett.* **92**, 212301 (2004); R. L. Thews, *Eur. Phys. J. C* **43**, 97 (2005).

23. F. Karsch, D. Kharzeev and H. Satz, *Phys. Lett. B* **637**, 75 (2006).

24. Y. Aoki, G. Endrodi, Z. Fodor, S. D. Katz and K. K. Szabo, *Nature* **443**, 675 (2006); K. K. Szabo, this conference.

25. D. Gabor, this conference.

26. B. Kampfer, this conference; M. Bluhm and B. Kampfer, hep-ph/0511015.

27. D. Antonov, this conference.

28. G. Krein, this conference.

29. G. Kozlov, this conference.

30. P. Christakoglou, this conference.

31. S. A. Bass, P. Danielewicz and S. Pratt, *Phys. Rev. Lett.* **85**, 2689 (2000).

32. C. Alt et al., NA49 Collaboration, *Phys. Rev. C* **71**, 034903 (2005).

33. D. Jiaxin, L. Na and L. Lianshou, nucl-th/0606062.

34. V. Koch, M. Bleicher and S. Jeon, *Nucl. Phys. A* **698**, 261 (2002).

35. B. Lungwitz, this conference.

36. M. Bleicher, this conference.

37. E. G. Ferreiro, this conference; L. Cunqueiro, E. G. Ferreiro, F. del Moral and C. Pajares, *Phys. Rev. C* **72**, 024907 (2005).

38. E. G. Ferreiro, F. del Moral and C. Pajares, *Phys. Rev. C* **69**, 034901 (2004).

39. M. A. Braun, N. Armesto, E. G. Ferreiro and C. Pajares, *Phys. Rev. Lett* **77**, 3736 (1996); M. Nardi and H. Satz, *Phys. Lett. B* **442**, 14 (1998).

40. N. A. Amelin, et al., *Phys. Rev. Lett.* **73**, 2813 (1984); S. Hanssler, M. Abdel-Aziz and M. Bleicher, nucl-th/0608075.

41. T. Tarnowsky, STAR Collaboration, nucl-ex/0606018.

42. E. Iancu, this conference.

43. N. Armesto, L. McLerran and C. Pajares, hep-ph/0607245 *Nucl. Phys. A* to appear.

44. D. Kharzeev, this conference; D. Kharzeev, E. Levin, K. Tuchin, hep-ph/0602063; D. Kharzeev, K. Tuchin, *Nucl. Phys. A* **753**, 316 (2005).

45. T. Lappi and L. McLerran, *Nucl. Phys. A* **722**, 200 (2006).

46. J. Dias de Deus and C. Pajares, *Phys. Lett. B* **642**, 455 (2006).

# QCD, New Physics and Experiment

## Giuseppe Nardulli

*Department of Physics, University of Bari and INFN-Bari*
*Via E. Orabona 4, 70126, Bari, Italy*
*E-mail: giuseppe.nardulli@ba.infn.it*

**Abstract.** I give a summary of Section E of the seventh edition of the Conference *Quark confinement and the hadron spectrum*. Papers were presented on different subjects, from spectroscopy, including pentaquarks and hadron structure, to the quest for physics beyond the standard model.

**Keywords:** Tests of the Standard Model, Physics beyond the Standard Model, Hadron Structure, Heavy Quarks, Pentaquark.
**PACS:** 12.60.Cn,12.40.Yx,14.40.Lb

## SEARCH FOR NEW PHYSICS

Before discussing how and where one should look for new physics, a preliminary analysis of the status of the Standard Model (SM) is needed. In Section E this was the task of T. Dorigo, who presented a summary of SM tests at the Tevatron and P. Taras, who discussed the BaBar experiment on the muon anomalous magnetic moment.

As reported by Dorigo, at present the top quark mass has been measured with an accuracy of 1.2%. The average result from Tevatron Run I/II is $m_{top} = 171.4 \pm 2.1$ GeV. As a result of the increased accuracy on $m_{top}$ the available parameter space for the Higgs mass within the SM has been sharpened. The result of the Tevatron studies is that the Higgs boson mass cannot be much larger than the present limit of 114.4 GeV [1]. The latest LEPEWWG results (summer 2006) are in fact $m_{higgs} = 85^{+39}_{-28}$ GeV, and $m_{higgs} < 166$ GeV at 95% C.L.

Apart from indirect hints from radiative corrections, the Higgs particle has been hunted in many different channels. Dorigo presented some of these results. For example in $WH \to l\nu bb$ searches D0 extracts a limit for the cross section of 2.4 pb (at 95%C.L.) for $m_{higgs} = 115$ GeV with 380 pb$^{-1}$ of data; CDF excludes cross sections above 3.4 pb with 1 fb$^{-1}$. These limits are above SM cross sections and therefore there is no exclusion region in $m_{higgs}$ yet. To give another example, in D0 searches for $H \to WW$ by selecting two high-$P_t$, isolated leptons ($ee$, $e\mu$, $\mu\mu$), with significant missing $E_t$, and little jet activity, in $ee$ 11 events are seen (11.4 expected); in $e\mu$: 18 seen (28.1 expected), in $\mu\mu$: 10 seen (10.5 expected). The expected SM Higgs signal is small and the limits are dominated by systematic effects. Though statistics is not yet sufficient to exclude definite $m_{higgs}$ regions, Tevatron Run II preliminary data are getting closer to this result day by day. If the Higgs boson exists and is light, it might be therefore discovered within two or three years at Fermilab.

CP892, *Quark Confinement and the Hadron Spectrum VII*
edited by J. E. F. T. Ribeiro
© 2007 American Institute of Physics 978-0-7354-0396-3/07/$23.00

The results presented by Taras were obtained by the BaBar collaboration measuring the ratio ($\sqrt{s}$ is the c.m. energy)

$$R(s) = \frac{\sigma(e^+e^- \to \text{hadrons})}{\sigma(e^+e^- \to \mu^+\mu^-)} \qquad (1)$$

using initial state radiation, in the context of the study of the anomalous muon magnetic moment $a_\mu$. It is defined by the relation $g - 2 = 2a_\mu$ and comprises three components: the QED part $a_\mu^{QED}$, the electroweak part $a_\mu^{EW}$, and the hadronic contribution $a_\mu^{had}$. The components $a_\mu^{QED}$ and $a_\mu^{EW}$ are computed with high precision (5 and 2 loops), so that the main source of theoretical uncertainties is from $a_\mu^{had}$ and this quantity is dominated by the integral

$$a_\mu^{had,Lo} = \left(\frac{\alpha m_\mu}{3\pi}\right)^2 \int_{4m_\pi^2}^\infty ds \, K(s) \frac{R(s)}{s^2} \qquad (2)$$

(apart from $a_\mu^{had,Lo}$ there are other smaller contributions to $a_\mu^{had,Lo}$ that tend to cancel out). In (2) $K(s)$ takes values in the interval $(0.63, 1)$ and $R(s)$ is obtained experimentally summing up several channels, both non resonant ($2\pi$) and resonant ($\omega, \phi, J/\Psi$, etc.). The result quoted by Paras is $a_\mu^{had,Lo} = (690.9 \pm 3.9_{exp} \pm 1.9_{rad} \pm 0.7_{QCD})^{-10}$. Existing experimental data on $e^+e^- \to \pi^+\pi^-$ show some disagreement (KLOE data differ appreciably from SND, CMD-2 data); moreover also measurements of spectral functions obtained by $\tau$ decay ($\tau^- \to \pi^0\pi^-\nu_\tau$) and $e^+e^-$ show some discrepancy, which must be eliminated. This makes the independent determination with BaBar initial state radiation especially interesting. It must be noted the BaBar offers a unique opportunity to get precision of 2-4% or even better than 1% for the two charged pions mode (the dominant mode, contributing by $\sim 73\%$ to $a_\mu^{had,Lo}$). If one compares $a_\mu^{exp}$ obtained by SND, CMD-2 and BaBar, with the result in the Standard Model $a_\mu^{SM}$:

$$\begin{aligned} a_\mu^{exp} &= (11659208.0 \pm 6.3) \times 10^{-10}, \\ a_\mu^{SM} &= (11659180.5 \pm 5.6) \times 10^{-10}, \end{aligned} \qquad (3)$$

one finds a discrepancy of 3.3 standard deviations:

$$a_\mu^{exp} - a_\mu^{SM} = (27.5 \pm 8.4) \times 10^{-10}. \qquad (4)$$

This might be taken as a signal of new physics, but, on the basis of recent lattice QCD and $\tau$-based calculations, it is fair to see that the theoretical error has been probably underestimated (see Rubakov's talk in [2]).

One of the talks (J.Ulbricht) presented in Section E was devoted to tests of non pointlike behavior of fermions. Measurements from various experiments (LEP, TRISTAN, CDF, D0, UA2) have been used to search for such a non point-like behaviour. In this way the group responsible for this analysis (I. Dymnikova, U. Burch, J.Ulbricht, C.H. LIN, S. Sakharov, J. Wu and J. Zhao) was able to put limits on the energy scale $\Lambda$ of the direct contact interaction. Ulbricht presented also models with excited fermions, contact interaction and compositeness and he put constraints on the mass of excited electron

144

$m^*$, and on $\Lambda$. In more detail, excited quarks with a mass between 80 and 570 GeV are excluded at 95 % confidence level. Using the UA2 data, according to his results one can exclude excited $u^*$ and $d^*$ quarks with masses smaller than 288 GeV at 90% CL. For EM interactions one gets limit on the mass of a heavy electron: $m^* = 308 \pm 56$ GeV and for the finite size of the electron a limit of $\Lambda = 1253.2 \pm 226$ GeV, corresponding to a size $r \approx 16 \times 10^{-18}$cm . For EW interaction the most stringent limits for the quarks are $r_q < 2.2 \times 10^{-18}$cm, for the leptons $r_l < 0.9 \times 10^{-18}$ cm, and the form factor puts a limit on the electron size of $r_e < 28 \times 10^{-18}$cm. Finally a scheme to describe all fundamental particles as extended objects of a finite geometrical size was presented.

Search for physics beyond SM is one of the missions of the future Large Hadron Collider at CERN. R. Mackeprang presented a talk on the quest for supersymmetry at Atlas. As a matter of fact this detector is sensitive to a broad spectrum of SUSY phenomenology, which strongly qualifies it for these studies. He presented some examples of ATLAS analyses with different scenarios and concluded that at 10 fb$^{-1}$ this experiment would be sensitive to a SUSY scale not larger than 2 TeV. Needless to say that also in this case a knowledge of the SM background as precise as possible will be of invaluable help in the identification of signals of SUSY, if they will be there. SUSY searches were mentioned also by H. Fox who presented a survey of results from the D0 experiment concerning new physics. Besides results on supersymmetry he presented limits on the masses of extra gauge bosons $W'$ and $Z'$. The results obtained are $m_{W'} > 965$ GeV and $m_{Z'} > 850$ GeV, both at 95% CL. He also presented results for physics beyond SM with extra dimensions. Actually the analysis, inconclusive so far, was limited to large extra dimensions studied in the context of Russel-Sundrum models. Search for Extra Dimensions will be pursued in the future with the ATLAS and CMS Detectors at the LHC , see e.g. [3].

I wish to make here a digression on extra dimensions and QCD. The linkage between the two is especially interesting for an understanding of Quantum ChromoDynamics in the strong coupling regime. Though the topic was not explicitly presented, it was part of the discussion and deserves a mention. These developments are related to the gauge/gravity correspondence [4]. In the last few years it has been realized that such a correspondence can be used to get information on QCD in the nonperturbative regime. In particular by the term AdS/QCD one identifies the mapping between D=4 strongly coupled gauge theories and gravitational theories in D=5 with an anti-de Sitter gravity background. This is a fast growing field of study (for a review talk see [5]) in which one can distinguish two different approaches. In the up-bottom approach one starts from a string theory with an appropriate background chosen so that some fundamental properties of QCD are reproduced. By the bottom-up approach, starting from QCD one tries to constrain the dual theory using the gauge/gravity correspondence. To give just a few recent examples, these methods have been used to shed light on the parameters of low energy effective theory of Goldstone bosons [6], linearity of Regge trajectories [7], the heavy quark potential [8], the thermal phase transition [9] or the BFKL Pomeron [10]. These few examples suffice to show the interest of this approach and convince the reader that more results will be certainly obtained in the near future.

# HADRON STRUCTURE, HEAVY QUARK PHYSICS AND THE PENTAQUARK

Hadron structure is also a lively field of study; more than 50 papers were presented at International Conference on High Energy Physics at Moscow, August 2006. The related arguments discussed in Section E were various. In particular ample space was given to structure functions and parton density functions from neutral and charged currents and jets. The link to new physics stems from the fact that parton distribution functions measured at the existing accelerators are an essential piece of information for LHC. These items were covered by Osipenko's and Cwiok's talks. Related aspects are the nucleon spin structure (Livingston's talk) and improvements by the inclusion of the Polyakov loop in the description of low energy QCD by effective field theories (Megias).

Another topic discussed in the Section was the status of the recently found charmonium-like states, i.e. the states X,Y and Z. Let us start with the state X(3872). It was reviewed by S. Ricciardi from BaBar and by A. Zupanc from the Belle collaboration. The average mass of this state (also observed at CDF and D0) is $3871.2 \pm 0.5$ MeV, with a width $\Gamma < 2.3$ MeV. Its quantum numbers were also established: $J^{PC} = 1^{++}$. This assignment follows from the following considerations. First, since the decay $X \to \gamma J/\Psi$ is observed, then the $X$ state must have $C = +1$. Second, the decay $X \to \pi^+\pi^- J/\Psi$ is also observed. The part of the $2\pi$ invariant mass spectrum that can be ascribed to a $\rho^0$ decay is consistent with S-wave decay of the $X$ state. From this the assignment $P = +1$ follows. Finally the angular distribution in this channel is incompatible with $J = 0$ and therefore the only remaining possibility is $J = 1$ or $J = 2$. However if the peak in the $D^0\bar{D}^0\pi^0$ decay channel at 3875.4 MeV, i.e. at only $2\sigma$ from the mass of $X(3872)$, is interpreted as due to our state, the $J = 2$ should be excluded and the only remaining possibility is $J = 1$.

Several interpretations of this state have been advanced in the literature. One possibility is that it is a $\chi'_{c1}$ state, but this is unlikely because of this result

$$\frac{\mathscr{B}(X \to \gamma J/\Psi)}{\mathscr{B}(X \to \pi^+\pi^- J/\Psi)} = 0.19 \pm 0.07 , \tag{5}$$

which is an average of BaBar and Belle results and is too small to be compatible with this identification. Another possibility is that this state comprises four quarks, more exactly two diquarks [11]. If this interpretation is correct, then the existence of additional states can be predicted. At present these new states have not been seen. Moreover the mass difference between the two neutral states is larger by two $\sigma$ than experimental data [12]. While it is too early to get definite conclusions, it is fair to say that a more economical interpretation is that in terms of a molecular $D^0\bar{D}^{*0}$ state [13, 14, 15, 16].

Let us now consider the state Y(3940), observed by Belle in the decay mode $B \to K \omega J/\Psi$. Its reported mass is $M = 3943 \pm 11 \pm 13$ MeV, with a width $\Gamma = 87 \pm 22 \pm 26$ MeV. Its interpretation as a charmonium $c\bar{c}$ state is possible, but should be corroborated by the observation of the decay mode $Y \to D^{(*)}\bar{D}^{(*)}$, which has not yet been seen (on the contrary the decay mode $Y \to \omega J/\Psi$ has a large branching ratio). It could be a $c\bar{c}$-gluon hybrid since in this case the decay mode $Y \to D^{(*)}\bar{D}^{(*)}$ would be suppressed, but the difficulty is in the predicted mass of such a state, around 4.3-4.5

GeV from lattice QCD computations [17, 18], significantly larger than the measured value. One can therefore conclude that more data are needed before an identification of this state as a $c\bar{c}g$ can be made. Zupanc also discussed the new state $X(3940)$ found in double charm production, probably different from $Y(3940)$ and the state $Z(3930)$, whose possible interpretation is $\chi'_{c2}$. I refer the interested reader to his talk, as reported in these proceedings, for a detailed discussion of these results.

S. Ricciardi discussed other two charmonium-like states. The first one is the state $Y(4260)$ observed by BaBar in $e^+e^- \rightarrow (\gamma)Y(4260) \rightarrow J/\Psi\pi^+\pi^-$ (with no $\gamma$ detection). BaBar finds for this state $M = 4259 \pm 8^{+2}_{-6}$ MeV and $\Gamma = 88 \pm 23^{+6}_{-4}$ MeV. This state is confirmed by Cleo-III and Belle (the mass measured by Belle is however $2.5\sigma$, i.e. 36 MeV, higher). Data show a large coupling to $J/\Psi\pi\pi$, which renders puzzling its interpretation as charmonium-like. Other interpretations, in terms of a tetraquark state [19] or a hybrid, should be seriously considered. The other structure is seen in $e^+e^- \rightarrow (\gamma)\psi(2S)\pi^+\pi^-$ at a mass 4.35 GeV. Data are not sufficient to draw conclusions about the nature of the structure or its consistency with the previously observed Y(4260).

Among the various topics discussed by Ricciardi, let me mention the problem of the state $D_s^*(2860)$. This charmed meson, seen through a fit to DK mass spectra, has mass and width as follows: $M = 2856.6 \pm 1.5 \pm 5.0$ MeV, and $\Gamma = 47 \pm 7 \pm 10$ MeV. Since it decays into two pseudoscalar mesons, its spin-parity assignment can be $J^P = 0^+, 1^-, 2^+, 3^-$. At present the decay mode $D^*K$ has not been observed and this means, according to the authors in [20], that the assignments $1^-, 2^+$ are not favored. Between the remaining alternatives $0^+, 3^-$ the latter seems more likely because in the former case the decays would occur in $S$−wave and the resonance would be broad. On the other hand the existence of a narrow state with $\Gamma = 35 - 140$ MeV, Strangeness=0 and $M$ around 2.8 GeV was predicted already in 2000 [21]. Its strange partner would have therefore a mass compatible with the new state. The calculation in [21] was based on QCD sum rules in the framework of the Heavy Quark Effective Theory (HQET) as applied to mesons comprising one heavy quark [22]. HQET classifies these states according to $s_\ell^P$, the total angular momentum of the light degrees of freedom. For $s_\ell^P = 3/2^-$ one has $J^P = 1^-, J^P = 2^-$; there are other two narrow states with $s_\ell^P = 5/2^-$ and therefore with $J^P = 2^-, J^P = 3^-$. For a $3^-$ state, the decay width into $DK$ is predicted to be small because it occurs as $F$−wave; on the other hand missing observation of the $2^-$ partner might be explained by its likely mixing with the broad state. Other talks on heavy quarks were on further results from BELLE, in particular on the $Y(4260)$ state (G.Pakhlova).

Another topic discussed in Section E was the negative result from CLAS in the search for pentaquark with increased statistics (R. Gothe's talk). Pentaquarks were one of the highlights of the Section E of the 2004 Conference on Quark Confinement and the Hadron Spectrum [23]. In photoproduction experiment off proton no significant signal for $\Theta^+$ or $\Theta^{++}$ is seen by CLAS. Also for photoproduction off deuteron: $\gamma d \rightarrow \Theta^+K^-$, followed by $\Theta^+ \rightarrow nK^+$, CLAS has now negative results, differently from their previously released analysis. In particular previous peak could not be reproduced under similar circumstances. Therefore the statistical significance of the old peak is reduced from $5.2\sigma$ to $3.1\sigma$ when new data are used as background. While the existence of the pentaquark cannot be excluded, stringent upper limits on total and differential cross sections were set. In particular, present data, together with phenomenological models, put an upper limit on the cross section for $\gamma n \rightarrow \Theta^+ K^-$ of around 3 nb (at 95% CL). Let me note that pentaquark searches were reported elsewhere at this Conference, with mixed results from other experiments, see e.g. K. Daum's talk given in Section B.

In conclusion, waiting for LHC, still a lot of physics can be done and is actually done. In Section E reports were presented on several active and promising areas: Heavy quarks, supersymmetry, compositeness, excited states and the role of extra dimensions. As to string physics, the new developments on AdS/QCD, if correct, would represent a further step in the long and complicated journey from strong interactions to gravity and back.

## ACKNOWLEDGMENTS

I thank P. Colangelo for most useful comments on the spectroscopy of charmed states,

## REFERENCES

1. W. -M. Yao, et al.,"Review of Particle Physics", *Journal of Physics G* **33**, 1 (2006).
2. V. Rubakov, "Conclusions and Outlook", summary talk at the XXXIII International Conference on High Energy Physics, July 26 - August 2, 2006, Moscow, Russia.
3. S. Shmatov, "Search for Extra Dimensions with ATLAS and CMS Detectors at the LHC", talk presented at the XXXIII International Conference on High Energy Physics, July 26 - August 2, 2006, Moscow, Russia.
4. J. M. Maldacena, *Adv. Theor. Math. Phys.* **2**, 231 (1998).
5. G. Marchesini, "QCD review", talk presented at the XXXIII International Conference on High Energy Physics, July 26 - August 2, 2006, Moscow, Russia.
6. J. Erlich. E. Katz, D. T. Son, and M. A. Stephanov, *Phys. Rev. Lett.* **95**, 261602 (2005).
7. A. Karch, E. Katz, D. T. Son, and M. A. Stephanov, *Phys. Rev. D* **74**, 015005 (2006) [arXiv:hep-ph/0602229].
8. O. Andreev, and V. I. Zakharov, *Phys. Rev. D* **74**, 025023 (2006) [arXiv:hep-ph/0604204].
9. O. Andreev, and V. I. Zacharov, arXiv:hep-ph/0607026.
10. R. C. Brower, J. Polchinski, M. J. Strassler, and C.-I Tan, arXiv:hep-th/0603115.
11. L. Maiani, F. Piccinini, A. D. Polosa, and V. Riquer, *Phys. Rev. D* **71**, 014028 (2005) [arXiv:hep-ph/0412098].
12. E. S. Swanson, *Phys. Rept.* **429**, 243 (2006) [arXiv:hep-ph/0601110].
13. N. A. Törnqvist, arXiv:hep-ph/0308277.
14. F. E. Close, and P. R. Page, *Phys. Lett. B* **578**, 119 (2004) [arXiv:hep-ph/0309253].
15. C. Y. Wong, *Phys. Rev. C* **69**, 055202 (2004) [arXiv:hep-ph/0311088].
16. S. Pakvasa, and M. Suzuki, *Phys. Lett. B* **579**, 67 (2004) [arXiv:hep-ph/0309294].
17. C. W. Bernard, *et al.* [MILC Collaboration], *Phys. Rev. D* **56**, 7039 (1997) [arXiv:hep-lat/9707008].
18. Z. H. Mei, and X. Q. Luo, *Int. J. Mod. Phys. A* **18**, 5713 (2003) [arXiv:hep-lat/0206012].
19. L. Maiani, V. Riquer, F. Piccinini, and A. D. Polosa, *Phys. Rev. D* **72**, 031502 (2005) [arXiv:hep-ph/0507062].
20. P. Colangelo, F. De Fazio, and S. Nicotri, *Phys. Lett. B* **642**, 48 (2006) [arXiv:hep-ph/0607245].
21. P. Colangelo, F. De Fazio, and G. Nardulli, *Phys. Lett. B* **478**, 408 (2000) [arXiv:hep-ph/0001200].
22. R. Casalbuoni, A. Deandrea, N. Di Bartolomeo, R. Gatto, F. Feruglio, and G. Nardulli, *Phys. Rept.* **281**, 145 (1997) [arXiv:hep-ph/9605342].
23. G. Nardulli, "QCD, hadrons and beyond", edited by U. D'Alesio et al., AIP Conference Proceedings 756, American Institute of Physics, New York, 2005, pp. 228-235 (2005) [arXiv:hep-ph/0411294].

# Summary of Section F: QCD in Nuclear Physics and Astrophysics

Thomas D. Cohen

*Department of Physics*
*University of Mayland*
*College Park, MD 20742*

**Abstract.** This talk summarizes the Section F of "Quark Confinement and the Hadron Spectrum VII". This section, newly added to the conference, deals with issues of QCD in nuclear physics and astrophysics

**Keywords:** QCD, nuclear physics, astrophysics
**PACS:** 12.38.-t , 20. , 97.

## OVERVIEW

Section F which deals with QCD in nuclear physics and astrophysics is a new section for the "Quark Confinement and the Hadron Spectrum" series of conferences. A cynic might say that there was a good reason why the section was not present in the past. If one required the talks to focus on theory in which computations directly connect QCD to the observed nuclear or astrophysical phenomena, the section would be short---indeed of infinitesimal length. In a certain sense the cynics are correct: the talks in this section involve theory which only indirectly connects to QCD (*eg.,* QCD inspired models), only indirectly connects to nuclear/astrophysical phenomenon (*eg.,* phases of QCD), or both. Despite this, this section dealt with a considerable amount of very interesting nuclear physics and astrophysics related to QCD.

There were three main themes of the talks in the section:
- *Quark matter in its many manifestations.*
  - ➤ A key focus of these talks was on such matter as cores of possible exotic compact stars.
- *Nuclear and neutron matter.*
  - ➤ These talks focused on either the many-body physics needed to calculate the equations of state or the physics of neutron stars.
- *Other nuclear phenomenon.*
  - ➤ These talks focused on issues such as nucleon-nucleon forces or effects on hadrons in the hadronic medium

CP892, *Quark Confinement and the Hadron Spectrum VII*
edited by J. E. F. T. Ribeiro
© 2007 American Institute of Physics 978-0-7354-0396-3/07/$23.00

These topics are interconnected in various ways. A simple illustration of this can be seen in Fig. 1. The dashed lines represent moderately direct connections while the dotted lines are more tenuous.

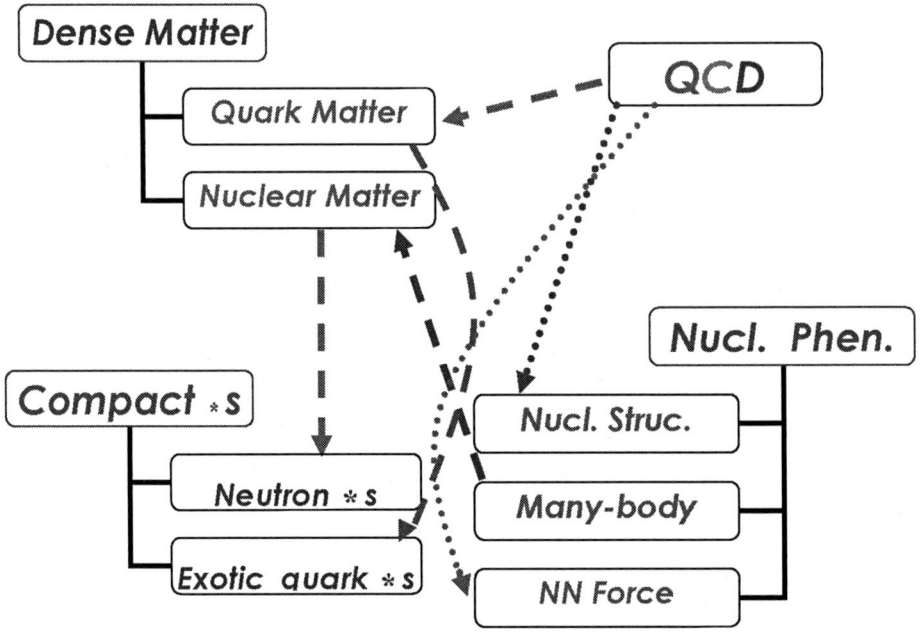

**FIGURE 1.** Interconnections of the various subjects in Section F.

My goal in this summary is to give a sense of the richness of the field. Given the great diversity of topics covered it is difficult or impossible to really summarize the intellectual content of this section. Instead what I will do here is to *very* briefly discuss each of the talks in the session to give some sense of the flavor of the field. For some talks I will also include a figure illustrating the issues. These figures are included to give a feel for the issues and not to convey much technical information; I refer you to the write-ups of the original talks for a more detailed description.

Of course, it is impossible to catch the essence of a serious scientific talk in a few choice lines. The spirit in which I am undertaking this is similar to that of the Reduced Shakespeare Company's production: *The Complete Works of William Shakespeare (abridged)* which goes through all 37 of Shakespeare's plays in 97 minutes. Clearly, in doing this all of the subtlety gets lost---but perhaps something of the original is captured. It is also important to keep in mind that the talks in this section fall naturally into three categories:

    i.    Those talks I understood.

    ii.   Those talks that I thought I understood, but in fact did not.

    iii.  Those talks which I did not understand, but which for the purpose of this summary will have to pretend that I did.

# THE TALKS OF SECTION F

**Ariel Zhitnitsky:** *"Gluon Vacuum Energy in Dense Matter."* A beautiful formal result on gluon condensate at finite chemical potential was discussed. The case of two QCD with two colors was studied as it is tractable. The result for low densities is that the condensate drops due to nucleon contribution as one deduces from naïve reasoning. Surprisingly, as shown below at slightly larger densities it grows and becomes larger than its vacuum value. The implication of this result for the real world of three colors remains unclear.

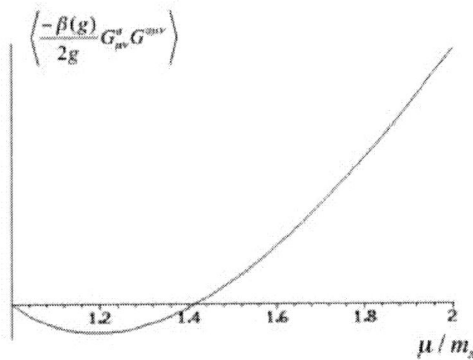

**FIGURE 2.** The gluon condensate as a function of chemical potential from the talk of Ariel Zhitnitsky.

*Charles Horowitz: "Neutron Rich Matter in Heaven and Earth."* This talk focused on two issues. One fundamental issue was raised in Gordon Baym's plenary talk: namely, whether neutron star data constrain the equation of state to be inconsistent with those of "exotic" matter (*eg.*, a color superconductor). In this talk the notion of quark/hadron duality was invoked to argue that the core of a neutron star could be a density in which it could be described either by strongly correlated nucleon *or* strongly correlated quarks. The second issue in this talk was the nearly universal behavior of low density neutron matter. One focus was on the extent to which real neutron matter with a large but finite scattering length ($a \approx 19\ fm$) can be approximated by the unitary limit of an infinite scattering length. The viral expansion may be used to study the system and the second viral coefficient was shown to be model independent.

*Andrew Steiner: "Crusts and Strange Quark Stars."* The key observation of this talk was that while the idea of a "quark star" is typically based on a quark matter core, the crust of such a star can contain "quark nuggets" depending on the parameters of the theory. The curves in the figure at the left with minima correspond to such quark nuggets. Possible signatures of quark stars were discussed.

*Fridolin Weber: "Neutron Star Interiors and the Equation of State of Super Dense Matter."* This talk was a whirlwind tour of many branches of this subject both from a theoretical and a phenomenological perspective. While it was nearly impossible to summarize most talks in this section in a few lines while retaining any flavor of the original, with this talk it was truly impossible.

*Sanjay Reddy:* *"First-Order Phase Transitions Generically Results in Heterogeneous States of Dense Matter."* General considerations of mixed phases in situations where electric neutrality plays a central role were discussed in this talk. In these problems the Debeye screening length, $\lambda_D = \left(4\pi\,\alpha_{em}\chi_Q\right)$ sets a natural length scale. As seen in the figure below this can have important consequences to the mixed phase between nuclear matter and a CFL phase.

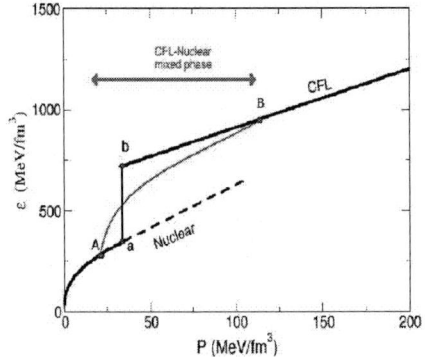

**FIGURE 3.** The effect of the mixed phase on the transition from nuclear matter to CFL from the talk of Sanjay Reddy.

*Achim Schwenk:* *"Superfluidity in Cold Atoms and Neutron Stars."* This talk reported on studies using many-body methods based on the renormalization group. This analysis showed that induced interactions suppress the s-wave gap. P-wave gaps were studied with these techniques. It was shown that the cooling of neutron stars is sensitive to these gaps and from this, large gaps were ruled out.

*Rémi Huguet:* *"Saturation Properties of Nuclear Matter in a Relativistic Mean Field Model Constrained by Quark Dynamics."* This talk reported on a hybrid approach to nuclear saturation. The problem was formulated in the context of a Walecka type relativistic nuclear model. However, the parameters were altered in medium due to physics at the quark level which were computed in the framework of an NJL type model. By tuning parameters, reasonable fits to saturation properties could be obtained.

*Jacek Rozynek:* *"The Nucleon Parton Distribution for Finite Densities."* There have been many proposed approaches to explain the EMC effect over the years. This talk reviewed the principle approaches on the market. It was shown that in deep inelastic scattering the magnitude of the nuclear Fermi motion effect is sensitive to the residual interaction between partons. This influenced both the nucleon structure function and the nucleon mass in the nuclear medium. Connections to Drell-Yan were explored as were connections to in-medium condensates.

*Laura Tolos:* *"Neutron Matter at Finite Temperature from Low-Momentum Interactions."* It has been known for several years that when nuclear potentials which fit the phase shifts are evolved to low scales using the renormalization group, the effective potential is universal---all starting potentials reduce to the same effective potential. This talk explored the use of such effective potentials in the many-body context. One key point is that the low scale of the interactions "tames" the hard core and makes systematic calculations tractable for the case of low density, low temperature neutron matter.

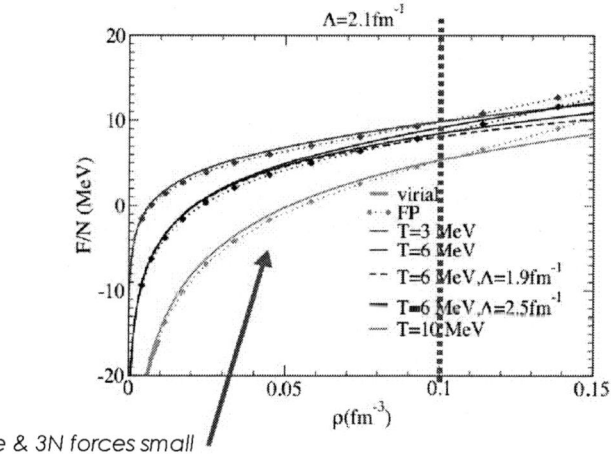

*Cutoff independence & 3N forces small*

**FIGURE 4.** The free energy of neutron matter from the talk of Laura Tolos.

**Vladimir Kukulin:** *"Dressed Dibaryon as Carrier of NN and 3N Interactions."* Traditional approaches to nuclear forces are based on exchanges of mesons on the s-channel. This talk reported on an alternative approach based on t-channel exchange of dibaryons. The approach yields good fits to the NN phase shifts with few free parameters. The approach naturally yields a strong scalar three-body force which *greatly* alters the description of nuclear matter relative to traditional approaches: of order ½ the nuclear matter binding comes from the three-body force in this approach.

**Hossein Malekzadeh:** *"Gluon Self-Energy in the Color-Flavor-Locked Phase of Color Superconductivity."* In a color superconducting state, the quarks in a very different configuration from the vacuums. Thus, the gluon's self-energy which includes the effects of quark loops is modified from the vacuums. This talk reported on calculations of the gluon self-energy based on a set of integral equations which included the modification to the quark propagators due to the color superconducting phase.

**Eduardo Fragga:** *"Nonzero Quark Masses and QCD Thermodynamics."* This talk discussed the role of the strange quark mass on the equation of state. There is a naïve expectation that due to approximate SU(3) flavor symmetry the effect of the strange quark mass should be small. Calculations based on pQCD were reported which illustrate the strong sensitivity to $m_s$.

**Paulo A. Faria da Veiga:** *"The Particle Spectrum for Lattice QCD Models at Strong Coupling."* A truncation scheme for QCD calculations was reported which preserved certain key symmetries. Meson and Baryon states computed. Depending on the details of the calculation two-nucleon bound states (*eg.* the deuteron) were found.

*Michaell Buballa:* *"Phase Diagram of Quark Matter Under Compact Star Conditions"* This talk reported on research on calculations of the phases of dense quark matter. As QCD is intractable in the regime of interest for compact stars, these calculations were based on a simple model of the NJL type. A rich phase structure was observed but it was highly dependent on the model details. Work based on the Schwinger-Dyson equations was also reported.

*Andreas Schmitt:* *"Stressed Pairing in Conventional Color Superconductors is Unavoidable."* Color and electric neutrality constraints along with SU(3) flavor breaking due to the strange quark mass move the fermi surfaces for the different types of quarks. As seen in the figure below, this "stresses" the pairing which can lead to breakdown of the color-flavor locking type of color superconductivity. This talk discussed the question of whether this effect can be evaded by a clever choice of pairing partners, thus yielding a more robust regime of color superconductivity. By considering all possibilities, it was shown that this is not possible: stressed paring is unavoidable.

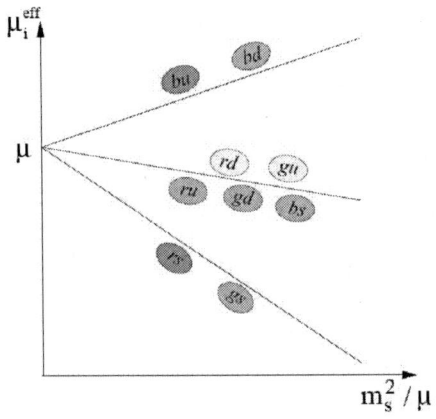

**FIGURE 5.** The phenomenon of stressed pairs from the talk of Andreas Schmitt..

*Marco Ruggieri:* *"The Three Flavor Crystalline Color Superconductive Phase of QCD."* The LOFF phase is a spatially inhomogeneous superconducting phase with spatial periodicity. This talk reported on model calculations of the phase of dense matter and found a relatively small window in chemical potential where the LOFF state appears to be the ground state. These calculations were based on a plane wave crystal structure. It remains possible that a larger window for which the LOFF phase is the ground state might emerge for more complicated crystal structures.

# Understanding Confinement From Deconfinement

M. Baker

*Dept of Physics, University of Washington*
*P. O. Box 351650, Seattle WA 98195, USA*

### Abstract

We show that a dual $SU(N)$ effective theory of the confined phase of $SU(N)$ Yang-Mills theory can be used in the temperature interval $T_c < T < 4T_c$ to calculate non-perturbative magnetic properties of the deconfined phase, including the magnetic screening mass and the spatial string tension. We suggest that evidence for the quanta of the effective theory should be sought in analyses of both lattice simulations of Yang-Mills theory and experiments on heavy ion collisions.

## 1 Introduction

In this talk we first review an effective dual $SU(N)$ gauge theory of the confined phase of $SU(N)$ Yang-Mills theory. The effective theory describes distances greater than the flux tube radius $R_{FT} \approx \frac{\sqrt{2}}{M} \approx 0.3fm$ and has a corresponding ultraviolet cutoff $M \approx 950MeV$. Since ([1]) the deconfinement temperature $T_c \sim 260MeV$ for $SU(N)$ groups with $N \geq 3$, the scale of the effective theory $M \sim 4T_c$. There is then a window of temperatures $T_c < T < 4T_c$ where the theory should be applicable also in the deconfined phase.

We use the effective theory, with parameters fixed by fits of heavy quark potentials in the confined phase, to calculate the magnetic screening mass and spatial string tensions in the deconfined phase. In this way the non-perturbative magnetic sector of finite temperature $SU(N)$ Yang-Mills theory is unified with the physics of confinement.

## 2 Effective Theory of the Confined Phase

The effective theory describing the low energy excitations of SU(N) Yang-Mills theory is a long distance dual $SU(N)$ Yang-Mills theory coupling non-Abelian magnetic SU(N) gauge potentials $\mathbf{C}_\mu$ to 3 scalar fields $\phi_i$ each in the adjoint representation of the magnetic gauge group. The Lagrangian $L_{eff}$ has the form [2]

$$L_{eff} = 2tr\left[-\frac{1}{4}\mathbf{G}^{\mu\nu}\mathbf{G}_{\mu\nu} + \frac{1}{2}(D_\mu\phi_i)^2\right] - V(\phi_i), \qquad (1)$$

CP892, *Quark Confinement and the Hadron Spectrum VII*
edited by J. E. F. T. Ribeiro
© 2007 American Institute of Physics 978-0-7354-0396-3/07/$23.00

where

$$\mathbf{G}_{\mu\nu} = \partial_\mu \mathbf{C}_\mu - \partial_\nu \mathbf{C}_\mu - ig_m [\mathbf{C}_\mu, \mathbf{C}_\nu], \tag{2}$$

$$D_\mu \phi_i = \partial_\mu \phi_i - ig_m [C_\mu, \phi_i], \tag{3}$$

$V(\phi_i)$ is a Higgs potential which has a minimum at nonzero values of $\phi_i$, and $g_m$ is the magnetic gauge coupling constant. This effective Lagrangian gives a concrete realization of the dual superconductor mechanism of confinement. [3]

In the confined phase the magnetic gauge symmetry is completely broken via a dual Higgs mechanism in which all particles become massive. (At least 3 adjoint scalars are necessary to completely break the symmetry.) The value, $\phi_0 \sim \Lambda_{QCD}$ of the magnetic Higgs condensate is fixed by the location of the minimum in the Higgs potential, and the magnetic gluon acquires a mass

$$M \approx g_m \phi_0, \tag{4}$$

via the dual Higgs mechanism. Since the effective theory describes fluctuations only at energy scales less than $M$, there is no physical excitation with this mass.

The vacuum condensate $\langle \phi_i \rangle \equiv \phi_{i0}$ has the color structure ([2])

$$\phi_{10} = \phi_0 J_x, \ \phi_{20} = \phi_0 J_y, \ \phi_{30} = \phi_0 J_z, \tag{5}$$

where $J_x$, $J_y$, and $J_z$ are the three generators of the N dimensional irreducible representation of the three dimensional rotation group corresponding to angular momentum $J = \frac{N-1}{2}$. Since any matrix which commutes with all three generators must be a multiple of the unit matrix, there is no $SU(N)$ transformation which leaves all three $\phi_i$ invariant and the dual gauge symmetry is completely broken.

The effective theory has classical flux tube solutions in which $Z_N$ electric flux is confined to narrow tubes of radius $R_{FT} \approx \frac{\sqrt{2}}{M}$, at the center of which the Higgs condensate vanishes. [4] The long wavelength fluctuations of the axis of these flux tubes give rise to an effective bosonic string theory. [5] These are the low energy excitations of the effective theory.

The effective theory has two parameters $g_m$ and $M$. Their values, $M \approx 950 MeV$ and $g_m \approx 3.91$, were determined [6] by comparing the predicted $SU(3)$ heavy quark potential with lattice simulations, which, for distances $R > 0.3 fm$, is well represented by the sum of a linear potential and a $\frac{1}{R}$ potential, $\frac{A_{lattice}}{R}$. [7] The value of $g_m \approx 3.91$ is fixed by writing the lattice $\frac{1}{R}$ potential in an effective Coulomb form:

$$\frac{A_{lattice}}{R} = -\frac{4}{3} \frac{\pi}{g_m^2} \frac{1}{R}. \tag{6}$$

The RHS of equation (6) is the potential obtained by coupling magnetic gluons to a Dirac string connecting a quark-antiquark pair with a strength $\frac{2\pi}{g_m}$.

The value of $g_m$ obtained from equation (6) includes the energy, $-\frac{\pi}{12R}$, of the long wave length oscillations of the axis of the flux tube, [5, 8]. Short distance fluctuations at energy scales greater than $M$ do not enter in the effective theory, and $g_m$ is a fixed coupling constant defined at the scale $M$. The value of $g_m$ is close to 4, so that the main contribution to $g_m$ comes from renormalization due to string fluctuations. At finite temperature the location of the minimum in the effective potential for the Higgs fields moves to $\phi_0 = 0$ when $T \sim \phi_0$, so that the effective theory has a deconfining transition at a temperature $T_c \sim \phi_0$ [9]. There is then a separation between the deconfinement temperature $T_c$ and the scale $M = g_m \phi_0 \approx g_m T_c$ of the effective theory. With $g_m \sim 4$ , we obtain the empirically observed separation between $T_c$ and $M \sim 4T_c$ which makes it possible to use the effective theory in the deconfined phase.

## 3   The Effective Theory in the Deconfined Phase

Above the deconfinement temperature the Higgs condensate vanishes, so that the magnetic gluon becomes massless. Furthermore since the deconfinement transition is first order, ([10]), the Higgs particles remain massive in the deconfined phase. The massless sector of the effective theory then reduces to a pure SU(N) Yang-Mills theory of magnetic gauge potentials $\mathbf{C}_\mu$ having the same form as the microscopic electric theory, but with a fixed gauge coupling constant $g_m$ and with a fixed ultraviolet cutoff $M$.

Thus, above $T_c$ the magnetic gluons, which at $T = 0$ confine Z(N) electric flux, become the physical degrees of freedom of the effective theory. These quanta are strongly interacting ($g_m \approx 3.91$), but their interaction is cut off at distances less than $0.3 fm$. Because of the duality between the microscopic electric $SU(N)$ Yang-Mills theory and the effective long distance magnetic $SU(N)$ gauge theory, calculations in the magnetic sector of the effective theory correspond to perturbative calculations of electric quantities in the microscopic theory. The effective theory will then provide access in the deconfined phase to physical quantities that are outside the perturbative realm of finite temperature Yang-Mills theory.

## 4   The Spatial Wilson Loop Calculated in the Dual Theory.

To test the idea of using the effective theory to calculate magnetic quantities in the deconfined phase we calculate spatial Wilson loops measuring magnetic flux with $Z(N)$ quantum number $k$ passing through a loop $L$. The spatial Wilson loop of Yang-Mills theory is the partition function of the dual theory in the presence of a current of $k$ quarks circulating around the loop $L$ (a closed Dirac string). This current is the source of a color magnetic field $\vec{B}$, the magnetic analogue of the color electric field $\vec{E}$ generated in the confined phase by a Dirac string connecting a static quark-antiquark pair [2] .

The operator creating the closed Dirac string is a singular magnetic gauge transformation which changes by a factor $e^{2\pi i \frac{k}{N}}$ when it encircles a curve linking the loop $L$. It is the dual of the spatial 't Hooft loop operator that creates a closed line of magnetic flux along a loop $L$ in Yang-Mills theory [11, 12].

## 4.1 The Effective Potential $U(\mathbf{C}_0)$

To use the effective theory to calculate spatial Wilson loops in the deconfined phase where there is no classical potential, requires evaluating the one loop effective potential $U(\mathbf{C}_0)$ in the background of a static magnetic scalar potential $\mathbf{C}_0$. This effective potential (in correspondence with the properties of the effective potential $U(\mathbf{A}_0)$ of Yang-Mills theory [13]) is a periodic function of the eigenvalues of $\mathbf{C}_0$ in the adjoint representation with period $2\pi/T$, having minima at the inequivalent $Z(N)$ vacua of the magnetic theory.

We have calculated $U(\mathbf{C}_0)$, integrating over the massless gauge modes of the magnetic theory and introducing a Pauli-Villars regulator mass M to account for the short distance cutoff of the dual theory. This calculation, aside from the presence of the regulator, mimics the calculation of the one loop effective potential $U(\mathbf{A}_0)$ [13, 14] used to evaluate the spatial 't Hooft loop [14, 15, 16] measuring electric $Z_N$ flux in Yang-Mills theory. We make the color ansatz $\mathbf{C}_0(x) = C_0(\vec{x})\mathbf{Y}_k$ , where $\mathbf{Y}_k$ is a diagonal matrix having the property that $e^{2\pi i Y_k} = e^{2\pi i \frac{k}{N}}$ , so that the corresponding background magnetic field, $\vec{\mathbf{B}}(\vec{x}) = -\vec{\nabla}C_0(\vec{x})\mathbf{Y}_k$, and then minimize the resulting effective action in the presence of a large current loop $L$.

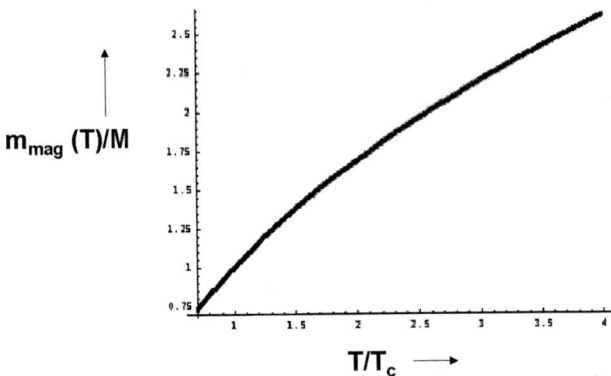

Figure 1: Ratio of magnetic screening mass to ultraviolet cutoff $M$ as a function of $T/T_c$.

The minimization yields a dual scalar potential $C_0(z)$ , which is a function only of the distance $z$ from the loop. The resulting color magnetic field profile $\vec{B}(z)$ is perpendicular to the loop and falls off exponentially at large distances from the loop. This exponential falloff determines the magnetic screening mass : $m_{mag}^2(T) = \frac{1}{2}d^2U(C_0)/dC_0^2$, evaluated at a $C_0$ vacuum.

## 4.2   The Magnetic Screening Mass $m_{mag}(T)$

The result for $m_{mag}(T)$ is equal to $\sqrt{\frac{N}{3}}g_m T$ multiplied by a function of $T/M$ accounting for the suppression of short distance contributions to $U(C_0)$. Using the values $M = 950 MeV$, $T_c = 260 MeV$, we plot in Fig. (1) $m_{mag}(T)/M$ (for $N = 3$) as a function of $T/T_c$ . We see from the figure that $m_{mag}(T_c) \approx M = 950 MeV$. Thus as the temperature is lowered toward $T_c$ the magnetic screening mass, generated from the fluctuations of the massless quanta of the effective theory in the deconfined phase, decreases and at $T \approx T_c$ becomes equal to the mass $M$, generated by the dual Higgs mechanism in the confined phase.

## 4.3   The Magnetic Energy Density Profile $\vec{B}^2(z)$

In Figure (2) we plot the profile of $\vec{B}^2(z)$ at $T = T_c$ as a function of the distance $z$ from the current loop. For $T > T_c$ where $m_{mag} > M$, the minimal wavelength $\sim \frac{1}{M}$ of the

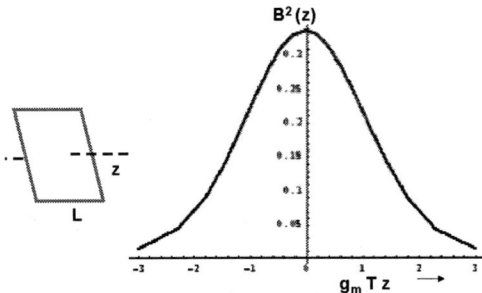

Figure 2· Magnetic energy density profile $\vec{B}^2(z)$ at $T = T_c$ as a function of distance $z$ from $L$

fluctuations present in the effective theory is larger than the width $\sim \frac{1}{m_{mag}}$ of the magnetic energy profile $\vec{B}^2(z)$. There are then no small scale fluctuations present in the theory to disturb the large scale structure of the classical solution. However, as $T$ is lowered

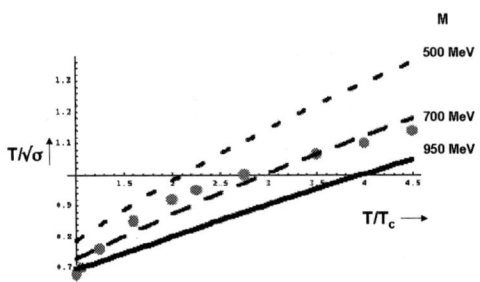

Figure 3: Comparison of 4d lattice data (dots) [17] for the spatial string tension $\sigma_k(T)$ with equation (7) obtained from the effective magnetic Yang-Mills theory. The value of the coupling constant $g_m = 3.91$ used in (7) was obtained from fitting lattice data for the heavy quark potential at $T = 0$. Solid curve uses a Pauli-Villars regulator mass $M = 950MeV$; dashed curve uses $M = 700MeV$; short dashed curve uses $M = 500MeV$.

below $T_c$ the flux tube width grows, becoming larger than the minimum wavelength of the fluctuations. We suggest that such fluctuations destabilize the distribution of magnetic energy around the loop $L$, facilitating the transition to the confined phase .

## 4.4   Spatial String Tension: Comparison with $SU(3)$ Lattice Simulations

The integral over the magnetic energy profile $\vec{B}^2(z)$, (the one loop effective action evaluated at the "classical" solution $\vec{B}(z)$), yields the spatial string tension $\sigma_k(T)$ :

$$\frac{\sigma_k(T)}{T^2} = \frac{4\pi^2 k(N-k)F(\frac{T}{M})}{3g_m\sqrt{3N}}, \tag{7}$$

where $F(\frac{T}{M})$ is the ratio of the action with regulator mass $M$ to the unregulated action. It determines the temperature dependence of $\frac{\sigma_k(T)}{T^2}$, and varies from 0.92 at $T = T_c$ to 0.44 at $T = 4T_c$. In Fig.(3) we plot $\frac{T}{\sqrt{\sigma_k(T)}}$ as a function of $\frac{T}{T_c}$ and compare with the results of 4d lattice simulations [17]. The predicted value 0.72 of $\frac{T}{\sqrt{\sigma_k(T)}}$ at $T = T_c$ lies close to the lattice result. (To test the sensitivity of the spatial string tension to the Pauli-Villars mass, we also plot curves using $M = 700MeV$ and $M = 500MeV$.)

160

## 4.5   Spatial String Tension $\sigma_k(T)$: Casimir Scaling

We note from Eq.(7) that $\sigma_k(T)$ is proportional to $k(N-k)$ (Casimir Scaling). This dependence of spatial string tensions above $T_c$ on the quantum number $k$ is consistent the results of lattice simulations of $SU(4)$, $SU(6)$, and $SU(8)$ gauge theories [10, 16].

On the other hand, from the point of view of the magnetic effective theory, such approximate Casimir scaling is not expected for the $T = 0$ string of the presence of the Higgs condensate in the confined phase.

## 5   Summary

We show that a dual $SU(N)$ effective theory of the confined phase of $SU(N)$ Yang-Mills theory can be used in the one loop approximation to calculate long distance magnetic properties of the deconfined phase, in analogy to the use of the effective theory in the classical approximation to describe the confined phase . Calculating the one loop effective potential for $C_0$ with an ultraviolet cutoff $M$, we predict :

- The magnetic screening mass $m_{mag}(T)$.

  - At $T = T_c$ it is approximately equal to the scale $M = 950MeV$, the ultraviolet cutoff of the effective theory, so that the width of the magnetic energy profile (Fig.(2)) is approximately equal to the radius of the $T = 0$ electric flux tube.
  - For $T > T_c$ it increases somewhat more slowly than linearly with $T$ (Fig.(1)).

- The spatial string tensions $\sigma_k(T)$ .

  - The predicted result for $SU(3)$ is compatible with lattice simulations (Fig.(3)).
  - For SU(N) groups with $N \geq 3$ the string tensions satisfy Casimir scaling (while Casimir scaling is not expected in the confined phase).

## 6   Further Tests and Investigations

- The effective theory predicts there is no $1/L$ correction to the area law behavior of spatial Wilson loops in the deconfined phase, in contrast with the predicted correction to the area law behavior of $T = 0$ temporal Wilson loops. Lattice simulations of spatial Wilson loops would test this prediction.

- Evidence for the magnetic quanta of the effective theory should be sought in analyses of both lattice simulations of Yang-Mills theory and experiments on heavy ion collisions which explore the temperature interval $T_c < T < 4T_c$ , where the theory is applicable.

# Acknowledgments

I would like to thank Ph. de Forcrand, C. P. Korthals Altes and A. Vuorinen for their valuable help, and J. E. Ribeiro and N. Brambilla for their hospitality and encouragement that made this work possible.

# References

[1] B. Lucini, M. Teper and U. Wenger, *Phys. Lett.* **B 545**, 197 (2002) (hep-lat/0206029).

[2] M. Baker, J. S. Ball and F. Zachariasen, *Phys. Rev.* **D44**, 3328 (1991).

[3] S. Mandelstam, *Phys. Rep.* **23C**, 245 (1976); G. 't Hooft, in *High Energy Physics, Proceedings of the European Physical Society Conference, Palermo, 1975*, ed. A. Zichichi (Editrice Compositori, Bologna, 1976).

[4] M. Baker, J. S. Ball and F. Zachariasen, *Phys. Rev.* **D41**, 2612 (1990).

[5] M. Baker and R. Steinke, *Phys. Rev.* **D63**, 094013 (2001); **D65**, 114042 (2002).

[6] M. Baker, J. S. Ball and F. Zachariasen, *Phys. Rev.* **D56**, 4400 (1997).

[7] O. Kaczmarek and F. Zantow, *Phys. Rev.* **D71**, 114510 (2005)( hep-lat/0503017).

[8] M. Lüscher, *Nucl. Phys.* **B180**, 317 (1981).

[9] M. Baker, J. S. Ball and F. Zachariasen, *Phys. Rev. Lett.* **61**, 521 (1988).

[10] B. Lucini, M. Teper and U. Wenger, hep-lat/052003.

[11] G.'t Hooft *Nucl. Phys.*, **B138**, 1 (1978); **B153**, 141 (1979).

[12] Ph. de Forcrand , M.d'Elia and M. Pepe, *Phys. Rev. Lett.* **86**, 1438 (2001); Ph. de Forcrand, B. Lucini and D. North, hep-lat/0510081 .

[13] D. Gross, R. D. Pisarski and L. Yaffe, *Rev. Mod. Phys.* **53**, 43 (1981).

[14] T. Bhattacharya, A.Gocksch, C. P. Korthals Altes and R. D. Pisarski, *Nucl. Phys.* **B 383**, 497 (1992); *Phys. Rev. Lett.* **66**, 998 (1991).

[15] C. P. Korthals Altes, A. Kovner and M. Stephanov, *Phys. Lett.* **B 346**, 94 (1999).

[16] C. P. Korthals Altes, H. Meyer, (2005), hep-ph/0509018.

[17] G. Boyd, J. Engels, F. Karsch, E. Laermann, C. Legeland, M. Lütgemeier and B. Petersson, *Nucl. Phys.* **B469**, 419 (1996), hep-lat/9602007; Y. Schröder and M. Laine, hep-lat/0509104; T. Umeda for the RBC-Bielefeld Collaboration, hep-lat/0610019.

# On the Noncommutativity Approach to Supersymmetry on the Lattice

Falk Bruckmann* and Mark de Kok†

*Institut für Theoretische Physik, Universität Regensburg, D-93040 Regensburg (speaker)
†Instituut-Lorentz, Universiteit Leiden, P.O. Box 9506, NL-2300 RA Leiden

**Abstract.** The noncommutativity approach to SUSY on the lattice is shown to be inconsistent and a similar inconsistency is displayed for the link approach.

Supersymmetry (SUSY) is a celebrated symmetry relating bosons and fermions which may be realized in particle physics and has many interesting theoretical features. It would be nice to be able to investigate SUSY numerically on the lattice. To this end, an exact SUSY invariance of the lattice action would be very useful (in analogy to exact lattice gauge invariance). As will be shown below, this is deeply related to keeping the Leibniz rule on the lattice, which is the aim of the noncommutativity approach.

The simplest example of a SUSY field theory is the 1D **supersymmetric quantum mechanics** (SUSYQM, see e.g. [1]). In the continuum it is described by the algebra

$$\{Q_1, Q_1\} = \{Q_2, Q_2\} = 2H, \quad \{Q_1, Q_2\} = 0 \tag{1}$$

where $H = \partial_t$ is the (only) generator of the (Euclidean) Poincaré algebra. The simplest multiplet consists of a real boson $\phi$, two Majorana fermions $\psi_{1,2}$ and an auxiliary boson $D$. The action is taken to be

$$S = \int dt \left[ \frac{1}{2}(\partial_t \phi)^2 - \frac{1}{2}D^2 - \frac{1}{2}(\psi_1 \partial_t \psi_1 + \psi_2 \partial_t \psi_2) - i(m + 3g\phi^2)\psi_1\psi_2 - D(m\phi + g\phi^3) \right] \tag{2}$$

Upon integrating out the nondynamical field $D$, mass and interaction terms for $\phi$ are generated. This action is invariant, $S(\phi, \psi_1, \ldots) = S(\phi + \delta_i\phi, \psi_1 + \delta_i\psi_1, \ldots)$, under the following variations $\delta_i = \varepsilon^i Q_i$, $i = 1, 2$ (no sum, $\varepsilon_i$ fermionic):

| $\Phi$ | $\phi$ | $\psi_1$ | $\psi_2$ | $D$ |
|---|---|---|---|---|
| $\delta_1\Phi$ | $i\varepsilon^1\psi_1$ | $i\varepsilon^1\partial_t\phi$ | $\varepsilon^1 D$ | $-\varepsilon^1\partial_t\psi_2$ |
| $\delta_2\Phi$ | $i\varepsilon^2\psi_2$ | $-\varepsilon^2 D$ | $i\varepsilon^2\partial_t\phi$ | $\varepsilon^2\partial_t\psi_1$ |

Several aspects become clearer in the superfield formalism where the fields are organized into components of a Hermitian superfield

$$\Phi(t, \theta^1, \theta^2) = \phi(t) + i\theta^1\psi_1(t) + i\theta^2\psi_2(t) + i\theta^2\theta^1 D(t) \tag{3}$$

The $\theta$'s are Grassmann coordinates and the supercharges can be represented as $Q_i = \partial_{\theta^i} + \theta^i\partial_t$ in close analogy to $H = \partial_t$. The action is constructed as

$$S = S_{\text{kin}} + S_{\text{pot}}, \quad S_{\text{kin}} = \int dt\, d^2\theta \frac{1}{2}D_2\Phi D_1\Phi, \quad S_{\text{pot}} = \int dt\, d^2\theta\, iF(\Phi), \tag{4}$$

CP892, *Quark Confinement and the Hadron Spectrum VII*
edited by J. E. F. T. Ribeiro
© 2007 American Institute of Physics 978-0-7354-0396-3/07/$23.00

with $D_i$ the superderivatives. Choosing $F(\Phi) = \frac{1}{2}m\Phi^2 + \frac{1}{4}g\Phi^4$ and integrating out the $\theta$'s gives the action in the form of Eq. (2).

Why $S$ is invariant will be demonstrated now for the kinetic term (skipping indices $i$; the potential term is invariant for the same reasons):

$$\delta S_{\text{kin}} \equiv S_{\text{kin}}[\Phi + \delta\Phi] - S_{\text{kin}}[\Phi] = \int dt\, d^2\theta\, \frac{1}{2}[D_2(\varepsilon Q\Phi)D_1\Phi + D_2\Phi D_1(\varepsilon Q\Phi)] \quad (5)$$

$$= \int dt\, d^2\theta\, \frac{1}{2}[(\varepsilon Q)D_2\Phi \cdot D_1\Phi + D_2\Phi \cdot (\varepsilon Q)D_1\Phi]$$

We used that the $D_i$ anticommute with $Q_j$ (and $\varepsilon^j$). The crucial step is the following:

$$\delta S_{\text{kin}} = \int dt\, d^2\theta\, \frac{1}{2}(\varepsilon Q)[D_2\Phi D_1\Phi] = 0, \qquad \varepsilon Q = \varepsilon\partial_\theta + \varepsilon\theta\partial_t \quad (6)$$

where we used **the Leibniz rule of $\varepsilon Q$** as a derivative operator. $S$ is invariant because of the total derivatives in $t$ and $\theta$ as integrands (just like for time translations).

The **problem on the lattice** comes from the discretizing the derivative $\partial_t$ to $\Delta_t$, the forward difference (here defined over two lattice spacings $a$; the backward difference works in an analogous way). This operator fulfills

$$\Delta_t[f(t)g(t)] = [\Delta_t f(t)]g(t) + f(t+2a)[\Delta_t g(t)] \quad (7)$$

Hence the Leibniz rule is violated on the lattice. Consequently, the naive application of continuum methods fails and the question is how to write down lattice actions, especially interacting theories, with exact SUSY invariance.

Most alternative attempts (see [2] for a review) have part of the SUSY implemented exactly at finite lattice spacing and hope (sometimes prove) to keep typical SUSY phenomena without fine-tuning problems in the continuum limit.

Now we come to the **noncommutativity approach** suggested by D'Adda et al. [3]. Its idea is to have an ordinary Leibniz rule for $\delta = \varepsilon Q$ by turning the modified Leibniz rule (7) of the lattice difference $\Delta_t$ into an ordinary Leibniz rule for $\varepsilon\theta\Delta_t$:

$$\varepsilon\theta\Delta_t[f(t)g(t)] = [\varepsilon\theta\Delta_t f(t)]g(t) + f(t)[\varepsilon\theta\Delta_t g(t)] \quad (8)$$

and keeping this rule for $\varepsilon\partial_\theta$. This can be done through the noncommutativities

$$[t,\theta^i] = a\theta^i, \quad [t,\partial_{\theta^i}] = -a\partial_{\theta^i}, \quad [t,\varepsilon^i] = a\varepsilon^i \quad (9)$$

As a consequence of the nc approach specific shifts appear in the action (by bringing $\theta$'s to the left to be integrated out). For example the mass terms in SUSYQM read [4]:

$$S_m = a\sum_t -m[i\psi_1(t+a)\psi_2(t) - i\psi_2(t+a)\psi_1(t) + D(t)\{\phi(t) + \phi(t+2a)\}/2] \quad (10)$$

This action is supposed to be invariant under *all* variations $\delta_i = \varepsilon^i Q_i$ (see the table on the first page) when using the noncommutativity of $t$ and $\varepsilon^i$, Eq. (9), in products of fields. The authors of [3] claim the nc approach works for theories with $N = D = 2$ and $N = D = 4$, whereas above we have applied it to the simplest case, SUSYQM [4].

However, there is an **inconsistency** in the nc approach: the SUSY variations of a product of fields depends on their order [4]. On the one hand, two fields are varied as

$$fg \rightarrow (f + \varepsilon Qf)(g + \varepsilon Qg) = fg + \varepsilon(Qf(t) \cdot g(t) + f(t+a) \cdot Qg(t)) \tag{11}$$

Interchanging the order (for simplicity restricting to bosonic fields $f$ and $g$) gives

$$gf \rightarrow (g + \varepsilon Qg)(f + \varepsilon Qf) = gf + \varepsilon(Qg(t) \cdot f(t) + g(t+a) \cdot Qf(t))$$
$$= fg + \varepsilon(Qf(t) \cdot g(t+a) + f(t) \cdot Qg(t)) \tag{12}$$

These variations, Eq.s (11) and (12), do not agree due to the different shifts, but they have to as they come from the product of two commuting fields!

As a consequence, when checking the invariance of the action, the expression $D(t)\phi(t) - i\psi_2(t+a)\psi_1(t)$ (see (10)) gives a total derivative under the variations $\varepsilon^1 Q_1$ (just like in the continuum), but $\phi(t)D(t) - i\psi_2(t+a)\psi_1(t)$ does not: because of the different shifts cancellations have been destroyed. Hence the two equivalent forms of this term give different answers concerning the SUSY invariance of it!

In [4] we have shown this inconsistency to be a generic feature of the noncommutativity approach. The rationale of this inconsistency lies in the fact that the noncommutativity $[t, \varepsilon] \neq 0$ forbids to treat $t$ as a number. Correspondingly, the component fields cannot be ordinary functions and this theory cannot be simulated numerically.

We would like to add a related finding. The **link approach** to lattice SUSY, emanating from the nc approach and proposed by the same authors in [5], suffers from **a similar inconsistency**. In [5] the SUSY transformations $s_A$ are defined via the (anti)commutator with a fermionic link $\nabla_A$. This link nature gives $s_A$ a modified Leibniz rule, much like in Eq. (11). For example, the trace over two bosonic site variables $f$ and $g$ transforms as

$$s_A[\text{tr}\, f_{x,x} g_{x,x}] = \text{tr}\,(s_A f)_{x+a_A,x} g_{x,x} + \text{tr}\, f_{x+a_A,x+a_A} (s_A g)_{x+a_A,x} \tag{13}$$

In order to arrive at the analogue of Eq. (12) we consider the inverse order

$$s_A[\text{tr}\, g_{x,x} f_{x,x}] = \text{tr}\,(s_A g)_{x+a_A,x} f_{x,x} + \text{tr}\, g_{x+a_A,x+a_A} (s_A f)_{x+a_A,x} \tag{14}$$

Because of the traces, the (matrix-valued) fields on the l.h.s.s can be interchanged. But the results on the r.h.s.s differ by shifts again! As will be shown in a forthcoming publication [6], this ambiguity also plaques the action of the link approach, which therefore is inconsistent, too.

FB likes to thank the organizers for a very nice conference. We are grateful to Simon Catterall, Joel Giedt, David Kaplan and the Jena group for helpful discussions.

## REFERENCES

1. F. Cooper, and B. Freedman, *Ann. Phys.* **146**, 262 (1983).
2. J. Giedt, *PoS* **LAT2006**, 008 (2006).
3. A. D'Adda, I. Kanamori, N. Kawamoto, and K. Nagata, *Nucl. Phys.* **B707**, 100–144 (2005).
4. F. Bruckmann, and M. de Kok, *Phys. Rev.* **D73**, 074511 (2006).
5. A. D'Adda, I. Kanamori, N. Kawamoto, and K. Nagata, *Phys. Lett.* **B633**, 645–652 (2006).
6. F. Bruckmann, S. Catterall, and M. de Kok, in preparation.

# Testing the QCD string at large $N_c$ from the thermodynamics of the hadronic phase

Thomas D. Cohen

*Department of Physics*
*University of Maryland*
*College Park, Maryland 2042*
*USA*

**Abstract.** It is generally believed that in the limit of a large number of colors ($N_c$) the description of confinement via flux tubes becomes valid and QCD can be modeled accurately via a hadronic string theory–at least for highly excited states. QCD at large $N_c$ also has a well-defined deconfinement transition at a temperature $T_c$. In this talk it is shown how the thermodyanmics of the metastable hadronic phase of QCD (above $T_c$) at large $N_C$ can be related directly to properties of the effective QCD string. The key points in the derivation is the weakly interacting nature of hadrons at large $N_c$ and the existence of a Hagedorn temperature $T_H$ for the effective string theory. From this it can be seen at large $N_c$ and near $T_H$, the energy density and pressure of the hadronic phase scale as $\mathscr{E} \sim (T_H - T)^{-(D_\perp - 6)/2}$ (for $D_\perp < 6$) and $P \sim (T_H - T)^{-(D_\perp - 4)/2}$ (for $D_\perp < 4$) where $D_\perp$ is the effective number of transverse dimensions of the string theory. This behavior for $D_\perp < 6$ is qualitatively different from typical models in statistical mechanics and if observed on the lattice would provide a direct test of the stringy nature of large $N_c$ QCD. However since it can be seen that $T_H > T_c$ this behavior is of relevance only to the metastable phase. The prospect of using this result to extract $D_\perp$ via lattice simulations of the metastable hadronic phase at moderately large $N_c$ is discussed.

**Keywords:** Large $N_c$ QCD, QCD String
**PACS:**

## BACKGROUND

This talk explores what thermodynamics of the large $N_c$ limit of QCD can teach us about the nature of confinement, particularly in the context of the QCD string. The ultimate goal here is to establish how practical lattice calculations may be able to shed light on the QCD string. Underlying this approach are some basic features of the thermodynamics of QCD at large $N_c$. These basic features were recognized long ago by Charles Thorn[1]. In this talk, these basic ideas will be used to gain insight on the QCD string and on large $N_c$ continuum reduction. The talk is based largely on work published in refs. [2] and [3].

The key physical inputs are that couplings between light hadrons (those with mass of order $N_c^0$)—*i.e.* the mesons and glueballs—become weak at large $N_c$ while the baryon masses scale as $N_c^1$ and hence diverge at large $N_c$ [4]. Thus the couplings between light hadrons go to zero as some power of $1/N_c$ and the states become narrow in the large $N_c$ limit. Consider the thermodynamics of large $N_c$ QCD in the hadronic phase and temperatures in the regime $T \sim N_c^0$, in light of these scaling rules. Baryons are thermally suppressed in this regime by an exponential factor because they are heavy. Thus, the system will look like a gas of mesons and glueballs which interact weakly (and, indeed

CP892, *Quark Confinement and the Hadron Spectrum VII*
edited by J. E. F. T. Ribeiro

are noninteracting in the large $N_c$ limit). In this limit, the thermodynamic quantities, such as energy density and pressure, are given by a free hadron gas model:

$$\mathscr{E}(T) = \int_0^\infty dm \rho(m) \varepsilon(m,T) + \mathscr{O}(1/N_c^2)$$

$$\mathscr{P}(T) = \int_0^\infty dm \rho(m) P(m,T) + \mathscr{O}(1/N_c^2)$$

$$\text{where } \varepsilon(m_k,T) \equiv \int \frac{d^3 p}{(2\pi)^3} \frac{\sqrt{p^2 + m_k^2}}{e^{\sqrt{p^2 + m_k^2}/T} - 1} \rightarrow \frac{T^{3/2} m_k^{5/2} e^{-m_k/T}}{(2\pi)^{3/2}} = m_k n_k,$$

$$P(m_k,T) \equiv \left(\frac{m_k^2 T^2}{2\pi^2}\right) \sum_{n=1}^{\infty} K_2\left(\frac{nm_k}{T}\right) \rightarrow \frac{m_k^{3/2} T^{5/2}}{2^{3/2} \pi^{3/2}} e^{-m_k/T} \tag{1}$$

where $\rho(m)$ is the density of hadrons and the arrow indicates the asymptotic behavior for large $m_k$.

One immediate consequence of this is that at Large $N_c$ QCD must have a phase transition. At high temperatures, asymptotic freedom implies the system will look like a gas of noninteracting gluons whose energy density scales as $N_c^2$ while in the hadronic phase, it has an energy density which scales as $N_c^0$. The mismatch implies a phase transition of some sort must occur.

A second consequence is that many relevant physical intensive quantities—those which diverge at large $N_c$—are completely insensitive to temperature in the hadronic phase at large $N_c$. This is important since typical quantities of interest, such as the chiral condensate ($\sim N_c^1$) or the gluon condensate ($\sim N_c^2$), are in this set. This behavior is easy to understand: in a hadronic gas, the change of such a quantity is the spatially integrated change per hadron times the density of hadrons. Simple $N_c$ counting gives both of order $N_c^0$ which is an infinitesimal change. This fact has profound consequences. It clarifies the physical content[2] of large $N_c$ continuum reduction[5]. This subject will not be discussed further here due to space limitations.

## THE QCD STRING

The idea of strings for the strong interaction is quite old, predating QCD, and has had some real phenomenological and theoretical success. The modern view is that the picture emerges from QCD as flux tubes for highly excited states: open strings correspond to mesons (with a quark and antiquark on the ends) while glueballs correspond to closed strings. It should be apparent that the string picture of QCD is only clean at large $N_c$. Clearly string breaking destroys the simple picture, but it is suppressed by $1/N_c$. Phenomenological predictions of string theory, such as the Regge trajectories or the Hagedorn spectrum also depend on large $N_c$. These are predictions about the masses of excited hadrons and as such are only well defined to the extent that the masses are unambiguous. However, hadron masses *are* ambiguous due to decay widths. At large $N_c$ the widths go to zero and the ambiguity vanishes.

While it is easy to verify in lattice simulations that QCD—or at least its pure gauge version relevant at large $N_c$– has a static flux tube as one would expect in a string picture;

it is much harder to test the guts of the QCD string which concerns the excitations of highly excited states. However, there is a practical way to use the thermodynamics of the metastable hadronic phase of large $N_c$ QCD to test an essential aspect of the string dynamics—the Hagedorn spectrum[6]. In a string theory the density of meson and glueball states is given asymptotically by[7]

$$\rho(m) \to \rho_{asy}(m) = A\,(m/T_H)^{-(D_\perp+3)/2}\,e^{m/T_H} \tag{2}$$

where $T_H$ is the Hagedorn temperature, $A$ is a constant and $D_\perp$ is the number of transverse dimensions of the string ($D_\perp = 2$ in a simple Nambu-Goto string).
>From the Hagedorn spectrum of Eq. (2) and the thermodynamics of a hadronic gas of Eq. (1)

$$\mathcal{E} = (2\pi)^{-3/2} A\, T^{(9-D_\perp)/2}\, T_H^{7/2}\,(T_H-T)^{(D_\perp-6)/2}\, \Gamma\left(\frac{6-D_\perp}{2}\right) \quad (D_\perp < 6)$$

$$P = (2\pi)^{-3/2} A\, T^{(9-D_\perp)/2}\, T_H^{5/2}\,(T_H-T)^{(D_\perp-4)/2}\, \Gamma\left(\frac{4-D_\perp}{2}\right) \quad (D_\perp < 4) \tag{3}$$

as $T - T_H$ in the hadronic phase of large $N_c$ QCD. Note that for $D_\perp < 4$, both the energy density and pressure diverge. For $D_\perp = 2$ as one would have in a simple string picture:

$$\mathcal{E} \sim \frac{1}{(T_H-T)^2} \qquad P \sim \frac{1}{(T_H-T)}\ . \tag{4}$$

This means that if this picture is correct, $T_H$ is an unobtainable temperature for the hadronic phase of QCD. This is is a concrete realization of Hagedorn's old idea of an unreachable hadronic temperature. Note, however, this scenario is only well posed at large $N_c$.

Clearly numerical evidence of the diverging energy density and pressure as $T_H$ for lattice simulations of QCD with many colors (extrapolated to the infinite color limit) would be strong evidence of the Hagedorn spectrum and, hence, of the stringy dynamics. Note that the behavior as $T \to T_H$ is characteristically different from what one has in conventional statistical mechanics models and would be quite compelling. Conventional models have a specific heat which diverges a critical point as $(T_c - T)^\alpha$, with the critical exponent $\alpha \leq 1$. In contrast, the string picture with $D_\perp = 2$, $\alpha = 3$.

However, as one increases the temperature starting from the low temperature hadronic phase, one expects that the system undergoes a first order phase transition before reaching the regime near $T_H$. The fact that $T_H$ is unreachable in the hadronic phase at large $N_c$ says that something must happen before one gets to $T_H$. The specific heat in such a transition goes as $N_c^2$. Lattice studies by the Oxford group for large but finite $N_c$ confirm this scenario[8].

Given the existence of the phase transition, can one observe the Hagedorn behavior in the thermodynamics to verify the stringy behavior? The previous arguments are irrelevant if this is not the case. Fortunately, it is possible with sufficient numerical resources. The fact that the QCD transition at large $N_c$ is first order implies that a metastable phase above the phase transition exists. In fact, if the string theory is correct the metastable phase is highly unusual. A metastable phase is one which is locally stable but globally unstable. Typically there is an upper bound for the temperature for such a phase: the spinodal point where the phase becomes locally unstable. Near a spinodal point, thermodynamic quantities behave essentially as they do near usual second order phase transitions. In contrast, if a QCD string is correct the metastable hadronic phase at large $N_c$ does not have a spinodal point—rather it has a temperature which is dynamically unreachable due to a divergent energy.

Generically it is possible to do lattice simulations for a metastable phase. Indeed, lattice studies have been done for QCD with large but finite $N_c$ in the metastable hadronic phase[9]. While, the lattice studies to date have focused on the string tension and not on thermodynamics, nothing in principle prevents thermodynamic studies. As discussed in ref. [3], care must be taken in extrapolations to large $N_c$ and $T_H$ since the limits $T \rightarrow T_H$ and $N_c \rightarrow \infty$ do not commute. If such simulations do demonstrate the Hagedorn behavior, it will be the first direct evidence for the dynamics of a QCD string.

Finally, it should be noted that the lattice studies at large $N_c$ of the string tension in the hadronic phase fit the data quite well by assuming that $T_H$ *is* a spinodal point and the string tension had a critical exponent associated with a standard statistical mechanics model[9]. At blush, that is inconsistent with the picture above where no spinodal point exists and might seem to be evidence *against* the QCD string. However, this is not the case. The string tension is *not* an intensive thermodynamical observable associated with a local operator. Thus, there is no thermodynamic argument to prevent it from having some power law behavior reminiscent of critical behavior near a spinodal point. The key question is not the behavior of the string tension but rather of the e thermodynamic observables.

## ACKNOWLEDGMENTS

The work reported in this talk was supported in part by the US Department of Energy.

## REFERENCES

1. C.B.Thorn, *Phys. Lett.* **B99**,458 (1981) .
2. T.D. Cohen, *Phys. Rev. Lett.* **93**, 201601 (2004).
3. T.D. Cohen, *Phys. Lett.* **B637** 81 (2006).
4. G. t'Hooft, Nucl. Phys. **B72** (1974) 461; E. Witten, Nucl. Phys. **B160** (1979) 57.
5. J. Kiskis, R. Narayanan and H. Neuberger, Phys. Rev. **D66**, 025019 (2002); *Phys. Lett.* **B574**, 65 (2003); R. Narayanan and H. Neuberger, *Phys. Rev. Lett.* **91**, 081601 (2003).
6. R. Hagedorn, *Nuo. Cimen.* **56A**, 1027. (1968).
7. K.R. Dienes and J.-R. Cudell, *Phys. Rev. Lett* **72**, 187 (1994).
8. B. Lucini, M. Teper and U. Wenger, *JHEP* **0502**, 033 (2005).
9. B. Bringoltz and M. Teper, *Phys. Rev.* **D73**, 014517 (2006).

# Dual superconductivity and typology of the QCD vacuum

A. D'Alessandro[*,†], M. D'Elia[†] and L. Tagliacozzo[**]

[*]Speaker at the conference
[†]Dipartimento di Fisica, Università di Genova and INFN, I-16146 Genova, Italy
[**]Departament ECM, Universitat de Barcelona, 08028 Barcelona, Spain.

**Abstract.** Within the dual superconductor model for the QCD vacuum we compare the field penetration depth $\lambda$ with the correlation length $\xi$ of the Higgs condensate. The comparison places the vacuum marginally on the type II side. A qualitative check of the interaction between two parallel flux tubes gives also consistent indications for a weak repulsive behaviour.

**Keywords:** dual superconductor, QCD vacuum, color confinement
**PACS:** 11.15.Ha, 12.38.Aw

## Introduction

The dual superconductivity model of color confinement [1] assimilates the properties of the QCD vacuum to those of a superconductor where the roles of the electric and magnetic field are exchanged: the chromoelectric field of a $q\bar{q}$ couple is compelled into narrow flux tubes by the dual Meissner effect yielding a linearly rising potential.

In the framework of this model we may ask whether the QCD vacuum is a type-I or a type-II superconductor. Superconductors are classified into these two categories according to whether the penetration lenght $\lambda$ of an external field is less (type I) or greater (type II) than the correlation length $\xi$ of the Higgs condensate.

The determination of the typology of the QCD vacuum is done first from a direct numerical analysis of the two parameters $\lambda$ and $\xi$, the latter obtained from the observation of the temporal correlator of a magnetically charged operator, and then from a qualitative study of the behavior of parallel flux tubes. We study the $SU(2)$ pure gauge case.

## Numerical determination of $\lambda$

The penetration length $\lambda$ can be determined directly from its definition by observing the shape of the flux tube generated by a static quark-antiquark couple. The $q\bar{q}$ charges are created at time 0 and destroyed at time $T$ with a Wilson loop $W$ while the plaquette $\Pi_{0i}$ is used as a probe to detect $F_{0i}$ at a certain site: actually we observe

$$E_i = \frac{< \operatorname{tr}\left(W^{AbPr}\, \Pi_{0i}^{AbPr}\right) >}{< \operatorname{tr}\left(W^{AbPr}\right) >} - \frac{< \operatorname{tr}\left(W^{AbPr}\right) \operatorname{tr}\Pi_{0i}^{AbPr} >}{2 < \operatorname{tr}\left(W^{AbPr}\right) >}$$

where the superscript $^{AbPr}$ reminds us that all the quantities are projected [2] onto $U(1)$.

CP892, *Quark Confinement and the Hadron Spectrum VII*
edited by J. E. F. T. Ribeiro
© 2007 American Institute of Physics 978-0-7354-0396-3/07/$23.00

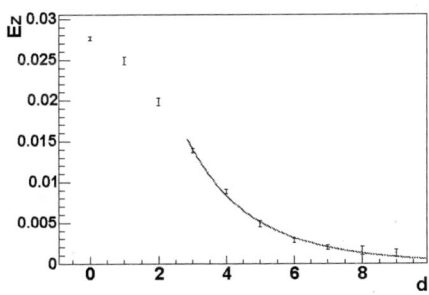 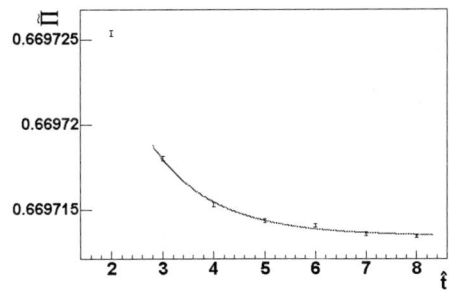

**FIGURE 1.** $E_z(d)$ fitted with $K_0(\frac{d}{\lambda})$ (left) and mean modified plaquette $\hat{\Pi}$ (right) fitted according to (1).

According to London equations the transversal shape of the flux tube is proportional to $K_0(\frac{d}{\lambda})$ where $d$ is the distance from the axis and that is the functional form we use for the fit (fig.1 left). At $\beta = 2.5115$ a check against finite size effects is provided by comparing the value for a $12^3 \times 20$ lattice ($\hat{\lambda} = 1.24(21)$) with the one for a $16^3 \times 20$ lattice ($\hat{\lambda} = 1.28(25)$): no evidence of finite size is seen. At $\beta = 2.6$ we obtain $\hat{\lambda} = 2.58(12)$ in agreement with previous literature [5].

## Numerical determination of $\xi$

We measure $\xi$ from the temporal correlator of the magnetically charged operator $\mu$ developed by the Pisa group [3]: this operator shifts an eigenstate of the vector potential operator $\vec{A}$ by the field $\vec{b}_\perp$ whose rotor gives a coulombian magnetic field. As the temporal correlator $\langle \bar{\mu}(t,\vec{x})\mu(0,\vec{x})\rangle$ itself is a very noisy quantity we are looking for

$$\rho(\hat{t}) = \frac{d}{d\beta}\ln\langle\bar{\mu}\mu\rangle = \langle S\rangle_S - \langle\tilde{S}(0,\hat{t})\rangle_{\tilde{S}(0,\hat{t})} \overset{\hat{t}\to\infty}{\simeq} \frac{d\ln <\mu>^2}{d\beta} + A\frac{e^{-\hat{t}/\hat{\xi}}}{\hat{t}^{1/2}} \qquad (1)$$

$$\rho'(\hat{t}) = \frac{d}{d\hat{t}}\ln\langle\bar{\mu}\mu\rangle \simeq -\frac{\beta}{2}\langle\tilde{S}(0,\hat{t}+1) - \tilde{S}(0,\hat{t}-1)\rangle_{\tilde{S}(0,\hat{t})} \overset{\hat{t}\to\infty}{\simeq} A'\frac{e^{-\hat{t}/\hat{\xi}}}{\hat{t}^{3/2}} \qquad (2)$$

where $\tilde{S}(\hat{t}_1,\hat{t}_2)$ is the action with a monopole inserted at time $\hat{t}_1$ and an antimonopole inserted at time $\hat{t}_2$ [3, 4]. At $\beta = 2.6$ we extract $\hat{\xi} = 1.52(18)$ from $\rho$ and the compatible $\hat{\xi} = 1.32(25)$ from $\rho'$. The value for $\hat{\xi}$ is just below (fig.2 left) $\hat{\lambda}(\beta = 2.6) = 2.58(12)$: this gives indications for weak type-II superconductivity.

## Qualitative behavior of two parallel tubes

We know that two parallel flux tubes attract in a type-I superconductor while they repel in a type-II: by putting two Wilson loops aside and observing their repulsive or

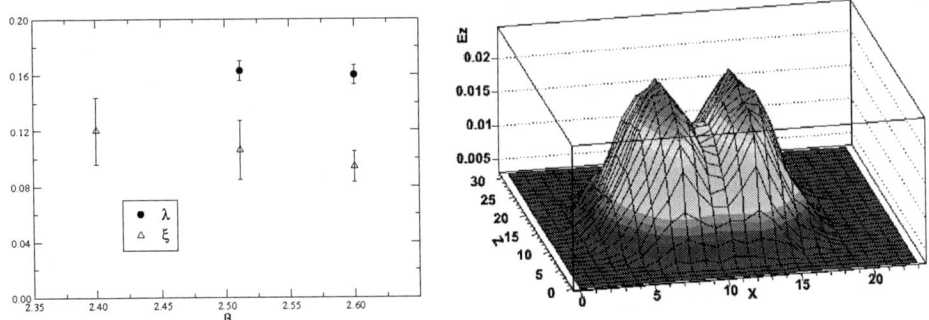

**FIGURE 2.** $\xi$ and $\lambda$ in fm (left) and qualitative behavior of two parallel flux tubes (right).

attractive behavior we may have another (qualitative) indication. We observe

$$E_i = \frac{< \mathrm{tr}\left(W_1^{AbPr} W_2^{AbPr} \Pi_{0i}^{AbPr}\right) >}{< \mathrm{tr}\left(W_1^{AbPr} W_2^{AbPr}\right) >} - \frac{< \mathrm{tr}\left(W_1^{AbPr} W_2^{AbPr}\right) \mathrm{tr}\Pi_{0i}^{AbPr} >}{2 < \mathrm{tr}\left(W_1^{AbPr} W_2^{AbPr}\right) >}$$

No visible sign of attraction or repulsion is seen (fig.2 right) but from a numerical analysis we notice that two tubes whose axes are at distance $D = 4$ repel more (mean deviation=0.44(7) lattice spacings at $\beta = 2.6$) than two tubes at distance $D = 5$ (mean deviation=0.22(14)). This is a sign of repulsive behavior.

## Conclusions

By a numerical analysis of the penetration lenght $\lambda$ and the correlation lenght $\xi$ we have classified the QCD vacuum as a weak type-II superconductor. This evidence has been confirmed by a qualitative analysis of the behavior of two parallel tubes; closer tubes provide a stronger repulsion (type-II behavior) but anyway weak and not clearly visible. A complete account of our results can be found in [6].

## REFERENCES

1. G. 't Hooft, in "High Energy Physics", EPS International Conference, Palermo 1975, ed. A. Zichichi; S. Mandelstam, *Phys. Rept.* **23**, 245 (1976).
2. G. 't Hooft, *Nucl. Phys.* **B190**, 455 (1981).
3. A. Di Giacomo, L. Del Debbio, G. Paffuti, *Phys. Lett.* **B 349**, 513 (1995); A. Di Giacomo and G. Paffuti, *Phys. Rev. D* **56**, 6816 (1997).
4. L. Tagliacozzo, *arXiv:hep-lat/0603022*.
5. G. S. Bali, C. Schlichter, K. Schilling, *Prog. Theor. Phys. Suppl.* **131**, 645 (1998); R. W. Haymaker and T. Matsuki, *arXiv:hep-lat/0505019*.
6. A. D'Alessandro, M. D'Elia, L. Tagliacozzo *hep-lat/0607014* (2006)

# $SU(4)$ potentials with two vortices

Sedigheh Deldar* and Shahnoosh Rafibakhsh†

*Department of Physics, University of Tehran, P.O. Box 14395/547, Tehran 1439955961, Iran [1]
†Department of Physics, University of Tehran, P.O. Box 14395/547, Tehran 1439955961, Iran

**Abstract.** There are three vortices for $SU(4)$ gauge group where two of them are independent. Using both of these vortices in calculating the induced potentials between static sources, we show that the second vortex only affects the potentials at large distances and does not have a significant effect on the intermediate distance potentials and their slopes. The ratio of the probabilities of piercing a plaquette by the two type of vortices may be fixed by the ratio of diquark string tension to the string tension of the quarks in the fundamental representation.

Thick center vortices model [1] is one of the phenomenological models which has been fairly successful in explaining the linear behavior of the potential between the static sources in various gauge groups [1], [2], [3]. Based on this model, the static sources are confined as the result of interaction between the Wilson loops and the vortices which are topological line-like or space-like field configurations. Each $SU(N)$ gauge group has $(N-1)$ types of vortices corresponding to the non-trivial center elements. However, vortices of type $n$ and $N-n$ are the same, except that their magnetic fluxes are in the opposite directions and therefore are not independent. In addition, there exist $int(\frac{N}{2})$ independent asymptotic string tensions for each $SU(N)$ gauge group. Therefore, studying $SU(N > 3)$ gauge groups which have more than one asymptotic string tension is interesting. The values of these string tensions will constrain the details of the confinement models.

In our previous calculation [3], we have studied the potentials between static sources by the thick center vortices model, using only the first non-trivial vortex of the $SU(4)$ gauge group. As expected, for $SU(4)$ gauge group, two asymptotic string tensions have been obtained. However, to get the lattice results [5], [4], one has to use both types of vortices. In this article, we report the preliminary results of our new calculations using both vortices of the $SU(4)$ gauge group.

The induced potential from the thick center vortices model which gives the intermediate and asymptotic $SU(4)$ string tensions is:

$$V(R) = \sum_{m=-\infty}^{m=\infty} \ln\left\{1 - 2f_1\left(1 - \mathrm{Re}\,\mathscr{G}_r\left[\vec{\alpha}_C^1(x)\right]\right) - f_2\left(1 - \mathrm{Re}\,\mathscr{G}_r\left[\vec{\alpha}_C^2(x)\right]\right)\right\}. \quad (1)$$

---

[1] Talk presented by S. Deldar

CP892, *Quark Confinement and the Hadron Spectrum VII*
edited by J. E. F. T. Ribeiro
© 2007 American Institute of Physics 978-0-7354-0396-3/07/$23.00

Where $f_1$ and $f_2$ are the probabilities of piercing the plaquettes by vortex number one and two, respectively. $\mathscr{G}_r$ is defined as:

$$\mathscr{G}_r[\vec{\alpha}] = \frac{1}{d_r}\mathrm{Trexp}[i\vec{\alpha}.\vec{H}] \tag{2}$$

Where $d_r$ is the dimension of the representation and $\{H_i, i = 1,2,3\}$ are the generators spanning the Cartan subalgebra. The parameter $\alpha_c(x)$ describes the flux distribution of the vortex. It depends on that fraction of the vortex core which is enclosed by the Wilson loop and thus depends on the position of the center of the vortex and the shape of the loop. We have used the vortex profile of reference [1] for both of the vortex types.

From the model, zero 4-ality representations are supposed to be screened and all representations with the same class of 4-ality should get the same slope at large distances. The $SU(4)$ asymptotic string tensions are obtained to be [6]: $\sigma_{fund.} = 2f_1 + 2f_2$, $\sigma_6 = 4f_1$ and $\sigma_{adj.} = 0$. Hence, the ratio of the asymptotic string tension of representations with 4-ality=2 to the string tension of the fundamental representation with 4-ality=1 depends on the values of $f_1$ and $f_2$:

$$\frac{\sigma_6}{\sigma_f} = \frac{2f_1}{f_1 + f_2} \tag{3}$$

On the other hand, asymptotic string tensions may be found from lattice calculations. We have used the data of B. Luicini [4] and S. Ohta [5] for the ratio of the diquark string tension to that of the fundamental one. Using these lattice data and equation (3), one may find the ratio of $\frac{f_2}{f_1}$. Fixing $f_1$ to an arbitrary value which leads to the physical results and reasonable potentials in agreement with lattice data, $f_2$ may be determined. We have chosen $f_1 = 0.1$, then $f_2 = 0.045$ and $0.054$ based on the data of Lucini [4] and Ohta [5], respectively. Table 1 shows the ratio of intermediate string tensions of different representations to that of the fundamental representation for fixed $f_1$ and different $f_2$'s. The last column shows the ratios of the diquark asymptotic string tension to the string tension of the fundamental representation, determined from equation (3). As the table shows, the intermediate string tensions are not that much sensitive to the value of $f_2$. However, to obtain the asymptotic string tensions which are in agreement with lattice calculations, one has to use the second vortex of the gauge group. The last two rows of the table represents the Casimir ratios and the number of fundamental fluxes in each representation which are qualitatively in agreement with the intermediate string tensions ratios.

Using only vortex type two in the calculations, one gets a screened potential at large distances (figure 1). This is because representation 6 is constructed of two quarks and at large distances where the Wilson loop is large enough to contain the vortex core, each Wilson loop multiplies by the square of the second center elements, $Z_2^2$, which is equal to one and the vortex does not have any effect on the Wilson loop. Thus, no confinement is supposed and the potential is screened. The ratio of $\frac{f_1^2}{f_2}$ would give some information about the interaction of vortices. Using lattice data in equation (3) gives $f_2 > f_1^2$. This is suggesting that there is an interaction between the vortex fluxes type one which are constructing vortex of type two. For non-interacting vortices, one would expect naively to have $f_2 = f_1^2$. If there is a repulsion force between vortices, one would expect to

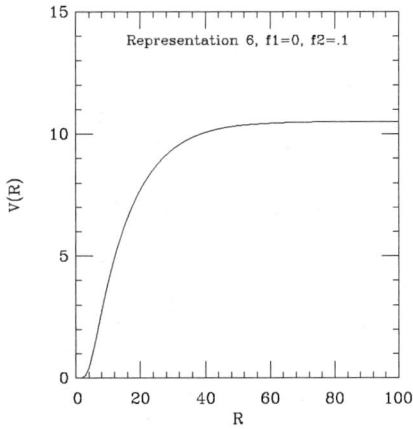

**FIGURE 1.** Potentials between static sources in representation 6 with $f_1 = 0.0$ and $f_2 = 0.1$. The potential is screened at large distance. However, to obtain the correct potential, one has to use both vortices.

**TABLE 1.** Using different $f_1$ and $f_2$, the ratio of string tension of each representation to the fundamental representation has been calculated. Number of fundamental fluxes of each representation and Casimir ratios are also shown in the table. At intermediate region, a qualitative agreement with Casimir scaling and flux tube counting is observed for all values of $f_1$ and $f_2$ we have used.

| ratios | $\frac{\sigma_6}{\sigma_f}$ | $\frac{\sigma_{15}}{\sigma_f}$ | $\frac{\sigma_{10}}{\sigma_f}$ | $\frac{\sigma_{20}}{\sigma_f}$ | $\frac{\sigma_{35}}{\sigma_f}$ | $\frac{\sigma_6}{\sigma_f}$ (asymptotic) |
|---|---|---|---|---|---|---|
| $f_2 = 0, f_1 = .1$ | 1.51 | 1.56 | 1.76 | 2.31 | 2.66 | 2 |
| $f_2 = 0.045, f_1 = .1$ | 1.61 | 1.57 | 1.86 | 2.53 | 3.20 | 1.377 |
| $f_2 = 0.054, f_1 = .1$ | 1.62 | 1.56 | 1.84 | 2.50 | 3.15 | 1.3 |
| $f_2 = .1\, f_1 = .1$ | 1.68 | 1.51 | 1.77 | 2.41 | 2.93 | 1 |
| Casimir ratio | 1.33 | 2.13 | 2.4 | 4.2 | 6.4 | |
| no. of fund. fluxes | 2 | 2 | 2 | 3 | 4 | |

get $f_2 < f_1^2$. We would like to thank the research council of the University of Tehran for supporting this work. We wish to thank M. Faber, J. Greensite, M. Engelhardt, M. Shifman, E.M. Ilgenfritz and M. Polikarpov for the nice discussions. Many thanks to the organizers for making such an enjoyable scientific atmosphere.

## REFERENCES

1. M. Faber, J. Greensite, S. Olejník, Phys. Rev. D57, (1998) 2603.
2. S. Deldar, JHEP 0101, (2001) 013.
3. S. Deldar, S. Rafibakhsh, Eur. Phys. J. C42, (2005) 319.
4. B. Lucini, M. Teper, Phys. Lett. B501, (2001)128.
5. S. Ohta, M. Wingate, Nucl. Phys. 73, (Proc. Suppl.), (1999) 435.
6. S. Deldar, S. Rafibakhsh, To be published.

# Confinement and center vortex dynamics in different gauge groups

## M. Engelhardt and B. Sperisen

*Department of Physics, New Mexico State University, Las Cruces, NM 88003, USA*

**Abstract.** The random vortex world-surface model is extended to the gauge groups $SU(4)$ and $Sp(2)$. Compared to the $SU(2)$ and $SU(3)$ models studied previously, which reproduce the infrared properties of the corresponding Yang-Mills theories on the basis of a simple vortex world-surface curvature action, new dynamical characteristics become important. In the $SU(4)$ case, an explicit dependence of the vortex effective action on the configuration of the Abelian magnetic monopoles residing on the vortices emerges; in the $Sp(2)$ case, a new "stickiness" contribution to the vortex action serves to drive the deconfinement phase transition towards the correct first-order behavior.

**Keywords:** Center vortices, infrared effective theory, confinement
**PACS:** 12.38.Aw, 12.38.Mh, 12.40.-y

## INTRODUCTION

The random vortex world-surface model is a concrete realization of the center vortex picture of the strong interaction vacuum [1–7], i.e., the notion that the relevant infrared gluonic degrees of freedom of the strong interaction are closed tubes of quantized chromomagnetic flux. The random vortex world-surface model was initially investigated for $SU(2)$ Yang-Mills theory [8–10], and in this simplest case, the main characteristics of the strongly interacting vacuum were reproduced. Both a confining low-temperature phase as well as a deconfined high-temperature phase are found [8], separated by a second-order deconfinement phase transition; furthermore, the topological susceptibility [9, 11–13] and the (quenched) chiral condensate [10] match the ones extracted from $SU(2)$ lattice Yang-Mills theory quantitatively. Extending the investigation to the $SU(3)$ gauge group [14–16], the deconfinement phase transition exhibits weakly first-order behavior [14], and a Y-law for the baryonic static potential is found [15], again matching the corresponding characteristics of $SU(3)$ lattice Yang-Mills theory. Studies of the topological and chiral properties in the $SU(3)$ case are pending.

The aforementioned successes of the $SU(2)$ and $SU(3)$ random vortex world-surface models are obtained on the basis of very simple vortex dynamics, with the action determined purely by world-surface curvature. Accordingly, these models only contain one dimensionless coupling parameter, which is adjusted in practice to reproduce the ratio of the deconfinement temperature $T_c$ to the square root of the zero-temperature string tension $\sigma$ found in the corresponding Yang-Mills theory. Recent efforts have focused on the question of how far this simple picture carries as the gauge group is varied. There are two systematic ways of extending the Yang-Mills gauge group beyond the cases discussed above: On the one hand, one may increase the number of colors $N$ determining the $SU(N)$ gauge symmetry; on the other hand [17–19], the $SU(2)$ group

CP892, *Quark Confinement and the Hadron Spectrum VII*
edited by J. E. F. T. Ribeiro

can alternatively be considered as the smallest symplectic group $Sp(1)$, and the $Sp(N)$ sequence can also be used to generalize $SU(2) = Sp(1)$. Accordingly, random vortex world-surface models for the infrared sectors of both $SU(4)$ and $Sp(2)$ Yang-Mills theory have been constructed [20, 21] and are presented in the following. In both cases, new dynamical characteristics emerge.

The concrete modeling methodology used in the random vortex world-surface model is discussed in detail in [8, 14, 20]. Vortex world-surfaces are composed of elementary squares on a hypercubic space-time lattice. The lattice spacing is a fixed physical quantity related to the transverse thickness of vortices; it represents the ultraviolet cut-off inherent in any infrared effective description. An ensemble of closed vortex world-surfaces is generated by Monte Carlo update. For different underlying gauge groups, random vortex world-surface models differ in two respects: On the one hand, the quantization of vortex flux is determined by the center of the gauge group. When encircling a vortex, a Wilson loop acquires a phase given by one of the nontrivial center elements. Accordingly, in general, several species of center vortices, corresponding to the different nontrivial center elements, can exist. They can merge and disassociate into one another. On the other hand, the models can differ in the effective action governing the vortices.

## $SU(4)$ **VORTEX MODEL**

The $SU(4)$ group contains the nontrivial center elements $\{i, -i, -1\}$. Vortex fluxes associated with the elements $i$ and $-i$ are related by an inversion of space-time orientation; therefore, there are altogether only two physically distinct types of center vortices. A vortex generating a phase factor $-1$ when linked to a Wilson loop can branch into two vortices associated with a phase factor $\pm i$ and vice versa.

Correspondingly, $SU(4)$ Yang-Mills theory [22–24] also induces two distinct string tensions, the quark string tension $\sigma_1$ and the diquark string tension $\sigma_2$. The $SU(4)$ Yang-Mills confinement properties thus are characterized by the ratios $\sigma_2/\sigma_1$ and $T_c/\sqrt{\sigma_1}$, as well as the behavior at the deconfinement phase transition, which is about twice as strongly first-order as the one of $SU(3)$ Yang-Mills theory.

To properly model these characteristics, it is necessary to use an effective vortex action which is more complicated than for $SU(2)$ or $SU(3)$, and which can be symbolically represented as

$$ S = c_i c_j \times \begin{array}{c} j \\ \boxed{\phantom{xx}} \\ i \end{array} - b \times \diagup\!\!\!\square \qquad . \tag{1} $$

The first term is the curvature term already used in the $SU(2)$ and the $SU(3)$ models. It penalizes configurations in which two vortex squares share a link without lying in the same plane by an action increment $c_i c_j$ depending on the types of vortices participating; in the $SU(4)$ case, there are two vortex types, $i, j \in \{1, 2\}$. Studying a model based only on this term led to the conclusion that it cannot faithfully reproduce the confinement properties of $SU(4)$. When $\sigma_2/\sigma_1$ is tuned to the correct value, the deconfinement phase transition is second-order. Consequently, additional dynamics must be introduced, embodied in the second term in (1), the branching term. It facilitates vortex branchings by weighting links at which 3 or 5 vortex squares meet with an action decrement $b$.

Using this action, agreement with the $SU(4)$ Yang-Mills confinement characteristics is reached at the physical point [20]

$$c_1 = 0.45 \qquad c_2 = 0.80 \qquad b = 0.71 . \tag{2}$$

It should be noted that, in Abelian gauges, vortex branching can be associated with Abelian magnetic monopoles [20]. Thus, the above result can be interpreted as implying that a realistic vortex model for $SU(4)$ Yang-Mills theory is only achieved by including a dependence of the vortex dynamics on the configuration of the Abelian magnetic monopoles which reside on the vortices in Abelian gauges [9, 11, 12]. This confirms a corresponding expectation formulated in [25], that Abelian magnetic monopoles begin to play a role in infrared Yang-Mills vortex dynamics as the number of colors $N$ is raised. Note that Abelian magnetic monopoles are also present in $SU(2)$ and $SU(3)$ vortex configurations; however, there is no signature for an independent dynamical role of these monopoles. Their distribution appears to be essentially determined by the dynamics of the vortices on which they reside.

## $Sp(2)$ **VORTEX MODEL**

A remarkable property of the $Sp(N)$ sequence of groups is that all members have the same center, $Z(2)$, and allow for the same set of center vortex degrees of freedom. There is only one nontrivial center element, $-1$, and therefore only one type of vortex flux. Nevertheless, the effective vortex actions can be very different for different underlying $Sp(N)$ groups; after all, different cosets would be integrated out in each case if one were to derive the effective vortex action from first principles. Therefore, vortex models for different $Sp(N)$ Yang-Mills theories are by no means forced to display similar confinement characteristics, such as deconfinement transitions of the same order.

Indeed, while $SU(2) = Sp(1)$ Yang-Mills theory exhibits a second-order deconfinement phase transition, the deconfinement transition of $Sp(2)$ Yang-Mills theory is strongly first-order [17, 18]. As above, in order to generate such behavior, new dynamics must be introduced compared to the $SU(2)$ vortex model. The confinement characteristics of $Sp(2)$ Yang-Mills theory can be reproduced using an effective vortex action of the symbolic form

$$S = c \times \boxed{\quad} + s \times \boxed{\quad} . \tag{3}$$

The first term is the curvature term already discussed above. The second term can be interpreted in terms of a "stickiness" of vortices: When 4 (or even 6) vortex squares meet at a link, this corresponds to 2 (or even 3) intersecting vortex fluxes maintaining contact to one another for a finite space-time length instead of intersecting only at one space-time point. Enhancing such behavior by choosing a negative value for $s$ means that vortices become stickier. Indeed, a first-order deconfinement phase transition of the proper strength, together with the correct value of $T_c/\sqrt{\sigma}$ is achieved at the physical point [21]

$$c = 0.479 \qquad s = -1.745 . \tag{4}$$

# CONCLUSIONS

Extending the Yang-Mills gauge group to $SU(4)$ and $Sp(2)$, new dynamics emerge in the corresponding infrared effective vortex descriptions. The $SU(4)$ case exhibits clear signatures of Abelian magnetic monopoles (which are intrinsically present in vortex configurations cast in Abelian gauges) attaining a dynamical significance of their own as the number of colors is raised. This corroborates related arguments put forward in [25]. In the $Sp(2)$ case, a new "stickiness" term in the effective action serves to drive the deconfinement transition towards the correct first-order behavior. While $SU(2) = Sp(1)$ and $Sp(2)$ Yang-Mills theory contain the same center vortex degrees of freedom, the vortex effective actions in the two cases differ and thus naturally lead to different behavior at the deconfinement transition.

Having determined the physical points (2) and (4) of the $SU(4)$ and $Sp(2)$ infrared effective vortex models, the behavior of the spatial string tensions at high temperatures can be predicted [20, 21]. As discussed further in [20, 21], comparison with measurements in the corresponding full lattice Yang-Mills theories can be used to test the validity of the model constructions presented here.

# ACKNOWLEDGMENTS

This work was supported by the U.S. DOE under grants DE-FG03-95ER40965 (M.E.) and DE-FG02-94ER40847 (B.S.).

# REFERENCES

1. G. 't Hooft, *Nucl. Phys.* **B138**, 1 (1978).
2. H. B. Nielsen and P. Olesen, *Nucl. Phys.* **B160**, 380 (1979).
3. L. Del Debbio, M. Faber, J. Giedt, J. Greensite and Š. Olejník, *Phys. Rev. D* **58**, 094501 (1998).
4. T. G. Kovács and E. T. Tomboulis, *Phys. Rev. D* **57**, 4054 (1998).
5. P. de Forcrand and M. D'Elia, *Phys. Rev. Lett.* **82**, 4582 (1999).
6. M. Engelhardt, K. Langfeld, H. Reinhardt and O. Tennert, *Phys. Rev. D* **61**, 054504 (2000).
7. J. Greensite, *Prog. Part. Nucl. Phys.* **51**, 1 (2003).
8. M. Engelhardt and H. Reinhardt, *Nucl. Phys.* **B585**, 591 (2000).
9. M. Engelhardt, *Nucl. Phys.* **B585**, 614 (2000).
10. M. Engelhardt, *Nucl. Phys.* **B638**, 81 (2002).
11. J. M. Cornwall, *Phys. Rev. D* **61**, 085012 (2000).
12. M. Engelhardt and H. Reinhardt, *Nucl. Phys.* **B567**, 249 (2000).
13. F. Bruckmann and M. Engelhardt, *Phys. Rev. D* **68**, 105011 (2003).
14. M. Engelhardt, M. Quandt and H. Reinhardt, *Nucl. Phys.* **B685**, 227 (2004).
15. M. Engelhardt, *Phys. Rev. D* **70**, 074004 (2004).
16. M. Quandt, H. Reinhardt and M. Engelhardt, *Phys. Rev. D* **71**, 054026 (2005).
17. K. Holland, M. Pepe, and U.-J. Wiese, *Nucl. Phys.* **B694**, 35 (2004).
18. M. Pepe, *Nucl. Phys. Proc. Suppl.* **141**, 238 (2005).
19. M. Pepe, *PoS* **LAT2005**, 017 (2005).
20. M. Engelhardt, *Phys. Rev. D* **73**, 034015 (2006).
21. M. Engelhardt and B. Sperisen, hep-lat/0610074.
22. B. Lucini, M. Teper and U. Wenger, *JHEP* **0401**, 061 (2004).
23. B. Lucini, M. Teper and U. Wenger, *JHEP* **0406**, 012 (2004).
24. B. Lucini, M. Teper and U. Wenger, *JHEP* **0502**, 033 (2005).
25. J. Greensite and Š. Olejník, *JHEP* **0209**, 039 (2002).

# Extended Double Lattice BRST, Curci-Ferrari Mass and the Neuberger Problem

M. Ghiotti, L. von Smekal and A. G. Williams

*Centre for the Subatomic Structure of Matter, University of Adelaide, SA 5005, Australia*

**Abstract.** We present Extended Double BRST on the lattice and extend the Neuberger problem to include the ghost/anti-ghost symmetric formulation of the non-linear covariant Curci-Ferrari (CF) gauges. We then show how a CF mass regulates the 0/0 indeterminate form of physical observables, as observed by Neuberger, and discuss the gauge parameter and mass dependence of the model.

**Keywords:** Lattice BRST, TFT, Neuberger Problem, Curci-Ferrari Mass
**PACS:** 11.15.Ha, 11.30.Ly, 11.30.Pb

## INTRODUCTION

In the covariant continuum formulation of gauge theories one has to deal with the redundant degrees of freedom due to gauge invariance. Within the language of local quantum field theory, the machinery for this is based on the Becchi-Rouet-Stora-Tyutin (BRST) symmetry which can be considered the quantum version of local gauge invariance. Beyond perturbation theory one faces the famous Gribov ambiguity: the existence copies of gauge-configurations that satisfy the Lorentz condition (or any other local gauge fixing condition) but are related by gauge transformations, and are thus physically equivalent. As a result, the usual definitions of a BRST charge fail to be globally valid. A rigorous non-perturbative framework is provided by lattice gauge theory. Its strength and beauty derives from the fact that gauge-fixing is not required. However, in order to arrive at a non-perturbative definition of non-Abelian gauge theories in the continuum, from a lattice formulation, we need to be able to perform the continuum limit in a formally watertight way. The same ambiguity then shows in another form when attempting to fix a gauge via BRST formulations on the lattice. There it is known as the Neuberger problem which asserts that the expectation value of any gauge invariant (and thus physical) observable in a lattice BRST formulation will always be of the indefinite form 0/0 [1].

In this talk we present the ghost/anti-ghost symmetric Curci-Ferrari gauges with double BRST on the lattice. We show how Neuberger's argument can be extended to include these non-linear covariant gauges, and how the indeterminate form $0/0$ of expectation values is regulated by CF mass term [2] thereby decontracting the double BRST algebra to its extended version. Finally, we discuss how the gauge-parameter $\xi$ dependence of the model can be compensated by adjusting the CF mass with $\xi$.

In pure $SU(N)$ lattice gauge theory, the gauge transformation of link $U_{ij}$ is defined as $U_{ij}^g = g_i U_{ij} g_j^\dagger$. BRST and anti-BRST transformations $s$ and $\bar{s}$ in the topological setting do not act on the link variables $U$ directly, but on the gauge transformations $g_i$ like infinitesimal left translations in the gauge group with real ghost and anti-ghost

Grassmann fields $c_i^a$, $\bar{c}_i^a$ as parameters, $sg_i = c_i g_i$ and $\bar{s}g_i = \bar{c}_i g_i$, where $c_i \equiv c_i^a X^a$ and $\bar{c}_i \equiv \bar{c}_i^a X^a$. For the normalization of the anti-Hermitian generators $X^a$ in the fundamental representation we use $\mathrm{tr} X^a X^b = -\delta^{ab}/2$. The action of the topological lattice model for gauge fixing $a$ $la$ Faddeev-Popov with double BRST invariance can then be written as

$$S_{\mathrm{CF}} = i s \bar{s} \Big( V[U^g] + i \xi \sum_i \mathrm{tr}\, \bar{c}_i c_i \Big) = \sum_i \Big( i b_i^a F_i^a[U^g] - \frac{i}{2} \bar{c}_i^a M_i^a[U^g, c] + \frac{\xi}{2}(b_i^a)^2 + \frac{\xi}{8}(\bar{c}_i \times c_i)^2 \Big),$$

where $V[U^g] = -\sum_i \sum_{j \sim i} \mathrm{tr}\, U_{ij}^g = -2 \sum_{x,\mu} \mathrm{Re}\, \mathrm{tr}\, U_{x,\mu}^g$ is the gauge fixing functional of covariant gauges which here assumes the role of a Morse potential on a gauge orbit. $F_i^a[U^g] = 0$ is the gauge-fixing condition and $M_i^a[U^g, c] = \sum_j M_{ij}^{ab} c_j^b$ defines the Faddeev-Popov operator of the ghost/anti-ghost symmetric gauges. Note the occurrence of quartic ghost self-interactions $\propto (\bar{c}_i \times c_i)^2 \equiv (f^{abc} \bar{c}_i^b c_i^c)^2$ which make the Neuberger problem somewhat less obvious in these gauges. Details will be presented elsewhere.

## REGULARISATION OF $0/0$ AND $\xi$ INDEPENDENCE

Following Neuberger, we introduce an auxiliary parameter $t$ upfront the Morse potential, to write the Euclidean partition function used as the gauge-fixing device, with double BRST,

$$Z_{\mathrm{CF}}(t) = \int d[g, b, \bar{c}, c] \, \exp\Big\{ -i s \bar{s} \Big( t V[U^g] + i \xi \sum_i \mathrm{tr}\, \bar{c}_i c_i \Big) \Big\} \tag{1}$$

which is independent of $\{U\}$ and $\xi$. Because a derivative w.r.t. $t$ produces a BRST-exact operator in the integrand, it is in fact independent of $t$ also, $i.e.$, $Z_{\mathrm{CF}}'(t) = 0$. For $t = 0$ on the other hand we find that $Z_{\mathrm{CF}}(0) = 0$; and this is the reason for the indeterminate form of $0/0$ for all observables first derived for the standard linear covariant gauges in [1]. The fact that this conclusion holds also in the ghost/anti-ghost symmetric formulation with its quartic self-interactions directly relates to the topological interpretation [3] of the Neuberger zero: $Z_{\mathrm{CF}}$ can be viewed as the partition function of a Witten-type TQFT which computes the Euler character $\chi$ of the gauge group. On the lattice the gauge group is a direct product of $SU(N)$'s per site, and $Z_{\mathrm{CF}} = \chi(SU(N)^{\#\mathrm{sites}}) = \chi(SU(N))^{\#\mathrm{sites}} = 0^{\#\mathrm{sites}}$. For $t = 0$ the action decouples from the link-field configuration and $Z_{\mathrm{CF}}(0)$, albeit computing the same topological invariant, has no effect in terms of fixing a gauge. In the present formulation, $Z_{\mathrm{CF}}(0)$ factorises into independent Grassmann integrations per site of the quartic term containing the curvature of $SU(N)$, each of which computes the vanishing Euler character of $SU(N)$ via the Gauss-Bonnet theorem [4].

As proposed in [5], this zero can be regularised, however, by introducing a Curci-Ferrari mass $m$, such that the gauge-fixing action $S_{\mathrm{CF}}$ is replaced by

$$S_{\mathrm{mCF}}(t) = i(s \bar{s} - i m^2)\Big( t V_U[g] + i \xi \sum_i \mathrm{tr}\, \bar{c}_i c_i \Big). \tag{2}$$

The corresponding partition function $Z_{\mathrm{mCF}}(t)$ no-longer vanishes at $t = 0$, and this part in Neuberger's disastrous conclusion is thus avoided. We have explicitly calculated $Z_{\mathrm{mCF}}(0)$, which is polynomial in $\xi m^4$, for $SU(2)$ and $SU(3)$. The original zero is obtained

181

for $m^2 \to 0$ which corresponds to a Wigner-Ionu contraction of the so-called extended double BRST superalgebra. While a non-vanishing $m^2$ thereby breaks the nilpotency of BRST and anti-BRST transformations, which is known to result in a loss of unitarity, it also serves to regulate the $0/0$ indeterminate form of expectation values in lattice BRST formulations, and to obtain finite results for $m^2 \to 0$ via l'Hospital's rule.

For gauge fixing we need to have $t \neq 0$. The partition function $Z_{\mathrm{mCF}}(t)$ of the massive CF model is no-longer $t$-independent because $s$ and $\bar{s}$ are no-longer nilpotent and the simple argument above fails, i.e., $Z_{\mathrm{mCF}}'(t) \neq 0$ for $m^2 \neq 0$. However, the existence of 3 independent parameters $t$, $\xi$ and $m^2$ is an illusion. A change in $t$ can always be compensated by changing the gauge parameter $\xi$ and $m^2$. In fact, simple scaling arguments and explicit calculations show that $Z_{\mathrm{mCF}}$ only depends on 2 combinations of the 3, we can parametrise $Z_{\mathrm{mCF}} = f(t^2/\xi, \xi m^4) \equiv f(x^2, \widehat{m}^4)$, where we defined $x^2 \equiv t^2/\xi$ and $\widehat{m}^4 \equiv \xi m^4$. Our explicit calculations for $t = 0$ yield $f(0, \widehat{m}^4)$. Independence of $t$ then comes together with gauge parameter $\xi$ independence. To achieve this, we allow $\widehat{m}^2 \equiv \widehat{m}^2(x)$ so that $Z_{\mathrm{mCF}} = f(x, \widehat{m}^4(x))$. This means that we adjust the CF mass $\widehat{m}^2$ with $x$ such that our $x = 0$ results remain unchanged. In particular, we must have

$$\frac{\mathrm{d}}{\mathrm{d}x} Z_{\mathrm{mCF}} = \left( \frac{\partial}{\partial x} + \frac{\mathrm{d}\widehat{m}^2}{\mathrm{d}x} \frac{\partial}{\partial \widehat{m}^2} \right) Z_{\mathrm{mCF}} = 0 \tag{3}$$

which can be used to determine the derivative of $\widehat{m}^2(x)$. This is always possible. The crucial question at this point is whether it can be done independent of the link configuration $\{U\}$. As our explicit calculations are restricted to $x = 0$ we have explicitly verified that $\widehat{m}^{2\prime}(0)$ is finite and independent of $\{U\}$. While this is merely necessary, but not sufficient, it demonstrates that we can get away from $x = 0$, at least infinitesimally. This is of qualitative importance as a non-zero value of $x = t/\sqrt{\xi}$, no matter how small, corresponds to a large but finite $\xi$ at $t = 1$ and thus eliminates the gauge freedom.

## CONCLUSIONS

The massive Curci-Ferrari model with extended double BRST symmetry can be formulated on the lattice without the $0/0$ problem. The parameter $m^2$ is not interpreted as a physical mass but to meaningfully define a limit $m^2 \to 0$ in the spirit of l'Hospital's rule. At finite $m^2$ the topological nature of the gauge-fixing partition function seems lost. It is possible, however, to tune the CF mass with the gauge parameter $\xi$ so that the limit $m^2 \to 0$ can be defined along a certain trajectory in parameter space independent of $\xi$. An interesting open question might then be the topological interpretation of the model within the extended double BRST superalgebra framework.

## REFERENCES

1. H. Neuberger, *Phys. Lett.* **B175**, 69 (1986); *ibid.* **B183**, 337 (1987).
2. G. Curci, and R. Ferrari, *Phys. Lett.* **B63**, 91 (1976); *Nuovo Cim.* **A35**, 1 (1976).
3. B. Sharpe, J. Math. Phys. **25** (1984) 3324; L. Baulieu and M. Schaden, *Int. J. Mod. Phys.* **A13**, 985 (1998); M. Schaden, *Phys. Rev.* **D59**, 014508 (1999).
4. D. Birmingham, M. Blau, M. Rakowski, and G. Thompson, *Phys. Rept.* **209**, 129 (1991).
5. A. C. Kalloniatis, L. von Smekal, and A. G. Williams, *Phys. Lett.* **B609**, 424 (2005).

# K-strings and minimal surfaces

Ferdinando Gliozzi

*Dipartimento di Fisica Teorica, Università di Torino and INFN, Sezione di Torino*
*via P.Giuria 1, 110125 Torino, Italy*

**Abstract.**
Some general properties of the confining strings joining higher rank sources in SU(N) gauge theories (i.e. k-strings) are compared with those of the confining string attached to sources in the fundamental representation. Some aspects reflect features of the minimal surfaces associated with the string world-sheet.

**Keywords:** confining strings, higher representations
**PACS:** 11.15-q 11.15.Ha 11.25.Tq

$SU(N)$ gauge theories are characterised by an infinite set of irreducible representations, however it is usual to study uniquely the properties of the colour sources belonging to the fundamental (or anti-fundamental) representations, being the one where lie the quarks.

The question naturally arises: what are the properties of the colour fields generated by sources in higher representations? A first important notion is the $N-$ality $k_{\mathscr{R}} \equiv j \bmod N$ of a representation $\mathscr{R}$, i.e. the number $j$ modulo $N$ of copies of the fundamental representation $f$ needed to generate $\mathscr{R}$, namely, $\mathscr{R} = f \otimes f \otimes \dots$ ($j$ times). The sources with non-vanishing $N-$ality give rise to a confining potential: a pair of point-like sources belonging to the representations $\mathscr{R}$ and $\bar{\mathscr{R}}$ experiences at large distance a static potential of the form

$$V_{\mathscr{R}} = \sigma_{\mathscr{R}} r + \mu_{\mathscr{R}} + O(1/r) \tag{1}$$

where $\sigma_{\mathscr{R}}$ is the string tension associated to the representation $\mathscr{R}$. Most of these strings are expected to be unstable, because all representations with same $k$ can be converted into each other by the emission of a proper number of soft gluons. Heavier $\mathscr{R}-$strings are expected to decay into the strings with smallest string tension within the same $N$-ality class, called $k-$strings.

How does $\sigma_{\mathscr{R}}$ depend on $\mathscr{R}$? According to an old conjecture, known as Casimir scaling [1], one should have

$$\sigma_{\mathscr{R}} = C_{\mathscr{R}} \, \sigma \tag{2}$$

where $C_{\mathscr{R}}$ is the quadratic Casimir invariant and $\sigma = \sigma_f$ is the tension of the string in the fundamental representation. This seems approximately supported by *unstable $\mathscr{R}$* strings. Assuming this formula leads to conclusion that stable $k-$ strings belong to the totally anti-symmetric representation with $k$ indices, because it corresponds to the minimal value of the Casimir within the given $N-$ality class. According to Casimir scaling the tension of the $k$ string is given by

$$\sigma_k^{(c)} = \sigma \frac{k(N-k)}{N-1} \; . \tag{3}$$

CP892, *Quark Confinement and the Hadron Spectrum VII*
edited by J. E. F. T. Ribeiro

Although the way how $\sigma_k$ scales with $k$ is still debated, the Casimir scaling for *stable* $k-$strings seems excluded.

Much insight comes from SUSY $SU(N)$: some string-inspired gauge models allow to evaluate, at least in the $N \to \infty$ limit, the string tension of stable $k-$ strings. The $\mathcal{N} = 2$ supersymmetric $SU(N)$, softly broken to $\mathcal{N} = 1$ by an adjoint multiplet yields [2] the so called *sine law*

$$\sigma_k^{(s)} = \sigma \frac{\sin(k\pi/N)}{\sin(\pi/N)} . \tag{4}$$

This formula was also derived in the M theory description of $\mathcal{N} = 1$ supersymmetric gauge theory, called MQCD [3]. Similarly, and in the context of gauge/gravity duality it has been found the exact sine law for the so called Maldacena-Nuñez (MN) background (which in the IR limit behaves as a $\mathcal{N} = 1$ $SU(N)$ gauge theory), while for the Klebanov-Strassler (KS) background it results a slightly smaller value $\sigma_k^{KS} < \sigma_k^{(s)}$. More recently a generalised background was found in which the MN and the KS settings are particular cases.

In pure $SU(N)$ gauge theories there is no complete agreement between the numerical data of different groups: although some dedicated studies favour the sine law [5], other simulations with $N = 4, 5, 6, 8$ found string tension lying partway between the Casimir scaling and the sine law [6]. There are also some analytic results (or conjectures) in the large $N$ limit which describe the IR properties of pure $SU(N)$ gauge theories as duals of AdS Schwarzchild black holes. It turns out that [7] in $D = 3$ $\sigma_k \propto \sin^3 \theta$ with $\frac{k\pi}{N} = \theta - \sin\theta\cos\theta$; in $D = 4$ $\sigma_k = \sigma_k^{(c)}$ and finally in $D = 5$ $\sigma_k = \sigma_k^{(s)}$.

A curious fact: in all the known cases, in supersymmetric as well as in the pure gauge theories the sine law seems to be an upper bound on the $k-$string tensions: [1]

$$\sigma_k \leq \sigma_k^{(s)} . \tag{5}$$

In this talk I want to point out two general properties of the $k-$strings that might be related to the above observation. First, it turns out that mentioned range (5) corresponds exactly to the spontaneously broken symmetry phase of a spatial $\mathbb{Z}_N$ symmetry associated to the baryonic vertex. Secondly, I argue that the logarithmic broadening of the flux tube, a well-known property of the confining string, should be also true in the case of the $k-$strings, provided that $\sigma_k$ lies in the above range (5). Both arguments are based on simple properties of minimal surfaces.

A baryonic vertex is a gauge-invariant coupling of $N$ multiplets in the fundamental representation which gives rise to finite energy configurations with $N$ external quarks. When the separations among these quarks is large, one expects that $N$ strings of chromo-electric flux form, which meet at a common junction; their world-sheet forms a $N-$bladed surface with a common intersection.

When $N > 3$ one has to envisage the possibility that $k$ neighbouring strings of the baryon vertex, i.e. $k$ blades of the mentioned surface, coalesce into a single $k-$string.

---

[1] This consideration does not apply to the abelian gauge theories, where it is possible to vary with continuity the value of the $k-$string tension.

Assume for instance that $k_1$ neighbouring strings coalesce into a $k_1$−string, $k_2$ strings in a $k_2$−sting, and so on, with $N = k_1 + k_2 + \ldots k_m$. In order to find the state of minimal energy we have to compare the $\mathbb{Z}_N$ symmetric configuration, which gives rise to the potential $V_N$, with the broken-symmetry configuration, associated with the potential $V_{\{k_1,k_2,\ldots k_m\}}$, namely,

$$V_N = N\sigma R + O\left(\frac{1}{R}\right) \quad , \quad V_{\{k_1,k_2,\ldots k_m\}} = \sigma R \sum_{a=1}^{m} f_{k_a}(\rho_a) + O\left(\frac{1}{R}\right) , \tag{6}$$

with $f_k(\rho) = \rho\frac{\sigma_k}{\sigma} + \sum_{j=0}^{k-1}\left|\rho - e^{i\frac{2\pi j}{N} - i\alpha}\right|$ , where $\rho R$ is the length of the $k$-string, $\alpha = \frac{k-1}{N}\pi$ and $Re^{i\frac{2\pi j}{N}}$ $(j = 1,2\ldots N)$ are the coordinates of the polygon vertices.

In order not to break the $\mathbb{Z}_N$ symmetry the configuration of minimal energy should be characterised by $\rho_{k_a} = 0$ for all $a = 1,\ldots,m$. Expanding $f_k(\rho)$ around $\rho = 0$ yields

$$f_k(\rho) = k + \rho\frac{\sigma_k - \sigma_k^{(s)}}{\sigma} + \frac{\rho^2}{4}\left(k - \frac{\sin\frac{2\pi k}{N}}{\sin\frac{2\pi}{N}}\right) + \cdots \tag{7}$$

Inserting this expression in (6) and comparing it with the symmetric solution $V_N$ leads to conclusion that when $\sigma_k \geq \sigma_k^{(s)}$ the $\mathbb{Z}_N$-symmetric baryon vertex should be stable against the formation of $k$-strings, while for $\sigma_k < \sigma_k^{(s)}$ this is no longer true and the system breaks up into less symmetric configurations made with $k$-strings. In other terms, the sine law is the critical threshold below which the $\mathbb{Z}_N$ symmetry of the baryon vertex is spontaneously broken [8].

Another feature of the $k$−string physics that can be described in terms of minimal surfaces is related to the logarithmic broadening of the colour flux tube as a function of the quark separation. This quantum effect, which actually is very difficult to be observed, is due to the fluctuations of the underlying confining string. A particularly simple derivation of this phenomenon has been described by Lüscher, Münster and Weisz [9]. The starting point is the quantity $P(h) = [\langle W(C)W(c)\rangle - \langle W(C)\rangle\langle W(c)\rangle]/\langle W(C)\rangle$, where $W(C)$ and $W(c)$ are Wilson operators for parallel, coaxial circles of radii $R$ and $r$ placed at a mutual distance $h$ . $P(h)$ can be considered as the a density of the flux generated by the source $C$ and revealed by the probe $c$. The insight of [9] was to observe that $P(h)$ can be described in the effective string picture as $P(h) = \exp[-\sigma A(R,r,h)]$, where denotes the area of the minimal surface with $C$ and $c$ as boundaries. It is well-known that such a surface is a catenoid $y(x) = \frac{1}{\omega}\cosh\omega(x - x_o)$ with $R = \frac{1}{\omega}\cosh\omega x_o$ and $r = \frac{1}{\omega}\cosh\omega(h - x_o)$. The minimal area is

$$A(R,r,h) = \pi\left(\frac{h}{\omega} + R^2\sqrt{1 - 1/(\omega R)^2} - r^2\sqrt{1 - 1/(\omega r)^2}\right) . \tag{8}$$

In the limit $\log(R/r) \gg h/r$ one gets $\omega \sim \frac{1}{h}\log(R/r)$, which leads to the gaussian distribution $P(h) \propto \exp[-\sigma\pi h^2/\log(R/r)]$. As a consequence the mean square width, defined as $w^2 = \int h^2 P(h)\,dh / \int P(h)\,dh$, fulfils the universal logarithmic broadening

$$\sigma w^2 = \frac{1}{2\pi}\log h\Lambda , \tag{9}$$

185

with $\Lambda$ a suitable scale.

It is interesting to observe that putting the source $C$ in an arbitrary representation $\mathscr{R}$ of $N-$ality $k$ keeping the probe $c$ in the fundamental leads exactly to the same formula, the reason being that the world-sheet of the $k-$string can be thought as a stack of $k$ sheets in the fundamental representation, and the probe can feel only one of them. Thus we can conclude that the logarithmic growing of the flux tube as a function of the source separation does depend on the representation $\mathscr{R}$ only through the scale $\Lambda$. In other terms we are led to generalise (9) as

$$\sigma w^2_{\mathscr{R}} = \frac{1}{2\pi} \log h \Lambda_{\mathscr{R}} , \tag{10}$$

where $\sigma$ is the string tension of the <u>fundamental</u> representation. A nice check of such a prediction has been recently obtained in the case of $k = 2$ string in a 3D $\mathbb{Z}_4$ gauge model [10].

In some special cases, for a very narrow range of $h$, there is, beside the above general solution, another minimal surface made by a first catenoid composed by the $k-$string. At a suitable distance $d$ splits into a disk orthogonal to the symmetry axis made by the world-sheet of the $k - 1$ string and a second catenoid in the fundamental representation which reaches the probe $c$. The position of the intermediate disk is not arbitrary, but it is dynamically determined by the balance of the tensions at the string junction. When these special configurations are kinematically allowed the logarithmic broadening gets modified at large distance. In the $N = 4$ $k = 2$ case it is easy to prove that in the strong coupling phase (5) these special configurations are forbidden.

More generally, a configuration in which a $k-$string propagating along a catenoid decays into a $k - \ell$ string forming a disk and a $\ell-$string forming another catenoid is permitted only if the kinematic constraint $\sigma_k^2 \geq \sigma_{k-\ell}^2 + \sigma_\ell^2$ holds, showing that exceptional configurations are allowed only if the binding energy of the $k$-string is sufficiently small.

## REFERENCES

1. J. Ambjørn, P.Olesen and C. Peterson, *Nucl Phys. B* **240**,186 (1984; Nucl. Phys. B **240**, 533 (1984).
2. M.R. Douglas and S.H. Shenker, Nucl. Phys. B **481**, 513 (1995).
3. A. Hanany, M.J. Strassler and A. Zaffaroni, *Nucl Phys. B* **513**, 87 (1998).
4. C.P. Herzog and I. R. Klebanov, *Phys. Lett. B* **526**, 388 (2002).
5. L. Del Debbio, H. Panagopoulos, P. Rossi and E. Vicari, *Phys.Rev. D* **65**, 021501 (2002); *JHEP* **01** , 009 (2002).
6. B. Lucini and M. Teper, *Phys. Rev. D* **64**,105019 (2001); B. Lucini, M. Teper and U. Wengler, *JHEP***0406**, 012 (2004 ).
7. Y. Imamura, *Prog. Theor. Phys.* **115** (2006) 815.
8. F. Gliozzi, *Phys. Rev. D* **72** (2005). [arXiv:hep-th/0504105].
9. M.Lüscer, G. Münster and P. Weisz, *Nucl.Phys. B* **180**,1 (1981)
10. P. Giudice, F. Gliozzi and S. Lottini, arXiv:hep-lat/0609060.

# Vacuum structure as seen by overlap fermions

E.-M. Ilgenfritz*, K. Koller†, Y. Koma**, G. Schierholz‡,§, T. Streuer¶ and V. Weinberg‖

*Humboldt-Universität zu Berlin, Institut für Physik, 12489 Berlin, Germany
†Sektion Physik, Universität München, 80333 München, Germany
**Institut für Kernphysik, Johannes-Gutenberg Universität Mainz, 55099 Mainz, Germany
‡Deutsches Elektronen-Synchrotron DESY, 22603 Hamburg, Germany
§John von Neumann-Institut für Computing NIC, 15738 Zeuthen, Germany
¶Department of Physics and Astronomy, University of Kentucky, Lexington, KY 40506-0055, USA
‖Deutsches Elektronen-Synchrotron DESY, 15738 Zeuthen, Germany

**Abstract.** Three complementary views on the QCD vacuum structure, all based on eigenmodes of the overlap operator, are reported in their interrelation. (i) spectral density, localization and chiral properties of the modes, (ii) the possibility of filtering the field strength with the aim to detect selfdual and antiselfdual domains and (iii) the various faces of the topological charge density, with and without a cutoff $\lambda_{cut} = O(\Lambda_{QCD})$. The techniques are tested on quenched $SU(3)$ configurations.

**Keywords:** lattice fermions, localization, topological charge
**PACS:** 11.15.-q. 11.15.Ha, 12.38.Aw

## INTRODUCTION

Attempts to catch the structure of the QCD vacuum on the lattice were mostly inspired by the instanton/caloron picture, and appropriate lumps were searched for in the gluonic lattice topological charge density $q(x)$. With the advent of overlap fermions, the fermionic definition of the topological charge density has become viable, from the scale of the lattice spacing $a$ up to the infrared. This is possible since the Ginsparg-Wilson approach provides lattice fermions with perfect chiral properties. Results concerning the vacuum structure, based on the eigenmodes of the overlap Dirac operator that have been collected in the QCDSF collaboration [1, 2], are described in this contribution.

## PROPERTIES OF THE LOWEST EIGENMODES

We analyze a large set of overlap eigenmodes, $O(150)$ per lattice configuration, for five ensembles of quenched Yang-Mills theory generated with the Lüscher-Weisz action. This action is important to avoid dislocations. The analysis is performed on $12^3 \times 24$, $16^3 \times 32$ and $24^3 \times 48$ lattices at $\beta = 8.45$, on a $12^3 \times 24$ lattice at $\beta = 8.10$ and on a $16^3 \times 32$ lattice at $\beta = 8.00$. In Fig. 1 (left) the spectral densities for the three volumes at $\beta = 8.45$ are presented. The fit to quenched chiral perturbation theory [3] requires to know the distribution $w(Q)$ of topological charge, easily obtained from the number and chirality $\pm 1$ of the zeromodes, $Q = N_- - N_+$, per configuration. Fig. 1 shows the fit of

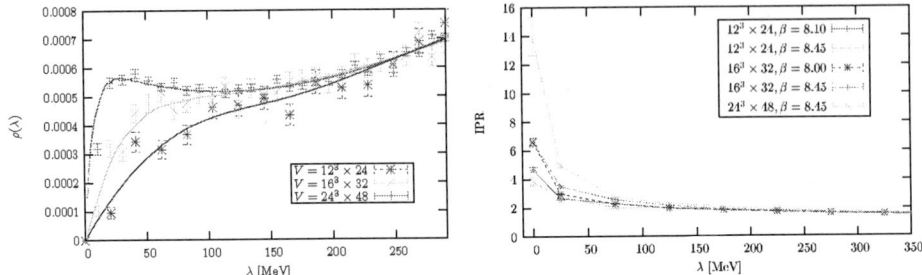

**FIGURE 1.** Left: The spectral density of non-zeromodes for three lattice sizes at $\beta = 8.45$, together with a simultaneous fit using the finite volume prediction of quenched chiral perturbation theory. Right: The average IPR for zeromodes and for non-zeromodes in $\Delta\lambda = 50$ MeV bins for all five ensembles.

the spectral densities according to

$$\rho(\lambda, V) = \Sigma_{\text{eff}}(V, \lambda) \sum_Q w(Q) \rho_Q(\lambda\, V\, \Sigma_{\text{eff}}(V, \lambda)) , \tag{1}$$

with the microscopic spectral density $\rho_Q(x) = (x/2)(J_{|Q|}^2(x) - J_{|Q|+1}(x)J_{|Q|-1}(x))$ in the fixed $Q$ sector, leading to an estimate of the quark condensate, $\Sigma^{\overline{\text{MS}}}(2\,\text{GeV}) = (231(20)\,\text{MeV})^3$.

In Fig. 1 (right) the average inverse participation ratio (IPR) of the zeromodes and non-zeromodes is presented. In the spectral region of strong volume dependence the modes are highly localized, the spread of the IPRs is large and the level compressibility $\alpha = \langle (N - \langle N \rangle)^2 \rangle / \langle N \rangle << 1$ (for $N$ modes falling in some interval) is small. In the theory of the metal-insulator transition this situation is called "critical", and the modes are multifractal. Fitting the $V$ dependence of the IPRs, one can assign a dimension $d^*(p = 2) \approx 2$ to the zeromodes and $d^*(p = 2) \approx 3 - 3.5$ to the non-zeromodes below 100 MeV (see Fig. 2 (left), upper curve). Similar results of the MILC collaboration [4] have been obtained with Asqtad fermions. The IPRs of the higher modes are consistent with $d^*(2) \approx 4$. Generalized IPRs based on higher moments $I_\lambda^{(p)} = \sum_x p_\lambda(x)^p$ of the scalar density $p_\lambda(x) = \psi_\lambda^\dagger(x)\psi_\lambda(x)$ localize the peaks of $p_\lambda(x)$ in regions of lower dimension, $d^*(p > 2) < d^*(p = 2)$. We find $d^*(p = 10) < 1$ for zeromodes and $d^*(p = 10) \sim 1$ for non-zeromodes with $\lambda < 100$ MeV. The multifractality evidenced in Fig. 2 (left) urges a more detailed study of the lowest eigenmodes, in particular closer to the continuum limit. Case studies [5] show that they might be pinned-down on low-dimension defects of the gauge field (monopoles and vortices, the candidates to create confinement).

Different from the high temperature phase [6], the distribution of local chirality of the lowest non-zeromodes peaks at non-zero values of the chirality variable

$$X_\lambda(x) = (4/\pi) \arctan \sqrt{p_{\lambda+}(x)/p_{\lambda-}(x)} - 1 \in [-1, +1] . \tag{2}$$

expressing the ratio between the density of the chirality components, $p_{\lambda\pm}(x) = \frac{1}{2}\psi_\lambda^\dagger(x)(1 \pm \gamma_5)\psi_\lambda(x)$. Focussing on the highest-ranking 1 % of lattice sites according

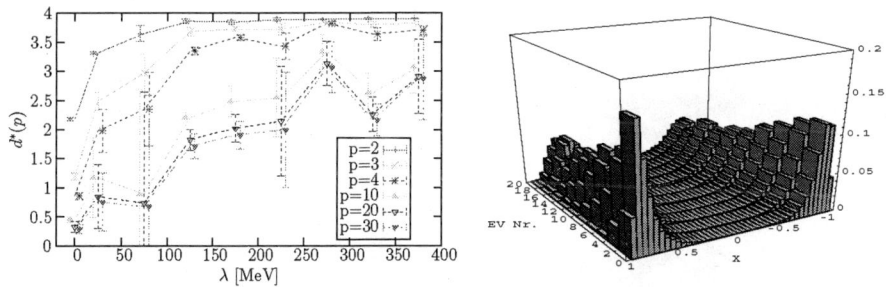

**FIGURE 2.** Left: The multifractal dimensions of zeromodes and of non-zeromodes, from the volume dependence of average $I_\lambda^{(p)}$ for the three ensembles at $\beta = 8.45$. Right: The histograms of local chirality $X_\lambda(x)$ for the 10 lowest pairs of non-zeromodes over 1 % of sites with highest scalar density.

to the scalar density $p_\lambda(x)$, the histograms for the 10 lowest pairs $\pm\lambda$ are shown in Fig. 2 (right).

## TOPOLOGICAL DENSITY AND A MEASURE OF SELFDUALITY

The *all-scale* topological charge density in terms of the overlap operator $D_N$ is

$$q(x) = -\,\mathrm{Tr}_{color,spinor}\left[\gamma_5\left(1 - \frac{a}{2}\,D_N(x,x)\right)\right],\qquad (3)$$

while the UV-filtered version $q_{\lambda_{cut}}(x)$ gets contributions to the trace only from modes with $|\lambda| < \lambda_{cut} \sim \Lambda_{QCD}$.

The density $q(x)$ satisfies the negativity of the topological correlator $C_q(r)$ [7] for $r \neq 0$ due to the diverging multiplicity of clusters at large enough $\beta$ or $a \lesssim 0.1$ fm. At $\beta = 8.45$ one finds $\approx 75$ such clusters per fm$^4$ that percolate at $|q(x)| = 0.25\ q_{max}$. Clusters with different fractal dimensions $d^* < 3$ are visible at $|q(x)|$ above the percolation threshold. The global 3D structure discovered by Horvath *et al.* [8] appears below that level. The UV-filtered density with $\lambda_{cut} = 200$ MeV forms only one cluster per fm$^4$ percolating at $|q_{\lambda_{cut}}(x)| = 0.05\ q_{\lambda_{cut}max}$.

Gattringer [9] proposed an UV filter for the field strength tensor using eigenmodes $\psi_\lambda(x)$ of the Dirac operator ($T^a$ denotes a color generator in the fundamental representation)

$$F_{\mu\nu}^a(x) \propto \sum_j \lambda_j^2 f_{\mu\nu}^a(x|j)\,,\ \text{with}\ f_{\mu\nu}^a(x|j) = -(i/2)\psi_{\lambda_j}^\dagger(x)\gamma_\mu\gamma_\nu T^a\psi_{\lambda_j}(x)\,.\qquad (4)$$

This prescription yields a filtered field strength up to an (undetermined) normalization when only low-lying modes are included. With a filtered action density $\tilde{s}(x) = \mathrm{Tr}\,F_{\mu\nu}(x)F_{\mu\nu}(x)$ and topological charge density $\tilde{q}(x) = \mathrm{Tr}\,F_{\mu\nu}(x)\tilde{F}_{\mu\nu}(x)$, one gets an estimator for the local (anti-)selfduality analogous to Eq. (2),

$$R(x) = (4/\pi)\arctan\sqrt{(\tilde{s}(x) - \tilde{q}(x))/(\tilde{s}(x) + \tilde{q}(x))} - 1 \in [-1, +1]\,.\qquad (5)$$

189

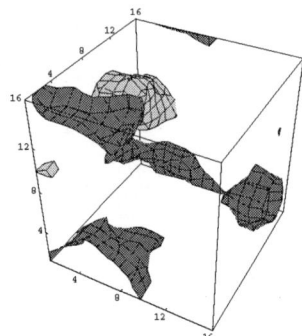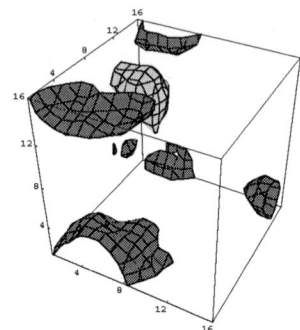

**FIGURE 3.** Comparison of the $R$-clusters for $R_{cut} = 0.99$ with the $q_{\lambda_{cut}}$-clusters for $\lambda_{cut} = 100$ MeV (at 1/10 of the maximal density) for one timeslice of a $16^3 \times 32$ lattice configuration with $Q = 0$ generated at $\beta = 8.45$. The similarity is the same in all timeslices.

This observable has good cluster properties if a cut separates (anti-)selfdual domains from the rest of the lattice. For $R_{cut} \lesssim 1$ the number of such domains is practically stable under changing the cutoff. It depends, however, on the number of modes used in the filter. $R(x)$ is correlated with the local chirality $X_\lambda(x) \approx \pm 1$ of the low eigenmodes, and the $R$-cluster structure corresponds to the cluster structure of $q_{\lambda_{cut}}$, as demonstrated in Fig. 3, if both are tuned to the relevant number of modes ($\lambda_{cut} \sim \Lambda_{QCD}$).

## CONCLUSION

The main conclusion is that overlap fermions can probe the lattice vacuum both in the ultraviolet and in the infrared. To see the phenomenologically known structure of 4D selfdual domains, some UV filtering by applying $\lambda_{cut} \sim \Lambda_{QCD}$ is necessary without the need of smoothing the gauge field. The *all-scale* topological charge density is multi-fractal, too, within the global 3D structure. It forms lower dimensional clusters above the percolation threshold. The phenomenological significance of the low-dimensional (singular) structures is still speculative.

## ACKNOWLEDGMENTS

E.-M. I. is grateful to the organizers for the opportunity to present this work. He is supported by the DFG Forschergruppe FOR 465 "Gitter-Hadronen-Phänomenologie".

## REFERENCES

1. E.-M. Ilgenfritz, et al. (QCDSF collaboration), in preparation.
2. V. Weinberg, et al. (2006), hep-lat/0610087.
3. P. H. Damgaard, *Nucl. Phys.* **B608**, 162–176 (2001), hep-lat/0105010.
4. C. Aubin, et al., *Nucl. Phys. Proc. Suppl.* **140**, 626–628 (2005), hep-lat/0410024.
5. J. Gattnar, et al., *Nucl. Phys.* **B716**, 105–127 (2005), hep-lat/0412032.
6. V. Weinberg, et al., (DIK collaboration), in preparation.
7. Y. Koma, et al., *PoS* **LAT2005**, 300 (2006), hep-lat/0509164.
8. I. Horvath, et al., *Phys. Rev.* **D68**, 114505 (2003), hep-lat/0302009.
9. C. Gattringer, *Phys. Rev. Lett.* **88**, 221601 (2002), hep-lat/0202002.

# Bosonic and Baryonic String Theory in Quantum Chromodynamics

## Julius Kuti

*Department of Physics, University of California, San Diego*
*La Jolla, California 92093, United States*

**Abstract.** Bosonic string formation in gauge theories is reviewed with particular attention to the confining flux in lattice QCD and its effective string theory description. Recent results on the Casimir energy of the ground state and the string excitation spectrum are analyzed in the Dirichlet string limit of large separation between static sources. The closed string-soliton (torelon) with electric flux winding around a compact dimension is discussed and a new bound state tower spectrum at baryon string junctions is presented.

**Keywords:** quark confinement, lattice QCD
**PACS:** 12.38Aw,12.38.Gc

## EFFECTIVE STRING THEORY

Polchinski and Strominger proposed a new way to construct an effective string theory of long flux lines emerging from quantum field theories [1]. They introduced the effective world sheet string action which is defined in light-cone coordinates as

$$S_{eff} = \frac{1}{4\pi} \int d\tau^+ d\tau^- \left[ \frac{1}{\alpha'/2} \partial_+ X^\mu \partial_- X_\mu + \beta \frac{\partial_+^2 X \partial_- X \partial_+ X \partial_-^2 X}{(\partial_+ X \partial_- X)^2} \right] + ..., \qquad (1)$$

where $\beta = \beta_c = \frac{D-26}{12}$ is required for the correct anomaly free theory. The action has manifest D-dimensional Poincare invariance and conformal symmetry. It is an effective theory only, because additional higher order terms may be required when the string action is expanded in derivative powers on large scales. The obvious example is the stiffness, or extrinsic curvature term, which will require added terms in Eq. 1. In the discussion below, I will use this effective string theory approach which is almost identical to the expansion of Lüscher and Weisz [2] which has one more free parameter to the same order of the expansion. Otherwise the two descriptions lead to identical prediction to the tested orders on large string scale of the infrared spectrum. References and a more detailed description of new developments is provided in a recent review [3]. Most of my work reported here has been done in collaboration with Jimmy Juge, Colin Morningstar, and Kieran Holland.

CP892, *Quark Confinement and the Hadron Spectrum VII*
edited by J. E. F. T. Ribeiro
© 2007 American Institute of Physics 978-0-7354-0396-3/07/$23.00

# CASIMIR ENERGY

The effective string prediction for the ground state energy of two static sources at separation $R$ is given asymptotically by

$$E_0(R) = \sigma R - \frac{\pi(D-2)}{24R} - \frac{\pi^2(D-2)^2}{1152\sigma R^3} , \qquad (2)$$

where the model dependent correction $\mathcal{O}(R^{-5})$ and a cutoff dependent constant, not contributing to $E_0'(R)$, are not included. There is some evidence, without rigorous proof, that the three terms of $E_0(R)$ are universal for bosonic strings[3]. Corrections to $E_0(R)$, starting at $\mathcal{O}(R^{-5})$, are not predictable without new terms in the effective string action. New physics, like string stiffness, will be coded in those higher order terms. Deviations are certainly expected from the simple Nambu-Goto form which has no new free parameters but its quantization becomes inconsistent on the $\mathcal{O}(R^{-5})$ scale in non-critical physical dimensions.

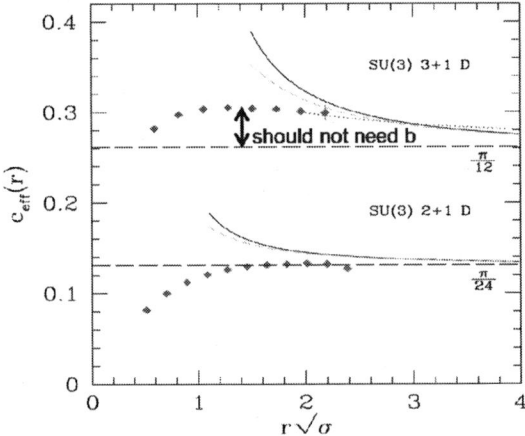

**FIGURE 1.** The solid lines are the NG predictions. The dashed lines, merging with the NG curves, designate $E_0(R)$ of Eq. 2. The diamond data points were obtained with high precision and the dotted line includes boundary operator corrections [4].

The effective Casimir term, defined by $C_{eff}(R) = -\frac{1}{2}R^3 E_0''(R)$, has the asymptotic behavior $C_{eff}(R) = \frac{\pi(D-2)}{24}(1 + \frac{\pi(D-2)}{8\sigma R^2} + ...)$ which isolates the Casimir energy with leading $R^{-2}$ correction. The definitive lattice simulation of $C_{eff}(R)$ came from an efficient method to measure the ground state energy with high precision [4] as shown in Fig. 1. The results had the popular interpretation of bosonic string formation on the scale $R \lesssim 1$ fm. In D=3 dimensions the agreement seemed to be a perfect match to $\pi/24$, but boundary operators had to be added to explain deviations from $\pi/12$ when D=4 [4]. In fact, the interpretation should have been the reverse. The universal $O(R^{-2})$ asymptotic behavior, represented by the dashed line in Fig. 1, is a measurable correction to $\pi/12$ and apparently reached by the data for D=4. Therefore the D=4 result does not require any boundary terms. It is in D=3 dimensions that the lattice results remain noticeably below the universal dashed line signaling strong boundary effects not being close enough to the asymptotic string description [5].

## DIRICHLET STRING SPECTRUM

In the 4D SU(3) Yang-Mills gauge model, which is the gluon sector of QCD, three exact quantum numbers determine the classification scheme of the gluon excitation spectrum in the presence of a static $q\bar{q}$ pair [6]. Restricted to the $R = 0.2 - 2$ fm range of selected

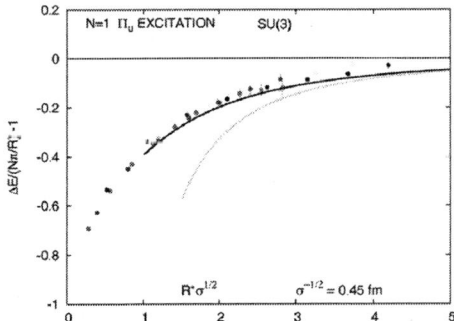

**FIGURE 2.** The lowest Dirichlet string excitation is displayed in the 4D SU(3) gauge model with reach to R=2 fm. Solid black line shows the full NG prediction which will break down at small $R$ values. The data points correspond to several gauge couplings, demonstrating scaling behavior in the spectrum. The dashed line represents universal prediction on the $\mathcal{O}(R^{-3})$ scale of the effective string action, as it was discussed earlier.

simulations, the energy gap $\Delta E$ above the ground state is compared to string predictions for the first excited state in Fig. 2. The quantity $R \cdot \Delta E / N\pi - 1$ is plotted to show percentage deviations from the asymptotic string levels. In sharp contrast, for $R \ll 1$ fm it was shown that the multipole expansion of the gluon field can be applied with success in the short-distance operator product expansion of gluon excitations around static color sources.[6] The observed short distance level ordering is very different from the string spectrum at short distances. For 0.5 fm $< R <$ 2 fm, a dramatic crossover of the energy levels toward a string-like spectrum is observed as $R$ increases. For example, the N=3 $\Sigma_u^-$ state breaks rapidly away from its N=1 $\Pi_u$ short-distance $O(3)$ degeneracy partner to approach the ordering and degeneracies expected from bosonic string theory [3, 6].

## TORELON STRING SPECTRUM

The spectrum of the closed string with unit winding number (torelon) was discussed earlier [3]. Fig. 3 shows two torelon states with the same principal quantum number, $N + \tilde{N} = 2$, but different momenta along the winding direction. A single right mover with $N = 2, \tilde{N} = 0$ and running with two units of momentum in the compactified direction has positive $\mathcal{O}(R^{-3})$ energy correction to the asymptotic string level and split from the composite state of back to back right and left movers with zero total momentum in the compact dimension. The dramatic fine structure is in agreement with lattice simulations [3] both for the Z(2) gauge model in three dimensions and the 4D SU(3) torelon excitation spectrum as well [7].

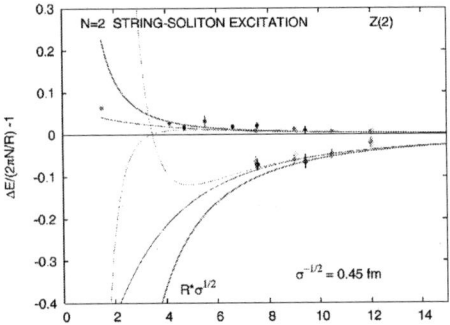

**FIGURE 3.** The spectra correspond to two different torelon excitations with $w = 1, N + \tilde{N} = 2$. Above the null line the spectrum of the $n = 2, N = 2, \tilde{N} = 0$ state is depicted with $p_n = 4\pi/R$ momentum in the compact dimension. Below the null line, the spectrum of the state $N = 1, \tilde{N} = 1, p_n = 0$ is shown. There are two pairs of solid lines. The thin lines correspond to the spectrum which coincides with the exact NG prediction. The thick lines show the universal $\mathscr{O}(R^{-3})$ predictions of the effective string action after the square root is expanded in inverse powers of $R$. The dotted lines are $\mathscr{O}(R^{-5})$ NG terms which were left in the plot only to indicate the divergence of the expansion as the tachion singularity is approached.

## BARYON STRING JUNCTION

When three static sources are inserted in a baryon configuration, a three-string junction with Y-shape is expected to form asymptotically. Two questions stand out: (1) Is the asymptotic baryon string really Y-shaped asymptotically? (2) What is the spectrum and Casimir energy of the Y-shape three-string from first principles string theory calculation and lattice simulations? In collaboration with Holland, I looked into the space-time picture of baryon string formation in the 3D Z(3) gauge model [8]. We find convincing evidence for baryon three-string formation with Y-junction in Fig. 4, continuing the investigation with the baryon string spectrum and the related baryon Casimir energy. The upper left part of the figure depicts the ground state of the Y-shape baryon string in the 3D Z(3)gauge model. The upper right part shows a typical string excitation which runs on the segments of the Y-shape Dirichlet string. The two lower parts are unexpected and very interesting. A tower excitation spectrum of localized bound states is found at the baryon junction, and the first two junction states are shown. Their significance within the context of effective 3-string theory remains to be understood.

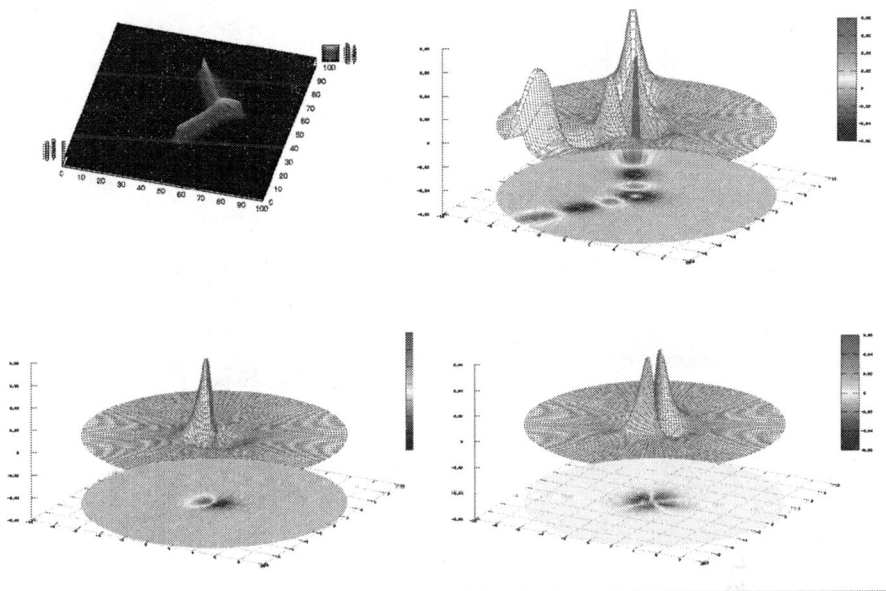

**FIGURE 4.** Baryon junction and its excitation spectrum.

## ACKNOWLEDGEMENTS

I wish to thank the organizers of QCHSVII for their hospitality. This work was supported by the DOE, Grant No. DE-FG03-97ER40546.

## REFERENCES

1. J. Polchinski and A. Strominger, *Effective string theory*, Phys. Rev. Lett. **67** (1991) 1681.
2. M. Lüscher and P. Weisz, *String excitation energies in SU(N) gauge theories beyond the free-string approximation*, JHEP **0407** (2004) 014.
3. J. Kuti, *Lattice QCD and String Theory*, Proceedings of Lattice 2005, PoS(LAT2005)001 [hep-lat/0511023].
4. M. Lüscher and P. Weisz, *Quark confinement and the bosonic string*, JHEP **0207** (2002) 049.
5. K.J. Juge, J. Kuti, C.J. Morningstar, *QCD string formation and the Casimir energy*, Proceedings of *Color confinement and hadrons in quantum chromodynamics*, Wako, Japan (2004)233 [hep-lat/0401032].
6. K.J. Juge, J. Kuti, C.J. Morningstar, *The fine structure of the QCD string spectrum*, Phys. Rev. Lett. **90** (2003) 161601.
7. K.J. Juge, J. Kuti, F. Maresca, C.J. Morningstar, M. Peardon, *Excitations of torelon*, Nucl. Phys. Proc. Suppl. **129** (2004) 703.
8. K. Holland and J. Kuti, *unpublished*.

# The Spectrum of Yang-Mills Theory in 2+1 Dimensions, Analytically

R.G. Leigh*, D. Minic† and A. Yelnykov†

*Department of Physics, University of Illinois, Urbana IL 61822, USA.
†Department of Physics, Virginia Tech, Blacksburg VA 24061, U.S.A.

**Abstract.** We review our recent work on the glueball spectrum of pure Yang-Mills theory in 2+1 dimensions. The calculations make use of Karabali-Nair corner variables in the Hamiltonian formalism, and involve a determination of the leading form of the ground-state wavefunctional.

**Keywords:** Yang-Mills, glueballs
**PACS:** 12.38.Lg, 11.15.Pg, 11.15.Tk

The understanding of the non-perturbative dynamics of Yang-Mills theory is one of the grand problems of theoretical physics. The $2+1$-dimensional theory is expected on many grounds to share the essential features of its $3+1$ dimensional cousin, such as asymptotic freedom and confinement, yet is distinguished by the existence of a dimensionful coupling constant. This simple fact has important consequences.

Here, we describe the determination of the ground state wave-functional in a specific approximation, the knowledge of which enables us to determine the mass gap, string tension and the glueball mass spectrum. The results are in excellent agreement with the lattice data in the planar limit [1]. Further details of this work, as well as a full set of references, may be found in [2, 3]. Here we will concentrate on the main ideas and outstanding issues.

We will describe the pure $SU(N)$ Yang-Mills theory by transforming to *corner variables* [4]. One advantage of these variables is that the passage to a lattice regulator is straightforward. This change of variables may be done exactly, the Jacobian of the path integration measure being computable. Given a local coordinatization of space, we introduce the straight Wilson lines $M_i(x) = Pexp\left[-\int_{-\infty}^{x} dx_i A_i\right]$ or equivalently $A_i = -\partial_i M_i \, M_i^{-1}, \forall i$. Gauge transformations act linearly $M_i \mapsto gM_i$ and local gauge invariant variables may be constructed, $H_{ij}(x) = M_i^{-1}(x)M_j(x)$. These are the (generally constrained) corner variables. In two spatial dimensions, a complex spatial coordinatization may be used, leading to the Karabali-Kim-Nair parameterization[5, 6] $A = -\partial M \, M^{-1}, \bar{A} = M^{\dagger-1}\bar{\partial}M^\dagger, H = M^\dagger M$. These variables possess a new local *holomorphic invariance* $M(z,\bar{z}) \mapsto M(z,\bar{z})h^\dagger(\bar{z}), M^\dagger(z,\bar{z}) \mapsto h(z)M^\dagger(z,\bar{z}), H(z,\bar{z}) \mapsto h(z)H(z,\bar{z})h^\dagger(\bar{z})$ which must be preserved. A holomorphic connection for this symmetry is $J = \partial H \, H^{-1}, J \mapsto hJh^{-1} + \partial h \, h^{-1}$ and the corresponding covariant derivative $D = \partial - J = M^\dagger \nabla M^{\dagger-1}$, with $\nabla = \partial + A$ the usual gauge-covariant derivative. The path integral measure may be written $d\mu[A] \sim d\mu[H]e^{2NS_{wzw}[H]}$. The Hamiltonian, written as a functional differential operator in $J$'s, was constructed by Karabali-Nair and takes the

CP892, *Quark Confinement and the Hadron Spectrum VII*
edited by J. E. F. T Ribeiro
© 2007 American Institute of Physics 978-0-7354-0396-3/07/$23.00

collective field form

$$\mathcal{H}_{KN}[J] = T + V = m \left( \int_x J^a(x) \frac{\delta}{\delta J^a(x)} + \int_{x,y} \Omega_{ab}(x,y) \frac{\delta}{\delta J^a(x)} \frac{\delta}{\delta J^b(y)} \right) + \frac{2}{g^2} \int_x \bar{\partial} J^a \partial J^a \quad (1)$$

where $m$ is the 't Hooft coupling $m = g^2 N/2\pi$. In principle, wavefunctionals, expressed as functionals of $J$, may be found by solving the functional Schrödinger equation. Note that at weak coupling, the Hamiltonian is dominated by the trailing potential term, while at strong coupling, it is dominated by the kinetic operator. By looking at the action of parity and charge conjugation, it becomes clear that holomorphic invariant wavefunctionals may be constructed from $\bar{\partial} J = [D, \bar{\partial}]$ and $\Delta = \{D, \bar{\partial}\}$, $\Psi = \Psi[\bar{\partial} J/m^2, L = \Delta/2m^2]$. As we shall see, in considering the ground state wavefunctional, it appears to be useful to restrict our attention to the gauge and holomorphic invariant form

$$\Psi_0 \simeq exp \left[ -\frac{N}{2\pi m^2} \int tr \, \bar{\partial} J \, K(L) \bar{\partial} J + \dots \right]. \quad (2)$$

We note that although this is not the most general functional (terms quartic in $\bar{\partial} J$, etc., are allowed by symmetries), it is highly non-trivial in the sense that the kernel $K(L)$ is allowed to be an arbitrary function, whose exact form will be determined by the Schrödinger equation, and its asymptotic features must be consistent with asymptotic freedom *and* confinement. We are assuming large $N$ here, which implies a single trace but does not further simplify the exponent. The real importance of large $N$ is in the expectation that interactions amongst gauge invariant states are eliminated and that the states are stable. Note further that the form (2) is not equivalent to a simple (covariant) derivative expansion, as all orders in derivatives are included. Since the magnetic field $B = M^\dagger \bar{\partial} J \, M^{\dagger -1}$, it appears that this may be thought of as a sort of gauge curvature expansion. In a sense, $\bar{\partial} J$ may be thought of as an adjoint constituent, with glueball states, which are normally identified with vibrating closed strings, modeled as configurations of a pair of constituents connected by strings. Whether or not this picture is borne out, is a matter for experiment to decide. As we detail below, we are able to compute glueball masses which agree remarkably well with the available lattice data. The form (2) may not be sufficient for the calculation of other observables.

The first term in the kinetic operator counts the number of $J$'s in a functional; through detailed computations, it appears that the second term in $T$ acts essentially to restore holomorphic invariance. This implies that, defining $\mathcal{O}_n \equiv tr \, \bar{\partial} J L^n \bar{\partial} J$,

$$T \mathcal{O}_n \simeq (2 + n) \mathcal{O}_n + \dots \quad (3)$$

The ellipsis contains mixing with operators containing more factors of $\bar{\partial} J$ which will not concern us. Eq. (3), crucial to the results that follow, may be explicitly demonstrated for low values of $n$, but has not been firmly established in general. The required calculations rely on a holomorphic invariant regulator and the calculations are tedious; better calculational methods remain to be found.

Now, given the assumed form of the ground-state wavefunctional (2), the Hamiltonian (1) and the result (3), we find that the Schrödinger equation takes the form

$$\mathscr{H}_{KN}\Psi_0 = E_0\Psi_0 = \left[\ldots + \int tr\, \bar{\partial} J\, \mathscr{R}\, \bar{\partial} J + \ldots\right]\Psi_0,\qquad(4)$$

The first ellipsis contains (divergent) terms contributing to the vacuum energy, while the trailing ellipsis contains terms of higher order in $\bar{\partial} J$. Given the assumed form of the vacuum wavefunctional, this truncation of the Schrödinger equation is appropriate and consistent. The quantity $\mathscr{R}$ may be found by regarding $K$ as a power series in $L$, and one finds

$$\mathscr{R} = -K(L) - \frac{L}{2}\frac{d}{dL}K(L) + LK^2(L) + 1\qquad(5)$$

The Schrödinger equation requires that we set this to zero, resulting in a differential equation for $K$ of the Riccati type. Through a series of redefinitions, this may be cast as a Bessel equation, and one obtains $K(L) = \frac{1}{\sqrt{L}}\frac{J_2(4\sqrt{L})}{J_1(4\sqrt{L})}$. Although there are other solutions to the differential equation, no other is normalizable in the given path integral measure. We note that although this is a very complicated function (which we now regard as a function of momentum), it has the asymptotics $K(p) \to 2m/p$ as $p \to \infty$ (asymptotic freedom) and $K(p) \to 1$ as $p \to 0$ (confinement). We have thus found that *the only normalizable ground state wavefunctional is consistent with both confinement and asymptotic freedom*. As argued by Karabali and Nair, the infrared limit of our wavefunctional $K$ implies a string tension $\sqrt{\sigma} \simeq \frac{g^2 N}{\sqrt{8\pi}}$ obtained by a dimensional reduction argument. This result agrees precisely with lattice data. However, it is not possible by this line of reasoning to detect different string tensions appropriate to different gauge representations. It is believed that this is consistent, being an artifact of the large $N$ approximation known as Casimir scaling [7] (in particular, this result would be inconsistent at finite $N$). In effect, there is an order of limits problem at large $N$ and large distance.

The ratio of Bessel functions has a rich analytic structure and encodes the mass spectrum of the theory. By Fourier transforming, we find $K^{-1}(|x-y|) = -\frac{1}{4\sqrt{2\pi}|x-y|}\sum_{n=1}^{\infty}(M_n)^{3/2}e^{-M_n|x-y|}$ where $M_n = \gamma_{2,n}m/2$ and $J_2(\gamma_{2,n}) = 0$. As we will see, this result has direct consequences for the mass spectrum. To probe the mass spectrum, we consider pair correlation functions of gauge invariant operators with definite spin, parity and charge conjugation quantum numbers. The simplest $0^{++}$ probe operator is $tr\bar{\partial} J\bar{\partial} J$, and we wish to compute $\langle tr\bar{\partial} J\bar{\partial} J(x)tr\bar{\partial} J\bar{\partial} J(y)\rangle = \int d\mu[H]e^{2NS_{WZW}[H]}|\Psi_0|^2\, tr\bar{\partial} J\bar{\partial} J(x)tr\bar{\partial} J\bar{\partial} J(y)$ at large spatial separation. To proceed, we first rewrite the measure as an integral over $J$. Note that if we introduce a variable $\bar{J} = \bar{\partial} HH^{-1}$, then we have the 'reality condition' $\bar{\partial} J = [D,\bar{J}]$. Given the gauge transformations $\delta A \simeq [\nabla, \delta MM^{-1}], \delta\bar{A} \simeq [\bar{\nabla}, M^{-\dagger}\delta M^{\dagger}]$, the measure is

$$\frac{d\mu[A]}{Vol\, G} = det\nabla\bar{\nabla}\,\frac{d\mu(M,M^{\dagger})}{Vol\, G} = det\nabla\bar{\nabla}\, d\mu[H]$$

with $d\mu[H]$ the measure corresponding to $ds_{inv}^2 = \int Tr(\delta HH^{-1})^2$. We now transform this to the $J$ variables, namely, we would like to find the measure corresponding to the distance $ds_J^2 = \int Tr(\delta \bar{\partial} J)^2$. Formally, this may be found by first complexifying $H$

$$d\mu[H] \to d\mu[H, H^\dagger] \delta(H^\dagger - H) = \frac{d\mu[\bar{\partial}J, [D, \bar{J}]]}{Det^2 \bar{\partial}D} \frac{\delta(\bar{\partial}J - [D, \bar{J}])}{Det^{-1}\bar{\partial}D} \equiv \frac{1}{Det\bar{\partial}D} d\mu[\bar{\partial}J] \quad (6)$$

So we have

$$\frac{d\mu[A]}{Vol\ G} = \frac{det\nabla\bar{\nabla}}{Det\bar{\partial}D} d\mu[\bar{\partial}J] \quad (7)$$

Since $D = M^\dagger \nabla M^{-\dagger}$ and $\bar{\partial} = M^\dagger \bar{\nabla} M^{-\dagger}$, the determinant factor cancels exactly, and we obtain

$$\langle tr\bar{\partial}J\bar{\partial}J(x) tr\bar{\partial}J\bar{\partial}J(y) \rangle = \int d\mu[\bar{\partial}J] |\Psi_0|^2 tr\bar{\partial}J\bar{\partial}J(x) tr\bar{\partial}J\bar{\partial}J(y) \quad (8)$$

To the approximation in which we ignore interactions (that is, take $K(L)$ to be a function of momentum), we may thus regard $\bar{\partial}J$ as a 'constituent,' with correlators determined by $K(p)$. Consequently, we find

$$\langle tr\bar{\partial}J\bar{\partial}J(x) tr\bar{\partial}J\bar{\partial}J(y) \rangle \simeq K^{-2}(|x-y|) = \sum_{m,n} \frac{\#}{|x-y|} e^{-(M_n+M_m)|x-y|} \quad (9)$$

a form consistent with single particle $0^{++}$ poles of mass $m_{m,n} = M_m + M_n$. We note that $K^{-1}$ is not the propagator of a physical mode, but it does determine physical propagators. Resulting masses for $0^{++}$ states, with lattice comparisons, [1] are collected in the table.

| State | Lattice, $N \to \infty$ | Our prediction | Diff, % |
|-------|-------------------------|----------------|---------|
| $0^{++}$ | $4.065 \pm 0.055$ | 4.098 | 0.8 |
| $0^{++*}$ | $6.18 \pm 0.13$ | 5.407 | -- |
| $0^{++**}$ | $6.18 \pm 0.13$ | 6.716 | -- |
| $0^{++***}$ | $7.99 \pm 0.22$ | 7.994 | 0.05 |
| $0^{++****}$ | $9.44 \pm 0.38$ | 9.214 | 2.4 |

We see very good agreement, apart from the second and third states. We note though that the average of these two states coincides with the second lattice state, and we may take this as a prediction that these two states were not resolved in the lattice studies. Similar results may be obtained for other states using suitable probe operators and results are equally encouraging. The resulting spectrum appears to have an exponentially rising density of states, which may be taken as a manifestation of the effective QCD string.

# REFERENCES

1. M. Teper, Phys. Rev. D **59** (1999) 014512;B. Lucini and M. Teper, Phys. Rev. D **66** (2002) 097502 .
2. R. G. Leigh, D. Minic and A. Yelnikov, Phys. Rev. Lett. **96** (2006) 222001.
3. R. G. Leigh, D. Minic and A. Yelnikov, hep-th/0604060.
4. I. Bars, Phys. Rev. Lett. **40**, 688 (1978).
5. D. Karabali and V. P. Nair, Nucl. Phys. B **464** (1996) 135; Phys. Lett. B **379** (1996) 141.
6. D. Karabali, C. J. Kim and V. P. Nair, Nucl. Phys. B **524** (1998) 661 ; Phys. Lett. B **434** (1998) 103 ; Phys. Rev. D **64** (2001) 025011 and references therein.
7. J. Greensite, Prog. Part. Nucl. Phys. **51** (2003) 1.

# Split-quaternionic representation of SDYM SU$(1,1)$ instantons in $S^2_- \times S^2_+$

Sungwook Lee* and Khin Maung Maung†

*Department of Mathematics, University of Southern Mississippi, Hattiesburg, MS 39406, USA
†Department of Physics and Astronomy, University of Southern Mississippi, Hattiesburg, MS 39406, USA

**Abstract.** Using split-quaternions, we find explicit SDYM SU$(1,1)$ instanton solutions in $S^2_- \times S^2_+$ which is the conformal compactification of the semi-Euclidean 4-spacetime $\mathbb{R}^{2+2}$ of split-signature $(-,-,+,+)$. It is also shown that SDYM and ASDYM fields in $S^2_- \times S^2_+$ can be described as simple split-quaternionic 2-forms.

The first explicit solution to non-Abelian gauge theory was found by Belavin, Polyakov, Schwarz and Tyupkin (BPST). [1] Now the natural question to ask is whether explicit SDYM or ASDYM instanton solutions can be found with other gauge groups and other manifolds. This can be answered by looking at the identity $**\sigma = (-1)^{p(n-p)} s\sigma$ where $\sigma$ is a $p$-form defined on an $n$-dimensional manifold endowed with a metric $g$ and $s \equiv \det g$. In the case of $n = 4$ and $p = 2$, there are only two 4-dimensional spaces that satisfy the Hodge duality $*^2 = id$. One is 4-dimensional Euclidean space $\mathbb{R}^4$ and the other is the semi-Euclidean spacetime $\mathbb{R}^{2+2}$ of split-signature $(-,-,+,+)$. In this paper, we find explicit SDYM instanton solutions in $\mathbb{R}^{2+2}$ with the gauge group SU$(1,1)$. Let $\mathbb{R}^{2+2}$ denote $\mathbb{R}^4$ with rectangular coordinates $x_0, x_1, x_2, x_3$ and the semi-Riemannian metric $ds^2 = -dx_0^2 - dx_1^2 + dx_2^2 + dx_3^2$ of split-signature $(-,-,+,+)$. One can show that the conformal compactification of $\mathbb{R}^{2+2}$ is $S^2 \times S^2$. To be more accurate $S^2_- \times S^2_+$ is a double cover of the conformal compactification of $\mathbb{R}^{2+2}$. [2] A vector $\underline{x} = (x_0, x_1, x_2, x_3)$ in $\mathbb{R}^{2+2}$ can be identified with

$$\underline{x} = x_0 \mathbf{1} + x_1 \mathbf{i} + x_2 \mathbf{j}' + x_3 \mathbf{k}'.$$

Where the basis set given by $\{\mathbf{1}, \mathbf{i}, \mathbf{j}', \mathbf{k}'\}$ forms the standard basis for the algebra of split-quaternions, i.e, $\mathbf{i}^2 = -1$, $\mathbf{j}'^2 = \mathbf{k}'^2 = 1$, $\mathbf{ij}' = -\mathbf{j}'\mathbf{i} = \mathbf{k}'$, $\mathbf{j}'\mathbf{k}' = -\mathbf{k}'\mathbf{j}' = -\mathbf{i}, \mathbf{k}'\mathbf{i} = -\mathbf{ik}' = \mathbf{j}'$.

Therefore, there is a natural correspondence between a vector $\underline{x} = (x_0, x_1, x_2, x_3)$ in $\mathbb{R}^{2+2}$ and a split-quaternion $x = x_0 \mathbf{1} + x_1 \mathbf{i} + x_2 \mathbf{j}' + x_3 \mathbf{k}'$. [3] We note that the conjugate of $x$ is given by $\bar{x} = x_0 \mathbf{1} - x_1 \mathbf{i} - x_2 \mathbf{j}' - x_3 \mathbf{k}'$. The space of time-like unit split-quaternions is identified with the noncompact Lie group SU$(1,1)$, i.e.

$$\mathrm{SU}(1,1) = \{g \in \mathrm{SL}(2,C) : g\sigma_3 g^* = \sigma_3\}$$

where $\sigma_3$ is the third Pauli matrix in the standard representation. Now, we are ready to find explicit solutions for the Yang-Mills equations $dF = d^*F = 0$ subject to self-dual

CP892, Quark Confinement and the Hadron Spectrum VII
edited by J. E. F. T. Ribeiro

condition $*F = F$ in $\mathbb{R}^{2+2}$. First, we define open sets covering
$S_-^2 \times S_+^2$ by $U_1 = (S_-^2 \setminus N) \times (S_+^2 \setminus N)$, $U_2 = (S_-^2 \setminus N) \times (S_+^2 \setminus S)$, $U_3 = (S_-^2 \setminus S) \times (S_+^2 \setminus N)$ and $U_4 = (S_-^2 \setminus S) \times (S_+^2 \setminus S)$ where $N$ and $S$ are the north and the south poles of 2-sphere $S^2$. Then $\mathcal{U} = \{U_\alpha : \alpha = 1, 2, 3, 4\}$ is an open cover of $S_-^2 \times S_+^2$. If we were to construct a principal G-bundle over $S_-^2 \times S_+^2$ with respect to $\mathcal{U}$, the transition functions would be defined as $g_{\alpha\beta} : U_\alpha \cap U_\beta \to G$. These transition functions fall into homotopy classes and $U_\alpha \cap U_\beta$ is homotopy equivalent to $S^1 \times \mathbb{R}^2$ or $S^1 \times S^1$. Note that $S^1 \times S^1 \subset S^1 \times \mathbb{R}^2$ and $S^1 \times \mathbb{R}^2$ is topologically equivalent to AdS$_3$. As is well-known, AdS$_3$ has a Lie group structure and is identified with the Lie group SU$(1,1)$. Hence, for our problem, we can naturally choose an appropriate gauge group to be SU$(1,1)$. In the principal SU$(1,1)$ bundle over $S_-^2 \times S_+^2$, the transition functions will be defined as $\tilde{g}_{\alpha\beta} : S^1 \times \mathbb{R}^2$ (or $S^1 \times S^1$) $\longrightarrow$ SU$(1,1)$.

On the boundary $S^1 \times \mathbb{R}^2 \cong$ AdS$_3$ of $\mathbb{R}^{2+2}$, the connection 1-form $A_\mu$ has the pure gauge so that the action will be finite. Hence, we consider connection 1-form $A_\mu$ in $\mathbb{R}^{2+2}$ in the form.

$$A_\mu = f(|x|^2)g^{-1}(x)\partial_\mu g(x),$$

where $f(|x|^2)$ is some function that approaches to 1 as $|x|^2 \to \pm\infty$. Note that unlike in the case of Euclidean space, we have to consider $|x|^2 \to \pm\infty$ because $x$ can be a space-like or a time-like vector in $\mathbb{R}^{2+2}$. Next, we take $g(x)$ to be a SU$(1,1)$-valued function. So, we can write $g(x) = g_\alpha(|x|^2)\tau^\alpha$, where $\tau^0 = 1$, $\tau^1 = \mathbf{i}'$, $\tau^2 = \mathbf{j}'$, $\tau^3 = \mathbf{k}'$. Here, the unknown function $g(|x|^2)$ is to be found. Now by standard procedure [1,4] we can find the explicit solutions for the self-dual connection $A_\mu$ as

$$A_\mu(|x|^2) = \frac{1}{\pm\lambda^2 - |x|^2}(x_\alpha \sigma_3 (\tau^\alpha)^\dagger \sigma_3 \tau_\mu - x_\mu \tau^0) \quad \text{if } \mu = 0, 1$$

$$A_\mu(|x|^2) = \frac{1}{\pm\lambda^2 - |x|^2}(x_\alpha \sigma_3 (\tau^\alpha)^\dagger \sigma_3 \tau_\mu + x_\mu \tau^0) \quad \text{if } \mu = 2, 3,$$

where $+$ sign is for $|x|^2 < 0$ and $-$ sign for $|x|^2 > 0$.

We are now ready to describe self-dual Yang-Mills fields in $S_-^2 \times S_+^2$ in terms of split-quaternions. Let $\mathscr{A}$ be an SU$(1,1)$ connection, i.e., $\mathscr{A} \in$ su$(1,1)$. Note that the Lie algebra su$(1,1)$ is isomorphic to Minkowski 3-space $\mathbb{R}^{2+1}$ which coincides with the space of pure imaginary split-quaternions. Hence, $\mathscr{A}$ can be written as

$$\mathscr{A} = \text{Im}A(x)dx = A_\mu dx_\mu,$$

where $A(x) = \tilde{A}_0(x)\mathbf{1} + \tilde{A}_1(x)\mathbf{i}\tilde{A}_2(x)\mathbf{j}' \mid \tilde{A}_3(x)\mathbf{k}'$ and $dx = dx_0\mathbf{1} + dx_1\mathbf{i}' + dx_2\mathbf{j}' + dx_3\mathbf{k}'$. Therefore we can write

$$\mathscr{A} = (\tilde{A}_1\mathbf{i} + \tilde{A}_2\mathbf{j}' + \tilde{A}_3\mathbf{k}')dx_0 + (\tilde{A}_0\mathbf{i} + \tilde{A}_3\mathbf{j}' - \tilde{A}_2\mathbf{k}')dx_1$$

$$+ (\tilde{A}_3\mathbf{i} + \tilde{A}_0\mathbf{j}' + \tilde{A}_1\mathbf{k}')dx_2 + (-\tilde{A}_2\mathbf{i} - \tilde{A}_1\mathbf{j}' + \tilde{A}_0\mathbf{k}')dx_3.$$

For brevity we assume that $x$ is a time-like vector in $\mathbb{R}^{2+2}$. Comparing this last expression with the previous expression we get

$$\mathscr{A} = \operatorname{Im} \frac{\bar{x}dx}{\lambda^2 - |x|^2} = \frac{1}{2} \frac{\bar{x}dx - xd\bar{x}}{\lambda^2 - |x|^2}.$$

Hence, the curvature 2-form $F = d\mathscr{A} + \mathscr{A} \wedge \mathscr{A}$ can be written as

$$F = \frac{\pm\lambda^2}{(\lambda^2 - |x|^2)^2} d\bar{x} \wedge dx$$

Note that $\frac{\pm\lambda^2}{(\lambda^2-|x|^2)^2} d\bar{x} \wedge dx$ is pure imaginary. Therefore, the SDYM field $F$ is given by

$$F = -\frac{\lambda^2}{(\lambda^2 + |x|^2)^2} d\bar{x} \wedge dx \text{ if } \text{x is space} - \text{like}$$

$$F = \frac{\lambda^2}{(\lambda^2 - |x|^2)^2} d\bar{x} \wedge dx \text{ if } \text{x is time} - \text{like}.$$

For ASDYM fields, one simply replace $x$ by its conjugate $\bar{x}$.

We found explicit SDYM instanton solutions in $S_- \times S_+$, where $S_- \times S_+$ is regarded as the conformal compactification of the semi-Euclidean 4-spacetime $\mathbb{R}^{2+2}$ of split-signature $(-,-,+,+)$. The noncompact gauge group $SU(1,1)$ is naturally introduced as an appropriate gauge group due to the geometric nature of $\mathbb{R}^{2+2}$. We have seen that the semi-Euclidean 4-spacetime $\mathbb{R}^{2+2}$ is identified with the algebra of split-quaternions and they played a crucial role to find explicit SDYM $SU(1,1)$ instanton solutions in $S_- \times S_+$.

## ACKNOWLEDGMENTS

The first named author wishes to thank Dr. Gueo Grantcharov and Dr. Jun-ichi Inoguchi for informing him about Mason's work on ASDYM instanton solutions in split-signature. He also wishes to thank Dr. Jun-ichi Inoguchi and Prof. Lionel Mason for helpful comments. The second named author would like to thank Jose L. Goity for many useful conversations.

## References

[1] A. A. Belavin, A. M. Polyakov, A. S. Schwarz, Yu. S. Tyuokin, *Pseudoparticle solutons of the Yang-Mills equations*, Phys. Lett. **B 59** (1975) 233.

[2] L. J. Mason and N. M. J. Woodhouse, Integrability,Self-Duality and Twister Theory, Oxford University Press, 1996. L. J. Mason.

[3] A. Fujioka and J. Inoguchi, *Spacelike Surfaces with harmonic inverse mean curvature*, J. Math. Sci. Univ. Tokyo 7 (2000), no.4,657-698., J. Inoguchi and M. Toda, *Timelike minimal surfaces via loop groups*, Acta Appl. Math. **83** (2004),313-335.

[4] C. Nash and T. Sen, Geometry and Topology for Physicists, Academic Press, 1992.

# Infrared Maximally Abelian Gauge

Tereza Mendes, Attilio Cucchieri and Antonio Mihara

*Instituto de Física de São Carlos, Universidade de São Paulo, Caixa Postal 369,*
*13560-970 São Carlos, SP, Brazil*

**Abstract.** The confinement scenario in Maximally Abelian gauge (MAG) is based on the concepts of Abelian dominance and of dual superconductivity. Recently, several groups pointed out the possible existence in MAG of ghost and gluon condensates with mass dimension 2, which in turn should influence the infrared behavior of ghost and gluon propagators. We present preliminary results for the first lattice numerical study of the ghost propagator and of ghost condensation for pure $SU(2)$ theory in the MAG.

**Keywords:** Yang-Mills theory; Green's functions; Confinement; Abelian projection
**PACS:** 11.15.Ha 12.38.Aw

The study of the infra-red (IR) limit of QCD is of central importance for understanding the mechanism of confinement. Despite being non-gauge-invariant, gluon and ghost propagators are powerful tools in the (non-perturbative) investigation of this limit. In recent years, (gauge-dependent) condensates of mass dimension two have received considerable attention. An example of such objects is the ghost condensate [1, 2], related to the breakdown of a global $SL(2,R)$ symmetry. In particular, in MAG the diagonal and off-diagonal components of the ghost propagators are expected to be modified by ghost condensation. In this paper we present preliminary results of lattice studies of the Faddeev–Popov (FP) matrix for pure $SU(2)$ theory in the MAG. We consider the ghost propagator, the ghost condensate and the smallest eigenvalue of the FP matrix.

On the lattice, for the $SU(2)$ case, the MAG is obtained (see e.g. [3]) by minimizing the functional

$$S = -\frac{1}{2dV}\sum_{x,\mu} Tr\left[\sigma_3 U_\mu(x)\sigma_3 U_\mu^\dagger(x)\right].\tag{1}$$

At any local minimum one has that the Faddeev-Popov matrix, defined as

$$\sum_{by} M^{ab}(x,y)\gamma^b(y) = \sum_\mu \gamma^a(x)[V_\mu(x)+V_\mu(x-e_\mu)] + 2\{\gamma^a(x-e_\mu)[1-2(U_\mu^0(x))^2]$$

$$-2\sum_b \gamma^b(x-e_\mu)[\varepsilon_{ab}U_\mu^0(x)U_\mu^3(x) + \sum_{cd}\varepsilon_{ad}\varepsilon_{bc}U_\mu^d(x)U_\mu^c(x)]\},\tag{2}$$

is positive-definite. Here the color indices take values $1,2$ and we follow the notation $U_\mu(x)=U_\mu^0(x)\mathbb{1}+i\,\sigma^a U_\mu^a(x)$ and $V_\mu(x)=(U_\mu^0(x))^2+(U_\mu^3(x))^2-(U_\mu^1(x))^2-(U_\mu^2(x))^2$, where $\sigma^a$ are the 3 Pauli matrices. Notice that (as in Landau gauge [7]) this matrix is symmetric under the simultaneous exchange of color and space-time indices. Using the relation $U_\mu(x)=\exp[-iag_0 A_\mu(x)]$ one finds (in the formal continuum limit $a\to 0$) the standard continuum results [4] for the stationary conditions above and for $M^{ab}(x,y)$.

CP892, *Quark Confinement and the Hadron Spectrum VII*
edited by J. E. F. T. Ribeiro
© 2007 American Institute of Physics 978-0-7354-0396-3/07/$23.00

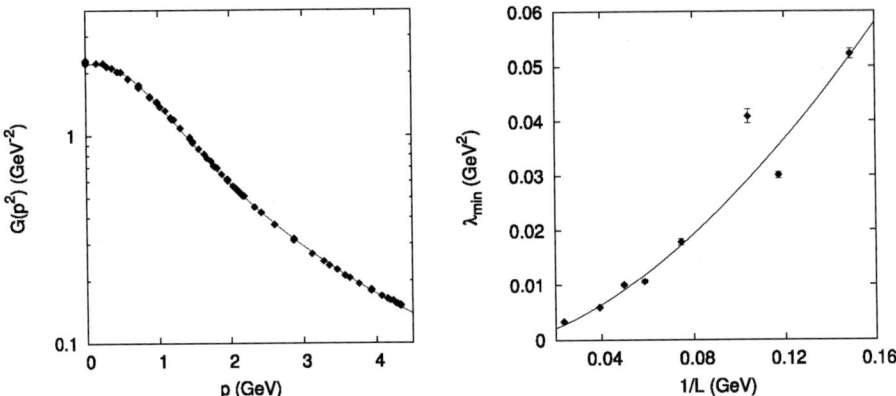

**FIGURE 1.** Left: plot of $G(p^2)$ as a function of improved $p$ for lattice volumes $V = 16^4$, $24^4$, $40^4$ and $\beta = 2.2$. Right: plot of the smallest eigenvalue of the FP operator, as a function of the inverse linear size of the system.

We have considered four values of $\beta$ (2.2, 2.3, 2.4, 2.512) and lattice volumes up to $40^4$. Our results for the gluon propagators are in agreement with the study by Bornyakov et al. [3]: we see a clear suppression of the off-diagonal propagators compared to the diagonal (transverse) one, supporting Abelian dominance. We have fitted our data for the various gluon propagators (at all values of $V$ and $\beta = 2.2$), obtaining the following behaviors. For $D(p^2)$ (transverse) diagonal, our data favor a Stingl-Gribov form

$$D(p^2) = \frac{1 + d\,p^2}{a + b\,p^2 + c\,p^4},\tag{3}$$

with a mass $m = \sqrt{a/b} \approx 0.72\,GeV$. Note that the above equation corresponds to a pair of complex conjugate poles $z$ and $z^*$. We can thus write $z = x + iy$ with $x = b/(2c) \approx 0.32\,GeV^2$ and $y = \sqrt{a/c - x^2} \approx 0.47\,GeV^2$. Let us recall that in the case of a Gribov-like propagator these two poles are purely imaginary. For $D(p^2)$ transverse off-diagonal our best fit is of Yukawa type, i.e. $D(p^2) = 1/(a + b\,p^2)$, with a mass $m = \sqrt{a/b} \approx 0.97\,GeV$. Finally, the longitudinal off-diagonal gluon propagator is best fitted by $D(p^2) = 1/(a + b\,p^2 + c\,p^4)$ (i.e. also of Yukawa type) with a mass $m = \sqrt{a/b} \approx 1.25\,GeV$. As expected from Abelian dominance, the mass is larger in the off-diagonal case.

In Fig. 1 (left) we show our data for the ghost propagator $G(p^2)$, as a function of an improved momentum $p$ (see Ref. [5]). The data show little volume dependence at small $p$. (Note that, contrary to Landau gauge, here we can evaluate the ghost propagator at zero momentum.) We see no sign of an enhanced IR propagator. We have fitted our data (at $\beta = 2.2$), obtaining a behavior of the type (3) above with $a = 0.45(1)\,GeV^2$, $b = 1.1(3)$, $c = 0.73(30)\,GeV^{-2}$, $d = 2.1(9)\,GeV^{-2}$. Thus, we see a Stingl-Gribov fit with mass $m \approx 0.6\,GeV$ and complex poles given by $x \approx 0.75$, $y \approx 0.22$.

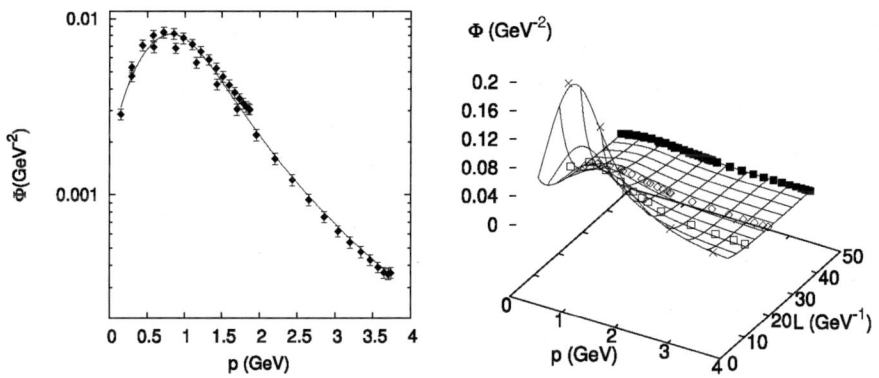

**FIGURE 2.** Left: plot of the quantity $\Phi(p^2) = L^2/\cos\left(\pi\,\tilde{p}_\mu\,a/L\right)\langle\,|\,\varepsilon_{ab}G^{ab}(p^2)/2\,|\,\rangle$ as a function of $p$ for lattice volumes $V = 8^4$, $16^4$, $24^4$, $40^4$ and $\beta = 2.2$. Right: plot of $\Phi(p^2)$ as a function of $p$ and $L$.

We next consider (see Fig. 1, right) the smallest eigenvalue of the FP matrix. We have looked at $\lambda_{min}$ for several lattice volumes and values of $\beta$ as a function of $1/L$. The data are fitted to $a\,(1/L)^b$ with $b = 1.6(1)$, showing that $\lambda_{min}$ vanishes more slowly than $(1/L)^2$ (Laplacian). This may explain why we do not see a diverging ghost propagator at zero momentum even at rather large lattice volumes [6].

Following the analysis done in Landau gauge [7], we consider the anti-symmetric off-diagonal ghost propagator $\langle\,|\,\varepsilon_{ab}G^{ab}(p^2)/2\,|\,\rangle$ rescaled by $L^2/\cos\left(\pi\,\tilde{p}_\mu\,a/L\right)$, as a function of the (unimproved) momentum $p$ for all lattice volumes and $\beta$ values considered. The data show nice scaling for all cases considered. The data at $V = 40^4$ and $\beta = 2.2$ can be fitted by $\Phi(p) = (a + b\,p/L^2)(p^4 + v^2)$ with $a = 0.0026(7)\,GeV^2$, $b = 32.6(7)\,GeV^{-1}$ and $v^2 = 1.7(1)\,GeV^4$. We thus have a rather large ghost condensate $v \approx 1.3\,GeV^2$, but we cannot be sure that it survives in the infinite-volume limit, since the overall constant $a$ might be null. We can also fit data at several $V$'s and $\beta$'s for $\Phi(p^2)$ as a function of $p$ and $L$ (see Fig. 2, right). We obtain $\Phi(p) = (a + b\,p/L^2)(p^4 + v^2)$ with $a = 0.0033(6)\,GeV^2$, $b = 35.8(5)\,GeV^{-1}$ and $v^2 = 1.87(8)\,GeV^4$. We note that the fit parameters change little with the (physical) lattice volume.

We are currently investigating the effects of Gribov copies on our results.

We thank M.I. Polikarpov and M. Schaden for helpful discussions. A. C. and T. M. were supported by FAPESP and CNPq. A. M. was supported by FAPESP.

## REFERENCES

1. M. Schaden, arXiv:hep-th/9909011.
2. M. A. L. Capri et al., Phys. Rev. D **72**, 085021 (2005).
3. V. G. Bornyakov et al., Phys. Lett. B **559**, 214 (2003).
4. F. Bruckmann, T. Heinzl, A. Wipf and T. Tok, Nucl. Phys. B **584**, 589 (2000).
5. J. P. Ma, Mod. Phys. Lett. A **15**, 229 (2000).
6. A. Cucchieri, these proceedings.
7. A. Cucchieri, T. Mendes and A. Mihara, Phys. Rev. D **72**, 094505 (2005).

# Confinement in $(2+1)$-Dimensional Gauge Theories at Weak Coupling

Peter Orland

*Physics Program, The Graduate School and University Center, The City University of New York, 365 Fifth Avenue, New York, NY 10016, U.S.A. and Department of Natural Sciences, Baruch College, The City University of New York, 17 Lexington Avenue, New York, NY 10010, U.S.A.*

**Abstract.** In axial gauge, the $(2+1)$-dimensional SU($N$) Yang-Mills theory is equivalent to a set of $(1+1)$-dimensional integrable models with a non-local coupling between charge densities. This fact makes it possible to determine the static potential between charges at weak coupling in an anisotropic version of the theory and understand features of the spectrum. We briefly mention a few open problems.

**Keywords:** Confinement, QCD, Yang-Mills theory, 2+1 dimensions, integrable models, k-string tensions
**PACS:** 11.15.Ha, 12.38.Mh

Many pictures and models have been proposed for confinement in QCD. Some sort of magnetic condensation clearly occurs, but no one knows why. Our grasp of the basic mechanism of confinement, at weak bare coupling, is little better than it was thirty years ago. In the opinion of the author, strong-coupling and variational methods can guide us toward a better understanding, but are not substitutes for first-principles weak-coupling calculations. We discuss such calculations here for a more modest theory. This is a $(2+1)$-dimensional SU($N$) gauge theory with two coupling constants [1], [2], [3].

The action is of Yang-Mills type: $\int d^3 \mathscr{L}$, where the Lagrangian is $\mathscr{L} = \frac{1}{2e'^2}\text{Tr}F_{01}^2 + \frac{1}{2e^2}\text{Tr}F_{02}^2 - \frac{1}{2e^2}\text{Tr}F_{12}^2$, where $A_0, A_1$ and $A_1$ are SU($N$)-Lie-algebra-valued components of the gauge field, and the field strength is $F_{\mu\nu} = \partial_\mu A_\nu - \partial_\nu A_\mu - i[A_\mu, A_\nu]$. The gauge transformation is $A_\mu(x) \to ig(x)^{-1}[\partial_\mu - iA_\mu(x)]g(x)$, where $g(x)$ is an SU($N$)-valued scalar field. To study this model, we will take $e' \ll e$; by doing this we lose rotation invariance.

The next step is to discretize the 2-direction, so that $x^2 = a, 2a, 3a\ldots$, where $a$ is a lattice spacing. All fields will be considered functions of $x = (x^0, x^1, x^2)$. We define the unit vector $\hat{2} = (0, 0, 1)$. We replace $A_2(x)$ by a field $U(x)$ lying in SU($N$), via $U(x) \approx \exp{-iaA_2(x)}$. There is a natural discrete covariant-derivative operator: $\mathscr{D}_\mu \mathfrak{U}(x) = \partial_\mu \mathfrak{U}(x) - iA_\mu(x)\mathfrak{U}(x) + i\mathfrak{U}(x)A_\mu(x+\hat{2}a)$, $\mu = 0, 1$, for any $N \times N$ complex matrix field $\mathfrak{U}(x)$. The action is $S = \int dx^0 \int dx^1 \sum_{x^2} a\mathscr{L}$ where

$$\mathscr{L} = \frac{1}{2e'^2}\text{Tr}F_{01}^2 + \frac{1}{2g_0^2}\text{Tr}[\mathscr{D}_0 U(x)]^\dagger \mathscr{D}_0 U(x) - \frac{1}{2g_0^2}\text{Tr}[\mathscr{D}_1 U(x)]^\dagger \mathscr{D}_1 U(x)\,, \qquad (1)$$

and $g_0^2 = e_0^2 a$. The Lagrangian (1) is invariant under the gauge transformation: $A_\mu(x) \to$

CP892, *Quark Confinement and the Hadron Spectrum VII*
edited by J. E. F. T. Ribeiro
© 2007 American Institute of Physics 978-0-7354-0396-3/07/$23.00

$ig(x)^{-1}[\partial_\mu - iA_\mu(x)]g(x)$ and $U(x) \to g(x)^{-1}U(x)g(x+\hat{2}a)$ where again, $g(x) \in SU(N)$ and $\mu$ is restricted to 0 or 1. Notice that the quantity $g_0$ is dimensionless. In the limit $a \to 0$ , (1) yields the anisotropic continuum action.. The action (1) is a collection of parallel $(1+1)$-dimensional $SU(N) \times SU(N)$ sigma models, each of which couples to the gauge fields $A_0$, $A_1$. The sigma model field is $U(x^0,x^1,x^2)$, and each discrete $x^2$ corresponds to a different sigma model. The sigma-model self-interaction is the dimensionless number $g_0$.

The left-handed and right-handed currents are, $j_\mu^L(x)_b = iTrt_b \partial_\mu U(x)U(x)^\dagger$ and $j_\mu^R(x)_b = iTrt_b U(x)^\dagger \partial_\mu U(x)$, respectively, where $\mu = 0,1$. The Hamiltonian obtained from (1) is $H_0 + H_1$, where

$$H_0 = \sum_{x^2} \int dx^1 \frac{1}{2g_0^2}\{[j_0^L(x)_b]^2 + [j_1^L(x)_b]^2\} \,, \tag{2}$$

and

$$
\begin{aligned}
H_1 &= \sum_{x^2} \int dx^1 \frac{(g_0')^2 a^2}{4} \partial_1 \Phi(x^1,x^2)\partial_1 \Phi(x^1,x^2) \\
&\quad - \left(\frac{g_0'}{g_0}\right)^2 \sum_{x^2=0}^{L^2-a} \int dx^1 [j_0^L(x^1,x^2)\Phi(x^1,x^2) - j_0^R(x^1,x^2)\Phi(x^1,x^2+a)] \\
&\quad + (g_0')^2 q_b \Phi(u^1,u^2)_b - (g_0')^2 q_b' \Phi(v^1,v^2)_b \,, \tag{3}
\end{aligned}
$$

where $-\Phi_b = A_{0\,b}$ is the temporal gauge field, $g_0'^2 = e'^2 a$, and where in the last term we have inserted two color charges - a quark with charge $q$ at site $u$ and an anti-quark with charge $q'$ at site $v$. There is some gauge invariance left over after the axial gauge fixing, namely that for each $x^2$

$$\left\{\int dx^1 [j_0^L(x^1,x^2)_b - j_0^R(x^1,x^2-a)_b] - g_0^2 Q(x^2)_b\right\} \Psi = 0 \,, \tag{4}$$

where $Q(x^2)_b$ is the total color charge from quarks at $x^2$ and $\Psi$ is any physical state.

The Hamiltonian (2), (3) can be derived more carefully, by starting with the Kogut-Susskind lattice formulation [1], [2] and assuming the lattice spacing is small in the $x^1$-direction. In any case, we assume that $H_1$ is suitably regularized.

From (3) we see that the left-handed charge of the sigma model at $x^2$ is coupled to the electrostatic potential at $x^2$. The right-handed charge of the sigma model is coupled to the electrostatic potential at $x^2 + a$. The excitations of $H_0$, which we call Fadeev-Zamoldochikov or FZ particles, behave like solitons, though they do not correspond to classical configurations. Some of these FZ particles are elementary and others are bound states of the elementary FZ particles. An elementary FZ particle has an adjoint charge and mass $m_1$. An elementary FZ particle state is a superposition of color-dipole states, with a quark charge at $x^1,x^2$ and an anti-quark charge at $x^1,x^2 + a$. The interaction $H_1$ produces a linear potential between color charges with the same value of $x^2$. Residual gauge invariance (4) requires that at each value of $x^2$, the total color charge is zero. If there are no quarks, the total right-handed charge of FZ particles in the sigma model at $x^2 - a$ is equal to the total left-handed charge of FZ particles in the sigma model at $x^2$.

The presence of a mass gap and the lack of magnetization in the $SU(N) \times SU(N)$ sigma model implies confinement in the (2+1)-dimensional $SU(N)$ gauge theory in an anisotropic weak-coupling approximation $g_0 \gg g_0'$ [1]. A formal weak-coupling perturbation theory in $g_0'$ around the sigma-model states can be considered in principle. Unfortunately, it is very hard to carry out this perturbation theory in practice, except for gauge group SU(2) [2].

The principal chiral sigma model spectrum is described by particles, each of which has a label $n$ which has the values $n = 1, \ldots, N - 1$ [4], [5]. Each particle of label $n$ has an antiparticle of the same mass, with label $N - n$. The masses are given by

$$m_n = m_1 \frac{\sin \frac{n\pi}{N}}{\sin \frac{\pi}{N}}, \quad m_1 = \frac{C}{a}(g_0^2 N)^{1/2} e^{-\frac{4\pi}{g_0^2 N}} + \text{non-universal corrections}, \tag{5}$$

where $C$ is a non-universal constant.

Lorentz invariance in each $x^0, x^1$ plane is manifest; hence the linear potential is not the only effect of $H_1$. The interaction also creates and destroys pairs of elementary FZ particles. This effect is unimportant, however, provided the interaction is small enough. This specifically means that the square of the $1 + 1$ string tension in the $x^1$-direction coming from $H_1$ is small compared to the square of the mass of fundamental FZ particle.

A rough picture of a gauge-invariant state for the gauge group SU(2) with no quarks is:

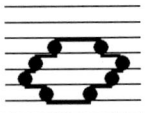

The horizontal coordinate is $x^1$ and the vertical coordinate is $x^2$. The FZ particles are the bullets joined by horizontal electric strings. The vertical electric flux consists of the FZ particles themselves. The lightest glueball is a pair of FZ particles with the same value of $x^2$. For small enough $g_0'$, its mass is $2m_1$.

The leading-order vertical $k$-string tension is just the energy of the bound state of $k$ fundamental FZ particles, divided by the lattice spacing [3]. This yields a sine law $\sigma_V^k = \frac{m_k}{a} \propto \sin \pi k / N$. The leading-order horizontal $k$-string tension is found by a simple argument [3]. If the coupling $g_0'$ is sufficiently small, then the mass gap in the principal chiral sigma models at $x^2 = u^2$ and $x^2 = u^2 - a$ forces electric flux along the line from $(u^1, u^2)$ to $(v^1, u^2)$. Thus the potential is just that of the $(1+1)$-dimensional $SU(N)$ Yang-Mills theory, and the horizontal string tension should behave as $\sigma_H^k = \left(\frac{g_0'}{a}\right)^2 C_k$, where $C_k$ is the quadratic Casimir operator. Adjoint sources are not confined [3].

These naive results for the string tension have further corrections in $g_0'$, which were determined for the horizontal string tension for SU(2) [2]:

$$\sigma_H = \frac{3}{2}\left(\frac{g_0'}{a}\right)^2 \left[1 + \frac{4}{3}\frac{0.7296}{C^2\pi^2}\frac{(g_0')^2}{g_0^2}e^{4\pi/g_0^2}\right]^{-1}. \tag{6}$$

This calculation was done using the exact form factor for sigma model currents obtained by Karowski and Weisz [6]. Particle states have rapidity $\theta$ and four internal states, labeled by $j$. The 2-particle current form factor is

$$\langle 0 | j \, {}_{0}^{L,R}(x)_b | \theta_2, j_2, \theta_1, j_1 \rangle = i\sqrt{2} \left( \delta_{j_1 4} \delta_{j_2 b} - \delta_{j_2 4} \delta_{j_1 b} \pm \varepsilon_{b j_1 j_2} \right) m(\cosh \theta_1 - \cosh \theta_2)$$
$$\times \exp\{-im[x^0(\cosh \theta_1 + \cosh \theta_2) - x^1(\sinh \theta_1 + \sinh \theta_2)]\} F(\theta_2 - \theta_1) , \qquad (7)$$

where the plus or minus sign corresponds to the left-handed ($L$) or right-handed ($R$) current, respectively, and

$$F(\theta) = \exp 2 \int_0^\infty \frac{d\xi}{\xi} \frac{e^{-\xi} - 1}{e^\xi + 1} \frac{\sin^2 \frac{\xi(\pi i - \theta)}{2\pi}}{\sinh \xi} = \exp - \int_0^\infty \frac{d\xi}{\xi} \frac{e^{-\xi}}{\cosh^2 \frac{\xi}{2}} \sin^2 \frac{\xi(\pi i - \theta)}{2\pi}. \qquad (8)$$

Other two-particle form factors can be obtained by crossing.

Our results for the mass gap and the string tension are not of the form one would expect for the isotropic theory. If $g_0' = g_0 = e\sqrt{a}$, naive dimensional arguments imply that the gap is proportional to $e^2$ and the string tension is proportional to $e^4$. In fact, general arguments imply a crossover should occur as $g_0'$ is increased, so this behavior is a real possibility as isotropy is approached [2]. If this is so, there is no possibility of extracting the isotropic gap and string tension from our results. Physical quantities cannot be expected to have a part which is analytic in both $e$ and $e'$ (to see this, try to do standard perturbation theory in both $e$ and $e'$).

We have shown there is confinement in the region $g_0$ and $g_0'$ small, provided $g_0' \ll g_0$. This leaves no doubt that that confinement persists over the entire phase diagram of $g_0$ and $g_0'$, even for the isotropic case. Though we have little to say about the specifics of the isotropic theory, the anisotropic theory is perhaps more interesting, as it is not simply finite, but asymptotically free.

There are more problems to be investigated for anisotropic gauge theories. Corrections need to be found in $g_0'$ for the vertical string tension and the mass gap. The former problem can probably only be solved easily for SU(2). The theory should be studied at non-zero temperature; it should be possible to see a phase transition to a deconfined phase. Matter fields can be introduced and the spectrum of mesons and baryons should be examined. Finally, there is a striking mathematical question. Our weak-coupling analysis looks very much like the strong-coupling picture of a lattice gauge theory; is there a duality present?

# REFERENCES

1. P. Orland, Phys. Rev. **D71** (2005) 054503, **hep-lat/0501026**.
2. P. Orland, Phys. Rev. **D74** (2006) 085001, **hep-th/0607013**.
3. P. Orland, submitted to Phys. Rev. D, **hep-th/0608067**.
4. A.M. Polyakov and P. Wiegmann, Phys. Lett. **B131** (1983) 121; P. Wiegmann, Phys. Lett. **B141B** (1984) 217.
5. E. Abdalla, M.C.B. Abdalla and A. Lima-Santos, Phys. Lett. **B140** (1984) 71; P. Wiegmann; Phys. Lett. **B142** (1984) 173.
6. M. Karowski and P. Weisz, Nucl. Phys. **B139** (1978) 455.

# Perturbative gauge theory in a background[1]

Dennis D. Dietrich\*, Paul Hoyer†, Matti Järvinen† and Stéphane Peigné\*\*

\*Niels Bohr Institute, Blegdamsvej 17, DK-2100 Copenhagen, Denmark
†Department of Physical Sciences and Helsinki Institute of Physics, POB 64, FIN-00014, Finland
\*\*SUBATECH, UMR 6457, Université de Nantes, IN2P3/CNRS, F-44307 Nantes, France

**Abstract.** We study the perturbative expansion of a gauge theory around a non-empty state, by considering the vacuum to consist of gauge bosons with zero four-momentum. The latter prescription allows to preserve Poincaré invariance, and to calculate dressed Green functions to all orders in the condensate. Those Green functions have the standard perturbative behaviour at short distance, but exhibit interesting features at long distance. Although we consider the case of an abelian gauge theory, this might give some insight on the Green functions of elementary fields in a confining theory.

**Keywords:** QCD, Nonperturbative Effects
**PACS:** 12.38.Aw, 12.38.Lg

## MOTIVATION AND MODEL

The perturbative expansion of a field theory is determined by its lagrangian, and by the perturbative vacuum $|\Omega\rangle$ around which the expansion is made. The perturbative vacuum is usually chosen to be the empty state $|0\rangle$. In QED this leads to a perturbative expansion which agrees with the data to an amazing precision, suggesting that the true QED ground state is quite close to the empty state. On the contrary, perturbative QCD fails at long distances. In particular the elementary fields of the QCD lagrangian - quarks and gluons - are not the asymptotic states of QCD. Since the QCD ground state is known to be a quark and gluon condensate, we intuitively understand that an expansion around $|0\rangle$ is inadequate to describe quark and gluon propagation on long distances. Since there are indications [1] that the strong coupling saturates, at low energy, at a value which is not large, $\alpha_s(Q^2 \to 0) \lesssim 1$, the situation in QCD might be similar to electron propagation in matter: although the electromagnetic coupling is small, long distance propagation is suppressed by multiple scattering and associated energy loss.

It is thus interesting to study the perturbative expansion of a gauge theory around a vacuum $|\Omega\rangle \neq |0\rangle$ and to address in such a framework the following questions:

- What are the features of Green functions of elementary fields?
- How to get an analytic and unitary $S$-matrix in terms of true asymptotic states?

---

[1] Talk presented by S. Peigné.

CP892, Quark Confinement and the Hadron Spectrum VII
edited by J. E. F. T. Ribeiro
© 2007 American Institute of Physics 978-0-7354-0396-3/07/$23.00

As a simple model for a non-empty perturbative vacuum $|\Omega\rangle$, we consider a modified gauge boson propagator (in Feynman gauge),

$$D^{\mu\nu}(p) = -g^{\mu\nu}\left[\frac{i}{p^2+i\varepsilon} + \Lambda^2(2\pi)^4\delta^4(p)\right] , \qquad (1)$$

which has an additional term $\propto \delta^4(p)$, signalling the presence of zero-momentum gluons in $|\Omega\rangle$ [2]. The modification (1) of the bare gauge boson propagator can be obtained by assuming a *constant background field* $\Phi_\mu^a$ and averaging over the Lorentz and color components with the gaussian weight [3, 4]

$$\left(\Pi \int d\Phi_\mu^a\right)\exp\left[\frac{1}{2\Lambda^2}\Phi_\nu^b\Phi_b^\nu\right] , \qquad (2)$$

where $\Lambda$ has the dimension of mass and is related to the strength of the background field. Poincaré invariance is preserved, and the term $\propto \delta^4(p)$ in (1) allows to 'dress' any Green function (at a given order in $\alpha_s$) to all orders in $\mu \equiv g\Lambda$. This dressing defines a modified perturbative expansion, where Green functions tend to the standard ones in the short distance limit ($|p^2| \gg \mu^2$) but exhibit new features at long distance ($|p^2| \lesssim \mu^2$). In a previous work [3] we studied the effect of the modification (1) in a non-abelian gauge theory in the large $N_c$ limit. Resumming planar diagrams, it was shown that the dressing removes quarks and gluons from the asymptotic states of the theory. In the following we summarize our study in the *abelian* case [2], where a different method is used to sum up the larger number of (planar and non-planar) diagrams.

## DRESSING GREEN FUNCTIONS

As an illustration we first consider the dressed tree (abelian) 'quark' propagator, given in Fig. 1, where a dashed line denotes the term $\propto \delta^4(p)$ in the 'gluon' propagator (1). For convenience we rescale momenta and the quark mass $m$ by $\mu \equiv g\Lambda$, namely $p/\mu \to p$, $m/\mu \to m$. Due to the identity $S_0(p)(-ig\gamma_\alpha)S_0(p) = -\frac{g}{\mu}\frac{\partial}{\partial p^\alpha}S_0(p)$, where $S_0(p) = i/[\mu(\not p - m)]$, attaching a zero-momentum gluon line to an internal quark propagator amounts to differentiating with respect to $p$. The dressed quark propagator $S(p)$ is then easily shown to satisfy the Dyson-Schwinger (DS) differential equation

$$(\not p - m - \not\partial_p)S(p) = i/\mu \qquad ; \qquad \not\partial_p \equiv \gamma^\alpha\partial/\partial p^\alpha . \qquad (3)$$

In coordinate space the DS equation reads $(i\not\partial_x - m - \frac{g}{\mu}\not A(x))S(x) = i\mu^3\delta^4(x)$, with $gA^\nu(x) = i\mu x^\nu = i\mu\partial^\nu(\frac{1}{2}x^2)$, showing that the propagator's dressing is formally equivalent to a pure gauge transformation with an *imaginary* gauge parameter[2]. In order to find $S(p)$, instead of solving directly the differential equation (3) (with the boundary

---

[2] The dressed propagator $S(p)$ thus differs from the bare propagator $S_0(p)$ in magnitude. Hence the on-shell modification (1) of the gluon propagator does influence the physics.

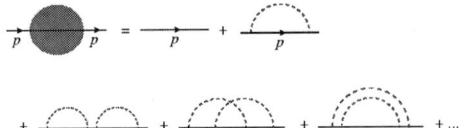

**FIGURE 1.** Dressed tree quark propagator to all orders in $\mu \equiv g\Lambda$.

condition $S(p) \to i/(\mu\!\!\!\slash{p})$ when $p^2 \to \infty$), an alternative method [2] making use of the latter formal equivalence proves to be more efficient. We find

$$S(p) = \frac{i}{2\mu}(\slash{p} + m - \slash{\partial}_p) \int_0^\infty dt \exp\left[ -\frac{t}{2}\left( p^2 - \frac{m^2}{1+t} \right) \right] ,\qquad (4)$$

which satisfies the DS equation (3) and has the standard limit $S_0(p)$ when $p^2 \to \infty$.

The expression (4) is well-defined for $\mathrm{Re}\, p^2 > 0$. Its continuation to the whole $p^2$ plane [2] can be shown to reduce to the standard result at short distance in all directions of the complex plane (*i.e.* for $|p^2| \to \infty$). It also exhibits non-trivial features at long distance:

- $S(p)$ is regular at $p^2 = m^2$ for any $\mu > 0$. The 'quark' is thus removed from the asymptotic states of the theory.
- $S(p)$ has an exponentially behaved discontinuity along the negative real axis.

A similar calculation can be done for the 'quark-antiquark' propagator [2]. When dressed with zero-momentum lines, the bare $q\bar{q}$ propagator

$$iG_{0,0}^{\alpha\beta,\rho\sigma}(k,\bar{k}) = \left( \frac{i}{\mu} \frac{\slash{k} + m}{k^2 - m^2} \right)^{\alpha\beta} \left( \frac{i}{\mu} \frac{\slash{\bar{k}} + m}{\bar{k}^2 - m^2} \right)^{\rho\sigma} \qquad (5)$$

satisfies the DS equation $(\slash{k} - m - \slash{\partial}_k)\, G(k,\bar{k}) = \frac{1}{\mu} S(\bar{k})$, where $k$ and $-\bar{k} = p - k$ are the quark and antiquark momenta. The solution $G(k,\bar{k})$ [2] has the following properties:

- It is consistent with the Ward identity for the dressed $gq\bar{q}$ vertex.
- Its short distance limit ($k^2, \bar{k}^2 \to \infty$) is the propagator (5).
- In the $m = 0$ case, we find that $G(k,\bar{k})$ is singular at $\sqrt{p^2} = \sqrt{k^2} + \sqrt{\bar{k}^2}$, *i.e.* at threshold, possibly indicating the existence, in our model, of *non-relativistic* bound states with constituent masses $\sqrt{k^2}$, $\sqrt{\bar{k}^2}$.

Our method [2] a priori allows calculating any dressed $n$-point Green function. True asymptotic states in the present framework remain to be identified.

## REFERENCES

1. Y. L. Dokshitzer, hep-ph/9812252; S. J. Brodsky et al., *Phys. Rev. D* **67** (2003) 055008; Y. L. Dokshitzer and D. E. Kharzeev, *Ann. Rev. Nucl. Part. Sci.* **54** (2004) 487.
2. D. D. Dietrich, P. Hoyer, M. Järvinen and S. Peigné, hep-ph/0608075.
3. P. Hoyer and S. Peigné, *JHEP* **0412** (2004) 051; S. Peigné, *AIP Conf. Proc.* **756** (2005) 296.
4. D. D. Dietrich and S. Hofmann, *Phys. Lett. B* **632** (2006) 439; D. D. Dietrich, hep-ph/0507112.

# Topology and confinement at $T \neq 0$: calorons with non-trivial holonomy[1]

P. Gerhold*, E.-M. Ilgenfritz*, M. Müller-Preussker*, B.V. Martemyanov[†] and A.I. Veselov[†]

*Humboldt-Universität zu Berlin, Institut für Physik, Newtonstr. 15, 12489 Berlin, Germany
[†]Institute for Theoretical and Experimental Physics, B. Cheremushkinskaya 25, Moscow 117259, Russia

**Abstract.** In this talk, relying on experience with various lattice filter techniques, we argue that the semiclassical structure of finite temperature gauge fields for $T < T_c$ is dominated by calorons with non-trivial holonomy. By simulating a dilute gas of calorons with identical holonomy, superposed in the algebraic gauge, we are able to reproduce the confining properties below $T_c$ up to distances $r = O(4\text{fm}) >> \rho$ (the caloron size). We compute Polyakov loop correlators as well as space-like Wilson loops for the fundamental and adjoint representation. The model parameters, including the holonomy, can be inferred from lattice results as functions of the temperature.

**Keywords:** Yang-Mills theory, lattice, caloron, non-trivial holonomy, confinement
**PACS:** 11.15.Ha, 11.10.Wx, 12.38.Gc

Instanton or caloron models of QCD successfully describe many non-perturbative features of hadron physics, in particular chiral symmetry breaking and the $U_A(1)$ anomaly (for reviews see [1, 2]). However, they fail to describe confinement unless they are endowed with long-range correlations. This is the case for instantons in the regular gauge [3] and has been discussed also at this conference [4]. An attractive alternative, at least for non-zero temperature $T$, is based on new caloron solutions with non-trivial holonomy [5, 6, 7, 8], worked out in various aspects by Kraan and van Baal. At this symposium we were happy to listen a review talk [9] by Pierre van Baal after his recovery.

The new caloron solutions - we call them KvBLL calorons - have characteristic properties which distinguish them from the old BPST instantons [10] or Harrington-Shepard (HS) calorons [11]. Like the latter, the new calorons are (anti)selfdual with integer topological charge and periodic in Euclidean time with the period $1/T$. The difference is a non-trivial asymptotic behaviour of $A_4(x)$ such that the Polyakov loop

$$P(\vec{x}) = \hat{P} \exp \left( \imath \int_0^{1/T} A_4(\vec{x}, t) dt \right) \overset{|\vec{x}| \to \infty}{\Longrightarrow} \mathscr{P}_\infty \tag{1}$$

can take arbitrary fixed values $\mathscr{P}_\infty \notin Z(N_c)$ at spatial infinity. Each single KvBLL (anti)caloron consists of $N_c$ monopole constituents localized at positions where the Polyakov loop $P(\vec{x})$ has degenerated eigenvalues. For $SU(2)$ this means that $L(\vec{x}) =$

---

[1] Talk given by M. Müller-Preussker

CP892, *Quark Confinement and the Hadron Spectrum VII*
edited by J. E. F. T. Ribeiro

$\frac{1}{2}\mathrm{tr}P(\vec{x})$ takes opposite values $\pm 1$ at the positions of the two monopoles. The profile of the Polyakov loop field inside a KvBLL solution is the most significant feature of the new calorons irrespective whether the constituents or 'instanton quarks' are separated or not. If the constituents are far from each other the caloron dissociates into $N_c$ static non-Abelian monopoles. The topological charge of either lump then depends on the eigenvalue differences of $\mathscr{P}_\infty$. The zero modes of the Dirac operator are localized only at one of the constituents. When the fermionic boundary condition is smoothly changed the zero mode jumps from one constituent to another [12] if these are separated.

First we used the cooling method applying it to pure $SU(2)$ and $SU(3)$ lattice Monte Carlo gauge fields. We demonstrated that the lumps of topological charge observed in the plateau configurations have to be interpreted in terms of KvBLL calorons [13, 14]. Closer to the deconfinement transition or for a smaller aspect ratio we found an increasing frequency of dissociated monopole pairs in $SU(2)$.

More recently we have studied Monte Carlo lattice fields with the 4d smearing method at different temperatures [15]. We found many clusters of topological charge and classified them with respect to their Abelian monopole content (in the maximally Abelian gauge). Two limiting cases suggest an interpretation in terms of KvBLL constituents or calorons: (i) clusters containing a monopole loop winding around the lattice in time direction, taken as candidates for a single constituent; (ii) clusters containing a closed monopole loop, taken as candidates for undissociated calorons. For these cases we have estimated the topological charge of the cluster, $Q_{\text{cluster}}$, and the Polyakov loop averaged over the positions of time-like Abelian monopoles, $< PL >_{\text{cluster}}$. In the confinement phase - *i.e.* for maximally non-trivial holonomy - we would expect half-integer topological charges for isolated monopoles and integer charge for full calorons. The averaged Polyakov loop should be close to $\pm 1$ for isolated monopoles and near zero for calorons according to the "dipole" profile of the Polyakov loop inside the KvBLL caloron. What is really observed is seen in the scatter plots in the $(Q_{\text{cluster}}, < PL >_{\text{cluster}})$-plane of Fig. 1. Each entry corresponds to one of the selected cluster candidates, and the scatter plot is clustering into classes with the expected signatures. In the deconfined phase (not shown), for holonomies closer to the trivial one we would expect to find disbalanced constituents, one with small action and a complementary one with large action. Apart from few full calorons accounting for the topological charge of the configurations, we found many "single-constituent" clusters with static Abelian monopole loops but small topological charge, whereas constituents with topological charges close to $\pm 1$ were completely missing. We conclude that the model picture of KvBLL calorons may fail in the deconfinement phase.

One can also study the topological content without cooling or smearing techniques by applying purely fermionic methods to equilibrium fields. Such investigations have also provided indications for the presence of KvBLL monopole constituents [16, 17].

What are the consequences if HS calorons are replaced by KvBLL ones as building blocks in a random caloron gas model at finite $T$? Such a model, so far realized in the $SU(2)$ case [18], requires to start the superposition in the so-called algebraic gauge for which $A_4(x)$ decreases sufficiently fast. A non-periodic gauge transformation is applied in order to render the gauge field periodic. This restricts us to superpositions of calorons with identical holonomy. The positions of the calorons are chosen randomly, the sizes $\rho$ (*i.e.* the distances $d = \pi\rho^2/\beta$ between the constituents) are sampled for $T > T_c$

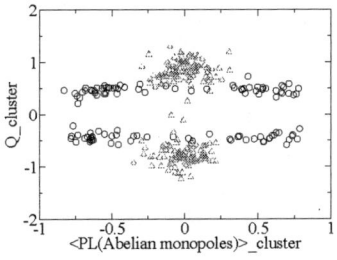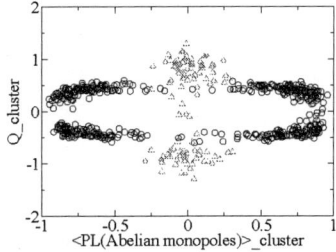

**FIGURE 1.** Scatter plots of topological charge versus averaged Polyakov loop for topological clusters of lattice fields produced with the Wilson action at $\beta = 2.3, 2.4$ and lattice size $24^3 \times 6$ (confinement). Triangles (circles) denote full caloron (isolated static monopole) candidates.

**TABLE 1.** Model parameters $n(T)$, $\omega(T)$, $\bar{\rho}(T)$, the lattice grid size $N_s \times N_\tau$, and the number of generated configurations # for selected temperature values $T/T_c$. Furthermore, the measured average Polyakov loop $< |L| >$ (together with the input value $cos(2\pi\omega)$) and the action surplus factor $\gamma = \frac{S_{tot}}{N_{caloron} \cdot S_{inst}}$ are given.

| $T/T_c$ | $N_s^3 \times N_\tau$ | $n^{\frac{1}{4}}$ [MeV] | $4\omega$ | $\bar{\rho}$ [fm] | # | $cos(2\pi\omega)$ | $< |L| >$ | $\gamma$ |
|---|---|---|---|---|---|---|---|---|
| 0.80 | $32^3 \times 10$ | 198 | 1.00 | 0.37 | 777 | 0.00 | 0.13(1) | 1.61(1) |
| 1.00 | $32^3 \times 8$ | 198 | 1.00 | 0.37 | 526 | 0.00 | 0.14(1) | 1.69(1) |
| 1.20 | $32^3 \times 8$ | 174 | 0.51 | 0.31 | 160 | 0.70 | 0.59(1) | 1.18(1) |

according to [19]

$$D(\rho, T) = A(T) \cdot \rho^{b-5} \cdot \exp(-\frac{4}{3}(\pi T \rho)^2), \quad b = 11N_c/3 = 22/3.  \qquad (2)$$

For $T < T_c$ temperature independence is postulated but keeping the suppression at $T_c$ fixed (see [20, 21]). For a statistically uncorrelated caloron gas the actual density $n(T)$ can be inferred from lattice computations of the topological susceptibility. The average size was fixed by comparison between model and lattice results for the spatial string tension in units of the critical temperature which then turned out $T_c \simeq 178$ MeV. The holonomy $\mathscr{P}_\infty \equiv \exp(2\pi i\omega\tau_3)$ was identified with the (renormalized) Polyakov loop. More details and references can be found in [18]. For some parameter sets see Table 1.

On a lattice grid we have computed spatial Wilson loops as well as Polyakov loop correlators, both within fundamental and adjoint representations. The spatial string tension is seen to drop at $T_c$, because the mechanism responsible for the observed rise are not the monopoles that are part of the calorons and suppressed at $T > T_c$. This problem corresponds to our observation, in the smeared configurations reported above, of many monopoles with low accompanying topological charge. The results for the free energy of a static quark-antiquark pair are shown in Fig. 2. We can follow a linear rise for distances up to $O(4$fm$)$ in the confinement phase, whereas it becomes screened above $T_c$. The free energy of adjoint charge pairs is screened in both phases.

We conclude that a semiclassical model for finite-temperature $SU(2)$ fields should start from calorons with generic holonomy. Such a model turns out to describe confine-

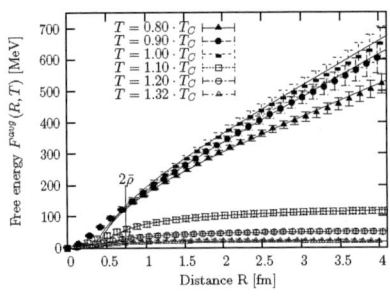 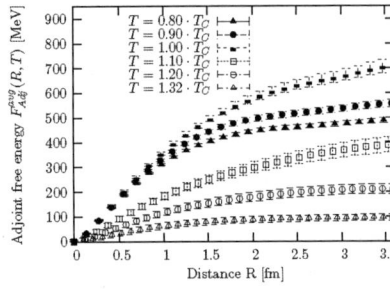

**FIGURE 2.** Colour averaged free energy versus distance $R$ at various temperatures for the fundamental (left) and adjoint (right) representations.

ment with parameters which are rather close to standard instanton model assumptions. Keeping all parameters fixed, merely changing the holonomy from $\omega = 1/4$ to $\omega = 0$ or $1/2$, removes the linear rise completely [18]. This underscores the rôle of non-trivial holonomy and the corresponding long-range nature of the caloron fields.

We are grateful to P. van Baal, F. Bruckmann, C. Gattringer, A. Schäfer and S. Solbrig for numerous useful discussions. We acknowledge financial support by the DFG through FOR 465 / Mu 932/2-4 and 436 RUS 113/739/0-2 as well as by RFBR grant 06-02-16309.

# REFERENCES

1. T. Schäfer, and E. V. Shuryak, *Rev. Mod. Phys.* **70**, 323 (1998), hep-ph/9610451.
2. D. Diakonov, *Prog. Part. Nucl. Phys.* **51**, 173 (2003), hep-ph/0212026.
3. J. W. Negele, F. Lenz, and M. Thies, *Nucl. Phys. Proc. Suppl.* **140**, 629 (2005), hep-lat/0409083.
4. M. Wagner (2006), hep-ph/0608090.
5. T. C. Kraan, and P. van Baal, *Phys. Lett.* **B428**, 268 (1998), hep-th/9802049.
6. T. C. Kraan, and P. van Baal, *Nucl. Phys.* **B533**, 627 (1998), hep-th/9805168.
7. T. C. Kraan, and P. van Baal, *Phys. Lett.* **B435**, 389–395 (1998), hep-th/9806034.
8. K.-M. Lee, and C.-H. Lu, *Phys. Rev.* **D58**, 025011 (1998), hep-th/9802108.
9. P. van Baal (2006), hep-ph/0610409.
10. A. A. Belavin, A. M. Polyakov, A. S. Shvarts, and Y. S. Tyupkin, *Phys. Lett.* **B59**, 85 (1975).
11. B. J. Harrington, and H. K. Shepard, *Phys. Rev.* **D17**, 2122 (1978).
12. M. N. Chernodub, T. C. Kraan, and P. van Baal, *Nucl. Phys. Proc. Suppl.* **83**, 556–558 (2000), hep-lat/9907001.
13. E.-M. Ilgenfritz, B. V. Martemyanov, M. Müller-Preussker, S. Shcheredin, and A. I. Veselov, *Phys. Rev.* **D66**, 074503 (2002), hep-lat/0206004.
14. E. M. Ilgenfritz, M. Muller-Preussker, and D. Peschka, *Phys. Rev.* **D71**, 116003 (2005), hep-lat/0503020.
15. E. M. Ilgenfritz, B. V. Martemyanov, M. Müller-Preussker, and A. I. Veselov, *Phys. Rev.* **D73**, 094509 (2006), hep-lat/0602002.
16. C. Gattringer, and S. Schäfer, *Nucl. Phys.* **B654**, 30 (2003), hep-lat/0212029.
17. C. Gattringer, and R. Pullirsch, *Phys. Rev.* **D69**, 094510 (2004), hep-lat/0402008.
18. P. Gerhold, E. M. Ilgenfritz, and M. Müller-Preussker (2006), hep-ph/0607315.
19. D. J. Gross, R. D. Pisarski, and L. G. Yaffe, *Rev. Mod. Phys.* **53**, 43 (1981).
20. E.-M. Ilgenfritz, and M. Müller-Preussker, *Nucl. Phys.* **B184**, 443 (1981).
21. D. Diakonov, and V. Y. Petrov, *Nucl. Phys.* **B245**, 259 (1984).

# Leutwyler-Smilga sum rules in the Schwinger model

L. Shifrin*,† and J. J. M. Verbaarschot*

*Department of Physics and Astronomy, Suny Stony Brook, Stony Brook, New York 11794, USA
†School of Information Systems, Computing and Mathematics, John Crank Building, Brunel University, Uxbridge, Middlesex UB8 3PH, United Kingdom

**Abstract.** We outline the microscopic derivation of the subset of Leutwyler-Smilga spectral sum rules in the 2-dimensional Schwinger model. As a side result, we obtain an expression for the sum rule in any external gauge field, for an arbitrary topological sector. This result generalizes the one due to Smilga, and may in principle be checked in lattice simulations of the model.

**Keywords:** CHIRAL SYMMETRY BREAKING, DIRAC OPERATOR, 2 DIMENSIONS
**PACS:** 12.38.Aw,11.15.Tk,11.30.Rd

## INTRODUCTION

Spontaneous breaking of Chiral Symmetry plays an important role in the low-energy dynamics of QCD. In the language of Dirac operator spectrum, it is reflected in universal relations such as Banks-Casher relation [1] and Leutwyler-Smilga (LS) spectral sum rules [2]. In an attempt to better understand the microscopic origin of the LS sum rules, we consider the Schwinger model [3], QED in 1+1 dimensions. It contains non-perturbative features that are also found in QCD, and was often used as a toy model for QCD. In particular, the Dirac eigenvalues in the Schwinger model satisfy the same LS sum rules as in $N_f = 1$ QCD [4]. The simplest of them has also been previously derived microscopically [4] in the trivial topological sector $\nu = 0$. We extend this result to the case of arbitrary $\nu$, starting from the standard fermionic description of the model.

## DERIVATION OF THE RESULT

### Leutwyler-Smilga sum rules in $N_f = 1$ QCD

The eigenvalues of the anti-hermitian Euclidean Dirac operator are defined by

$$D\!\!\!/\,\phi_k = i\lambda_k \phi_k. \tag{1}$$

In the topological sector $\nu$ the Dirac operator has exactly $\nu$ zero eigenvalues that are not paired. All other eigenvalues occur in pairs $\pm\lambda_k$. Noticing that the quark mass $m$ and the vacuum angle $\theta$ only enter in a combination $me^{i\theta}$, and expanding the partition function in powers of $m$, Leutwyler and Smilga [2] obtained a family of gauge field-averaged sum

CP892, *Quark Confinement and the Hadron Spectrum VII*
edited by J. E. F. T. Ribeiro
© 2007 American Institute of Physics 978-0-7354-0396-3/07/$23.00

rules ($\Sigma$ being the chiral condensate, V - the system's volume):

$$\left\langle\!\!\left\langle \sum_{n_1\neq\cdots\neq n_l} \frac{1}{\lambda_{n_1}^2\cdot\ldots\cdot\lambda_{n_l}^2}\right\rangle\!\!\right\rangle^v = \frac{|v|!}{2^l l!(|v|+l)!}\left(\Sigma V\right)^{2l}. \tag{2}$$

## Schwinger model

In this section we give a brief review of the Schwinger Model [3] which is massless QED in two spacetime dimensions. The Euclidean Lagrangian is defined by

$$\mathscr{L} = \frac{1}{4}F_{\mu\nu}^2 - \bar{\psi}[i\,\partial\!\!\!/ + g\,A\!\!\!/]\psi. \tag{3}$$

The Lagrangian of this model has a chiral symmetry which is broken by the $U(1)$ axial anomaly. Due to Index theorem, the Dirac operator in the topological sector $v$ has $|v|$ normalizable zero modes. The vector potential $A_\mu$ can be decomposed as

$$A_\mu = -\varepsilon_{\mu\nu}\partial^\nu\phi + \partial_\mu\lambda, \tag{4}$$

$\lambda$ being a pure gauge. In 2D, the Dirac operator has an important representation:

$$\partial\!\!\!/_\phi = e^{g\phi\gamma_5}\,\partial\!\!\!/\,e^{g\phi\gamma_5}. \tag{5}$$

Therefore, the model is solved by a local chiral rotation. The non-invariance of the fermionic measure [5] results in the anomalous term in the effective action [6, 7, 8]:

$$\det'\partial\!\!\!/ = \prod_{n>0}\lambda_n^2 = \mathscr{C}\det\mathscr{N}\exp\left(\frac{g^2}{2\pi}\int d^2x\phi(x)\Delta\phi(x)\right), \tag{6}$$

$\mathscr{N}$ being the norm matrix of the fermionic zero modes, which are (for $v > 0$):

$$\psi_p(x) = \frac{1}{\sqrt{2\pi}}(x^+)^P e^{-g\phi(x)}\begin{pmatrix}1\\0\end{pmatrix}\ ,\quad p=0\cdots v-1,\qquad x^\pm = x_0\pm ix_1. \tag{7}$$

More details on the model and further references can be found in e.g.[9].

## Simplest LS sum rule in the Schwinger model for arbitrary $v$

It follows from the spectral representation of the fermionic Green's function $G^v$ that

$$\sum_{\lambda_n\neq 0}\frac{1}{\lambda_n^2} = -\text{Tr}\left[G^{v2}\right], \tag{8}$$

where the trace is taken over both spinor indices and spatial coordinates. This was used in [4] to derive the sum rule for $v = 0$. For the non-zero (positive) $v$, $G^v$ satisfies

$$\partial\!\!\!/_x G^v(x,y) = \delta(x-y) - \gamma_5^+ P^v(x,y), \tag{9}$$

with $\gamma_5^+ \equiv \frac{1}{2}(1 + \gamma_5)$ and $\gamma_5^+ P^\nu(x,y)$ being the projection density on the zero mode subspace. The explicit solution is (in operator form):

$$G^\nu = (1 - \gamma_5^+ P^\nu)e^{-g\phi\gamma_5}G_0 e^{-g\phi\gamma_5}(1 - \gamma_5^+ P^\nu), \qquad (10)$$

with $G_0$ being a free 2D Dirac propagator. Using this in (8) one gets [9]:

$$\sum_{\lambda_n \neq 0} \frac{1}{\lambda_n^2} = \frac{1}{(2\pi)^2} \int d^2x d^2y d^2z P^\nu(x,y) e^{-g\phi(y)-g\phi(x)+2g\phi(z)} \times \qquad (11)$$

$$\times \frac{(x-y)^2 - 2i\varepsilon_{\mu\nu}(x-z)_\mu(z-y)_\nu}{(x-z)^2(z-y)^2}.$$

Computing the projector from (7) and inserting it into (11) leads to:

$$\sum_{\lambda_n \neq 0} \frac{1}{\lambda_n^2} = \frac{2}{(2\pi)^{\nu+2}(\nu+1)!\det\mathcal{N}} \int d^2z \int d^2x_1 \cdots d^2x_{\nu+1} \times$$

$$\times \frac{\prod_{i<j}^{\nu+1} |x_i - x_j|^2}{\prod_{p=1}^{\nu+1} |x_p - z|^2} e^{-2g(\phi(x_1)+\cdots+\phi(x_{\nu+1}))+2g\phi(z)}. \qquad (12)$$

The integration over the gauge field $\phi$ with the proper weight gives in the large volume limit [9]:

$$\left\langle\!\!\left\langle \sum_{\lambda_n \neq 0} \frac{1}{\lambda_n^2} \right\rangle\!\!\right\rangle_\nu = \frac{1}{2(|\nu|+1)}\Sigma^2 V^2, \qquad (13)$$

which is the LS sum rule (for a proper definition of the gauge averaging see [9]; the case of negative $\nu$ gives the same result). The last two equations are our main results. In the Schwinger model, all the LS sum rules can also be derived using 2D bosonization [9].

## ACKNOWLEDGMENTS

We would like to thank Andreas Ludwig, Andrei Smilga and Pierre van Baal for useful discussions. This work was supported in part by US DOE Grant DE-FG-88ER40388.

## REFERENCES

1. T. Banks and A. Casher, Nucl. Phys. B **169**, 103 (1980).
2. H. Leutwyler and A. Smilga, Phys. Rev. D **46**, 5607 (1992).
3. J. S. Schwinger, Phys. Rev. **128**, 2425 (1962).
4. A. V. Smilga, Phys. Rev. D **46**, 5598 (1992).
5. K. Fujikawa, Phys. Rev. D **21**, 2848 (1980) K. Fujikawa, Phys. Rev. Lett. **42**, 1195 (1979).
6. M. Hortacsu, K. D. Rothe and B. Schroer, Phys. Rev. D **20**, 3203 (1979).
7. R. Roskies and F. Schaposnik, Phys. Rev. D **23**, 558 (1981).
8. I. Sachs and A. Wipf, Helv. Phys. Acta **65**, 652 (1992).
9. L. Shifrin and J. J. M. Verbaarschot, Phys. Rev. D **73**, 074008 (2006) [arXiv:hep-th/0507220].

# Studying the infrared behaviour of gluon and ghost propagators using large asymmetric lattices

P. J. Silva and O. Oliveira

*Centro de Física Computacional, Departamento de Física, Universidade de Coimbra, P-3004-516 Coimbra, Portugal*

**Abstract.** We report on the infrared limit of the quenched lattice Landau gauge gluon propagator computed from large asymmetric lattices. In particular, the compatibility of the pure power law infrared solution $(q^2)^{2\kappa}$ of the Dyson-Schwinger equations is investigated and the exponent $\kappa$ is measured. Some results for the ghost propagator and for the running coupling constant will also be shown.

**Keywords:** lattice QCD; Landau gauge; confinement; gluon propagator; ghost propagator; strong coupling constant
**PACS:** 12.38.-t; 11.15.Ha; 12.38.Gc; 12.38.Aw; 14.70.Dj; 14.80.-j

Despite the success of Quantum Chromodynamics (QCD) as *the* theory of the strong interaction, a full understanding of the confinement mechanism is still missing. One line of research, very active in the last years, consists in the study of the QCD propagators for low momenta. Indeed, some works (for details, see [1] and references therein) relate the infrared behaviour of the gluon and ghost propagators in the Landau gauge with gluon confinement. In particular, the Zwanziger horizon condition implies a null zero momentum gluon propagator $D(q^2)$, and the Kugo-Ojima confinement mechanism requires an infinite zero momentum ghost propagator $G(q^2)$.

An investigation of the infrared behaviour of the gluon and ghost propagators should be done in a non-perturbative framework. At the moment, two first principles approaches are available for such a task, namely Dyson-Schwinger equations (DSE) and lattice QCD methods. Given the different nature of such approaches, a comparison between the results of the two methods is necessary.

A solution of the DSE [2] predicting pure power laws for gluon and ghost dressing functions,

$$Z_{gluon}(q^2) \sim (q^2)^{2\kappa}, \ Z_{ghost}(q^2) \sim (q^2)^{-\kappa}, \tag{1}$$

with $\kappa \sim 0.595$, has been extensively used in subsequent works (see [3] for a recent review). As shown in figure 1 of [11], these power laws are only valid for very low momenta, $q < 200 MeV$ (see also [4]). To test this solution of the DSE with lattice QCD, using a symmetric lattice, it would require a lattice volume much larger than a typical present day simulation (see, for example, [5]).

Large asymmetric lattices, in the form $L_s^3 \times L_t$, with $L_t \gg L_s$, give us a possibility to test these power laws on the lattice. In this paper we briefly report on the results [6, 7, 8, 9, 10, 11, 12] obtained by us, considering large asymmetric lattices with

CP892, *Quark Confinement and the Hadron Spectrum VII*
edited by J. E. F. T. Ribeiro
© 2007 American Institute of Physics 978-0-7354-0396-3/07/$23.00

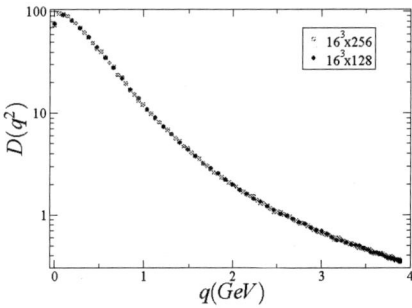 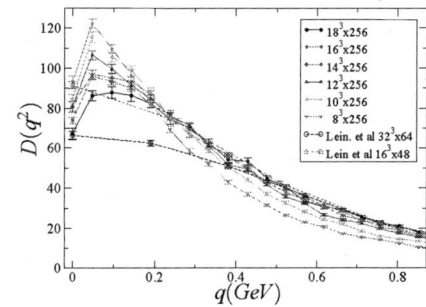

**FIGURE 1.** On the left, the gluon propagator for $16^3 \times 128$ and $16^3 \times 256$ lattices, considering only pure temporal momenta. Note the logarithmic scale in the vertical axis. On the right, the gluon propagator for all lattices $L_s^3 \times 256$. For comparisation, we also show the $16^3 \times 48$ and $32^3 \times 64$ propagators computed in [14].

$L_s = 8, 10, \ldots, 18$ and $L_t = 256$, about the infrared behaviour of the gluon and ghost propagators and the strong running coupling defined from these propagators. Despite the finite volume effects caused by the small spatial extension, the large temporal size of these lattices allow to access to momenta as low as 48 MeV.

In what concerns the gluon propagator, our results [9] show that the propagator dependence on the spatial volume is smooth. Indeed, for the smallest momenta, the bare gluon propagator decreases with the lattice volume, and increases for higher momenta. The values of the infrared exponent extracted from our lattices increase with the lattice volume. Although almost all values of $\kappa$ are below 0.5 (see table 1 in [9]), we obtain, by extrapolating the $\kappa$ values to the infinite volume, $\kappa$ values above 0.5, with a weigthed mean of the various estimations giving $\overline{\kappa}_\infty = 0.5246(46)$.

Considering the gluon propagator as a function of the spatial volume, we can also extrapolate it, and fit the obtained propagator to a pure power law. Going this way, we get values for $\kappa \in [0.49, 0.53]$. Note that the lattice data favours the values in the right hand side of this interval.

The reader should be also aware that fits to our data considering higher momenta and other model functions give higher values for $\kappa$ [11, 13].

Similarly to other studies, it is possible to use our gluon data to verify the positivity violation for the gluon propagator [12, 13].

We have also computed the ghost propagator and the strong coupling constant $\alpha_S(q^2)$ defined from these propagators, for our smallest lattices [10]. Our lattice data for these quantities also show sizeable dependence on the spatial volume of the lattices involved in our calculations. We also have found visible Gribov copy effects in the ghost propagator as well as in the strong coupling constant.

Concerning the infrared behaviour of the ghost propagator, we were unable to extract an infrared exponent from our results. Possible reasons for this negative result can be either the finite volume effects associated to the small spatial volume of the lattices involved in the computation, or the lack of lattice data in the infrared region — remember that the DSE ghost power law lacks validity well below 200 MeV.

In the infrared region, $\alpha_S(q^2)$ shows a decreasing behaviour for the smallest momenta,

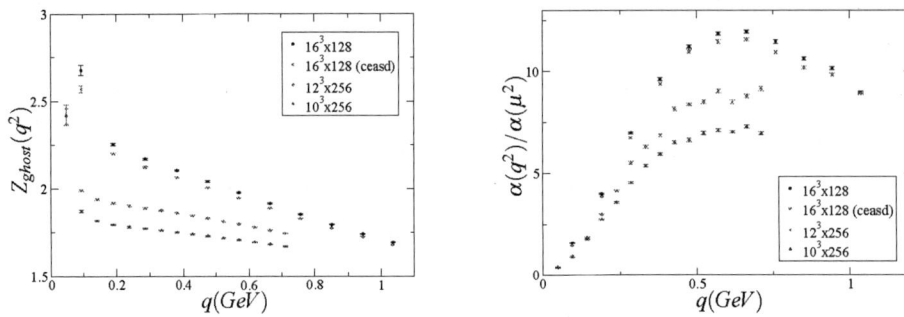

**FIGURE 2.** On the left, the bare ghost dressing function in the infrared region computed from a plane wave source. On the right, the strong coupling constant. Here, we only consider pure temporal momenta.

in apparent contradiction with the continuum DSE prediction — an infrared fixed point, but in agreement with other lattice studies [5] and the solution of DSE on a torus [15]. However, the reader should be aware that $\alpha_S(q^2)$, for the smallest momenta, seems to increase with the volume.

In a near future, we will improve the statistics for our larger lattices and the extrapolations to the infinite volume limit. We also plan to perform simulations with larger lattices.

## ACKNOWLEDGMENTS

This work was supported by FCT via grant SFRH/BD/10740/2002, and project POCI/FP/63436/2005.

## REFERENCES

1. R. Alkofer, L. von Smekal, *Phys. Rept.* **353** (2001) 281 [hep-ph/0007355].
2. C. Lerche, L. von Smekal, *Phys. Rev.* **D65** (2002) 125006 [hep-ph/0202194].
3. C. S. Fischer, *J.Phys.* **G32** (2006) R253-R291 [hep-ph/0605173].
4. C. S. Fischer, J. M. Pawlowski, hep-th/0609009.
5. A. Sternbeck, E.-M. Ilgenfritz, M. Müller-Preussker, A. Schiller, I. L. Bogolubsky, PoS(LAT2006)076 [hep-lat/0610053].
6. O. Oliveira, P. J. Silva, *AIP Conf. Proc.* **756** (2005) 290 [hep-lat/0410048].
7. P. J. Silva, O. Oliveira, PoS(LAT2005)286 [hep-lat/0509034].
8. O. Oliveira, P. J. Silva, PoS(LAT2005)287 [hep-lat/0509037].
9. P. J. Silva, O. Oliveira, *Phys. Rev.* **D74** (2006) 034513 [hep-lat/0511043].
10. O. Oliveira, P. J. Silva, hep-lat/0609027.
11. O. Oliveira, P. J. Silva, hep-lat/0609036, to appear in Brazilian Journal of Physics.
12. P. J. Silva, O. Oliveira, PoS(LAT2006)075 [hep-lat/0609069].
13. P. J. Silva, O. Oliveira, "Fitting the lattice gluon propagator and the question of positivity violation", these proceedings.
14. D. B. Leinweber, J. I. Skullerud, A. G. Williams, and C. Parrinello, *Phys. Rev.* **D60** (1999) 094507; Phys. Rev. **D61** (2000) 079901 [hep-lat/9811027].
15. C. S. Fischer, B. Gruter, R. Alkofer, *Annals Phys.* **321** (2006) 1918 [hep-ph/0506053].

# Topological Charge and the Laminar Structure of the QCD Vacuum

## H. B. Thacker

*University of Virginia, Charlottesville, VA 22904*

**Abstract.** Monte Carlo studies of pure glue $SU(3)$ gauge theory using the overlap-based topological charge operator have revealed a laminar structure in the QCD vacuum consisting of extended, thin, coherent, locally 3-dimensional sheets of topological charge embedded in 4D space, with opposite sign sheets interleaved. In this talk I discuss the interpretation of these Monte Carlo results in terms of our current theoretical understanding of theta-dependence and topological structure in asymptotically free gauge theories.

**Keywords:** QCD, topological charge, D-branes
**PACS:** 11.15.Ha, 11.30.Rd, 12.38.Gc, 12.39.Fe

## <A SECTION>

Recent studies of topological charge using the overlap-based topological charge density operator on the lattice have revealed a type of long-range structure that is profoundly different from what might have been expected in an instanton-based model of the QCD vacuum. These studies [1] (see also [2]) produced the surprising result that the $q(x)$ distribution in a typical Monte Carlo gauge configuration is dominated by extended, coherent, thin 3-dimensional sheets of topological charge. In each configuration, sheets of opposite sign are juxtaposed and are everywhere close together in what can roughly be described as a dipole layer which is spread throughout the 4-dimensional Euclidean space (with various folds and wrinkles). The vacuum is thus permeated with what is locally a laminar structure consisting of alternating sign sheets or membranes of topological charge. The thickness of these membranes is typically a few lattice spacings, independent of the physical mass scale, and thus the membranes apparently become infinitely thin in the continuum limit. This kind of "subdimensional" ordering, where coherence takes place on manifolds of lower dimensionality than spacetime itself, is closely related to the appearance of a contact term in the two-point topological charge correlator. The Euclidean correlator $G(x) = \langle q(x)q(0) \rangle$ *is required by spectral considerations to be negative for any nonzero separation* $|x| > 0$. In practice (i.e. in Monte Carlo calculations), this requirement places severe restrictions on what type of topological charge fluctuations can be dominant. For example, the negativity of the correlator rules out the dominance of bulk-coherent lumps of topological charge (e.g. finite size instantons), as this would lead to a positive correlator over distances smaller than the instanton size. Since the topological susceptibility can be obtained by integrating the 2-point correlator over all $x$, a positive susceptibility can only arise from a delta-function contact term at the origin. The observed arrangement of thin, nearby layers of $q(x)$ with opposite sign builds up a positive contact term at $x = 0$ while maintaining the required

CP892, *Quark Confinement and the Hadron Spectrum VII*
edited by J. E. F. T. Ribeiro
© 2007 American Institute of Physics 978-0-7354-0396-3/07/$23.00

negativity of the correlator for finite separation. The range of this contact term on the lattice is associated with the thickness of the sheets, both of these being a few lattice spacings and approaching zero in physical units. This thickness is consistent with the range of non-ultralocality of the overlap Dirac operator.

Since the initial discovery of coherent topological charge sheets in QCD, similar methods have been applied to the study of 2-dimensional $CP^{N-1}$ sigma models [3]. For $N > 3$, the topological charge distribution was found to be dominated by thin 1-dimensionally coherent membrane-like structures with interleaved membranes of opposite sign. This interleaved arrangement is exactly what one would expect as the analog of the observed 3-dimensional structures in 4D gauge theory. In both cases, the coherent structure has codimension 1, i.e. the dimensionality of a domain wall.

To interpret these Monte Carlo results, we recall the arguments of Luscher [4] that nonzero topological susceptibility implies the presence of a massless pole in the two-point correlator of the Chern-Simons current. Let us define the *abelian* 3-index Chern-Simons tensor

$$A_{\mu\nu\rho} = -Tr\left(B_\mu B_\nu B_\rho + \frac{3}{2}B_{[\mu}\partial_\nu B_{\rho]}\right) \tag{1}$$

where $B_\mu$ is the Yang-Mills gauge potential. We consider the Chern-Simons current that is dual to this tensor,

$$j_\mu^{CS} = \varepsilon_{\mu\nu\rho\sigma}A_{\nu\rho\sigma} . \tag{2}$$

Although $j_\mu^{CS}$ is not gauge invariant, its divergence is the gauge invariant topological charge density

$$\partial_\mu j_\mu^{CS} = Tr F\tilde{F} = 32\pi^2 q(x) . \tag{3}$$

Choosing a covariant gauge, $\partial_\mu A_{\mu\nu\rho} = 0$, the correlator of two Chern-Simons currents has the form

$$\langle j_\mu^{CS}(x) j_\nu^{CS}(0)\rangle = \int \frac{d^4p}{(2\pi)^4} e^{-ip\cdot x} \frac{p_\mu p_\nu}{p^2} G(p^2) . \tag{4}$$

From (3) we see that $G(p^2)$ must have a $p^2 = 0$ pole whose residue is the topological susceptibility,

$$G(p^2) \sim \frac{\chi_t}{p^2} . \tag{5}$$

This long-range correlation constitutes a "secret long-range order" of gauge fields associated with their topological charge fluctuations. Since the CS current is not gauge invariant, the presence of a $q^2 = 0$ pole does not imply the existence of a massless particle. On the other hand, the pole has a gauge invariant residue ($\propto \chi_t$) and cannot be transformed away. So it characterizes a physically significant long range coherence in the gauge field associated with topological charge fluctuations.

Luscher's analysis of QCD topological structure in terms of Wilson bags can be understood as a generalization of the analysis of similar properties in the 2-dimensional $CP^{N-1}$ sigma models. These models provide a quite detailed 2D analog of the coherent structure observed in 4D QCD. The $CP^{N-1}$ models have a $U(1)$ gauge invariance and have classical instanton solutions which come in all sizes. (Just like pure-glue QCD these models are classically scale invariant and acquire a mass scale via a conformal anomaly.) In 2D U(1) gauge theories like $CP^{N-1}$, the topological charge density in the continuum is just $(1/2\pi)\varepsilon_{\mu\nu}\partial_\mu A_\nu$ and the Chern-Simons current is just the dual of the gauge potential,

$$j_\mu^{CS} = \frac{1}{2\pi}\varepsilon_{\mu\nu}A_\nu \tag{6}$$

Just as in 4D QCD, nonzero topological susceptibility implies the presence of a $q^2 = 0$ pole in the CS current correlator. But in the two-dimensional case, this same pole appears in the $A_\mu$ correlator and is responsible for confinement of $U(1)$ charge via a linear coulomb potential. Thus, in 2-dimensional $U(1)$ theories, topological susceptibility and confinement of $U(1)$ charge are equivalent phenomena. An instructive way to illustrate this is to introduce a nonzero $\theta$ term in the action over a two-volume $V$ enclosed by a boundary $C = \partial V$ with $\theta = 0$ outside the boundary. After integration by parts, the theta term is equivalent to a Wilson loop around the boundary carrying a charge $\theta/2\pi$:

$$\exp\left[\frac{i}{2\pi}\int d^2x\theta(x)\varepsilon_{\mu\nu}F^{\mu\nu}\right] = \exp\left[\frac{i\theta}{2\pi}\oint_C A \cdot dx\right] \tag{7}$$

If the topological susceptibility is nonzero

$$\chi_t = \frac{\partial^2 E(\theta)}{\partial\theta^2}\Big|_{\theta=0} > 0 \tag{8}$$

then for small nonzero $\theta$, the vacuum energy density $E(\theta)$ inside the Wilson loop will be greater than that outside the loop, so it will obey an area law,

$$\langle W(C)\rangle \propto \exp\left[-(E(\theta)-E(0))V\right] \tag{9}$$

In a Hamiltonian framework, this corresponds to applying a background electric field $\theta$ by putting opposite charges at either end of the 1-dimensional spatial box. The topological susceptibility is just the vacuum polarizability with respect to this field. Periodicity in $\theta \to \theta + 2\pi$ arises via a discontinuous process of charged pair production at $\theta = \pi$, which screens a unit of electric flux. At $\theta = 2\pi$ the confining force is completely screened. This is essentially Coleman's original interpretation of $\theta$-dependence in the massive Schwinger model [5].

In specifying the analogy between 2D U(1) theories and 4D SU(N) gauge theories, we take the Chern-Simons currents (2) and (6) to be directly analogous. This means that the gauge field $A_\mu$ in the 2D theory should be identified *not* with the 4-dimensional gauge field, but with the abelian 3-index Chern-Simons tensor (1). Like the gauge field $A_\mu$ in 2 dimensions, this is dual to the Chern-Simons current (2). Similarly, the Wilson loop or line excitations in the 2-dimensional $U(1)$ models correspond not to Wilson loops in 4D, but to "Wilson bags," i.e. integrals of the Chern-Simons tensor over a 3-surface $\Sigma$.

$$B(\Sigma) = \exp\left[i(\theta/2\pi)\int_\Sigma A_{\mu\nu\lambda}dx_\mu dx_\nu dx_\lambda\right] \tag{10}$$

This is the analog of a Wilson loop in 2D $U(1)$ in the sense that, if $\Sigma$ is a closed 3-surface that forms the boundary of a 4-dimensional volume $V$, inserting the Wilson bag factor

(10) in the gauge field path integral is equivalent to including a $\theta$-term in the gauge action over the 4-volume $V$. The discussion of what happens as we vary $\theta$ from 0 to $2\pi$ is completely analogous to the screening of the 2D Wilson loop. For a fractional bag charge $\theta/2\pi$, with $0 < \theta < 2\pi$, the vacuum inside the bag will have a higher energy than the $\theta = 0$ vacuum outside. The expectation of the Wilson bag integral thus satisfies a volume law analogous to the area law for the Wilson loop in 2D, and there will be "bag confinement," a confining force between the walls of a fractionally charged bag. At $\theta = 2\pi$, the confining force between bag walls disappears. Integer charged bag surfaces are thus free to percolate throughout the vacuum. The topological charge is the curl of the Chern-Simons tensor, so for a uniform $A_{\mu\nu\lambda}$ which is nonzero on a flat bag surface, the topological charge distribution is a dipole layer consisting of thin, coherent positive and negative layers on either side of the bag surface. Like the Wilson line excitations in the $CP^{N-1}$ models [3], a vacuum full of unit-charged Wilson bags provides a reasonable model for interpreting the topological charge structure observed in lattice Monte Carlo simulations.

The role of Wilson bag excitations in QCD is greatly illuminated by string/gauge duality, exploiting the dual relationship between topological charge in gauge theory and Ramond-Ramond charge in IIA string theory [7]. As Witten showed [6], the string/gauge correspondence nicely confirms the k-vacuum/Wilson bag/domain wall scenario originally suggested by large-$N_c$ chiral Lagrangian arguments [8]. It also points to the correct candidate for the string theory analog of a Wilson bag in gauge theory. In IIA string theory, the Wilson bag can be interpreted as the holographic image of a D6-brane (which is wrapped around a compact $S_4$ and therefore appears as a 2-brane or membrane in 3+1 dimensions). In particular, the defining property of the Wilson bag, namely that the value of $\theta$ jumps by $\pm 2\pi$ when crossing the surface, is in fact nothing but the statement of quantization of Ramond-Ramond charge on a D6-brane.

This work was supported in part by the Department of Energy under grant DE-FG02-97ER41027.

## REFERENCES

1. I. Horvath et al, Phys. Rev. D68, 114505 (2003).
2. E.-M. Ilgenfritz et al., Nucl. Phys. Proc. Suppl. 153: 328 (2006); and hep-lat/0610087.
3. S. Ahmad, J. T. Lenaghan, H. B. Thacker, Phys. Rev. D72: 114511 (2005); Y. Lian and H. B. Thacker, hep-lat/0607026.
4. M. Luscher, Phys. Lett. B78, 465 (1978).
5. S. Coleman, Ann. Phys. 101: 239 (1976).
6. E. Witten, Phys. Rev. Lett. 81: 2862 (1998).
7. For a discussion, see H. Thacker, hep-lat/0610049.
8. E. Witten, Ann. Phys. 128: 363 (1980).

# String/Flux Tube Duality

## Charles B. Thorn

*Department of Physics, University of Florida, Gainesville FL 32611*

**Abstract.** We describe Field/String duality as applied to the response of gauge fields to separated quark and antiquark sources.

**Keywords:** Confinement, Gauge/String duality
**PACS:** 11.15.Pg, 11.25.Tq

## INTRODUCTION

The physics of quark confinement is generally believed to involve two conjectured properties of quantum Yang-Mills gauge theory: (1) that there is a mass gap $m_G$ (the lightest glueball mass) and (2) that the gauge field responds to a fixed $Q$ source separated from a fixed $\bar{Q}$ source by a distance $L$ by forming a gluonic flux tube (or gluon chain) between $Q$ and $\bar{Q}$ with energy $U(L) \simeq T_0 L$ for large $L$.

Although both of these facets of quark confinement are firmly established numerically, an analytic understanding of either is so far unattained. Indeed just proving the mass gap is one of the Clay Institute millenium problems. My message here is that a less daunting path to such an analytic understanding may be the reformulation of Yang-Mills as a String Theory (Field/String Duality) [1, 2, 3]. This reformulation can proceed without solving the theory or even without establishing a mass gap. It might well provide both a setting, in which the physics of confinement can be understood by using string variables to construct a tractable model of the gluonic flux tube (gluon chain), and a vehicle for a self-consistent determination of a mass gap.

The AdS/CFT correspondence shows that a string reformulation of QFT is quite independent of confinement and also of the existence of a mass gap. Indeed, the best understood case of Field/String duality is the equivalence of $\mathcal{N} = 4$ supersymmetric SU(N) Yang Mills to IIB superstring theory on $AdS_5 \times S_5$ [1]. In this case the finiteness of the $\mathcal{N} = 4$ theory implies conformal invariance which in turn implies a vanishing mass gap and zero string tension. On the string side these features are consequences of the curved AdS background. The string interpretation is particularly transparent when $N \to \infty$ because $g_{\text{string}} \sim 1/N$, so it is a limit in which strings do not break or interact.

From this point of view string theory offers something much more tangible to theoretical physics than a nebulous and quasi-religious "theory of everything". I believe that it provides a practical theoretical framework for resolving some of the most intriguing but difficult conundra of quantum field theory. In this regard, I offer a new definition of string theory by way of an analogy:

String Theory : $\sum$ (Planar Diagrams) :: Bethe-Salpeter Equation :$\sum$ (Ladder Diagrams)

Just as the sum of ladder diagrams gives a zeroth order Bethe-Salpeter equation so

CP892, *Quark Confinement and the Hadron Spectrum VII*
edited by J. E. F. T. Ribeiro
© 2007 American Institute of Physics 978-0-7354-0396-3/07/$23.00

does the sum of planar diagrams give a zeroth order (noninteracting) string theory. In both cases the full QFT can be regained by systematic corrections. For string theory the nonplanar corrections are neatly handled via 't Hooft's $1/N$ expansion [4], since the planar approximation is exact in the large $N$ limit with $\lambda = N\alpha_s/\pi$ fixed.

In this talk I shall explain how stringy features appear in the conformally invariant $\mathscr{N} = 4$ case. Though this theory lacks a mass gap and quark confinement, it nonetheless produces a stringy gluonic flux tube between separated color sources. Moreover, we can easily reach interesting conclusions about this flux tube's physical properties, especially in the strong 't Hooft coupling limit when the string can be treated semi-classically. This limit makes sense here because the coupling does not depend on the scale. In a string theory formulation of QCD such a semiclassical limit is not possible because there is no tunable coupling. Nonetheless, the conformal case is an important example because it shows a limit in which the planar sum can actually be done. Although there should be a strikingly different outcome for the sum of planar diagrams in QCD, the technical difficulties in the two problems are quite comparable. Solving one should teach us a great deal about solving the other.

## SEPARATED $Q\bar{Q}$ SOURCES

At $N = \infty$ the response of the $\mathscr{N} = 4$ theory to separated static color sources is very interesting: a flux tube forms with an excitation spectrum that becomes string-like in the limit of strong 't Hooft coupling $\lambda \to \infty$. A convenient probe that reveals this excitation spectrum is the expectation of a rectangular $L \times T$ Wilson Loop, $\langle W(L,T) \rangle \sim \sum_n w_n \exp(-TE_n(L))$ as $T \to \infty$. The $L$ dependence of the ground state energy tests the presence of confinement, a test the $\mathscr{N} = 4$ theory fails because it is conformal $E_G(L) \sim -c/L$ at large $L$. But the excitation spectrum $E_n(L)$ at fixed $L$ can look stringy and does at strong 't Hooft coupling.

But first we consider the the weak coupling limit of this $Q\bar{Q}$ system, $\lambda \ll 1$. In this limit in Coulomb gauge the planar approximation to the Wilson loop, in pure Yang-Mills theory, is given by multiple instantaneous Coulomb exchanges (See Fig. 1). The

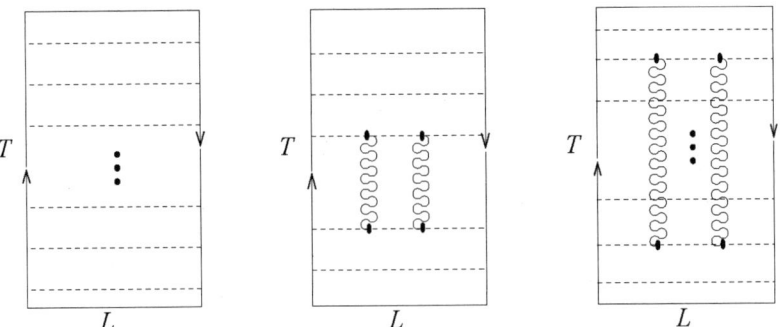

**FIGURE 1.** Wilson loop with only Coulomb exchange (left), a planar radiative correction (middle), and a nonplanar correction (right).

Wilson loop ensures $Q\bar{Q}$ are in a singlet. We can therefore read off the singlet energy as the coefficient of $-T$ in the exponential behavior at large $T$. In the case of the sum of diagrams with only Coulomb exchange, this shows $E_{singlet} = -\pi\lambda/2L$, and this is the only eigenstate that couples. The resolvent shows only a single pole

$$R_0(E,L) \equiv \int_0^\infty dTe^{ET}W_0 \sim \frac{\rho_0'}{-E - \pi\lambda/2L}$$

Note that in the $\mathcal{N} = 4$ conformal theory the Wilson loop is modified to include coupling to the scalar fields. This (1) doubles the effect of Coulomb exchange, so we have to understand $\lambda \to 2\lambda$ in this formula, and (2) introduces a continuum for $E > 0$ at lowest order, because the scalar propagator is not instantaneous. A very interesting feature of the planar radiative corrections (e. g. the middle diagram) is that, since planarity forbids gluon exchanges between parts of the diagram separated by propagating transverse gluons, there is a gap between the discrete ground state and the continuum, summarized by the following form for the planar resolvent (now for the $\mathcal{N} = 4$ theory)

$$R_{\text{planar}}(E,L) \sim \frac{\rho_0}{-E - \pi[\lambda + O(\lambda^2)]/L} + \int_0^\infty dE' \frac{\rho_1(E')}{E' - E}$$

In contrast nonplanar corrections, suppressed by powers of $1/N$, show no such gap:

$$R_{\text{nonplanar}}(E,L) \sim \frac{1}{N^2} \int_{-\pi\lambda/2L}^\infty dE' \frac{\rho_2(E')}{E' - E}$$

Thus at $N = \infty$ and arbitrarily weak coupling, there is a gap in the $Q\bar{Q}$ system [5].

At strong coupling, $\lambda \gg 1$, the AdS/CFT correspondence gives the excitation spectrum of the $Q\bar{Q}$ system in $\mathcal{N} = 4$ supersymmetric Yang-Mills theory as the semiclassical quantization of a IIB superstring connecting the two sources on the boundary of $AdS_5$. The static solution [6] has energy $E_0(L) = -(2\pi)^3 \sqrt{\lambda}/(\Gamma(1/4)^4 L)$. $E_0$ decreases as $1/L$ with separation, showing the absence of a confining force. Nonetheless, this stretched string has an infinite number of stringy excitations.

Semi-classical quantization of the small oscillations about this static solution [7] gives string like modes with discrete levels just above $E_0$:

$$E_{N_n} - E_0 = \sum N_n \omega_n; \quad \omega_n = \frac{(2\pi)^{3/2}}{\Gamma(1/4)^2 L} \xi_n; \quad \xi_n \sqrt{\xi_n^4 - 1} \int_0^1 \frac{t^2 dt}{[1 + \xi_n^2 t^2]\sqrt{1 - t^4}} = \frac{n\pi}{2}$$

Where $n = 1,2,\ldots$. For large $n$, $\omega_n \sim (2\pi)^3(n+1)/(\Gamma(1/4)^4 L)$ [8], typical of normal modes of a string with effective tension $T_{eff} \sim 1/L^2$. We see that $\mathcal{N} = 4$ is teetering on the brink of quark confinement: a stringy object is there for $L$ finite, but a mechanism to prevent $T_{eff}$ from dropping to 0 when $L \to \infty$ is lacking. Finally, we note that near threshold ($E = 0$) discrete levels accumulate [5] $E_{n+1}/E_n \sim e^{-\pi/\sqrt{4\lambda}}$. This is shown by a semiclassical quantization of the motion of the midpoint of string stretched a distance $D >> L$ from the boundary of AdS, where the string ends reside.

To summarize we show the energy level diagram for the $\mathcal{N} = 4$ $Q\bar{Q}$ system at $N = \infty$ for weak and strong 't Hooft coupling (see Fig. 2). The transition from weak to strong

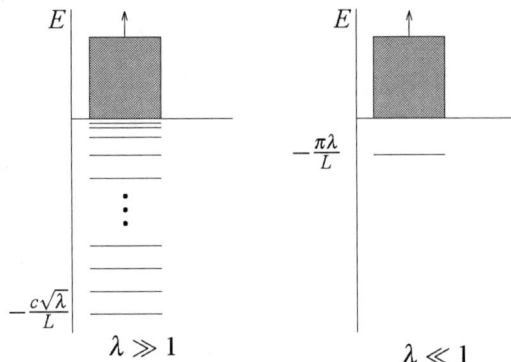

**FIGURE 2.** Energy spectrum of the $\mathcal{N} = 4$ $Q\bar{Q}$ system for strong and weak coupling at $N = \infty$.

coupling is rather mundane: a stronger binding with a deepening gap that eventually supports excited discrete levels that peel off the continuum and move into the gap. $N = \infty$ is essential here: otherwise the continuum goes all the way down to $E_0$. At large finite $N$ this continuum would be dominated by very narrow resonances.

Surprisingly a simple Feynman gauge ladder diagram model [9] shows qualitatively similar physics. The authors show that the weak coupling energy is recovered and at strong coupling the ground energy $\propto -\sqrt{\lambda}/L$ with a different numerical coefficient than the AdS string. At intermediate coupling [8] this model shows no more bound states for $\lambda < 1/4$. For $\lambda > 1/4$ infinite number of bound states appear with threshold behavior $E_{n+1}/E_n \sim e^{-\pi/\sqrt{4\lambda-1}}$, *with the same strong coupling behavior as* $\mathcal{N} = 4$. The big qualitative difference from the exact strong coupling behavior is that the ladder model shows only a single mode of small oscillations instead of an infinite number of stringy modes $\omega_n$. This defect is due to the neglect of all the non-ladder planar diagrams. An accurate treatment of intermediate coupling requires either the proper quantization and solution of IIB superstring on $AdS_5 \times S_5$, or a way to sum *all* planar diagrams.

# REFERENCES

1. J. Maldacena, Adv. Theor. Math. Phys. **2**, 231 (1998), hep-th/9711200
2. S.S. Gubser, I.R. Klebanov and A.M. Polyakov, Phys. Lett. **B428**, 105 (1998), hep-th/9802109;
   E. Witten, Adv. Theor. Math. Phys. **2**, 253 (1998), hep-th/9802150.
3. K. Bardakci and C. B. Thorn, Nucl. Phys. B **626** (2002) 287 [arXiv:hep-th/0110301].
   C. B. Thorn, Nucl. Phys. B **637** (2002) 272 [Erratum-ibid. B **648** (2003) 457] [arXiv:hep-th/0203167];
   S. Gudmundsson, C. B. Thorn and T. A. Tran, Nucl. Phys. B **649** (2003) 3 [arXiv:hep-th/0209102].
4. G. 't Hooft, Nucl. Phys. B **72**, 461 (1974).
5. I. R. Klebanov, J. M. Maldacena and C. B. Thorn, JHEP **0604**, 024 (2006) [arXiv:hep-th/0602255].
6. J. Maldacena, Phys. Rev. Lett. **80**, 4859 (1998), hep-th/9803002.
   S.-J. Rey and J. Yee, hep-th/9803001;
7. C. G. Callan and A. Guijosa, Nucl. Phys. B **565** (2000) 157 [arXiv:hep-th/9906153].
8. R. C. Brower, C. I. Tan and C. B. Thorn, Phys. Rev. D **73**, 124037 (2006) [arXiv:hep-th/0603256].
9. J. K. Erickson, G. W. Semenoff, R. J. Szabo and K. Zarembo, Phys. Rev. D **61** (2000) 105006 [arXiv:hep-th/9911088].

# Properties of confining gauge field configurations in the pseudoparticle approach

Marc Wagner

*Institute for Theoretical Physics III, University of Erlangen-Nürnberg, Staudtstraße 7, 91058 Erlangen, Germany*

**Abstract.** The pseudoparticle approach is a numerical method to approximate path integrals in SU(2) Yang-Mills theory. Path integrals are computed by summing over all gauge field configurations, which can be represented by a linear superposition of a small number of pseudoparticles with amplitudes and color orientations as degrees of freedom. By comparing different pseudoparticle ensembles we determine properties of confining gauge field configurations. Our results indicate the importance of long range interactions between pseudoparticles and of non trivial topological properties.

**Keywords:** SU(2) Yang-Mills theory, pseudoparticles, confinement.
**PACS:** 11.15.-q.

## 1. THE BASIC PRINCIPLE OF THE PSEUDOPARTICLE APPROACH

The pseudoparticle approach [1, 2, 3] is a numerical technique to approximate Euclidean path integrals. In this work we consider pure SU(2) Yang-Mills theory, where the expectation value of a quantity $\mathscr{O}$ is given by

$$\left\langle \mathscr{O} \right\rangle = \frac{1}{Z} \int DA\, \mathscr{O}[A] e^{-S[A]} \quad , \quad S[A] = \frac{1}{4g^2} \int d^4x F^a_{\mu\nu} F^a_{\mu\nu} \tag{1}$$

with $F^a_{\mu\nu} = \partial_\mu A^a_\nu - \partial_\nu A^a_\mu + \varepsilon^{abc} A^b_\mu A^c_\nu$. Furthermore, the pseudoparticle approach is a tool to analyze the importance of certain classes of gauge field configurations with respect to confinement (c.f. section 2).

The basic idea of the pseudoparticle approach is to consider only those gauge field configurations, which can be written as a sum of a fixed number ($\approx 400$) of pseudoparticles:

$$A^a_\mu(x) = \sum_i \mathscr{A}(i)\mathscr{C}^{ab}(i)a^b_{\mu,\text{inst.}}(x-z(i)) + \sum_j \mathscr{A}(j)\mathscr{C}^{ab}(j)a^b_{\mu,\text{antiinst.}}(x-z(j)) +$$
$$\sum_k \mathscr{A}(k)\mathscr{C}^{ab}(k)a^b_{\mu,\text{akyron}}(x-z(k)) \tag{2}$$

with amplitudes $\mathscr{A}(i) \in \mathbb{R}$, color orientation matrices $\mathscr{C}^{ab}(i) \in \text{SO}(3)$ and positions $z(i) \in \mathbb{R}^4$ as degrees of freedom. Our standard choice of pseudoparticles are "regular gauge instantons", "regular gauge antiinstantons" and akyrons:

$$a^b_{\mu,\text{inst.}}(x) = \eta^b_{\mu\nu} x_\nu \frac{1}{x^2 + \lambda^2} \quad , \quad a^b_{\mu,\text{antiinst.}}(x) = \bar{\eta}^b_{\mu\nu} x_\nu \frac{1}{x^2 + \lambda^2} \quad ,$$

CP892, *Quark Confinement and the Hadron Spectrum VII*
edited by J. E. F. T. Ribeiro

$$a_{\mu,\text{akyron}}^b(x) = \delta^{b1} x_\mu \frac{1}{x^2 + \lambda^2} \tag{3}$$

with $\eta_{\mu\nu}^b = \varepsilon_{b\mu\nu} + \delta_{b\mu}\delta_{0\nu} - \delta_{b\nu}\delta_{0\mu}$ and $\bar{\eta}_{\mu\nu}^b = \varepsilon_{b\mu\nu} - \delta_{b\mu}\delta_{0\nu} + \delta_{b\nu}\delta_{0\mu}$. Note, however, that instead of these pseudoparticles any set of localized gauge field configurations can be used (an example, Gaussian localized pseudoparticles, are discussed in section 2). The path integral (1) is approximated by an integration over amplitudes and color orientation matrices:

$$\langle \mathscr{O} \rangle = \frac{1}{Z} \int \left( \prod_i d\mathscr{A}(i)\, d\mathscr{C}(i) \right) \mathscr{O}(\mathscr{A}(i), \mathscr{C}(i)) e^{-S(\mathscr{A}(i),\mathscr{C}(i))}. \tag{4}$$

To be more concrete, we put 400 pseudoparticles (pseudoparticle size $\lambda = 0.5$) with randomly chosen positions $z(i)$ inside a hyperspherical spacetime region with radius 3.0 (this amounts to a pseudoparticle density of 1.0) and compute the multidimensional integrals (4) via Monte-Carlo sampling. Boundary effects are excluded by "measuring" observables sufficiently far away from the boundary.

In [2, 3] it has been shown in detail that the pseudoparticle approach applied with around 400 instantons, antiinstantons and akyrons is able to reproduce many essential features of SU(2) Yang-Mills theory. For example the static quark antiquark potential is linear for large separations (c.f. Figure 1a), i.e. there is confinement. Like in lattice gauge theory the string tension $\sigma$ is an increasing function of the coupling constant $g$. Therefore, when the scale is set by identifying the numerical value of $\sigma$ with its physical value, e.g. $\sigma = 4.2/\text{fm}^2$, the physical size of the hyperspherical spacetime region can be adjusted by choosing appropriate values for $g$. Furthermore, the string tension $\sigma$, the topological susceptibility $\chi$ and the critical temperature of the confinement deconfinement phase transition $T_{\text{critical}}$ scale consistently with $g$, i.e. the dimensionless ratios $\chi^{1/4}/\sigma^{1/2}$ and $T_{\text{critical}}/\sigma^{1/2}$ are constant with respect to $g$. They are also of the right order of magnitude when compared to lattice results.

Note that there are significant differences between our method and well-established instanton gas and instanton liquid models [4]: (i) in general, our gauge field configurations (2) are not even close to classical solutions, i.e. we do not consider a semiclassical limit, but try to approximate full quantum physics; (ii) the latter is related to the fact that our "regular gauge pseudoparticles" (3) interact over large distances; this is so, be-

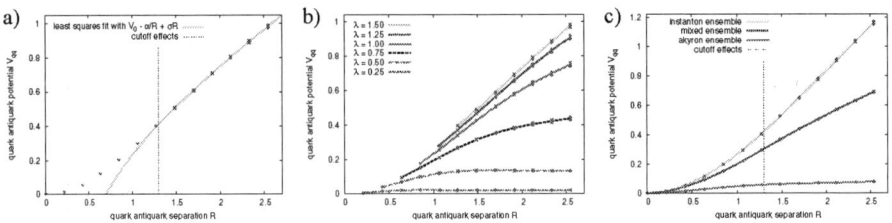

**FIGURE 1.** The static quark antiquark potential plotted against the separation. **a)** Standard choice of pseudoparticles (the data points have been fitted with $V_{q\bar{q}}(R) = V_0 - \alpha/R + \sigma R$). **b)** Gaussian localized pseudoparticles of different size. **c)** Akyron ensemble, instanton ensemble and mixed ensemble.

cause their gauge fields decrease like $1/|x|$ for large $|x|$ in contrast to the $1/|x|^3$-behavior of singular gauge instantons and antiinstantons; as we will demonstrate in section 2, these long range interactions are intimately connected to confinement; (iii) because we include amplitudes as degrees of freedom, it is also possible to model small quantum fluctuations; furthermore, in the limit of infinitely many pseudoparticles any gauge field configuration can be represented according to (2), i.e. in this limit the pseudoparticle approach is identical to full SU(2) Yang-Mills theory (c.f. [2]).

## 2. PROPERTIES OF CONFINING GAUGE FIELD CONFIGURATIONS

In the following we apply the pseudoparticle approach with different types of pseudoparticles to study the effect of different classes of gauge field configurations on confinement.

*Pseudoparticles of different size.* We have compared ensembles with different pseudoparticle size $\lambda$. Note that $\lambda$ strongly affects the shape of a pseudoparticle near its center, but has essentially no effect on the $1/|x|$ long range behavior (c.f. (3)). For $\lambda = 0.2, \dots, 1.1$ the static quark antiquark potential is essentially unaffected by the pseudoparticle size $\lambda$. This indicates that confinement is a consequence of the $1/|x|$ long range behavior of the pseudoparticles, which is the same for all values of $\lambda$. Typical gauge field configurations for $\lambda = 0.2$ and for $\lambda = 1.1$ are shown in Figure 2a. For both values of $\lambda$ the global structure is the same, but for $\lambda = 0.2$ there are additional local ultraviolet fluctuations. Apparently, these ultraviolet fluctuations have no effect on confinement and the string tension.

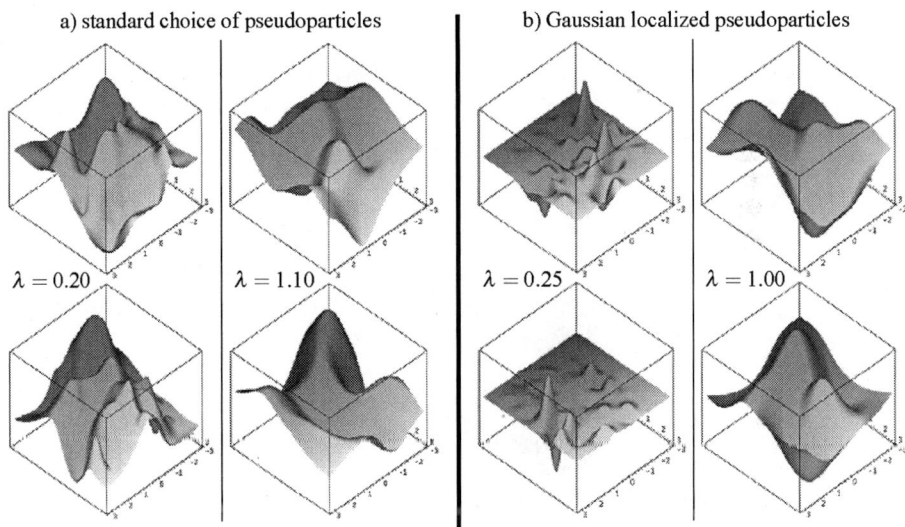

a) standard choice of pseudoparticles      b) Gaussian localized pseudoparticles

$\lambda = 0.20$    $\lambda = 1.10$    $\lambda = 0.25$    $\lambda = 1.00$

**FIGURE 2.** Typical gauge field configurations: one of the gauge field components $A_\mu^a$ plotted against two of the spacetime coordinates $x_\mu$ and $x_\nu$.

*Pseudoparticles with a limited range of interaction.* We have compared ensembles of Gaussian localized pseudoparticles with different size $\lambda$. These Gaussian localized pseudoparticles arise by replacing $1/(x^2 + \lambda^2)$ in (3) by $(1/\lambda^2)\exp(-x^2/2\lambda^2)$. They have a limited range of interaction, which is proportional to $\lambda$. For short range pseudoparticles ($\lambda \leq 0.50$) there is little overlap between neighboring pseudoparticles and no confinement. For long range pseudoparticles ($\lambda \geq 1.00$) there is significant overlap and the static quark antiquark potential is confining (c.f. Figure 1b). To put it another way, there is a close relation between "pseudoparticle percolation" and confinement. Typical gauge field configurations for $\lambda = 0.25$ and for $\lambda = 1.00$ are shown in Figure 2b. For $\lambda = 0.25$ (no confinement) there are only local ultraviolet fluctuations, while for $\lambda = 1.00$ (confinement) there are global excitations. It seems that gauge field configurations responsible for confinement contain extended structures and large area excitations.

*Pseudoparticles without topological charge.* We have compared a pure akyron ensemble (400 akyrons), a pure instanton ensemble (200 instantons, 200 antiinstantons) and a "mixed ensemble" (150 instantons, 150 antiinstantons, 100 akyrons). There is no confinement in the akyron ensemble (c.f. Figure 1c). That demonstrates that akyrons alone are not suited to reproduce correct Yang-Mills physics. Since superpositions of akyrons always have vanishing topological charge density (c.f. [3]), it also supports the common expectation that confinement and topological charge are closely related. Both in the instanton and in the mixed ensemble there is confinement. However, comparing physically meaningful quantities in the instanton ensemble and in the mixed ensemble (e.g. $(\chi^{1/4}/\sigma^{1/2})_{instanton} = 0.26$, $(\chi^{1/4}/\sigma^{1/2})_{mixed} = 0.35$) with results from lattice calculations ($(\chi^{1/4}/\sigma^{1/2})_{lattice} = 0.49$ [5]) indicates that using akyrons is beneficial with respect to quantitative results.

*Summary.* We have presented evidence that confining gauge field configurations contain extended structures and large area excitations. Topological charge also seems to play an important role. In contrast to that, confinement is not affected by ultraviolet fluctuations.

## ACKNOWLEDGMENTS

I would like to thank M. Faber, J. Greensite and M. Polikarpov for inviting me to the "Quark Confinement and the Hadron Spectrum VII"-conference to give this talk, as well as J. E. Ribeiro for his support. Furthermore, it is a pleasure to thank F. Lenz for many interesting and fruitful discussions.

## REFERENCES

1. M. Wagner and F. Lenz, PoS **LAT2005** (2005) 315 [arXiv:hep-lat/0510083].
2. M. Wagner, PhD thesis, Institute for Theoretical Physics III, University of Erlangen-Nürnberg (2006).
3. M. Wagner, arXiv:hep-ph/0608090.
4. T. Schäfer and E. V. Shuryak, Rev. Mod. Phys. **70** (1998) 323 [arXiv:hep-ph/9610451].
5. M. J. Teper, arXiv:hep-th/9812187.

# Monopole Condensate Stability in three- and four-colour QCD

## M.L. Walker

*Department of Physics, Chiba University, Chiba, 263-8522, Japan*

**Abstract.** I demonstrate, to one-loop order, that the ground state of three- and four-colour QCD contains a monopole condensate, necessary for the dual Meissner effect to be the mechanism of confinement, and establish its stability to arbitrary loop order by showing that it gives the off-diagonal gluons an effective mass sufficient to remove the unstable ground state mode.

**Keywords:** QCD, confinement, magnetic condensate, magnetic monopoles
**PACS:** 12.38.-t, 12.38.Aw, 14.80.Hv

The proof of colour confinement is one of the most important, long-running problems in quantum field theory today. Particularly promising is the dual Meissner effect, in which chromomagnetic monopoles exclude the chromoelectric field in the same way that Cooper pairs in a superconductor exclude the magnetic field. The parallels between the QCD vacuum and superconductors in this model are laid out in Table 1). Although a non-

**TABLE 1.** QCD vacuum vs type II superconductor.

| QCD vacuum | | Superconductor |
|---|---|---|
| (Pairs of) Monopoles | $\leftrightarrow$ | Cooper Pairs |
| Colour Confinement | $\leftrightarrow$ | Exclusion of Magnetism |

zero magnetic background is a lower energy state than the perturbative vacuum, there remained the twin difficulties of specifying the magnetic condensate in a gauge-invariant manner that ensured it was due to monopoles, and removing the apparent instability found by Nielsen and Olesen [1] in the zero-point gluon fluctuations in two-colour QCD in the presence of a magnetic condensate.

Although t'Hooft's Maximal Abelian Gauge (MAG) is more widely known, the former of these difficulties is more elegantly solved using the Cho-Faddeev-Niemi (CFN) decomposition [2]. Begin by specifying the internal directions corresponding to the $SU(3)$ Abelian generators $\lambda^{(3)}, \lambda^{(8)}$ with the unit isovectors $\hat{n}, \hat{m}$ respectively, and then define a field $\vec{C}_\mu$ such that

$$g\vec{C}_\mu \times \hat{n} = -\partial_\mu \hat{n}; \ g\vec{C}_\mu \times \hat{m} = -\partial_\mu \hat{m}$$
$$\Rightarrow \vec{C}_\mu = A_\mu \hat{n} + B_\mu \hat{m} + \vec{B}_\mu, \tag{1}$$

where

$$\vec{B}_\mu = g^{-1}[\partial_\mu \hat{n} \times \hat{n} + \partial_\mu \hat{m} \times \hat{m}]. \tag{2}$$

$\hat{n}, \hat{m}$ exhibit the full homotopy class of mapping

$$\pi_2(SU(3)/U(1) \otimes U(1)) \rightarrow \pi_1(U(1) \otimes U(1)). \tag{3}$$

CP892, *Quark Confinement and the Hadron Spectrum VII*
edited by J. E. F. T. Ribeiro
© 2007 American Institute of Physics 978-0-7354-0396-3/07/$23.00

which gives $\vec{B}_\mu$ the monopole field configuration.

We define

$$\hat{D}_\mu = \partial_\mu + g\vec{C}_\mu \times,$$
$$\hat{D}_\mu \hat{n} = 0 = \hat{D}_\mu \hat{m}, \tag{4}$$

The gluon field then decomposes into

$$\begin{aligned} \vec{A}_\mu &= \vec{C}_\mu + \vec{X}_\mu \\ &= A_\mu \hat{n} + B_\mu \hat{m} + \vec{B}_\mu + \vec{X}_\mu, \end{aligned} \tag{5}$$

where

$$\begin{aligned} \vec{C}_\mu &= A_\mu \hat{n} + B_\mu \hat{m} + \vec{B}_\mu \\ \vec{X}_\mu &\perp \hat{n}, \hat{m} \\ \vec{X}_\mu &= g^{-1}[\hat{n} \times \vec{D}_\mu \hat{n} + \hat{m} \times \vec{D}_\mu \hat{m}], \\ \vec{D}_\mu &= \partial_\mu + g\vec{A}_\mu \times. \end{aligned} \tag{6}$$

The extension to higher $N$ is straightforward

The monopole field strength

$$\begin{aligned} \vec{H}_{\mu\nu} &= \partial_\mu \vec{B}_\nu - \partial_\nu \vec{B}_\mu + g\vec{B}_\mu \times \vec{B}_\nu \\ &= -g\vec{B}_\mu \times \vec{B}_\nu, \end{aligned} \tag{7}$$

has only $\hat{n}, \hat{m}$ components,

$$(\hat{n}\hat{n} \cdot + \hat{m}\hat{m} \cdot)\vec{H}_{\mu\nu} = \vec{H}_{\mu\nu}, \tag{8}$$

Stability was an issue because a non-zero magnetic background

$$\|\vec{H}^{(\alpha)}\| \neq 0, \tag{9}$$

regardless of whether or not it is due to monopoles, will add an imaginary part to the ground-state energy of $\vec{X}^{(\alpha)}$ via

$$\sqrt{\vec{k}^2 + \left(n - \frac{1}{2}\right) 2g\|\vec{H}^{(\alpha)}\|}, \tag{10}$$

when $n = 0$. However, Kondo [3] argued in the two-color theory that if the gluons gain an effective mass this becomes

$$\sqrt{\vec{k}^2 + M_X^2 + \left(n - \frac{1}{2}\right) 2g\|\vec{H}^{(\alpha)}\|}. \tag{11}$$

Repeating his approach in three-color QCD, the monopole condensate gives the off-diagonal gluons an effective mass via the term

$$\frac{1}{2}(\hat{\mathbf{D}}_\mu \vec{X}_\nu - \hat{\mathbf{D}}_\nu \vec{X}_\mu)^2 \xrightarrow{IBP} (\vec{X}_\mu \hat{\mathbf{D}}_\nu) \cdot (\hat{\mathbf{D}}_\mu \vec{X}_\nu) - (\vec{X}_\mu \hat{\mathbf{D}}_\nu) \cdot (\hat{\mathbf{D}}_\nu \vec{X}_\mu). \tag{12}$$

The latter term gives

$$g^2 B_\rho^D X_\mu^E B_\rho^B X_\mu^C f_{ABC} f_{ADE} \tag{13}$$

which provides the effective gluon mass matrix

$$M_{EC}^2 = g^2 B_\rho^D B_\rho^B f_{ABC} f_{ADE} \tag{14}$$

This is too hard to diagonalise conventionally (even using mathematica!), but we need simply observe that the sum of the mass eigenvalues is the trace of the mass matrix, $3g^2 \vec{B} \cdot \vec{B}$. Since there are six physical internal directions the average effective mass squared is

$$M_X^2 = \frac{1}{2} g^2 \vec{B} \cdot \vec{B}. \tag{15}$$

Since all physical masses are equal by the isotropy of the condensate and the gauge invariance of the mass term (13), (15) is the valence gluons' effective mass.

Starting from

$$\vec{H}_{\mu\nu}^{(\alpha)} = \partial_\mu \mathbf{B}_\nu^{(\alpha)} - \partial_\nu \mathbf{B}_\mu^{(\alpha)} + g\mathbf{B}_\mu^{(\alpha)} \times \mathbf{B}_\nu^{(\alpha)}, \tag{16}$$

a lot of dry, technical mathematics (see [4] for details) can show that

$$g\|\vec{H}^{(\alpha)}\| < \frac{1}{2} \vec{B} \cdot \vec{B} = M_X^2, \tag{17}$$

which is sufficient to ensure that the argument under the square root sign in the zero-point calculation (eq. 11) is always positive. Hence the imaginary part is removed and the monopole background is stable.

The corresponding result is similarly shown for $SU(4)$ QCD [4] but QCD with more than four colors require a different analysis [5].

In summary, we have applied the CFN decomposition to three- and four-color QCD, and found that there is indeed a monopole condensate. Although it naively appears to generate instability, an associated gluon mass generation takes place and can be shown (although the details were skipped here) to be sufficient to restore stability.

## REFERENCES

1. N. K. Nielsen and P. Olesen. An unstable yang-mills field mode. *Nucl. Phys.*, B144:376, 1978.
2. Y. M. Cho. A restricted gauge theory. *Phys. Rev.*, D21:1080, 1980.
3. K.-I. Kondo. Magnetic condensation, abelian dominance and instability of savvidy vacuum. *Phys. Lett.*, B600:287–296, 2004.
4. M.L. Walker. Stability of the magnetic monopole condensate in three- and four-colour qcd. 2006. hep-th/0605103.
5. H. Flyvbjerg. Improved qcd vacuum for gauge groups su(3) and su(4). *Nucl. Phys.*, B176:379, 1980.

# Dyson-Schwinger Equations and Coulomb Gauge Yang-Mills Theory

P. Watson and H. Reinhardt

*Institut für Theoretische Physik, Auf der Morgenstelle 14, D-72076 Tübingen, Germany*

**Abstract.** Coulomb gauge Yang-Mills theory is considered within the first order formalism. It is shown that the action is invariant under both the standard BRS transform and an additional component. The Ward-Takahashi identity arising from this non-standard transform is shown to be automatically satisfied by the equations of motion.

Dyson-Schwinger equations [DSEs] are a natural tool for studying the continuum properties of field theories. Applied to Quantum Chromodynamics [QCD] in the Landau gauge, significant progress has been made in understanding the nature of confinement, the infrared behaviour of propagators and the light hadron spectrum with good agreement, where pertinent, with lattice studies (see for example the recent review [1] and references therein). As with any such calculation, the question naturally arises whether the relevant observables are properly gauge invariant and this leads to the desire for a DSE study of Coulomb gauge QCD. In Coulomb gauge there are many different existing studies, though perhaps the most closely related are those based on the Hamiltonian approach [2], however, a true DSE study has not yet been attempted. Here we outline some of the basic formalism for such a study of Yang-Mills theory within the first order formalism and highlight an interesting feature – that there exists more than one BRS-type symmetry but that the 'extra' invariance is equivalent to the equations of motion. A more complete description of this work can be found in [3].

Let us consider the following functional integral (working in Minkowski space):

$$Z = \int \mathscr{D}\Phi \exp\{\imath\mathscr{S}\} = \int \mathscr{D}\Phi \exp\left\{\imath \int d^4x \left[\frac{1}{2}E^2 - \frac{1}{2}B^2\right]\right\}$$

where $\vec{E}^a = -\partial_t\vec{A}^a - \vec{D}^{ab}A_0^b$ and $\vec{B}^a$ are the chromo-electric and -magnetic components of the field strength tensor $F_{\mu\nu}$ ($A$ is the gauge field, $\vec{D}$ is the covariant derivative and $\Phi$ denotes the collection of all fields). If we fix to Coulomb gauge ($\vec{\nabla}\cdot\vec{A} = 0$) with a Lagrange-multiplier field $\lambda^a$ using the Faddeev-Popov technique then the action gets the additional term

$$\mathscr{S}_{fp} = \int d^4x \left[-\lambda^a\vec{\nabla}\cdot\vec{A}^a - \bar{c}^a\vec{\nabla}\cdot\vec{D}^{ab}c^b\right].$$

where the fields $\bar{c}$ and $c$ are the ghost fields. The ghost term can also be written as a functional determinant $\text{Det}(-\vec{\nabla}\cdot\vec{D})$ in $Z$. The action as it is so far written, is invariant under the standard BRS transform [4]. To convert to the first order formalism [5], we

CP892, *Quark Confinement and the Hadron Spectrum VII*
edited by J. E. F. T. Ribeiro
© 2007 American Institute of Physics 978-0-7354-0396-3/07/$23.00

introduce $\vec{\pi}$-fields (classically, momentum conjugate to the $\vec{A}$-fields) via the following identity

$$\exp\left\{\imath \int \frac{1}{2}E^2\right\} = \int \mathscr{D}\pi \exp\left\{\imath \int \left[-\frac{1}{2}\pi^2 - \vec{\pi}^a \cdot \vec{E}^a\right]\right\}$$

and to maintain BRS invariance (given that under an infinitessimal local transform parameterised by $\theta_x^a = c_x^a \delta\lambda$, the variation of $\vec{E}^a$ is $\delta\vec{E}^a = f^{abc}\theta_x^b\vec{E}^c$) we require that

$$\delta\pi^a = f^{abc}\theta_x^b\left[(1-\alpha)\vec{\pi}^c - \alpha\vec{E}^c\right],$$

where $\alpha$ is some constant. Since $\alpha$ is unconstrained we have *two* invariances. To complete, we split the $\vec{\pi}$-field into transverse and longitudinal parts with

$$\text{const} = \int \mathscr{D}\phi\mathscr{D}\tau \exp\left\{-\imath \int \tau^a \left(\vec{\nabla}\cdot\vec{\pi}^a + \nabla^2\phi^a\right)\right\}$$

and translate $\vec{\pi} \to \vec{\pi} - \vec{\nabla}\phi$ to finally give

$$Z = \int \mathscr{D}\Phi\exp\{\imath\mathscr{S}_B + \imath\mathscr{S}_{fp} + \imath\mathscr{S}_\pi\},$$

$$\mathscr{S}_\pi = \int \left[-\tau^a\vec{\nabla}\cdot\vec{\pi}^a - \frac{1}{2}\left(\vec{\pi} - \vec{\nabla}\phi\right)^2 + \frac{1}{2}\left(\vec{\pi}^a - \vec{\nabla}\phi^a\right)\cdot\left(\partial_t\vec{A}^a + \vec{D}^{ab}A_0^b\right)\right].$$

The (non-standard) $\alpha$-dependent part of the BRS transform reads ($\vec{X}^c = \vec{\pi}^c - \vec{\nabla}\phi^c - \partial_t\vec{A}^c - \vec{D}^{cd}A_0^d$)

$$\delta\pi^a = f^{abc}\theta_x^b\vec{X}^c + \vec{\nabla}\delta\phi^c, \quad \delta\phi^a = f^{abc}\frac{\vec{\nabla}}{(-\nabla^2)}\cdot\vec{X}^c\theta_x^b,$$

with all other fields unaltered.

At this stage, it is pertinent to motivate the use of the first order formalism and this lies in the ability, at least formally, to reduce to transverse $\vec{A}$ and $\vec{\pi}$ degrees of freedom [5]. Those fields occuring linearly in the action may be integrated out, leaving functional $\delta$-functions, one of which implements Gauß' law:

$$\delta(-\vec{\nabla}\cdot\vec{D}^{ab}\phi^b - gf^{ade}\vec{A}^d\cdot\vec{\pi}^e).$$

Defining the inverse Faddeev-Popov operator $M$ via $-\vec{\nabla}\cdot\vec{D}^{ab}M^{bc} = \delta^{ac}$, the $\delta$-function now reads

$$\text{Det}^{-1}(-\vec{\nabla}\cdot\vec{D})\delta(\phi^a - M^{ac}gf^{cde}\vec{A}^d\cdot\vec{\pi}^e)$$

and the inverse determinant *cancels* the Faddeev-Popov determinant exactly. The energy divergences associated with the static ghost propagators in Coulomb gauge are automatically removed [5]. We are left with

$$Z = \int \mathscr{D}\Phi\delta\left(\vec{\nabla}\cdot\vec{A}\right)\delta\left(\vec{\nabla}\cdot\vec{\pi}\right)\exp\{\imath\mathscr{S}_B + \imath\mathscr{S}_\pi\},$$

$$\mathscr{S}_\pi \sim \int \left[-\frac{1}{2}\pi^2 - \frac{1}{2}\left(\vec{A}\cdot\vec{\pi}\right)M(-\nabla^2)M\left(\vec{A}\cdot\vec{\pi}\right) + \vec{\pi}\cdot\partial_t\vec{A}\right]$$

which expresses the theory in terms of transverse gluon degrees of freedom. Since the action is now non-local these manipulations can be regarded as merely formal but do give insight into the previous, local formulation.

Each continuous transform of the theory may be regarded as a change of variables in the functional integral and that the action is invariant leads to a Ward-Takahashi identity. In the case of the $\alpha$-dependent part of the BRS transform the Jacobian factor is trivial [3] and we have that

$$0 = \int \mathscr{D}\Phi f^{abc} \vec{X}^c \cdot \left[ \vec{K}^a - \frac{\vec{\nabla}}{(-\nabla^2)} \left( \kappa^a - \vec{\nabla} \cdot \vec{K}^a \right) \right] \exp\{\imath\mathscr{S}\} \qquad (1)$$

where $\vec{K}$ and $\kappa$ are sources introduced for the $\vec{\pi}$ and $\phi$ fields respectively. In addition, for each field, from the observation that the integral of a total derivative vanishes (up to boundary terms which are here assumed to vanish) we have equations of motion – the Dyson-Schwinger equations are functional derivatives of such equations. For the $\vec{\pi}$ and $\phi$ fields respectively, these equations are:

$$\vec{K}^a Z\left[\vec{K}, \kappa\right] = -\int \mathscr{D}\Phi \left[\vec{\nabla}\tau^a - \vec{X}^a\right] \exp\{\imath\mathscr{S}\} \qquad (2)$$

$$\kappa^a Z\left[\vec{K}, \kappa\right] = \int \mathscr{D}\Phi \vec{\nabla}\cdot\vec{X}^a \exp\{\imath\mathscr{S}\} \qquad (3)$$

where $Z\left[\vec{K}, \kappa\right]$ is now the generating functional (the functional integral $Z$ from before in the presence of sources). Inserting Eqs. (2) and (3) into the right-hand side of Eq. (1) and using the antisymmetry property of the structure constant $f^{abc}$ it is immediately clear that the equation is automatically satisfied. The two equations of motion and the $\alpha$-dependent part of the BRS transform give rise to exactly the same constraints on the functional integral and thus to any functional derivatives (i.e., Green's functions).

## ACKNOWLEDGMENTS

Work supported by the DFG under contracts no. Re856/6-1 and Re856/6-2.

## REFERENCES

1. C. S. Fischer, "Infrared properties of QCD from Dyson-Schwinger equations," J. Phys. G **32** (2006) R253 [arXiv:hep-ph/0605173].
2. C. Feuchter and H. Reinhardt, "Variational solution of the Yang-Mills Schroedinger equation in Coulomb gauge," Phys. Rev. D **70** (2004) 105021 [arXiv:hep-th/0408236] and references therein.
3. P. Watson and H. Reinhardt, in preparation.
4. W. J. Marciano and H. Pagels, "Quantum Chromodynamics," Phys. Rep. 36C ( 1978) 137-276.
5. D. Zwanziger, "Renormalization in the Coulomb gauge and order parameter for confinement in QCD," Nucl. Phys. B **518** (1998) 237.

# A Review of Instanton Quarks and Confinement

Pierre van Baal

*Instituut-Lorentz for Theoretical Physics, University of Leiden,*
*P.O.Box 9506, NL-2300 RA Leiden, The Netherlands*

**Abstract.** We review the recent progress made in understanding instantons at finite temperature (calorons) with non-trivial holonomy, and their monopole constituents as relevant degrees of freedom for the confined phase.

**Keywords:** Calorons, Confinement, QCD
**PACS:** 11.10.Wx, 12.38.Aw, 14.80.Hv

## INTRODUCTION

New instantons (also called calorons) have been obtained recently, where the Polyakov loop at spatial infinity (the so-called holonomy) is non-trivial [1, 2]. Trivial holonomy, i.e. with values in the center of the gauge group, is typical for the deconfined phase [3, 4]. Non-trivial holonomy is therefore expected to play a role in the confined phase (i.e. for $T < T_c$) where the trace of the Polyakov loop fluctuates around small values.

The Polyakov loop plays the role of the Higgs field, $P(t,\vec{x}) = \text{Pexp}\left(\int_0^\beta A_0(t+s,\vec{x})ds\right)$, where $\beta = 1/kT$ is the period in the imaginary time direction. For $SU(n)$, finite action requires this to tend to

$$\mathscr{P}_\infty = \lim_{|\vec{x}|\to\infty} P(0,\vec{x}) = g^\dagger \exp(2\pi i\text{diag}(\mu_1,\mu_2,\ldots,\mu_n))g, \tag{1}$$

where $g$ is chosen to bring $\mathscr{P}_\infty$ to its diagonal form, with the $n$ eigenvalues being ordered according to $\sum_{i=1}^n \mu_i = 0$ and $\mu_1 \leq \mu_2 \leq \ldots \leq \mu_n \leq \mu_{n+1} \equiv 1+\mu_1$. One can recognize $8\pi^2 v_m/\beta$ (with $v_m = \mu_{m+1} - \mu_m$) as being the monopole mass.

Monopoles as constituents are close to the picture of instanton quarks, which was already introduced more than 25 years ago [5]. The only difference is that instanton quarks were pointlike, whereas here we have to work in terms of monopole degrees of freedom. We will investigate in how far this plays a role in describing confinement.

Caloron solutions are such that the total magnetic charge vanishes. The "force" stability of these solutions in terms of its constituent monopoles is based, as for exact BPS multi-monopole solutions, on balancing the electromagnetic with the scalar (Higgs) force [6], except that for calorons repulsive and attractive forces are interchanged as compared to multi-monopoles. A single caloron with topological charge one contains $n-1$ monopoles with a unit magnetic charge in the $i$-th $U(1)$ subgroup, which are compensated by the $n$-th monopole of so-called type $(1,1,\ldots,1)$, having a magnetic charge in each of these subgroups. At topological charge $k$ there are $kn$ constituents,

CP892, *Quark Confinement and the Hadron Spectrum VII*
edited by J. E. F. T. Ribeiro
© 2007 American Institute of Physics 978-0-7354-0396-3/07/$23.00

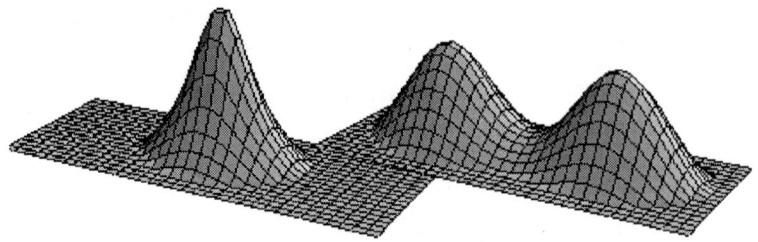

**FIGURE 1.** Profiles for a caloron at $\omega = 1/4$ with $\rho \ll T$ (left) and $\rho = T$ (right), where vertically the logarithm of the action density is plotted, cutoff below $1/(2e)$.

$k$ monopoles of each of the $n$ types. The sum rule $\sum_{j=1}^{n} v_j = 1$ guarantees the correct action, $8\pi^2 k$, for calorons with topological charge $k$.

## ONE-LOOP CORRECTIONS

Prior to their explicit construction, calorons with non-trivial holonomy were considered irrelevant [4], because the one-loop correction gives rise to an infinite action barrier. However, the infinity simply arises due to the integration over the finite energy density induced by the perturbative fluctuations in the background of a non-trivial Polyakov loop [7]. The non-perturbative contribution of calorons (with a given asymptotic value of the Polyakov loop) to this energy density as the relevant quantity to be considered, was first calculated in supersymmetric theories [8], where the perturbative contribution vanishes. It has a minimum where the trace of the Polyakov loop vanishes, i.e. at maximal non-trivial holonomy. Recently the calculation of the non-perturbative contribution was performed in ordinary gauge theory at high temperatures [9]. When added to the perturbative contribution with its minima at center elements, these minima turn unstable for decreasing temperature right around the expected value of $T_c$. This lends some support to monopole constituents being the relevant degrees of freedom which drive the transition from a phase in which the center symmetry is broken at high temperatures to one in which the center symmetry is restored at low temperatures.

## A CALORON GAS MODEL FOR CONFINEMENT

A caloron gas model has been constructed recently for SU(2) [10], where one solves for overlapping instantons approximately. One takes

$$A_\mu^{\text{per}}(x) = e^{-2\pi i \vec{\omega} \cdot \vec{\tau}} \sum_i A_\mu^{(i),\text{alg}}(x) e^{2\pi i \vec{\omega} \cdot \vec{\tau}} + 2\pi \vec{\omega} \cdot \vec{\tau} \delta_{\mu 4} \tag{2}$$

to be valid when the density is of the order of 1 fm$^{-4}$ and size $\rho$ is roughly 0.33 fm. In other words, one adds the caloron gauge fields (with the *same* $\mathscr{P}_\infty = e^{2\pi i \vec{\omega} \cdot \vec{\tau}}$) in the algebraic gauge $A_\mu^{\text{alg}}(x + \beta) = \mathscr{P}_\infty A_\mu^{\text{alg}}(x) \mathscr{P}_\infty^{-1}$ in order not to change the boundary

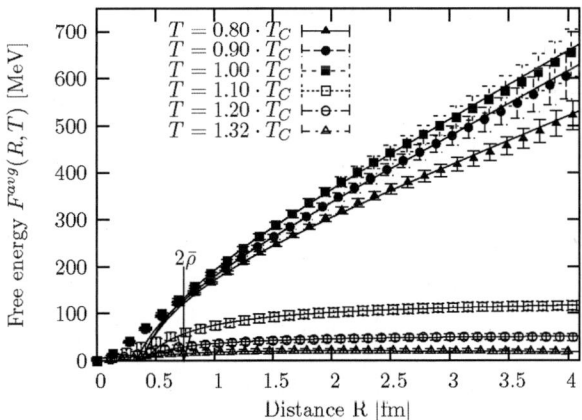

**FIGURE 2.** Free energy versus distance $R$ at different temperatures $T/T_C = 0.8, 0.9, 1.0$ for the confined and at $T/T_C = 1.10, 1.20, 1.32$ for the deconfined phase in the fundamental representation.

conditions. Only at the end one transforms to the periodic gauge. This has been shown to be exact for multi-calorons [11], but for the above parameters it is a good approximation for a superposition of (anti)calorons.

Remarkably this seems to give confinement for $T < T_c$ and deconfinement for $T > T_c$. In the confining phase one imposes $\omega = |\vec{\omega}| = 1/4$ and $\text{Tr}\mathscr{P}_\infty = 0$, whereas in the deconfining phase one tends to find $\omega = 0$ (or $1/2$) and $\text{Tr}\mathscr{P}_\infty = 2$ (one takes into account that $\omega$ only gradually becomes 0 or 1/2 with increasing temperature, but we will ignore this here). In figure 1 the caloron is shown for $\omega = 1/4$, where we contrast $\rho \ll T$ and $\rho = T$. Of course $\rho$ is somewhere in between, but it clearly gives a confining force over the distances probed.

To show this they have solved for

$$D_1(\rho,T) = A_1\rho^{b-5}\exp(-c\rho^2) \quad \text{and} \quad D_2(\rho,T) = A_2\rho^{b-5}\exp(-4[\pi\rho T]^2/3), \quad (3)$$

where in the first case $\bar{\rho}$ is fixed, $T \leq T_c$ and $\omega = 1/4$ (which means $\nu = 1/2$), and in the second case $\bar{\rho}$ is running, $T \geq T_c$ and $\omega = \nu = 0$. Finally one requires $\bar{\rho}(T_c)_{\text{conf}} = \bar{\rho}(T_c)_{\text{deconf}} = 0.37$ fm, which determines $c$. With $b = (11n - 2n_f)/3 = 22/3$ ($n_f = 0$) and $\int D_{1,2}(\rho,T)d\rho = 1$ this gives the model. Determining $\bar{\rho}(T < T_c)$, they have also fixed $T_c \approx 178$ MeV and $\sigma(0) \approx 318$ MeV/fm.

In figure 2 the free energy versus the distances at different temperatures is given and although the string tension should go to zero as one approaches $T_c$ from below, it is true that for $T < T_c$ the string tension is finite and becomes zero for $T > T_c$. This model, in a sense, assumes weak coupling. Also in the spatial Wilson loops one finds an area law.

## DENSE MATTER

There has been yet another development that introduces instanton quarks to describe confinement [12], which has been summarized in [13]. At low energies and large chemical potential the $\eta'$ interactions are determined by ordinary instantons, with a periodicity of $\theta$ which is $2\pi$. But at small chemical potential (and temperature) one finds for $\eta' = \phi = \text{Tr}(U)$, where $U$ is the chiral matrix, that (ignoring the mass corrections)

$$L_{\eta'} = f^2(\partial_\mu \phi)^2 + \lambda \cos([\phi - \theta]/n). \qquad (4)$$

Now the topological charge is $Q_a = \pm 1/n$, but with the sum $Q = \Sigma_a Q_a$ an integer. The conjecture is that in the confined phase instanton quarks can be far apart, but remain strongly correlated, requiring large and overlapping instantons. One has to see if it is strongly interacting and if the constituents are line like (the constituent monopoles), instead of point like (at least semi-classically). The conclusions are nevertheless interesting.

In conclusion instanton quarks seem to play a role in the confined phase. The interpretation is of course different than what was assumed in [5], where now the time coordinate is replaced in a sense by a phase. What remains true is, however, that charge $k$ SU($n$) solutions are described by $kn$ lumps of charge $1/n$.

## ACKNOWLEDGMENTS

I thank Manfried Faber, Misha Polikarpov and above all Jeff Greensite for inviting me to this wonderful place. I thank Michael Ilgenfritz and Michael Müller-Preussker for explaining their work with Phillip Gerhold, and Falk Bruckmann for taking over when I was away, and from which I am determined to come back, if only for the warmly felt wishes of many participants.

## REFERENCES

1. T.C. Kraan and P. van Baal, Phys. Lett. B428 (1998) 268 [hep-th/9802049]; Nucl. Phys. B533 (1998) 627 [hep-th/9805168]; Phys. Lett. B435 (1998) 389 [hep-th/9806034].
2. K. Lee, Phys. Lett B426 (1998) 323 [hep-th/9802012]; K. Lee and C. Lu, Phys. Rev. D58 (1998) 025011 [hep-th/9802108].
3. B.J. Harrington and H.K. Shepard, Phys. Rev. D17 (1978) 2122; Phys. Rev. D18 (1978) 2990.
4. D.J. Gross, R.D. Pisarski and L.G. Yaffe, Rev. Mod. Phys. 53 (1981) 43.
5. A.A. Belavin, V.A. Fateev, A.S. Schwarz and Yu.S. Tyupkin, Phys. Lett. B83 (1979) 317.
6. N.S. Manton, Nucl. Phys. B126 (1977) 525; C. Montonen and D. Olive, Phys. Lett. 72B (1977) 117.
7. N. Weiss, Phys. Rev. D24 (1981) 475.
8. N.M. Davies, T.J. Hollowood, V.V. Khoze and M.P. Mattis, Nucl. Phys. B559 (1999) 123 [hep-th/9905015].
9. D. Diakonov, N. Gromov, V. Petrov and S. Slizovskiy, Phys. Rev. D70 (2004) 036003 [hep-th/0404042]; D. Diakonov and N. Gromov, Phys. Rev. D72 (2005) 025003 [hep-th/0502132].
10. P. Gerhold, E.-M. Ilgenfrizt and M. Müller-Preussker, An $SU(2)$ KvBLL caloron gas model and confinement, hep-ph/0607315.
11. F. Bruckmann and P. van Baal, Nucl. Phys. B645 (2002) 105 [hep-th/0209010]; F. Bruckmann, D. Nógrádi and P. van Baal, Nucl. Phys. B666 (2003) 197 [hep-th/0305063]; Nucl. Phys. B698 (2004) 233 [hep-th/0404210].
12. D.T. Son, M.A. Stephanov and A.R. Zhitnitsky, Phys. Lett. B510 (2001) 167 [hep-ph/0103099]; D. Toublan and A.R. Zhitnitsky, Phys. Rev. D73 (2006) 034009 [hep-ph/0503256].
13. A.R. Zhitnitsky, Confinement-Deconfinement Phase Transition and Fractional Instanton Quarks in Dense Matter, hep-ph/0601057.

# Gauge-invariant masses through
# Schwinger-Dyson equations

## A. Bashir[*,†] and A. Raya[*]

*Instituto de Física y Matemáticas, Universidad Michoacana de San Nicolás de Hidalgo, Apartado Postal 2-82, Morelia, Michoacán 58040, México
†Institute for Particle Physics Phenomenology, University of Durham, Durham DH1 3LE, U.K.

**Abstract.**
   Schwinger-Dyson equations (SDEs) are an ideal framework to study non-perturbative phenomena such as dynamical chiral symmetry breaking (DCSB). A reliable truncation of these equations leading to gauge invariant results is a challenging problem. Constraints imposed by Landau-Khalatnikov-Fradkin transformations (LKFT) can play an important role in the hunt for physically acceptable truncations. We present these constrains in the context of dynamical mass generation in QED in 2 + 1-dimensions.

**Keywords:** Schwinger-Dyson equations, Landau-Khalatnikov-Fradkin transformations, Non-perturbative phenomena, Dynamical chiral symmetry breaking, QED3.
**PACS:** 11.15.Tk,12.20.-m

Solving Quantum Chromodynamics (QCD) in the non-perturbative domain of its coupling has been elusive for decades. Careless truncations can incur gauge dependence. It is hard to pin point the source of such problem in QCD. Its non-abelian nature makes the corresponding Schwinger-Dyson equations (SDEs) prohibitively complicated and several simplifying assumptions have to be employed to render them solvable. Gribov ambiguities provide an additional set back. Slavnov Taylor identities and generalized Landau-Khalatnikov-Fradkin transformations (LKFT) are more complicated in form than their counterpart in QED. Therefore, addressing the issue of gauge invariance in QCD is highly non-trivial. Due to its mathematically simpler and more revealing structure, we address this issue in parity conserving 4-component formalism of 2 + 1-dimensional QED (QED3) in the light of LKFT, [1].

## GAUGE COVARIANCE OF THE FERMION PROPAGATOR

We write out the Euclidean space fermion propagator in momentum and coordinate spaces, respectively, in their most general forms as :

$$S(p;\xi) \equiv \frac{F(p;\xi)}{i\not{p} - \mathcal{M}(p;\xi)} , \qquad S(x;\xi) \equiv \not{x} X(x;\xi) + Y(x;\xi) . \qquad (1)$$

$F$ is often referred to as the fermion wavefunction renormalization and $\mathcal{M}$ as the mass function. Expressions in Eq. (1) are related through a Fourier transformation. The LKFT relating the coordinate space fermion propagator in the Landau gauge to the one in an arbitrary covariant gauge is $S(x;\xi) = S(x;0)e^{-ax}$, where $a = \alpha\xi/2$ with

CP892, *Quark Confinement and the Hadron Spectrum VII*
edited by J. E. F. T. Ribeiro
© 2007 American Institute of Physics 978-0-7354-0396-3/07/$23.00

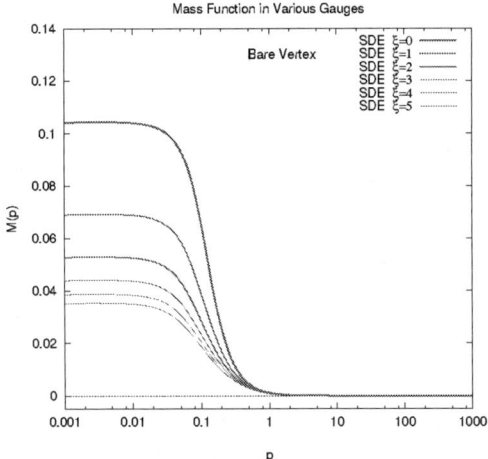

**FIGURE 1.** Solution for the mass function generated through SDEs.

$\alpha = e^2/(4\pi)$, $e^2$ being the dimensionful electromagnetic coupling of QED3 and $\xi$ the covariant gauge parameter. These transformations guarantee the gauge-independence of the chiral condensate which is defined as $\langle \bar{\psi}\psi \rangle = -\mathrm{Tr}S(x=0;\xi)$. Therefore, they can impose vital constraints on the truncations of SDEs. Such constraints have been studied in some detail in [2, 3, 4]. In the momentum space, these transformations translate as [5, 6].

$$\frac{F(p;\xi)}{p^2+\mathcal{M}^2(p;\xi)} = \frac{a}{\pi p^2}\int_0^\infty dk\,k^2\frac{F(k;0)}{k^2+\mathcal{M}^2(k;0)}\left[\frac{1}{\lambda^-}+\frac{1}{\lambda^+}+\frac{1}{2kp}\ln\left|\frac{\lambda^-}{\lambda^+}\right|\right],$$

$$\frac{F(p;\xi)\mathcal{M}(p;\xi)}{p^2+\mathcal{M}^2(p;\xi)} = \frac{a}{\pi p}\int_0^\infty dk\,k\frac{F(k;0)\mathcal{M}(k;0)}{k^2+\mathcal{M}^2(k;0)}\left[\frac{1}{\lambda^-}-\frac{1}{\lambda^+}\right], \qquad (2)$$

where $\lambda^\pm = a^2 + (k\pm p)^2$. Thus the knowledge of the fermion propagator in one gauge, i.e., $S(p;0)$, is the input required to obtain the same in an arbitrary covariant gauge. It is easy to verify that these relations continue to hold chiral condensate gauge invariant.

## CONSTRAINTS ON TRUNCATIONS

SDE for the fermion propagator can be written as follows for QED3 :

$$S^{-1}(p;\xi) = S_0^{-1}(p)+e^2\int\frac{d^3k}{(2\pi)^3}\Gamma_\nu(k,p)S(k;\xi)\gamma_\mu\Delta_{\mu\nu}(q), \qquad (3)$$

where $q = k - p$, $\Delta_{\mu\nu}(q)$ is the photon propagator and $\Gamma_\mu(k,p)$ the full 3-point vertex. For the sake of simplicity, let us take $\Gamma_\mu(k,p) = \gamma_\mu$, i.e., the bare vertex.

Figures 1 and 2 compare the mass function $\mathcal{M}$ generated from solving SDEs in various gauges versus the same function as obtained through LKFT once the Landau

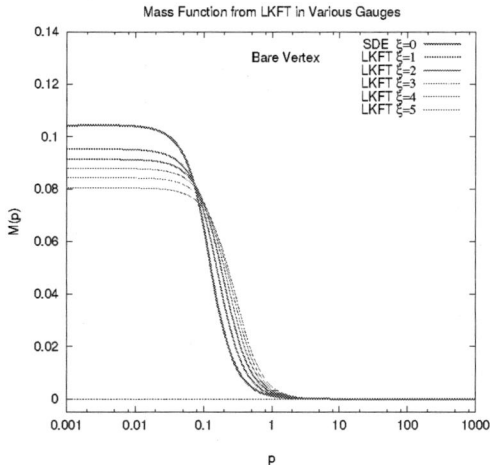

**FIGURE 2.** Solution for the mass function generated through LKFT.

gauge solution is obtained via SDEs. Note that the qualitative behaviour of the curves is different. LKFT requires a turning over of the mass function in various gauges for the bare vertex truncation employed. It cannot be a correct truncation in all gauges as Figure 1 does not exhibit the turning over of the mass function, crucial for the gauge invariance of the chiral condensate. In this way, we can check the validity of a vertex ansatz through the behaviour of the mass function under the LKFT.

## ACKNOWLEDGMENTS

We thank M. R. Pennington and R. Williams for helpful discussions. AB wishes to acknowledge a short term visitor grant by a joint scheme of The Royal Society, U.K, and The Mexican Academy of Sciences, Mexico. Support has also been provided by CIC and CONACyT (grants 4.10 and 46614-I).

## REFERENCES

1.  L.D. Landau and I.M. Khalatnikov, Zh. Eksp. Teor. Fiz. **29** 89 (1956); L.D. Landau and I.M. Khalatnikov, Sov. Phys. JETP **2** 69 (1956); E.S. Fradkin, Sov. Phys. JETP **2** 361 (1956); K. Johnson and B. Zumino, Phys. Rev. Lett. **3** 351 (1959); B. Zumino, J. Math. Phys. **1** 1 (1960). S. Okubo, Nuovo Cim. **15** 949 (1960). I Bialynicki-Birula. Nuovo Cim. **17** 951 (1960).
2.  A. Bashir, Phys. Lett. **B491** 280 (2000).
3.  A. Bashir and A. Raya, Phys. Rev. **D66** 105005 (2002);
4.  A. Bashir and R. Delbourgo, J. Phys. **A37** 6587 (2004).
5.  A. Bashir and A. Raya, *"Gauge Independent Chiral Condensate in QED3"* hep-ph/0511291.
6.  A. Bashir and A. Raya, Nucl. Phys. **B709** 307 (2005); Nucl. Phys. Proc. Suppl. **B141** 259 (2005); proceedings of "2004 International Workshop on Dynamical Symmetry Breaking", Nagoya University, Nagoya, Japan, Dec. 21-22, 257-261 (2004).

# The polarized EMC effect

W. Bentz*, I. C. Cloet[†,**] and A. W. Thomas**

*Department of Physics, Tokai University, Hiratsuka-shi, Kanagawa 259-1292, Japan
†Special Research Centre for the Subatomic Structure of Matter and
Department of Physics and Mathematical Physics, University of Adelaide, SA 5005, Australia
**Jefferson Lab, 12000 Jefferson Avenue, Newport News, VA 23606, U.S.A.

**Abstract.** We calculate both the spin independent and spin dependent nuclear structure functions in an effective quark theory. The nucleon is described as a composite quark-diquark state, and the nucleus is treated in the mean field approximation. We predict a sizable polarized EMC effect, which could be confirmed in future experiments.

**Keywords:** Spin-dependence, medium modifications, structure functions
**PACS:** 12.39Fe, 14.20Dh, 25.30Fj

In this paper we determine the structure functions for both unpolarized and polarized deep inelastic scattering of leptons on nuclear targets. The basic quantities are the light-cone quark distributions in a nucleus, which have the following form for the polarized case: [1]

$$\Delta q_A^H(x_A) = \frac{P_-}{A} \int \frac{d\omega^-}{2\pi} e^{iP_- x_A \omega^-/A} \langle A,P,H|\overline{\psi}_q(0)\,\gamma^+\gamma_5\,\psi_q(\omega^-)|A,P,H\rangle, \qquad (1)$$

where $\psi_q$ is the quark field (flavor $q$) and $A$, $P^\mu$, $H$ are the mass number, 4-momentum, and helicity (along the direction of the incoming lepton momentum) of the nucleus with spin $J$. We evaluate these distributions using the convolution formalism.

In our model, we describe the nucleon as a bound state of a quark and a diquark (scalar and axial vector) in the Nambu-Jona-Lasinio model[2]. The nucleus is treated in the mean field approximation, where we assume scalar and vector potentials of Woods-Saxon shape and depth parameters given by our earlier self consistent nuclear matter calculations[3]. Using the resulting Dirac spinors for the nucleons, we calculate the light cone momentum distribution of nucleons in the nucleus, and convolute them with the quark distributions in the bound nucleon. The essential point in our calculation is that these quark distributions in the bound nucleon are calculated in the presence of the nuclear scalar and vector mean fields, i.e., they respond to the nuclear environment.

By using the QCD evolution equations up to the next-to-leading order[4], we can obtain the nuclear structure functions $F_{2A}$ and $g_{1A}^H$, where for the latter we indicate the dependence on the helicity $H$ of the nucleus. Alternatively, we can express the spin dependent structure functions in terms of $K$-multipoles $g_{1A}^{(K)}$, which are linear combinations of the helicity structure functions[1]. We present our results for the structure functions in terms of the following EMC ratios:

$$R_A = \frac{F_{2A}}{ZF_{2p} + NF_{2n}}, \qquad R_{As}^H = \frac{g_{1A}^H}{P_p^H g_{1p} + P_n^H g_{1n}} \qquad (2)$$

CP892, *Quark Confinement and the Hadron Spectrum VII*
edited by J. E. F. T. Ribeiro
© 2007 American Institute of Physics 978-0-7354-0396-3/07/$23.00

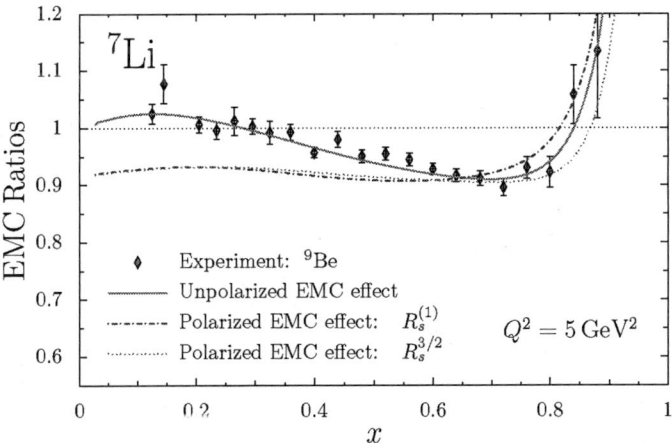

**FIGURE 1.** The solid line is the unpolarized EMC ratio for $^7$Li, which is compared to experimental data[6]. The dash-dotted line is the prediction of the polarized EMC ratio for the dominant $K = 1$ multipole, and the dotted line refers to helicity $H = 3/2$ of the nucleus.

Here $P_p^H \equiv <J,H|2S_z^p|J,H>$ is the polarization factor for protons, and similar for the neutrons. These EMC ratios are such that in the extreme nonrelativistic limit, with no medium modifications, they are unity. For the spin dependent case, we will denote the EMC ratio for the leading multipole ($K = 1$) by $R_{As}^{(1)}$.

Our results for the EMC ratios for $^7$Li, $^{11}$Be and $^{27}$Al are shown in Figs.1 to 3[5]. We have chosen these nuclei because their polarization is determined mainly by the protons, which avoids uncertainties associated with $g_{1n}$ (see Eq.(2)), and because they are not too heavy, which is desirable in view of the approximate $1/A$ suppression of $g_{1A}$ relative to $F_{2A}$.

Because our calculation includes the important medium modifications of the single nucleon structure functions, we are readily able to explain the unpolarized EMC data. These medium modifications also lead to a decrease of the fraction of the nucleon spin carried by the quarks, i.e., some part of the spin is converted into orbital angular momentum. This leads to a decrease of the spin dependent light cone momentum distributions and structure functions in the medium, and the resulting polarized EMC effect is larger than, or at least comparable to, the unpolarized one. Experimental confirmation would give important insights into in-medium quark dynamics, thereby helping to quantify the role of quark degrees of freedom in the nuclear environment.

This work was supported by the Australian Research Council and DOE contract DE-AC05-84150, under which JSA operates Jefferson Lab, and by the Grant in Aid for Scientific Research of the Japanese Ministry of Education, Culture, Sports, Science and Technology, Project No. C2-16540267.

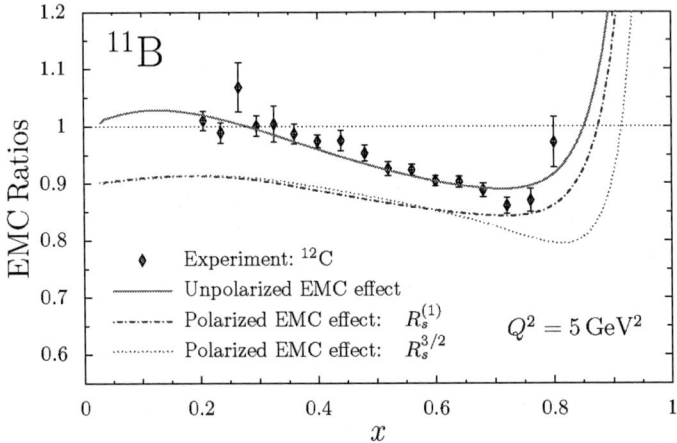

**FIGURE 2.** Same as Fig.1 for the nucleus $^{11}$B.

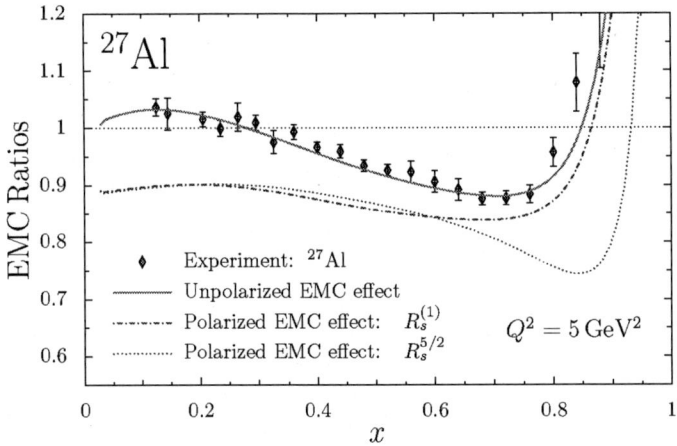

**FIGURE 3.** Same as Fig.1 for the nucleus $^{27}$Al. Here the dotted line refers to the helicity $H = 5/2$.

# REFERENCES

1. R. L. Jaffe and A. Manohar, Nucl. Phys. B **321**, 343 (1989).
2. N. Ishii, W. Bentz and K. Yazaki, Nucl. Phys. A **587**, 617 (1995).
3. I. C. Cloet, W. Bentz and A. W. Thomas, Phys. Rev. Lett. **95**, 052302 (2005).
4. M. Hirai, S. Kumano and M. Miyama, Comput. Phys. Commun. **108**, 38 (1998).
5. I. C. Cloet, W. Bentz and A. W. Thomas, to be published in Phys. Lett. **B**.
6. J. Gomez, et et., Phys. Rev. D **49**, 4348 (1994).

# Chiral Approach to $\phi$ radiative decays

Deirdre Black*, Masayasu Harada[†] and Joseph Schechter**

*Department of Physics, Cavendish Laboratory, J.J. Thomson Avenue, Cambridge CB3 OHE, UK
[†]Department of Physics, Nagoya University, Nagoya 464-8602, Japan
**Department of Physics, Syracuse University, Syracuse, New York 13244-1130 USA

**Abstract.** Rare decays of the $\phi$ vector meson, to $\pi\pi\gamma$ and $\pi\eta\gamma$, have been measured in the last few years. We give some background about why these decays are of interest in that they may provide some insights about the puzzling light scalar mesons. We then present our approach to studying strong and radiative decays involving light scalar mesons more generally, pointing out that it has the advantage of potentially relating many radiative decays at tree-level. Finally we discuss the rare radiative $\phi$ decays and comparison with recent experimental data in more detail.

**Keywords:** Chiral symmetry, vector meson dominance, radiative meson decays, spectroscopy
**PACS:** 14.40.-n, 11.30.Rd, 12.40.Vv, 13.20.-v

## INTRODUCTION

### Motivation for studying rare radiative $\phi$ decays

In the late 1980s it was realized that certain rare radiative decays of the $\phi$ vector meson, to two pseudoscalar mesons and a photon, could possibly shed some light on the scalar mesons $a_0(980)$ and $f_0(980)$. The main interest was in studying CP violation in the kaon system at $\phi$ factories such as Daphne at Frascati, but it was realized that the process $\phi \to K\bar{K}\gamma$ would be one of the backgrounds to the main decay $\phi \to K\bar{K}$. Initial calculations of this background gave branching ratios varying between about $10^{-6}$ and $10^{-4}$ [1]. In these calculations the radiative decays occur primarily through charged kaon loops like the ones shown in Fig. 2. The kaons in the loop couple to a scalar meson ($S$) which in turn couples to two pseudoscalar mesons (say $P$ and $P'$) in the final state. Therefore estimates for the branching ratio for $\phi \to PP'\gamma$ depend on the $SPP'$ and $SK\bar{K}$ couplings, meaning that measuring such radiative $\phi$ decays can potentially give information about the scalar mesons involved. Two sets of experiments (see [2] and references therein) have recently reported measurements of the decays $\phi \to \pi\pi\gamma$ and $\phi \to \pi\eta\gamma$ and many authors have studied these decays from a theoretical perspective (see [3] for references).

### Light scalar mesons in pseudoscalar meson scattering and decays

The light scalar mesons are an old puzzle from the point of view of quark model spectroscopy. For example, their masses are much lower than one would expect for p-wave quark-antiquark states. Also, assuming that $f_0(600)$ or $\sigma$, $K_0^*(800)$ or $\kappa$, $a_0(980)$

CP892, *Quark Confinement and the Hadron Spectrum VII*
edited by J. E. F. T. Ribeiro
© 2007 American Institute of Physics 978-0-7354-0396-3/07/$23.00

and $f_0(980)$ form an SU(3) nonet, it is surprising that the I=0 state $a_0(980)$, which would contain no strange quarks if it is a conventional meson, is heavier than the strange state. Some possibilities are that these scalar mesons are multiquark states, meson-meson molecules or dynamically generated states, possibly mixing with conventional and glueball states having the same quantum numbers (see also talks at this conference by B. Hiller, H. Leutwyler, J. Pelaez, N.A. Tornqvist and H. Zheng).

In our approach (see [4] and references therein for more detail) a nonet of scalar mesons is added in a chiral invariant way to the lowest order nonlinear chiral Lagrangian for pseudoscalar mesons, giving the following trilinear interaction:

$$
\begin{aligned}
\mathscr{L}_{SPP} &= A\varepsilon^{abc}\varepsilon_{def}N_a^d\partial_\mu P_b^e\partial_\mu P_c^f + B\mathrm{Tr}\,[N]\,\mathrm{Tr}\,[\partial_\mu P\partial_\mu P] \\
&+ C\mathrm{Tr}\,[N\partial_\mu P]\,\mathrm{Tr}\,[\partial_\mu P] + D\mathrm{Tr}\,[N]\,\mathrm{Tr}\,[\partial_\mu P]\,\mathrm{Tr}\,[\partial_\mu P]\,.
\end{aligned} \tag{1}
$$

Here $N$ is the putative nonet of scalar mesons and P is the usual pseudoscalar multiplet. We found that including the light scalar mesons in this way gave a good fit to $\pi\pi$ and $\pi K$ scattering data, up to energies beyond the $f_0(980)$ and the $K_0^*(1430)$ regions respectively, as well as $\eta' \to \eta\pi\pi$ decay data. In addition to neatly explaining the mass ordering and general pattern of decays of the scalar states below 1 GeV, a multiquark interpretation for these states is also suggested by the value of the scalar meson octet-singlet mixing angle, which was a parameter fixed by our fits. Our best fit was about $-20^o$ which, in our mixing convention, would be close to ideal mixing for a "dual" diquark-antidiquark nonet.

## RADIATIVE DECAYS INVOLVING LIGHT SCALAR MESONS

A phenomenologically successful non-linear chiral Lagrangian treatment of both pseudoscalar and vector mesons is given in [5] (which is equivalent at tree level to that based on Hidden Local Symmetry [6]). It turns out that in this chiral invariant formulation, vector meson dominance emerges naturally in the sense that the direct photon coupling to two pseudoscalar mesons vanishes in the limit where the KSRF relation holds. In [7] we extended our framework to also include chiral invariant interactions of the vector and scalar mesons. An attractive feature of this approach is that many processes, amongst them the radiative decays of the $\phi$, are related. The hope is that using most of the scalar parameters as fixed from fits to scattering and strong decays, we get many predictions. We only introduce four new parameters, which appear as coupling constants in a new effective Scalar-Vector-Vector interaction given by:

$$
\begin{aligned}
\mathscr{L}_{SVV} &= \beta_A\,\varepsilon_{abc}\varepsilon^{a'b'c'}\left[F_{\mu\nu}(\rho)\right]_{a'}^a\left[F_{\mu\nu}(\rho)\right]_{b'}^b N_{c'}^c + \beta_B\,\mathrm{Tr}\,[N]\,\mathrm{Tr}\,[F_{\mu\nu}(\rho)F_{\mu\nu}(\rho)] \\
&+ \beta_C\,\mathrm{Tr}\,[NF_{\mu\nu}(\rho)]\,\mathrm{Tr}\,[F_{\mu\nu}(\rho)] + \beta_D\,\mathrm{Tr}\,[N]\,\mathrm{Tr}\,[F_{\mu\nu}(\rho)]\,\mathrm{Tr}\,[F_{\mu\nu}(\rho)]\,.
\end{aligned} \tag{2}
$$

Here $\rho$ is the vector meson multiplet and $F_{\mu\nu}(\rho) = \partial_\mu\rho_\nu - \partial_\nu\rho_\mu - i\tilde{g}\,[\rho_\mu,\rho_\nu]$ (see [5] for more detail). Using this effective interaction the radiative decays occur at tree-level as shown in Fig. 1 for the example of $\phi \to \pi\eta\gamma$. In [7] we fit to the experimental values for the total widths $\Gamma(a_0 \to \gamma\gamma)$, $\Gamma(f_0 \to \gamma\gamma)$ and $\Gamma(\phi \to a_0\gamma)$ and predicted nine other branching ratios, including those for $\phi \to f_0\gamma$ and $\sigma \to \gamma\gamma$. Assuming that $\phi \to \pi\eta\gamma$

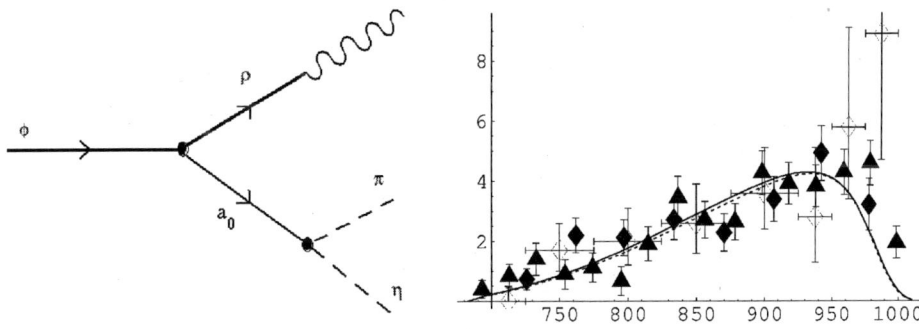

**FIGURE 1.** Left: Tree-level decay $\phi \to \pi\eta\gamma$. The $a_0\rho\phi$ coupling follows from Eq. (2) and the $a_0\eta\pi$ coupling is in principle determined elsewhere. Right: Best fit to $\phi \to \pi\eta\gamma$ data in tree-level model. Plot shows the $dB(\phi \to \pi^0\eta\gamma)/dq \times 10^7$ (in units of MeV$^{-1}$) as a function of the $\pi\eta$ invariant mass $q = m_{\pi\eta}$. Experimental data are from SND (white diamonds) and KLOE (filled diamonds and triangles.

and $\phi \to \pi\pi\gamma$ are dominated by $\phi \to a_0\gamma$ and $\phi \to f_0\gamma$ respectively, our prediction for $\Gamma(f_0 \to \gamma\gamma)$ was too small when compared with [2].

We considered [8] the effect of changing the scalar isoscalar mixing angle from the small angle $\sim -20^o$ (suggestive of a multiquark interpretation for the scalar mesons), found in our scattering fits, to a large angle $\sim -90^o$ which is consistent with the scalar meson masses and would be natural for conventional $q\bar{q}$ mesons. We found that the predictions for these two scenarios did not differ significantly in our tree-level model. We also did a preliminary investigation of the effect of mixing between a light, probably non-$q\bar{q}$, scalar nonet $N$ and a heavier $q\bar{q}$-type nonet $N'$ described by a mixing Lagrangian $\mathscr{L}_{\text{mix}} = \gamma\text{Tr}(NN')$. We take the mixing parameter $\gamma = 0.33\text{GeV}^4$ from a fit to the properties of the $I = 1$ and $I = \frac{1}{2}$ scalar mesons in [9]. Considering only the simplest OZI-type couplings for each nonet, given by

$$\mathscr{L} = \alpha\varepsilon_{abc}\varepsilon^{a'b'c'} \left[ F_{\mu\nu}(\rho) \right]^a_{a'} \left[ F_{\mu\nu}(\rho) \right]^b_{b'} N^c_{c'} + \beta\text{Tr}\left[ N'F_{\mu\nu}(\rho)F_{\mu\nu}(\rho) \right], \qquad (3)$$

and fitting to the measured two-photon widths of $a_0$ and $f_0$ we found predictions for the widths for $\phi \to a_0\gamma$ and $\phi \to f_0\gamma$ which were still too small. In fact mixing between scalar mesons is likely to be more complicated than just $\mathscr{L}_{mix}$, especially for the $I = 0$ states (see for example [10]), and we could of course include more general interactions of both scalar nonets with the vector mesons than Eq. (3).

In [3] we used our simple model to study the radiative $\phi$ decay spectra in detail. For example in Fig. 1 we show our best fits for the partial branching fraction for $\phi \to \pi\eta\gamma$. We see that the fit is quite good, except towards the higher end of the spectrum, where there is also a larger spread in the central values obtained by the different experimental groups. We note that in our non-linear chiral Lagrangian approach there is derivative coupling between the scalar and pseudoscalar mesons. The resultant extra momentum factor in the decay width counteracts falling phase space towards the end of the spectrum and we found that this gives a substantially better fit than a model where the scalar and pseudoscalar mesons couple non-derivatively. We also calculated the contribution of the

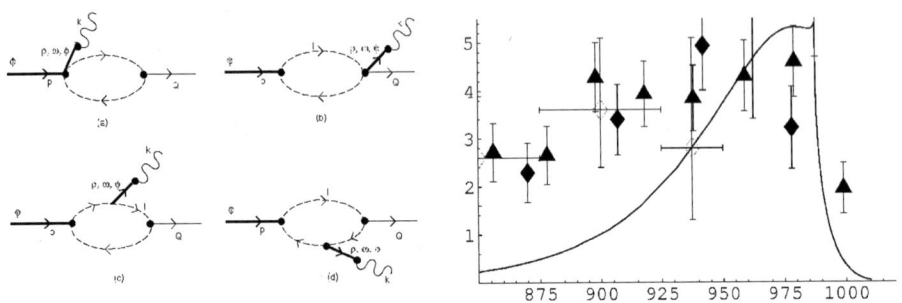

**FIGURE 2.** Left: Kaon loop diagrams contributing to decays of $\phi \to S\gamma$ where $S$ is a scalar meson which gives two pseudoscalar mesons in the final state. Right: Fit of the kaon loop contribution to upper end of $\phi \to \pi\eta\gamma$ decay spectrum. Data and axes are as in Fig. 1

charged kaon loop diagrams in our model. In Fig 2. we show our prediction for the kaon loop contribution to the $\phi \to \pi\eta\gamma$ partial branching ratio in the $a_0(980)$ resonance region towards the high end of the spectrum. A next step is to combine the tree and one-loop contributions, also including known non-resonant background contributions.

## ACKNOWLEDGMENTS

D. B. (speaker) is supported by the Royal Society, UK. M. H. is supported in part by the Daiko Foundation #9099, the 21st Century COE Program of Nagoya University provided by Japan Society for the Promotion of Science (15COEG01), and the JSPS Grant-in-Aid for Scientific Research (c) (2) 16540241. The work of J. S. is supported in part by the U. S. DOE under contract No. DE-FG-02-85ER 40231.

## REFERENCES

1. N. N. Achasov and V. N. Ivanchenko, Nucl Phys **B315**, 465 (1989), F. E. Close, N. Isgur and S. Kumano, Nucl. Phys. **B389** 513 (1993).
2. M. N. Achasov *et al.* (SND Collaboration), Phys. Lett. B **479**, 53 (2000); A. Aloisio *et al.* (KLOE Collaboration), Phys. Lett. B **537**, 21 (2002); *ibid* **536**, 209 (2002).
3. D. Black, M. Harada and J. Schechter, Phys. Rev. D **73**, 054017 (2006).
4. D. Black, A. H. Fariborz, F. Sannino and J. Schechter, Phys. Rev. D **59** 074026 (1999).
5. M. Harada and J. Schechter, Phys. Rev. D **54**, 3394 (1996).
6. M. Harada and K. Yamawaki, Phys. Rep. **381**, 1 (2003).
7. D. Black, M. Harada and J. Schechter, Phys. Rev. Lett. **88**, 181603 (2002).
8. M. Harada, hep-ph/0408189. Talk given at YITP workshop on "Multi-quark Hadrons; four, five and more?", February 17-19, 2004, Yukawa Insitute, Kyoto, Japan
9. D. Black, A. H. Fariborz and J. Schechter, Phys. Rev. **D** 61, 074001 (2000).
10. T. Teshima, I. Kitamura, N. Morisita, J. Phys. G28 (2002), 1391-1402, A. H. Fariborz, Int. J. Mod. Phys. A19(2004) 2095-2112, A. H. Fariborz, R. Jora and J. Schechter, Phys. Rev. **D** 72, 034001 (2005),

# Behavior of physical observables in the vicinity of the QCD critical end point

## Pedro Costa

*Centro de Física Teórica, Departamento de Física, Universidade, P3004-516 Coimbra, Portugal*

**Abstract.** Using the SU(3) Nambu-Jona-Lasinio (NJL) model, we study the chiral phase transition at finite $T$ and $\mu_B$. Special attention is given to the QCD critical end point (CEP): the study of physical quantities, as the pressure, the entropy, the baryon number susceptibility and the specific heat near the CEP, will provide complementary information concerning the order of the phase transition. We also analyze the information provided by the study of the critical exponents around the CEP.

**Keywords:** NJL model, chiral phase transition, critical end point
**PACS:** 11.30.Rd, 11.55.Fv, 14.40.Aq

The existence of the CEP in QCD was suggested in the end of the eighties, and its properties have been studied since then (for a general review see Refs. [1, 2]). The most recent lattice results for the study of dynamical QCD with $N_f = 2 + 1$ staggered quarks of physical masses indicate the location of the CEP at $T^{CEP} = 162 \pm 2\text{MeV}$, $\mu^{CEP} = 360 \pm 40\text{MeV}$ [3], however its exact location is not yet known once the location of the CEP depends strongly of the mass of the strange quark. At the CEP the phase transition is of second order, belonging to the three-dimensional Ising universality class, and this kind of phase transitions are characterized by long-wavelength fluctuations of the order parameter.

As pointed out in [4, 5], the critical region around the CEP is not pointlike but has a very rich structure. The vicinity of the CEP is a privileged region to study the influence of different type phase transitions in the physical observables, namely, the pressure, the entropy, the baryon number susceptibility, $\chi_B$, and the specific heat, $C_V$.

We perform our calculations in the framework of the three–flavor NJL model, including the determinantal 't Hooft interaction that breaks the $U_A(1)$ symmetry, which has the following Lagrangian:

$$\mathcal{L} = \bar{q}(i\partial \cdot \gamma - \hat{m})q + \frac{g_S}{2}\sum_{a=0}^{8}\left[(\bar{q}\lambda^a q)^2 + (\bar{q}(i\gamma_5)\lambda^a q)^2\right]$$
$$+ g_D\left[\det\left[\bar{q}(1+\gamma_5)q\right] + \det\left[\bar{q}(1-\gamma_5)q\right]\right]. \tag{1}$$

By using a standard hadronization procedure, an effective meson action is obtained, leading to the gap equations for the constituent quark masses from which several observables are calculated (we follow the methodology presented in detail in Refs. [6, 7]).

CP892, *Quark Confinement and the Hadron Spectrum VII*
edited by J. E. F. T. Ribeiro
© 2007 American Institute of Physics 978-0-7354-0396-3/07/$23.00

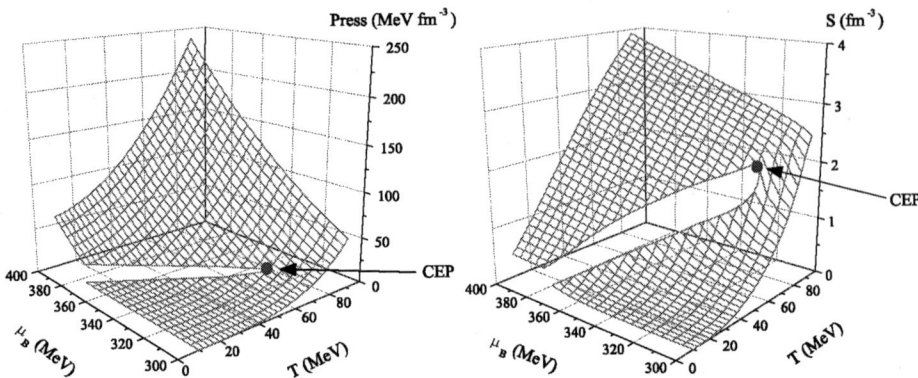

**FIGURE 1.** The pressure (left panel) and the entropy (right panel) as functions of the temperature and the baryonic chemical potential.

The fundamental relation is provided by the baryonic thermodynamic potential

$$\Omega(\mu_i, T) = E - TS - \sum_{i=u,d,s} \mu_i N_i, \tag{2}$$

from which the relevant observables as the pressure $P$, the entropy $S$ and the particle number $N_i$ can be calculated as usually (the expressions are given in Ref. [6]).

The nature of the chiral phase transition in NJL type models at finite $T$ and/ or $\mu_B$ has been discussed by different authors [6]. For zero temperature the transition is of first order. As the temperature increases, the first order transition line persists up to the CEP. In the CEP the chiral transition becomes of second order. At higher temperatures there is a smooth crossover.

In the vicinity of the CEP, the nature of the chiral phase transition influence strongly the behaviour of the physical observables. In the first order phase transition region ($T < T^{CEP}, \mu_B > \mu_B^{CEP}$) both, pressure and entropy, show a discontinuity which end at the CEP. In the crossover region ($T > T^{CEP}, \mu_B < \mu_B^{CEP}$), the discontinuities of $P$ and $S$ vanish, and both quantities change gradually in a continuous way (see Fig. 1). The same situation can be seen in the behaviour of $\chi_B$: in the first order phase transition region $\chi_B$ has a discontinuity (left panel of Fig. 2). At the CEP, the phase transition is of second order, and the slope of the baryonic density tends to infinity which implies a diverging $\chi_B$. In the crossover region, $\chi_B$ changes gradually in a continuous way. A similar behavior is found for the specific heat for three different chemical potentials around the CEP, as we can observe from the right panel of Fig. 2.

Focusing our attention on the critical behavior of the baryon number susceptibility in the vicinity of the CEP, we verify that $\chi_B$ diverges with a certain critical exponent. Considering a path parallel to the $\mu_B$-axis in the ($T, \mu_B$)-plane from lower $\mu_B$ towards the critical $\mu_B^{CEP} = 318.5$ MeV at fixed temperature $T^{CEP} = 67.7$ MeV, one can calculate the critical exponent $\varepsilon$ using the linear logarithmic fit $\ln \chi_B = -\varepsilon \ln |\mu_B - \mu_B^{CEP}| + const$, where the term *const* is independent of $\mu_B$. The result that we obtain is $\varepsilon = 0.67 \pm 0.01$, which is consistent with the mean field theory prediction: $\varepsilon = 2/3$. Since there is no

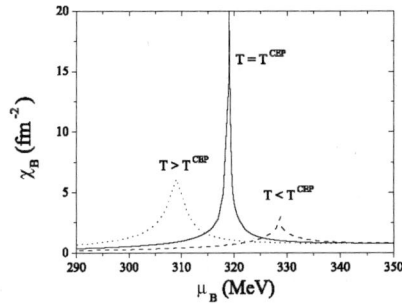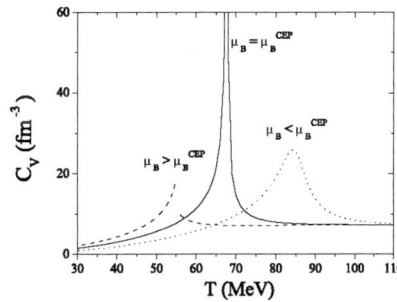

**FIGURE 2.** Left panel: baryon number susceptibility as a function of $\mu_B$ for different temperatures around the CEP: $T^{CEP} = 67.7$ MeV and $T = T^{CEP} \pm 10$ MeV. Right panel: specific heat as a function of $T$ for different values of $\mu_B$ around the CEP: $\mu_B^{CEP} = 318.5$ MeV and $\mu_B = \mu_B^{CEP} \pm 10$ MeV.

reason why the critical exponent should be equal for both regions, below and above $\mu_B^{CEP}$, we also study $\chi_B$ from higher $\mu_B$ towards the critical $\mu_B^{CEP}$. Using again a logarithmic fit, the result is $\varepsilon' = 0.68 \pm 0.01$ which is very near the value of $\varepsilon$. This means that the size of the region we observe is approximately the same independently of the direction we choose in the path parallel to the $\mu_B$-axis. Complementary information is also obtained from the study of the critical exponent of $C_V$ [8].

Summarizing our discussion, we have analyzed the vicinity of the QCD critical end point in the SU(3) NJL model. We conclude that, in this region, the physical observables are strongly influenced by the nature of the phase transition. Around the CEP we have studied the baryon number susceptibility which is related with event-by-event fluctuations of $\mu$ in heavy-ion collisions. The study of the specific heat is also important once it is related with event-by-event fluctuations of $T$ in heavy-ion collisions [9]. We also conclude that the critical exponents of $\chi_B$ obtained in our model are consistent with the mean field value $\varepsilon \simeq \varepsilon' \simeq 2/3$.

Work supported by grant SFRH/BPD/23252/2005 from F.C.T. and Centro de Física Teórica.

# REFERENCES

1. M. A. Stephanov, Prog. Theor. Phys. Suppl. 153 (2004) 139; Int. J. Mod. Phys. A 20 (2005) 4387.
2. R. Casalbuoni, hep-ph/0610179.
3. Z. Fodor, S. D. Katz, J. High Energy Phys. 0204 (2004) 050.
4. Y. Hatta, T. Ikeda, Phys. Rev. D 67 (2003) 014028.
5. B.-J. Schaefer, J. Wambach, hep-ph/0603256.
6. P. Costa, M. C. Ruivo, Y. L. Kalinovsky, C. A. de Sousa, Phys. Rev. C 70 (2004) 025204.
7. P. Costa, M. C. Ruivo, Yu. L. Kalinovsky, Phys. Lett. B 560 (2003) 171.
8. P. Costa, *et al.*, in preparation.
9. M. Stephanov, K. Rajagopal, E. Shuryak, Phys. Rev. Lett. 81 (1998) 4816.

# The narrow pentaquark

Dmitri Diakonov

*Petersburg Nuclear Physics Institute, Gatchina, 188 300, St. Petersburg, Russia*

**Abstract.** The experimental status of the pentaquark searches is briefly reviewed. Recent null results by the CLAS collaboration are commented, and new strong evidence of a very narrow $\Theta^+$ resonance by the DIANA collaboration is presented. On the theory side, I revisit the argument against the existence of the pentaquark – that of Callan and Klebanov – and show that actually a strong resonance is predicted in that approach, however its width is grossly overestimated. A recent calculation gives 2 MeV for the pentaquark width, and this number is probably still an upper bound [1].

The original claim for the discovery of a narrow exotic baryon resonance in two independent experiments by T. Nakano *et al.* [2] and A. Dolgolenko *et al.* [3], announced in the end of 2002 [1], was followed in 2003-04 by a dozen experiments confirming the resonance and about the same amount of non-sighting experiments. In 2005 the results of the two CLAS high-statistics experiments were announced, which didn't see a statistically significant signal of the $\Theta^+$ resonance in the $\gamma d$ and $\gamma p$ reactions and gave upper bounds for its production cross sections. Although those upper bounds didn't seem to contradict the theoretical estimates [1] many people in the community jumped to the conclusion that "pentaquarks do not exist". In 2005-06 new results became available [5, 6] partly based on new data, confirming seeing the $\Theta^+$, see Fig. 1.

We have to keep in mind that there are numerous and so far uncontested observations of a $KN$ resonance at 1.53 GeV in neutrino-, photon- and proton-induced reactions. The analysis of old $K^+d$ data by Gibbs calls for the exotic resonance with the width $0.9 \pm 0.3$ MeV. An anomaly in $K^+$ scattering off nuclei needs an "additional reactivity" as compared to the usual optical potential scattering. Last but not least, the GRAAL collaboration reports a possible narrow $N^*(1675)$ resonance in the $\gamma n \to \eta n$ reaction (but not in the $\gamma p \to \eta p$) [7] which is consistent with the resonance being the antidecuplet partner of the $\Theta^+$.

Given a small $KN\Theta$ coupling constant (since the width is very small) and a small $K^*N\Theta$ coupling (since the transition magnetic moment is small [8]), it is difficult to arrange for a sizable production of the $\Theta^+$. Future progress can be obtained a) by performing a high-flux $KN$ direct formation experiment (planned at J-PARC), b) by learning to make reliable estimates for the production cross sections, such that the comparison with the data becomes meaningful, and c) by inventing clever new methods of searching $\Theta^+$ taking into account that all its couplings to normal hadrons are small.

---

[1] They were totally independent as both groups didn't know about the work of one another and made a tedious re-analysis of data taken long before, however both searches were triggered off by the authors of Ref. [4] where the resonance at $\sim 1530$ MeV and width less than 15 MeV had been predicted.

CP892, *Quark Confinement and the Hadron Spectrum VII*
edited by J. E. F. T. Ribeiro
© 2007 American Institute of Physics 978-0-7354-0396-3/07/$23.00

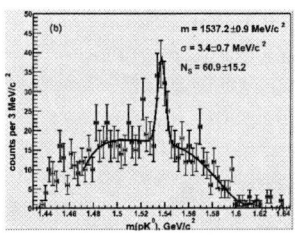

**FIGURE 1.** $\Theta^+$ as seen in the $K^+n(\mathrm{Xe}) \to K^0p$ reaction: there are 60 events above the estimated background from rescattering; $M_\Theta = 1537 \pm 2\,\mathrm{MeV}$, $\Gamma_\Theta = 0.36 \pm 0.11\,\mathrm{MeV}$ [3].

Probably the only theoretical argument against the existence of exotic baryons is due to Callan and Klebanov [9] [2]. It relies on the academic limit of large number of colours $N_c$ when baryons can be considered in the mean field approximation with quarks bound by the self-consistent pion field, the "soliton" (*à la* large-Z Thomas–Fermi atom or the large-A shell model for nuclei). The Skyrme model is a popular realization of this idea, although not a too realistic one [11].

At large $N_c$ the existence or non-existence of the $\Theta^+$ can be studied by considering kaon scattering off a 'classical' nucleon (which I shall generically call the 'Skyrmion') and that is what Callan and Klebanov did. After the discovery of the $\Theta^+$, the study has been repeated in more detail in Ref. [13]. One has to solve a Schrödinger-Klein-Fock equation but with a Wess-Zumino-Witten term linear in the time derivative, and find the scattering phases for given quantum numbers. The resulting phase in the strangeness +1, spin $1/2^+$ channel is plotted in Fig. 2a.

The point made in Refs. [9, 13] is that the $K^+n$ phase shift in Fig. 2a does not pass through $90^o$ as it should be for an isolated Breit–Wigner resonance, and therefore there is no exotic resonance, at least in the large-$N_c$ limit. However, if there is both a resonance *and* a potential scattering, the phase shift needs not go through $\pi/2$.

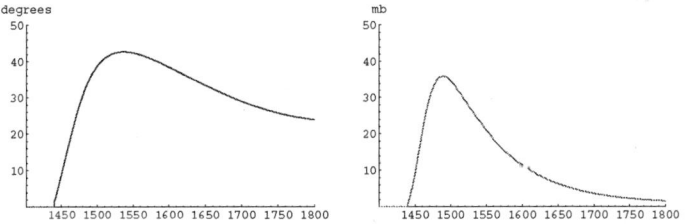

**FIGURE 2.** The $K^+n$ scattering phase [13] (left) and the ensuing $K^+n$ scattering cross section (right) as function of the invariant $K^+n$ mass in the Skyrme model (in the academic limit $N_c \to \infty$). Note that at the maximum the cross section is as large as 35 mb! Courtesy V. Petrov.

---

[2] T. Cohen gave an additional argument [10] why the Callan–Klebanov approach to the exotic baryon must be correct at large number of colours.

It is instructive to solve the Callan–Klebanov $K^+n$ scattering equation in the complex energy plane, simultaneously varying the coefficient of the Wess–Zumino–Witten term [14]. When it is zero, there is exact zero-energy solution corresponding to the rotation of the soliton as a whole in the flavour space. It was on the base of the quantization of this rotation that the light and narrow $\Theta^+$ was predicted [4]. As one increases the coefficient of the Wess–Zumino–Witten term towards its physical value, the would-be zero energy level moves up but obtains an imaginary part. With the standard Skyrme model parameters used by Klebanov *et al.*, the pole position of the $\Theta^+$ resonance is at $1510 - \frac{i}{2} \cdot 120 \, \text{MeV}$. Indeed, had Klebanov *et al.* [13] plotted the $K^+n$ cross section from their phase shift according to the well-known formula $\sigma = (4\pi/k^2)(2j+1)\sin^2\delta$, they would have observed a very strong resonance, see Fig. 2b.

Thus, the prediction of the Skyrme model is not that there is no exotic resonance but just the opposite: **there is a very strong resonance**, at least when the number of colours is taken to infinity! Varying the parameters of the Skyrme model or modifying it can make the exotic resonance narrower or broader but one cannot get rid of it. The reason is very general: energy levels do not disappear as one varies the parameters but move into the complex plane.

However, the Callan–Klebanov large-$N_c$ logic in general and the concrete Skyrme model in particular grossly overestimate the resonance width. When the limit $N_c \to \infty$ is used the width is increased at least by a factor of 5 [1]. At $N_c = 3$ the resonance becomes very narrow, and that is why it is so difficult to observe it. Had $N_c$ been 300 instead of 3, $\Theta^+$ could be as broad as any other well-established baryon resonance. It would have been produced in abundance in hadron collisions.

If $N_c$ is not infinitely large, one can use the Relativistic Mean Field Approximation [12] (alias the Chiral Quark Soliton Model [11]). Being a relativistic field-theoretic model, it allows to account for quark pair creation and annihilation in a consistent way, and that is what we need here.

In the mean field approximation *all* quark wave functions inside *all* baryons belonging to the octet, decuplet and exotic antidecuplet are known for *all* their Fock, i.e. $3Q, 5Q, 7Q, \ldots$ components [12]. The leading component in the ordinary octet and decuplet baryons is naturally the $3Q$ one but there is a sizable ($\sim 30\%$) addition of the $5Q$ component. For some baryon observables the $5Q$ component gives a mild correction (and that is why the primitive $3Q$ constituent quark models are not so bad as one would naively expect) but in some other observables higher components are critical to obtain agreement with the experiment. About 30% of the time nucleons are pentaquarks!

As to the exotic $\Theta^+$ and other members of the antidecuplet, their *lowest* Fock component is the $5Q$ one. The extra $Q\bar{Q}$ pair in the $\Theta^+$ is a known [12] mixture of $0^+, 0^-, 1^+$ and $1^-$ waves corresponding to scalar, pseudoscalar, axial and vector mesons which, however, are deep inside baryons and do not form 'molecules'.

To evaluate the width of the $\Theta^+ \to K^+n$ decay one has to compute the transition matrix element of the strange axial current, $< \Theta^+ | \bar{s} \gamma_\mu \gamma_5 u | n >$. The important point is that there are, generally, two contributions: the "fall apart" $Q\bar{Q}$ annihilation process and the "5-to-5" process where $\Theta^+$ decays into the $5Q$ component of the nucleon.

Each of those decay amplitudes are not Lorentz-invariant, only their sum is. A convenient way to evaluate the sum of two amplitudes is to go to the infinite momentum frame where only the "5-to-5" amplitude survives, as axial (and vector) currents with a finite momentum transfer do not create or annihilate quarks with infinite momenta. The baryon matrix elements are thus non-zero only between Fock components with *equal number* of quarks and antiquarks.

The transition matrix element of the strange axial charge $< \Theta^+ | \bar{s} \gamma_5 u | n >$ was evaluated in [12, 15] with the resulting width $\Gamma_\Theta \approx 2\,\text{MeV}$, assuming the chiral limit for the kaon and zero momentum transfer. One would expect a further formfactor suppression of this estimate such that $\Gamma_\Theta$ may well end up at the sub-MeV level. As stressed in our first publication [4], in the imaginary non-relativistic limit when ordinary baryons are made of three quarks with no admixture of $Q\bar{Q}$ pairs the $\Theta^+$ width tends to zero strictly.

To summarize: The very small width of $\Theta^+$ is natural; the present estimate $\Gamma_\Theta \approx 2\,\text{MeV}$ will probably go down when formfactor suppression is included. We have revisited the theoretical argument of Callan and Klebanov against the exotics and found that actually it is the opposite: the Skyrme model at large $N_c$ predicts a too strong resonance. A broad resonance, however, is a very-large-$N_c$ artifact. On the experimental side, there is new strong evidence of an extremely narrow $\Theta^+$ from DIANA, a very significant new evidence from LEPS, and other older evidence which is difficult to brush aside. The null results from the new round of CLAS experiments are compatible with what one should expect based on the estimates of production cross sections.

I thank A. Dolgolenko, V. Guzey, A. Hosaka, T. Nakano, A. Titov and especially V. Petrov and M. Polyakov for discussions. The contribution has been written up during the visit to Bochum University, sponsored by the A.v.Humboldt Award. This work is supported in part by Russian Government grants 1124.2003.2 and RFBR 06-02-16786. I would like to thank cordially Jose Emilio Ribeiro for the fantastic organization of the meeting, and for hospitality.

# REFERENCES

1. A more detailed version of this contribution is: D. Diakonov, arXive:hep-ph/0610166.
2. T. Nakano [LEPS Collaboration], Talk at the PANIC 2002 (Oct. 3, 2002, Osaka); T. Nakano *et al.*, Phys. Rev. Lett. **91**, 012002 (2003).
3. V.A. Shebanov [DIANA Collaboration], Talk at the Session of the Nuclear Physics Division of the Russian Academy of Sciences (Dec. 3, 2002, Moscow); V.V. Barmin *et al.*, Phys. Atom. Nucl. **66**, 1715 (2003).
4. D. Diakonov, V. Petrov and M. Polyakov, Z. Phys. **A359**, 305 (1997).
5. T. Nakano, talk at the Bochum workshop on $\eta$ photoproduction (Feb. 23-25, 2006); talk at the Internat. Conf. on Strangeness in Quark Matter (UCLA, March 26-31, 2006) and other presentations.
6. V.V. Barmin *et al.* [DIANA Collaboration], arXive:hep-ex/0603017.
7. V. Kuznetsov [for GRAAL Collaboration], arXive:hep-ex/0606065.
8. M.V. Polyakov and A. Rathke, Eur. Phys.J. **A18**, 691 (2003); H.-C. Kim *et al.*, Phys. Rev. **D71**, 094023 (2005).
9. C. Callan and I. Klebanov, Nucl Phys. **B262**, 365 (1985).
10. T. Cohen, Phys. Lett. **B581**, 175 (2004).
11. D. Diakonov and V. Petrov, in: *At the frontier of particle physics*, M. Shifman (ed.), World Scientific, Singapore, vol. 1, pp. 359-415 [arXive:hep-ph/0009006].
12. D. Diakonov and V. Petrov, Phys. Rev. **D72**, 074009 (2005).
13. N. Itzhaki, I.R. Klebanov, P. Ouyang and L. Rastelli, Nucl. Phys. **B684**, 264 (2004).
14. D. Diakonov and V. Petrov, in preparation.
15. C. Lorcé, Phys. Rev. **D74**, 054019 (2006) [arXive:hep-ph/0603231].

# Search for medium effects on light vector mesons

C. Djalali [a], R. Nasseripour [a], D. P. Weygand [b], M. H. Wood [c], and CLAS Collaboration

[a] University of South Carolina, Columbia, SC 29208
[b] Thomas Jefferson Accelerator Facility, Newport News, VA 23606
[c] University of Massachusetts, Amherst, MA 01003

**Abstract.** The photoproduction of vector mesons on various nuclei has been studied using the Cebaf Large Acceptance Spectrometer (CLAS) at Jefferson Laboratory. The $\rho$, $\omega$, and $\phi$ mesons are observed via their decay to $e^+e^-$. The $\rho$ spectral function is extracted from the data on carbon, iron, and titanium. We observe no effects on the mass of the $\rho$ meson, some widening in titanium and iron is observed consistent with standard collisional broadening.

**Keywords:** Medium modifications, vector mesons, di-lepton decay.
**PACS:** 25.20.Lj, 13.20.-v, 13.60.Le, 14.40.-n.

## INTRODUCTION

Hadron masses, for example the proton at $\sim 1$ GeV/c$^2$, are much larger than the summed masses of their constituent quarks, which are a few MeV/c$^2$, indicating that much of the hadron mass is generated dynamically. Hadron masses are somewhat effected by the spontaneous breaking of chiral symmetry. At high temperature or pressure, chiral symmetry may likely be restored. At normal nuclear densities, partial restoration of chiral symmetry may effect the properties of hadrons, in particular masses and widths. [1-6]

The first evidence of a medium-effected $\rho$ mass came from CERN in 1995.[7,8] Theorists were able to account for the observations by assuming a decrease in the mass of the $\rho$ meson.[9] Relativistic heavy-ion results are integrated over a wide range of densities and temperatures. Theoretical predictions of in-medium effects by the different models are so large that they should have observable consequences already at normal nuclear density in $\gamma$ or $\pi$-induced reactions.

Hatsuda and Lee [4], based on QCD sum rule calculations, obtain spectral changes of the vector mesons in the nuclear medium. Their calculations result in a linear decrease of the masses as a function of density:

$$\frac{m_{VM}(\rho)}{m_{VM}(\rho=0)} = 1 - \alpha \frac{\rho}{\rho_0}, \quad \alpha = 0.16 \pm 0.06 \tag{1}$$

CP892, *Quark Confinement and the Hadron Spectrum VII*
edited by J. E. F. T. Ribeiro
© 2007 American Institute of Physics 978-0-7354-0396-3/07/$23.00

Models based on nuclear many-body effects predict a broadening in the width of the ρ meson with increasing density.[5,6] An observation of a medium-modified vector meson invariant mass decrease has been claimed by a KEK-PS collaboration.[10] Very recently, the Crystal Barrel/TAPS collaboration has reported a downward shift in the mass of the ω.[11] All these experiments are yielding results complementary to each other, but no clear consensus has yet emerged between the various analyzes.

## EXPERIMENTAL SETUP

The data for this study were taken in 2002 using the CEBAF accelerator and the CLAS detector located in the Hall-B of the Jefferson Laboratory.[12,13] CLAS is a nearly 4π-detector which was designed to track charged particles with momenta greater than 200 MeV/c. The detector is made of 3 regions of drift chambers, time-of-flight scintillators, Cerenkov counters (CC) and electromagnetic calorimeters (EC). The $e^+e^-$ event selection and the rejection of the very large $\pi^+\pi^-$ background were done through cuts on the EC and the CC.

Lepton pair production has a background of random combinations of pairs due to the uncorrelated sources. We have treated this background using the combinatorial method that has successfully been used in the past for measurements involving opposite-sign pairs of pions or muons.[14,15]

## RESULTS AND DISCUSSION

To simulate each physics process, the events were generated using a code based on a semi-classical Boltzmann-Uehling-Uhlenbeck (BUU) transport model developed by the group of U. Mosel at the University of Giessen.[16,17]

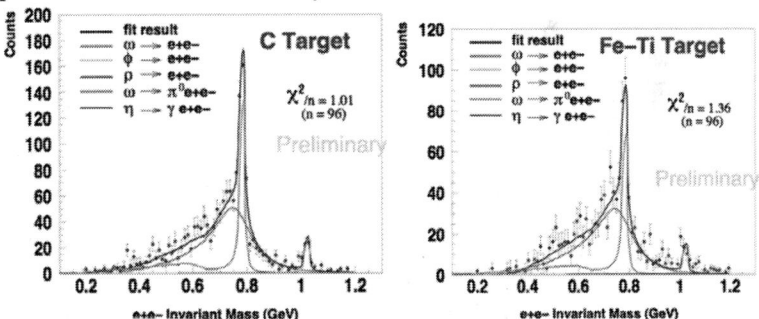

**FIGURE 1.** Results of the fit to the $e^+e^-$ invariant mass obtained for C (left) and Fe-Ti (right) data. Curves are Monte-Carlo calculations by the BUU model [18,19] for various $e^+e^-$ channels.

The combinatorial background distributions are subtracted from the $e^+e^-$ spectra. The shape of the narrow ω and φ vector mesons, and the ω Dalitz channel are well described by BUU model. These fits for C and Fe/Ti are shown in Fig.1. The extracted ρ mass distributions and the ratio to the deuterium data are then simultaneously fit with the suggested functional form of $1/m^3$ times a Breit-Wigner function.[20-22] The result of the fits are tabulate in Table 1. The fits describe the data very well. The masses are consistent with the PDG values and the widths are consistent with the collisional broadening. We don't observe the doubling of the ρ width reported by NA60.[23,24] Our results do not favor the prediction of Brown and Rho for the mass shift (20%) or Hatsuda and Lee ($\alpha$ =0.16 ± 0.06).

TABLE 1. Mass and width of the ρ meson obtained from fits to the mass spectra.(Preliminary results)

| Target | Mass (MeV) g7 data | Width (MeV) g7 data | Mass (MeV) BUU | Width (MeV) BUU |
|--------|--------------------|---------------------|----------------|-----------------|
| $D_2$ | 770.3 ± 3.2 | 185.2 ± 8.6 | No BUU | No BUU |
| C | 762.5 ± 3.7 | 176.4 ± 9.5 | 769.2 ± 2.0 | 160.3 ± 3.0 |
| Fe | 779.0 ± 5.7 | 217.7 ± 14.5 | 764.0 ± 2.5 | 186.5 ± 5.0 |

# ACKNOWLEDGMENTS

The authors would like to thank U. Mosel, P. Muehlich, J. Weil, O. Buss, and A. Afanasev for providing us with theoretical support during this work. The U.S. Department of Energy and the National Science Foundation supported this work. The Jefferson Science Associates, LLC, operates the Thomas Jefferson National Accelerator Facility for the United States Department of Energy under contract No. DE-AC05-06OR23177.

# REFERENCES

1. V. Bernard and U. G. Meissner, *Nucl. Phys.* **A489,** Issue 4, 647 (1988).
2. S. Klimt *et al., Phys. Lett.* **B249,** 386 (1990).
3. Brown and Rho, *et al., Phys. Rev Lett.* **66,** 2720 (1991).
4. Hatsuda and Lee, *Phys. Rev.* **C46,** R34 (1992).
5. M. Herrman *et al., Nucl. Phys.* **A545,** 267c (1992).
6. R. Rapp *et al., Nucl. Phys.* **A617,** 472 (1997).
7. G. Agakichiev *et al., Phys. Rev. Lett.* **75,** 1272 (1995).
8. M. Massera *et al., Nucl. Phys.* **A590,** 93c (1995).
9. G. Q. Li *et al., Phys. Rev. Lett.* **75,** 4007 (1995).
10. M. Naruki *et al., Phys. Rev. Lett.* **96,** 092301 (2006).
11. D. Trnka *et al., Phys. Rev. Lett.* **94,** 192303, (2005).
12. C. W. Leemann, D. R. Douglas, and G. A. Krafft, *Annu. Rev. Nucl. Part. Sci.* **51,** 413 (2001)
13. B. A. Mecking *et al., Nucl. Instr. Methods* **A503,** 513 (2003).
14. G. Jancso *et al., Nucl. Phys.* **B124,** 1 (1977)
15. B. D. Jouan *et al., IPNO-DR-02.015* (2002).
16. P. Muehlich, T. Falter, C. Greiner, J. Lehr, M. Post and U. Mosel, arXiv:nucl-th/0210079 (2002).
17. M. Effenberger, E. L. Bratkovskaya and U. Mosel, *Phys. Rev.* **C60,** 044614 (1999).
18. M. Effenberger and U. Mosel, *Phys. Rev.* **C62,** 014605 (2000).
19. M. Effenberger, E. L. Bratkovskaya, W. Cassing, and U. Mosel, *Phys. Rev.* **C60,** 027601 (1999).
20. Guo-Qiang Li *et al.,* arXiv:nucl-th/9611037 v1 Nov. (1996).
21. M. Effenberger *et al.,* arXiv:nucl-th/9903026 v2 Aug. (1999).
22. H. B. O'Connell *et al., Prog. Part. Nucl. Phys.* **39,** 201 (1997)
23. S. Damjanovic *et al.,* for the NA60 collaboration, Quark Matter (2005).
24. R. Arnaldi *et al., Phys. Rev. Lett.* **96,** 162302 (2006).

# Photo-Production of Proton Antiproton Pairs

Paul Eugenio and Burnham Stokes

*Department of Physics, Florida State University, Tallahassee, FL USA*
*for the CLAS Collaboration*

**Abstract.** Results are reported on the reaction $\gamma p \to pp\bar{p}$. A high statistic data set was obtained at the Thomas Jefferson National Accelerator Facility utilizing the CLAS detector and a tagged photon beam of 4.8 to 5.2 GeV incident on a liquid hydrogen target. The focus of this study was to search for possible intermediate resonances which decay to proton-antiproton. Both final state protons were detected in the CLAS apparatus whereas the antiproton was identified via missing mass. General features of the data are presented along with results on narrow and broad resonance studies.

**Keywords:** baryonia
**PACS:** 13.60.Rj

## Introduction

The proton-antiproton system has had a rich history spanning more than thirty years. Initially, the $p\bar{p}$ system had much interest due to theoretical predictions of exotic matter. These predictions included: nucleon-antinucleon states that are loosely bound in a molecule-like structure called quasi-nuclear baryonium, and tightly-bound multi-quark baryonium ($qq - \bar{q}\bar{q}$) which have favored decays to nucleon-antinucleon final states.

Around 1970, there were claims of a unusually-narrow meson resonance with a mass of 1.93 $GeV/c^2$ [1], and it was believed that this particle was not an ordinary meson and that it would couple to the proton-antiproton system. There were then claims that experiments found the narrow resonance in proton-antiproton scattering experiments [2]. Also, in the late 1970s there were claims of additional higher mass narrow resonances at 2.02 and 2.20 $GeV/c^2$ in the proton-antiproton system [3]. However follow up experiments did not make such claims[4]. And until recently, the debate had died out.

In 1997, CERN refuted their own earlier claims of the 1.93 and 2.02 $GeV/c^2$ resonances. Yet in 1999, a reanalysis of the CERN data confirmed the existence of the 2.02 and 2.2 $GeV/c^2$ resonances. Presently, the only well-known particle that decays to proton-antiproton is the $J/\psi$ particle, with a mass of 3.097 $GeV/c^2$ [4]. Most of the past experiments involved proton-antiproton scattering or pion production. Recently Jefferson Laboratory has provided the first look at the proton-antiproton system through photoproduction. In 2001 JLAB E01-017, observed nearly 20,000 $p\bar{p}$ events, and it is this data that is the main focus of the present work.

CP892, *Quark Confinement and the Hadron Spectrum VII*
edited by J. E. F. T. Ribeiro

# Experimental Results

The data were obtained using a photon beam incident on a liquid hydrogen target in the CEBAF Large Angle Spectrometer (CLAS). The particle detection system consists of drift chambers to determine the trajectories of charged particles, gas Čerenkov detectors for particle identification, scintillation counters for measuring time-of-flight (TOF) and particle identification, and electromagnetic calorimeters to detect neutral particles [5]. These detectors are designed to provide as much coverage of the $4\pi$ solid angle as possible.

For this experiment, the trigger required that a photon of an energy between 4.8 and 5.5 GeV be detected, it required that at least two of the timing counters surrounding the target measure hits, and it required that at least two of the six downstream TOF sectors measure hits.

After filtering events by particle identification, initial cuts were applied to the data set. These selections include beam energy, vertex position, and timing requirements. The photon energy was determined using the electron beam tagger. While the trigger required that a photon with an energy in the range of 4.8 to 5.5 GeV be identified, additional low energy photons could also be measured during the time window allowed to acquire the event. This would lead to an ambiguity in which photon beam particle was associated with the event measured in CLAS. Tight timing requirements as well as energy conservation cuts were used to take this ambiguity into account. Events were required to have a beam energy in the range of 4.8 to 5.5 GeV, and events below this energy are excluded from further analysis.

Nearly five thousand exclusive events were observed where all final state particles were identified in the CLAS spectrometer. However in CLAS, there are detector regions where particles can go unmeasured. For example, the CLAS toroidal magnetic field bends negatively charged particles back toward the beam. Quite often, these particles end up going back into the beam-line, and are lost. To increase the exclusive data yield, the anti-proton was allowed to be identified via the missing mass.

a prominent peak at a mass squared of $0.880 (GeV/c^2)^2$, which is consistent with a missing antiproton. Selecting the events consistent with a missing antiproton $(0.85(GeV/c^2)^2 \leq MM^2 \leq 0.91(GeV/c^2)^2)$ yields approximately 17,100 $\gamma p \rightarrow pp(\bar{p})$ events. Yet not all of these events are $\gamma p \rightarrow pp\bar{p}$ events as seen by the nearly flat background of non-antiprotons in the missing-mass squared distribution. Efforts are underway to further clean up and understand the background events under the antiproton signal.

Possible production mechanisms which describe the photoproduction of a proton-antiproton pair are diffraction/meson exchange, baryon exchange, and antibaryon exchange. In each process, an intermediate resonance may be produced. In meson exchange the photon transfers very little momentum to the target, but interacts, causing the photon to produce a resonance that decays to a fast forward-going proton-antiproton pair. In baryon exchange, the photon interacts with an exchange baryon converting it to a fast forward-going proton leaving behind a slow moving resonance at the target vertex which decays to a proton-antiproton pair. For antibaryon exchange, the photon interacts with an exchange antibaryon converting it to a fast forward-going antiproton, leaving behind a resonance at the target vertex which decays to two protons.

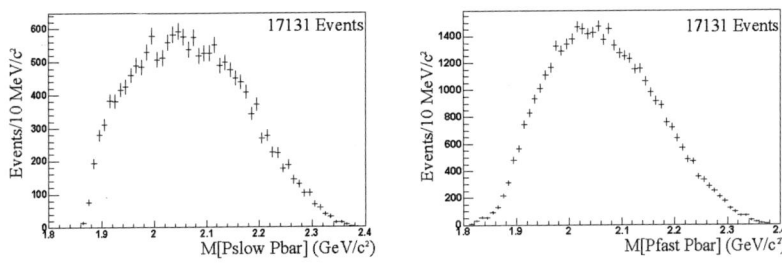

**FIGURE 1.** The invariant mass of the slow proton with the antiproton(left) and the invariant mass of the fast proton with the antiproton(right).

The distinction of meson exchange and baryon exchange production is clouded by the two identical protons. Without information identifying which is which, the two mechanisms are nearly indistinguishable. For antibaryon exchange it does not matter since both protons are at the same decay vertex. No obvious peaks or features are observed in the two proton invariant mass .

We label the protons by sorting on momentum. The proton with the greatest magnitude of momentum is defined as the fast proton, and the other proton is defined as the slow proton.

The invariant mass of $p_{slow}\bar{p}$ is shown in Fig. 1. The distribution has some statistical fluctuation, with a sharp rise at threshold and a possible narrow peak or dip near 2.0 GeV and broader peak at 2.04 GeV. In the invariant mass distribution of $p_{fast}\bar{p}$ there are no obvious structures.

In both distributions, no narrow resonant peaks are obvious in the proton antiproton invariant mass distributions. Preliminary results for the upper limit of the claimed resonance at 2.02 GeV is placed at 0.3 nb. Our results contradict results from CERN and DESY. Monte Carlo simulations show that the data is well described by a mixture of 75% meson exchange (exponential slope of 3.0 (GeV)$^{-2}$) and 25% baryon exchange (exponential slope of 0.9 (GeV)$^{-2}$). A moments analysis of the decay angular distributions show that the data is isotropic in nature. Since the final proton antiproton polarizations are not measured, both $J^{PC} = 0^{-+}$ and $J^{PC} = 1^{--}$ are consistent with the $J^{PC}$ of the produced proton antiproton pairs.

## REFERENCES

1. M.N. Focacci *et. al.*, Phys. Rev. Lett. <u>17</u>, 890(1966); D. Cline *et. al.*, Phys. Rev. Lett. <u>17</u>, 1268(1968).
2. A.S, Carroll *et. al.*, Phys. Rev. Lett. <u>32</u>, 247(1974). T.E. Kalogeropoulos and G.S. Tzanakos Phys. Rev. Lett. <u>34</u>, 1047(1975) V. Chaloupka *et. al.*, Phys. Lett. <u>61 B</u>, 487(1976). P. Benkheiri *et. al.*, Phys. Lett. <u>68 B</u>, 483(1977).
3. J. Bodenkamp *et. al.*, Phys. Lett. <u>133 B</u>, 275(1983). B.G. Gibbard *et. al.*, Phys. Rev. Lett. <u>42</u>, 1593(1979). R. Bizzarri *et. al.*, Phys. Rev. D <u>6</u>, 160(1972).
4. J. Bensinger *et. al.*, Phys. Rev. D <u>23</u>, 1417(1983). M.W. Eaton *et. al.*, Phys. Rev. D <u>29</u>, 805(1984).
5. B.A. Mecking *et. al.*, NIM <u>A503</u>, 513(2003).

# Dispersive approach in Sudakov resummation

Georges Grunberg

Centre de Physique Théorique, École polytechnique, CNRS,
91128 Palaiseau, France
E-mail: grunberg@cpht.polytechnique.fr

**Abstract.** The dispersive approach to power corrections is given a precise implementation, valid beyond single gluon exchange, in the framework of Sudakov resummation for deep inelastic scattering and the Drell-Yan process. It is shown that the assumption of infrared finite Sudakov effective couplings implies the universality of the corresponding infrared fixed points. This property is closely tied to the universality of the virtual contributions to space-like and time-like processes, encapsulated in the second logarithmic derivative of the quark form factor.

The infrared (IR) finite coupling ("dispersive") approach to power corrections [1] provides an attractive framework where the issue of universality can be meaningfully raised. This approach however seems to be tied in an essential way to the single gluon exchange approximation. In this talk I show that it can actually find a precise implementation in the framework of Sudakov resummation, and that its validity extends beyond single gluon exchange.

Consider first the scaling violation in deep inelastic scattering (DIS) in Mellin space at large $N$. One can show [2] that Sudakov resummation takes in this case the very simple form

$$\frac{d\ln F_2(Q^2, N)}{d\ln Q^2} = 4C_F \int_0^{Q^2} \frac{dk^2}{k^2} G(Nk^2/Q^2) A_{\mathcal{S}}(k^2) + 4C_F H(Q^2) + \mathcal{O}(1/N), \quad (1)$$

where the "Sudakov effective coupling" $A_{\mathcal{S}}(k^2)$, as well as $H(Q^2)$, are given as power series in $\alpha_s$ with $N$-independent coefficients. In the standard resummation framework one has $A_{\mathcal{S}}(k^2) = A_{\mathcal{S}}^{stan}(k^2)$ with $4C_F A_{\mathcal{S}}^{stan}(k^2) = A(\alpha_s(k^2)) + dB(\alpha_s(k^2))/d\ln k^2$ where $A$ (the universal "cusp" anomalous dimension) and $B$ are the standard Sudakov anomalous dimensions relevant to DIS, and $G(Nk^2/Q^2) = G_{stan}(Nk^2/Q^2) \equiv \exp(-Nk^2/Q^2) - 1$. It was further observed in [2] that the separation between the constant terms contained in the Sudakov integral on the right hand side of eq.(1) and the "leftover" constant terms contained in $H(Q^2)$ is arbitrary, yielding a variety of Sudakov resummation procedures, different choices leading to a different "Sudakov distribution function" $G(Nk^2/Q^2)$ and effective coupling $A_{\mathcal{S}}(k^2)$, as well as to a different function $H(Q^2)$. This freedom of selecting the constant terms actually disappears by taking one more derivative, namely

CP892, *Quark Confinement and the Hadron Spectrum VII*
edited by J. E. F. T. Ribeiro
© 2007 American Institute of Physics 978-0-7354-0396-3/07/$23.00

$$\frac{d^2 \ln F_2(Q^2, N)}{(d \ln Q^2)^2} = 4C_F \int_0^\infty \frac{dk^2}{k^2} \dot{G}(\frac{Nk^2}{Q^2}) A_S(k^2) + 4C_F [\frac{dH}{d \ln Q^2} - A_S(Q^2)] + \mathcal{O}(\frac{1}{N}),$$
(2)

where $\dot{G} = -dG/d \ln k^2$. The point is that the Sudakov integral $\mathcal{S}(Q^2, N)$ on the right hand side of eq.(2) being UV convergent, all the large $N$ logarithmic terms are now determined by the $\mathcal{O}(N^0)$ terms contained in the integral, which therefore cannot be fixed arbitrarily anymore. Thus $\mathcal{S}(Q^2, N)$ is uniquely determined. This observation implies in turn that the combination $dH/d \ln Q^2 - A_S(Q^2)$, which represents the "leftover" constant terms not included in $\mathcal{S}(Q^2, N)$, is also fixed. In fact, I conjecture that it is related to the space-like on-shell electromagnetic quark form factor [3] $\mathcal{F}_q(Q^2)$ by

$$4C_F \left( \frac{dH}{d \ln Q^2} - A_S(Q^2) \right) = \frac{d^2 \ln(\mathcal{F}_q(Q^2))^2}{(d \ln Q^2)^2}.$$
(3)

Eq.(3) has been checked [4] to $\mathcal{O}(\alpha_s^4)$. For the short distance Drell-Yan cross section, the analogue of (3) is $4C_F \left( \frac{dH_{DY}}{d \ln Q^2} - A_{S,DY}(Q^2) \right) = \frac{d^2 \ln |\mathcal{F}_q(-Q^2)|^2}{(d \ln Q^2)^2}$ where $\mathcal{F}_q(-Q^2)$ is the time-like quark form factor.

However, although $\mathcal{S}(Q^2, N)$ is uniquely determined, the Sudakov distribution function and effective coupling are still *not*. We deal with an infinite variety of different representations of $\mathcal{S}(Q^2, N)$, all equivalent for the purpose of resumming Sudakov logarithms. A prescription to single out the correct representation relevant for the issue of power corrections in the IR finite coupling approach is needed. In absence of the appropriate criterion, predictions such as existence of an $\mathcal{O}(1/Q)$ linear power correction [5] in Drell-Yan, or logarithmically-enhanced power corrections [2], which follow from particular choices of $G$, cannot be a priori dismissed. In the last paper of [2], it was suggested to select the correct $G$ by requiring the corresponding $A_S$ to be identical at large $N_f$ to the *Minkowskian* coupling defined as the time-like (integrated) discontinuity of the *Euclidean* one-loop coupling (the so-called "V-scheme" coupling) associated to the dressed gluon propagator: $A_{\mathcal{S},\infty}^{Mink}(k^2) = \frac{1}{\beta_0} \left[ \frac{1}{2} - \frac{1}{\pi} \arctan(t/\pi) \right]$ with $t = \ln(k^2/\Lambda_V^2)$ (where $\Lambda_V$ is the V-scheme scale parameter). As shown in [2], this ansatz fixes the corresponding "Minkowskian" Sudakov distribution function (which one could also call "characteristic function" following [1]) to be given in the DIS case by $G_{Mink}(Nk^2/Q^2) = \ddot{\mathcal{G}}(\epsilon)$, with $\epsilon = Nk^2/Q^2$. The function $\ddot{\mathcal{G}}(\epsilon)$ is obtained from the finite $N$ characteristic function [1] $\mathcal{F}(\lambda^2/Q^2, N)$ (where $\lambda$ is the "gluon mass") by defining $\mathcal{G}(y, N) \equiv \mathcal{F}(\lambda^2/Q^2, N)$ with $y \equiv N\lambda^2/Q^2$, and taking the $N \to \infty$ limit at *fixed* $y$: $\ddot{\mathcal{G}}(y, N) \to \ddot{\mathcal{G}}(y, \infty) \equiv \ddot{\mathcal{G}}(y)$ (where $\ddot{\mathcal{G}} = -d\mathcal{G}/d \ln Q^2$). In the Drell-Yan case, the same requirement yields instead $G_{Mink}^{DY}(Nk/Q) = \ddot{\mathcal{G}}_{DY}(\epsilon_{DY})$, with $\epsilon_{DY} = Nk/Q$. Similarly, $\ddot{\mathcal{G}}_{DY}(y_{DY})$ is obtained by taking the large $N$ limit at fixed $y_{DY} = N^2\lambda^2/Q^2$ of the finite $N$ characteristic function [1] $\mathcal{F}_{DY}(\lambda^2/Q^2, N)$. The same Sudakov distribution function $G_{Mink}^{DY}(Nk/Q)$ also follows from the resummation formalism (not tied to the single gluon approximation) of [6], which therefore uses an implicitly Minkowskian framework in the above sense.

Since they are $N_f$-independent, the *same* Minkowskian Sudakov distribution functions, now fixed through the large $N_f$ identification of the Sudakov effective couplings, can then be used to determine the corresponding effective couplings at *finite* $N_f$ in the usual way, requiring the large $N$ logarithmic terms on the left hand side of eq.(1) (with $G = G_{Mink}$) to be correctly reproduced order by order in perturbation theory. This proposal is equivalent to generalize the basic equation of the dispersive approach [1] to *all orders* in $\alpha_s$ in the large $N$ limit to the statement that $\frac{d\ln F_2(Q^2, N)}{d\ln Q^2} = 4C_F \int_0^\infty \frac{d\lambda^2}{\lambda^2} \ddot{\mathcal{F}}(\frac{\lambda^2}{Q^2}, N) A_{\mathcal{S}}^{Mink}(\lambda^2) + 4C_F \Delta H_{Mink}(Q^2) + \mathcal{O}(\frac{1}{N})$ (a "left-over" $\Delta H_{Mink}(Q^2)$ contribution is still expected at finite $N_f$). It is natural to keep referring to the resulting $A_{\mathcal{S}}^{Mink}(\lambda^2)$ coupling as Minkowskian even at *finite* $N_f$, where identification to a dressed gluon propagator is no longer possible. In the Minkowskian formalism, only *non-analytic* terms in the small "gluon mass" expansion of the characteristic function do contribute [1] to the IR power corrections through their discontinuities, and power corrections are given by low-energy moments of the corresponding [2] Euclidean coupling $A_{\mathcal{S}}^{Eucl}(k^2)$. Thus consistency with IR renormalons expectations is guaranteed.

Universality issues: the IR finite coupling approach allows some statement on the universality of power corrections to various processes. Indeed, at large $N_f$ there is universality to all orders in $\alpha_s$ between $A_{\mathcal{S}}^{Eucl}(k^2)$ and $A_{\mathcal{S},DY}^{Eucl}(k^2)$, which in the present framework are both prescribed to be equal to the one-loop V-scheme coupling in this limit. At finite $N_f$ however it easy to check that universality in the ultraviolet region holds only up to next to leading order in $\alpha_s$, where the DIS and Drell-Yan Euclidean Sudakov effective couplings actually coincide (up to a $1/4C_F$ factor) with the "cusp" anomalous dimension $A(k^2)$, but is lost beyond that order. On the other hand, an interesting universality property holds in the IR region at finite $N_f$, *assuming* the Sudakov effective couplings reach non-trivial IR fixed points at zero momentum. Indeed, eq.(3) shows that for a given selection of "left-over" constant terms (contained in the renormalization group invariant function $H(Q^2)$), the corresponding Sudakov effective coupling $A_{\mathcal{S}}(Q^2)$ differs from the *universal* quantity (for space-like processes) $A_{\mathcal{S}}^{all}(Q^2) \equiv -\frac{1}{4C_F} \frac{d^2 \ln(\mathcal{F}_q(Q^2))^2}{(d\ln Q^2)^2}$ only by the total derivative $dH/d\ln Q^2$. The latter is again expected to vanish at zero momentum if one assumes $H(Q^2)$ also reaches a non-trivial IR fixed point $H(0)$. A similar argument applies to $A_{\mathcal{S},DY}(Q^2)$, with $A_{\mathcal{S}}^{all}(Q^2)$ replaced by its time-like counterpart $Re[A_{\mathcal{S}}^{all}(-Q^2)] = -\frac{1}{4C_F} \frac{d^2 \ln |\mathcal{F}_q(-Q^2)|^2}{(d\ln Q^2)^2}$. Thus we expect, for *any* resummation procedure (assuming the IR fixed points exist) $A_{\mathcal{S}}(0) = A_{\mathcal{S},DY}(0) = A_{\mathcal{S}}^{all}(0)$.

# REFERENCES

1. Yu.L. Dokshitzer, G. Marchesini and B.R. Webber, *Nucl. Phys.* **B469** (1996) 93, and references therein; Yu.L. Dokshitzer and B.R. Webber, Phys.Lett. **B404** (1997) 321.
2. G. Grunberg, hep-ph/0601140; *Phys.Rev.* **D73** (2006) 091901(R); hep-ph/0609309.
3. S. Moch, J.A.M. Vermaseren and A. Vogt, *JHEP* **0508** (2005) 049, and references therein.
4. S. Friot and G. Grunberg, in preparation.
5. G.P. Korchemsky and G. Sterman, *Nucl. Phys.* **B437** (1995) 415.
6. E. Laenen, G. Sterman and W. Vogelsang, hep-ph/0010183; *Phys. Rev.* **D63** (2001) 114018.

# Stable Multiquark Interactions

B. Hiller*, A.A. Osipov*,†, A.H. Blin* and J. da Providência*

*Centro de Física Teórica, Departamento de Física da Universidade de Coimbra, 3004-516
Coimbra, Portugal
†Joint Institute of Nuclear Research, Laboratory of Nuclear Problems, 141980 Dubna, Moscow
Region, Russia

**Abstract.** The necessity of adding higher order multiquark interactions to the three flavor NJL model with $U_A(1)$ breaking, in order to stabilize its vacuum, is discussed.

**Keywords:** Hadronic vacuum, stability, multiquark interactions, meson spectra
**PACS:** 12.39.Fe, 11.30Rd, 11.30Qc

It has recently been shown [1] that the $SU(3)_L \times SU(3)_R$ chiral symmetric Lagrangian with axial $U_A(1)$ breaking composed of the $4q$ ($q$ =quark) Nambu-Jona-Lasinio (NJL) [2] and $2N_f = 6q$ determinantal 't Hooft [4] terms, for the u,d,s quarks, has no stable ground state. Global stability is implemented [3] by the addition of chiral invariant and OZI violating eight quark interactions. The multiquark picture is present in several approaches to low energy hadron dynamics: i) The instanton vacuum provides evidence in favour of $2N_f$-quark interactions (in the zero mode approximation). In leading $1/N_c$ order they are given by the 't Hooft determinant [4], which breaks the axial $U_A(1)$ symmetry and is a source of OZI-violating effects. ii) Non zero modes play an important role as well [5] , to comply with the Banks and Casher result for chiral symmetry breaking [6]. The effective quark Lagrangian derived from the instanton gas model, considered beyond the zero mode approximation, predicts $4q, 6q, \ldots, 2nq, \ldots$ interactions, all equally weighted at large $N_c$ [7]. iii) Lattice results for gluon correlators [8] reveal a hierarchy with dominance of the lowest ones; they could trigger a similar hierarchy in terms of the multiquark interactions, after integrating out the gluonic degrees of freedom. Within our model the hierarchy in multiquark interactions is made possible in the large $N_c$ counting scheme. The minimal constellations which support the symmetry principles of low energy QCD and have a globally stable vacuum involve $4q$, $6q$ and $8q$ interactions. We suppose that these interactions are localized in the interval $\Lambda_{conf} < \Lambda < \Lambda_{\chi SB}$ of confining and chiral symmetry breaking scales. The effects of multiquark terms on the vacuum are easily illustrated by calculating the scalar effective potential in the $SU(3)$ chiral limit (the general case $m_u \neq m_d \neq m_s$ is derived in [3]). Results are displayed in fig. 1, as function of the order parameter for chiral symmetry breaking. The upper panel represents the Wigner-Weyl, the lower panel the phase of spontaneous chiral symmetry breaking, the corresponding curvature of the potential at the origin is not altered by inclusion of higher multiquark interactions, see details in caption and [9]. Spontaneous breakdown of chiral symmetry can also occur superimposed to the Wigner-Weyl phase, induced by the 't Hooft interactions (figure at far right).

CP892, *Quark Confinement and the Hadron Spectrum VII*
edited by J. E. F. T. Ribeiro
© 2007 American Institute of Physics 978-0-7354-0396-3/07/$23.00

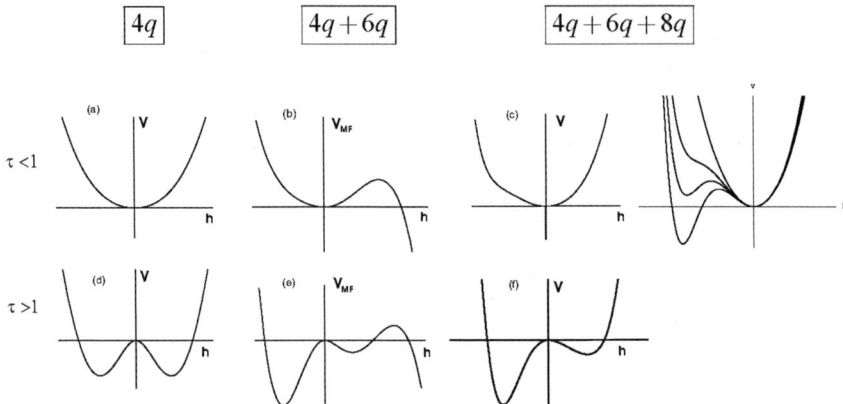

**FIGURE 1.** Effective potential V in the $SU(3)$ limit, calculated in the stationary phase approximation (SPA), $h \propto$ quark condensate, $\tau \propto$ curvature of V at origin. Each panel shows the typical form of the potential when one adds successively to the $4q$ case (a,d) the $6q$ (b,e) and $8q$ (c,f) terms. In (b,d) it is shown that the 't Hooft interaction $\sim 6q$ renders the NJL vacuum of (a,d) metastable in the mean field approximation, $V_{MF}$, (SPA leads to a vacuum without any local minimum; $V_{MF} = V$ only for stable configurations). Global stabilization is achieved by adding the $8q$ terms, figs. (c,f). Upper far right:a closer view of fig. 1c. A further mechanism of chiral symmetry breaking is at play at some critical value of $\kappa$. This phase coexists with the Wigner-Weyl phase.

The $N_c$ counting rules for the different couplings of the model are $G \sim 1/N_c$, $\kappa \sim 1/N_c^3$, $1/N_c^5 \leq g_1 \leq 1/N_c^4$, related with the $4q, 6q$ and $8q$ interactions respectively.

We now address the effects of the $8q$ forces on the low lying spin zero meson mass spectra. Here we discuss the trends, for numerical results see [9]. The effect on pseudoscalars is small and observed in the $\eta - \eta'$ mass splitting,

$$m_\eta^2 = m_0^2 - \frac{8(m_K^2 - m_\pi^2)^2 + 3c_q}{9(m_{\eta'}^2 - m_0^2)}, \tag{1}$$

where $m_0^2 = \frac{1}{3}(4m_K^2 - m_\pi^2)$ is the Gell-Mann–Okubo result for the $\eta$-mass. Numerically $m_0 = 565\,\text{MeV}$ is just a bit larger than the phenomenological value $m_\eta = 547.30 \pm 0.12\,\text{MeV}$. The remainder originates in the repulsion of $\eta$ and $\eta'$ and is a $SU(3)$ breaking effect of second order. The Witten-Veneziano correction, i.e., the second term $\sim (m_K^2 - m_\pi^2)^2$, is related to the topological susceptibility and is about four times larger than required [10], leading to a too low value for the $\eta$ mass. The coefficient $c_q$ extends the Veneziano result by including corrections from 't Hooft and $8q$ interactions [9]. They yield an additional small correction $\sim 2\%$ to the Witten-Veneziano term. In the large $N_c$ limit the $\eta, \eta'$ masses coincide with the result of Veneziano and Witten and $8q$ effects are absent in the leading order result for the topological susceptibility, although formally they could contribute [9],

$$\chi(0)|_{YM} = \frac{\kappa}{4} \left( \frac{M}{2G} \right)^3 \qquad (\text{large } N_c), \tag{2}$$

where $M$ stands for the constituint quark mass in the $SU(3)$ limit.

In the scalar sector we get the following hierarchy within the nonet ($f_0^-$, $f_0^+$ stand for the singlet-octet mixed states) :

$$m_{f_0^-} < m_{a_0} < m_{K_0^*} < m_{f_0^+}$$

which can be understood already at leading order of $N_c$, [9]. So within the model restrictions (no confinement, no renormalizability) the scalar mesons built from one-loop quark-antiquark interactions do not conform with the empirical ordering of their masses. If the "ab initio" calculated $l_i$ of the model reproduce the values obtained in a recent investigation within unitarized CHPT, in which an imposed large $N_c$ limit helps to explore the quark-antiquark versus a more intricate meson structure [11], the present Lagrangian yields further evidence in favour of a more complex structure.

A significant effect of $8q$ interactions is present in the low lying $\sigma$ meson ($f_0^-$), seen from the approximate sum rule [12] in a large $N_c$ estimate

$$m_{\eta'}^2 + m_{\eta}^2 - 2m_K^2 + m_{f_0^+}^2 + m_{f_0^-}^2 - 2m_{K_0^*}^2 = -6E_1^{LO} + \mathcal{O}\left(\frac{1}{N_c^2}\right). \tag{3}$$

The $E_1$ term on the RHS has its origin in an $8q$ interaction term stemming from the mass relations $f_0^{-,+}$, and contributes at $1/N_c$ order, if $g_1 \sim 1/N_c^4$. This term has a negative sign, decreasing the sum $m_{f_0^-}^2 + m_{f_0^+}^2$. With increasing $g_1$ mainly the value of $m_{f_0^-}$ is lowered [9] and the octet-singlet splitting grows in the scalar nonet. This sum rule is a good illustration of the possible impact of the eight-quark OZI violating forces on the scalar mesons.

Research supported by FCT, Unidade I&D 535, POCI/FP/63930/2005 and the EU integrated infrastructure initiative Hadron Physics project No.RII3-CT-2004-506078. A.A. Osipov gratefully acknowledges Fundação Calouste Gulbenkian for financial support.

## REFERENCES

1. A.A. Osipov, B. Hiller, V. Bernard, A.H. Blin, *Annals of Physics (N.Y.)* **321**, pp. 2504–2534 (2006), [ArXiv:hep-ph/0507226]
2. Y.Nambu, G. Jona-Lasinio, *Physical Review* **122**, pp. 345–358 (1961).
3. A.A. Osipov, B. Hiller, J. da Providência, *Physics Letters B* **634**, pp. 48–54 (2006) [ArXiv:hep-ph/0508058].
4. G.'t Hooft, *Physical Review* D**14**, pp. 3432–3450 (1976).
5. B.O. Kerbikov, D.S. Kuzmenko, Yu.A. Simonov, *JETP Letters* **65**, pp. 128 (1997), [ArXiv:hep-ph/9609440].
6. T. Banks, A. Casher, *Nuclear Physics* B**169**, pp. 103–125 (1980).
7. Yu. A. Simonov, *Physical Review* D**65**, pp. 094018 1–10 (2002), [ArXiv:hep-ph/0201170].
8. G.S. Bali, *Physics Reports* **343**, pp. 1–136 (2001), [ArXiv:hep-ph/0001312].
9. A.A. Osipov, B. Hiller, A.H. Blin, J. da Providência, *Annals of Physics (N.Y.)* **in print**,(2006), [ArXiv:hep-ph/0607066].
10. E. Witten, *Nuclear Physics* B**156**, pp. 269–283 (1979); G. Veneziano, *Nuclear Physics* B**159**, pp. 213–224 (1979).
11. J. Pelaez, these proceedings.
12. V. Dmitrasinović, *Physical Review* C**53**, pp. 1383–1396 (1996).

# Impact of Four-Quark Condensates on In-Medium Effects of Hadrons[1]

R. Thomas*, T. Hilger†, S. Zschocke** and B. Kämpfer*,†

*Forschungszentrum Dresden-Rossendorf, PF 510119, 01314 Dresden, Germany
†Institut für Theoretische Physik, TU Dresden, 01062 Dresden, Germany
**TU Dresden, Lohrmann-Observatorium, 01062 Dresden, Germany

**Abstract.** Spectral properties of hadrons in nuclear matter are treated in the framework of QCD sum rules. The influence of the ambient strongly interacting medium is encoded in various condensates. Especially, the structure of different four-quark condensates and their density dependencies in light quark systems are exemplified for the $\omega$ meson and the nucleon.

**Keywords:** Medium modifications, Four-quark condensates, QCD sum rules
**PACS:** 24.85.+p, 12.38.Lg, 12.40.Yx

## INTRODUCTION

Strongly interacting matter, being subject of QCD, appears in different phases, depending on the temperature and the chemical potential, which characterize a thermalized system. Thereby, distinct regions of the corresponding QCD phase diagram can exhibit rich structures, especially in the areas of high temperature and density, where deconfinement is expected. However, also for moderate temperature and density in the hadronic phase one expects a change in properties of matter and its constituents, the hadrons. In photoproduction off nuclei (CB-TAPS [1], TAGX [2], LEPS [3], CLAS [4]) or by measurements of the dilepton channel in $C+C$ collisions (HADES [5]), $p+A$ reactions (KEK [6]) or heavy-ion collisions (e.g. CERES [7], NA60 [8], STAR [9]) the study of medium-induced modifications for light mesons is pursued. Since hadrons are confined composite objects of the fundamental degrees of freedom of QCD, excited from a ground state, it becomes possible to probe changes of the QCD vacuum at various conditions. A quantitative description of hadron properties can be linked, within the QCD sum rule approach, to expectation values of quark and/or gluon operators — the QCD condensates. We consider here the four-quark condensates and relate their density dependencies to spectral modifications of $\omega$ meson and nucleon in cold nuclear matter.

## FOUR-QUARK CONDENSATES IN QCD SUM RULES

QCD sum rules [10] match non-perturbative QCD condensates to hadronic quantities. Therefore each hadron considered is represented by an interpolating current $j_\mu$ built out of quarks and gluons. The correlation function

---

[1] The work is supported by BMBF and GSI.

CP892, *Quark Confinement and the Hadron Spectrum VII*
edited by J. E. F. T. Ribeiro
© 2007 American Institute of Physics 978-0-7354-0396-3/07/$23.00

$$\Pi_{\mu\nu}(q) = i \int d^4 x\, e^{iqx} \langle \Psi | T[j_\mu(x) j_\nu(0)] | \Psi \rangle \qquad (1)$$

is evaluated, on the one side for time-like momenta $q$, where the physical hadrons are realized, and on the other side in terms of quarks and gluons via an operator product expansion (OPE) for large space-like euclidian momenta. Both representations for $\Pi$ are related using subtracted dispersion relations. Thus, integrals over spectral densities on the hadronic side correspond to QCD condensates, which enter as expectation values of the OPE terms. The leading condensates are the chiral condensate $m_q \langle \bar{q} q \rangle$, a measure for chiral symmetry breaking, the gluon condensate $\langle \frac{\alpha_s}{\pi} G^2 \rangle$, related by the trace anomaly to a breaking of scale invariance, the mixed quark-gluon condensate $\langle \bar{q} g_s \sigma G q \rangle$, and condensates of mass dimension 6: the triple gluon condensate $\langle g_s^3 G^3 \rangle$ and four-quark condensates $\langle \bar{q} \Gamma q \bar{q} \Gamma' q \rangle$ ($\Gamma$ is symbolic for all possible structures specified below). The extension to finite temperatures and densities induces on the OPE side modification of the condensates. For cold nuclear matter, modelled in leading order as Fermi gas of non-interacting nucleons, this dependence $\langle \mathcal{O} \rangle = \langle \mathcal{O} \rangle_0 + \frac{n}{2M_N} \langle \mathcal{O} \rangle_N$ is linear in the baryon density $n$ and dictated by nucleon matrix elements $\langle \mathcal{O} \rangle_N \equiv \langle N | \mathcal{O} | N \rangle$.

**TABLE 1.** Left part: List of all four-quark condensates in vacuum for one flavor $q$ ($\Gamma \in \{1, \gamma_\alpha, \sigma_{\alpha\beta}, \gamma_5 \gamma_\alpha, \gamma_5\}$ are elements of the Clifford algebra, $\lambda^A$ Gell-Mann matrices). The columns correspond to two color singlet structures which can, for identical flavor, be transformed into each other by Fierz relations. Right part: The four-quark condensates which arise additionally in medium.

| | | | |
|---|---|---|---|
| $\langle \bar{q} q \bar{q} q \rangle$ | $\langle \bar{q} \lambda^A q \bar{q} \lambda^A q \rangle$ | $\langle \bar{q} \slashed{v} q \bar{q} \slashed{v} q / v^2 \rangle$ | $\langle \bar{q} \slashed{v} \lambda^A q \bar{q} \slashed{v} \lambda^A q / v^2 \rangle$ |
| $\langle \bar{q} \gamma_\alpha q \bar{q} \gamma^\alpha q \rangle$ | $\langle \bar{q} \gamma_\alpha \lambda^A q \bar{q} \gamma^\alpha \lambda^A q \rangle$ | $\langle \bar{q} \sigma_{\alpha\beta} v^\beta q \bar{q} \sigma^{\alpha\gamma} v_\gamma q / v^2 \rangle$ | $\langle \bar{q} \sigma_{\alpha\beta} v^\beta \lambda^A q \bar{q} \sigma^{\alpha\gamma} v_\gamma \lambda^A q / v^2 \rangle$ |
| $\langle \bar{q} \sigma_{\alpha\beta} q \bar{q} \sigma^{\alpha\beta} q \rangle$ | $\langle \bar{q} \sigma_{\alpha\beta} \lambda^A q \bar{q} \sigma^{\alpha\beta} \lambda^A q \rangle$ | $\langle \bar{q} \gamma_5 \slashed{v} q \bar{q} \gamma_5 \slashed{v} q / v^2 \rangle$ | $\langle \bar{q} \gamma_5 \slashed{v} \lambda^A q \bar{q} \gamma_5 \slashed{v} \lambda^A q / v^2 \rangle$ |
| $\langle \bar{q} \gamma_5 \gamma_\alpha q \bar{q} \gamma_5 \gamma^\alpha q \rangle$ | $\langle \bar{q} \gamma_5 \gamma_\alpha \lambda^A q \bar{q} \gamma_5 \gamma^\alpha \lambda^A q \rangle$ | $\langle \bar{q} \slashed{v} q \bar{q} q \rangle$ | $\langle \bar{q} \slashed{v} \lambda^A q \bar{q} \lambda^A q \rangle$ |
| $\langle \bar{q} \gamma_5 q \bar{q} \gamma_5 q \rangle$ | $\langle \bar{q} \gamma_5 \lambda^A q \bar{q} \gamma_5 \lambda^A q \rangle$ | $\langle \bar{q} \gamma_5 \gamma_\alpha q \bar{q} \gamma_5 \sigma^{\alpha\beta} v_\beta q \rangle$ | $\langle \bar{q} \gamma_5 \gamma_\alpha \lambda^A q \bar{q} \gamma_5 \sigma^{\alpha\beta} v_\beta \lambda^A q \rangle$ |

For the following discussion it is important to distinguish the different four-quark condensate structures. A list of all possible four-quark condensates follows from the demanded Lorentz invariance, invariance with respect to parity and time reversal and symmetry under $SU(3)_{color}$ of the normal ordered expectation values $\langle \Psi | : \bar{q} \Gamma q \bar{q} \Gamma' q : | \Psi \rangle$. Flavor symmetry will further reduce the number of independent four-quark condensates. The left part of Tab. 1 shows these condensates for $n_f = 1$ flavors in vacuum. In a medium with four-velocity $v_\mu$ they become density dependent and additional Lorentz scalars (right part of Tab. 1) can be formed giving rise to additional structures. In each part of Tab. 1 the two possible color singlets are listed which transform into each other by Fierz relations in the case of identical flavors presented here. Note that this four-quark condensate catalog increases when considered for more flavor degrees of freedom, also because flavor mixing terms are realized.

In particular sum rule calculations, distinct linear combinations of four-quark condensates occur and their relevance will be case-specific. This prevents from an extraction of a particular four-quark condensate although the QCD condensates are generic for the whole hadronic spectrum. For quantitative evaluations the four-quark condensates, even in vacuum not well-known, are often expressed by the squared chiral condensate, i.e. a factorization ansatz motivated by large-$N_c$ arguments [11]. In linearized form w.r.t.

the density, $\langle \bar{q}q \rangle^2$ provides a first guess for the four-quark condensates and their density dependencies. Deviations of this ansatz are parameterized as $\langle \bar{q}_{f_1} \Gamma_1 q_{f_1} \bar{q}_{f_2} \Gamma_2 q_{f_2} \rangle = A\kappa^{vac} + B\kappa^{med}n$, where $A$, $B$ are individual constants for each structure $\Gamma_{1,2}$ tabulated (and generalized to flavors $f_{1,2}$; $B$ contains also the density dependence of $\langle \bar{q}q \rangle$) and $\kappa^{med}$ allows different assumptions for the behavior at finite density (e.g. $\kappa^{med} = 1$ is the factorization limit, $\kappa^{med} = 0$ means no density dependence at all).

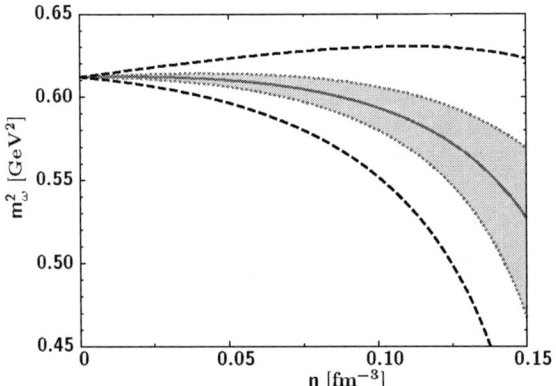

**FIGURE 1.** The $\omega$ meson mass parameter $m_\omega^2$ as a function of the baryon density $n$ from a full QCD sum rule evaluation [12] for $\kappa^{med} = 4$ and condensates up to mass dimension 6 (solid curve). Dimension 8 condensates are globally accumulated in a coefficient $c_4 = c_4^{(0)} + c_4^{(1)}n$ [12]. The effect of such a term is exhibited by the shaded area for $|c_4^{(1)}| \leq 5 \times 10^{-5} n_0^{-1}$ GeV$^8$; the inclusion of $c_4^{(0)} \neq 0$ requires a readjustment of $\kappa_\omega^{vac}$ to recover the vacuum value of $m_\omega^2$. The upper (lower) dashed curve is for $\kappa_\omega^{med} = 3.5$ ($\kappa_\omega^{med} = 4.5$).

*$\bar{q}q$-sector: $\omega$ Meson.* In the photoproduction off $Nb$ the CB-TAPS collaboration has observed additional strength of the $\omega$ meson at lower invariant mass compared to a test reaction with a $LH_2$ (proton) target [1]; this is interpreted as experimental evidence for a medium-modified spectral distribution of the $\omega$ meson. One may use this finding to constrain the density dependence of a linear combination of particular four-quark condensates, $\langle \bar{u}\gamma^\mu \lambda_A u d\gamma_\mu \lambda_A d \rangle$, $\langle \bar{u}\gamma_5 \gamma^\mu \lambda_A u d\gamma_5 \gamma_\mu \lambda_A d \rangle$, $\langle \bar{q}\gamma^\mu \lambda_A q \bar{q}\gamma_\mu \lambda_A q \rangle$ and $\langle \bar{q}\gamma_5 \gamma^\mu \lambda_A q \bar{q}\gamma_5 \gamma_\mu \lambda_A q \rangle$ ($q \equiv u,d$). For the center of the $\omega$ spectral distribution not increasing, as indicated by the data [1], one concludes a strong decrease of the special four-quark condensate combination entering the QCD sum rule for the $\omega$ meson in the coefficient $c_3 = c_3^{(0)}(\kappa_\omega^{vac}) + c_3^{(1)}(\kappa_\omega^{med})n$: $\kappa_\omega^{med} \gtrsim 4$. The limiting situation with a constant spectral moment $m_\omega^2$ at small $n$ is depicted in Fig. 1.

*$qqq$-sector: Nucleon.* The situation becomes more involved for the nucleon, where 3 connected sum rule equations arise [13], each including another four-quark condensate combination which requires the three independent parameters $\kappa_s^{med}$, $\kappa_q^{med}$ and $\kappa_v^{med}$ governing their density dependencies. Fig. 2 exhibits qualitatively their expected behavior reproduced in a sum rule evaluation using $\kappa^{med}$ values derived from a perturbative chiral quark model [14]. Especially $\kappa_q^{med} = 0$, which corresponds to a combination remaining constant while density increases, signals a deviation from the factorization ansatz.

The color structure of four-quark condensates in baryon sum rules, a mixture of both columns in Tab. 1, prevents their combinations to be equated with those in the $\omega$ sum rule.

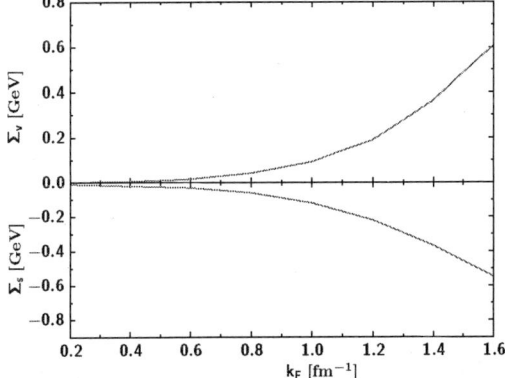

**FIGURE 2.** The scalar ($\Sigma_s$) and vector ($\Sigma_v$) self-energies of the nucleon with momentum $p = k_F$ at density $n(k_F)$ as functions of the Fermi momentum $k_F$ for the specific choice of density dependencies of the combined four-quark condensates: $\kappa_s^{med} = 1, \kappa_q^{med} = 0, \kappa_v^{med} = 0$. Note the similarity with advanced nuclear matter calculations [15].

## SUMMARY

The research in modified hadronic properties opens access to an understanding of the QCD ground state and its response to finite density and temperature. The description of these effects via condensates can constrain particular combinations, as in the case of the $\omega$ meson or the nucleon, where four-quark condensates differently combined from an extensive catalog of possible structures tend to have either strong or weak density dependencies, respectively. In systems containing heavy quarks, like $D$ mesons — envisaged within the CBM and PANDA projects at FAIR — conclusions for other dominating condensates could be drawn.

## REFERENCES

1.  D. Trnka, et al., *Phys. Rev. Lett.* **94**, 192303 (2005), nucl-ex/0504010.
2.  G. M. Huber, et al., *Phys. Rev.* **C68**, 065202 (2003), nucl-ex/0310011.
3.  T. Ishikawa, et al., *Phys. Lett.* **B608**, 215–222 (2005), nucl-ex/0411016.
4.  D. P. Weygand, et al., JLAB proposal PR-06-016.
5.  P. Salabura, et al., *Nucl. Phys.* **A749**, 150–159 (2005). G. Agakichiev, et al., nucl-ex/0608031.
6.  M. Naruki, et al., *Phys. Rev. Lett.* **96**, 092301 (2006), nucl-ex/0504016.
7.  D. Adamova, et al., *Phys. Rev. Lett.* **91**, 042301 (2003), nucl-ex/0209024.
8.  S. Damjanovic, et al., *Nucl. Phys.* **A774**, 715–718 (2006), nucl-ex/0510044.
9.  J. Adams, et al., *Phys. Rev. Lett.* **92**, 092301 (2004), nucl-ex/0307023.
10. M. A. Shifman, A. I. Vainshtein, and V. I. Zakharov, *Nucl. Phys.* **B147**, 385–447 (1979).
11. S. Leupold, *Phys. Lett.* **B616**, 203–207 (2005), hep-ph/0502061.
12. R. Thomas, S. Zschocke, and B. Kämpfer, *Phys. Rev. Lett.* **95**, 232301 (2005), hep-ph/0510156.
13. R. J. Furnstahl, D. K. Griegel, and T. D. Cohen, *Phys. Rev.* **C46**, 1507–1527 (1992).
14. E. G. Drukarev, et al., *Phys. Rev.* **D68**, 054021 (2003), hep-ph/0306132.
15. O. Plohl, and C. Fuchs, *Phys. Rev.* **C74**, 034325 (2006), nucl-th/0607053.

# Exploring backward pion electroproduction in the scaling regime

J.P. Lansberg[a,b], B. Pire[a], L. Szymanowski[a,b,c]

[a] Centre de Physique Théorique, École polytechnique, CNRS, 91128 Palaiseau, France
[b] Physique théorique fondamentale, Université de Liège, B-4000 Liège 1, Belgium
[c] Soltan Institute for Nuclear Studies, Warsaw, Poland
E-mail: Jean-Philippe.Lansberg@cpht.polytechnique.fr

**Abstract.** We use general relations between the Transition Distribution Amplitudes (TDAs), entering the description of the $p \to \pi^0$ transition, and the proton Distribution Amplitudes (DAs) in the soft-pion limit to estimate the size of the amplitude for backward electroproduction of $\pi^0$ at large $Q^2$.

We have recently [1] shown that factorisation theorems [2] for exclusive processes apply to $\pi^- \pi^+ \to \gamma^\star \gamma$ in the kinematical regime where the virtual photon is highly virtual but at small $t$. We also advocated the extension of this approach to $P\bar{P} \to \gamma^\star \gamma$, to backward VCS $\gamma^\star P \to P'\gamma$ [3], to backward pion electroproduction $\gamma^\star P \to P'\pi$ and to $P\bar{P} \to \gamma^\star \pi$ in the near forward region and for large virtual $Q^2$, which may be studied in detail at GSI.

For the $\gamma^\star$ to $\rho$ transition, a perturbative limit of the TDA may be obtained [5]. For $\gamma \to \pi$ one, where there are only four leading-twist TDAs [1] related to $\langle \gamma | \bar{q}_\alpha(z_1 n) [z_1; z_0] q_\beta(z_0 n) | \pi \rangle$, where $[z_1; z_0]$ denotes the Wilson line we have recently shown [6] that experimental analysis of e.g. $\gamma^\star \gamma \to \rho\pi$ and $\gamma^\star \gamma \to \pi\pi$ could be carried out since the Bremsstrahlung contribution is small and rates are sizable at present $e^+ e^-$ facilities. Whereas in the pion case, models used for GPDs (see [7] and references therein) could be applied to TDAs, this is not obvious for baryonic ones, for which the soft limit considered here is therefore very interesting.

In Ref. [4], we have defined the leading-twist proton to pion $P \to \pi$ transition distribution amplitudes from the Fourier transform of the matrix element

$$\langle \pi | \epsilon^{ijk} q^i_\alpha(z_1 n) [z_1; z_0] q^j_\beta(z_2 n) [z_2; z_0] q^k_\gamma(z_3 n) [z_3; z_0] | P \rangle, \qquad (1)$$

We define here the leading-twist TDAs for the $P \to \pi^0$ transition at $\Delta_T = 0$ as[1]:

$$4\mathcal{F}\left( \langle \pi^0(p_\pi) | \epsilon^{ijk} u^i_\alpha(z_1 n)[z_1; z_0] u^j_\beta(z_2 n)[z_2; z_0] d^k_\gamma(z_3 n)[z_3; z_0] | P(p_1, s_1) \rangle \right) \qquad (2)$$

$$= i\frac{f_N}{f_\pi}\left[ V_1^{p\pi^0}(\not{p}C)_{\alpha\beta}(N^+)_\gamma + A_1^{p\pi^0}(\not{p}\gamma^5 C)_{\alpha\beta}(\gamma^5 N^+)_\gamma + T_1^{p\pi^0}(\sigma_{p\mu}C)_{\alpha\beta}(\gamma^\mu N^+)_\gamma \right],$$

---

[1] In the following, we shall use the notation $\mathcal{F} \equiv (p.n)^3 \int_{-\infty}^{\infty} dz_i e^{\Sigma_i x_i z_i p.n}$.

CP892, *Quark Confinement and the Hadron Spectrum VII*
edited by J. E. F. T. Ribeiro
© 2007 American Institute of Physics 978-0-7354-0396-3/07/$23.00

where $\sigma^{\mu\nu} = 1/2[\gamma^\mu, \gamma^\nu]$, $C$ is the charge conjugation matrix and $N^+$ is the large component of the nucleon spinor $(N = (\not p\not n + \not n\not p)N = N^- + N^-$ with $N^+ \sim \sqrt{p_1^+}$ and $N^- \sim \sqrt{1/p_1^+})$. $f_\pi$ is the pion decay constant $(f_\pi = 133$ MeV$)$ and $f_N$ has been estimated through QCD sum rules to be of order $5.2 \cdot 10^{-3}$ GeV$^2$ [8]. All the TDAs $V_i$, $A_i$ and $T_i$ are dimensionless.

Now, we shall derive the general limit of these three contributing TDAs at $\Delta_T = 0$ when $\xi$ gets close to 1. In that limit, the soft-meson theorems [9] derived from current algebra apply [10], which allow us to express these 3 TDAs in terms of the 3 Distribution Amplitudes (DAs) of the corresponding baryon. In the case of the proton DA [8], $V^p(x_i)$, $A^p(x_i)$, $T^p(x_i)$ are defined such as

$$4\mathcal{F}\Big(\langle 0|\epsilon_{ijk}u_\alpha^i(z_1 n)u_\beta^j(z_2 n)d_\gamma^k(z_3 n)|p(p,s)\rangle\Big) = f_N \times \qquad (3)$$

$$\Big[V^p(x_i)(\not p C)_{\alpha\beta}(\gamma^5 N^+)_\gamma + A^p(x_i)(\not p\gamma^5 C)_{\alpha\beta}N_\gamma^+ + T^p(x_i)(\sigma_{p\mu}C)_{\alpha\beta}(\gamma^\mu\gamma^5 N^+)_\gamma\Big].$$

We use the general soft pion theorem [9] to write:

$$\langle \pi^a(p_\pi)|\mathcal{O}|P(p_1, s_1)\rangle = -\frac{i}{f_\pi}\langle 0|[Q_5^a, \mathcal{O}]|P(p_1, s_1)\rangle + \text{ pole term} \qquad (4)$$

The second term, which takes care of the nucleon pole term, does not contribute at threshold and will not be considered in the following.

For the transition $P \to \pi^0$, $Q_5^a = Q_5^3$ and the flavour content of $\mathcal{O}$ is $u_\alpha u_\beta d_\gamma$. Since the commutator of the chiral charge $Q_5$ with the quark field $\psi$ ($\tau^a$ being the isospin matrix) is $[Q_5^a, \psi] = -\frac{\tau^a}{2}\gamma^5\psi$, the first term in the rhs of Eq. (4) gives three terms from $(\gamma^5 u)_\alpha u_\beta d_\gamma$, $u_\alpha(\gamma^5 u)_\beta d_\gamma$ and $u_\alpha u_\beta(\gamma^5 d)_\gamma$. The corresponding multiplication by $\gamma^5$ (or $(\gamma^5)^T$ when it acts on the index $\beta$) on the vector and axial structures of the DA (Eq. (3)) gives two terms which cancel each other and the third one, which remains, is the same as the one for the TDA, up to the modification that in the DA decomposition $p$ is the proton momentum, whereas for the TDA one, $p$ is the light-cone projection of $P \equiv (p_1 + p_\pi)/2$, i.e. half the proton momentum if one neglects $p_\pi$. This introduces a factor 2 in the relation between the DA $V^p$ ($V^p$) and the TDA $V_1^{p\pi^0}$ ($A_1^{p\pi^0}$), which cancels the factor $1/2$ from $[Q_5^a, \psi]$. To what concerns the tensorial structure multiplying $T^p$, the three terms are identical at leading-twist accuracy and yield a factor 3 in $T_1$.

We eventually have the soft limit[2] for our three TDAs at $\Delta_T = 0$:

$$V_1^{p\pi^0}(x_i, \xi, t) \to V^p\Big(\frac{x_i}{2}\Big), \quad A_1^{p\pi^0}(x_i, \xi, t) \to A^p\Big(\frac{x_i}{2}\Big), \quad T_1^{p\pi^0}(x_i, \xi, t) \to 3T^p\Big(\frac{x_i}{2}\Big). \qquad (5)$$

---

[2] The factor $\frac{1}{2}$ in the argument of the DA in Eq. (5) comes from the fact that for the TDAs, the $x_i$ are defined with respect to $p$ ( see e.g. $\mathcal{F} \equiv (p.n)^3 \int_{-\infty}^\infty dz_i e^{\Sigma_i x_i z_i p.n}$) and $p \to \frac{p_1}{2}$ when $\xi \to 1$. Therefore, they vary within the interval $[0:2]$, whereas for the DAs, the momentum fraction are defined with respect to the proton momentum $p_1$ and vary between 0 and 1.

At leading order in $\alpha_s$, the amplitude for $\gamma^*(q)P(p_1,s_1) \to P'(p_2,s_2)\pi^0(p_\pi)$ is

$$\mathcal{M}^\mu = -ieF^{p\pi^0}(Q^2,\xi,t)\bar{u}(p_2)\gamma^\mu\gamma^5 u(p_1), F^{p\pi^0} = \frac{Cf_N^2}{f_\pi Q^4} \int_{-1+\xi}^{1+\xi} d^3x \int_0^1 d^3y \sum_{\alpha=1}^{14} T'_\alpha(x_i,y_j), \quad (6)$$

to be compared with the leading amplitude for the baryonic form factor [8]

$$\mathcal{M}^\mu = -ieF_1^p(Q^2)\bar{u}(p_2)\gamma^\mu u(p_1), F_1^p = \frac{Cf_N^2}{Q^4} \int_0^1 d^3x \int_0^1 d^3y \sum_{\alpha=1}^{14} T_\alpha(x_i,y_j). \quad (7)$$

Considering, for now, only the contribution from the ERBL region $x_i > 0$, the integration between $-1+\xi$ and $1+\xi$ can be converted into one between 0 and 1 by a change of variable. Since the expressions of $T'_\alpha$ and $T_\alpha$ are identical up to the 3 replacements the initial-state DAs by the $P \to \pi^0$ TDAs, they would in fact differ only by the factor 3 in the last relation of Eq. (5) extrapolating the $\xi \to 1$ limit to the ERBL region,.

Due to this factor, whereas the asymptotic choice [8], 120 $x_1 x_2 x_3$, for the DAs gives a vanishing result for $F_1^p$ or $G_M^p$, the result is nonzero for $F^{p\pi^0}$. This lets us therefore hope that the onset of the dominance of the perturbative contribution to $\gamma^*P \to P'\pi^0$ with TDAs may happen at much lower $Q^2$ than for the proton form factor. Quantitative results to be compared with the measurement of [11] will be presented soon.

## ACKNOWLEDGMENTS

This work is partially supported by the scientific agreement Polonium, the Polish Grant 1 P03B 028 28, the EU program I3HP, contract RII3-CT-2004-506078 and the FNRS (Belgium). L.Sz. is a Visiting Fellow of the FNRS (Belgium).

## REFERENCES

1. B. Pire and L. Szymanowski, Phys. Rev. D **71** (2005) 111501 [arXiv:hep-ph/0411387].
2. J. C. Collins *et al.*, Phys. Rev. D **56**, 2982 (1997).
3. J. P. Lansberg, *et al.* , arXiv:hep-ph/0607130.
4. B. Pire and L. Szymanowski, Phys. Lett. B **622** (2005) 83 [arXiv:hep-ph/0504255].
5. B. Pire, *et al.* , Phys. Lett. B **639** (2006) 642 [arXiv:hep-ph/0605320].
6. J. P. Lansberg, *et al.* , Phys. Rev. D **73** (2006) 074014 [arXiv:hep-ph/0602195].
7. F. Bissey, *et al.* , Phys. Lett. B **587** (2004) 189 [arXiv:hep-ph/0310184].
8. V. L. Chernyak and A. R. Zhitnitsky, Phys. Rept. **112**, 173 (1984).
9. S. Adler and R. Dashen, *Current Algebras*, Benjamin, New York, 1968.
10. P. V. Pobylitsa, M. V. Polyakov and M. Strikman, Phys. Rev. Lett. **87**, 022001 (2001).
11. G. Laveissiere *et al.* [JLab Hall A Collaboration], Phys. Rev. C **69** (2004) 045203 [arXiv:nucl-ex/0308009].

# Chiral properties of the constituent quark model

Wolfgang Lucha[*], Dmitri Melikhov[*,†] and Silvano Simula [**]

[*]HEPHY, Austrian Academy of Sciences, Nikolsdorfergasse 18, A-1050 Vienna, Austria
[†]Institute of Nuclear Physics, Moscow State University, 119992, Moscow, Russia
[**]INFN, Sezione di Roma 3, Via della Vasca Navale 84, I-00146, Roma, Italy

**Abstract.** We show that, in a model based exclusively on constituent-quark degrees of freedom interacting via a potential, the full axial current is conserved *if* the spectrum of $\bar{Q}Q$ states contains a massless pseudoscalar. The current conservation emerges nonperturbatively if the model satisfies certain constraints on (i) the axial coupling $g_A$ of the constituent quark and (ii) the $\bar{Q}Q$ potential at large distances. We define the chiral point of the constituent quark model as that set of values of the parameters (such as the masses of the constituent quarks and the couplings in the $\bar{Q}Q$ potential) for which the mass of the lowest pseudoscalar $\bar{Q}Q$ bound state vanishes. At the chiral point the main signatures of the spontaneously broken chiral symmetry are shown to be present, namely: the axial current is conserved, the decay constants of the excited pseudoscalar bound states vanish, and the pion decay constant has a nonzero value.

**Keywords:** Spontaneously broken chiral symmetry, constituent quarks
**PACS:** 11.30.Rd,12.39.Ki,11.40.Ha

Chiral symmetry is a basic symmetry of massless QCD which, apart from the axial anomaly in the flavor-singlet channel, entails the conservation of the axial-vector current. The masses of the light $u$ and $d$ quarks are small compared to the confinement scale, and consequently the chiral limit serves as a good approximation for the light-quark sector of QCD. Chiral symmetry in QCD is spontaneously broken, and is thus not a symmetry of the hadron spectrum: except for the existence of the octet of light pseudoscalar mesons, the lowest-energy part of the hadron spectrum shows no trace of chiral symmetry.

Because of confinement, the calculation of the hadron mass spectrum directly from the QCD Lagrangian is a very challenging task, which requires a nonperturbative approach. QCD-inspired constituent quark models (i.e., models based on constituent-quark degrees of freedom in which mesons appear as $\bar{Q}Q$ bound states in a potential) proved to be quite successful for the description of the mass spectrum of hadrons and their interactions at low momentum transfers [1, 2, 3]. Because of the proper description of the hadron mass spectrum, the Lagrangian of the constituent quark model cannot be chirally invariant: it would produce a chirally invariant spectrum of hadron states. Consequently, the Noether axial current found in such models is not conserved but satisfies the divergence equation

$$\partial^\mu [\bar{Q}(x)\gamma_\mu \gamma_5 Q(x)] = 2m_Q \bar{Q}(x) i\gamma_5 Q(x). \tag{1}$$

In a recent paper [4] we have shown that, nevertheless, taking into account the infinite number of diagrams describing the $\bar{Q}Q$ interactions, leads to the full axial current of the constituent quarks, which turns out to have the structure

$$\langle 0|\bar{q}\gamma_\mu \gamma_5 q|\bar{Q}Q\rangle = g_A(p^2)\left\{ \bar{Q}\gamma_\mu \gamma_5 Q + 2m_Q \frac{p_\mu}{p^2}\bar{Q}\gamma_5 Q \right\} + g_A(p^2)\frac{p_\mu}{p^2}\bar{Q}\gamma_5 Q \frac{2m_Q O(M_\pi^2)}{p^2 - M_\pi^2}. \tag{2}$$

CP892, Quark Confinement and the Hadron Spectrum VII
edited by J. E. F. T. Ribeiro
© 2007 American Institute of Physics 978-0-7354-0396-3/07/$23.00

Obviously, the term in curly brackets is transverse by virtue of Eq. (1). Therefore, the full "constituent-quark axial current" is conserved if the mass $M_\pi$ of the pion, the lowest $Q\bar{Q}$ pseudoscalar bound state, vanishes. As has been demonstrated in Ref. [4], to guarantee the axial current conservation up to terms of order $O(M_\pi^2)$ requires that the axial coupling $g_A$ of the constituent quarks is not constant but that it is related to the pion wave function $\Psi_\pi(s)$ by

$$g_A(s) = \eta_A(s - M_\pi^2)\Psi_\pi(s) + O(M_\pi^2), \quad \eta_A = \text{const.} \tag{3}$$

It should be recalled here that the spontaneous breaking of chiral symmetry requires not only the conservation of the axial current, but also the nonvanishing of the coupling of a massive fermion (such as a nucleon or a constituent quark) to the pion, i.e., $g_A(s)$ should be nonzero at $s = M_\pi^2$. Consequently, to be compatible with the spontaneous breaking of chiral symmetry, the potential model should generate a light pseudoscalar bound state for which $\Psi_\pi(s = M_\pi^2)$ has a pole at $s = M_\pi^2$ [4].

In Ref. [4] it is shown that the behavior of $\Psi_\pi(s)$ at $s = M_\pi^2$ is related to the behavior of the potential of the $Q\bar{Q}$ interaction at large separations $r$. More precisely, $\Psi_\pi(s)$ exhibits a pole at $s = M_\pi^2$ only if the potential saturates at large $r$:

$$V(r \to \infty) = \text{const} < \infty. \tag{4}$$

In this case the nearly massless pion is a strongly bound $Q\bar{Q}$ state with binding energy $\varepsilon \simeq 2m$.

The observed conservation of the axial current allows us to define the *chiral point of the constituent quark model* as exactly that set of values of the parameters which leads to a massless lowest pseudoscalar $Q\bar{Q}$ bound state.[1] In the following, let us consider certain properties of the constituent quark model at the chiral point.

## Decay constants of pseudoscalar mesons

Making use of the relation (3) between the axial coupling of the constituent quark, $g_A(s)$, and the pion wave function, the standard quark-model expression for the decay constant of the $n$-th excitation of a pseudoscalar meson [3] takes the form [4]

$$f_P(n) = 2m_Q\eta_A\sqrt{N_c}\int ds\,\Psi_0(s)\Psi_n(s)\rho(s,m_Q^2,m_Q^2)\frac{s - M_\pi^2}{s} + O(M_\pi^2). \tag{5}$$

The wave functions $\Psi_n(s)$ of the pseudoscalar states satisfy the orthogonality condition

$$\int ds\,\Psi_n(s)\Psi_m(s)\rho(s,m_Q^2,m_Q^2) = \delta_{mn}. \tag{6}$$

For the ground state, $n = 0$, the decay constant $f_P(0) \equiv f_\pi$ is clearly finite in the chiral limit. With the help of Eq. (6), we obtain the relation [4]

$$f_\pi = 2m_Q\eta_A\sqrt{N_c} + O(M_\pi^2). \tag{7}$$

---

[1] Precisely, we consider a given potential (i.e., with fixed couplings) and adjust the constituent mass of the light quark until $M_\pi$ vanishes.

For excited states, $n \neq 0$, Eq. (5) implies [4], by virtue of the orthogonality condition (6),

$$f_P(n \neq 0) = -2m_Q \eta_A M_\pi^2 \int ds\, \Psi_0(s) \Psi_n(s) \frac{\rho(s, m_Q^2, m_Q^2)}{s} + O(M_\pi^2). \tag{8}$$

This decay constant is proportional to $M_\pi^2$ and therefore vanishes in the chiral limit, in accordance with the equations of motion in QCD. Also, beyond the chiral limit the decay constants of the excited pseudoscalars are expected to be strongly suppressed compared to the pionic decay constant $f_\pi$ [5]. However, all more accurate predictions for the decay constants of the excited pseudoscalars require a better knowledge of the details of $g_A(s)$, since in this case the unknown terms of the order $O(M_\pi^2)$ are of the same order as the contribution given by the main term in $g_A(s)$.

## Pionic coupling of hadrons

The result (2) for the full axial current contains an explicit pion pole, thus providing the possibility to extract the amplitude $A(h_1 \rightarrow h_2 \pi)$ for pionic decays $h_1 \rightarrow h_2 + \pi$ [4]:

$$p_\mu A(h_1 \rightarrow h_2 \pi) = \lim_{p^2 \rightarrow M_\pi^2} \frac{p^2 - M_\pi^2}{f_\pi} \langle h_2 | j_\mu^5 | h_1 \rangle = p_\mu \frac{2m_Q}{f_\pi} \langle h_2 | \bar{Q} \gamma_5 Q | h_1 \rangle. \tag{9}$$

It is understood that the amplitude $\langle h_2 | \bar{Q} \gamma_5 Q | h_1 \rangle$ is calculated in terms of the constituent quark description of the hadrons $h_1$ and $h_2$. The expression (9) for the amplitude has been successfully applied to pionic decays of charmed mesons [6].

## The chiral constituent quark mass

Clearly, the constituent quark mass does not vanish in the chiral point. We give now an estimate for the constituent quark mass corresponding to the chiral limit, $m_Q^0$, making use of the following relation between the constituent quark mass $m_Q$ and the current quark mass $m$ at the chiral-symmetry breaking scale $\mu_\chi \simeq 1$ GeV [7]:

$$\langle \bar{q}q \rangle = \frac{N_c}{\pi^2} \int_0^\infty dk\, k^2 \exp(-k^2/\beta_\infty^2) \left\{ \frac{m}{\sqrt{m^2 + k^2}} - \frac{m_Q}{\sqrt{m_Q^2 + k^2}} \right\}, \tag{10}$$

with $\beta_\infty \simeq 0.7$ GeV [7]. We now have to take into account the dependence of the quark condensate on the value of the current quark mass. For the physical value of the quark condensate, corresponding to the current quark mass $m = 6$ MeV, we use $\langle \bar{q}q \rangle = -(240 \pm 15$ MeV$)^3$. Eq. (10) then gives $m_Q = 220$ MeV, a typical value of the $u$ and $d$ constituent quark mass [2]. In order to consider the chiral limit, $m \rightarrow 0$, the dependence of the quark condensate on the current quark mass should be taken into account. Setting $m = 0$, and making use of the chiral quark condensate $\langle \bar{q}q \rangle_{m=0} \simeq -(230 \pm 15$ MeV$)^3$, Eq. (10) gives the chiral constituent quark mass $m_Q^0 = 180$ MeV. Let us notice that this is precisely the value of the chiral constituent quark mass of the Godfrey–Isgur model [2].

In summary, we have demonstrated that the relativistic quark picture based exclusively on constituent-quark degrees of freedom is fully compatible with the (well-known) chiral properties of QCD *if* it encompasses the following features:

- The axial coupling $g_A$ of the constituent quarks is a momentum-dependent quantity, $g_A = \underline{g}_A(s)$, and is related to the pion $Q\bar{Q}$ wave function.
- The $Q\bar{Q}$ potential $V(r)$ saturates at large interquark separations: $V(r \to \infty) \to$ const.

Under the above conditions, a summation of the infinite number of diagrams describing constituent-quark soft interactions leads to the full axial current of the constituent quarks which is then conserved up to terms of order $O(M_\pi^2)$.

We defined the chiral point of the constituent quark model as that set of values of the parameters of the model (masses of the constituent quarks and couplings in the quark potential) for which the mass of the lowest pseudoscalar $Q\bar{Q}$ bound state, $M_\pi$, vanishes. Although the constituent quark mass clearly does not vanish at the chiral point, we claim that the chiral point of the constituent quark model corresponds to the spontaneously broken chiral limit of QCD for the following three reasons. (i) At the chiral point the full nonperturbative axial current of the constituent quarks is conserved (without the explicit introduction of Goldstone degrees of freedom). (ii) The lowest-energy part of the hadron spectrum has no other traces of chiral symmetry except for a massless pseudoscalar. (iii) Two important signatures of the spontaneously-broken chiral symmetry can be seen: the decay constant $f_\pi$ of the massless pion is finite, that means, nonvanishing, whereas all the decay constants of the excited massive pseudoscalars vanish.

We emphasize that the nonperturbative emergence of chiral symmetry in a model with merely constituent-quark degrees of freedom [4] is qualitatively different from the chiral symmetry of models which explicitly contain Goldstones along with constituent quarks: the latter may be rendered chirally invariant for any value of the constituent quark mass, whereas in our approach chirally symmetry is present only for a definite (nonvanishing) value of the constituent quark mass which leads to a massless ground-state pseudoscalar.

*Acknowledgments.* D. M. was supported by the Austrian Science Fund (FWF) under project No. P17692.

## REFERENCES

1. W. Lucha, F. F. Schöberl, and D. Gromes, *Phys. Rep.* **200**, 127 (1991).
2. S. Godfrey and N. Isgur, *Phys. Rev. D* **32**, 189 (1985).
3. V. V. Anisovich *et al.*, *Nucl. Phys. A* **544**, 747 (1992); F. Cardarelli *et al.*, *Phys. Lett. B* **332**, 1 (1994); D. Melikhov, *Phys. Rev. D* **53**, 2460 (1996); F. Cardarelli *et al.*, *Phys. Rev. D* **53**, 6682 (1996); D. Melikhov and B. Stech, *Phys. Rev. D* **62**, 014006 (2000).
4. W. Lucha, D. Melikhov, and S. Simula, *Phys. Rev. D* **74**, 054004 (2006).
5. M. A. Shifman, A. I. Vainshtein, and V. I. Zakharov, *Nucl. Phys. B* **147**, 385 (1979); A. Höll, A. Krassnigg, and C. D. Roberts, *Phys. Rev. C* **70**, 042203(R) (2004); W. Lucha and D. Melikhov, *Phys. Rev. D* **73**, 054009 (2006).
6. D. Melikhov and O. Pene, *Phys. Lett. B* **446**, 336 (1999); D. Melikhov and M. Beyer, *Phys. Lett. B* **452**, 121 (1999); D. Melikhov and B. Stech, *Phys. Rev. D* **74**, 034022 (2006).
7. D. Melikhov and S. Simula, *Eur. Phys. J. C* **37**, 437 (2004).

# The $[70,1^-]$ baryon multiplet in the $1/N_c$ expansion revisited

## N. Matagne and Fl. Stancu

*University of Liège, Physics Department,*
*Institute of Physics, B.5,*
*Sart Tilman, B-4000 Liège 1, Belgium*
*E-mail: nmatagne@ulg.ac.be, fstancu@ulg.ac.be*

**Abstract.**
The mass splittings of the baryons belonging to the $[70,1^-]$-plet are derived by using a simple group theoretical approach to the matrix elements of the mass formula. The basic conclusion is that the first order correction to the baryon masses is of order $1/N_c$ instead of order $N_c^0$, as previously found. The conceptual difference between the ground state and the excited states is therefore removed.

**Keywords:** baryon spectroscopy, large $N_c$ QCD
**PACS:** 12.39.-x,11.15.Pg,11.30.Hv

*Introduction.* The $1/N_c$ expansion of QCD [1] is a powerful theoretical tool which allows to systematically analyze baryon properties. The success of the method stems from the discovery that the ground state baryons have an exact contracted $SU(2N_f)$ symmetry when $N_c \to \infty$ [2, 3], $N_f$ being the number of flavors. A considerable amount of work has been devoted to the ground state baryons, summarized in several review papers as, e. g., [4, 5]. For $N_c \to \infty$ the ground state baryons are degenerate. For large $N_c$ the mass splitting starts at order $1/N_c$. The applicability of the approach to excited states is a subject of current investigation. The experimental facts indicate a small breaking of $SU(2N_f)$ which make the $1/N_c$ studies of excited states plausible. When the $SU(N_f)$ symmetry is exact, the baryon mass operator is a linear combination of terms

$$M = \sum_i c_i O_i, \tag{1}$$

with the operators $O_i$ having the general form

$$O_i = \frac{1}{N_c^{n-1}} O_\ell^{(k)} \cdot O_{SF}^{(k)}, \tag{2}$$

where $O_\ell^{(k)}$ is a $k$-rank tensor in $SO(3)$ and $O_{SF}^{(k)}$ a $k$-rank tensor in $SU(2)$, but invariant in $SU(N_f)$. The latter is expressed in terms of $SU(N_f)$ generators. For the ground state one has $k = 0$. The first factor gives the order $\mathcal{O}(1/N_c)$ of the operator in the series expansion. The lower index $i$ represents a specific combination of generators, see Table 1. In Eq. (1), each $O_i$ is multiplied by an unknown coefficient $c_i$ which is a reduced matrix element. All these coefficients encode the QCD dynamics

CP892, *Quark Confinement and the Hadron Spectrum VII*
edited by J. E. F. T. Ribeiro
© 2007 American Institute of Physics 978-0-7354-0396-3/07/$23.00

and are obtained from a fit to the existing data. Additional terms are needed if $SU(N_f)$ is broken.

*Excited states.* The excited states can be grouped into excitation bands with N = 1, 2, 3, etc. units of excitation energy. Among these, the states belonging to the N = 1 band, described by the $[\mathbf{70}, 1^-]$-plet, have been most extensively studied, either for $N_f = 2$, see *e.g.* [6, 7, 8] or for $N_f = 3$ [9]. The conclusion was that the splitting starts at order $N_c^0$.

The method has been applied to the N = 2 band multiplets $[\mathbf{56}', 0^+]$ for $N_f = 2$ [10], $[\mathbf{56}, 2^+]$ for $N_f = 3$ [11] and $[\mathbf{70}, \ell^+]$ for $N_f = 2$ [12] and $N_f = 3$ [13] and also to $N_f = 3$ baryons [14] of the $[\mathbf{56}, 4^+]$-plet (N = 4 band).

The excited states belonging to $[\mathbf{56}, \ell]$ multiplets are rather simple and can be studied by analogy to the ground state. Naturally the splitting starts at order $1/N_c$ [11, 14].

**TABLE 1.** List of operators and coefficients of the mass operator (1) for $[\mathbf{70}, 1^-]$.

| Operator | | Fit 1 (MeV) | Fit 2 (MeV) | Fit 3 (Mev) | Fit 4 (MeV) | Fit 5 (MeV) |
|---|---|---|---|---|---|---|
| $O_1 = N_c \mathbf{1}$ | $c_1 =$ | $481 \pm 5$ | $482 \pm 5$ | $484 \pm 4$ | $484 \pm 4$ | $498 \pm 3$ |
| $O_2 = \frac{1}{N_c} \ell^i s^i$ | $c_2 =$ | $-47 \pm 39$ | $-30 \pm 34$ | $-31 \pm 20$ | $8 \pm 15$ | $38 \pm 34$ |
| $O_3 = \frac{1}{N_c} s^i s^i$ | $c_3 =$ | $161 \pm 16$ | $149 \pm 11$ | $159 \pm 16$ | $149 \pm 11$ | $156 \pm 16$ |
| $O_4 = \frac{1}{N_c} T^a T^a$ | $c_4 =$ | $169 \pm 36$ | $170 \pm 36$ | $138 \pm 27$ | $142 \pm 27$ | |
| $O_5 = \frac{3}{N_c^2} \ell^{(2)ij} G^{ia} G^{ja}$ | $c_5 =$ | $-443 \pm 459$ | | $-371 \pm 456$ | | $-514 \pm 458$ |
| $O_6 = \frac{1}{N_c^2} \ell^i T^a G^{ia}$ | $c_6 =$ | $473 \pm 355$ | $433 \pm 353$ | | | $-606 \pm 273$ |
| $\chi_{\text{dof}}^2$ | | 0.43 | 0.68 | 1.1 | 0.96 | 11.5 |

The states belonging to $[\mathbf{70}, \ell]$-plets are more difficult due to the presence of a mixed symmetry. So far, in calculating the mass spectrum, the general practice was to split the baryon into an excited quark and a symmetric core, in this way reducing the problem to the well known ground state. There are two drawbacks in this procedure. One is that each generator of $SU(2N_f)$ is written as a sum of two terms, one acting on the excited quark and the other on the core. As a consequence, the number of linearly independent operators to be used in Eq. (2) increases tremendously and the number of coefficients to be determined becomes larger or much larger than the experimental data available. For example, for the $[\mathbf{70}, 1^-]$ multiplet with $N_f = 2$ one has 12 linearly independent operators up to order $1/N_c$ [7], instead of 6 as in the present approach (see Table 1). We recall that there are only 7 nonstrange resonances belonging to this band. They are given in Table 2. Consequently, in selecting the most dominant operators one has to make an arbitrary choice [7]. The second drawback is due to the truncation of the wave function containing the orbital and the spin-flavor parts. The exact wave function is given by a linear combination of terms of equal weight where each term corresponds to a given Young tableau of mixed symmetry denoted by $[N_c - 1, 1]$. In the above procedure only the term where the last quark is in the second row was kept, the other $N_c - 2$ terms being ignored. As a consequence the order of the spin-orbit operator became $N_c^0$.

In this practice the matrix elements of the excited quark are straightforward, as being described by single-particle operators. The matrix elements of the core operators $S_c^i$ and $T_c^a$ are also simple to calculate, while $G_c^{ia}$ are more involved. Analytic group theoretical formulas for several types of matrix elements of the SU(4) generators have been derived in the late sixtieths [15], in the context of nuclear physics. Every matrix element is factorized according to a generalized Wigner-Eckart theorem into a reduced matrix element and an SU(4) Clebsch-Gordan coefficient. Recently we have extended the approach to calculate isoscalar factors needed for the matrix elements of SU(6) generators between symmetric $[N_c]$ states [16].

*The $[\mathbf{70},1^-]$-plet.* Here we propose a new method where the splitting into an excited quark and a core is unnecessary. Details can be found in Ref. [17]. All one needs to know are the matrix elements of the $SU(2N_f)$ generators between mixed symmetric states $[N_c - 1, 1]$. For $N_f = 2$ these are provided by the work of Hecht and Pang [15]. To our knowledge such matrix elements are yet unknown for $N_f = 3$. Thus our work deals with nonstrange baryon resonances only. The list of operators $O_i$ contributing to the mass operator (1) is shown in Table 1 together with the coefficients $c_i$. The first column gives all linearly independent operators of type (2) up to order $1/N_c$. The other columns indicate the values obtained for the dynamical coefficients $c_i$ from various fits. Fit 1 contains all operators and gives the best $\chi^2_{\text{dof}}$. In the other columns one or two operators have been removed in order to understand their role in the fit. One can see that the removal of $O_5$ or of $O_6$ is not so dramatic but that the removal of $O_4$ badly deteriorates the fit. The data are from Ref. [18]. It is important to note that here the angular momentum operator of components $\ell_i$ ($i = 1,2,3$) is an intrinsic operator, acting on the entire system, not on the $N_c$-th quark only, as in previous studies, which means it was taken relative to a fixed center of mass [7]. In other words here we do not have a center of mass problem which may complicate the $N_c$ counting.

**TABLE 2.** Partial contributions and total mass (MeV) of $[\mathbf{70},1^-]$ resonances predicted by the $1/N_c$ expansion. The last two columns reproduce the experimental masses and the status of resonances [18].

| | Part. contrib. (MeV) | | | | | | Total (MeV) | Exp. (MeV) | Name, status |
|---|---|---|---|---|---|---|---|---|---|
| | $c_1O_1$ | $c_2O_2$ | $c_3O_3$ | $c_4O_4$ | $c_5O_5$ | $c_6O_6$ | | | |
| $^2N_{\frac{1}{2}}$ | 1444 | -16 | 40 | 42 | 0 | -13 | $1529 \pm 11$ | $1538 \pm 18$ | $S_{11}(1535)$**** |
| $^4N_{\frac{1}{2}}$ | 1444 | 39 | 201 | 42 | -31 | -33 | $1663 \pm 20$ | $1660 \pm 20$ | $S_{11}(1650)$**** |
| $^2N_{\frac{3}{2}}$ | 1444 | -8 | 40 | 42 | 0 | 7 | $1528 \pm 8$ | $1523 \pm 8$ | $D_{13}(1520)$**** |
| $^4N_{\frac{3}{2}}$ | 1444 | 16 | 201 | 42 | 25 | -13 | $1714 \pm 45$ | $1700 \pm 50$ | $D_{13}(1700)$*** |
| $^4N_{\frac{5}{2}}$ | 1444 | -24 | 201 | 42 | -6 | 20 | $1677 \pm 8$ | $1678 \pm 8$ | $D_{15}(1675)$**** |
| $^2\Delta_{\frac{1}{2}}$ | 1444 | 16 | 40 | 211 | 0 | -66 | $1645 \pm 30$ | $1645 \pm 30$ | $S_{31}(1620)$**** |
| $^2\Delta_{\frac{3}{2}}$ | 1444 | -8 | 40 | 211 | 0 | -33 | $1720 \pm 50$ | $1720 \pm 50$ | $D_{33}(1700)$**** |

The partial contributions and the total mass predicted by the $1/N_c$ expansion are given in Table 2. One can see that the contributions of all terms containing angular momentum, and in particular that of the spin-orbit, are small. This can be viewed as a dynamical effect. An entirely new quantitative result is that the isospin-isospin term $O_4$ brings a dominant contribution to $\Delta$ resonances, of the same order as the spin-spin term $O_3$ brings to $N$ resonances.

*Conclusions.* These results shed a new light into the description of the baryon multiplet $[\mathbf{70}, 1^-]$ in the $1/N_c$ expansion. The main findings are:

- In the mass formula the expansion starts at order $1/N_c$, as for the ground state, instead of $N_c^0$ as previously concluded.
- The isospin operator $O_4 = \frac{1}{N_c} T^a T^a$ is crucial in the fit to the existing data and its contribution is as important as that of the spin term $O_3 = \frac{1}{N_c} S^i S^i$.

It would be interesting to reconsider the study of higher excited baryons, for example those belonging to $[\mathbf{70}, \ell^+]$ multiplets, in the spirit of the present approach. Based on group theoretical arguments it is expected that the mass splitting starts at order $1/N_c$, as a general rule.

# REFERENCES

1. G. 't Hooft, Nucl. Phys. **72** (1974) 461; E. Witten, Nucl. Phys. **B160** (1979) 57.
2. J. L. Gervais and B. Sakita, Phys. Rev. Lett. **52** (1984) 87; Phys. Rev. D **30** (1984) 1795.
3. R. Dashen and A. V. Manohar, Phys. Lett. **B315** (1993) 425; ibid **B315** (1993) 438.
4. R. Dashen, E. Jenkins, and A. V. Manohar, Phys. Rev. **D51** (1995) 3697.
5. E. Jenkins, Ann. Rev. Nucl. Part. Sci. **48** (1998) 81
6. J. L. Goity, Phys. Lett. **B414** (1997) 140.
7. C. E. Carlson, C. D. Carone, J. L. Goity and R. F. Lebed, Phys. Rev. **D59** (1999) 114008.
8. C. E. Carlson and C. D. Carone, Phys. Rev. **D58** (1998) 053005.
9. J. L. Goity, C. L. Schat and N. N. Scoccola, Phys. Rev. **D66** (2002) 114014.
10. C. E. Carlson and C. D. Carone, Phys. Lett. **B484** (2000) 260.
11. J. L. Goity, C. L. Schat and N. N. Scoccola, Phys. Lett. **B564** (2003) 83.
12. N. Matagne and Fl. Stancu, Phys. Lett **B631** (2005) 7.
13. N. Matagne and Fl. Stancu, Phys. Rev. **D74** (2006) 034014
14. N. Matagne and Fl. Stancu, Phys. Rev. **D71** (2005) 014010.
15. K. T. Hecht and S. C. Pang, J. Math. Phys. **10** (1969) 1571.
16. N. Matagne and Fl. Stancu, Phys. Rev. **D73** (2006) 114025
17. N. Matagne and Fl. Stancu, arXiv:hep-ph/0610099.
18. W.-M. Yao *et al.* [Particle Data Group], J. Phys. **G33** (2006) 1.

# Parity doublers in chiral potential quark models

Yu. S. Kalashnikova*, A. V. Nefediev* and J. E. F. T. Ribeiro†

*Institute of Theoretical and Experimental Physics, 117218, B.Cheremushkinskaya 25, Moscow, Russia
†Centro de Física das Interacções Fundamentais (CFIF), Departamento de Física, Instituto Superior Técnico, Av. Rovisco Pais, P-1049-001 Lisboa, Portugal

**Abstract.** The effect of spontaneous breaking of chiral symmetry over the spectrum of highly excited hadrons is addressed in the framework of a microscopic chiral potential quark model (Generalised Nambu-Jona-Lasinio model) with a vectorial instantaneous quark kernel of a generic form. A heavy-light quark-antiquark bound system is considered, as an example, and the Lorentz nature of the effective light-quark potential is identified to be a pure Lorentz-scalar, for low-lying states in the spectrum, and to become a pure spatial Lorentz vector, for highly excited states. Consequently, the splitting between the partners in chiral doublets is demonstrated to decrease fast in the upper part of the spectrum so that neighboring states of an opposite parity become almost degenerate. A detailed microscopic picture of such a "chiral symmetry restoration" in the spectrum of highly excited hadrons is drawn and the corresponding scale of restoration is estimated.

Chiral symmetry is known to be broken spontaneously in QCD, and this phenomenon plays an important role for low–lying hadrons. The (almost) massless chiral pion constitutes the most prominent example of chiral symmetry breaking manifestation in the hadronic spectrum. In the meantime, considering quite general quantum–mechanical principles [1], one can argue that this manifestation should asymptotically disappear for highly excited states [2]. Although this property was studied before using various effective and phenomenological approaches [3, 4, 5], the microscopic picture of such an effective chiral symmetry restoration in highly excited hadrons was not disclosed so far. In the meantime, a model exists which meets all the requirements necessary to reproduce this phenomenon [6, 7]. This is the Generalised Nambu-Jona-Lasinio (GNJL) model for QCD [8, 9, 10]. Indeed, the model is chirally symmetric (in the chiral limit) and is able to describe microscopically the phenomenon of spontaneous breaking of chiral symmetry in the vacuum. It is intrinsically relativistic and contains confinement, so it should be able to address the problem of highly excited hadrons. The model is described by the Hamiltonian:

$$H = \int d^3x q^\dagger(x)(-i\vec{\alpha}\vec{\nabla})q(x) + \int d^3x d^3y \left[ q^\dagger(x)\frac{\lambda^a}{2}q(x) \right] V(\vec{x}-\vec{y}) \left[ q^\dagger(y)\frac{\lambda^a}{2}q(y) \right], \quad (1)$$

where, for the sake of simplicity, we stick to the simplest form of the Hamiltonian (1) compatible with the requiremets of confinement and chiral symmetry breaking. The standard approach used in this kind of models is the Bogoliubov–Valatin transformation from bare to dressed quarks parametrised with the help of the chiral angle $\varphi_p$ [10]. The mass–gap equation which defines the profile of the chiral angle appears from the requirement that the quadratic part of the normally ordered Hamiltonian (1) is diagonal, $: H_2 :\propto b^\dagger b + d^\dagger d$. Then two phases of the theory can be identified: the unbroken phase

CP892, *Quark Confinement and the Hadron Spectrum VII*
edited by J. E. F. T. Ribeiro
© 2007 American Institute of Physics 978-0-7354-0396-3/07/$23.00

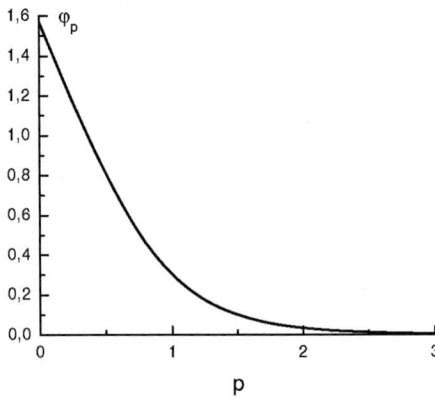

**FIGURE 1.** The typical profile of the chiral angle — solution to the mass–gap equation

with $\varphi_p \equiv 0$, and the broken phase, with the chiral angle taking the form depicted in Fig. 1. The broken phase possesses a lower vacuum energy and is therefore the true vacuum state of the theory.

Every mesonic state in the model is described by a pair of wave functions: one responsible for the time–forward and the other — for the time–backward motion of the quark–antiquark pair in the meson [8, 9, 10, 11]. To make things simpler, we consider one quark infinitely heavy, so that the heavy–light meson w.f. becomes one–component and it obeys the Schrödinger-like equation, in momentum space, [6]

$$E_p \psi(\vec{p}) + \int \frac{d^3k}{(2\pi)^3} V(\vec{p}-\vec{k}) \left[ C_p C_k + (\vec{\sigma}\hat{\vec{p}})(\vec{\sigma}\hat{\vec{k}}) S_p S_k \right] \psi(\vec{k}) = E\psi(\vec{p}), \qquad (2)$$

where $E_p$ is the dressed–quark dispersive law; $C_p = \cos\frac{1}{2}\left(\frac{\pi}{2} - \varphi_p\right)$, $S_p = \sin\frac{1}{2}\left(\frac{\pi}{2} - \varphi_p\right)$. The effect of chiral symmetry restoration in highly excited hadrons can be quite naturally exemplified by the behaviours of the splitting between the opposite–parity states $\psi(\vec{p})$ and $\psi'(\vec{p})$ as the excitation number increases (in this work, we keep the radial excitation number fixed and increase the light–quark angular momentum). It is easy to notice that, for highly excited mesons, the mean interquark momentum is large, so that $C_p \approx S_p \approx \frac{1}{\sqrt{2}}$ and the resulting bound–state equation becomes symmetric with respect to the substitution $\psi(\vec{p}) \to \psi'(\vec{p}) = (\vec{\sigma}\hat{\vec{p}})\psi(\vec{p})$. Thus these opposite–parity states become degenerate. It is easy to trace the origin of such a degeneracy. Indeed, the splitting between $\psi(\vec{p})$ and $\psi'(\vec{p})$ comes from the difference $C_p^2 - S_p^2 = \sin\varphi_p$, so that this is the large–momentum behaviour of the chiral angle to play a decisive role for the symmetry restoration — see Fig. 1. It is instructive to approach the same problem from the point of view of the Lorentz nature of the effective interquark interaction. Thus we perform Foldy counter–rotation of Eq. (2) to arrive at the Dirac–like equation for the light quark:

$$(\vec{\alpha}\vec{p} + \beta m)\Psi(\vec{x}) + \frac{1}{2} \int d^3z\, U(\vec{x}-\vec{z})[V(\vec{x}) + V(\vec{z}) - V(\vec{x}-\vec{z})]\Psi(\vec{z}) = E\Psi(\vec{x}). \qquad (3)$$

The Lorentz nature of confinement in this equation follows from the matrix structure of $U$ which takes the form, in momentum space,

$$U(\vec{p}) = \beta \sin \varphi_p + (\vec{\alpha}\hat{\vec{p}}) \cos \varphi_p. \tag{4}$$

We conclude therefore that, for low–lying states, the effective scalar interaction dominates. It is chirally nonsymmetric and thus chiral symmetry breaking manifests itself in this part of the spectrum (this regime was studied in detail in [12]). On the contrary, for highly excited states, the effective interquark interaction (asymptotically) becomes purely vectorial, so that chiral symmetry is an approximate symmetry of the interaction and thus approximate chiral multiplets appear in the spectrum of excited states. Notice that the transition between the two regimes is governed by $\sin \varphi_p$ which is known to be exactly the quantity "responsible"' for chiral symmetry breaking in this class of models.

Below we exemplify our qualitative conclusions drawn above with exact calculations performed for the harmonic oscillator potential case, $V(r) = K_0^3 r^2$, with $K_0$ being the only dimensional parameter of the model [8, 9, 10]. For this potential, the mass–gap equation and the bound–state euqation become ordinary second–order differential equations. As soon as the mass–gap equation is solved, we fix the value of the scale $K_0$ by evaluating the chiral condensate, $\langle \bar{q}q \rangle = -\frac{3}{\pi^2} \int_0^\infty dp\, p^2 \sin \varphi_p \approx -(0.51 K_0)^3$, and setting it equal to the standard value of $-(250 MeV)^3$ (for future references we call this scale the BCS scale, $\Lambda_{BCS} = 250 MeV$). This gives $K_0 = 490 MeV$. The bound–state Eq. (2) can be written now, in momentum space, as

$$-K_0^3 u'' + V(p)u = Eu, \quad \psi(\vec{p}) = \Omega_{jlm}(\hat{\vec{p}}) \frac{u(p)}{p}, \tag{5}$$

with the effective potential

$$V(p) = E_p + K_0^3 \left[ \frac{1}{4} \varphi_p'^2 + \frac{(j+1/2)^2}{p^2} + \frac{\kappa}{p^2} \sin \varphi_p \right], \quad \kappa = \pm (j+1/2). \tag{6}$$

This allows us to extract the difference between the potentials operative for the opposite–parity states:

$$\Delta V = -\frac{2(j+1/2)K_0^3}{p^2} \sin \varphi_p, \tag{7}$$

and which is responsible for the splitting between such states. As it was anticipated before, this potential is directly proportional to $\sin \varphi_p$ which vanishes as $p \to \infty$, and so does the splitting [6]. It is instructive to compare this situation with the Salpeter equation $[\sqrt{p^2 + m^2} + K_0^3 r^2] \psi(\vec{x}) = E \psi(\vec{x})$ which leads to the splitting potential $\Delta V = -\frac{2(l+1)K_0^3}{p^2}$. Results of numerical calculations are plotted in the form of Regge trajectories, in Fig. 2 (left plot) for the radial quantum number $n = 0$. It is clearly seen from this plot that, for the full bound–state equation the trajectories corresponding to the parity doublers merge thus reflecting the effect of chiral symmetry restoration.

Finally, we estimate the restoration scale [6]. To this end we notice that the splitting in chiral doublets appears of order of the BCS scale in the lowest part of the spectrum

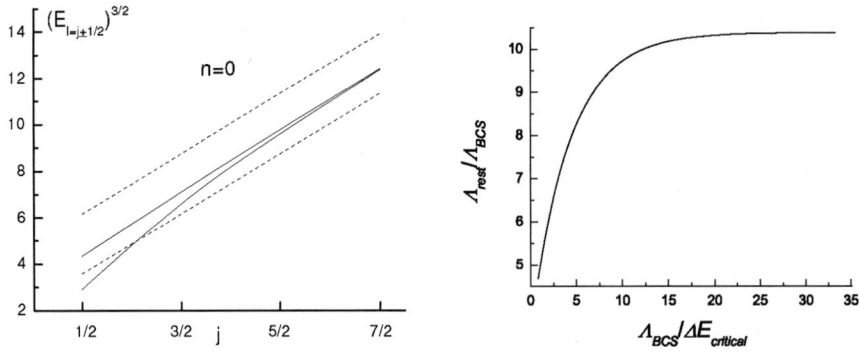

**FIGURE 2.** Left plot: Regge trajectories for the full bound–state Eq. (5) (solid line) and for the Salpeter equation (dashed line). Right plot: the restoration scale against the splitting.

and, above some scale — the restoration scale $\Lambda_{rest}$, it is much smaller than $\Lambda_{BCS}$. We use this as the definition of $\Lambda_{rest}$. Then, from the right plot at Fig. 2, one can conclude that $\Lambda_{rest} \approx 10\Lambda_{BCS} \approx 2.5\,GeV$.

Therefore we observe that the Generalised Nambu–Jona-Lasinio model provides a clear microscopical pattern of chiral symmetry restoration in the spectrum of highly excited hadrons. This effective asymtotic restoration happens rather fast and the model is able to predict a reasonable value of the restoration scale in agreement with other estimates found in the literature [5].

The research of A.N. and Yu.K. was supported by the grants RFFI-05-02-04012-NNIOa, DFG-436 RUS 113/820/0-1(R), NSh-843.2006.2, by the Federal Programme of the Russian Ministry of Industry, Science, and Technology No. 40.052.1.1.1112, and by the Russian Governmental Agreement N 02.434.11.7091.

# REFERENCES

1. L. Ya. Glozman, *Int. J. Mod. Phys.* **A21**, 475 (2006) 475.
2. L. Ya. Glozman, *Phys. Lett.* **B475**, 329 (2000); T. D. Cohen and L. Ya. Glozman, *Int. J. Mod. Phys.* **A17**, 1327 (2002); *Phys. Rev.* **D65**, 016006 (2002).
3. W. A. Bardeen, E. J. Eichten, and C. T. Hill, *Phys. Rev.* **D68**, 054024 (2003).
4. M. Nowak, M. Rho, and I. Zahed, *Acta Phys. Polon.* **B35**, 2377 (2004).
5. E. S. Swanson, *Phys. Lett.* **B582**, 167 (2004).
6. Yu. S. Kalashnikova, A. V. Nefediev, and J. E. F. T. Ribeiro, *Phys. Rev.* **D72**, 034020 (2005).
7. L. Ya. Glozman, A. V. Nefediev, and J. E. F. T. Ribeiro, *Phys. Rev.* **D72**, 094002 (2005).
8. A. Amer, A. Le Yaouanc, L. Oliver, O. Pene, and J.-C. Raynal, *Phys. Rev. Lett.* **50**, 87 (1983); A. Le Yaouanc, L. Oliver, O. Pene, and J.-C. Raynal, Phys. Lett. B **134**, 249 (1984); *Phys. Rev.* **D29**, 1233 (1984).
9. A. Le Yaouanc, L. Oliver, S. Ono, O. Pene and J.-C. Raynal, *Phys. Rev.* **D31**, 137 (1985).
10. P. Bicudo and J. E. Ribeiro, *Phys. Rev.* **D42**, 1611 (1990); *ibid.*, 1625 (1990); *ibid.*, 1635 (1990).
11. A. V. Nefediev and J. E. F. T. Ribeiro, *Phys. Rev.* **D70**, 094020 (2004).
12. Yu. A. Simonov, *Yad. Fiz.* **60**, 2252 (1997) [*Phys. Atom. Nucl.* **60**, 2069 (1997)].

# Pion and Kaon Masses and Pion Form Factors from Dynamical Chiral-Symmetry Breaking with Light Constituent Quarks

Michael D. Scadron*, Frieder Kleefeld[1][†] and George Rupp[†]

*Physics Department, University of Arizona, Tucson, AZ 85721, USA
†Centro de Física das Interacções Fundamentais, Instituto Superior Técnico, Edifício Ciência,
P-1049-001 Lisboa, Portugal

**Abstract.** Light constituent quark masses and the corresponding dynamical quark masses are determined by data, the quark-level linear $\sigma$ model, and infrared QCD. This allows to define effective nonstrange and strange current quark masses, which reproduce the experimental pion and kaon masses very accurately, by simple additivity. In contrast, the usual nonstrange and strange current quarks employed by the Particle Data Group and Chiral Perturbation Theory do not allow a straightforward quantitative explanation of the pion and kaon masses.

**Keywords:** Light constituent quarks, dynamical quark mass, effective current quark mass, pion and kaon masses, pion form factors, dynamical chiral-symmetry breaking, quark-level linear $\sigma$ model
**PACS:** 14.65.Bt, 14.40.Aq, 13.40.Gp, 11.30.Rd

## INTRODUCTION

The pion is commonly accepted to be massless in the chiral limit (CL). Not only is its physical mass a good measure of chiral-symmetry breaking (ChSB), but also the related nonstrange constituent quark mass $\hat{m}$. In the present short note, we shall show that $\hat{m}$ can be additively decomposed into a bulk part called dynamical quark mass ($m_{\mathrm{dyn}}$), associated with chiral-symmetric strong interactions, and a smaller part called current quark mass ($m_{\mathrm{cur}}$), which arises from ChSB in the electroweak sector. This *effective* $m_{\mathrm{cur}}$ turns out to be precisely half the pion mass. Moreover, the strange constituent quark mass allows a similar decompostition as well, with the kaon mass being the simple sum of the effective nonstrange and strange current quark masses.

## QUARK-MASS DIFFERENCE $m_d - m_u \approx 4$ MeV

A simple estimate of the mass difference between a down and an up quark gives

$$\left. \begin{array}{c} m_{K^0} - m_{K^+} \\ m_{\Sigma^-} - m_{\Sigma^+} \end{array} \right\} \implies m_d - m_u \approx 4\,\mathrm{MeV}\,. \tag{1}$$

---

[1] Home address: Pfisterstr. 31, D-90762 Fuerth, Germany

CP892, *Quark Confinement and the Hadron Spectrum VII*
edited by J. E. F. T. Ribeiro
© 2007 American Institute of Physics 978-0-7354-0396-3/07/$23.00

Note that this holds for both current *and* constituent quarks. On the other hand, from the proton magnetic moment we can derive [1] an average constituent quark mass as

$$\hat{m} = (m_u + m_d)/2 = 337.5 \, \text{MeV} \,. \tag{2}$$

Using Eq. (1), this yields the constituent masses

$$m_u \approx 335.5 \, \text{MeV} \quad, \quad m_d \approx 339.5 \, \text{MeV} \,. \tag{3}$$

We can also obtain $\hat{m}$ in the context of the quark-level linear $\sigma$ model (QLL$\sigma$M), via the Goldberger-Treiman relation [1, 2]

$$\hat{m} \approx f_\pi g = 93 \, \text{MeV} \times \frac{2\pi}{\sqrt{3}} \approx 337.4 \, \text{MeV} \,. \tag{4}$$

The agreement with the value in Eq. (2) is remarkable.

## DYNAMICAL QUARK MASS $m_{\text{dyn}}$

A bulk dynamical CL nonstrange quark mass can be estimated as

$$m_{\text{dyn}} \approx \frac{m_N}{3} = 313 \, \text{MeV} \,. \tag{5}$$

A check via the CL pion charge radius gives

$$m_{\text{dyn}} \approx \frac{\hbar c}{r_\pi^{\text{CL}}} = \frac{197.3 \, \text{MeV·fm}}{0.63 \, \text{fm}} = 313 \, \text{MeV} \,, \tag{6}$$

employing vector-meson dominance or the QLL$\sigma$M to predict

$$r_\pi^{\text{CL}} = 0.63 \, \text{fm} \,. \tag{7}$$

Alternatively, using infrared QCD, with $\alpha_s \approx 0.5$ at a 1 GeV cutoff, we get

$$m_{\text{dyn}} = \left[ \frac{4\pi}{3} \alpha_s \langle -\bar{q}q \rangle \right]^{\frac{1}{3}} \approx 313 \, \text{MeV} \,, \tag{8}$$

for the commonly accepted value of the quark condensate

$$\langle -\bar{q}q \rangle \approx (245 \, \text{MeV})^3 \,. \tag{9}$$

## EFFECTIVE CURRENT QUARK MASS VIA QCD

Away from the CL, we define the effective current quark mass as

$$\hat{m}_{\text{cur}} = \hat{m} - m_{\text{dyn}} \,, \tag{10}$$

where $\hat{m}$ is the constituent quark mass, and the dynamical mass $m_{\text{dyn}}$ runs as

$$m_{\text{dyn}}(p^2) \sim p^{-2} \tag{11}$$

according to QCD. On the $\hat{m} = 337.5$ MeV mass shell, selfconsistency then requires

$$m_{\mathrm{dyn}}(p^2 = \hat{m}^2) = \frac{m_{\mathrm{dyn}}^3}{\hat{m}^2} = \frac{(313)^3}{(337.5)^2} \, \mathrm{MeV} = 269.2 \, \mathrm{MeV} \,. \tag{12}$$

This yields
$$\hat{m}_{\mathrm{cur}} = (337.5 - 269.2) \, \mathrm{MeV} = 68.3 \, \mathrm{MeV} \,, \tag{13}$$

near the pion-nucleon sigma term

$$\sigma_{\pi N} = (55 \pm 13) \, \mathrm{MeV} \, [3] \,, \quad \sigma_{\pi N} = (66 \pm 9) \, \mathrm{MeV} \, [4] \,, \quad \sigma_{\pi N} = (64 \pm 8) \, \mathrm{MeV} \, [5] \,. \tag{14}$$

Note that both $\sigma_{\pi N}$ and $\hat{m}_{\mathrm{cur}}$ vanish in the CL.

## PION $\bar{q}q$ MASS

With the effective current quark mass derived in Eq. (13), we get a $\bar{q}q$ pion mass

$$m_\pi = 2\hat{m}_{\mathrm{cur}} = 136.6 \, \mathrm{MeV} \,, \tag{15}$$

almost midway between the observed $m_{\pi^0} = 134.98$ MeV and $m_{\pi^+} = 139.57$ MeV!

## PION FORM-FACTOR RATIO

In the QLL$\sigma$M, the conserved-vector-current pion form-factor ratio is predicted as

$$\frac{F_A^\pi(0)}{F_V^\pi(0)} = 1 - \frac{1}{3} = \frac{2}{3} \,, \tag{16}$$

for $q^2 = 0$. Here, the terms 1 and 1/3 are due to quark and meson loops, respectively, and the relative minus sign is a Feynman rule. The result is very near data [6] at

$$\frac{(0.0116 \pm 0.0016)}{(0.017 \pm 0.008)} = 0.68 \pm 0.33 \,. \tag{17}$$

## KAON $\bar{q}q$ MASSES

Chiral-symmetry-breaking experimental kaon masses (given $m_d - m_u \approx 4$ MeV as above) are computed as (neglecting small experimental errors)

$$m_{K^+(\bar{s}u)} = m_{s,\mathrm{cur}} + m_{u,\mathrm{cur}} = \hat{m}_{\mathrm{cur}} \left[ 1 + \left( \frac{m_s}{\hat{m}} \right)_{\mathrm{cur}} \right] - 2 \, \mathrm{MeV} = 493.677 \, \mathrm{MeV} \,, \tag{18}$$

$$m_{K^0(\bar{s}d)} = m_{s,\mathrm{cur}} + m_{d,\mathrm{cur}} = \hat{m}_{\mathrm{cur}} \left[ 1 + \left( \frac{m_s}{\hat{m}} \right)_{\mathrm{cur}} \right] + 2 \, \mathrm{MeV} = 497.648 \, \mathrm{MeV} \,. \tag{19}$$

For $\hat{m}_{cur} \approx 68.3$ MeV (see Eq. (13)), this gives *in both cases*

$$\left(\frac{m_s}{\hat{m}}\right)_{cur} \approx 6.257 , \tag{20}$$

which compares well to the light-plane result [7]

$$\left(\frac{m_s}{\hat{m}}\right)_{cur} \approx 6-7 , \tag{21}$$

and to other theory work [8], but *not* to the chiral-perturbation-theory (ChPT) predictions

$$\left(\frac{m_s}{\hat{m}}\right)_{cur} \approx 25-30 \text{ and } \hat{m}_{cur} \sim 5 \text{ MeV} . \tag{22}$$

## ChPT ALTERNATIVE

The value for $(m_s/\hat{m})_{cur}$ presently adopted by the Particle Data Group [6] and the ChPT [9] research community is as large as 25–30, with the small nonstrange current-mass value $\hat{m}_{cur} = 2.5$–5.5 MeV being strongly biased by the ChPT estimate

$$\hat{m}_{cur} = \frac{(f_\pi m_\pi)^2}{2\langle -\bar{q}q \rangle} \approx 5.5 \text{ MeV} , \tag{23}$$

for $f_\pi \approx 93$ MeV. On the basis of these scales, the quantitative explanation of $m_\pi$ and $m_K$ appears, however, rather cumbersome and "unnatural", contrary to what we observe in our aforementioned scheme resulting from the QLL$\sigma$M and dynamically broken QCD.

## ACKNOWLEDGMENTS

This work was supported by the *Fundação para a Ciência e a Tecnologia* of the *Ministério da Ciência, Tecnologia e Ensino Superior* of Portugal, under contract POCI/FP/63437/2005, and by the Czech project LC06002.

## REFERENCES

1. M. D. Scadron, R. Delbourgo and G. Rupp, J. Phys. G **32**, 735 (2006) [hep-ph/0603196].
2. R. Delbourgo, D. S. Liu and M. D. Scadron, Phys. Rev. D **59**, 113006 (1999) [hep-ph/9808253]; M. D. Scadron, F. Kleefeld and G. Rupp, hep-ph/0601196.
3. G. Hoehler, H. P. Jakob and R. Strauss, Phys. Lett. B **35**, 445 (1971).
4. H. B. Nielsen and G. C. Oades, Nucl. Phys. B **72**, 310 (1974).
5. R. Koch, Z. Phys. C **15**, 161 (1982).
6. W. M. Yao *et al.* [Particle Data Group], J. Phys. G **33**, 1 (2006).
7. H. Sazdjian and J. Stern, Nucl. Phys. B **94**, 163 (1975).
8. J. F. Gunion, P. C. McNamee and M. D. Scadron, Phys. Lett. B **63**, 81 (1976); J. F. Gunion, P. C. McNamee and M. D. Scadron, Nucl. Phys. B **123**, 445 (1977); N. H. Fuchs and M. D. Scadron, Phys. Rev. D **20**, 2421 (1979); N. H. Fuchs, H. Sazdjian and J. Stern, Phys. Lett. B **238**, 380 (1990).
9. J. Gasser and H. Leutwyler, Phys. Rept. **87**, 77 (1982).

# The Scalar Mesons and $Z(3)$ Symmetry

Nils A. Törnqvist

*Department of Physical Sciences, University of Helsinki, POB 64, FIN–00014*

**Abstract.**
It is pointed out that the $\det\Sigma + \det\Sigma^\dagger$ term, which resolves the $U_A(1)$ problem in effective theories, gives rise to three classical minima along the $U_A(1)$ circle when $N_f = 3$. The three minima are related to the center $Z(3)$ of $SU(3)$. This $Z(3)$ symmetry can be retained if the $SU(3)_L \times SU(3)_R$ symmetry breaking is assumed to be trilinear in the fields. The three vacua suggests a connection to the strong $CP$ problem and confinement.

**Keywords:** Scalar Mesons, Z(3). $U_A(1)$, Symmetry Breaking
**PACS:** 11.15.Ex, 11.30.-j, 11.30.Rd, 12.39.Fe, 14.65.Bt

This conference talk is a short version of my recent paper[1] on the possibility that the $Z(3)$ subgroup of the chiral $SU(3)_L \times SU(3)_R$ can be retained in the light scalar spectrum.

In effective theories for scalar and pseudoscalar mesons one models the global $U(3)_L \times U(3)_R$ symmetry by potential terms. One usually writes

$$V_{U3U3} = \frac{\mu^2}{2}\mathrm{Tr}[\Sigma\Sigma^\dagger] + \lambda\,\mathrm{Tr}[\Sigma\Sigma^\dagger\Sigma\Sigma^\dagger] + \lambda'(\mathrm{Tr}[\Sigma\Sigma^\dagger])^2, \tag{1}$$

where $\Sigma$ is the usual $3 \times 3$ matrix containing the scalar ($s$) and pseudoscalar ($p$) nonets.

To have a realistic zeroth order $SU(3)_L \times SU(3)_R$ model, one must break the axial $U_A(1)$ symmetry in eq.(1) explicitly. The simplest way to do it[2] is by adding a determinant term to the Lagrangian,

$$V_{SU3SU3} = V_{U3U3} + \beta[\det(e^{i\theta}\Sigma) + \det(e^{i\theta}\Sigma)^*]. \tag{2}$$

The addition of the complex conjugate term is required by parity, and also by $C$ parity, since a trilinear coupling of three $C = +$ mesons must by Bose statistics be symmetric under interchange of two mesons. We have included a $U_A(1)$ phase factor given by the angle $\theta$, which naïvely breaks strong $CP$. To give the pseudoscalar octet members mass (and the $\eta'$ a small extra mass) one conventionally assumes a symmetry breaking term which is linear in the fields $\propto (\mathrm{Tr}[\Sigma M_q] + h.c.)$, where $M_q$ is a diagonal matrix containing the chiral light quark masses.

Thereby one obtains essentially the $SU(3)$ version of the linear sigma model, by which one can model the basic global symmetries of QCD and their zeroth order breaking with the nonperturbative instanton term. In its first formulations it has been with us for almost 50 years[4, 6]. Our main point of this note is to show that determinant term in the symmetry limit gives rise to three classical minima. These can be related to color symmetry by spin statistics.

CP892, *Quark Confinement and the Hadron Spectrum VII*
edited by J. E. F. T. Ribeiro
© 2007 American Institute of Physics 978-0-7354-0396-3/07/\$23.00

There is a well known identity for a $3 \times 3$ determinant, which is useful for our purpose

$$6\det\Sigma = (\text{Tr}\Sigma)^3 + 2\text{Tr}(\Sigma^3) - 3\text{Tr}(\Sigma^2)\text{Tr}(\Sigma) . \tag{3}$$

In this expression each term has less symmetry $(SU(3)_F)$ than the sum $SU(3)_L \times SU(3)_R$. Another identity for a determinant $\det\Sigma_{ij} = \det(\bar{q}q_j)$ comes directly from its basic definition

$$\det\Sigma = \det(\bar{q}q_j) = \varepsilon_{ijk}\,\bar{q}_1 q_i\,\bar{q}_2 q_j\,\bar{q}_3 q_k \tag{4}$$

$$= \overline{uu}\,\overline{dd}\,\overline{ss} - \overline{uu}\,\overline{ds}\,\overline{sd} + \overline{ud}\,\overline{ds}\,\overline{su} - \overline{ud}\,\overline{du}\,\overline{ss} + \overline{us}\,\overline{du}\,\overline{sd} - \overline{us}\,\overline{dd}\,\overline{su}. \tag{5}$$

The most important physics properties of these determinant forms are (a) The determinant is completely antisymmetric with respect to flavor. (b) In each term one has 3 quarks and 3 anti-quarks, and any quark flavor occurs only once, and similarly any anti-quark flavor occurs only once. (c) It is a flavor singlet both in the three quarks and in the three anti-quarks, and as already noted invariant under an $SU(3)$ transformation from both the left as well as from the right of $\Sigma$. (d) A $U_A(1)$ transformation is just a simple phase transformation $e^{i\varphi}$ from the left and from the right, whereby only the phase of $\Sigma$ changes by $e^{2i\varphi}$. Because of this we have the freedom in choosing $\theta$ in eq.(2).

In particular, note that because the three quarks or three anti-quarks involved form a flavor singlet, any diquark subsystem must be in the $\mathbf{3}_F$ representation of $SU(3)_F$. In fact, many years ago Jaffe[7] found that in the bag model the strongest bound diquarks are those, which are in the antisymmetric $\mathbf{3}_F$ $SU(3)_F$ representation, have antisymmetric spin $S = 0$, symmetric space (S-wave) and are antisymmetric in color $\mathbf{3}_C$. Therefore he suggested a diquark model for the lightest scalar nonet, which would have an "inverted" mass spectrum (compared to the vector mesons), where the $\sigma(600)$ is the lightest, followed by a $\kappa$ near 800 MeV and the $a_0(980), f_0(980)$. In fact, the model described by eq.(2) predicts a very broad, light sigma and the determinant term (when including $s - d$ quark mass splitting) shifts the $\kappa$ down from the $a_0$ by the same amount as the K is shifted up from the $\pi$ .

The light and broad sigma, the $\sigma(600)$, is now a well established resonance. Also an extremely broad $\kappa$ pole, which has been claimed in experiments as well as in phenomenology has very recently[8] been determined to a remarkable accuracy by Roy-Steiner constraints involving crossing symmetry, analyticity and unitarity.

The connection between Jaffe's diquark model and the determinant term is clear. It is natural to expect the lowest diquarks to have spin 0 and to be in an S-wave. Since the determinant requires any diquark to be in the $\mathbf{3}_F$ they must also be in the antisymmetric $\mathbf{3}_C$ by spin-statistics. Thus if one wants to include color, then the determinant term should be multiplied by a similar factor, but now with color replacing flavor in the indices. This shows a flavor-color connection through Fermi-Dirac statistics within the scalar mesons, in a analogous way as the color factor is needed for the proton wave function.

Now in the flavour symmetric limit the pure $SU(3)$ singlet states are equal superposition of $\overline{u}u, \overline{d}d, \overline{s}s$. They are thus represented by the complex matrix $\Phi = \phi \cdot \mathbf{1}/\sqrt{3}$, where $\phi = (s_0 + ip_0)$ and where $\mathbf{1}$ is the $3 \times 3$ unit matrix. First neglect the phase angle $\theta$ in eq.(2). There is then a real minimum of the potential eq.(2) i.e. a non zero vacuum value. (For $\mu = 0$ this is $v = \frac{1}{\sqrt{3}}\phi^{min} = -\beta/(2\lambda + 6\lambda').$)

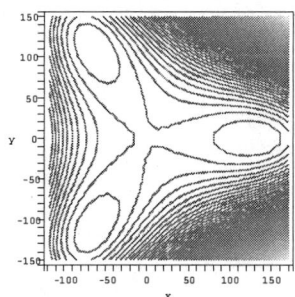

**FIGURE 1.** An illustrative example of the potential $V(\phi)$ of eq. (6) as a contour plot in the complex $\phi$ plane. The three minima are here at $|\phi_{min}| \approx 130$ MeV. (This corresponds to an average $f_\pi$ and $f_K$ decay constant of $130\sqrt{(2/3)}$ MeV $\approx 106$ MeV.) The parameters in eq.(6) are chosen in this illustation as $\mu = 0$, $\beta = -1700$ MeV and $\lambda + 3\lambda' = 11.5$. The masses of the $SU(3)$ singlet pseudoscalar and singlet scalar states are given by the second derivatives at any of the three minima.

But, in fact, there are three minima in the effective potential defining 3 vacuum expectation values! Substituting $\Phi$ into $\Sigma$ of eq.(6) one finds, now including the phase $\theta$ in eq.(2):

$$V(\phi) = \frac{\mu^2}{2}|\phi|^2 + \frac{\lambda + 3\lambda'}{3}|\phi|^4 + \frac{2\beta}{3\sqrt{3}}|\phi|^3 \cos[3\theta + 3\arg(\phi)]. \tag{6}$$

The cosine factor in the $\beta$ term makes this potential different from the usual "Mexican hat" potential. As an illustrative example it is shown in fig. 1 as a contour plot in the complex $\phi$ plane near parameter values found in Ref.[5]. It has three "hills" and three valleys in between. Most importantly, provided $\mu^2$ is not too large and positive, it has three minima defining three vacuum expectation values in the downhill directions of the steepest hills $v_j = ve^{i(2j\pi/3-\theta)}$, $j = 1, 2, 3$. Thus the $U_A(1)$ symmetry is broken by the determinant, but the $Z(3)$ center symmetry of $SU(3)$ is retained.

One should expect that instantons in QCD can tunnel between these vacua and, in fact, 't Hooft[2] motivated the determinant term because of instantons. The inclusion of the $\theta$ angle shows that the three minima are all on the same footing. In the $SU(3)_F$ limit, i.e. if one neglects weak interactions and chiral quark masses, one has the freedom to chose this chiral angle $\theta$ to be a multiple of $2\pi/3$, such that this choice ($\theta = 0, 2\pi/3$ or $4\pi/3$) makes any of the three minima real and $> 0$. Reality of $v_i e^{i\theta}$ is required by $CP$, at least as long as weak interactions are neglected. Expanding the meson fields around any of these vacua $\Sigma \to \Sigma + v_i \mathbf{1}$ one finds a singlet $\eta'$ mass, $m_{p_0}^2 = -6\beta|v| = 12(\lambda + 3\lambda')|v|^2$, from the second derivative in the angular variable ($\arg(\phi) \propto p_0$) of the potential (6).

The scalar singlet mass is similarly obtained $m_{s_0}^2 = 4(\lambda + 3\lambda')v^2 = m_{p_0}^2/3$, or 553 MeV for a 958 MeV $p_0$, from the second derivative in the radial direction $|\phi|$ of the same potential (6). The $0^{-+}$ octet remain massless while the scalar octet mass is given by $m_{s_{1..8}}^2 = 16(\lambda + 3/2\lambda')v^2 = 4/3m_{p_0}^2 - 8\lambda'v^2$, which means in the region of 1 GeV.

What is the significance of these 3 minima? The threefold symmetry, together with $CP$, is related to the center $Z(3)$ of the axial $SU(3)$ symmetry in $SU_L(3) \times SU_R(3)$. Above we showed how the determinant connects flavor and three colors because of Fermi-Dirac statistics. This makes three flavors special for scalar mesons, and $N_f = 3$ is also special because $SU(3)_F$ remains approximate after symmetry breaking from the small chiral quark masses. The symmetry breaking is small compared to $\Lambda_{QCD}$ or, here perhaps better, compared to the $\eta'$ or proton mass.

Thus for the meson spectrum it does not matter which of the three $v_i$'s is chosen in the shift, $\Sigma \to \Sigma + v_i \mathbf{1}$. The meson masses remain the same since they depend on $|v_i|^2 = v^2$, but for fermions a problem appears because of the possible phase of $v_i$. A constituent quark can get mass, $m_q^{const} = gv_i$, through Yukawa couplings to the vacuum as in the original linear sigma model[4]: $g\,\bar{q}_L \Sigma q_R + h.c. \to gv_i \bar{q}_L q_R + h.c.$, where $g$ is a pion quark coupling. Here $v_i$ must be chosen chosen real for each quark. A phase of $v_i$ violates parity and charge conjugation, by which one could argue that such single free quarks are forbidden not only by color but also $CP$.

The three minima in fig.1 are puzzling, Are these connected to the longstanding strong $CP$ problem[9] and perhaps confinement? The axial $U_A(1)$ current is, of course, well known not to be conserved, because of the triangle quark graph and the gluon chiral anomaly. In the strong $CP$ problem one also derives from the anomaly many different vacua connected by "large" gauge transformations and winding numbers in the same $U_A(1)$ degree of freedom as discussed here.

Now, chiral and flavor symmetry can be broken in a way that maintains the permutation $Z(3)$ symmetry of the three vacua. To get a finite pseudoscalar octet mass one can, instead of a conventional term $\propto m_q \text{Tr}\Sigma + h.c.$, introduce a small term $\propto m_q (\text{Tr}\Sigma)^3 + h.c.$, which retains the $Z(3)$ symmetry (like the terms on the r.h.s. of eq.(3)). Similarly, instead of a conventional term $\propto \text{Tr}(\Sigma M_q) + h.c.$, which breaks $SU(3)_F$, one can introduce e.g. a term $\propto (\text{Tr}\Sigma)^2 \text{Tr}(\Sigma M_q) + h.c.$, which also retains the $Z(3)$ symmetry, i.e. one still has the three equal minima as in fig.1. In conclusion the puzzling fact that this well known effective model can have three vacua, which are illustrated in fig. 1, opens many interesting questions. A better understanding should illuminate the long standing strong $CP$ and confinement problems.

# REFERENCES

1. Nils. A. Törnqvist hep-ph/0606041 and work in progress.
2. G. 't Hooft, Phys.Rev.Lett. **37** (1976) 8; ăG. 't Hooft, Phys.Rep. **142** (1986) 357; G. 't Hooft, "The physics of instantons in the pseudoscalar and vector meson mixing", hep-th/9903189 (unpublished).
3. S. Weinberg, Phys.Rev. **D11** (1975) 3583.
4. J. Schwinger, Ann.Phys. **2** (1957) 407; M. GellŨ-Mann, M. Levy, Nuovo Cim. **XVI** (1960) 705; B.W. Lee, Nucl. Phys. **B9** (1969) 64.
5. N.A. Törnqvist, Eur.Phys.J. **C11** 359 (1999); M. Napsuciale, A.Wirzba, M.Kirchbach, Nucl.Phys. A **703** (2002), 306 and ref. [1].
6. J.Schechter, Y.Ueda, Phys.Rev. **D3** (1971) 2874; S. Gasiorowicz, D.A.Geffen, Rev.Mod.Phys. **41** (1969) 531; T.Hatsuda, T.Kunihiro, Phys.Rep. **247** (1994) 223 S.P. Klevansky, Rev.Mod.Phys. **64** (1992) 643.
7. R.L.Jaffe, Phys.Rev. D **15** (1977) 267, 281; R.L.Jaffe, F.E.Low, Phys.Rev. D **19** (1979) 2105; R.L.Jaffe hep-ph/0001123 (unpublished);
8. S.Descotes-Genon, B.Moussallam, LPT-Orsay/06-43, hep-ph/0607133.
9. For a recent review see: R.D. Peccei "The strong CP problem and axions", hep-ph/0607268.

# Exclusive $\rho\rho$ production in $\gamma\gamma$ interaction at LEP

Igor Vorobiev
(on behalf of the L3 Collaboration)

*Carnegie Mellon University, Pittsburgh, PA 15213*

**Abstract.** Exclusive $\rho\rho$ production in two-photon collisions is studied at LEP for quasi-real photons ($\gamma\gamma$, centre-of-mass energies 161 GeV $\leq \sqrt{s} \leq$ 209 GeV, total integrated luminosity L=698 $pb^{-1}$) and one virtual photon ($\gamma\gamma^*$, 89 GeV $\leq \sqrt{s} \leq$ 209 GeV, L=855 $pb^{-1}$). The cross sections of the $\rho\rho$ production processes are determined as a function of the photon virtuality, $Q^2$, and the two-photon centre-of-mass energy, $W_{\gamma\gamma}$, in the kinematic region: $Q^2 \leq 30$ GeV$^2$ and 1 GeV $\leq W_{\gamma\gamma} \leq$ 3 GeV.

**Keywords:** LEP, two-photon physics, exclusive meson pair production, QCD model verification.
**PACS:** 13.60.Le

The L3 Collaboration performed a series of measurements of neutral and charged $\rho$-meson pair production in interactions involving two quasi-real photons

$$e^+e^- \to e^+e^- \, \gamma\gamma \to e^+e^- \, \rho\rho \,, \tag{1}$$

or one photon of higher virtuality

$$e^+e^- \to e^{\pm}e^{\mp}_{tag} \, \gamma\gamma^* \to e^{\pm}e^{\mp}_{tag} \, \rho\rho \,. \tag{2}$$

In reaction (1) both electron and positron escape detection (untagged events), while in reaction (2) one electron/positron is tagged either by the Luminosity Monitor (LUMI) or Very Small Angle Tagger (VSAT) ($\rho^0\rho^0$ with the LUMI, $\rho^+\rho^-$ with the LUMI, $\rho^0\rho^0$ with the VSAT, $\rho^+\rho^-$ with the VSAT, $\rho$-pairs in untagged events - see papers in [1]). The LUMI covers the $Q^2$ range $1.2 - 8.5$ GeV$^2$ at LEPI, $8.8 - 30$ GeV$^2$ at LEPII, and the VSAT (available only at LEPII) - $0.2 - 0.85$ GeV$^2$. This is the first measurement of tagged exclusive $\rho^+\rho^-$ production in $\gamma\gamma$ interactions and more precise measurement of tagged $\rho^0\rho^0$ production. The untagged $\rho\rho$ measurement is performed with an order of magnitude higher statistics than previous measurements.

For untagged events a spin-parity-helicity analysis was employed. The $\rho\rho$ production was considered in different spin-parity and helicity states ($J^P, J_Z$) together with an isotropic nonresonant production of four pions. It was checked that the inclusion of the $\rho\pi\pi$ state practically didn't change the results on $\rho\rho$ production.

In order to determine the differential $\rho\rho$ production rates in tagged events, a maximum likelihood fit was performed in intervals of $Q^2$ and $W_{\gamma\gamma}$ using a box method. Data were fit to the sum of non-interfering contributions from the processes, generated by a Monte Carlo in a simple model of isotropic production and phase space decay. For the data sample with four charged pions the processes used in fit were: $\gamma\gamma^* \to \rho^0\rho^0$, $\gamma\gamma^* \to \rho^0\pi^+\pi^-$, $\gamma\gamma^* \to \pi^+\pi^-\pi^+\pi^-$ (nonresonant); and for two charged and two neutral pions: $\gamma\gamma^* \to \rho^+\rho^-$, $\gamma\gamma^* \to \rho^{\pm}\pi^{\mp}\pi^0$, $\gamma\gamma^* \to \pi^+\pi^-\pi^0\pi^0$ (nonresonant). The inputs

CP892, *Quark Confinement and the Hadron Spectrum VII*
edited by J. E. F. T. Ribeiro

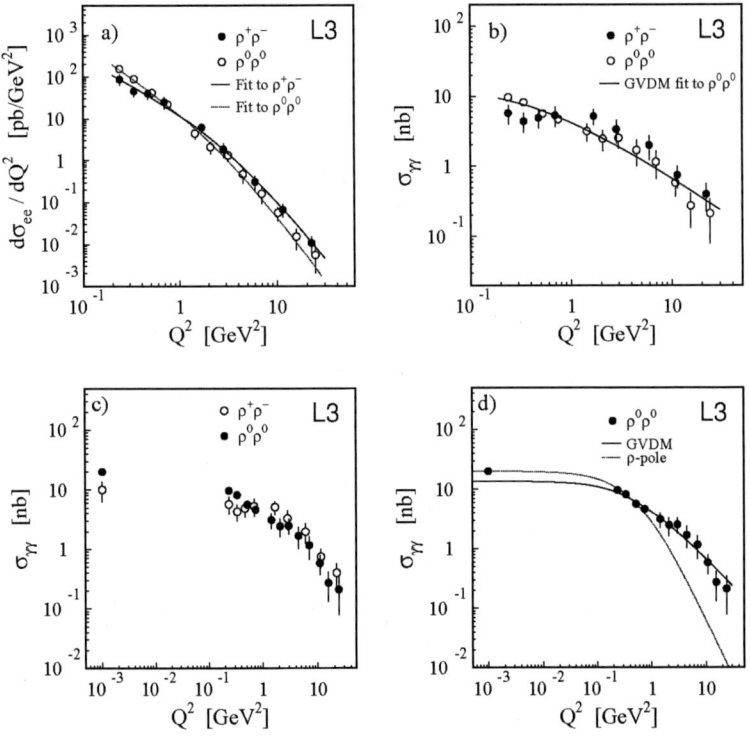

**FIGURE 1.** The $\rho\rho$ production cross section as a function of $Q^2$: a) differential cross section of the process $e^+e^- \to e^{\pm}e^{\mp}_{tag}\rho\rho$, b) cross section of the process $\gamma\gamma^* \to \rho\rho$, c) $\sigma_{\gamma\gamma}$ cross section comparison for $\rho^+\rho^-$ and $\rho^0\rho^0$ for the whole range of $Q^2$, d) $\sigma_{\gamma\gamma}$ for $\rho^0\rho^0$ with $\rho$-pole and GVDM fit.

to the fit were the six possible two-pion mass combinations in an event. The fit provides a good description of all mass and angular distributions. From the numbers of $\rho\rho$ events, obtained by the fit in intervals of $Q^2$ and $W_{\gamma\gamma}$ and corrected for efficiencies and background, the cross sections of $\rho\rho$ pair production were calculated. The cross section dependences on $Q^2$ and $W_{\gamma\gamma}$ for both $\rho^0\rho^0$ and $\rho^+\rho^-$ production are shown in Figures 1 and 2. The $Q^2$ dependence of the differential cross section can be well described by the approximate formula

$$d\sigma_{ee}/dQ^2 \sim 1/Q^n(Q^2 + <W_{\gamma\gamma}>^2)^2, \qquad (3)$$

following QCD-based calculations [2], where $n$ is expected to equal 2. In the range $Q^2 > 1.2\ \mathrm{GeV}^2$, the fit gives values $n = 2.4 \pm 0.3$ for $\rho^0\rho^0$ and $n = 2.5 \pm 0.4$ for $\rho^+\rho^-$, which are compatible with the expected value 2. The $\rho^0\rho^0$ production at $Q^2 >$

1.2 GeV$^2$ was analyzed in the first paper of [3]. The $Q^2$ dependence of the differential cross section is perfectly reproduced by the formulas with three phenomenological parameters, one of which, $C_1 = 1.20 \pm 0.23$ GeV$^2$, gives the normalization of the $\rho\rho$ generalized distribution amplitudes. In the $Q^2 > 0.2$ GeV$^2$ range, the fit to approximate formula (3) gives $n = 2.3 \pm 0.15$ for $\rho^+\rho^-$ and $n = 2.9 \pm 0.14$ for $\rho^0\rho^0$ (fit is shown in Figure 1a by the lines).

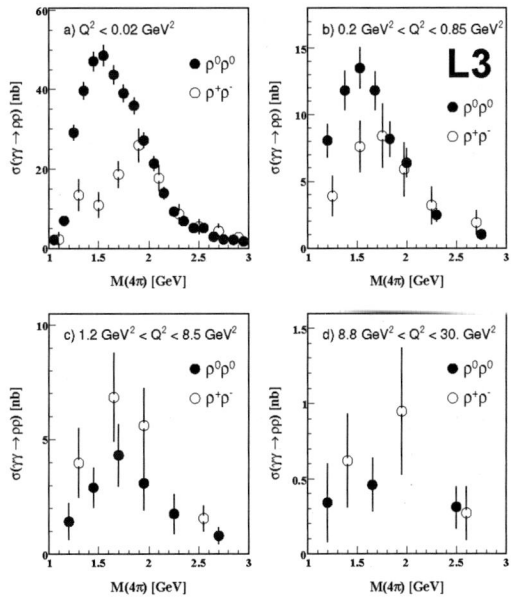

**FIGURE 2.** The $\rho\rho$ production cross section as a function of $W_{\gamma\gamma}$ in four $Q^2$ intervals.

The measured cross section of the process $\gamma\gamma^* \to \rho\rho$ as a function of $Q^2$ at $Q^2 > 0.2$ GeV$^2$ is shown in Figure 1b. It is well described by the GVDM model for $\rho^+\rho^-$ at $Q^2 > 1$ GeV$^2$, and for $\rho^0\rho^0$ - in the entire $Q^2$ range. The fit for $\rho^0\rho^0$ in the whole $Q^2$ range finds a cross section of $13.6 \pm 0.7$ nb at $Q^2 = 0$ (Figure 1d). A $\rho$-pole description is excluded both for $\rho^0\rho^0$ and $\rho^+\rho^-$ data.

The ratio of cross sections $R = \sigma_{e^+e^- \to e^+e^-\rho^+\rho^-} / \sigma_{e^+e^- \to e^+e^-\rho^0\rho^0}$ for $1.1$ GeV $\leq W_{\gamma\gamma} \leq 2.1$ GeV is $2.2 \pm 1.1 \pm 0.6$ in the $Q^2$ range $8.8$ GeV$^2 \leq Q^2 \leq 30$ GeV$^2$ (Figure 2d) and $1.81 \pm 0.47 \pm 0.22$ for $1.2$ GeV$^2 \leq Q^2 \leq 8.5$ GeV$^2$ (Figure 2c). This is very compatible with the factor 2, expected for an isospin $I = 0$ state.

Contrary to this the measurements at lower $Q^2$ show the $\rho^0\rho^0$ cross section to be higher than the $\rho^+\rho^-$ one, $R = 0.63 \pm 0.10 \pm 0.09$ for $0.2$ GeV$^2 \leq Q^2 \leq 0.85$ GeV$^2$ (Figure 2b) and $0.42 \pm 0.05 \pm 0.09$ for $Q^2 \leq 0.02$ GeV$^2$ (Figure 2a). The change of the relative magnitude of $\rho^+\rho^-$ and $\rho^0\rho^0$ production in the vicinity of $Q^2 \approx 1$ GeV$^2$ is clearly seen on Figures 1a,1c suggesting different $\rho$-pair production mechanisms at low and high $Q^2$.

If to assume the production of an isospin $I = 2$ exotic state at small $Q^2$, the whole ensemble of data can be well described for $Q^2 > 0.2$ GeV$^2$ (both $Q^2$ and $W_{\gamma\gamma}$ dependencies for $\rho^0\rho^0$ and $\rho^+\rho^-$, and change of relative amplitude in vicinity of $Q^2 \approx 1$ GeV$^2$) - see the second paper of [3].

The presented data lay new experimental grounds for obtaining information about QCD in going from nonperturbative to the perturbative regime.

## REFERENCES

1. L3 Coll., P. Achard *et al*, *Phys. Lett.* **B 568**, 11 (2003); *Phys. Lett.* **B 597**, 26 (2004); *Phys. Lett.* **B 604**, 48 (2004); *Phys. Lett.* **B 615**, 19 (2005); *Phys. Lett.* **B 638**, 128 (2006).
2. M. Diehl, T. Gousset and B. Pire, *Phys. Rev.* **D 62**, 073014 (2000).
3. I.V. Anikin, B. Pire and O.V. Teryaev, *Phys. Rev.* **D 69**, 014018 (2004); *Phys. Lett.* **B 626**, 86 (2005).

# The Olsson sum rule and the rho Regge pole

## F. J. Yndurain

*Departamento de Física Teórica, Universidad Autónoma de Madrid*

**Abstract.** We consider the Olsson sum rule, i.e., the forward dispersion relation for pion-pion scattering with exchange of isospin unity at threshold. We show that, if using the S0, S2 and P wave expressions of Colangelo, Gasser and Leutwyler, then either the sum rule is not satisfied or, if adjusting the residue of the rho exchange Regge amplitude to have the sum rule satisfied (as recently proposed by Caprini, Colangelo and Leutwyler) then the subsequent high energy amplitude is in disagreement with experimental pi-pi cross sections.

**Keywords:** Dispersion relations, Regge theory.
**PACS:** 11.55.Hx, 11.55.Jy

In a paper written some time ago by Peláez and myself[1] (see also ref. 2 for a full discussion of the Regge amplitudes) we remarked that, if in the so-called Olsson sum rule,

$$2a_0^{(0)} - 5a_0^{(2)} = D_{\text{Ol.}}, \quad D_{\text{Ol.}} \equiv 3M_\pi \int_{4M_\pi^2}^{\infty} ds \, \frac{\operatorname{Im} F^{(I_t=1)}(s,0)}{s(s-4M_\pi^2)}$$

we use for calculating the quantity $\operatorname{Im} F^{(I_t=1)}(s,0)$ the values of the S0, S2 and P wave phase shifts proposed by Colangelo, Gasser and Leutwyler [3] at low energy, together with experimental information for the other waves and for S0, S2 and P waves at intermediate energy; and, at high energy, we input the amplitude for rho exchange deduced in the previously quoted articles by Peláez and myself from factorization, then the Olsson sum rule is violated by a bit more than two standard deviations. Subsequently, Caprini, Colangelo, Gasser and Leutwyler[4] conceded that we were right, but stated that they were sure that one could find a Regge parametrization that would restore fulfillment of the Olsson relation. This was done recently, in a paper by Caprini, Colangelo and Leutwyler:[5] here, these authors determine the parameters for exchange of isospin 1 that make the Olsson sum rule satisfied, with the $\pi\pi$ amplitudes of Colangelo, Gasser and Leutwyler.[3] That is to say: they consider the Olsson sum rule given above, they substitute these low energy amplitudes and fix the residue $\beta_\rho$ of the rho trajectory, assumed to give the imaginary part of the amplitude at high energy,

$$\operatorname{Im} F^{(I_t=1)}(s,0) \simeq \beta_\rho (s/s_0)^{\alpha_\rho},$$

by requiring the Olsson sum rule to be verified. However, the existing high energy experimental data can be used to get the cross section for exchange of isospin unity. We can define the cross section $\sigma^{(I_t=1)}(s)$ terms of $\operatorname{Im} F^{(I_t=1)}(s,0)$:

$$\sigma^{(I_t=1)}(s) = \frac{4\pi^2}{\lambda^{1/2}(s, M_\pi^2, M_\pi^2)} \operatorname{Im} F^{(I_t=1)}(s,0); \quad \lambda(a,b,c) = a^2 + b^2 + c^2 - 2ab - 2ac - 2bc.$$

CP892, *Quark Confinement and the Hadron Spectrum VII*
edited by J. E. F. T. Ribeiro
© 2007 American Institute of Physics 978-0-7354-0396-3/07/$23.00

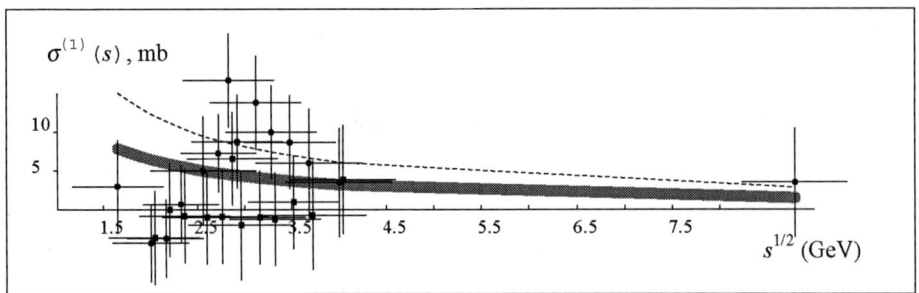

**FIGURE 1.** $\pi\pi$ cross section for Regge exchange of isospin unity $\sigma^{(I_t=1)} = \sigma_{\text{tot},\,\pi^+\pi^-} - \sigma_{\text{tot},\,\pi^-\pi^-} \cdots$
The experimental points have been obtained from [6]. We plot as a thick line the results of Peláez and Ynduráin. The dotted line is the central value of Caprini, Colangelo and Leutwyler.

The cross section $\sigma^{(I_t=1)}(s)$ may then be written as

$$\sigma^{(1)} \equiv \sigma^{(I_t=1)} = \sigma_{\text{tot},\,\pi^+\pi^-} - \sigma_{\text{tot},\,\pi^-\pi^-},$$

and one can take the $\sigma_{\text{tot},\,\pi^+\pi^-}$, $\sigma_{\text{tot},\,\pi^-\pi^-}$ from experiment.[6] The results are shown in the accompanying figure, where we plot (*thick line*) against experiment what was obtained by Peláez and Ynduráin (loc. cit.) either by direct fits to data or from factorization; this two methods agree, within the thickness of the line. Then, the *dotted line* is central value of Caprini, Colangelo and Leutwyler[5] that follows from assuming the Olsson sum rule satisfied with the low energy parameters of Golangelo, Gasser and Leutwyler.

The data are not very good, but it is clear that the central value that Caprini, Colangelo and Leutwyler get from the Olsson sum rule goes well above what one finds from factorization, and also above the majority of the experimental data points. One may conclude that if you want a rho Regge amplitude that leads to satisfaction of the Olsson sum rule with the Colangelo, Gasser and Leutwyler low energy partial waves, such amplitude will disagree with factorization and with most experimental information for $\pi\pi$ total cross sections.

This note contains only part of the contribution of the author to the *Quark Confinement and the Hadron Spectrum* Conference celebrated in the Azores, Portugal. The rest of the contribution was identical to that presented at the *IV International Conference on Quarks and Nuclear Physics* celebrated in Madrid between 5 and 10 June, 2006, by Kamiński, Peláez and myself.[7].

I am grateful to J. R. Peláez with whom the essentials of this work were done.

## ACKNOWLEDGMENTS

I am grateful to J. R. Pelaez with whom the essentials of this work were done.

# REFERENCES

1. J. R. Pelaez and F. J. Yndurain, Phys. Rev. D **68**, 074005 (2003)
2. J. R. Pelaez and F. J. Yndurain, Phys. Rev. D **69**, 114001 (2004)
3. G. Colangelo, J. Gasser and H. Leutwyler, Nucl. Phys. B **603**, 125 (2001)
4. I. Caprini, G. Colangelo, J. Gasser and H. Leutwyler, Phys. Rev. D **68**, 074006 (2003)
5. I. Caprini, G. Colangelo and H. Leutwyler, Int. J. Mod. Phys. A **21**, 954 (2006)
6. Biswas, N. N., et al., *Phys. Rev. Letters*, **18**, 273 (1967) $[\pi^-\pi^-, \pi^+\pi^-$ and $\pi^0\pi^-]$; Cohen, D. et al., *Phys. Rev.* **D7**, 661 (1973) $[\pi^-\pi^-]$; Robertson, W. J., Walker, W. D., and Davis, J. L., *Phys. Rev.* **D7**, 2554 (1973) $[\pi^+\pi^-]$; Hoogland, W., et al. *Nucl. Phys.*, **B126**, 109 (1977) $[\pi^-\pi^-]$; Hanlon, J., et al, *Phys. Rev. Letters*, **37**, 967 (1976) $[\pi^+\pi^-]$; Abramowicz, H., et al. *Nucl. Phys.*, **B166**, 62 (1980) $[\pi^+\pi^-]$
7. R. Kaminski, J. R. Pelaez and F. J. Yndurain, arXiv:hep-ph/0610315.

# Is the $\sigma$ meson dynamically generated?[1]

## Zhi-hui Guo, L. Y. Xiao and H. Q. Zheng

*Department of Physics, Peking University, Beijing 100871, P. R. China*

**Abstract.** We study the problem whether the $\sigma$ meson is generated 'dynamically'. A pedagogical analysis on the toy O(N) linear sigma model is performed and we find that the large $N_c$ limit and the $m_\sigma \to \infty$ limit does not commute. The sigma meson may not necessarily be described as a dynamically generated resonance. On the contrary, the sigma meson may be more appropriately described by considering it as an explicit degree of freedom in the effective lagrangian.

**Keywords:** Scalar meson, chiral symmetry, dispersion relations
**PACS:** 14.40.Cs, 13.85.Dz, 11.55.Bq, 11.30.Rd

The long standing debate on whether there exists a light and broad resonance (for the historical reason named as the '$\sigma$' meson [1]) has finally come to an end. The Roy equation analysis clearly indicates a light and broad resonance pole located just inside the analyticity domain established from Roy equations [2]. Despite that there may still exist some disputes at technical level on pole locations, Roy equation analysis starts from first principles of quantum field theory and the result on the existence of the $\sigma$ pole is robust. The very existence of the sigma pole can actually be understood in a simper, more intuitive and also rigorous way: one writes a dispersion relation for the analytic continuation of $\sin(2\delta_\pi)$ in the I,J=0,0 channel of elastic $\pi\pi$ scattering and use chiral perturbation theory to estimate the background (left hand cut) contribution. In this way one finds that the background contribution to the phase shift ($\sin(2\delta_\pi)$) is negative and concave whereas the experimental data is positive and convex [3]. The difference can only be made up by a pole contribution, according to the standard $S$ matrix theory principle. On the other side, the $\sigma$ pole location found from different dispersive analyses [2, 4] agree with each other, within error bars, hence proving convincingly the stability of the numerical outputs on the pole location from dispersive analyses. Also the results from dispersive analyses are also in fair agreement with recent experimental determinations [5]. On the other side, dispersive analyses also reveal, in a model independent way, the existence of a light and broad resonance named $\kappa$ [6, 7], which are again in agreement with the recent experimental determinations [8]. In the partial wave dispersion relation for $\pi K$ scattering, there exists the circular cut due to un-equal mass kinematics. In Ref. [6], the background contribution is estimated along the outer edge of the circular cut, hence the complicated cut structure inside the circular cut is avoided.

Having firmly established the existence of the light and broad $\sigma$ and $\kappa$, the next important question is what are these resonances? There exist different proposals to explain these resonances. For example, there are tetra quark model [9], linear sigma

---

[1] Talk presented by Zheng at "Quark Confinement and Hadron Spectrum VII", 2–7 Sept. 2006, Ponta Delgada, Acores, Portugal

CP892, *Quark Confinement and the Hadron Spectrum VII*
edited by J. E. F. T. Ribeiro
© 2007 American Institute of Physics 978-0-7354-0396-3/07/$23.00

model at hadron level [10], linear sigma model at quark level [11], and also the ENJL model [12]. Also there exists the approach to explain the light scalars as dynamically generated resonances [13]. In Ref. [14] it is observed that since the widths of $\sigma$ and $\kappa$ are very large the pole mass and and 'line shape' mass are very different quantities, one should be extremely careful when discussing the mass relations between scalars: the tree level mass relations have to be among those line shape masses rather than those pole masses. The mass relations obtained from the extended Nambu Jona-Lasinio lagrangian are discussed and it is argued that the lightest scalars, $\sigma$, $\kappa$, $a_0(980)$ and $f_0(980)$ form an nonet, as the chiral partners of the pseudo-goldstone bosons [14]. The mass relations are crude, but it is expected that they grasp the major characters of the physics underlined. The major difficulty in this approach is to explain the large mass of $f_0(980)$ [14], and more detailed dynamical analysis may be needed [15].

In the approach to consider the lightest scalars as chiral partners of the pesudo-goldstone bosons, one thing remains to be explained is how to understand the approach that the $\sigma$ and $\kappa$ are generated dynamically. Owing to the complexity of the problem, we in the following discuss the unitarization approximations to the solvable O(N) $\sigma$ model. As will be shown later, it will be helpful to understand several difficult issues.

The O(N) linear $\sigma$ model lagrangian is

$$\mathscr{L} = \frac{1}{2}\partial_\mu \Phi^T \partial^\mu \Phi - \frac{1}{2}m^2 \Phi^T \Phi - \frac{\lambda}{8N}(\Phi^T \Phi)^2 \tag{1}$$

where $\Phi = (\Phi_1, \Phi_2, \cdots, \Phi_N)^T$. The explicit symmetry breaking interaction is characterized by,

$$\mathscr{L}_{S.B.} = vm_\pi^2 \Phi_N . \tag{2}$$

Here we treat the lagrangian as a cutoff effective lagrangian. That is, in our calculation we make the following replacement:

$$\Gamma(\varepsilon) + \ln 4\pi + \ln \frac{\mu^2}{m_\pi^2} \Rightarrow \ln \frac{\Lambda^2}{m_\pi^2} . \tag{3}$$

It has been proved that in such a toy model the [n,n] Padé amplitudes reproduce the exact sigma pole location and the K matrix unitarizations are good approximations [16]. Nevertheless, such a nice property is not maintained if the pion fields are expressed in the non-linear representation, since for the latter the chiral expansion series has to be truncated. There are variants of Eq. (1). For example one may make a polar decomposition to the linear lagrangian and recast it into the following form:

$$\mathscr{L}_{polar} = \mathscr{L}^\sigma + \frac{1}{2}(1 + \frac{\sigma}{v})^2 (\partial_\mu \vec{\pi} \cdot \partial^\mu \vec{\pi} + \partial_\mu \sqrt{v^2 - \vec{\pi} \cdot \vec{\pi}} \partial^\mu \sqrt{v^2 - \vec{\pi} \cdot \vec{\pi}}) \tag{4}$$

where

$$\mathscr{L}^\sigma = \frac{1}{2}\partial_\mu \sigma \partial^\mu \sigma - \frac{1}{2}m^2(\sigma + v)^2 - \frac{\lambda}{8N}(\sigma + v)^4 . \tag{5}$$

One further expands the square root in Eq. (4) when calculating scattering amplitudes. Also one may completely neglect the sigma field in Eq. (4) to get the non-linear sigma

309

model,

$$\mathscr{L}_{NL} = \frac{1}{2}(\partial_\mu \vec{\pi} \cdot \partial^\mu \vec{\pi} + \partial_\mu \sqrt{v^2 - \vec{\pi} \cdot \vec{\pi}} \partial^\mu \sqrt{v^2 - \vec{\pi} \cdot \vec{\pi}}) \ . \tag{6}$$

Or one integrate out the sigma field at tree level to get the modified non-linear sigma model lagrangian,

$$\mathscr{L}_{\overline{NL}} = \mathscr{L}_{NL} + \frac{1}{2m_\sigma^2 v^2}[(\partial_\mu \vec{\pi} \cdot \partial^\mu \vec{\pi})^2 - m_\pi^2 \vec{\pi} \cdot \vec{\pi} \partial_\mu \vec{\pi} \cdot \partial^\mu \vec{\pi} + \frac{m_\pi^4}{4}(\vec{\pi} \cdot \vec{\pi})^2]. \tag{7}$$

We have tested various unitarization approximations and the details will be given elsewhere. Here we only briefly discuss the properties of [1,1] Padé amplitudes constructed using $\mathscr{L}_{polar}$, $\mathscr{L}_{NL}$, $\mathscr{L}_{\overline{NL}}$, respectively. Notice that here we work in the cutoff version of effective lagrangian and hence no counter term is needed when one make calculations at 1-loop level. In each amplitude, a pole is found close to or not far from the sigma pole of the original lagrangian Eq. (1). In the case of $\mathscr{L}_{polar}$, the pole found in the unitarized amplitude is not dynamical. For $\mathscr{L}_{NL}$, $\mathscr{L}_{\overline{NL}}$, the poles are called 'dynamical'. Except these '$\sigma$' poles being reproduced, there may exist other spurious poles. The spurious pole does not occur in the lagrangian with linearly realized chiral symmetry, hence one may find that the lagrangian with linearly realized chiral symmetry are better for the purpose of unitarization. Another lesson one may learn is that a 'dynamically generated' resonance may or may not be truly dynamical. For $\mathscr{L}_{NL}$, the '$\sigma$' pole is indeed dynamical, but for $\mathscr{L}_{\overline{NL}}$ the '$\sigma$' pole just regenerates the $\sigma$ particle being integrated out in the original lagrangian Eq. (1). For the latter case, the $\sigma$ is, of course, better (or more conveniently) described by explicitly including it in the lagrangian.

The dynamical poles generated from Eq. (6) and Eq. (7) have quite different dynamical properties, however. It is not difficult to check that the pole location of the the '$\sigma$' pole produced by Eq. (6) is $\sqrt{s_p} \propto v = f_\pi \propto \sqrt{N_c}$ and moves to infinity when $N_c \to \infty$, whereas the pole generated from Eq. (7) behaves as $\sqrt{s_p} \to m_\sigma$ when $N_c \to \infty$. Apparently only the latter is correct when simulating Eq. (1). The lesson one may learn from here is that the large $N_c$ limit and the $M_\sigma \to \infty$ limit do not commute.

O(N) model is only a simple toy model, comparing with the complicated structure of QCD. However one may still learn some useful lessons from above. The Eq. (6) simulates the current algebra non-linear sigma model in reality whereas Eq. (7) resembles $O(p^4)$ chiral perturbation theory lagrangian in reality. In the real situation, Actually similar things happen. The current algebra prediction to the $\sigma$ pole location [17]

$$\sqrt{s_\sigma} \simeq \sqrt{16i\pi f_\pi^2} \simeq 463 - 463i \ , \tag{8}$$

which moves to $\infty$ when $N_c \to \infty$, may receive important corrections:

$$s_\sigma \simeq \frac{16i\pi f_\pi^2}{1 + 16i\pi f_\pi^2 \triangle} \ , \tag{9}$$

where $\triangle = \frac{2}{3f_\pi^2}(22L_1 + 14L_2 + 11L_3) \propto O(N_c^0)$. The above expression is obtained from [1,1] Padé approximation in the chiral limit [18]. Hence it was not clear what approximation is made in obtaining Eq. (9). It is however also obtainable using the PKU parametrization form under two assumptions in the large $N_c$ and chiral limit [18]: 1) one pole (the '$\sigma$' pole) dominance in the $s$ channel, 2) neglecting all resonance exchanges in the crossed channels, which can also be at the leading order in $1/N_c$ expansion. Even though Eq. (9) is only a rough approximation, we expect it gives the correct $N_c$ dependence of the sigma pole, if the '$\sigma$' meson contributes to the low energy constants when $N_c$ is large [19]. For more detailed discussion related to the $\sigma$ pole location in the large $N_c$ limit one is referred to Ref. [18].

Since the '$\sigma$' meson found in the [1,1] Padé approximation finally falls down to the real axis in the large $N_c$ limit, it is suggested, through the analysis given above, that the '$\sigma$' meson is a true particle, i.e., the $\sigma$ meson being responsible for the spontaneous chiral symmetry breaking in the linear realization of chiral symmetry.

**Acknowledgement:** This work supported in part by China National Natural Science Foundation under grant number 10575002 and 10421503.

# REFERENCES

1. N. A. Tornqvist, Talk given at YITP Workshop on Possible Existence of the sigma meson and its Implications to Hadron Physics, Kyoto, Japan, 12-14 Jun 2000, hep-ph/0008135.
2. I. Caprini, G. Colangelo and H. Leutwyler, Phys. Rev. Lett. **96** (2006) 132001.
3. Z. G. Xiao and H. Q. Zheng, Nucl. Phys. **A695**, 273(2001).
4. Z. Y. Zhou et al., JHEP 0502(2005)043.
5. E. M. Aitala et al. (E791 Collaboration), Phys. Rev. Lett. **86**(2001)770; M. Ablikim et al. (BES Collaboration), Phys. Lett. **B598**(2004)149.
6. Z. Y. Zhou and H. Q. Zheng, Nucl. Phys. **A775**(2006)212; H. Q. Zheng et al., Nucl. Phys. **A733**(2004)235.
7. S. Descotes-Genon, B. Moussallam, hep-ph/0607133.
8. E. M. Aitala et al., Phys. Rev. Lett. **89**, 121801(2002); M. Ablikim et al. (BES Collaboration), Phys. Lett. **B633**(2006)681; D. V. Bugg, Eur. Phys. J. **A25**(2005)107, Erratum-ibid. **A26**(2005)151; D. V. Bugg, Phys. Lett. **B632**(2006)471.
9. R. L. Jaffe, Phys. Rev. **D15**(1977)267.
10. N. A. Tornqvist, Z. Phys. **C68**(1995)647; N. N. Achasov and G. N. Shestakov, Phys. Rev. **D49**(1994)5779; R. Kaminski, et al., Phys. Rev. **D50**(1994)3154; M. Ishida et al., Prog. Theor. Phys. **99**(1998)1031; D. Black et al., Phys. Rev. **D64** (2001) 014031.
11. M. D. Scadron, F. Kleefeld, G. Rupp, E. van Beveren, Nucl. Phys. **A724**(2003)391.
12. A. A. Osipov, H. Hansen, B. Hiller, Nucl. Phys. **A745**(2004)81 and reference therein; V. Dmitrasinovic, Phys. Rev. **C53** (1996) 1383.
13. A. Gomez Nicola, J. R. Pelaez, Phys. Rev. **D65**(2002)054009; J. A. Oller, E. Oset, J. R. Pelaez, Phys. Rev. **D59**(1999)074001, Erratum-ibid. **D60**(1999)099906.
14. L. Y. Xiao, H. Q. Zheng and Z. Y. Zhou, talk given at QCD06, Montpellier, France, July 3 – 7, 2006. hep-ph/0609009.
15. See for example V. Baru, J. Haidenbauer, C. Hanhart, Yu. Kalashnikova, A. E. Kudryavtsev, Phys. Lett. **B586**(2004)53.
16. S. Willenbrock, Phys. Rev. **D43**(1991)1710.
17. H. Leutwyler, contributions to MESON2006, hep-ph/0608218.
18. Z. X. Sun et al., hep-ph/0503195.
19. Z. G. Xiao and H. Q. Zheng, hep-ph/0502199, accepted for publication by MPLA.

# Quark model study of the semileptonic $B \to \pi$ decay

C. Albertus[*], J. M. Flynn[*], E. Hernández[†], J. Nieves[**] and J. M.
Verde–Velasco[†]

[*]School of Physics and Astronomy, University of Southampton, Southampton SO17 1BJ, United
Kingdom.
[†]Grupo de Física Nuclear, Departamento de Física Fundamental e IUFFyM, Universidad de
Salamacanca, E-37008 Salamanca, Spain.
[**]Departamento de Física Atómica, Molecular y Nuclear, Universidad de Granada, E-18071
Granada, Spain.

**Abstract.** The semileptonic decay $B \to \pi l \bar{v}_l$ is studied starting from a simple quark model and taking into account the effect of the $B^*$ resonance. A novel, multiply subtracted, Omnès dispersion relation has been implemented to extend the predictions of the quark model to all physical $q^2$ values. We find $|V_{ub}| = 0.0034 \pm 0.0003(\text{exp.}) \pm 0.0007(\text{theory})$, in good agreement with experiment.

**Keywords:** Decay of bottom mesons, nonrelativistic quark model, dispersion relations, Kobayashi-Maskawa matrix elements
**PACS:** 12.15.Hh, 11.55.Fv, 12.39.Jh, 13.20.He

## INTRODUCTION

The measurement of the exclusive semileptonic decay $B \to \pi l \bar{v}_l$ can be used to determine de Cabibbo-Kobayashi-Maskawa (CKM) matrix element $|V_{ub}|$. With no flavor symmetry constraining the hadronic matrix elements, the errors on $|V_{ub}|$ are currently dominated by theoretical uncertainties, being a determination of $|V_{ub}|$, with well understood uncertainties, a priority of heavy flavor physics. The application of Watson's theorem to the $B \to \pi$ semileptonic decay allows one to write a dispersion relation for each of the form factors entering in the hadronic matrix element. This leads to the so-called Omnès representation, which can be used to constrain the $q^2$ dependence of the form factors assuming some knowledge of the elastic $\pi B \to \pi B$ scattering amplitudes. The use of multiple subtractions will allow to combine predictions from various methods in different $q^2$ regions. In this talk, we show how this Omnès scheme can be used to combine the results at $q^2 = 0$ of the relevant hadron $B \to \pi l \bar{v}_l$ form factors from light cone sum rules (LCSR) calculations, with those obtained from a simple nonrelativistic constituent quark model (NRCQM) in its region of applicability, near zero recoil. In this way we end up, with an accurate description of the differential decay rate, except for $V_{ub}$, in the whole physically accessible $q^2$ range. We also use a Monte Carlo simulation to estimate the theoretical error bands of our procedure.

CP892, *Quark Confinement and the Hadron Spectrum VII*
edited by J. E. F. T. Ribeiro

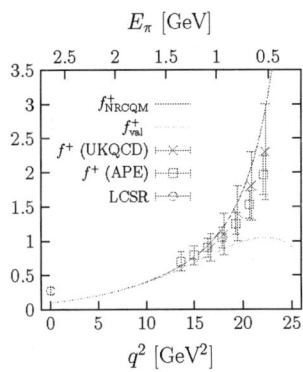

$E_\pi$ [GeV]

**FIGURE 1.** Valence quark (val), and valence quark plus $B^*$ contribution (NRCQM) to $f^+$. We also plot lattice QCD and LCSR $f^+$ results. See Ref. [1] for details.

## NRCQM: VALENCE QUARK, $B^*$ RESONANCE AND OMNÈS REPRESENTATION

The hadronic matrix element for $B^0 \rightarrow \pi^- l^+ v_l$ can be parametrized in terms of two dimensionless form factors, $f^+$ and $f^0$, of which only $f^+$ contributes for massless final leptons. In Figure 1 we show under "*val*" the NRCQM prediction for $f^+$ when considering only the valence quark contribution. The description fails in the whole $q^2$ range: from the region close to $q^2_{max}$, where a nonrelativistic model should work best, to the opposite end where the pion is ultrarelativistic and thus predictions from a nonrelativistic scheme are unreliable. As first pointed out in Ref. [2], near zero recoil the $B \rightarrow \pi l^+ v_l$ decay is dominated by the effects of the $B^*$ resonance, which is quite close to $q^2_{max}$. These effects of the $B^*$ resonance must be added as a distinct coherent contribution. We have consistently evaluated within our model the $B^*$ resonance contribution to the $f^+$ form factor (See Ref. [1] and references therein for details). The result is that the $B^*$ resonance plays a role only near $q^2_{max}$, being strongly suppressed by a soft hadronic vertex outside that region. The inclusion of the $B^*$ resonance contribution to the form factor ($f^+_{NRCQM}$ in the figure) improves the simple valence quark contribution down to values around 15 GeV$^2$. Below that the description is still poor.

We have now used the Omnès representation to combine the NRCQM predictions at high $q^2$ with the LCSR result at $q^2 = 0$. This representation requires as an input the elastic $\pi B \rightarrow \pi B$ phase shift $\delta(s)$ in the $J^P = 1^-$ and isospin $I = 1/2$ channel plus the form factor at different $q^2$ values below the $\pi B$ threshold where we will perform the subtractions. With a large enough number of subtractions only the phase shift at or near threshold is needed. We can then approximate $\delta(s) \approx \pi$ (Levinson's theorem) which renders our calculation analytic.

In Figure 2 we show with a solid line the form factor obtained using the Omnès representation. We have used as subtraction points five $q^2$ values between 18 GeV$^2$ and

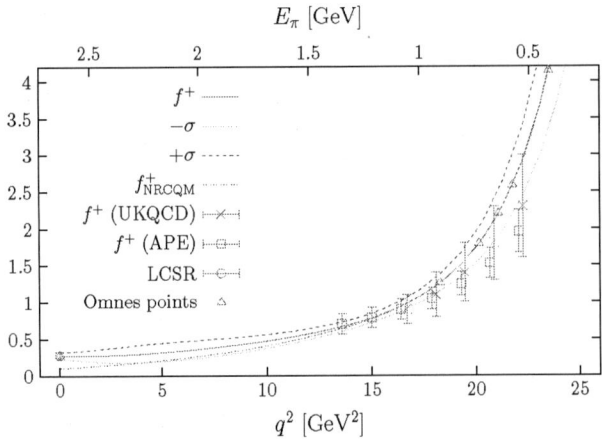

**FIGURE 2.** The solid line represents the Omnès improved form factor. The subtraction points are denoted by triangles. The $\pm\sigma$ lines show the theoretical uncertainty band on the Omnès form factor. We compare with previous lattice results.

$q^2_{\max}$, for which the NRCQM predictions (valence+pole) have been used, plus the LCSR prediction at $q^2 = 0$. We have paid special attention to the estimation of theoretical uncertainties that come from two main sources: (i) uncertainties in the quark–antiquark nonrelativistic interaction and (ii) uncertainties on the $[g_{B^*B\pi}f_{B^*}]$ product and on the input to the multiply subtracted Omnès representation. As a result, we obtain the 68% confidence level region enclosed between $\pm\sigma$ lines. See Ref. [1] for details. Comparing with the experimental decay width, we obtain $|V_{ub}| = 0.0034 \pm 0.0003\,(\text{exp.}) \pm 0.0007\,(\text{theo.})$ in good agreement with a recent experimental determination by the CLEO Collaboration [3].

## ACKNOWLEDGMENTS

This work was supported by DGI and FEDER funds, under Contracts No. FIS2005-00810, BFM2003-00856 and FPA2004-05616, by the Junta de Andalucía and Junta de Castilla y León under Contracts No. FQM0225 and No. SA104/04, and it is a part of the EU integrated infrastructure initiative Hadron Physics Project under Contract No. RII3-CT-2004-506078. J.M.V.-V. acknowledges an E.P.I.F contract with the University of Salamanca. C. A. acknowledges a research contract with the University of Granada.

## REFERENCES

1. C. Albertus, J. M. Flynn, E. Hernández, J. Nieves, and J. M. Verde-Velasco, Phys. Rev. D **72** (2005) 033002.
2. N. Isgur and M. B. Wise, Phys. Rev. D **41** (1990) 151.
3. B. Athar *et al.* (CLEO Collaboration), Phys. Rev. D **68** (2003) 072003.

# Relativistic Corrections to $e^+e^- \to J/\psi + \eta_c$ in a Potential Model[1]

Geoffrey T. Bodwin[2]*, Daekyoung Kang[†], Taewon Kim[†], Jungil Lee[†] and Chaehyun Yu[†]

*High Energy Physics Division, Argonne National Laboratory, 9700 South Cass Avenue, Argonne, Illinois 60517
[†]Department of Physics, Korea University, Seoul 136-701, Korea

**Abstract.** We compute relativistic corrections to the process $e^+e^- \to J/\psi + \eta_c$ and find that they resolve the discrepancy between theory and experiment.

The disagreement between theory and experiment for the exclusive double-charmonium process $e^+e^- \to J/\psi + \eta_c$ has, for a number of years, been one of the largest discrepancies in the standard model. The production cross section times the branching fraction into two or more charged tracks $\sigma(e^+e^- \to J/\psi + \eta_c) \times B_{>2}$, has been measured by the Belle Collaboration to be $25.6 \pm 2.8 \pm 3.4$ fb (Ref. [1]) and by the *BABAR* Collaboration to be $17.6 \pm 2.8^{+1.5}_{-2.1}$ fb (Ref. [2]). In contrast, nonrelativistic QCD (NRQCD) factorization [3] calculations at leading order in $\alpha_s$ predict cross sections of $3.78 \pm 1.26$ fb (Ref. [4]) and $5.5$ fb (Ref. [5]). The differences between these calculations arise from different choices of $m_c$, NRQCD matrix elements, and $\alpha_s$ and from the fact that the calculation of Ref. [4] includes QED effects, while that of Ref. [5] does not. An important recent development is the calculation of the corrections of next-to-leading order in $\alpha_s$ (Ref. [6]), which yield a $K$ factor of about 1.96. However, even if one includes this $K$ factor, a significant discrepancy remains.

It is known from the work of Ref. [4] that relativistic corrections to the process are potentially large. The first relativistic correction appears at order $v^2$, where $v$ is the heavy quark or antiquark velocity in the quarkonium rest frame. ($v^2 \approx 0.3$ for charmonium.) In Ref. [4], the order-$v^2$ corrections are estimated to give a $K$ factor $2.0^{+2.9}_{-1.1}$. The large uncertainties arise because of large uncertainties in the relevant nonperturbative NRQCD matrix element of relative order $v^2$. If the order-$v^2$ $K$ factor is indeed large, then this casts doubt on the convergence of the $v$ expansion of NRQCD. In what follows, we address the uncertainty in the order-$v^2$ matrix through a potential-model calculation and the convergence of the $v$ expansion through resummation.

---

[1] Talk presented by G. T. Bodwin.
[2] Work in the High Energy Physics Division at Argonne National Laboratory is supported by the U. S. Department of Energy, Division of High Energy Physics, under Contract No. W-31-109-ENG-38.

The NRQCD matrix element of leading-order in $v$ for production (or decay) of the $\eta_c$ is related to the wave function at the origin $\psi(0)$:

$$\psi(0) \equiv \int \frac{d^3 p}{(2\pi)^3} \tilde{\psi}(p) = \frac{1}{\sqrt{2N_c}} \langle 0|\chi^\dagger \psi|\eta_c\rangle. \tag{1}$$

Here $\psi$ annihilates a heavy quark, $\chi^\dagger$ annihilates a heavy antiquark, and $\psi(\mathbf{x})$ and $\tilde{\psi}(p)$ are coordinate-space and momentum-space Schrödinger wave functions, respectively. A similar matrix element appears in NRQCD expressions for $J/\psi$ production and decay. The corresponding NRQCD matrix elements of higher order in $v^2$ are given by

$$\psi^{(2n)}(0) \equiv \int \frac{d^3 p}{(2\pi)^3} p^{2n} \tilde{\psi}(p) = \frac{1}{\sqrt{2N_c}} \langle 0|\chi^\dagger (-\nabla^2)^n \psi|\eta_c\rangle. \tag{2}$$

We use the following notation for such matrix elements: $\langle p^{2n}\rangle \equiv \psi^{(2n)}(0)/\psi(0)$ and $\langle v^2\rangle = \langle p^2\rangle/m_c^2$. Previous attempts to determine $\psi^{(2)}(0)$ from phenomenology, from lattice measurements, and from the Gremm-Kapustin relation [7] have resulted in large uncertainties. Even the sign of $\psi^{(2)}(0)$ is not known with great confidence from these methods.

Our strategy is to use a potential-model calculation, with the Cornell potential [8], to determine $\psi^{(2)}(0)$. Details of this calculation can be found in Ref. [9]. If the model potential is the exact static $Q\bar{Q}$ potential, then the errors in the potential model are of relative order $v^2$ (Ref. [10]). With an appropriate choice of parameters, the Cornell potential provides a good fit to lattice data for the static $Q\bar{Q}$ potential.

The matrix element $\psi^{(2)}(0)$ contains a linear UV divergence and must be regulated. Because existing calculations in NRQCD of order $\alpha_s$ and higher make use of dimensional regularization, we ultimately want to obtain a dimensionally regulated matrix element. We introduce two methods to achieve this.

Method 1 requires knowledge only of the wave function. In this method, we first regulate using a simple, analytic momentum-space hard cutoff $\Lambda^2/(p^2 + \Lambda^2)$. Then, we calculate the difference between $\psi^{(2)}(0)$ in hard-cutoff and dimensional regularization, which we call $\Delta\psi^{(2)}(0)$. Since $\Delta\psi^{(2)}(0)$ gives the difference between UV regulators, it is dominated by large momenta and can be computed in perturbation theory. We subtract $\Delta\psi^{(2)}(0)$ from the hard-cutoff result. In order to obtain the dimensionally regulated result, we must extrapolate this subtracted expression to $\Lambda = \infty$ because our computation of $\Delta\psi^{(2)}(0)$ does not include contributions that are suppressed as $1/\Lambda$.

Method 2 is applicable only to potential models. It requires knowledge of the binding energy and the potential, in addition to the wave function. In this approach, we use the Bethe-Salpeter equation to expose an explicit loop from the wave function in the expression for the matrix element. Then, we regulate that loop dimensionally. The result is equal to that which would be obtained from the Gremm-Kapustin relation, but for the binding energy of the potential model.

The results from Method 1 and Method 2 agree well numerically and yield $\psi^{(2)}(0) = 0.118 \pm 0.024 \pm 0.035$ GeV$^{7/2}$ and $\langle p^2\rangle = 0.50 \pm 0.09 \pm 0.15$ GeV$^2$, which imply for $m_c = 1.4$ GeV that $\langle v^2\rangle \approx 0.25 \pm 0.05 \pm 0.08$. This last result is in good agreement with expectations from the NRQCD $v$-scaling rules. In all of these results, the first error bar reflects the uncertainty in the input potential-model parameters and the wave function at the origin, and the second error bar reflects relative-order-$v^2$ corrections that have been neglected. This is the first determination of $\psi^{(2)}(0)$ with small enough uncertainties to be useful phenomenologically.

We can extend Method 2 to matrix elements of higher order in $v$. By using the equation of motion, dimensional regularization, and the scalelessness of the individual terms in the Cornell potential, we obtain the simple relation

$$\langle p^{2n} \rangle = \langle p^2 \rangle^n. \tag{3}$$

This relation allows one to resum a class of the relativistic corrections to $S$-wave quarkonium decay and production amplitudes to all orders in $v$.

We now apply these results to the relativistic corrections to $\sigma[e^+e^- \to J/\psi + \eta_c]$. These corrections arise in two ways. First, they appear directly in the process $e^+e^- \to J/\psi + \eta_c$ itself. Second, they enter indirectly through $|\psi(0)|^2$, which appears as a factor in the contribution of leading order in $v^2$. The quantity $|\psi(0)|^2$ is determined from the experimental value for the width for $J/\psi \to e^+e^-$ and the theoretical expression for that process, which is affected by relativistic corrections.

Our preliminary results, including the effects of QED, as well as QCD, contributions are as follows. The resummed relativistic corrections to $\sigma[e^+e^- \to J/\psi + \eta_c]$ itself yield a $K$ factor 1.34, while the resummed relativistic corrections to $|\psi(0)|^2$ yield a $K$ factor 1.32. In both cases, the resummation of contributions beyond order $v^2$ has only about a 10% effect on the $K$ factors, which indicates that $v$ expansion of NRQCD converges well for these processes. Combining the resummed relativistic-correction $K$ factors with the relative-order-$\alpha_s$ $K$ factor 1.96, we obtain a complete $K$ factor 4.15. Applying this to the calculation of Ref. [4] and using the most recent value for $\Gamma[J/\psi \to e^+e^-]$ (Ref. [11]), we obtain a prediction $\sigma[e^+e^- \to J/\psi + \eta_c] = 17.5 \pm 5.7$ fb. The quoted uncertainty reflects only the uncertainties in the values of $m_c$ and $\langle p^2 \rangle$. Other theoretical uncertainties are large and need to be quantified. Nevertheless, our prediction is in agreement with the Belle and BABAR results, and it seems that the inclusion of relative-order-$\alpha_s$ and relativistic corrections resolves the discrepancy between theory and experiment at the present level of precision.

# REFERENCES

1. K. Abe *et al.* [Belle Collaboration], Phys. Rev. D **70**, 071102 (2004).
2. B. Aubert *et al.* [BABAR Collaboration], Phys. Rev. D **72**, 031101 (2005).
3. G. T. Bodwin, E. Braaten, and G. P. Lepage, *Phys. Rev.* **D51**, 1125 (1995); **55**, 5855(E) (1997).
4. E. Braaten and J. Lee, *Phys. Rev.* **D67**, 054007 (2003).
5. K. Y. Liu, Z. G. He, and K. T. Chao, *Phys. Lett.* **B557**, 45 (2003).
6. Y. J. Zhang, Y. j. Gao, and K. T. Chao, Phys. Rev. Lett. **96**, 092001 (2006).
7. M. Gremm and A. Kapustin, Phys. Lett. B **407**, 323 (1997).
8. E. Eichten, K. Gottfried, T. Kinoshita, K. D. Lane, and T. M. Yan, Phys. Rev. D **17**, 3090 (1978); **21**, 313(E) (1980).
9. G. T. Bodwin, D. Kang, and J. Lee, Phys. Rev. D **74**, 014014 (2006).
10. N. Brambilla, A. Pineda, J. Soto, and A. Vairo, Nucl. Phys. B **566**, 275 (2000).
11. W. M. Yao *et al.* [Particle Data Group], J. Phys. G **33**, 1 (2006).

# $X(3872)$ as a near-threshold state in the coupled-channel model for charmonia levels

Yu. S. Kalashnikova

*Institute of Theoretical and Experimental Physics, 117218, B.Cheremushkinskaya 25, Moscow, Russia*

**Abstract.** A coupled-channel approach for charmonia levels is formulated, based on the $^3P_0$ model for pair creation. Special attention is paid to the levels above the open–charm threshold. It is shown, in particular, that the coupling of the $2\,^3P_1$ $\bar{c}c$ state to the $DD^*$ channel generates a well-pronounced near-threshold state. Interrelation between this state and the recently discovered mesonic state $X(3872)$ is discussed.

The recently discovered mesonic state $X(3872)$ [1] is the most interesting member of the family of "homeless" charmonia, i.e. mesons which definitely contain the $\bar{c}c$ pair but do not fit the standard charmonium assignement. The Belle experiment has found the $X(3872)$ in the $B \to KJ/\psi\pi^+\pi^-$ reaction. While the $B \to KX$ decay is considered as good one for excited charmonia searches, the $X(3872)$ appears to have the mass too high for $1D$ levels, and too low for $2P$ ones [2]. The observation of the $J/\psi\pi^+\pi^-\pi^0$ ($J/\psi\omega$) and $J/\psi\gamma$ decay modes of the $X(3872)$ [3] points to the positive $C$–parity of the $X$, so that the dipion in the $\pi^+\pi^-J/\psi$ mode is $C$–odd, originating from the $\rho$. This means considerable isospin violation in the wavefunction of the $X$. At present, the quantum numbers of the $X(3872)$ are established to be either $1^{++}$ or $2^{-+}$, while other hypotheses are excluded by studies [4] of the dipion mass spectrum.

The most interesting feature of the $X(3872)$ is that it resides exactly at the $D^0\bar{D}^{0*}$ threshold,

$$M_X - M(D^0 D^{*0}) = -0.4 \pm 0.7 \; MeV. \tag{1}$$

This suggests that that some interplay between the resonance $X(3872)$ and the $D\bar{D}^*$ threshold should take place, so that the wavefunction of the $X$ contains considerable fraction of $D\bar{D}^*$ molecule. Due to about 8 MeV mass difference between charged and neutral $D\bar{D}^*$ thresholds, this fraction could generate considerable isospin mixing. So the molecular model for the $X(3872)$ has become very popular, and several mechanisms which could be responsible for the $D\bar{D}^*$ threshold attraction were suggested.

In fact, one of such mechanisms was discovered many years ago, [5, 6]. This is $t$–channel one-pion exchange, shown to be attractice in the $1^{++}$ $DD^*$ channel. The molecule is formed in full analogy with the deuteron, and was called "deuson" in [6]. Another mechanism, suggested in [7], operates via quark–rearrangement kernels responsible for the $D\bar{D}^* \leftrightarrow J/\psi\rho/\omega$ transitions, and, again, produces attaction in the $1^{++}$ partial wave. In terms of colourless states, this can be considered as $u$–channel exchange. The estimated strength of attraction caused by quark–rearrangement kernels was found to be not enough to bind $D\bar{D}^*$, so in actual calculations [7] both $t$– and $u$–

CP892, *Quark Confinement and the Hadron Spectrum VII*
edited by J. E. F. T. Ribeiro
© 2007 American Institute of Physics 978-0-7354-0396-3/07/$23.00

channel exchanges were employed.

It appears that the s–channel exchanges could be also responsible for threshold attraction in the $D\bar{D}^*$ $1^{++}$ channel [8]: account for $\bar{c}c \leftrightarrow D\bar{D}^{(*)}$ mixing can generate a structure at the $D\bar{D}^*$ threshold in the $1^{++}$ partial wave. The phenomenon of generating such state is rather peculiar: one and a same charmonium state gives rise both to the resonance well above the threshold and a near-threshold state. Due to large coupling of the bare quarkonium state to hadronic channel, the one–to–one correspondence between bare and physical states could be destroyed, as widely discussed in connection with light scalar mesons, see e.g. [9].

The key observation here is that the effect of virtual mesonic loops on charmonium spectrum is large near S–wave mesonic thresholds. In the charmonium sector it means that one should expect pronounced cusps and threshold effects to show up in the mass range of about $3.8 - 4.00$ GeV, i.e. where the S–wave $D\bar{D}^{(*)}$ thresholds start to open. The channels affected are going to be the ones coupled to $2P$ charmonia levels, with quantum numbers $1^{+-}$ and $J^{++}, J = 0,1,2$.

The microscopic coupled–channel model was developed in [8], where the dressing of bare $c\bar{c}$ levels due to coupling to $D\bar{D}^{(*)}$ and $D_s\bar{D}_s^{(*)}$ channels was taken into account. The details of the coupled–channel model formalism can be found in [11]. The nonrelativistic quark potential model was employed to describe the bare charmonia and $D$-mesons, and the $c\bar{c} \leftrightarrow D\bar{D}^{(*)}$ trnsitons were calculated with the $^3P_0$ model for quark pair creation. The quark model parameters and $^3P_0$ pair–creation strength were chosen to reproduce, with reasonable accuracy, the positions of charmonia levels below open charm threshold, and the $D\bar{D}$ width of the $\psi(3770)$. The predicted mass of the $1^{++}$ resonance is calculated to be 3990 MeV, with the width of about 27 MeV.

The relative strength of the S–wave threshold effects in various channels is defined by the values of the S–wave spin-recoupling coefficients which enters the $c\bar{c} \leftrightarrow D\bar{D}^{(*)}$ transition amplitude. In any reasonable light–quark pair–creation model, including the $^3P_0$ [8] and $^3S_1$ [10] models, these coefficients take the form

$$
\begin{aligned}
^3P_2 &\to D^*\bar{D}^* \\
^3P_1 &\to \frac{1}{\sqrt{2}}(D\bar{D}^* + \bar{D}D^*) \\
^3P_0 &\to \frac{\sqrt{3}}{2}D\bar{D} + \frac{1}{2}D^*\bar{D}^* \\
^1P_1 &\to \frac{1}{2}(D\bar{D}^* - \bar{D}D^*) + \frac{1}{\sqrt{2}}D^*\bar{D}^*.
\end{aligned}
\tag{2}
$$

It is clear from these equations that the strongest threshold effects are to be expected at the $D^*\bar{D}^*$ threshold in the $2^{++}$ partial wave, and at the $D\bar{D}^*$ threshold in the $1^{++}$ partial wave. This is confirmed by spectral density analysis.

The so–called spectral density $w(M)$, the probability to find the bare state in the mesonic continuum, arises naturally in the coupled–channel model. The spectral density has several important properties. First, it is a normalizable quantity, with its integral over continuum spectrum being equal to unity (if there is no bound state) [12]. Second, if the mesonic final state is produced in some reaction via intermediate $c\bar{c}$ state, one obtains that the cross-section is proportional to $w(M)$.

The spectral density of the bare $2^3P_1$ level is shown at Fig. 1. The resonance at 3990 MeV is clearly seen, and, in addition, there is a near-threshold peak, rising at the flat

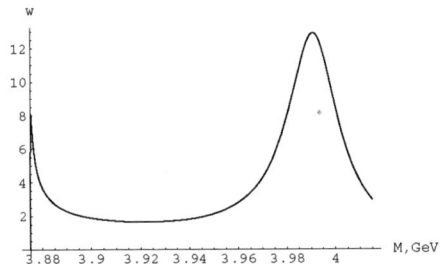

**FIGURE 1.** The spectral density of the $2^3P_1$ $c\bar{c}$ state.

background. This signals the presence of the near-threshold singularity in the $D\bar{D}^*$ $t$-matrix. The $D\bar{D}^*$ scattering length appears to be negative and large,

$$a_{D\bar{D}^*} = -8 \; fm, \tag{3}$$

so there is a virtual state very close to the $D\bar{D}^*$ threshold, with the energy $\varepsilon = 0.32$ MeV. Slight decrease of the $2^3P_1$ bare state mass or several per cent increase of the pair-creation strength could lead to moving the state to the physical sheet, i.e. to appearance of the bound state. It is tempting to idetify such virtual (or bound) state with the $X(3872)$.

As the spectral density is normalized to unity, the near-threshold fraction of the $2^3P_1$ bare charmonium state is small, so the state is essentially the $D\bar{D}^*$ molecule, with the properties not very distiguishable from the ones of the "deuson" molecule [5, 6].

This research was supported by the grants RFFI-05-02-04012-NNIOa, DFG-436 RUS 113/820/0-1(R), NSh-843.2006.2, by the Federal Programme of the Russian Ministry of Industry, Science, and Technology No. 40.052.1.1.1112, and by the Russian Governmental Agreement N 02.434.11.7091.

# REFERENCES

1. S. K. Choi *et al.* [Belle Collaboration], *Phys. Rev. Lett* **91**, 262001 (2003); D. Acosta *et al.* [CDF II Collaboration], *Phys. Rev. Lett* **93**, 072001 (2004).
2. T. Barnes ans S. Godfrey, *Phys. Rev.* **D69**, 054008 (2004).
3. K. Abe *et al.* [Belle Collaboration], BELLE-CONF-0540, arXiv:hep-ph/0505037.
4. A. Abulencia *et al.* [CDF Collaboration], *Phys. Rev. Lett.* **96**, 102002 (2006).
5. M. B. Voloshin and L. B. Okun, *JETP Lett.* **23**, 333 (1976).
6. N. A. Tornqvist, *Phys. Rev. Lett.* **67**, 556 (1991).
7. E. S. Swanson, *Phys. Lett.* **B588**, 189 (2004).
8. Yu. S. Kalashnikova, *Phys. Rev.* **D72**, 034010 (2005)
9. M. Boglione, M. R. Pennington, *Phys. Rev.* **D65**, 114010 (2002).
10. E. Eichten, K. Gottfried, T. Kinoshita, K. D. Lane, and T. M. Yan, *Phys. Rev.* **D17**, 3090 (1978).
11. V. Baru *et al Phys. Lett.* **B586**, 53 (2004).
12. L.N. Bogdanova, G.M. Hale, and V.E. Markushin, *Phys. Rev.* **C44**, 1289 (1991).

# Evidence for $B^+ \to \tau^+\nu_\tau$ decays and measurement of $f_B$ from Belle

Youngjoon Kwon[1]

*Dept. of Physics, Yonsei University, Seoul 120-749, KOREA*

**Abstract.** We present the first evidence for $B^+ \to \tau^+\nu_\tau$ decay, using 414 fb$^{-1}$ of $B$ meson decay event sample collected with the Belle detector at the KEKB $e^+e^-$ collider. To cope with large missing energy due to multiple neutrinos in the final state, events are tagged by fully reconstructing one of the $B$ mesons. We find the evidence for signal with a significance of 3.5 $\sigma$ including systematic uncertainties. The branching fraction is measured to be $\mathscr{B}(B^+ \to \tau^+\nu_\tau) = (1.79^{+0.56+0.46}_{-0.49-0.51}) \times 10^{-4}$. From this we obtain $f_B = 0.229^{+0.036+0.034}_{-0.031-0.037}$ GeV, the first direct determination of the $B$ meson decay constant.

**Keywords:** Leptonic decay, $B^+ \to \tau^+\nu_\tau$, $B$ meson decay constant $f_B$
**PACS:** PACS numbers: 13.20.-v, 13.25.Hw

## INTRODUCTION

The purely leptonic decays $B^+ \to \ell^+\nu_\ell$[1] are allowed in the SM via annihilation of initial-state quarks, $\bar{b}$ and $u$, into a virtual $W$ boson (Figure 1). It provides a direct determination of $f_B|V_{ub}|$ where $f_B$ is the $B$ meson decay constant. The branching fraction is given by

$$\mathscr{B}(B^+ \to \ell^+\nu_\ell) = \frac{G_F^2 m_B m_\ell^2}{8\pi} \left(1 - \frac{m_\ell^2}{m_B^2}\right)^2 f_B^2 |V_{ub}|^2 \tau_B.$$

Because of helicity suppression, the expected branching fraction to $\tau^+\nu_\tau$ mode is larger than the others by a few orders of magnitude. Physics beyond the SM, such as supersymmetry or two-Higgs doublet models, could modify $\mathscr{B}(B^+ \to \tau^+\nu_\tau)$ through the introduction of a charged Higgs boson [2]. Purely leptonic $B$ decays have not been observed in past experiments.

## ANALYSIS

To search for this decay we analyze $449 \times 10^6$ $B$ meson pairs collected with the Belle detector at the KEKB asymmetric-energy $e^+e^-$ collider operating at the $\Upsilon(4S)$ resonance. To improve signal purity we fully reconstruct one of the $B$ mesons in the event ($B_{\text{tag}}$), and compare properties of the remaining particle(s) ($B_{\text{sig}}$) to those expected for signal and background. This method allows us to suppress strongly the combinatorial

---

[1] Representing the Belle collaboration

CP892, *Quark Confinement and the Hadron Spectrum VII*
edited by J. E. F. T. Ribeiro
© 2007 American Institute of Physics 978-0-7354-0396-3/07/$23.00

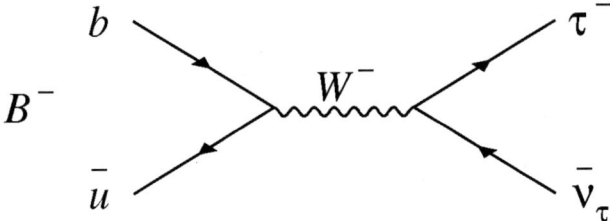

**FIGURE 1.** Purely leptonic $B$ decay proceeds via quark annihilation into a $W$ boson.

background. The $B_{\text{tag}}$ candidates are reconstructed in the hadronic $B$ decays modes. Using the variables $M_{\text{bc}}$ and $\Delta E$, we estimate the number of $B_{\text{tag}}$'s and their purity in the selected region to be $6.80 \times 10^5$ and 0.55, respectively.

In the events where a $B_{\text{tag}}$ is reconstructed, we search for decays of $B_{\text{sig}}$ into a $\tau$ and a neutrino. Candidate events are required to have one or three charged track(s) on the signal side with the total charge being opposite to that of $B_{\text{tag}}$. The $\tau$ lepton is identified in the five decay modes, $\mu^-\bar{\nu}_\mu\nu_\tau$, $e^-\bar{\nu}_e\nu_\tau$, $\pi^-\nu_\tau$, $\pi^-\pi^0\nu_\tau$ and $\pi^-\pi^+\pi^-\nu_\tau$, which taken together correspond to 81% of all $\tau$ decays.

The most powerful variable for separating signal and background is the remaining energy in the CsI electromagnetic calorimeter, denoted as $E_{\text{ECL}}$, which is sum of the energy of photons that are not associated with either the $B_{\text{tag}}$ or the $\pi^0$ candidate from the $\tau^- \to \pi^-\pi^0\nu_\tau$ decay. For signal events, $E_{\text{ECL}}$ must be either zero or a small value arising from beam background hits, therefore, signal events peak at low $E_{\text{ECL}}$. On the other hand background events are distributed toward higher $E_{\text{ECL}}$ due to the contribution from additional neutral clusters. We find a significant excess of events in the $E_{\text{ECL}}$ signal region.

The validity of $E_{\text{ECL}}$ simulation is tested on a control sample of double-tagged events, where the $B_{\text{tag}}$ is fully reconstructed and $B_{\text{sig}}$ is reconstructed in the decay chain, $B^- \to D^{*0}\ell^-\bar{\nu}$ ($D^{*0} \to D^0\pi^0$), followed by $D^0 \to K^-\pi^+$ or $K^-\pi^+\pi^+\pi^+$. The sources affecting the $E_{\text{ECL}}$ distribution in the control sample are similar to those affecting the distribution in the signal MC. The results of data and MC agree within the statistical uncertainty of the test sample.

Figure 2 shows the $E_{\text{ECL}}$ distribution for all $\tau$ decay modes combined. One can see a significant excess of events in the region below $E_{\text{ECL}} < 0.25$ GeV. For the events in this region, we verify that the distributions of the event selection variables other than $E_{\text{ECL}}$, such as $M_{\text{bc}}$ and $p_{\text{miss}}$, are consistent with those expected from MC.

We deduce the final results by fitting the obtained $E_{\text{ECL}}$ distributions to the sum of the expected signal and background shapes. Probability distribution functions (PDFs) for the signal $f_s$ ($E_{\text{ECL}}$) and for the background $f_b$ ($E_{\text{ECL}}$) are constructed for each $\tau$ decay mode from the MC simulation. The PDFs are combined into an extended likelihood function,

$$\mathscr{L} = \frac{e^{-(n_s+n_b)}}{N!} \prod_{i=1}^{N} (n_s f_s(E_i) + n_b f_b(E_i)) \;.$$

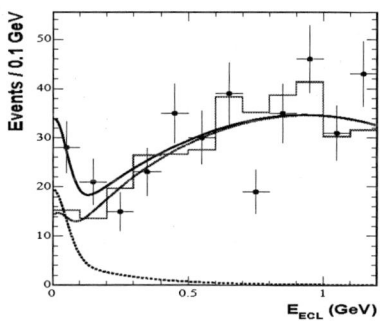

**FIGURE 2.** $E_{ECL}$ distributions. The data and background MC samples are represented by the points with error bar and the solid histogram, respectively. The solid curve shows the result of the fit with the sum of the signal shape (dashed) and background shape (dotted).

The branching fraction is calculated as $\mathscr{B} = N_s/(2 \cdot \varepsilon \cdot N_{B^+B^-})$. The branching fraction calculated from each $\tau$ decay mode is consistent with one another within statistical error. To deduce the combined result for all $\tau$ decay modes, the likelihood functions are multiplied to produce the combined likelihood (), and constrain the signal components of the five modes by a single branching fraction. The combined fit results in $24.1^{+7.6}_{-6.6}$ signal events in the fitted $E_{ECL}$ region. Including the systematic uncertainty, the branching fraction is obtained:

$$\mathscr{B}(B^+ \to \tau^+ \nu_\tau) = (1.79^{+0.56+0.46}_{-0.49-0.51}) \times 10^{-4} .$$

The significance of the signal, after including the systematics, is 3.5 $\sigma$. Using the HFAG average value for $|V_{ub}|$[3], we obtain $f_B = 0.229^{+0.036+0.034}_{-0.031-0.037}$ GeV, the first direct determination of the $B$ meson decay constant.

## ACKNOWLEDGMENTS

We thank the KEKB group, the KEK cryogenics group, and the KEK computer group for excellent support for this study. This work is supported in part by the CHEP program of KOSEF.

## REFERENCES

1. Charge conjugate states are implied throughout this paper.
2. W. S. Hou, Phys. Rev. D **48**, 2342 (1993).
3. Heavy Flavor Averaging Group, winter 2005 results, (http://www.slac.stanford.edu/xorg/hfag/).

# Off-shell and non-static contributions to heavy-quarkonium production

## J.P. Lansberg

*Centre de Physique Théorique, École polytechnique, CNRS,*
*91128 Palaiseau, France*
*and*
*Physique théorique fondamentale, Département de Physique, Université de Liège,*
*allée du 6 Août 17, bât. B5, B-4000 Liège 1, Belgium*
*E-mail: Jean-Philippe.Lansberg@cpht.polytechnique.fr*

**Abstract.** We have shown that if one relaxes the constraint that the quarks in a heavy quarkonium are at rest and on-shell, new contributions to the discontinuity of the production amplitude appear. These can be seen as a $s$-cut in the amplitude and are on the same footage as the classical cut of the Colour-Singlet Model (CSM), where the heavy quarks forming the quarkonium are put on-shell by hypothesis. We treat this cut in a gauge-invariant manner by introducing necessary new 4-point vertices, suggestive of the colour-octet mechanism. We have further shown that this cut contributes at least as much as the LO CSM at large $P_T$. However, the 4-point vertices cannot be totally constrained and an ambiguity remains to what concerns their actual contribution. Theoretical insights from meson photoproduction are discussed in that context.

**Keywords:** heavy-quarkonium production, vector-meson production, gauge invariance, relativistic effects, non-static extension
**PACS:** 14.40.Gx, 13.85.Ni, 11.10.St, 13.20.Gd

More than ten years ago, the CDF collaboration [1, 2] brought to light the "$\psi'$ anomaly", *i.e.* an excessively large experimental cross section for $\psi'$ (and $J/\psi$) production compared with theoretical expectations. No totally conclusive solution to this problem has been proposed so far (for recent reviews see [3, 4]). Even though the Colour-Octet Mechanism (COM), coming from the application of NRQCD to heavy quarkonium, looked as a very promising solution, it appears clearly that as long as fragmentation is the dominant production contribution and the velocity-scaling rules of NRQCD hold, it cannot accommodate the polarisation measurements of CDF [5], which show a non-polarised, if not slightly longitudinal, production.

In that context, we have felt the necessity to reconsider the appropriateness of the static and on-shell approximation of the Colour-Singlet Model (CSM) [6].

In order to study properly non-static and off-shell effects, we have used a vertex function as an input for the bound-state characteristics, whereas the Schrödinger wave function at the origin is used in the CSM and Long Distance Matrix Elements (LDME) of NRQCD enter the COM. We emphasise again that we probe all the internal phase space of the quarkonium, and thus need a function, where these two approaches simply need a constant factor.

In the case of $^3S_1$ quarkonium (noted $\mathcal{Q}$) production in high-energy hadronic

CP892, *Quark Confinement and the Hadron Spectrum VII*
edited by J. E. F. T. Ribeiro
© 2007 American Institute of Physics 978-0-7354-0396-3/07/$23.00

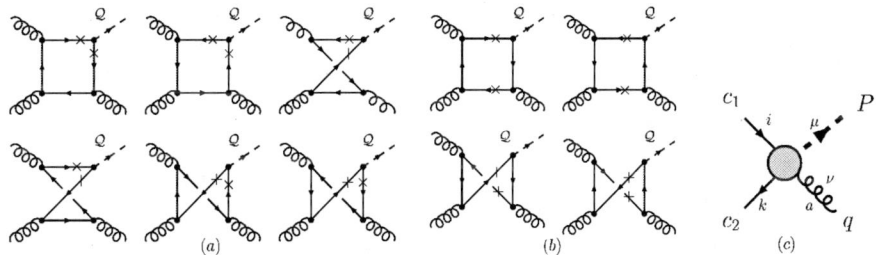

**FIGURE 1.** The first family (a) has 6 diagrams and the second family (b) 4 diagrams contributing the discontinuity of $gg \to {}^3S_1 g$ at LO in QCD.(c): the gauge-invariance restoring vertex, $\Gamma^{(4)}$.

collisions, we are to consider gluon fusion $gg \to \mathcal{Q}g$. Using the Landau equations [7], we have shown in [8] that there are two families of contributions (see Fig. 1 (a) and (b)): the first is the usual colour-singlet mechanism, where in the context of our model, we use a 3-point function $\Gamma_\mu^{(3)}(p,P) = \Gamma(p,P)\gamma_\mu$ at the $Q\bar{Q}\mathcal{Q}$ vertex; the second family was never considered before. To simplify the study, we set $m > M/2$ so that the first cut does not contribute. The functional form of $\Gamma(p,P)$ (gaussian or dipole) and its parameters have been discussed in details in [9].

In addition to the second family, one is driven – to preserve gauge invariance (GI) – to introduce new contributions arising from the presence of 4-point vertices. Beside restoring GI, these vertices have to satisfy specific constraints [8, 10, 11]. For the following simple choice for $\Gamma_{\mu\nu}^{(4)}(c_1, c_2, P, q)$

$$-ig_s T_{ki}^a (\Gamma_1 - \Gamma_2) \left[ \frac{c_1^\nu}{(c_1 - q)^2 - m^2} + \frac{c_2^\nu}{(c_2 + q)^2 - m^2} \right] \gamma^\mu, \tag{1}$$

where $\Gamma_1 \equiv \Gamma(2c_1 - P, P)$, $\Gamma_2 \equiv \Gamma(2c_2 - P, P)$, the momenta and indices are as in Fig. 1 (c), the results obtained for $J/\psi$ and $\psi'$ production at the Tevatron are exposed in [8]. For the $J/\psi$, we saw that the $s$-channel cut contributes at least as much as the classical cut of the CSM at large $P_T$. In the $\psi'$ case, we employed the ambiguity upon the vertex-function normalisation [9] due to the node position to show that agreement with the data at low $P_T$ was conceivable. We further noticed that the $P_T$ slope was only slightly different from that of the data. This is at variance with what is widely believed since COM fragmentation (with a typical $1/P_T^4$ behaviour) processes are in agreement with experimental measurements.

Another possible Ansatz for $\Gamma_{\mu\nu}^{(4)}(c_1, c_2, P, q)$ [12], inspired from studies of meson photoproduction [13, 14, 15] and which possesses a better behaviour at low momenta, is

$$-ig_s T_{ki}^a \left( \frac{(2c_1 - q)^\nu}{(c_1 - q)^2 - m^2}(\Gamma_2 - F) + \frac{(2c_2 + q)^\nu}{(c_2 + q)^2 - m^2}(\Gamma_1 - F) \right) \gamma^\mu \tag{2}$$

with $F = \Gamma_0 - h(\Gamma_0 - \Gamma_1)(\Gamma_0 - \Gamma_2)$ ($\Gamma_0$ is the value of the vertex function when $(c_1 - q)^2 = (c_2 + q)^2 = m^2$) and $h$ an arbitrary crossing-symmetric function of the momenta. This will be studied in a future work. Applications to $\eta_c$ and $\eta_c'$ decays could also be relevant in view of the possible anomaly of $\eta_c' \to \gamma\gamma$ decay [16].

However, there exist further other choices for the GI restoring vertex. Interesting results are indeed obtained by studying the effects of autonomous vertices, which link different suitable choices: they are GI alone and a priori unconstrained in normalisation. The latter can be fitted to described data from [1, 2, 17, 18] as shown in [10, 19].

In conclusion, we have shown that it is possible to go beyond the on-shell and static approximations of the CSM. It may also be possible to extend the COM in the same manner. This necessitates the introduction of 4-point vertices due to the non-local 3-point vertex relevant for the non-static and off-shell contributions.

By going deeper in the analysis, we have seen that the form of these 4-point vertices is not absolutely constrained, even after imposing necessary conditions to conserve crossing symmetry and the analytic structure of the amplitude. By exploiting this lack of constraint, we have been able to reproduce the direct-production cross section for the $J/\psi$, $\psi'$ and $\Upsilon(1S)$ as measured at the Tevatron by CDF (and also at RHIC by PHENIX for $J/\psi$).

# REFERENCES

1. F. Abe *et al.* [CDF Collaboration], Phys. Rev. Lett. **79** (1997) 572.
2. F. Abe *et al.* [CDF Collaboration], Phys. Rev. Lett. **79** (1997) 578.
3. N. Brambilla *et al.*, *Heavy quarkonium physics*, CERN Yellow Report, CERN-2005-005, 2005 Geneva : CERN, 487 pp [arXiv:hep-ph/0412158].
4. J. P. Lansberg, $J/\psi$, $\psi'$ and $\Upsilon$ *production at hadron colliders: A review*, Int. J. Mod. Phys. A **21** (2006) 3857 [arXiv:hep-ph/0602091].
5. T. Affolder *et al.* [CDF Collaboration], Phys. Rev. Lett. **85** (2000) 2886 [arXiv:hep-ex/0004027].
6. C-H. Chang, Nucl. Phys. **B 172** (1980) 425; R. Baier and R. Rückl, Phys. Lett. B **102** (1981) 364; R. Baier and R. Rückl, Z. Phys. **C 19** (1983) 251.
7. L. D. Landau, Nucl. Phys. **13** (1959) 181.
8. J. P. Lansberg, J. R. Cudell and Yu. L. Kalinovsky, Phys. Lett. B **633**, 301 (2006) [arXiv:hep-ph/0507060].
9. J. P. Lansberg, AIP Conf. Proc. **775** (2005) 11 [arXiv:hep-ph/0507184].
10. J. P. Lansberg, *Quarkonium Production at High-Energy Hadron Colliders*, Ph.D. Thesis, ULg, Liège, Belgium, 2005.
11. S. D. Drell and T. D. Lee, Phys. Rev. D **5** (1972) 1738.
12. H. Haberzettl, private communication.
13. H. Haberzettl, Phys. Rev. C **56** (1997) 2041 [arXiv:nucl-th/9704057].
14. H. Haberzettl *et al.*, Phys. Rev. C **58** (1998) 40 [arXiv:nucl-th/9804051].
15. R. M. Davidson and R. Workman, Phys. Rev. C **63** (2001) 025210 [arXiv:nucl-th/0101066].
16. J. P. Lansberg and T. N. Pham, Phys. Rev. D **74** (2006) 034001 [arXiv:hep-ph/0603113].
17. D. Acosta *et al.* [CDF Collaboration], Phys. Rev. Lett. **88** (2002) 161802.
18. S. S. Adler *et al.* [PHENIX Collaboration], Phys. Rev. Lett. **92** (2004) 051802 [arXiv:hep-ex/0307019].
19. J. P. Lansberg, AIP Conf. Proc. **792** (2005) 823 [arXiv:hep-ph/0507118].

# Non-leptonic B decays: tests of factorization, measurements of polarizations, strong phases, and final state interactions

Emmanuel Latour

*Laboratoire Leprince-Ringuet, CNRS/IN2P3, Ecole Polytechnique, F-91128 Palaiseau, FRANCE*

**Abstract.** Various factorization approaches can be challenged by measurements on a variety of quantities such as branching fractions, CP asymmetries, strong phases and polarization fractions. This is of crucial importance to better understand hadronic $B$ decays, in particular in unveiling the role of final state interactions. Some recent BaBar measurements concerning charm and charmless decays are presented here, mostly on $211 \, \text{fb}^{-1}$, and provide valuable information on factorization, its origin and breakdown.

**Keywords:** Hadronic decays, factorization, polarizations, strong phases, final state interactions
**PACS:** 12.38.Qk,12.39.St, 13.25.Hw, 13.88.+e

## Factorization tests with polarisations, strong phases and FSI

Hadronic $B$ decays are of fundamental interest as they can be used to extract CKM parameters or to search for New Physics. Unfortunatly, they involve strong interactions between particles in final states that cannot be computed theoretically. Factorization, based on color transparency, estimates the non-perturbative decay matrix elements involving local operators of OPE. Several approaches have been developped (naive factorization [1, 2], generalized factorization [3, 4], and the more sophisticated perturbative QCD [5, 6, 7], QCD factorization [8, 9], and SCET [10, 11]), which differ in many aspects such as power counting, treatment of non-factorizable contributions, or scheme/renormalization dependency.

FSI generate strong phases. Since they are treated in different ways among the various factorization schemes, measuring strong phases and comparing them with the various factorization predictions give precious information on FSI. This is fundamental as FSI can be responsible for branching fraction enhancement or polarization anomalies. In addition, factorization approaches provide many other predictions on polarization, branching ratios CP asymmetries that all can be used to test further those approaches and search for breakdowns.

## Experimental results

The variety of charmless decays provides an ideal framework to test extensively factorization predictions. It is not possible to cover them all, so only a few examples will be given. Several analyses of $B$ decays to final state involving an $\eta^{(\prime)}$ have recently been

CP892, *Quark Confinement and the Hadron Spectrum VII*
edited by J. E. F. T. Ribeiro
© 2007 American Institute of Physics 978-0-7354-0396-3/07/$23.00

performed. $B \to \eta^{(\prime)} K^*$ have shown a very good agreement between QCD factorization predictions and data both on branching ratios and CP asymmetries (See Table 1 and [12]). Other modes such as $B \to \eta^{(\prime)} \eta^{(\prime)}, \eta^{(\prime)} \phi, \eta^{(\prime)} \pi^0$, were not observed, but the upper limits on branching ratios ($\mathscr{O}(10^{-6})$) are now close to the range predicted by different factorization schemes. Thus those modes should soon be seen if factorization holds.

**TABLE 1.** Significance, branching ratios, CP asymmetries in $B \to \eta K^*$.

| | $S(\sigma)$ | $BR(10^{-6})$ | $\mathscr{A}_{CP}$ |
|---|---|---|---|
| $B^0 \to \eta K^{*0}(892)$ | 18.8 | $16.5 \pm 1.1 \pm 0.8$ | $0.21 \pm 0.06 \pm 0.02$ |
| $B^+ \to \eta K^{*+}(892)$ | 13.0 | $18.9 \pm 1.8 \pm 1.3$ | $0.01 \pm 0.08 \pm 0.02$ |
| $B^0 \to \eta (K\pi)_0^{*0}$ | 5.7 | $11.0 \pm 1.6 \pm 1.5$ | $0.06 \pm 0.13 \pm 0.02$ |
| $B^+ \to \eta (K\pi)_0^{*+}$ | 5.9 | $18.2 \pm 2.6 \pm 2.6$ | $0.05 \pm 0.13 \pm 0.02$ |
| $B^0 \to \eta K_2^{*0}(1430)$ | 5.3 | $9.6 \pm 1.8 \pm 1.1$ | $-0.7 \pm 0.19 \pm 0.02$ |

The longitudinal polarization anomaly of $B \to \phi K^*$ observed by BaBar and Belle [13][14] ($f_L \sim 0.5$ while theory predicted $f_L \sim 0.9$) can be investigated further by studying $SU(3)_F$ related modes. Two channels ($\rho^+ K^{*0}$ and $\rho^0 K^{*0}$) in the $B \to \rho K^*$ analysis have been observed with a significance of more than $5\sigma$, and the longitudinal polarizations, $f_L = 0.52 \pm 0.10 \pm 0.04$ and $f_L = 0.57 \pm 0.09 \pm 0.08$ respectively, show a similar behavior as for $B \to \phi K^*$. The $B \to \omega K^*$ analysis, which could also give some information, did not observe any signal. Only the $B \to \omega \rho^+$ was observed in this analysis, at a 5.7 $\sigma$ significance, and yielded a dominant longitudinal polarization fraction ($f_L = 0.82 \pm 0.11 \pm 0.02$) as could be expected from $B \to \rho \rho$.

In decays involving $b \to c$ transitions, two recent analyses enable nice tests of factorization. First, the $\bar{B}^0 \to D^{*+} \omega \pi^-$ analyses probes the origin of factorization. Following Ref.[15], factorization predicts:

$$\frac{d\Gamma(B \to D^* X)/dm_{X^2}}{d\Gamma(B \to D^* l\nu)/dm_{X^2}} = 3\pi \left( c_1(m_b) + \frac{c_2(m_b)}{3} \right) v(m_{X^2})(1 + \delta_{NF}) \qquad (1)$$

$v(m_{X^2})$ can be extracted from $\tau \to X\nu$ data. Depending on the factorization approach used, the non-factorizable contributions included in $\delta_{NF}$ will behave differently as a function of $m_{X^2}$. In the large $N_c$ limit it will be $\mathscr{O}(1/N_c)$, while corrections will be suppressed by powers of $m_X/E_X$ in perturbative QCD. Looking at the spectrum of $\bar{B}^0 \to D^{*+} \omega \pi^-$ in the region where $\tau \to X\nu$ data is available, factorization predictions agree both in shape and normalization with data (see Fig. 1), on the whole range of $m_{X^2}$, which favors approaches having no $m_{X^2}$ dependency.

In addition to this measurement, the analysis gives access to the longitudinal polarization fraction of $D^{*+}$. The result $f_L = 0.654 \pm 0.042 \pm 0.016$ is in perfect agreement with the generalized factorization predictions.

The full angular analysis of the Vector-Vector $B \to (c\bar{c}) K^*$, with $(c\bar{c}) = J/\psi, \psi(2S), \chi_{c_1}$, provides the measurement of the three polarization fraction $|A_0|^2$, $|A_\parallel|^2, |A_\perp|^2$ and strong phases $\delta_\parallel - \delta_0, \delta_\perp - \delta_0$ (See Table 2). Those decays being color-suppressed, color transparency does not hold, hence factorization may be questionable. The results reveal a clear breaking of factorization. First, $\delta_\parallel - \delta_\perp \neq 0$ at $8\sigma$. Furthermore, factorization approaches including non-factorizable contributions [17]

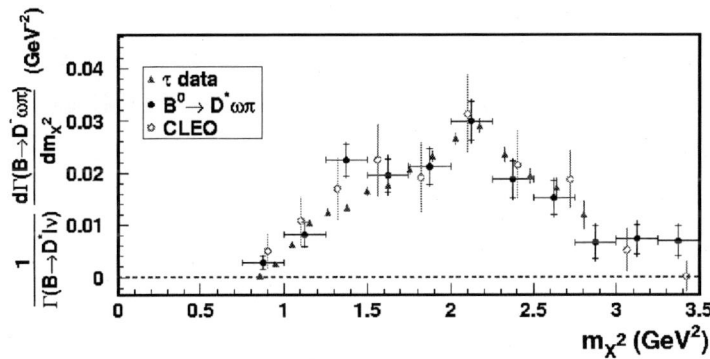

**FIGURE 1.** Differential decay rate as a function of $m_{X^2}$ for BaBar, CLEO, and as predicted by factorization. [16]

predict a larger longitudinal polarization for $J\psi K^*$ larger than for $\chi_{c_1} K^*$, which is not the observed. Finally, the fact that $\delta_\parallel \sim 0$ for $\chi_{c_1} K^*$ while $\delta_\parallel \sim -\pi$ for $J/\psi K^*/\psi(2S)K^*$ is not understood.

**TABLE 2.** Polarizations and strong phases in $B \to (c\bar{c})K^*$.

|  | $J/\psi K^*$ | $\psi(2S)K^*$ | $\chi_{c_1} K^*$ |
|---|---|---|---|
| $|A_0|^2$ | $0.556 \pm 0.009 \pm 0.010$ | $0.48 \pm 0.05 \pm 0.02$ | $0.77 \pm 0.07 \pm 0.04$ |
| $|A_\parallel|^2$ | $0.211 \pm 0.010 \pm 0.006$ | $0.22 \pm 0.06 \pm 0.02$ | $0.20 \pm 0.07 \pm 0.04$ |
| $|A_\perp|^2$ | $0.233 \pm 0.010 \pm 0.005$ | $0.30 \pm 0.06 \pm 0.02$ | $0.03 \pm 0.04 \pm 0.02$ |
| $\delta_\parallel - \delta_0$ | $-2.93 \pm 0.08 \pm 0.04$ | $-2.8 \pm 0.4 \pm 0.1$ | $0.0 \pm 0.3 \pm 0.1$ |
| $\delta_\perp - \delta_0$ | $2.91 \pm 0.05 \pm 0.03$ | $2.8 \pm 0.3 \pm 0.1$ | (-) |

# REFERENCES

1.  M. Wirbel, B. Stech and M. Bauer, Z. Phys. C **29**, 637 (1985).
2.  M. Bauer, B. Stech and M. Wirbel, Z. Phys. C **34**, 103 (1987).
3.  A. Ali, G. Kramer and C. D. Lu, Phys. Rev. D **58**, 094009 (1998)
4.  Y. H. Chen, H. Y. Cheng, B. Tseng and K. C. Yang, Phys. Rev. D **60**, 094014 (1999)
5.  H. n. Li, Phys. Rev. D **52**, 3958 (1995)
6.  C. H. Chang and H. n. Li, Phys. Rev. D **55**, 5577 (1997)
7.  T. W. Yeh and H. n. Li, Phys. Rev. D **56**, 1615 (1997)
8.  M. Beneke, G. Buchalla, M. Neubert and C. T. Sachrajda, Phys. Rev. Lett. **83**, 1914 (1999)
9.  M. Beneke, G. Buchalla, M. Neubert and C. T. Sachrajda, Nucl. Phys. B **591**, 313 (2000)
10. C. W. Bauer, S. Fleming, D. Pirjol and I. W. Stewart, Phys. Rev. D **63**, 114020 (2001)
11. C. W. Bauer, D. Pirjol and I. W. Stewart, Phys. Rev. D **65**, 054022 (2002)
12. M. Beneke and M. Neubert, Nucl. Phys. B **675**, 333 (2003) [arXiv:hep-ph/0308039].
13. B. Aubert *et al.* [BABAR Collaboration], Phys. Rev. Lett. **93**, 231804 (2004)
14. K. F. Chen *et al.* [BELLE Collaboration], Phys. Rev. Lett. **94**, 221804 (2005)
15. Z. Ligeti, M. E. Luke and M. B. Wise, Phys. Lett. B **507**, 142 (2001)
16. B. Aubert *et al.* [BABAR Collaboration], Phys. Rev. D **74**, 012001 (2006) [arXiv:hep-ex/0604009].
17. C. H. Chen and H. N. Li, Phys. Rev. D **71**, 114008 (2005)

# Charmed baryon spectroscopy with Belle

Tadeusz Lesiak [1]

*Institute of Nuclear Physics PAN*
*Radzikowskiego 152, 31-142 Kraków, Poland*

**Abstract.** Recent studies concerning charmed baryon spectroscopy, performed by the Belle collaboration, are briefly described. We report the first observation of two new baryons $\Xi_{cx}(2980)$ and $\Xi_{cx}(3077)$, a precise determination of the masses of $\Xi_c(2645)$ and $\Xi_c(2815)$, observation of the $\Lambda_c(2940)^+$ and experimental constraints on the possible spin-parity of the $\Lambda_c(2880)^+$. Observations of several exclusive decays of $B$ mesons to the final states containing charmed baryons are also briefly presented.

**Keywords:** hadron spectroscopy, charmed baryons
**PACS:** PACS numbers: 13.30.-a, 13.30.Rj, 14.20.Lq

## INTRODUCTION

In the last three years the Belle collaboration has provided evidence for several new hadrons. Among them are the states X(3872), X(3940), Y(3940) and Z(3931), discussed at this conference in a separate talk [1]. This paper is devoted to the recent studies by the Belle collaboration concerning charmed baryons. The Belle detector at the KEKB asymmetric $e^+e^-$ collider [2] is a general purpose spectrometer, described in detail in [3].

## OBSERVATION OF NEW STATES $\Xi_{CX}(2980)$ AND $\Xi_{CX}(3077)$

In the beginning of this year, the Belle collaboration using the data sample of $461.5 \text{ fb}^{-1}$, reported the first observation of two baryons [4], denoted as $\Xi_{cx}(2980)^+$ and $\Xi_{cx}(3077)^+$ and decaying into $\Lambda_c^+ K^- \pi^{+2}$ (Fig. 1(a)). Assuming that these states carry charm and strangeness, the above observation would comprise the first example of a baryonic decay in which the initial $c$ and $s$ quarks are carried away by two different final state particles. Most naturally, these two states would be interpreted as excited charm-strange baryons $\Xi_c$. This interpretation is strengthened by the positive results of the search for neutral isospin related partners of the above states, performed in $\Lambda_c^+ K_s^0 \pi^-$ final state. It yielded an evidence of the $\Xi_{cx}(3077)^0$ together with a broad enhancement near the threshold i.e. in the mass range corresponding to the $\Xi_{cx}(2980)^0$. The preliminary parameters of the states $\Xi_{cx}(2980)^+$ and $\Xi_{cx}(3077)^+$ are collected in table 1.

---

[1] partially supported by the KBN grant No. 2P03B 01324
[2] Charge-conjugate modes are included everywhere, unless otherwise stated

CP892, *Quark Confinement and the Hadron Spectrum VII*
edited by J. E. F. T. Ribeiro
© 2007 American Institute of Physics 978-0-7354-0396-3/07/$23.00

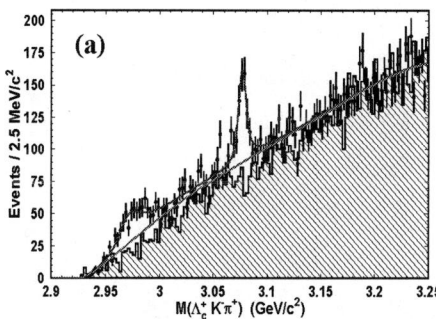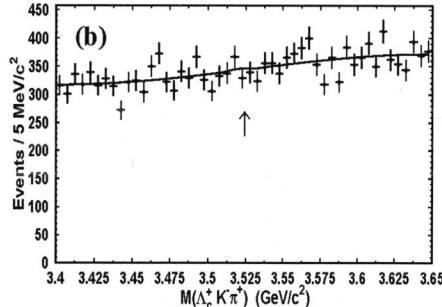

**FIGURE 1.** **(a)**: $M(\Lambda_c^+ K^- \pi^+)$ distribution (points with error bars) together with the fit (solid curve). The dashed region represents the background component corresponding to the wrong-sign combinations $\Lambda_c^+ K^+ \pi^-$. **(b)**: The distribution of $\Lambda_c^+ K^- \pi^+$ invariant mass in the region around the value of 3.52 GeV (marked with an arrow) corresponding to the SELEX evidence of the double-charmed baryon.

In the $\Lambda_c^+ K^- \pi^+$ final state, the SELEX collaboration [5] reported the observation of a double charmed baryon at 3520 MeV/$c^2$. The study by Belle shows no evidence for this state (Fig. 1(b)) and provides an upper limit on the ratio of production cross-sections $\sigma(\Xi_{cc}(3520)^+) \times \mathscr{B}(\Xi_{cc}^+ \to \Lambda_c^+ K^- \pi^+)/\sigma(\Lambda_c^+) < 1.5 \times 10^{-4}$ (estimated for $p^*(\Lambda_c) > 2.5$ GeV/c, where $p^*$ denotes the CMS momentum of the $\Lambda_c$).

**TABLE 1.** Parameters of the new charm-strange baryons $\Xi_{cx}(2980)^{+,0}$ and $\Xi_{cx}(3077)^{+,0}$

| State | Mass (MeV/c$^2$) | Width (MeV) | Yield (events) | Significance ($\sigma$) |
|---|---|---|---|---|
| $\Xi_{cx}(2980)^+$ | $2978.5 \pm 2.1 \pm 2.0$ | $43.5 \pm 7.5 \pm 7.0$ | $405.3 \pm 50.7$ | 5.7 |
| $\Xi_{cx}(3077)^+$ | $3076.7 \pm 0.9 \pm 0.5$ | $6.2 \pm 1.2 \pm 0.8$ | $326.0 \pm 39.6$ | 9.2 |
| $\Xi_{cx}(2980)^0$ | $2977.1 \pm 8.8 \pm 3.5$ | 43.5 (fixed) | $42.3 \pm 23.8$ | 1.5 |
| $\Xi_{cx}(3077)^0$ | $3082.8 \pm 1.8 \pm 1.5$ | $5.2 \pm 3.1 \pm 1.8$ | $67.1 \pm 19.9$ | 4.4 |

# MEASUREMENT OF MASSES OF $\Xi_C(2645)$ AND $\Xi_C(2815)$

The Belle collaboration, using the data sample of 414 fb$^{-1}$, studied the baryons $\Xi_c(2645)^{0,+}$, reconstructed in the $\Xi_c^{+,0} \pi^{-,+}$ decay modes and the $\Xi_c(2815)^{0,+}$, reconstructed in the decays to $\Xi_c(2645)^{+,0} \pi^{-,+}$ [6]. The following exclusive decays of the $\Xi_c$ hyperons are considered: $\Xi_c^+ \to \Xi^- \pi^+ \pi^-$, $\Xi_c^0 \to \Xi^- \pi^+$, $\Lambda^0 K^- \pi^+$, $pK^- K^- \pi^+$. The signal of $\Xi_c(2645)^{0,+}$, $(\Xi_c(2815)^{0,+})$ amounted to 2330 (172) events, respectively. The preliminary results of mass determination together with the mass difference of the charged and neutral state are the following:

$$m_{\Xi_c(2645)^+} = (2645.4 \pm 0.1(\text{stat}) \pm 0.8(\text{syst})) \text{ MeV}/c^2,$$

$$m_{\Xi_c(2645)^0} = (2645.6 \pm 0.2(\text{stat})^{+0.6}_{-0.7}(\text{syst})) \text{ MeV}/c^2,$$

$$m_{\Xi_c(2645)^+} - m_{\Xi_c(2645)^0} = (-0.2 \pm 0.3(\text{stat}) \pm 0.7(\text{syst})) \text{ MeV}/c^2,$$

$$m_{\Xi_c(2815)^+} = (2816.7 \pm 0.6(\text{stat})^{+0.7}_{-0.8}(\text{syst})) \text{ MeV}/c^2,$$

$$m_{\Xi_c(2815)^0} = (2819.7\pm0.8(\text{stat})\pm0.9(\text{syst}))\ \text{MeV}/c^2,$$
$$m_{\Xi_c(2815)^+} - m_{\Xi_c(2815)^0} = (-3.0\pm1.0(\text{stat})\pm0.8(\text{syst}))\ \text{MeV}/c^2.$$

Their precision is much better than the current world averages.

## OBSERVATION OF THE $\Lambda_C(2940)^+$ AND SPIN OF THE $\Lambda_C(2880)^+$

The baryon $\Lambda_c(2940)^+$ was observed first by the BaBar collaboration in the final state $pD^0$ [7]. The Belle collaboration, using a 553 fb$^{-1}$ data set, has recently reported the evidence for another decay mode $\Lambda_c(2940)^+ \to \Sigma_c(2455)^{0,++}\pi^\pm$ [8]. The mass of the $\Lambda_c(2940)^+$ is measured to be $(2937.9\pm1.0^{+1.8}_{-0.4})$ MeV/c$^2$ and its width is found to be $(10\pm4\pm5)$ MeV (Fig. 2(a)), in agreement with the determination of BaBar[7]. The angular analysis of $\Lambda_c(2880)^+ \to \Sigma_c(2455)^{0,++}\pi^\pm$ decays strongly favours a $\Lambda_c(2880)^+$ spin assignment of $\frac{5}{2}$ over $\frac{3}{2}$ or $\frac{1}{2}$ (Fig. 2(b)).

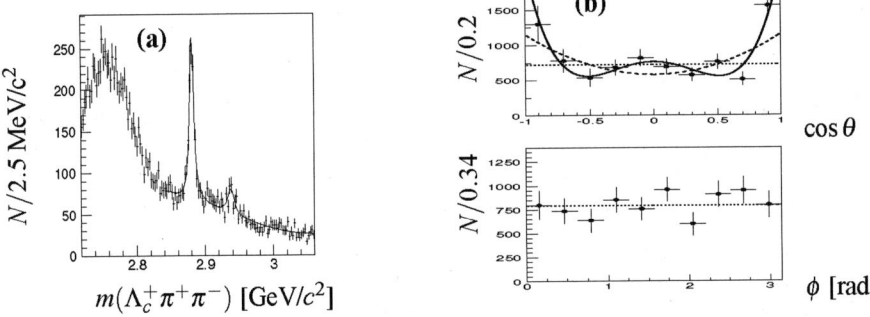

**FIGURE 2.** **(a)**: The invariant mass distribution of the $\Lambda_c^+\pi^+\pi^-$ combinations corresponding to the $\Sigma_c(2455)$ mass peak of the $\Lambda_c\pi^\pm$ combinations. The signals of $\Lambda_c(2765)^+$, $\Lambda_c(2880)^+$ and $\Lambda_c(2940)^+$ can be clearly distinguished. **(b)**: The yield of $\Lambda_c(2880)^+ \to \Sigma_c(2455)^{0,++}\pi^\pm$ decays as a function of $\cos\theta$ (upper plot) and $\phi$ (lower plot). Both angles are defined in the $\Lambda_c(2880)$ rest frame. The $\theta$ is defined as the one between the pion momentum and the boost direction of the $\Lambda_c(2880)^+$. The angle $\phi$ is defined in the plane perpendicular to the $\Lambda_c(2880)^+$ boost direction between the pion momentum and the $e^+e^- \to \Lambda_c(2880)^+X$ reaction plane. The solid, dashed and dotted curves in the upper plot correspond to the spin hypotheses 5/2, 3/2 and 1/2, respectively. The $\phi$ dependence, related to non-diagonal elements in the production density matrix, is uniform.

## $B$ MESON DECAYS TO CHARMED BARYONS

The Belle collaboration has recently reported the observation of several exclusive decays of $B$ mesons to the final state containing charm baryons. In particular the following branching fractions for three body decays $\overline{B^0} \to \Sigma_c^{++,0}\overline{p}\pi^\mp$ ($\Sigma_c^{++,0} \to \Lambda_c^+\pi^\pm$, $\Lambda_c \to pK\pi$) are measured [9]:

$$\mathscr{B}(\overline{B^0} \to \Sigma_c(2455)^{++}\overline{p}\pi^-) = (2.1\pm0.2\pm0.3\pm0.5)\times10^{-4}\ (13.1\sigma),$$
$$\mathscr{B}(\overline{B^0} \to \Sigma_c(2455)^0\overline{p}\pi^+) = (1.4\pm0.2\pm0.2\pm0.4)\times10^{-4}\ (9.4\sigma),$$

$$\mathscr{B}(\overline{B^0} \to \Sigma_c(2520)^{++}\overline{p}\pi^-) = (1.2 \pm 0.1 \pm 0.2 \pm 0.3) \times 10^{-4} \ (7.1\sigma),$$
$$\mathscr{B}(\overline{B^0} \to \Sigma_c(2520)^0 \overline{p}\pi^+) < 0.33 \times 10^{-4} \ (90\% \ \text{C.L.}) \ (1.3\sigma),$$

(here and below the third error is due to the uncertainty in $\mathscr{B}(\Lambda_c^+ \to pK\pi)$).

The Belle collaboration has also provided the observation of the decay $B^+ \to \overline{\Xi_c^0}\Lambda_c^+$ together with the $3.8\sigma$ evidence for $B^0 \to \overline{\Xi_c^-}\Lambda_c^+$ [10]. This is the first example of a two-body exclusive $B$ decay into two charmed baryons. The $\Lambda_c$ was reconstructed in the final state $pK\pi$, while for the $\Xi_c$ baryons the following decays are considered: $\Xi_c^0 \to \Xi^-\pi^+$, $\Xi_c \to \Lambda K^-\pi^+$ and $\Xi_c^+ \to \Xi^-\pi^+\pi^+$. The following branching fractions are evaluated:

$$\mathscr{B}(B^+ \to \overline{\Xi_c^0}\Lambda_c^+) \times \mathscr{B}(\overline{\Xi_c^0} \to \Xi^+\pi^-) = (5.6^{+1.9}_{-1.5} \pm 1.1 \pm 1.5) \times 10^{-5} \ (6.8\sigma),$$
$$\mathscr{B}(B^+ \to \overline{\Xi_c^0}\Lambda_c^+) \times \mathscr{B}(\overline{\Xi_c^0} \to \overline{\Lambda}K^+\pi^-) = (4.0^{+1.1}_{-0.9} \pm 0.9 \pm 1.0) \times 10^{-5} \ (5.9\sigma),$$
$$\mathscr{B}(B^0 \to \overline{\Xi_c^-}\Lambda_c^+) \times \mathscr{B}(\overline{\Xi_c^-} \to \Xi^+\pi^-\pi^-) = (9.3^{+3.7}_{-2.8} \pm 1.9 \pm 2.4) \times 10^{-5} \ (3.8\sigma).$$

Finally the Belle collaboration has also observed the decays $B^+ \to \Lambda_c^+\Lambda_c^-K^+$ and $B^0 \to \Lambda_c^+\Lambda_c^-K^0$ [11]. This study provides the first evidence for three body decays of $B$ mesons to the state containing a charmed baryon, a charmed anti-baryon and a strange meson. The following branching fractions are measured:

$$\mathscr{B}(B^+ \to \Lambda_c^+\Lambda_c^-K^+) = (6.5^{+1.0}_{-0.9} \pm 1.1 \pm 3.4) \times 10^{-4} \ (15.4\sigma),$$
$$\mathscr{B}(B^0 \to \Lambda_c^+\Lambda_c^-K^0) = (7.9^{+2.9}_{-2.3} \pm 1.2 \pm 4.1) \times 10^{-4} \ (6.6\sigma).$$

## SUMMARY

The Belle collaboration has recently observed two new charm-strange excited baryons $\Xi_{cx}(2980)$ and $\Xi_{cx}(3077)$ and confirmed the existence of the $\Lambda_c(2940)^+$. The study of angular distribution of the decay $\Lambda_c(2880)^+ \to \Sigma_c(2455)^{0,++}\pi^\pm$ strongly supports the assignment of spin-parity $\frac{5}{2}$ for the $\Lambda_c(2880)^+$. The masses of hyperons $\Xi_c(2645)$ and $\Xi_{(}2815)$ were determined with considerably improved precision. Several new decay modes of $B$ mesons with charmed baryons in the final state were also reconstructed.

## REFERENCES

1. A. Zupanc, these proceedings.
2. S. Kurokawa, E. Kikutani, *Nucl. Instrum. Methods A*, **499**, 1 (2003), and other papers in this vol.
3. Belle Collaboration, A. Abashian *et al*, *Nucl. Instrum. Methods A*, **479**, 117 (2002).
4. Belle Collaboration, R. Chistov *et al*, *Phys. Rev. Lett.* **97**, 162001, (2006).
5. SELEX Collaboration, M. Mattson *et al*, *Phys. Rev. Lett.*, **89**, 112001 (2002).
6. Belle Collaboration, K. Abe *et al*, preprint hep-ex/0608012 (2006).
7. BaBar Collaboration, D. Aubert *et al*, preprint hep-ex/0603052 (2006).
8. Belle Collaboration, K. Abe *et al*, preprint hep-ex/0608043 (2006).
9. Belle Collaboration, K. Abe *et al*, preprint hep-ex/0608025 (2006).
10. Belle Collaboration, K. Abe *et al*, preprint hep-ex/051074 (2005), subm. to Phys. Rev. Lett.
11. Belle Collaboration, K. Abe *et al*, preprint hep-ex/0508015 (2005), subm. to Phys. Rev. Lett.

# Measurements of $b \to s\gamma$ Decays at *BABAR*

Timofei Piatenko (on behalf of the *BABAR* Collaboration)

*California Institute of Technology, MC 356-48, Pasadena, CA 91125, USA*

**Abstract.** We present measurements of the Branching Fraction and photon energy spectrum in $B \to X_s\gamma$ decays in a sample of 89 million $B\bar{B}$ pairs collected at the *BABAR* detector at Stanford Linear Accelerator Center's PEP-II asymmetric B-factory. Results from a fully-inclusive and a sum of 38 exclusive final states techniques are presented and found to be consistent with the Standard Model calculations, as well as experimental results obtained from semileptonic $B \to X_c l \nu$ decays.

**Keywords:** *B* meson, radiative penguin decays, photon energy spectrum, semileptonic decays
**PACS:** 13.30.Ce, 13.25.Hw, 12.39.Hg, 12.38.Lg

## MOTIVATION

An overall goal of the *BABAR* experiment is to precisely measure and over-constrain parameters of the Cabbibo-Kobayashi-Maskawa (CKM) mixing matrix, which governs the weak couplings of quarks in the Standard Model (SM). The smallest element of the CKM matrix, $V_{ub}$, can be obtained from measurements of the Branching Fraction (BF) of semileptonic $B \to X_u l \nu$ decays that present a clean experimental signature. However, theoretical calculations of the decay amplitude are complicated by the Fermi motion of the *b* quark inside the *B* meson. While Operator Product Expansion (OPE) can be applied to deal with non-perturbative corrections to the quark-level calculations, the validity of this approach is limited by the kinematic restrictions imposed by experimental conditions. When the non-perturbative contributions are expanded in $1/m_b$ in what is known as Heavy Quark Expansion (HQE), the terms can be re-summed into a Shape Function, which cannot be calculated analytically. The decay rate is given by a convolution of the Shape Function and the perturbative part[1]. Since the Shape Function applies to all decays of *B* meson to light quarks, it can be measured in kinematically simple radiative penguin $B \to X_s\gamma$ decays by relating HQE parameters to moments of the $E_\gamma$ spectrum: $\langle E_\gamma \rangle \approx \frac{m_b}{2}, \langle E_\gamma^2 - \langle E_\gamma \rangle^2 \rangle \propto \mu_\pi^2$ ([2], [3], and [4]). Theoretically, there's less dependence on the heavy quark distribution at low $E_\gamma$, where different expansion schemes agree the best, while higher energy photons constitute a cleaner experimental signature.

## EXPERIMENTAL TECHNIQUE

Current next-to-leading-order theoretical calculations give, for example, $BF(B \to X_s\gamma, E_\gamma > 1.6\text{GeV}) = (3.61^{+0.37}_{-0.49}) \times 10^{-4}$[5], making the measurement challenging. At the *BABAR* detector (described in detail in [6]), excellent energy resolution of the Electromagnetic Calorimeter allows for rather clean detection of high-energy photons, while superior performance of the particle identification system allows for $\sim 4\sigma$ sep-

CP892, *Quark Confinement and the Hadron Spectrum VII*
edited by J. E. F. T. Ribeiro
© 2007 American Institute of Physics 978-0-7354-0396-3/07/$23.00

aration between $K$'s and $\pi$'s. This helps suppress the overwhelming background from continuum $e^+e^- \rightarrow q\bar{q}$ events, where $q$ is one of the lighter $u$, $d$, $s$, or $c$ quarks.

Two separate analyses, both based on 89 million $B\bar{B}$ pairs collected at BABAR at the $\Upsilon(4s)$ resonance, were carried out. The fully-inclusive analysis[7] reconstructs the signal photon, but not the hadron, avoiding the issue of final state fragmentation and $X_s$ modes missing from Monte Carlo simulation, problematic for the semi-inclusive method that uses a sum of 38 exclusive modes[8]. On the other hand, it suffers from a higher level of background and poorer $E_\gamma$ resolution. The semi-inclusive analysis also has the benefit of working entirely in the $B$ meson frame.

The fully-inclusive analysis applies a cut at 1.9 GeV on $E_\gamma^*$ in the $\Upsilon(4s)$ rest frame. The $q\bar{q}$ background is suppressed using a lepton tag of the other $B$ meson in the event, as well as event shape variables that take advantage of the fact that in the $\Upsilon(4s)$ frame, $B$'s are produced almost at rest and decay isotropically, while continuum events tend to be jet-like. Photons consistent with the decay of a $\pi^0$ or $\eta$ are vetoed. Data collected about 40 MeV below the $\Upsilon(4s)$ resonance is used to subtract remaining continuum background, while appropriate control samples are used to estimate the systematic effects of background resulting from non-signal decays of the $B$ meson.

In the semi-inclusive analysis, 38 fully-reconstructed decay modes to $\pi$'s, $K$'s, $\pi^0$'s, and $\eta$'s are combined. The decays are simulated using JETSET[9], which requires control sample studies to correct for missing modes. The BF, calculated for $E_\gamma > 1.9$ GeV and $0.6 < M(X_s) < 2.8$ GeV, is determined from a fit to beam energy substituted mass of the $B$ meson, $m_{ES} \equiv \sqrt{E_{Beam}^{*}{}^2 - p_B^{*2}}$, where the star refers to the $\Upsilon(4s)$ frame.

## RESULTS AND CONCLUSIONS

Both analyses carry out fits to the moments of the $E_\gamma$ distributions, shown in Figure 1. The fully-inclusive analyses obtains $\langle E_\gamma \rangle = (2.288 \pm 0.025 \pm 0.017 \pm 0.015)$ GeV and $\langle (E_\gamma - \langle E_\gamma \rangle)^2 \rangle = (0.0328 \pm 0.0040 \pm 0.0023 \pm 0.0036)$ GeV$^2$, while the semi-inclusive results are $\langle E_\gamma \rangle = (2.321 \pm 0.038^{+0.017}_{-0.038})$ GeV and $\langle (E_\gamma - \langle E_\gamma \rangle)^2 \rangle = (0.0253 \pm 0.0101^{+0.0041}_{-0.0028})$ GeV$^2$. In the Kinetic scheme[2], these numbers correspond to $m_b = (4.44 \pm 0.08 \pm 0.14)$ GeV and $\mu_\pi^2 = (0.64 \pm 0.13 \pm 0.24)$ GeV$^2$ for the fully-inlcusive and $m_b = (4.70^{+0.04}_{-0.08})$ GeV and $\mu_\pi^2 = (0.29^{+0.09}_{-0.04})$ GeV$^2$ for the semi-inclusive analyses. The errors are statistical and systematic, respectively, for the fully-inclusive result, and a combination of the two for the semi-inclusive.

The measured BF's for $E_\gamma^{(*)} > 1.9$ GeV are $BF(B \rightarrow X_s\gamma) = (3.67 \pm 0.29 \pm 0.34 \pm 0.29) \times 10^{-4}$ and $BF(B \rightarrow X_s\gamma) = (3.27 \pm 0.18^{+0.55+0.04}_{-0.40-0.09}) \times 10^{-4}$ for fully and semi-inclusive analyses, respectively. The errors are statistical, systematic, and due to the choice of the fit model. To compare BF results with theoretical calculations, one must choose a particular scheme and extrapolate the measurements down to $E_\gamma > 1.6$ GeV. For the fully-inclusive approach, this yields, in the Kinetic scheme, $BF(B \rightarrow X_s\gamma) = (3.94 \pm 0.31 \pm 0.36 \pm 0.21) \times 10^{-4}$. Similarly, the semi-inclusive analysis obtains $BF(B \rightarrow X_s\gamma) = (3.35 \pm 0.19^{+0.56+0.04}_{-0.41-0.09}) \times 10^{-4}$, except that here the Shape Function[3] and Kinetic schemes are averaged. The numbers agree well with the SM expectations.

Buchmüller and Flächer have recently combined all available measurements of the

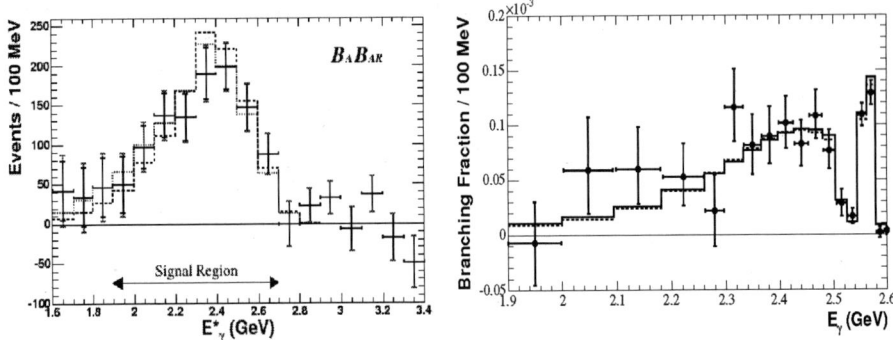

**FIGURE 1.** $E_\gamma$ spectra for fully-inclusive (left) and semi-inclusive (right) analyses. Data points are compared to Kinetic (dashed or solid line) and Shape Function (dotted or dashed line) schemes for the best-fit HQE parameters provided in the text.

$E_\gamma$ spectrum in $B \to X_s\gamma$ decays with lepton energy and hadron mass spectra from $B \to X_c l \nu$ decays[10]. Performing combined fits, they obtain, in Kinetic scheme, $m_b = (4.590 \pm 0.025_{exp} \pm 0.030_{HQE})\,\mathrm{GeV}$ and $\mu_\pi^2 = (0.401 \pm 0.019_{exp} \pm 0.035_{HQE})\,\mathrm{GeV}^2$, as well as a value for $|V_{cb}| = (41.96 \pm 0.23_{exp} \pm 0.35_{HQE} \pm 0.59_{\Gamma_{SL}}) \times 10^{-3}$. The first error is a combination of experimental statistical and systematic errors, the second accounts for theoretical uncertainties from HQE, and $\Gamma_{SL}$ is the semileptonic decay rate. The study also demonstrates good agreement between $B \to X_s\gamma$ and $B \to X_c l \nu$ decays, confirming the validity of universality assumption for the Shape Function approach to non-perturbative corrections in inclusive decays of the $B$ meson.

The *BABAR* collaboration is working on updating $B \to X_s\gamma$ results with much greater statistical precision. The current full dataset consists of about 350 million $B\bar{B}$ pairs, with plans to more than double this number by the end of 2008. Precision measurements of radiative $B \to X_s\gamma$ decays are very important for assessing the validity of the Standard Model of particle physics. The current agreement between theoretical calculations and experimental results stands at around 10%, and the aim is to lower both errors to a 5% level in the near future.

# REFERENCES

1.  G. Ricciardi (2006), hep-ph/0608220.
2.  D. Benson, I. I. Bigi, and N. Uraltsev, *Nucl. Phys.* **B710**, 371–401 (2005), hep-ph/0410080.
3.  S. W. Bosch, B. O. Lange, M. Neubert, and G. Paz, *Nucl. Phys.* **B699**, 335–386 (2004), hep-ph/0402094.
4.  A. L. Kagan, and M. Neubert, *Eur. Phys. J.* **C7**, 5–27 (1999), hep-ph/9805303.
5.  T. Hurth, E. Lunghi, and W. Porod, *Nucl. Phys.* **B704**, 56–74 (2005), hep-ph/0312260.
6.  B. Aubert, et al., *Nucl. Instrum. Meth.* **A479**, 1–116 (2002), hep-ex/0105044.
7.  B. Aubert, et al., *Phys. Rev. Lett.* **97**, 171803 (2006), hep-ex/0607071.
8.  B. Aubert, et al., *Phys. Rev.* **D72**, 052004 (2005), hep-ex/0508004.
9.  T. Sjostrand (1995), hep-ph/9508391.
10. O. Buchmuller, and H. Flacher, *Phys. Rev.* **D73**, 073008 (2006), hep-ph/0507253.

# Charmonium and Bottomonium from Classical $SU(3)$ Gauge Configurations

R. A. Coimbra and O. Oliveira

*Centro de Física Computacional, Universidade de Coimbra, 3004-516 Coimbra, Portugal*

**Abstract.** The charmonium and bottomonium spectra computed from a potential defined from a single gauge configuration, obtained from solving the classical field equations, is discussed. The theoretical spectra shows good agreement with the measured states.

A discussion of possible interpretations, within the same non-relativistic potential model, for the new charmonia states $X(3872)$, $\chi_{c1}(2P)$ and $Y(4260)$ is performed. In particular, we give predictions for electromagnetic E1 transitions for various scenarios.

**Keywords:** Charmonium, Bottomonium, Potential Models

**PACS:** 12.39.Pn Potential models - 12.38.Lg Other nonperturbative calculations - 12.38.-t Quantum chromodynamics

## INTRODUCTION AND MOTIVATION

In [1] it was proposed a generalized Cho-Faddeev-Niemi ansatz for the SU(3) gauge fields. For the simplest form of the ansatz, the classical field equations were solved and a potential for heavy quarkonia motivated. The single configuration potential is coulombic for short interquark distances and grows exponentially for large interquark distances. In this way, the potential provides quark confinement. Then, assuming that the quark interaction is a pure vectorial interaction, the spectra of charmonium was investigated.

As described in [1], the single configuration potential is able to describe the charmonium spectra with an error of less than 3% ($\sim 100$ MeV). However, the prediction for the $1S$ and $2S$ hyperfine splitting is about half of the experimental value. This result is probably due to the definition of the potential from a single gauge configuration. In the same work, the theoretical predictions for the leptonic widths of $1^{--}$ states where computed and the results are in-line with the experimental values.

In [2] the investigation was extended to include the E1 electromagnetic transitions together with an analysis of the bottomonium system. For $b\bar{b}$, the spectra is reproduced with an error of less than 1% ($\sim 100$ MeV) and the theoretical prediction for the $b\bar{b}$ leptonic widths reproduces the same level of accurary as found in charmonium.

In what concerns the electromagnetic transitions, the single configuration potential is able to reproduce well the quoted particle data book numbers [3].

## CHARMONIUM AND BOTTOMONIUM SPECTRA

In table 1 we show the theoretical $J^{PC} = 1^{--}$ spectra; see [1, 2] for details. The third column is the deviation of the theoretical mass to the experimental measured mass, in

CP892, *Quark Confinement and the Hadron Spectrum VII*
edited by J. E. F. T. Ribeiro

**TABLE 1.** $J^{PC} = 1^{--}$ charmonium and bottomnium spectra. The table shows the theoretical state, the mass prediction, the difference between the experimental measured mass and the theoretical prediction. All numbers are in MeV.

| | Charmonium | | | | Bottomonium | | |
|---|---|---|---|---|---|---|---|
| $1^3S_1$ | 3097 | 0 | $J/\psi(1S)$ | $1^3S_1$ | 9460 | 0 | $\Upsilon(1S)$ |
| $2^3S_1$ | 3659 | -27 | $\psi(2S)$ | $2^3S_1$ | 10023 | 0 | $\Upsilon(2S)$ |
| $1^3D_1$ | 3688 | -83 | $\psi(3770)$ | $1^3D_1$ | 10159 | - | |
| $3^3S_1$ | 4164 | -95 | $Y(4260)$ | $3^3S_1$ | 10385 | 30 | $\Upsilon(3S)$ |
| $2^3D_1$ | 4155 | 2 | $\psi(4160)$ | $2^3D_1$ | 10476 | - | |
| $4^3S_1$ | 4669 | - | | $4^3S_1$ | 10727 | 148 | $\Upsilon(4S)$ |
| $3^3D_1$ | 4636 | - | | $3^3D_1$ | 10796 | -69 | $\Upsilon(10860)$ |
| | | | | $5^3S_1$ | 11065 | 46 | $\Upsilon(11020)$ |

MeV. The overall agreement between theoretical and particle spectra is good. The only particles which don't fit well in the theoretical spectra are $\psi(4040)$ and $\psi(4415)$. In what concerns these two states, the experimental information is scarce and the particle data book comments that the "interpretation of these states as a single resonance is unclear because of the expectation of substantial threshold effects in this energy region". Curiously, both particle masses are essentially the sum of $J/\psi$ with light $J^{PC} = 0^{++}$ mesons.

The work reported in [1, 2] and summarized in table 1 is based on the nonrelativistic analysis of a confining potential derived from a single configuration. Our previous work suggests that either one should include the contribution from other configurations and/or one should perform a coupled-channel analysis. We are currently engaged in extending our previous studies in both ways. Anyway, in what concerns the spectra, the single channel analysis shows that the potential is able to explain the observed states if one allows for an error of $\sim 100$ MeV.

In the following we report on the predictions of the single channel analysis for the potential obtained in [1] for the new charmonium states $X(3872)$, $\chi_{c2}(2P)$ and $Y(4260)$ (we follow the particle data book notation). In order to be able to distinguish the possible quantum number assignements, when possible, we also report on our predictions for electromagnetic E1 transitions. We call the reader attention to the good agreement between theoretical predictions and experimental measures of charmonium and bottomonium electromagnetic widths - see tables 4, 5,6 of [2].

## $X(3872)$

In what concerns the quantum numbers of $X(3872)$, experimentaly only the parity, $C = +$, is known. Belle Collaboration [4] has performed an analysis of possible quantum numbers and conclude in favor of $J^{PC} = 1^{++}, 2^{++}$. In table 2, we report the charmonium states compatible with these quantum numbers and whose mass differs from the $X(3872)$ by 100 MeV, including the E1 electromagnetic widths for the assignement favoured by Belle data.

**TABLE 2.** $X(3872)$ possible interpretations and E1 electromagnetic transitions.

| | Mass (MeV) | $J^{PC}$ | E1 Electromagnetic Transition | (KeV) |
|---|---|---|---|---|
| $2^3P_1$ | 3938 | $1^{++}$ | $\longrightarrow \psi(2S) + \gamma$ | 63 |
| | | | $\longrightarrow J/\psi(1S) + \gamma$ | 48 |
| $1^3F_2$ | 3932 | $2^{++}$ | | |

## $\chi_{c2}(2P)$

According to the particle data book, this is a $J^{PC} = 2^{++}$ with a mass of $3929 \pm 5$ MeV. In our model, the $J = 2$ states around this mass value are $2^3P_2$, with $J^{PC} = 2^{++}$ and mass 4048 MeV; the $1^3F_2$, with $J^{PC} = 2^{++}$ and mass 3932 MeV.

If this state is a $2^3P_2$ state, it has a large E1 width of 140 KeV for the transition to $\psi'(2S)$ and a E1 width of 59 KeV to $J/\psi(1S)$.

## $Y(4260)$

This $c\bar{c}$ state is a $J^{PC} = 1^{--}$ with a mass of $4259^{+8}_{-10}$ MeV. According to the particle data book, the "interpretation as due to two interfering resonances is not excluded". In our study, possible candidates with a mass between $\sim 4160$ MeV and $\sim 4360$ MeV are: $3^3S_1$, 4155 MeV; $2^3D_2$, 4327 MeV; $2^3D_2$, 4230 MeV; $2^1D_2$, 4230 MeV; $1^3G_4$, 4260; $1^3G_3$, 4161 MeV; $3^3P_0$, 4300 MeV. Only the state $3^3S_1$ has $J^{PC} = 1^{--}$, as it should be for a state produced via initial state radiation.

If $Y(4260)$ is a $3^3S_1$ state, its larger E1 electromagnetic widths are transitions to $\chi(2P)$ states, namely: $2^3P_0$, $\Gamma = 358$ KeV; $2^3P_1$, $\Gamma = 475$ KeV; $2^3P_2$, $\Gamma = 233$ KeV. The corresponding widths for $\chi(1P)$ states being: $1^3P_0$, $\Gamma = 4$ KeV; $1^3P_1$, $\Gamma = 9$ KeV; $1^3P_2$, $\Gamma = 13$ KeV, making them hard to measure. Given the values for the various widths, it seems that a combine investigation of $Y(4260)$ and $\chi_c(2P)$ could be helpfull in understanding the nature of this particle.

## ACKNOWLEDGMENTS

R. A. C. acknowldges F.C.T. for financial support, grant SFRH/BD/8736/2002. This work was partly supported by F.C.T. under contract POCI/FP/63436/2005.

## REFERENCES

1. O. Oliveira, R. A. Coimbra, hep-ph/0603046.
2. R. A. Coimbra, O. Oliveira, hep-ph/0610142.
3. W.-M. Yao *el al.* (Particle Data Book), *J. Phys.* **G33**, 1 (2006).
4. K. Abe *et al.* (Belle Collaboration), hep-ex/0505038.

# Extracting infrared QCD coupling from meson spectrum

M. Baldicchi, G. M. Prosperi, C. Simolo

*Dipartimento di Fisica, Università di Milano*

*I.N.F.N., sezione di Milano*

*via Celoria 16, 120133 Milano, Italy*

**Abstract.** In the framework of the Bethe-Salpeter formalism used in previous papers to evaluate the quarkonium spectrum, here we reverse the point of view to extract an "experimental" running coupling $\alpha_s^{\exp}(Q^2)$ in the infrared (IR) region from the data. The values so obtained agree within the erros with the Shirkov-Solovtsov analytic coupling for 200 MeV $< Q <$ 1.2 GeV, thus giving a very satisfactory unifying description of high and low energy phenomena. Below 1 GeV however $\alpha_s^{\exp}(Q^2)$ seems to vanish as $Q \to 0$. The paper is based on a work in progress in collaboration with D. V. Shirkov.

**Keywords:** Running coupling, QCD, meson spectrum
**PACS:** 12.38.Aw, 11.10.St, 12.38.Lg, 12.39.Ki

As well known a very consistent picture of the high energy processes can be obtained by perturbative QCD if the running coupling $\alpha_s(Q^2)$, as derived from the renormalization group, is used and a good convergence is already attained at 3-loop level (see e.g. [1]). In the traditional $\overline{\text{MS}}$ renormalization scheme, however, $\alpha_s(Q^2)$ develops at any loop level unphysical singularities for $Q \sim \Lambda_{\text{QCD}}$ (Landau singularities) that make the expression useless in the IR region. This is a serious difficulty in any quark model where $Q$ should be identified with the momentum transfer taking values typically between few GeV and some hundred MeV.

Among the various attempts to eliminate Landau singularities (see e.g. [2]) we consider the proposal of Shirkov and Solovtsov, which consists in imposing analyticity on $\alpha_s(Q^2)$ [3]. At 1-loop the analytic coupling can be written explicitly

$$\alpha_{\text{an}}^{(1)}(Q^2) = \frac{1}{\beta_0} \left( \frac{1}{\ln(Q^2/\Lambda^2)} + \frac{\Lambda^2}{\Lambda^2 - Q^2} \right). \tag{1}$$

At 2- or 3-loop level $\alpha_{\text{an}}(Q^2)$ can be only numerically computed. At 3-loop an useful approximation [4] is however given by the "1-loop-like" model

$$\alpha_{\text{an}}^{(3)}(Q^2) = \frac{4\pi}{\beta_0} \left( \frac{1}{l} + \frac{1}{1 - e^l} \right), \quad \text{with} \quad l = \ln \frac{Q^2}{\Lambda^2} + \frac{\beta_1}{\beta_0^2} \ln \sqrt{\ln^2 \frac{Q^2}{\Lambda^2} + 2\pi^2} \tag{2}$$

and the proper $\overline{\text{MS}}$ value for $\Lambda$. For 1.5 GeV $< Q <$ 200 GeV eq. (2) differs from the exact expression by no more than 2%. Furthermore below 1 GeV $\alpha_{\text{an}}^{(3)}(Q^2)$ with $\Lambda_{n_f=3}^{(3)} = 375$ MeV differs even less from eq. (1) with $\Lambda_{n_f=3}^{(1)} = 206$ MeV.

On the other side in the last years we have developed a Bethe -Salpether formalism like [5] that was applied with a certain success to the calculation of the meson spectrum in the light and in the heavy quark sectors. The formalism was essentially derived from QCD first principles, making only an ansatz on the Wilson loop correlator $W$, which consists in writing $i \ln W$ as the sum of a one-gluon exchange and an area term encoding

CP892, *Quark Confinement and the Hadron Spectrum VII*
edited by J. E. F. T. Ribeiro
© 2007 American Institute of Physics 978-0-7354-0396-3/07/$23.00

(MeV)

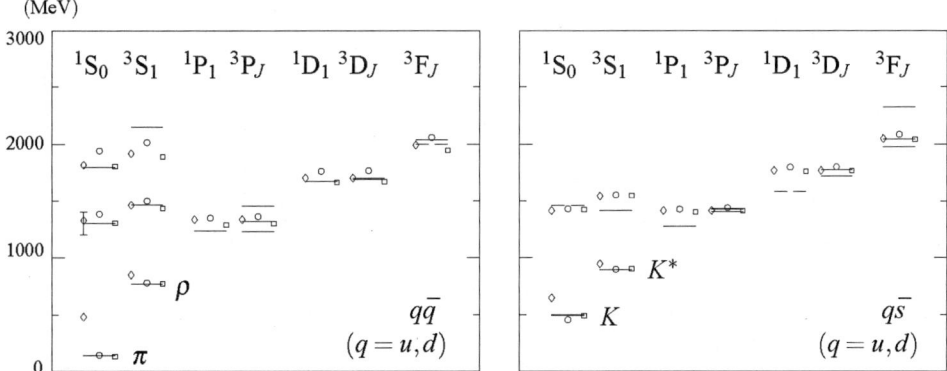

**FIGURE 1.** Quarkonium spectrum, three different calculations. Diamonds refer to the truncation prescription for $\alpha_s$, squares and circles refer to the 1-loop analytic coupling (1) and two different parametrizations for constituent masses of light quarks. Lines represent experimental data.

confinement, $i\ln W = (i\ln W)_{\text{OGE}} + \sigma S$. The resulting reduced Salpeter equation is then in the form of the eigenvalue equation for a squared bound state mass

$$M^2 = M_0^2 + U_{\text{OGE}} + U_{\text{CF}}, \tag{3}$$

where $M_0 = w_1 + w_2 = \sqrt{m_{P1}^2 + \mathbf{k}^2} + \sqrt{m_{P2}^2 + \mathbf{k}^2}$ and $U = U_{\text{OGE}} + U_{\text{CF}}$ the potential (see [5, 6] and references therein). By neglecting the spin orbit and the the tensorial term but including the hyperfine splitting term, $U_{\text{OGE}}$ has the form

$$\langle \mathbf{k}|U_{\text{OGE}}|\mathbf{k}'\rangle = \rho \, \frac{4}{3} \frac{\alpha_s(\mathbf{Q}^2)}{\pi^2}\left[-\frac{1}{\mathbf{Q}^2}\left(q_{10}q_{20} + \mathbf{q}^2 - \frac{(\mathbf{Q}\cdot\mathbf{q})^2}{\mathbf{Q}^2}\right) + \frac{1}{6}\sigma_1\cdot\sigma_2\right] \tag{4}$$

where $\rho$ is a kinematic factor. In [6] we have computed the meson masses by the equation $m_a^2 = \langle\phi_a|M_0^2|\phi_a\rangle + \langle\phi_a|U_{\text{OGE}}|\phi_a\rangle + \langle\phi_a|U_{\text{CF}}|\phi_a\rangle$, where $\phi_a$ is the zero-order wave function for the state $a$ obtained by solving the eigenvalue equation for the static limit Hamiltonian $H_{\text{CM}} = w_1 + w_2 - \frac{4}{3}\frac{\alpha_s}{r} + \sigma r$ by the Rayleigh-Ritz method.

Calculations have been performed by using both a truncation prescription for $\alpha_s(Q^2)$ and the 1-loop analytic coupling (1). The results of three sets of calculations are graphically reported in Fig. 1 for the light-light and light-strange sectors as an example. The key point is that, while the two different assumptions on $\alpha_s(Q^2)$ give similar results for the heavy-heavy quark states a correct reproduction of the $\pi$ and $K$ masses can be obtained, as it can be seen, only with the analytic coupling.

In this paper we focus our attention on the reversed point of view. $\Lambda^{(1)}_{n_f=3}$ and quark masses have been fixed by fitting $\pi, \rho, \phi, J/\psi$ and $\Upsilon$ mesons, while the string tension has been fixed a priori to the value $\sigma = 0.18\,\text{GeV}^2$ (see Fig. 2). For each state $a$ we then define a theoretical fixed coupling $\alpha_{s,a}^{\text{th}}$ which leads to the same theoretical mass as by using $\alpha_{\text{an}}^{(1)}(Q^2)$. Thus an effective momentum transfer $Q_a$ is assigned to each state by the equation $\alpha_{\text{an}}^{(1)}(Q_a^2) = \alpha_{s,a}^{\text{th}}$. We finally define $\alpha_s^{\text{exp}}(Q_a^2)$ as the value of the coupling to be inserted in (4) in order to exactly reproduce the experimental mass:

341

$$\langle \phi_a | M_0^2 | \phi_a \rangle + \alpha_s^{\text{exp}}(Q_a^2) \langle \phi_a | \mathscr{O}(\mathbf{q}; \mathbf{Q}) | \phi_a \rangle + \langle \phi_a | U_{\text{CF}} | \phi_a \rangle = m_{\text{exp}}^2 \quad (\mathscr{O}(\mathbf{q}; \mathbf{Q}) \text{ given by (4)}).$$

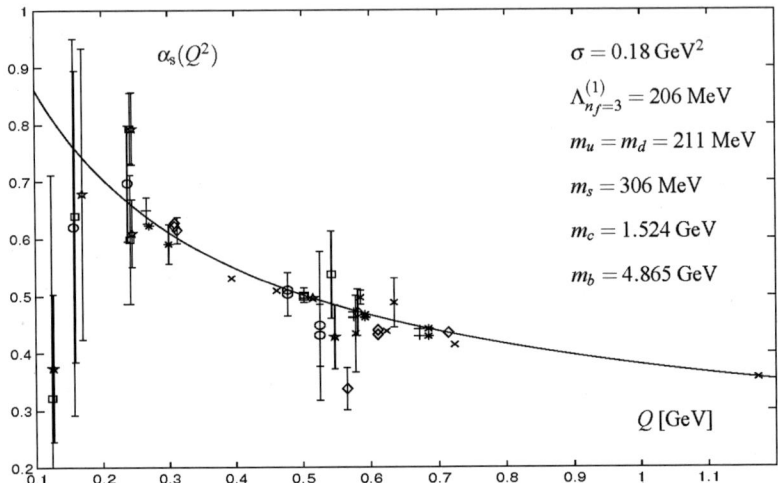

**FIGURE 2.** 1-loop analytic coupling with $\Lambda_{n_f=3}^{(1)} = 206$ MeV and $\alpha_s^{\text{exp}}$. Circles, stars and squares refer respectively to $q\bar{q}$, $s\bar{s}$ and $q\bar{s}$ with $q = u, d$; diamonds and crosses stay for $c\bar{c}$ and $b\bar{b}$, plus signes for $q\bar{c}$ and $q\bar{b}$, while asterisks for $s\bar{c}$ and $s\bar{b}$. Error bars are drawn only if relevant.

The results are given pictorially on Fig. 2; points representing $\alpha_s^{\text{exp}}(Q_a^2)$ are compared with the analytic curve (1) for $\Lambda_{n_f=3}^{(1)} = 206$ MeV. Error bars take into account the theoretical errors in the determination of the spectrum as well as the experimental ones when relevant. The theoretical incertitude expected in our procedure, that does not include coupling among different channels, is assumed to be roughly expressed by the half width of the state. As it can been seen the $\alpha_s^{\text{exp}}(Q_a^2)$ values agree rather well with the analytic coupling expression within the quoted errors for $200 \, \text{MeV} < Q < 1.2 \, \text{GeV}$. Below 200 MeV, however, there seems to exist a consitent tendency of $\alpha_s^{\text{exp}}(Q_a^2)$ to vanish rather than to approach a finite limit.

The present paper is based on a work in collaboration with D. V. Shirkov.

## REFERENCES

1.  W. M. Yao *et al.* [Particle Data Group], J. Phys. G **33** (2006) 1; S. Bethke, arXiv:hep-ex/0606035.
2.  G. M. Prosperi, M. Raciti and C. Simolo, arXiv:hep-ph/0607209.
3.  D. V. Shirkov and I. L. Solovtsov, JINR Rapid Comm. No.2[76]-96 (1996) 5, arXiv:hep-ph/9604363; Phys. Rev. Lett. **79** (1997) 1209; Phys. Lett. B **442** (1998) 344; D. V. Shirkov, Eur. Phys. J. C **22** (2001) 331.
4.  D. V. Shirkov and A. V. Zayakin, arXiv:hep-ph/0512325.
5.  N. Brambilla, E. Montaldi, G.M. Prosperi, *Phys. Rev.* **D 54** (1996) 3506; G.M. Prosperi, *Problems of Quantum Theory of Fields*, Pag. 381, B.M. Barbashov, G.V. Efimov, A.V. Efremov Eds. JINR Dubna 1999, hep-ph/9906237.
6.  M. Baldicchi and G. M. Prosperi, AIP Conf. Proc. **756** (2005) 152; *Color confinement and hadrons Quantum Chromodynamics*, Page. 183, H. Suganuma, *et al.* eds. World Scientific 2004, hep-ph/0310213; Phys. Rev. D **66** (2002) 074008; Phys. Rev. D **62** (2000) 114024; *Fizika* **B 8** (1999) 2, 251; Phys. Lett. B **436** (1998) 145.

# A dynamical approach to semi-inclusive B decays

## Giulia Ricciardi

*Dipartimento di Scienze Fisiche, Università di Napoli "Federico II" and I.N.F.N., Sezione di Napoli, Complesso Universitario di Monte Sant'Angelo, Via Cintia, 80126 Napoli, Italy.*

**Abstract.**
A dynamical approach to semi-inclusive decays based on an effective running coupling for the strong interactions is described.

**Keywords:** QCD, HeavyFlavour
**PACS:** 12.38.Bx, 13.25.Hw

## INTRODUCTION

Semi-leptonic and radiative semi-inclusive decays of heavy mesons are currently under intense investigation. Non perturbative physics seems to be more manageable in such decays, and there is hope for a reduction of theoretical assumptions and a more stringent comparison with experimental data. Several effective approaches to such decays are available. Most commonly, in order to calculate the decay rate, a series of operators is used, whose coefficients and matrix elements are weighted differently according to the theoretical assumptions. The matrix elements of the operators are not calculable within perturbation theory; the largest is the number of operators included in the calculation, the more accurate is the result. Another possible approach is based on an effective strong coupling, in turn based on perturbative threshold resumming and analyticity principles. [1, 2, 3, 4, 5] In such approach, an effective strong coupling is introduced and used in the resummation soft–gluon formulas.

## EFFECTIVE COUPLING

Long distance effects manifest themselves in perturbation theory in the form of series of large infrared logarithms, coming from "incomplete" cancellation of infrared divergencies in real and virtual diagrams in the threshold kinematical region. Such logarithms need to be resummed at all orders and resumming formulas are available within perturbation theory. Resumming requires integration over all possible kinematical domains, included low energy, order of $\Lambda_{QCD}$ ones; therefore, divergencies arise when the running coupling constant hits the Landau pole in the integrations. Several prescriptions can be used used in order to keep under control divergencies in the soft–gluon resumming formulas for the decay rates. An interesting possibility is to substitute an effective running coupling, with no Landau pole singularity, into the resumming formulas in place of the standard one. The non-physical Landau singularity can be removed by means of a

CP892, *Quark Confinement and the Hadron Spectrum VII*
edited by J. E. F. T. Ribeiro
© 2007 American Institute of Physics 978-0-7354-0396-3/07/$23.00

dispersion relation [5]

$$\bar{\alpha}(Q^2) = \frac{1}{2\pi i} \int_0^\infty \frac{ds}{s + Q^2} \operatorname{Disc}_s \alpha(-s); \tag{1}$$

$Q$ is the hard scale of the process. At lowest order

$$\bar{\alpha}_{lo}(Q^2) = \frac{1}{\beta_0} \left[ \frac{1}{\log Q^2/\Lambda^2} - \frac{\Lambda^2}{Q^2 - \Lambda^2} \right]. \tag{2}$$

Let us improve the effective coupling by adding the contributions of secondary emissions off the radiated gluons. The final effective coupling is given by the prescription

$$\tilde{\alpha}(k_\perp^2) = \frac{i}{2\pi} \int_0^{k_\perp^2} ds \operatorname{Disc}_s \frac{\bar{\alpha}(-s)}{s}. \tag{3}$$

If we neglect the $-i\pi$ terms in the integral over the discontinuity — i.e. the absorptive effects — the cascade coupling exactly reduces to the ghost-less one:

$$\tilde{\alpha}(k_\perp^2) \to \bar{\alpha}(k_\perp^2). \tag{4}$$

Let us notice that the dispersion relation (2) has automatically added a power term to the coupling. That leads to another assumption, that is that the effective coupling may have a role outside the perturbative contest where it has been introduced, by including long distance effects. The perturbative QCD formulas is extrapolated to a non-perturbative region by assuming that the relevant non-perturbative effects can be relegated into an effective coupling. By using an effective coupling to mimic also long distance effects, one can exploit the fact that resummation formulas have universal characteristics which do not depend on the single process. Of course, that implies that not all long distance effects can be accounted for. The description is assumed valid for bound state effects; they are due to the vibration of the $b$ quark inside the $B$ quark as a consequence of its interactions with light degrees of freedom (the so called Fermi motion). On the other side, other effects, like f.i. the $K^*$ peak which appears in the radiative hadron mass distribution or the $\pi$ and $\rho$ peaks which appear in the semileptonic one, cannot be accurately predicted in this approach.

## COMPARISON WITH DATA

Ultimately, the validity of the approach previously described relies on the comparison with the experimental data. The agreement is generally good [5]. F.i., in Fig. (1) and in Fig. (2) the invariant hadron mass distribution $d\Gamma_r/dm_X$ and the photon energy spectrum $d\Gamma_r/dt$ for the radiative decay $B \to X_s \gamma$ are compared, respectively, with experimental data given by the BaBar collaboration [6, 7]. There is good agreement also for the hadron spectrum in the semileptonic $B \to X_u l \nu$ decay, as shown f.i. in Fig. (3), where data from Babar Collaboraion are presented [8]. The electron spectrum in the semileptonic $B \to X_u l \nu$ decay is affected by a large background coming from the decay

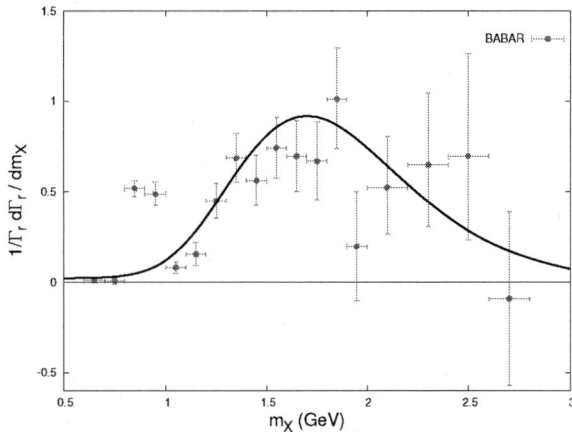

**FIGURE 1.** $B \to X_s\gamma$ invariant hadron mass distribution compared with BaBar experimental points for $\alpha_S(m_Z) = 0.123$.

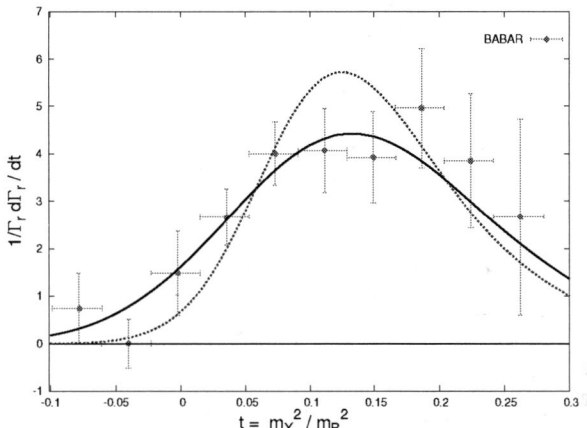

**FIGURE 2.** $B \to X_s\gamma$ photon spectrum from BaBar compared to the theoretical curve. The theoretical curve has been convoluted with a normal distribution with a standard deviation, $\sigma_\gamma$. Here $t \equiv 1 - \frac{2E_\gamma}{m_B}$. Dotted line (blue): $\alpha_S(m_Z) = 0.130$ and $\sigma_\gamma = 100$ MeV; continuous line (black): $\alpha_S(m_Z) = 0.129$ and $\sigma_\gamma = 200$ MeV.

$B \to X_c l \nu$. For the electron spectrum the agreement is not as good as in the previous cases, as can be seen in Fig. (4), where Belle data are shown [9]. However, before considering theory improvement, one has to investigate more sophisticated comparison with data, that are in principle possible; f.i., in this case, one could take into account more recent data for the charm background or a better analysis of the correlation of the systematics of the various bins.

345

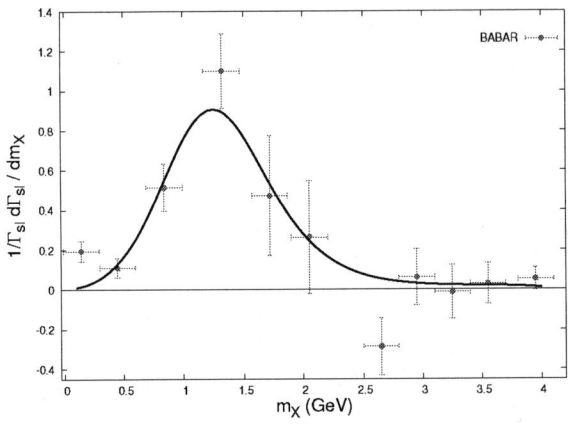

**FIGURE 3.** invariant hadron mass distribution in semileptonic decays from BaBar for $\alpha_S(m_Z) = 0.119$.

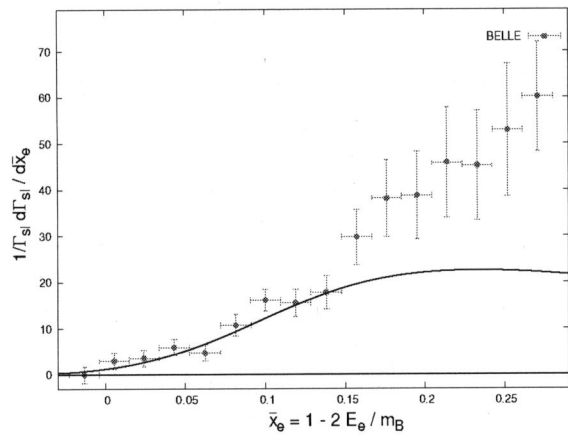

**FIGURE 4.** electron spectrum in semileptonic decay from Belle for $\alpha_S(m_Z) = 0.135$.

# REFERENCES

1. U. Aglietti and G. Ricciardi, *Phys. Rev. D* **70** 114008 (2004).
2. U. Aglietti, G. Ricciardi, and G. Ferrera, *Phys. Rev. D* **74** 034004 (2006).
3. U. Aglietti, G. Ricciardi, and G. Ferrera, *Phys. Rev. D* **74** 034005 (2006).
4. U. Aglietti, G. Ricciardi, and G. Ferrera, *Phys. Rev. D* **74** 034006 (2006).
5. U. Aglietti, G. Ferrera and G. Ricciardi, hep-ph/0608047.
6. B. Aubert *et al.* [BABAR Collaboration], *Phys. Rev. D* **72** 052004 (2005). [arXiv:hep-ex/0508004].
7. B. Aubert *et al.* [BABAR Collaboration], arXiv:hep-ex/0507001.
8. B. Aubert *et al.* [BABAR Collaboration], Phys. Rev. Lett. **96** (2006) 221801 [arXiv:hep-ex/0601046].
9. A. Limosani *et al.* [Belle Collaboration], Phys. Lett. B **621** (2005) 28 [arXiv:hep-ex/0504046].

# Charmonium and Charmonium–like (?) States

Kamal K. Seth

*Department of Physics and Astronomy, Northwestern University, Evanston, IL, 60208, USA*

**Abstract.** The last few years have witnessed a renaissance in the spectroscopy of heavy quarks. Several long elusive states have now been firmly identified, and several unexpected states have been reported by the high luminosity experiments at Belle, Babar, CLEO, and Fermilab. These discoveries have posed important theoretical questions for our understanding of QCD, and a variety of theoretical models have been proposed. These developments are critically discussed.

**Keywords:** Charmonium Spectroscopy
**PACS:** 13.25.Gv,14.40.Cs,13.65.+i

## INTRODUCTION

Until a few years ago, in my talks I used to point out how heavy–quark ($c$, $b$) spectroscopy is so much cleaner experimentally than the spectroscopy of light quarks ($u$, $d$, $s$) because of narrow and well-separated states, and so much more amenable to understanding in terms of QCD because of the smaller value of the strong coupling constant, $\alpha_S$, and the less drastic relativistic effects. As you will see, this was only true as long as we dealt only with bound states. It is not true now, as we have moved on to higher states.

## CHARMONIUM ($C\overline{C}$)

The spectrum of charmonium states is well known. Below the $D\overline{D}$ threshold at 3730 MeV, the bound states are $1^1S_0(\eta_c)$, $1^3S_1(J/\psi)$, $2^1S_0(\eta_c')$, $2^3S_1(\psi')$, $1^1P_1(h_c)$, and $1^3P_J(\chi_{c0,c1,c2})$. Despite thirty years of spectroscopy, two glaring holes have remained in the spectrum of the bound states of charmonium. Neither SLAC, nor Fermilab, nor BES were able to identify the two spin–singlet states, the $\eta_c'(2^1S_0)$ and $h_c(1^1P_1)$, both known to be bound states. A milestone in charmonium spectroscopy has now been reached. Both these states have now been firmly identified.

## $\eta_c'(2^1S_0)$   The Radial Excitation of the Charmonium Ground State

In 1982, the Crystal Ball Collaboration at SLAC claimed the identification of $\eta_c'$ in radiative transition from $\psi'$, with mass $M(\eta_c') = 3534(5)$ MeV. This corresponded to the 2S hyperfine splitting $\Delta M_{hf}(2S) \equiv M(\psi') - M(\eta_c') = 92(5)$ MeV. This was rather surprising because a 'model–independent' prediction based on $\Delta M_{hf}(1S) = 117(1)$ MeV [1], is that $\Delta M_{hf}(2S) = 62(2)$ MeV. Fortunately, the Crystal Ball observation was never

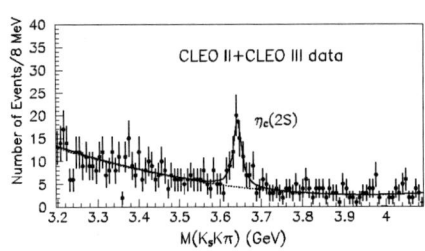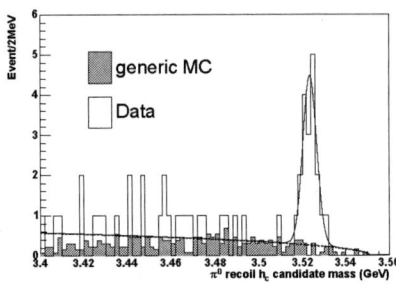

**FIGURE 1.** (left) Observation of $\eta_c'$ in the reaction $\gamma\gamma \to K_S K\pi$ by CLEO [3]. (right) Observation of $h_c$ in exclusive analysis of $\psi' \to \pi^0 h_c$, $h_c \to \gamma\eta_c$ by CLEO [5].

confirmed; $\eta_c'$ remained unidentified despite repeated attempts by the $\bar{p}p$ experiment E760/E835 at Fermilab, and the $e^+e^-$ measurements by DELPHI, L3, and CLEO.

The first observation of $\eta_c'$ was reported by Belle in $B$-decays [2]. This was followed by identifications by CLEO [3] and BaBar [4] in two–photon fusion, $\gamma\gamma \to \eta_c' \to K_S K\pi$, as illustrated in Fig. 1 (left). Since then, both Belle and BaBar have reported its observation in double charmonium production in $e^+e^-$ collisions in the $\Upsilon(4S)$ region. The weighted average of the observed masses is $M(\eta_c') = 3638(4)$ MeV [1], which leads to the 2S hyperfine splitting, $\Delta M_{hf}(2S) = 48(4)$ MeV. This is unexpectedly small compared to $\Delta M_{hf}(1S) = 117(1)$ MeV. Attempts have been made to explain this by invoking channel mixing, but it is fair to say that the observation remains a challenge to the theorists. The width of $\eta_c'$ remains essentially undetermined, so far. It is hoped that with the $\sim 30$ million $\psi'$ that CLEO-c expects to have soon, the direct M1 transition, $\psi' \to \gamma\eta_c'$ can be identified, and the width and mass of $\eta_c'$ can be determined with precision.

## $h_c(1^1P_1)$ – The Singlet $P$–state of Charmonium

This state has been the object of many frustrating searches since the early days of charmonium spectroscopy. The great interest in identifying it comes from the fact that if the $|q\bar{q}>$ confinement potential is a Lorenz scalar, as is generally assumed, there is no long-range spin-spin interaction, and for $P$–wave (and higher $L$–waves), the hyperfine splitting should be identically zero, i.e., $\Delta M_{hf}(1P) \equiv M(\langle^3P_J\rangle) - M(^3P_J) = 0$. The weighted average of the masses of the $^3P_J$ states, $\chi_{c0}$, $\chi_{c1}$, $\chi c2$, is very accurately known, being

$$M(\langle^3P_J\rangle) = [5M(\chi_{c2}) - 3M(\chi_{c1}) - M(\chi_{c0})] = 3525.36(6) \text{ MeV}$$

so that $M(^1P_1)$ should be exactly the same. However, speculations abound on how it could be up to 10–15 MeV different from this. A precision measurement of $M(h_c)$ is therefore mandatory.

348

The main difficulty in identifying $h_c$ is that its formation in radiative decay of $\psi'$ is forbidden by charge conjugation, as is its radiative decay to $J/\psi$. Attempts by the Fermilab E760/E835 to search for $h_c$ via the reaction $p\bar{p} \to h_c \to \pi^0 J/\psi$ were unsuccessful, as were earlier attempts by the Crystal Barrel Collaboration to search for it.

With a state-of-the-art detector and large luminosity, CLEO has returned to the isospin forbidden reaction $\psi' \to \pi^0 h_c$, $h_c \to \gamma\eta_c$, and successfully identified $h_c$ [5]. In the inclusive measurements, either the photon energy or the $M(\eta_c)$ were loosely constrained, and $h_c$ was identified as an enahancement in the $\pi^0$ recoil spectrum. In the exclusive measurement, shown in Fig. 1 (right), neither the photon energy nor $M(\eta_c)$ were constrained, but $\eta_c$ was identified in seven hadronic decays. The two measurements gave consistent results, their average being $M(h_c) = 3524.4 \pm 0.6 \pm 0.4$ MeV, which leads to $\Delta M_{hf}(1P) = +1.0 \pm 0.6 \pm 0.4$ MeV. This time the surprise is a pleasant one, in that the naive expectation of zero hyperfine splitting seems to be almost true. Once again, it is hoped that with the nearly ten times $\psi'$ which are expected to be soon available at CLEO, a better measurement of $M(h_c)$ and $\Gamma(h_c)$ will be forthcoming soon.

# THE SURPRISING AND UNEXPECTED CHARMONIUM–LIKE (?) STATES

During the last two years, unexpected states have been popping up all over. The first of these is X(3872), and the last is Y(4260). In between are three states X, Y, Z, all having masses near 3940 MeV. This proliferation is both exciting and rather baffling. It arises primarily from the fact that with huge integrated $e^+e^-$ luminosities available at the B–factories, the Belle and BaBar detectors are observing very weakly excited resonances. It is obvious that it will take a while before the dust settles down, and when it does, it is likely that not all the resonances will survive.

Of the alphabet soup, there are only two resonances which have been observed by more than one experiment. These are X(3872) and Y(4260). The other three, X, Y, Z(3940) have been reported only by Belle, and the silence from BaBar is deafening.

**X(3872):** First reported by Belle [6], this resonance, which decays primarily into $\pi^+\pi^- J/\psi$, has been confirmed by CDF, DØ, and BaBar. Its average mass is $M(X) = 3871.5 \pm 0.4$ MeV, and width $\langle\Gamma\rangle \leq 2.3$ MeV. A variety of theoretical explanations for X(3872) have been suggested, ranging from mixed charmonium to a $D\overline{D}^*$ molecule. A large number of decays of X(3872) have been investigated and angular correlations have been studied. My summary of these is that its spin $J^{PC} = 1^{++}$ or $2^{-+}$. It could be a displaced $1^1D_2(2^{-+})$ or $2^3P_1(1^{++})$ state of charmonium, or a $D\overline{D}^*$ molecule. If it is the latter, the binding energy of the molecule is $E_b = +0.61 \pm 0.62$ MeV [7]. This produces a big problem for the molecule, because Swanson [8] predicts the ratio $R \equiv \Gamma(X \to D\overline{D}\pi^0)/\Gamma(X \to \pi^+\pi^- J/\psi) \approx 1/20$, while Belle [9] has measured $R \approx 10$.

**Y(4260), or V(4260):** BaBar [10] has reported an enhancement in the $\pi^+\pi^- J/\psi$ invariant mass, labeled Y(4260), in $e^+e^-$ annihilation following initial state raditiation (ISR). They report $M(Y) = 4259 \pm 8^{+2}_{-6}$ MeV and $\Gamma(Y) = 88 \pm 23^{+6}_{-4}$ MeV. The production of this state in ISR would make it a vector (hence my suggestion to call it V(4260)). Unfortunately, all the charmonium vectors in the 3.8–4.4 GeV mass region are spo-

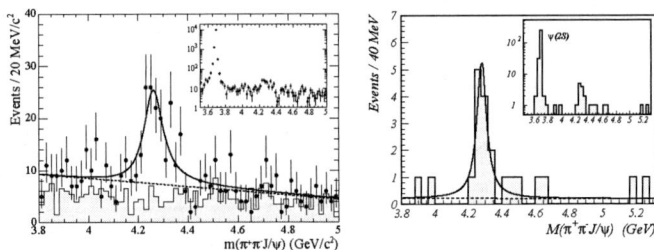

**FIGURE 2.** Observation of Y(4260) by BaBar (left) and CLEO (right) in the spectra of $M(\pi^+\pi^-J/\psi)$.

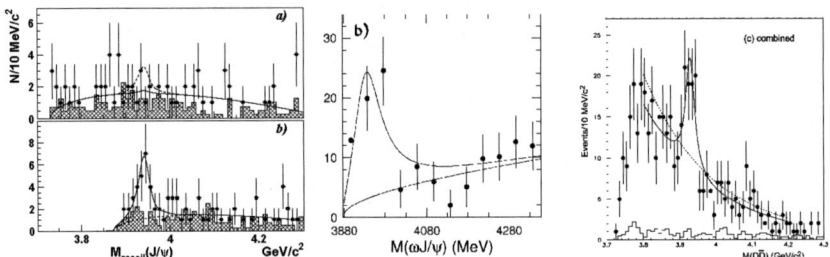

**FIGURE 3.** The three resonances of Belle: (left) X(3936), (middle) Y(3943), and (right) Z(3929).

**TABLE 1.** Summary of the Belle resonances X, Y, Z.

| | M (MeV) | $\mathscr{L}$ fb$^{-1}$ | N (evts) | $\Gamma$ (MeV) | Formed in/ Decays to | No decays to | Suggested? |
|---|---|---|---|---|---|---|---|
| X [13] | 3936(14) | 357 | 266(63) 25(7) | 39(26) 15(10) | $e^+e^- \to J/\psi(X)$ $X \to \bar{D}D^*$ | $X \nrightarrow \bar{D}D$ $X \nrightarrow \omega J/\psi$ | $\eta_c''(2^1S_0)$ |
| Y [14] | 3943(17) | 253 | 58(11) | 87(34) | $B \to KY$ $Y \to \omega J/\psi$ | $Y \nrightarrow \bar{D}D^*$ | $c\bar{c}$ hybrid |
| Z [15] | 3929(5) | 395 | 64(18) | 29(10) | $\gamma\gamma \to Z$ $Z \to D\bar{D}$ | $Z \nrightarrow \bar{D}D^*$ | $\chi'_{c2}(2^3P_2)$ |

ken for, and actually there is a deep minimum in $R \equiv \sigma(e^+e^- \to \text{hadrons})/\sigma(e^+e^- \to \mu^+\mu^-)$ [11]. thus, Y(4260) is extremely surprising and it sexistence had to be independently confirmed. CLEO [12] has now done that. Although CLEO statistics are much smaller, the much lower background enables it to make a firm confirmation of the Y(4260) resonance with $M(Y) = 4284^{+17}_{-16} \pm 4$ MeV and $\Gamma(Y) = 73^{+39}_{-25} \pm 5$ MeV. Fig. 2 shows the spectra obtained by both BaBar and CLEO. No theoretical understanding of Y(4260) exists so far. It is a truely mysterious state.

**X, Y, Z(3940):** Belle [13, 14, 15] has reported three different states, all having the same mass within $\pm 6$ MeV, but produced in different reactions and decaying into different final states (see Fig. 3). The observations are summarized in Table 1. The statistics of the observations are low, and there are open questions. Since Estia Eichten

[16] is talking about these states in a plenary talk, I will not go into details here. I do, however, want to note some of the problems with the favorite theoretical suggestions (listed in the last column of Table 1) for the possible nature of these states.

- **X(3940):** Can this be $\eta_c''(3^1S_0)$? $\psi(4040)$ is generally accepted as $\psi''(3^3S_1)$. If X(3940) is $\eta_c''$, $\Delta M_{hf}(3S) \approx 100$ MeV. With $\Delta M_{hf}(2S) = 48(4)$ MeV, can we have $\Delta M_{hf}(3S) \approx 100$ MeV? Further, is it possible that X(3940) is the same as Z(3929)? $D\overline{D}$ and $D\overline{D}^*$ decays of both need to be investigated carefully.
- **Y(3940):** Can this be a hybrid? The lowest $|c\overline{c}g >$ hybrid is predicted with mass $M = 4300 - 4500$ MeV. $D\overline{D}$ and $D\overline{D}^*$ decays need to be measured.
- **Z(3929):** Why is there yield only in $D^0\overline{D}^0$, and none in $D^+D^-$? Why no $D\overline{D}^*$?

My biased conclusion is that it is entirely likely that not all three X, Y, Z are seperate entities. We need consistent analysis of the full Belle data set, and for BaBar to weigh in with confirmations or refutations.

# REFERENCES

1. W. -M. Yao *et al.* (Particle Data Group), J.Phys. **G 33**, 1 (2006).
2. Belle Collaboration, S. K. Choi *et al.*, Phys. Rev. Lett. **89**, 142991 (2002).
3. CLEO Collaboration, D. M. Asner et al., Phys. Rev. Lett. **92**, 142001 (2004).
4. BaBar Collaboration, B. Aubert et al., Phys. Rev. Lett. **92**, 142002 (2004).
5. CLEO Collaboration, Phys. Rev. Lett. **95** (2005) 102003.
6. Belle Collaboration, S.–K. Choi et al., Phys. Rev. Lett. **91**, 262001 (2003).
7. Based on a recent precision measurement of $M(D^0)$ by CLEO. See P. Pakhove, ICHEP06.
8. E. S. Swanson, Phys. Lett. **B 588**, 189 (2004)
9. Belle Collaboration, G. Gokhroo *et al.*, Phys. Rev. Lett. **97**, 162002 (2006).
10. BaBar Collabortation, Phys. Rev. Lett. **95**, 142001 (2005).
11. K. K. Seth, Phys. Rev. **D 72**, 017501 (2005).
12. CLEO Collaboration, Q. He *et al.*, Phys. Rev. **D** (Rapid Communications), in press.
13. Belle Collaboration, K. Abe et al., hep-ex/0507019.
14. Belle Collaboration, S. K. Choi et al., Phys. Rev. Lett. **94**, 182002 (2005).
15. Belle Collaboration, S. Uehara et al., Phys. Rev. Lett. **96**, 082003 (2006).
16. E. Eichten, elsewhere in these proceedings.

# Relativistic spectra of bound fermions

Riccardo Giachetti* and Emanuele Sorace[†]

*Dipartimento di Fisica, Università di Firenze, Italy.
[†]INFN Sezione di Firenze, Italy

**Abstract.** A two fermion relativistic invariant wave equation is used for numerical calculations of the hyperfine shifts of the Positronium levels in a Breit interaction scheme. The results agree with known data up to the order $\alpha^4$.

**Keywords:** Bound states, Relativistic wave equations.
**PACS:** PACS 03.65.Pm, 03.65.Ge

In two recent papers [1] we presented a Lorentz invariant quantum mechanical formulation of two interacting Dirac particles in terms of a set of canonical coordinates where the relative time entered as a cyclic variable. We analyzed the nonperturbative solution of the spectral problem of a system with a purely scalar Coulomb interaction and performed the numerical computations getting the masses of bound states for any ratio of the two component masses $m_1$ and $m_2$. We then considered a system with $m_1 = m_2$ and an electromagnetic coupling of Breit type [2] : the levels were firstly calculated without approximations (despite the presence of divergences in the integration domain making the numerical procedure non trivial) and successively in a perturbative expansion, averaging the magnetic term over the relativistic Coulomb eigenstates. Denoting by $\alpha$ the fine structure constant, the eigenvalue equation for the system mass $\lambda$ reads:

$$\left[ q_a \left( \check{\gamma}_{(1)} \gamma_{(1)_a} - \check{\gamma}_{(2)} \gamma_{(2)_a} \right) + \check{\gamma}_{(1)} m_1 + \check{\gamma}_{(2)} m_2 \right.$$
$$\left. - \frac{\alpha}{r} \left( 1 - \varepsilon \left( \check{\gamma}_{(1)} \gamma_{(1)_a} \check{\gamma}_{(2)} \gamma_{(2)_a} + (\check{\gamma}_{(1)} \gamma_{(1)_a} \frac{r_a}{r}) (\check{\gamma}_{(2)} \gamma_{(2)_b} \frac{r_b}{r}) \right) \right) - \lambda \right] \Psi = 0 \quad (1)$$

where $(\check{\gamma}_{(1)}, \gamma_{(1)_a}) = (\check{\gamma}, \gamma_a) \otimes Id_4$, $(\check{\gamma}_{(2)}, \gamma_{(2)_a}) = Id_4 \otimes (\check{\gamma}, \gamma_a)$, $q_a$ is the relative momentum, $r_a$ is the relative distance. $q_a, r_a, \gamma_a$, $a = 1, 2, 3$ are Wigner vectors of spin one, so that their 3-dim inner products, as well as $\check{\gamma}_{(i)}$ are Lorentz scalars.

By reducing the 16-dim problem on states of given $(j, m)$ and total parity $((-)^j$ even; $(-)^{1+j}$ odd) one obtains a 4-dim system of O.D.E. for each parity sector states $\Psi_+$, $\Psi_-$:

$$\frac{dY(r)}{dr} + \mathscr{B} Y(r) = 0, \quad (2)$$

where $Y(r) = {}^t(y_1(r), y_2(r), y_3(r), y_4(r))$ and $\mathscr{B}$ is a matrix with general structure

$$\mathscr{B} = \begin{bmatrix} 0 & E_{\varepsilon=0}(r) & F_{\varepsilon=0}(r) & 0 \\ E_\varepsilon(r) & 1/r & 0 & F_\varepsilon(r) \\ G_{1,\varepsilon}(r) & 0 & 2/r & E_\varepsilon(r) \\ 0 & G_{2,\varepsilon}(r) & E_{\varepsilon=0}(r) & 1/r \end{bmatrix} . \quad (3)$$

CP892, *Quark Confinement and the Hadron Spectrum VII*
edited by J. E. F. T. Ribeiro
© 2007 American Institute of Physics 978-0-7354-0396-3/07/$23.00

In the even case, letting $h(r) = \lambda + \alpha/r$, $M = m_1 + m_2$, $\mu = (m_1 - m_2)/M$, the matrix elements of $\mathscr{B}$ read

$$E_\varepsilon(r) = \frac{\sqrt{j(j+1)}\,\mu}{rh(r) - 2\alpha\varepsilon}, \qquad F_\varepsilon(r) = \frac{(h^2(r) - \mu^2)r^2 - (2\alpha\varepsilon)^2}{2r(rh(r) - 2\alpha\varepsilon)}$$

$$G_{1,\varepsilon}(r) = \frac{h(r)}{2} + \frac{4\alpha\varepsilon + \dfrac{4j(j+1)}{2\alpha\varepsilon - rh(r)} + \dfrac{r^2 M^2}{4\alpha\varepsilon - rh(r)}}{2r}$$

$$G_{2,\varepsilon}(r) = \frac{2j(j+1)}{r^2 h(r)} + \frac{4\alpha^2\varepsilon^2 + (-h(r)^2 + M^2)r^2}{2r(rh(r) - 2\alpha\varepsilon)} \tag{4}$$

From the solution $Y(r)$ the complete vectors $\Psi_+$, $\Psi_-$ are reconstructed. The odd coefficients are obtained from the previous ones simply by changing $M \to -\mu$ and $\mu \to -M$. The pure Coulomb case corresponds to $\varepsilon = 0$. When $E_\varepsilon(r) = 0$, i.e. for $j = 0$ or $\mu = 0$ in even parity case, $(y_1, y_3)$ decouple from $(y_2, y_4)$. Besides the simplified situations, where the problem is of the second order, we have in general to solve a spectral problem of fourth order: we must therefore look for two acceptable asymptotic solutions both in zero and at infinity. We thus slightly modify the double shooting method, integrating numerically the system from zero and from infinity up to a chosen crossing point. Here the matching of the solutions, given by the vanishing of a $4 \times 4$ determinant, constitutes the spectral condition and provides the values of the mass $\lambda$.

The main features of the spectra we obtained in the Coulomb case for the ground and first excited states are the following: (*i*) the degeneracies of the singlet and triplet ground and first excited states for any ratios of the masses; (*ii*) the crossing of the $p$ and $s$ first excited states terms both for even (singlet) and odd (triplet); (*iii*) the difference between the numerical results and the analytical spectra calculated up to the order $\alpha^4$ that turns out to be of order $\alpha^4$ also.

We have then considered for the Positronium ($m_1 = m_2 = m_e$, the electron mass) the addition of the Breit term with $\varepsilon = 1/2$, that should correspond to the physical value, and with generic values of $\varepsilon$, allowing for considerable simplifications in the perturbative calculations that appear more suited to the problem. The numerical method we have used is the same as before, but its application is made somewhat more difficult by the presence of additional singularities appearing for positive values of $r$ and for $\varepsilon$ sufficiently close to $1/2$. Fortunately it happens that the new singularities fall in the categories classified by H. Weyl as "*limit cycle*", i.e. having a complete set of regular solutions in their neighborhood: these regular solutions can be used to bridge the singularity and to produce a regular eigenfunction in the whole $r$ domain. This procedure, however, is rather lengthy, due to the huge dependence of the solutions upon the initial data that makes impossible any use of the standard numerical integration codes. The calculations confirm the known fact that the complete Breit term gives physically wrong values, e.g. producing the mass of the ground singlet higher than that of the triplet – an erroneus prediction even omitting, as we do, the annihilation channel. On the other hand, a perturbative treatment of the Breit term leads to spectral values given by the relation

$$\lambda_{\text{Breit}} = \lambda_{\text{Coulomb}} + (1/2)\,d\lambda(\varepsilon)/d\varepsilon\big|_{\varepsilon=0} \tag{5}$$

that can be shown to be equivalent to adding the average of the magnetic term on the relativistic Coulomb states to the pure Coulomb eigenvalues. The derivative $d\lambda(\varepsilon)/d\varepsilon\big|_{\varepsilon=0}$

is obtained by solving the eigenvalue problem with different values of $\varepsilon$ and making a Lagrangian interpolation. The following table summarizes the comparison of our results and the numerical values obtained by the analytical semi-classical approximations [3]:

| State | $w_{num}$ | $w_{semi-classical}$ |
|-------|-----------|----------------------|
| $1^1s_0$ | -.5000349313 | $-\frac{1}{2} - \frac{21}{32}\alpha^2 = -.5000349462$ |
| $1^3s_1$ | -.4999994484 | $-\frac{1}{2} + \frac{1}{96}\alpha^2 = -.4999994453$ |
| $2^1s_0$ | -.1250055105 | $-\frac{1}{8} - \frac{53}{512}\alpha^2 = -.1250055123$ |
| $2^3s_1$ | -.1250010756 | $-\frac{1}{8} - \frac{31}{1536}\alpha^2 = -.1250010747$ |
| $2^1p_1$ | -.1250010747 | $-\frac{1}{8} - \frac{31}{1536}\alpha^2 = -.1250010747$ |
| $2^3p_1$ | -.1250016293 | $-\frac{1}{8} - \frac{47}{1536}\alpha^2 = -.1250016294$ |
| $2^3p_0$ | -.1250032935 | $-\frac{1}{8} - \frac{95}{1536}\alpha^2 = -.1250032935$ |
| $2^3p_2$ | -.1250002977 | $-\frac{1}{8} - \frac{43}{7680}\alpha^2 = -.1250002982$ |

where $w = (\lambda - 2m_e)/(\alpha^2 m_e/2)$.

We observe that the degeneracy of the states $2^3s_1$ and $2^1p_1$ is predicted from the perturbative expansion and that is completely confirmed from the numerical calculations. The existing difference of about $0.01\,\alpha^2$ for the pure Coulomb interaction disappears when introduncing the Breit term. The relativistic calculation we have presented provides a theoretical and mathematical instruments to investigate bound states without any semi-classical approximation or expansion in the fine structure constant: the only due perturbative treatment has been reserved to the magnetic interaction term. We have also proved that the new singularities we have found bear no serious consequences neither in the integration of the wave equations, nor in their spectral behavior. This, in a sense, can be considered an indirect test of the reliability of the approach to bound states through relativistic wave equations up to the quantum field theoretic corrections [4] and probably can have fruitful applications to hadron quark models [5], allowing a clearcut separation of kinematics from the dynamical approximations.

# REFERENCES

1. R.Giachetti. E.Sorace, J. Phys. A **38**, 1345, (2005), and arXiv hep-ph/0608216 (2006), J. Phys. A in press.
2. Breit, Phys. Rev. **34**, 553, (1929) and **36**, 383, (1930).
3. V.B. Berestetski, L.D. Landau, JEPT **19**, 673, (1949); T. Fulton, P.C. Martin, Phys. Rev. **93**, 904, (1955); S. Berko, H.N. Pendleton; Ann. Rev. Nuc. Part. Sci. **30**, 543, (1980).
4. A.A.Penin, arXiv hep-ph/0308204 (2003), S.G. Karshenboim, Phys. Reports **422**, 1, (2005).
5. H. Crater and P. Van Alstine, Phys. Rev. **D 70**, 034026, (2004).

# Unraveling the nature of heavy quarkonia through radiative decays

## Joan Soto

*Departament d'Estructura i Constituents de la Matèria, Universitat de Barcelona*

**Abstract.** It will be shown that precise data on the photon spectrum of radiative decays of heavy quarkonia provides important information about their nature. It may eventually tell us whether a given state is bound due to a Coulomb-like potential or to a confining one.

**Keywords:** heavy quarkonium, effective field theories, radiative decays
**PACS:** 13.20.Gd, 12.39.St, 14.40.Gx, 18.38.Bx, 18.38.Cy, 18.38.Lg

## INTRODUCTION

The understanding of heavy quarkonium systems (see [1] for an extensive review) from QCD has increased considerable during the last years. The extensive use of effective field theories (EFTs, see [2] for a review), following the pionering work of Caswell and Lepage [3], has allowed to work out model independent formulas for a number of processes, in particular for radiative decays to light hadrons which we will be concerned with here.

Heavy quarkonium systems enjoy the following hierarchies of physical scales $m \gg mv \gg mv^2$ and $m \gg \Lambda$, where $m$ is the mass of the heavy quark, $v$ the typical heavy quark velocity in the center of mass frame, $mv$ the typical relative momenta, $mv^2$ the typical binding energy and $\Lambda$ the typical hadronic scale. The hierarchy $m \gg mv, mv^2, \Lambda$ is most conveniently exploited using Non-Relativistic QCD (NRQCD) [4], whereas the hierarchy $mv \gg mv^2$ is most conveniently exploited using Potential NRQCD (pNRQCD) [5, 6]. The degrees of freedom of the last EFT depend on whether $\Lambda \lesssim mv^2$ (weak coupling regime) or $\Lambda \gg mv^2$ (strong coupling regime). In the weak coupling regime the binding is due to a Coulomb-like potential and the leading non-perturbative effects are of non-potential type [7, 8]. In the strong coupling regime the binding is due to a confining potential, and the EFT essentially reduces to a potential model [9, 10].

Since $m$, $v$ and $\Lambda$ are not directly observable, given a heavy quarkonium state it is not clear to which of the above regimes it must be assigned to. The leading non-perturbative ($\sim \Lambda$) corrections to the spectrum in the weak coupling regime scale as a large power of the principal quantum number [7, 8], which suggests that only the $n = 1$ states of bottomonium and charmonium may belong to this regime. However if one ignores this and proceeds with weak coupling calculations one finds, for instance, that renormalon based approaches at NNLO [11, 12] and the NNNLO calculation [13] give a reasonable description of the bottomonium spectrum up to $n = 3$. It would be desirable to have a theoretically sound procedure to infer the regime of a given state from data. This is not only an accademic issue: the counting of the EFT depends on the regime, and there are

CP892, *Quark Confinement and the Hadron Spectrum VII*
edited by J. E. F. T. Ribeiro
© 2007 American Institute of Physics 978-0-7354-0396-3/07/$23.00

observables, like the heavy quarkonium polarization in hadron colliders, which are very sensible to it (see [14] and references therein). It has recently been proposed that precise measurements of the photon spectra in radiative decays, as the ones carried out by CLEO [15], will clarify the assignments [16] (see also [17]), as I shall explain in the following.

## SEMI-INCLUSIVE RADIATIVE DECAYS

The contributions to the decay width of a state $n$ can be split into direct and fragmentation, $d\Gamma_n/dz = d\Gamma_n^{dir}/dz + d\Gamma_n^{frag}/dz$. Direct contributions are those in which the observed photon is emitted from the heavy quarks and fragmentation contributions those in which it is emitted from the decay products (light quarks). $z \in [0,1]$ is defined as $z = 2E_\gamma/M_n$ ($M_n$ is the mass of the heavy quarkonium state), namely the fraction of the maximum energy the photon may have in the heavy quarkonium rest frame. The approximations required to calculate the direct contributions are different in the lower end-point region ($z \to 0$), in the central region ($z \sim 0.5$) and in the upper end-point region ($z \to 1$) of the spectrum [18] . We shall restrict our discussion to $z$ in the central region, in which no further scale is introduced beyond those inherent of the non-relativistic system. Consequently, the photon spectrum can be expressed in terms of matrix elements of local NRQCD operators $\mathcal{Q}$ with matching coefficients $C[\mathcal{Q}](z)$ which depend on $m$ and $z$.

$$\frac{d\Gamma_n^{dir}}{dz} = \sum_{\mathcal{Q}} C[\mathcal{Q}](z)\frac{\langle \mathcal{Q} \rangle_n}{m^{\delta_{\mathcal{Q}}}} \tag{1}$$

$\delta_{\mathcal{Q}}$ is an integer which follows from the dimension of $\mathcal{Q}$, $\langle \mathcal{Q} \rangle_n := \langle V(nS)|\mathcal{Q}|V(nS)\rangle$ and $V(nS)$ stands for a vector $S$-wave state of principal quantum number $n$. The fragmentation contributions read [19]

$$\begin{aligned}
\frac{d\Gamma_n^{frag}}{dz} &= \sum_{a=q,\bar{q},g} \int_z^1 \frac{dx}{x} \sum_{\mathcal{Q}} C_a[\mathcal{Q}](x) D_{a\gamma}\left(\frac{z}{x}, m\right) \\
&:= \sum_{\mathcal{Q}} f_{\mathcal{Q}}(z)\frac{\langle \mathcal{Q} \rangle_n}{m^{\delta_{\mathcal{Q}}}}
\end{aligned} \tag{2}$$

$D_{a\gamma}(x,m)$ are the fragmentation functions and $C_a[\mathcal{Q}](x)$ the partonic kernels. It is important for what follows that the $f_{\mathcal{Q}}(z)$ are universal and do not depend on the specific bound state $n$. Due to the behavior of the fragmentation functions above, the fragmentation contributions are expected to dominate the spectrum in the lower $z$ region and to be negligible in the upper $z$ one. In the central region, in which we will focus on, they can be treated as a perturbation.

In the strong coupling regime, the NRQCD matrix elements can be further factorize into a wave function at the origin square (or derivatives of it), which contains all the dependence on the quantum numbers of the state, times universal non-perturbative parameters [20, 21, 22]. Due to this further factorization, one can work out the following parameter free formula, which holds at NLO,

$$\frac{\frac{d\Gamma_n}{dz}}{\frac{d\Gamma_r}{dz}} = \frac{\langle \mathcal{O}_1(^3S_1)\rangle_n}{\langle \mathcal{O}_1(^3S_1)\rangle_r} \left( 1 + \frac{C_1' \left[^3S_1\right](z)}{C_1 \left[^3S_1\right](z)} \frac{1}{m} (E_n - E_r) \right)$$

$$\frac{\langle \mathcal{O}_1(^3S_1)\rangle_n}{\langle \mathcal{O}_1(^3S_1)\rangle_r} = \frac{\Gamma\left(\Upsilon(n) \to e^+e^-\right)}{\Gamma\left(\Upsilon(r) \to e^+e^-\right)} \left[ 1 - \frac{\mathrm{Im} g_{ee}\left(^3S_1\right)}{\mathrm{Im} f_{ee}\left(^3S_1\right)} \frac{E_n - E_r}{m} \right]$$

$C_1'\left[^3S_1\right](z), C_1\left[^3S_1\right](z), \mathrm{Im} g_{ee}\left(^3S_1\right), \mathrm{Im} f_{ee}\left(^3S_1\right)$ are matching coefficients computable in perturbation theory, and $E_n - E_r$ the mass difference between the two states. If data follow this formula it will indicate that both $n$ and $r$ are in the strong coupling regime. For the $n = 1,2,3$ of bottomonia, current data, which is not very precise for $n = 3$, disfavor $n = 1$ in the strong coupling regime and is compatible with $n = 2,3$ in it.

## ACKNOWLEDGMENTS

I acknowledge financial support from MEC (Spain) grant CYT FPA 2004-04582-C02-01, the CIRIT (Catalonia) grant 2005SGR00564, and the RTNs Euridice HPRN-CT2002-00311 and Flavianet MRTN-CT-2006-035482 (EU).

## REFERENCES

1. N. Brambilla *et al.*, arXiv:hep-ph/0412158.
2. N. Brambilla, A. Pineda, J. Soto and A. Vairo, Rev. Mod. Phys. **77**, 1423 (2005) [arXiv:hep-ph/0410047].
3. W. E. Caswell and G. P. Lepage, Phys. Lett. B **167**, 437 (1986).
4. G. T. Bodwin, E. Braaten and G. P. Lepage, Phys. Rev. D **51**, 1125 (1995) [Erratum-ibid. D **55**, 5853 (1997)] [arXiv:hep-ph/9407339].
5. A. Pineda and J. Soto, Nucl. Phys. Proc. Suppl. **64**, 428 (1998) [arXiv:hep-ph/9707481].
6. N. Brambilla, A. Pineda, J. Soto and A. Vairo, Nucl. Phys. B **566**, 275 (2000) [arXiv:hep-ph/9907240].
7. M. B. Voloshin, Nucl. Phys. B **154** (1979) 365.
8. H. Leutwyler, Phys. Lett. B **98** (1981) 447.
9. N. Brambilla, A. Pineda, J. Soto and A. Vairo, Phys. Rev. D **63**, 014023 (2001) [arXiv:hep-ph/0002250].
10. A. Pineda and A. Vairo, Phys. Rev. D **63**, 054007 (2001) [Erratum-ibid. D **64**, 039902 (2001)]
11. N. Brambilla, Y. Sumino and A. Vairo, Phys. Lett. B **513** (2001) 381 [arXiv:hep-ph/0101305].
12. N. Brambilla, Y. Sumino and A. Vairo, Phys. Rev. D **65** (2002) 034001 [arXiv:hep-ph/0108084].
13. A. A. Penin and M. Steinhauser, Phys. Lett. B **538**, 335 (2002) [arXiv:hep-ph/0204290].
14. J. Soto, arXiv:nucl-th/0611055.
15. D. Besson *et al.* [CLEO Collaboration], Phys. Rev. D **74** (2006) 012003 [arXiv:hep-ex/0512061].
16. X. Garcia i Tormo and J. Soto, Phys. Rev. Lett. **96**, 111801 (2006) [arXiv:hep-ph/0511167].
17. X. Garcia i Tormo, arXiv:hep-ph/0610145.
18. X. Garcia i Tormo and J. Soto, Phys. Rev. D **72**, 054014 (2005) [arXiv:hep-ph/0507107].
19. F. Maltoni and A. Petrelli, Phys. Rev. D **59**, 074006 (1999) [arXiv:hep-ph/9806455].
20. N. Brambilla, D. Eiras, A. Pineda, J. Soto and A. Vairo, Phys. Rev. Lett. **88**, 012003 (2002)
21. N. Brambilla, D. Eiras, A. Pineda, J. Soto and A. Vairo, Phys. Rev. D **67**, 034018 (2003)
22. N. Brambilla, A. Pineda, J. Soto and A. Vairo, Phys. Lett. B **580**, 60 (2004) [arXiv:hep-ph/0307159].

# The Decay Constant $f_{D_s}$ and other Form Factor Measurements from BABAR

Jörg Stelzer

*Stanford Linear Accelerator Center,*
*2575 Sand Hill Road, MS 61,*
*Menlo Park, CA 94025, USA*
*and*
*CERN,*
*CH-1211 Genevé 23, Switzerland*
*Joerg.Stelzer@cern.ch*

**Abstract.** Since its start in 1999 the BABAR experiment has collected a vast amount of data, opening the doors for high precision measurements of decay constants and of semileptonic decay form-factors of heavy-light mesons, $B$ and $D$. In this article a number of such measurements is presented.

**Keywords:** Form Factors, fDs
**PACS:** 13.20.He, 14.40.Nd, 14.60.Fg

## INTRODUCTION

Long distance effects of QCD can not be calculated perturbatively by expansion in orders of the strong coupling constant, due to the increasing strength of the strong coupling constant at growing distances. The unknown effects of low-energy QCD are described by phenomenological form factors, which are functions of the momentum transfer and the helicity states of the physical process.

Lattice gauge theory can, potentially, provide calculations of hadronic matrix elements based on first principles. The theories success is founded by a remarkable agreement with experimental data, in particular for masses and decay constants of light mesons[1]. Yet, simulations in the regime of heavy-light mesons have not lined up in this list of success stories. Decay-constant simulations of heavy-light mesons, for instance, have still an error of more than 10%. The large separation of the scales that characterises hadronic bound states of heavy and light quarks requires large lattices to minimize simulation errors from heavy quarks discretization effects and finite volume effects from light quarks. Chiral Perturbation Theory and Heavy Quark Effective Theory have been developed to treat these problems. Combined with Lattice QCD they offer the most rigorous tools to control the hadronic physics in $B$ and $D$ meson decays today.

## MEASUREMENT OF THE DECAY CONSTANT $f_{D_s}$

The decay constants of the $B_d$ meson and the $B_s$ meson, $f_B$ and $f_{B_s}$ respectively, are needed to extract the CKM parameters $|V_{td}|$ and $|V_{ts}|$, using the relation

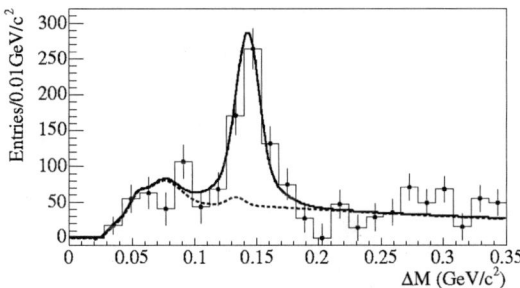

**FIGURE 1.** $\Delta M$ distribution after the tag sidebands and the electron sample are subtracted. The solid line is the fitted signal and background distribution, the dashed line is the background distribution alone.

$$\left|\frac{V_{td}}{V_{ts}}\right|^2 = \frac{\Delta M_d}{\Delta M_s}\frac{m_{B_s}}{m_{B_d}}\frac{f_{B_s}^2 B_{B_s}}{f_B^2 B_{B_d}}, \tag{1}$$

where $\Delta M_{d/s}$ are the mass differences between the heavy and the light mass-eigenstates of the neutral $B_{d/s}$ mesons. $\Delta M_d$ and $\Delta M_s$ are known from measurements of the flavor oscillations in the two neutral meson systems to a precision of 1.5% and 2.4%, respectively. The ratio of the bag parameters $B_{B_d}$ and $B_{B_s}$ has been simulated on the lattice with an error of 4%. The branching ratios of the decays $B^+ \to \mu^+ \nu_\mu$ and $B^+ \to \tau^+ \nu_\tau$ are $\sim 4 \times 10^{-7}$ and $1 \times 10^{-4}$, respectively, rendering a measurement of $f_B$ to better than $\sim 20\%$, even with the full dataset to be expected from the two B-factories, nearly impossible. The importance of the measurements of $f_{D_s}$ and $f_D$ lies in their use as calibration source for lattice simulations of decay constants of heavy-light mesons.

To measure the branching ratio of $D_s^+ \to \mu^+ \nu_\mu$, the decay chain $D_s^{*+} \to \gamma D_s^+, D_s^+ \to \mu^+ \nu_\mu$ is reconstructed, where the $D_s^{*+}$ meson is produced in the hard fragmentation of $e^+e^- \to c\bar{c}$ events. Signal candidates are required to lie in the recoil of a fully reconstructed $D$ meson (the "tag"), wherein the tag flavor, and hence the expected charge of the signal muon, is uniquely determined. The signal is reconstructed from a photon and a muon candidate opposite the tag candidate. The neutrino is estimated from the events missing four momentum, with a $D_s^+$ mass constraint on the muon-neutrino pair. The signature of the signal is a narrow peak in the mass difference $\Delta M = M(\mu\nu\gamma) - M(\mu\nu)$ at 143.5 MeV/$c^2$. Requirements on the photon energy, the missing energy, and the $D_s^{*+}$ momentum are made.

The tag mass sidebands are used to subtract background from events $e^+e^- \to q\bar{q}$ ($q = u, d, s$). Events $e^+e^- \to B\bar{B}$ are eliminated by a minimum tag momentum requirement above the kinematic limit from $B \to D$ decays. The contribution from decays $D \to X\mu\nu_\mu$ is subtracted using $D \to Xe\nu_e$ decays. Since the decay $D_s^+ \to e\nu_e$ is strongly helicity suppressed, signal events are not affected by this subtraction. The remaining background is estimated from simulation. The $\chi^2$-fit to the data yields $489 \pm 55_{\text{stat}}$ signal events (Fig. 1).

The number of produced $D_s^{*+}$ mesons in the dataset is determined by reconstructing $D_s^{*+} \to \gamma D_s^+ \to \gamma\phi\pi^+$ decays. The analysis of this decay yields 2093 events. Using the *BABAR* average for the branching ratio $\mathscr{B}(D_s^+ \to \phi \pi^+) = (4.71 \pm 0.46)\%$ [2][3], the branching fraction $\mathscr{B}(D_s^+ \to \mu^+ \nu_\mu) = (6.74 \pm 0.83 \pm 0.26 \pm 0.66) \times 10^{-3}$ and the decay constant $f_{D_s} = (283 \pm 17_{stat} \pm 7_{sys} \pm 14)$ MeV are obtained [4]. Here, the third error is the uncertainty from $\mathscr{B}(D_s^+ \to \phi \pi^+)$. The ratio of our value for $f_{D_s}$ to $f_D$ from the CLEO-c measurement, $f_{D_s}/f_D = 1.27 \pm 0.14$, is consistent with lattice QCD.[5]

## OTHER RECENT FORM FACTOR MEASUREMENTS OF SEMILEPTONIC *B* AND *D* DECAYS

Measurements of branching ratios of of semileptonic *B* and *D* decays always determine the product of a CKM matrix element and a hadronic form factor $f(q^2)$. While the form factor normalization $f(0)$ has to be calculated, the $q^2$ dependence can be measured testing different theories, and the extrapolation of $f$ to zero momentum transfer.

### $B^0 \to D^* l^+ \nu_l$ **Form Factors and** $|V_{cb}|$

The hadronic weak current of the semileptonic decay $B^0 \to D^* l^+ \nu_l$ is described by two axial and one vector form factor, $A_1(\omega)$, $A_2(\omega)$, and $V(\omega)$, where $\omega$ is the Lorentz boost of the $D^*$ meson in the $B$ meson rest frame. These are often given in terms of

$$h_{A_1}(\omega) = \frac{M_B + M_{D^*}}{\sqrt{M_B M_{D^*}}} \frac{1}{\omega + 1} A_1(\omega), \qquad (2)$$

and the HQET inspired form factor ratios $R_1(\omega)$ and $R_2(\omega)$

$$\frac{R_1(\omega)}{V(\omega)} = \frac{R_2(\omega)}{A_2(\omega)} = \frac{2M_B M_{D^*}}{(M_B + M_{D^*})^2} \frac{\omega + 1}{A_1(\omega)}. \qquad (3)$$

These functions can be parametrized by [6]

$$h_{A_1}(\omega) = h_{A_1}(\omega) \times \left\{ 1 - \rho^2 z + (53\rho^2 - 15)z^2 - (231\rho^2 - 91)z^3 \right\},$$
$$R_1(\omega) = R_1(1) - 0.12(\omega - 1) + 0.05(w - 1)^2 \quad \text{and} \qquad (4)$$
$$R_2(\omega) = R_2(1) + 0.11(\omega - 1) - 0.06(w - 1)^2,$$

where $z = (\sqrt{\omega + 1} - \sqrt{2})/(\sqrt{\omega + 1} + \sqrt{2})$. Since the dependence of $R_{1/2}$ on $\omega$ is small, $R_1$ and $R_2$ are assumed to be constant in this analysis.

The analysis [7] reconstructs events $B^0 \to D^{*+} l\nu_l$, $D^{*+} \to D^0 \pi^+$, with the $D^0$ decaying into $K^- \pi^+$, $K^- \pi^+ \pi^- \pi^+$, or $K^- \pi^+ \pi^0$. The full four-dimensional differential decay rate can be written as

$$\frac{d\Gamma}{d\omega d\cos\theta_l d\cos\theta_V d\chi} = F(\omega, \cos\theta_l, \cos\theta_V, \chi | \rho^2, R_1(1), R_2(1)), \qquad (5)$$

where, in addition to $\omega$, $\theta_l$, $\theta_V$, and $\chi$, (the polar angle of the $W^+ \to l^+ \nu_l$ decay, the polar angle of the $D^{*+} \to D^0 \pi^+$ decay, and the angle between the two decay planes, respectively), are the physical observable from which the form factors $\rho^2$, $R_1(1)$, and $R_2(1)$ are extracted.

The event selection starts with requirements on the lepton identification probability and the mass of the reconstructed $D^0$. The soft pion and the $D^{*+}$ momentum are required to be consistent with the signal decay. The sidebands of the $M_{D^{*+}} - M_{D^0}$ distribution are used to evaluate the background from randomly combined $D^{*+}$ candidates; data taken below the $\Upsilon(4S)$ resonance describe the background from non-$B\bar{B}$ events. Fake lepton contributions are estimated using a lepton veto. Most remaining backgrounds are estimated from a fit to the distribution of the angle between the reconstructed $B^0$ direction and the direction of the $(D^{*+}l^-)$ system. Less than 0.5% of the backgrounds are simulated.

A four-dimensional binned $\chi^2$-fit is performed to extract the form factors and $|V_{cb}f(\omega = 1)|$ simultaneously. The results are combined with those from a second BABAR analysis, that uses only electrons and reconstructs the $D^0$ only from $D^0 \rightarrow K^- \pi^+$, but performs a full four-dimensional unbinned maximum likelihood fit[8]. The combined result is

$$\rho^2 = 1.179 \pm 0.048_{\text{stat}} \pm 0.028_{\text{sys}},$$
$$R_1 = 1.417 \pm 0.061_{\text{stat}} \pm 0.044_{\text{sys}},$$
$$R_2 = 0.836 \pm 0.037_{\text{stat}} \pm 0.022_{\text{sys}} \quad \text{and}$$
$$|V_{cb}| = \left(37.74 \pm 0.35_{\text{stat}} \pm 1.25_{\text{sys}}{}^{+1.23}_{-1.44\text{f(1)}}\right) \times 10^{-3}, \tag{6}$$

where for the calculation of (6) the lattice result [9] for $f(\omega = 1)$ is used.

## Summary

The BABAR experiment has measured the decay constant of the $D_s^+$ meson to a very good precision, comparable with the result from the CLEO-c $D_s^+$-factory. It also widely opened the door into the realm of precision measurements of form factors of semileptonic $B$ decays. With this much progress made already, the prospects of the increased dataset of the BABAR $B$-factory are exciting.

## REFERENCES

1. C. T. H. Davies, et al., *Phys. Rev. Lett.* **92**, 022001 (2004).
2. B. Aubert, et al., *Phys. Rev. D* **71**, 091104 (2005).
3. B. Aubert, et al., *Phys. Rev.* **D74**, 031103 (2006).
4. B. Aubert, *hep-ex/0607094* .
5. C. Aubin, et al., *Phys. Rev. Lett.* **95**, 122002 (2005).
6. I. Caprini, L. Lellouch, and M. Neubert, *Nucl. Phys.* **B530**, 153–181 (1998).
7. B. Aubert, et al., *hep-ex/0607076* .
8. B. Aubert, et al., *hep-ex/0602023* .
9. S. Hashimoto, A. S. Kronfeld, P. B. Mackenzie, S. M. Ryan, and J. N. Simone, *Phys. Rev.* **D66**, 014503 (2002).

# A Canonical $D_S$ Spectrum

Eric S. Swanson

*Department of Physics and Astronomy, University of Pittsburgh, Pittsburgh PA 15260*

**Abstract.** Quark mass dependence induced by one loop corrections to the Breit-Fermi spin-dependent one gluon exchange potential permits an accurate determination of heavy-light meson masses. In this model, the $D_s(2317)$, $D_s(2460)$ and the newly discovered $D_s(2700)$ and $D_s(2860)$ are canonical $c\bar{s}$ mesons. The multiplet splitting relationship of chiral doublet models, $M(1^+) - M(1^-) = M(0^+) - M(0^-)$, holds to good accuracy in the $D$ and $D_s$ systems, but is accidental.

**Keywords:** meson spectroscopy
**PACS:** 12.39.Pn, 14.40.Ev, 13.20.Fc

BaBar's discovery of the $D_s(2317)$ state generated strong interest in heavy meson spectroscopy, chiefly due to its surprisingly low mass with respect to expectations[1]. These expectations are based on quark models or lattice gauge theory. Unfortunately, at present large lattice systematic errors do not allow a determination of the $D_s$ mass with a precision better than several hundred MeV. And, although quark models appear to be exceptionally accurate in describing charmonia, they are less constrained by experiment and on a weaker theoretical footing in the open charm sector. It is therefore imperative to examine reasonable alternative descriptions of the open charm sector.

It is natural to assume that the $D$ and $D_s$ spectra should be very similar once the strange and light quark mass difference has been accounted for. However, one finds that a shifted $D_s$ spectrum agrees quite well with the $D$ spectrum, except for the $D_{s0}$ and $D_{s1}$ states. One way to account for this discrepancy is to introduce mass-dependence into the constituent quark model. In fact, the quark model explanation of these states rests on P-wave mass splittings induced by spin-dependent interactions. A common model of spin-dependence is based on the Breit-Fermi reduction of the one-gluon-exchange interaction supplemented with the spin-dependence due to a scalar current confinement interaction. The general form of this potential has been computed by Eichten and Feinberg at tree level using Wilson loop methodology. The result is parameterised in terms of four nonperturbative matrix elements, $V_i$, which can be determined by electric and magnetic field insertions on quark lines in the Wilson loop. Subsequently, Gupta and Radford[2] performed a one-loop computation of the heavy quark interaction and showed that a fifth interaction, $V_5$ is present in the case of unequal quark masses. The net result is a quark-antiquark interaction that can be written as:

$$V_{q\bar{q}} = V_{conf} + V_{SD} \qquad (1)$$

CP892, *Quark Confinement and the Hadron Spectrum VII*
edited by J. E. F. T. Ribeiro
© 2007 American Institute of Physics 978-0-7354-0396-3/07/$23.00

where $V_{conf}$ is the standard Coulomb+linear scalar form:

$$V_{conf}(r) = -\frac{4}{3}\frac{\alpha_s}{r} + br \qquad (2)$$

and

$$
\begin{aligned}
V_{SD}(r) &= \left(\frac{\sigma_q}{4m_q^2} + \frac{\sigma_{\bar{q}}}{4m_{\bar{q}}^2}\right)\cdot \mathbf{L}\left(\frac{1}{r}\frac{dV_{conf}}{dr} + \frac{2}{r}\frac{dV_1}{dr}\right) + \left(\frac{\sigma_{\bar{q}}+\sigma_q}{2m_q m_{\bar{q}}}\right)\cdot \mathbf{L}\left(\frac{1}{r}\frac{dV_2}{dr}\right) \\
&\quad + \frac{1}{12 m_q m_{\bar{q}}}\left(3\sigma_q\cdot\hat{\mathbf{r}}\,\sigma_{\bar{q}}\cdot\hat{\mathbf{r}} - \sigma_q\cdot\sigma_{\bar{q}}\right)V_3 + \frac{1}{12 m_q m_{\bar{q}}}\sigma_q\cdot\sigma_{\bar{q}}V_4 \\
&\quad + \frac{1}{2}\left[\left(\frac{\sigma_q}{m_q^2} - \frac{\sigma_{\bar{q}}}{m_{\bar{q}}^2}\right)\cdot\mathbf{L} + \left(\frac{\sigma_q - \sigma_{\bar{q}}}{m_q m_{\bar{q}}}\right)\cdot\mathbf{L}\right]V_5. \qquad (3)
\end{aligned}
$$

The first four $V_i$ are order $\alpha_s$ in perturbation theory, while $V_5$ is order $\alpha_s^2$; for this reason $V_5$ has been largely ignored by quark modellers. In practice this is acceptable except in the case of unequal quark masses, where the additional spin-orbit interaction can play an important role [3, 4]. The model of Ref. [3, 4] can be described in terms of vector and scalar kernels defined by $V_{conf} = V + S$ where $V = -4\alpha_s/3r$ is the vector kernel and $S = br$ is the scalar kernel, and by the order $\alpha_s^2$ contributions to the $V_i$, denoted by $\delta V_i$. Expressions for the matrix elements of the spin-dependent interaction are then

$$
\begin{aligned}
V_1 &= -S + \delta V_1 \\
V_2 &= V + \delta V_2 \\
V_3 &= V'/r - V'' + \delta V_3 \\
V_4 &= 2\nabla^2 V + \delta V_4 \\
V_5 &= \delta V_5. \qquad (4)
\end{aligned}
$$

The important new element is $V_5$ given by $V_5(m_q, m_{\bar{q}}, r) = \frac{1}{4r^3}C_F C_A \frac{\alpha_s^2}{\pi}\ln\frac{m_{\bar{q}}}{m_q}$ where $C_F = 4/3$ and $C_A = 3$.

Predictions of the new model in the $D_s$ sector are summarised in Table 1. One sees that the $D_s(2317)$ and $D_s(2460)$ are described as conventional $c\bar{s}$ scalar and axial mesons. Furthermore, the new BaBar state, $D_s(2860)$ is identified with the excited $c\bar{s}$ scalar and the BaBar $D_s(2688)$ and Belle $D_s(2715)$ are identified with the excited vector meson. These identifications are supported by the strong decay widths and production properties (in B decays) [3, 6]. Note that we regard the SELEX state as unestablished[5].

The multiplet splittings prediction of chiral doublet theory, $M(1^+) - M(1^-) = M(0^+) - M(0^-)$ is explored in Table 2. One sees that this relationship holds remarkably well in the extended constituent quark model, and therefore must be regarded as an accident in this context.

The bottom flavoured meson spectra of Table 3 have been obtained with the 'average' extended model parameters and $m_b = 4.98$ GeV. As with the open charm spectra,

**TABLE 1.** $D_s$ Spectrum.

| state | mass (GeV) | expt[?] (GeV) |
|---|---|---|
| $D_s(1^1S_0)$ | 1.968 | 1.968 |
| $D_s(2^1S_0)$ | 2.637 | |
| $D_s(3^1S_0)$ | 3.097 | |
| $D_s^*(1^3S_1)$ | 2.112 | 2.112 |
| $D_s^*(2^3S_1)$ | 2.711 | 2.688/2715? |
| $D_s^*(3^3S_1)$ | 3.153 | |
| $D_s(1^3D_1)$ | 2.784 | |
| $D_{s0}(1^3P_0)$ | 2.329 | 2.317 |
| $D_{s0}(2^3P_0)$ | 2.817 | 2.857? |
| $D_{s0}(3^3P_0)$ | 3.219 | |
| $D_{s1}(1P)$ | 2.474 | 2.459 |
| $D_{s1}(2P)$ | 2.940 | |
| $D_{s1}(3P)$ | 3.332 | |
| $D_{s1}'(1P)$ | 2.526 | 2.535 |
| $D_{s1}'(2P)$ | 2.995 | |
| $D_{s1}'(3P)$ | 3.389 | |
| $D_{s2}(1^3P_2)$ | 2.577 | 2.573 |
| $D_{s2}(2^3P_2)$ | 3.041 | |
| $D_{s2}(3^3P_2)$ | 3.431 | |

**TABLE 2.** Chiral Multiplet Splittings (MeV).

| params | $M(1^+(1/2^+)) - M(1^-)$ | $M(0^+) - M(0^-)$ |
|---|---|---|
| $D$ low | 411 | 412 |
| $D$ avg | 391 | 389 |
| $D$ high | 366 | 368 |
| $D_s$ low | 384 | 380 |
| $D_s$ avg | 373 | 370 |
| $D_s$ high | 349 | 346 |

a flavour-dependent constant was fit to each pseudoscalar. The second row reports recently measured P-wave $B$ meson masses; these are in reasonable agreement with the predictions of the first row.

**TABLE 3.** Low Lying Bottom Meson Masses (MeV)

| flavour | $0^-$ | $1^-$ | $0^+$ | $1^+$ | $1^+$ | $2^+$ |
|---|---|---|---|---|---|---|
| $B$ | 5279 | 5322 | 5730 | 5752 | 5753 | 5759 |
| expt | 5279 | 5325 | – | $5724 \pm 4 \pm 7$ | – | $5748 \pm 12$ |
| $B_s$ | 5370 | 5416 | 5776 | 5803 | 5843 | 5852 |
| expt | 5369.6 | 5416.6 | – | – | – | – |
| $B_c$ | 6286 | 6333 | 6711 | 6746 | 6781 | 6797 |
| expt | 6286 | – | – | – | – | – |

Finally, the work presented here may explain the difficulty in accurately computing the mass of the $D_{s0}$ in lattice simulations. If the extended quark model is correct, it implies that important mass and spin-dependent interactions are present in the one-loop level one-gluon-exchange quark interaction. It is possible that current lattice computations are not sufficiently sensitive to the ultraviolet behaviour of QCD to capture this physics. The problem is exacerbated by the nearby, and presumably strongly coupled, $DK$ continuum; which requires simulations sensitive to the infrared behaviour of QCD. Thus heavy-light mesons probe a range of QCD scales and make an ideal laboratory for improving our understanding of the strong interaction.

## ACKNOWLEDGMENTS

I am grateful to my colleagues F.E. Close, C. Thomas, and O. Lakhina for fruitful collaboration on this topic. This work is supported by the U.S. Department of Energy under contract DE-FG02-00ER41135.

## REFERENCES

1. E. S. Swanson, Phys. Rept. **429**, 243 (2006) [arXiv:hep-ph/0601110].
2. S. N. Gupta and S. F. Radford, Phys. Rev. D **24**, 2309 (1981). J. T. Pantaleone, S. H. H. Tye and Y. J. Ng, Phys. Rev. D **33**, 777 (1986).
3. F. E. Close, C. E. Thomas, O. Lakhina and E. S. Swanson, arXiv:hep-ph/0608139.
4. O. Lakhina and E. S. Swanson, arXiv:hep-ph/0608011.
5. T. Barnes, F. E. Close, J. J. Dudek, S. Godfrey and E. S. Swanson, Phys. Lett. B **600**, 223 (2004) [arXiv:hep-ph/0407120].
6. F. E. Close and E. S. Swanson, Phys. Rev. D **72**, 094004 (2005) [arXiv:hep-ph/0505206].

# Strong one-pion decay of $\Sigma_c$, $\Sigma_c^*$ and $\Xi_c^*$

C. Albertus[*],  E. Hernández[†], J. Nieves[*] and  J. M. Verde-Velasco[†]

[*]*Departamento de Física Atómica, Molecular y Nuclear, Universidad de Granada, E-18071 Granada, Spain.*
[†]*Grupo de Física Nuclear, Departamento de Física Fundamental e IUFFyM, Facultad de Ciencias, E-37008 Salamanca, Spain.*

**Abstract.** Working in the framework of a nonrelativistic quark model we evaluate the widths for the strong one-pion decays $\Sigma_c \to \Lambda_c \pi$, $\Sigma_c^* \to \Lambda_c \pi$ and $\Xi_c^* \to \Xi_c \pi$. We take advantage of the constraints imposed by heavy quark symmetry to solve the three-body problem by means of a simple variational ansatz. We use partial conservation of the axial current hypothesis to get the strong vertices from weak axial current matrix elements. Our results are in good agreement with experimental data.

**Keywords:** Quark model, Charmed baryons, Strong decays
**PACS:** 11.40.Ha, 12.39.Jh, 13.30.Eg, 14.20.Lq

In this contribution we present results for the widths of the strong one-pion decay processes $\Sigma_c \to \Lambda_c \pi$, $\Sigma_c^* \to \Lambda_c \pi$ and $\Xi_c^* \to \Xi_c \pi$, obtained within a nonrelativistic quark model. To the best of our knowledge this is the first fully dynamical calculation of these observables done within a nonrelativistic approach. We use heavy quark symmetry constraints on baryons with a heavy quark to solve the three-body problem by means of a simple variational ansatz. The orbital wave functions thus obtained are simple and manageable. Their functional form and the corresponding variational parameters are given in Ref. [1]. In order to check the dependence of the results on the interquark interaction we have used five different quark-quark potentials that we took from Refs. [2, 3]. All the potentials include a confining term plus Coulomb and hyperfine terms coming from one-gluon exchange, while they differ in the power of the confining term and/or in the different regularization of the singular behaviour of the one-gluon terms at the origin. The pion emission amplitude is evaluated in a one-quark pion emission model (spectator approximation) in which we use partial conservation of the axial current to determine the strong couplings through the evaluation of axial matrix elements. Due to the limitations of space, we shall focus on the presentation of the results and their comparison with experimental data. Nevertheless our tables also show other theoretical results obtained using the constituent quark model (CQM), heavy hadron chiral perturbation theory (HHCPT), and relativistic quark models like the light-front quark model (LFQM) and the relativistic three-quark model (RTQM). Full details on our calculation are given in Ref. [4].

Our results for the $\Sigma_c$ one-pion decay widths are given in Table 1. They include two different classes of errors: the first one reflects the change of the results with the potential used, while the second is purely numerical. Our results are in very good agreement with experimental data by CLEO [5, 6] and in a reasonable agreement with data by FOCUS [7].

In Table 2 we present the results for the $\Sigma_c^*$ one-pion decay widths. Our central value

**TABLE 1.** Total decay widths for $\Gamma(\Sigma_c^{++} \to \Lambda_c^+ \pi^+)$, $\Gamma(\Sigma_c^+ \to \Lambda_c^+ \pi^0)$ and $\Gamma(\Sigma_c^0 \to \Lambda_c^+ \pi^-)$.

| | $\Gamma(\Sigma_c^{++} \to \Lambda_c^+ \pi^+)$ [MeV] | $\Gamma(\Sigma_c^+ \to \Lambda_c^+ \pi^0)$ [MeV] | $\Gamma(\Sigma_c^0 \to \Lambda_c^+ \pi^-)$ [MeV] |
|---|---|---|---|
| This work | $2.41 \pm 0.07 \pm 0.02$ | $2.79 \pm 0.08 \pm 0.02$ | $2.37 \pm 0.07 \pm 0.02$ |
| Experiment | $2.3 \pm 0.2 \pm 0.3$ [5] $2.05^{+0.41}_{-0.38} \pm 0.38$ [7] | $< 4.6$ (C.L.=90%) [6] | $2.5 \pm 0.2 \pm 0.3$ [5] $1.55^{+0.41}_{-0.37} \pm 0.38$ [7] |
| Theory CQM | $1.31 \pm 0.04$ [13] $2.025^{+1.134}_{-0.987}$ [14] | $1.31 \pm 0.04$ [13] | $1.31 \pm 0.04$ [13] $1.939^{+1.114}_{-0.954}$ [14] |
| HHCPT | $2.47, 4.38$ [8] $2.5$ [15] | $2.85, 5.06$ [8] $3.2$ [15] | $2.45, 4.35$ [8] $2.4$ [15] $1.94 \pm 0.57$ [16] |
| LFQM | $1.64$ [17] | $1.70$ [17] | $1.57$ [17] |
| RTQM | $2.85 \pm 0.19$ [9] | $3.63 \pm 0.27$ [9] | $2.65 \pm 0.19$ [9] |

for $\Gamma(\Sigma_c^{*++} \to \Lambda_c^+ \pi^+)$ is above the central value of the latest experimental data by CLEO [10]. We get results within experimental errors for the AP1 and AP2 potentials of Ref. [3]. For $\Gamma(\Sigma_c^{*+} \to \Lambda_c^+ \pi^0)$ our central value is slightly above the upper experimental bound obtained by CLEO [6], while for the AP1 and AP2 potentials we are below that bound. In the case of $\Gamma(\Sigma_c^{*0} \to \Lambda_c^+ \pi^-)$ decay we agree with experiment.

**TABLE 2.** Total decay widths for $\Gamma(\Sigma_c^{*++} \to \Lambda_c^+ \pi^+)$, $\Gamma(\Sigma_c^{*+} \to \Lambda_c^+ \pi^0)$ and $\Gamma(\Sigma_c^{*0} \to \Lambda_c^+ \pi^-)$.

| | $\Gamma(\Sigma_c^{*++} \to \Lambda_c^+ \pi^+)$ [MeV] | $\Gamma(\Sigma_c^{*+} \to \Lambda_c^+ \pi^0)$ [MeV] | $\Gamma(\Sigma_c^{*0} \to \Lambda_c^+ \pi^-)$ [MeV] |
|---|---|---|---|
| This work | $17.52 \pm 0.74 \pm 0.12$ | $17.31 \pm 0.73 \pm 0.12$ | $16.90 \pm 0.71 \pm 0.12$ |
| Experiment | $14.1^{+1.6}_{-1.5} \pm 1.4$ [10] | $< 17$ (C.L.=90%) [6] | $16.6^{+1.9}_{-1.7} \pm 1.4$ [10] |
| Theory CQM | $20$ [13] | $20$ [13] | $20$ [13] |
| HHCPT | $25$ [15] | $25$ [15] | $25$ [15] |
| LFQM | $12.84$ [17] | | $12.40$ [17] |
| RTQM | $21.99 \pm 0.87$ [9] | | $21.21 \pm 0.81$ [9] |

Finally in Table 3 we present results for partial and total $\Xi_c^*$ one-pion decay widths. Our central value for $\Gamma(\Xi_c^{*+} \to \Xi_c^0 \pi^+ + \Xi_c^+ \pi^0)$ is slightly above the experimental bound obtained by CLEO [11]. As before our results for the AP1 and AP2 potentials are below that bound. For $\Gamma(\Xi_c^{*0} \to \Xi_c^+ \pi^- + \Xi_c^0 \pi^0)$ our result is clearly smaller than the experimental upper bound determined by CLEO [12].

Our results are stable against the use of different potentials with variations at the level of $6 \sim 8\%$. They are in an overall good agreement with experimental data, in most cases in better agreement than predictions by other models.

TABLE 3. Decay widths for $\Gamma(\Xi_c^{*+} \to \Xi_c^0\pi^+)$, $\Gamma(\Xi_c^{*+} \to \Xi_c^+\pi^0)$, $\Gamma(\Xi_c^{*0} \to \Xi_c^+\pi^-)$ and $\Gamma(\Xi_c^{*0} \to \Xi_c^0\pi^0)$.

| | $\Gamma(\Xi_c^{*+} \to \Xi_c^0\pi^+)$ [MeV] | $\Gamma(\Xi_c^{*+} \to \Xi_c^+\pi^0)$ [MeV] | $\Gamma(\Xi_c^{*0} \to \Xi_c^+\pi^-)$ [MeV] | $\Gamma(\Xi_c^{*0} \to \Xi_c^0\pi^0)$ [MeV] |
|---|---|---|---|---|
| This work | $1.84 \pm 0.06 \pm 0.01$ | $1.34 \pm 0.04 \pm 0.01$ | $2.07 \pm 0.07 \pm 0.01$ | $0.956 \pm 0.030 \pm 0.007$ |
| Theory | | | | |
| LFQM | 1.12 [17] | 0.69 [17] | 1.16 [17] | 0.72 [17] |
| RTQM | $1.78 \pm 0.33$ [9] | $1.26 \pm 0.17$ [9] | $2.11 \pm 0.29$ [9] | $1.01 \pm 0.15$ [9] |

| | $\Gamma(\Xi_c^{*+} \to \Xi_c^0\pi^+ + \Xi_c^+\pi^0)$ [MeV] | $\Gamma(\Xi_c^{*0} \to \Xi_c^+\pi^- + \Xi_c^0\pi^0)$ [MeV] |
|---|---|---|
| This work | $3.18 \pm 0.10 \pm 0.01$ | $3.03 \pm 0.10 \pm 0.01$ |
| Experiment | $< 3.1$ (C.L.=90%) [11] | $< 5.5$ (C.L.=90%) [12] |
| Theory | | |
| CQM | $< 2.3 \pm 0.1$ [13] , $1.191 - 3.971$ [14] | $< 2.3 \pm 0.1$ [13] , $1.230 - 4.074$ [14] |
| HHCPT | $2.44 \pm 0.85$ [16] | $2.51 \pm 0.88$ [16] |
| LFQM | 1.81 [17] | 1.88 [17] |
| RTQM | $3.04 \pm 0.50$ [9] | $3.12 \pm 0.33$ [9] |

# ACKNOWLEDGMENTS

This research was supported by DGI and FEDER funds, under contracts FIS2005-00810, BFM2003-00856 and FPA2004-05616, by the Junta de Andalucía and Junta de Castilla y León under contracts FQM0225 and SA104/04, and it is part of the EU integrated infrastructure initiative Hadron Physics Project under contract number RII3-CT-2004-506078. C. A. wishes to acknowledge a research contract with Universidad de Granada. J. M. V.-V. acknowledges an E.P.I.F. contract with Universidad de Salamanca.

# REFERENCES

1. C. Albertus, J. E. Amaro, E. Hernández, J. Nieves, Nucl. Phys. A 740 (2004) 333.
2. R. K. Bhaduri, L. E. Cohler, Y. Nogami, Nuovo Cimento A 65 (1981) 376.
3. B. Silvestre-Brac, Few-Body Systems 20 (1996) 1; C. Semay and B. Silvestre-Brac, Z. Phys. C 61 (1994) 271.
4. C. Albertus, E. Hernández, J. Nieves, J. M.Verde-Velasco, Phys. Rev. D72 (2005) 094022.
5. M. Artuso et al. (CLEO Collaboration), Phys. Rev. D 65 (2002) 071101.
6. R. Ammar et al. (CLEO Collaboration), Phys. Rev. Lett. 86 (2001) 1167.
7. J. M. Link et al. (FOCUS Collaboration), Phys. Lett. B 525 (2002) 205.
8. T. M. Yan, H.-Y. Cheng, C.-Y. Cheung, G.-L. Lin, Y. C. Lin, H.-L. Yu, Phys. Rev. D 46 (1992) 1148.
9. M. A. Ivanov, J. G. Körner, V. E. Lyubovitskij, A. G. Rusetsky, Phys. Rev. D 60 (1999) 094002. Phys. Lett. B 442 (1998) 435.
10. S. B. Athar et al. (CLEO Collaboration), Phys. Rev. D 71 (2005) 051101.
11. L. Gibbons et al. (CLEO Collaboration), Phys. Rev. Lett. 77 (1996) 810.
12. P. Avery et al. (CLEO Collaboration), Phys. Rev. Lett. 75 (1995) 4364.
13. J. L. Rosner, Phys. Rev. D 52 (1995) 6461.
14. D. Pirjol, T. M. Yan, Phys. Rev. D 56 (1997) 5483.
15. M.-Q. Huang, Y.-B. Dai, C.-S. Huang, Phys. Rev. D 52 (1995) 3986.
16. H.-Y. Cheng, Phys. Lett. B 399 (1997) 281.
17. S. Tawfiq, P. J. O'Donnell, J. G. Körner, Phys. Rev. D 58 (1998) 054010.

# A Heavy $Q\bar{Q}$-Pair Below $T_c$

D. Antonov, S. Domdey and H.-J. Pirner

*Institut für Theoretische Physik, Universität Heidelberg,
Philosophenweg 19, D-69120 Heidelberg, Germany*

**Abstract.** Thermodynamics of a heavy quark-antiquark pair in SU(3)-QCD is studied below the deconfinement critical temperature, $T_c$. In the quenched case, a model of the string passing through heavy valence gluons yields a correct estimate of $T_c$ and a behavior of the string tension near $T_c$. For two light flavors, entropy and internal energy can be obtained from the partition function of heavy-light mesons and baryons. They are in a good qualitative agreement with the lattice results.

**Keywords:** Confinement-deconfinement phase transition, effective string models in QCD, relativistic quark model
**PACS:** 12.38.-t, 25.75.Nq, 12.39.-x

In the last few years, various indications appeared that the quark-gluon plasma is not a weakly coupled system even up to temperatures of the order of a few times the deconfinement one, $T_c$. In particular, a lot of information on the nonperturbative properties of the plasma comes from the lattice. Among the recent results obtained there and calling for a theoretical explanation, are the anomalously large (from the perturbation theory standpoint) maxima of the entropy and internal energy of the static quark-antiquark pair in unquenched QCD around $T_c$ [1], [2]. This work is aimed at a description of these data below $T_c$. The strategy of our analysis is the following. First, we will determine $T_c$ and the effective string tension $\sigma(T)$ in the quenched SU($N_c$) QCD within the so-called gluon-chain model. Then, with the use of $\sigma(T)$ adjusted to the unquenched ($N_c = 3, N_f = 2$)-case, we will calculate, within the relativistic quark model, the partition function of heavy-light mesons and baryons. Entropy and internal energy stemming from this partition function will further be compared to the corresponding lattice data.

When heavy $Q$ and $\bar{Q}$ are separated by large distances, the string joining them passes through numerous valence gluons. Such an object can naturally be called gluon chain. At low temperatures, the free energy of one string bit between two nearest gluons in the chain, being a constant, exceeds thermal mass of a gluon, which grows with $T$ linearly from zero on. However, at a certain temperature $T_0$ smaller than $T_c$, the gluon's thermal mass becomes larger than the free energy of one string bit. Therefore, at $T_0 < T < T_c$, a gluon chain becomes a sequence of static nodes with adjoint charges, connected by independently fluctuating string bits. At the moment of formation of such a chain, its end-point originating from the heavy $Q$ performs a random walk towards $\bar{Q}$ over the lattice of static nodes. Since color may alter from one node to another, the total number of states of the gluon chain is $N_c^{L/a}$, where $L$ is its length and $a$ is the length of one bit. Taking into account that the full free energy is the sum of the usual linear potential and the free energy of the random walk, we have for the effective string tension:

CP892, *Quark Confinement and the Hadron Spectrum VII*
edited by J. E. F. T. Ribeiro
© 2007 American Institute of Physics 978-0-7354-0396-3/07/$23.00

$\sigma(T) = \sigma - T \ln \frac{\mathscr{Z}(R,T)}{\mathscr{Z}(R,T_0)}\Big|_{R\to\infty}$, where $\mathscr{Z}(R,T)$ is the partition function of the random walk. At asymptotically large $R$'s of interest, one finds $\sigma(T) = \sigma + T[f(T) - f(T_0)]/\sqrt{a}$, where $f(T) = \sqrt{\frac{\sigma}{T} - \frac{\ln N_c}{a}}$. An estimate for $T_c$ stems from the equation $f(T_c) = 0$: $T_c|_{N_c>1} = \frac{\sigma a}{\ln N_c}$. Equating $T_c$ to the modern $N_c = 3$ lattice value [1], 270 MeV, we obtain an effective length of one string bit $a \simeq 0.31$ fm. This is larger than the minimal possible value of this quantity, $a = 0.22$ fm – the so-called vacuum correlation length [3], which defines the onset of a string-bit formation. The temperature $T_0$, below which the lattice of valence gluons does not exist, can be defined from the condition $\sigma(T_c) = 0$, which yields $T_0 = \frac{T_c}{\ln N_c + 1} \simeq 130$ MeV. An important finding of this model is the behavior $\sigma(T) \sim \sqrt{T_c - T}$ at $T \to T_c$. It is the same as the one which follows from the Nambu-Goto model for the two-point correlation function of Polyakov loops [4].

Let us now consider the unquenched case, where, at a certain distance, the $Q\bar{Q}$-string breaks due to the production of a light $q\bar{q}$-pair. The subsequent hadronization process leads to the formation of heavy-light mesons ($Qq$) and heavy-light-light baryons ($Qqq$), as well as their antiparticles. We will consider the ($N_c = 3, N_f = 2$)-case, with light $u$- and $d$-quarks, and use the value $T_c = 200$ MeV [2]. This yields $a = 0.23$ fm, that should be used in the function $\sigma(T)$. Then, for a heavy-light meson, the Hamiltonian of the relativistic quark model reads $H_{\bar{Q}q} = m_{\bar{Q}} + \sqrt{\mathbf{p}^2 + m_q^2} + V(r)$. Here, $V(r) = \sigma(T)r - (2 - \delta)\sqrt{\sigma(T)}$, $m_{\bar{Q}}$ is the mass of a heavy antiquark, whereas $m_q$ is the constituent mass of a light quark, $m_q \simeq 300$ MeV. The subtraction of $2\sqrt{\sigma(T)}$ in $V(r)$ is known to be important to reach an agreement between the predictions of the relativistic quark model to the phenomenology of meson spectroscopy [5]. An additional correction, $\delta\sqrt{\sigma(T)}$, is needed to obey the normalization condition $F \to 2(m_{D^0} - m_c)$ at $T \to 0$. Here, $F$ is the resulting free energy of two noninteracting mesons, $m_{D^0} = 1.864$ GeV is the $D^0$-meson mass, $m_c = 1.48$ GeV is the $c$-quark mass. The value of $\delta$ we find by this procedure is 0.344. Further strategy for the calculation of the partition function of a meson [5] consists of accounting for its ground state exactly and modeling its higher eigenenergies (together with their degeneracy factors) by those of a 3d harmonic oscillator. Similarly, one can treat $(Qqq)$-baryons, whose Hamiltonian reads $H_{Qqq} =$ $m_Q + \sum_{i=1}^{2}\left(\sqrt{\mathbf{p}_i^2 + m_q^2} + V(r_i)\right)$. Here, we have approximated the position of the baryon string-junction point by the position of $Q$, that is legitimate due to the heaviness of $Q$. The value of the parameter $\delta$, at which the free energy of a baryon goes to $m(\Lambda_c^+) - m_c$ at $T \to 0$, is 0.393. Here, $m(\Lambda_c^+) = 2.286$ GeV is the mass of the $\Lambda_c^+$-baryon. In the figures, we plot the entropy and the internal energy of mesons together with baryons, at the averaged value $\delta = 0.37$. In the same figures, we plot the corresponding lattice data for the $Q\bar{Q}$-pair [2].

In conclusion, we have analytically addressed thermodynamics of the static $Q\bar{Q}$-pair below the deconfinement phase transition, in quenched and unquenched QCD. In the quenched SU($N_c$)-case, within the so-called gluon-chain model, we have derived an effective temperature-dependent string tension. With this string tension adjusted to the unquenched ($N_c = 3, N_f = 2$)-case, we have calculated the entropies and internal energies of heavy-light mesons and baryons. In general, the results obtained are in a

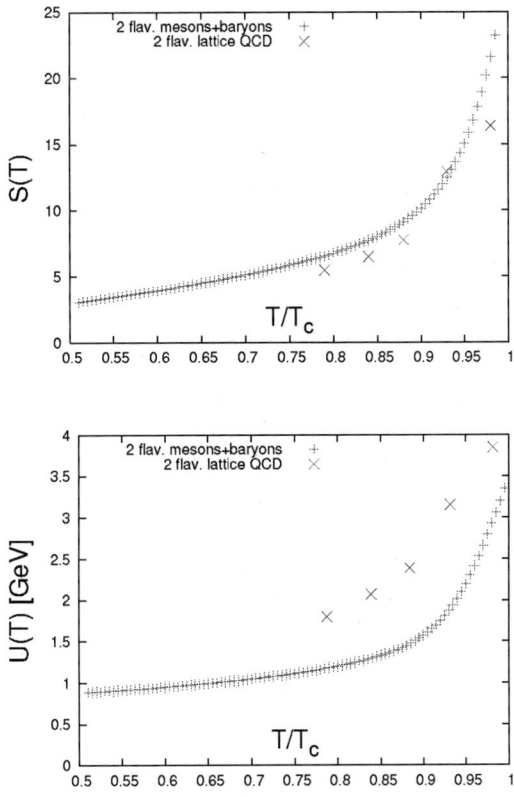

good agreement with the corresponding recent lattice data for the same quantities. Our analysis shows also that the entropy and the internal energy of heavy-light mesons alone are not enough to reproduce correctly the lattice data.

D.A. acknowledges the organizers of the conference 'QCHS7' (Ponta Delgada Açores, Portugal, 2-7 September, 2006) for an invitation and an opportunity to present these results in a stimulating atmosphere. The work of D.A. has been supported through the contract MEIF-CT-2005-024196. S.D. thanks P. Petreczky and F. Zantow for providing him with the details of the lattice data.

## REFERENCES

1. P. Petreczky, *Eur. Phys. J. C* **43**, 51–57 (2005).
2. O. Kaczmarek and F. Zantow, "Static quark anti-quark interactions at zero and finite temperature QCD. II: Quark anti-quark internal energy and entropy," preprint hep-lat/0506019 (unpublished).
3. A. Di Giacomo and H. Panagopoulos, *Phys. Lett. B* **285**, 133–136 (1992); G. S. Bali, N. Brambilla and A. Vairo, *Phys. Lett. B* **421**, 265–272 (1998).
4. R. D. Pisarski and O. Alvarez, *Phys. Rev. D* **26**, 3735–3737 (1982).
5. H. J. Pirner and M. Wachs, *Nucl. Phys. A* **617**, 395–413 (1997).

# Baryon-Strangeness Correlations from Hadron/String- and Quark-Dynamics

Stephane Haussler[*], Stefan Scherer[*] and Marcus Bleicher[†]

[*]*Frankfurt Institute for Advanced Studies (FIAS), Johann Wolfgang Goethe Universität, Max-von-Laue-Str. 1, 60438 Frankfurt am Main, Germany*
[†]*Institut für Theoretische Physik, Johann Wolfgang Goethe Universität, Max-von-Laue-Str. 1, 60438 Frankfurt am Main, Germany*

**Abstract.** Baryon-strangeness correlations ($C_{BS}$) are studied with a hadron/string transport approach (UrQMD) and a dynamical quark recombination model (quark molecular dynamics, qMD) for various energies from $E_{lab} = 4A$ GeV to $\sqrt{s_{NN}} = 200$ GeV. As expected, we find that the hadron/string dynamics shows correlations similar to a simple hadron gas. In case of the quark molecular dynamics, we find that initially the $C_{BS}$ correlation is that of a weakly interacting QGP but changes in the process of hadronization also to the value for a hadron gas. Therefore, we conclude that the hadronization process itself makes the initial baryon strangeness correlation unobservable. To make an experimental study of this observable more feasible, we also investigate how a restriction to only charged kaons and $\Lambda$'s (instead of all baryons and all strange particles) influences the theoretical result on $C_{BS}$. We find that a good approximation of the full result can be obtained in this limit in the present simulation.

**Keywords:** event-by-event, fluctuations, correlations
**PACS:** <http://www.aip..org/pacs/index.html>

A plasma of quarks and gluons is believed to be created in the course of the collision of two heavy nuclei travelling at ultra-relativistic speeds. Probes based on fluctuations have been proposed throughout the last decade to study the properties of QCD-matter close to the phase transition from hadronic to quark degrees of freedom [1, 2, 3, 4, 5, 6, 7, 8, 9, 10]. Even though they promised to be most adequate due to the strongly fluctuating energy density, initial temperature, isospin or particles density no experimental data up to now relying on event-by-event analyses could show a decisive signal for the production of quark-gluon matter (QGP).

A novel event-by-event observable has been introduced by Koch et al. [11], the baryon-strangeness correlation coefficient $C_{BS}$. This correlation is proposed as a tool to specify the nature (ideal QGP or strongly coupled QGP or hadronic matter) of the highly compressed and heated matter created in heavy ions collisions. The idea is that depending on the phase the system is in, the relation between baryon number and strangeness will be different: On the one hand, if one considers an ideal plasma of quarks and gluons, strangeness will be carried by freely moving strange and anti-strange quarks, carrying baryon number in strict proportions. This leads to a strong correlation between baryon number and strangeness. On the other hand, if the degrees of freedom are of hadronic nature, this correlation is different, because it is possible to carry strangeness without baryon number, e.g. in mesons or QGP bound states.

To quantify to which degree strangeness and baryon number are correlated, the following correlation coefficient has been proposed [11]:

CP892, *Quark Confinement and the Hadron Spectrum VII*
edited by J. E. F. T. Ribeiro
© 2007 American Institute of Physics 978-0-7354-0396-3/07/$23.00

$$C_{BS} = -3 \frac{\langle BS \rangle - \langle B \rangle \langle S \rangle}{\langle S^2 \rangle - \langle S \rangle^2} \quad , \tag{1}$$

where $B$ is the baryon charge and $S$ is the strangeness in a given event. If a QGP is created, the value of $C_{BS}$ will be unity as expected from lattice QCD, compatible with the ideal weakly coupled QGP. For a hadron gas, where the correlation is non trivial, this quantity has been evaluated in [11] to be $C_{BS} = 0.66$.

In this paper, we study the correlation coefficient $C_{BS}$ with the Ultra-relativistic Quantum Molecular Dynamics model (UrQMD v2.2) and the quark Molecular Dynamics model (qMD). The UrQMD is a non-equilibrium microscopic transport model that simulates the full space-time evolution of heavy ions collisions. It is valid from a few hundreds of MeV to several TeV per nucleon in the laboratory frame. It describes the rescattering of incoming and produced particles, the excitation and fragmentation of color strings and the formation and decay of resonances. This model has been used before to study event-by-event fluctuations rather successfully [3, 12, 13, 14, 15, 16] and yields a reasonable description of inclusive particle distributions. For a complete review of the model, the reader is referred to [17, 18]. Since the UrQMD is based on hadrons and strings it provides an estimate of the $C_{BS}$ value in the case where no QGP is created, however taking into account the rescattering and the non-equilibrium nature of the heavy ion reactions.

In contrast, the qMD model provides an out-of-equilibrium estimate of $C_{BS}$ with an explicit phase transition from QGP to hadronic matter. It describes the dynamics and the hadronization through an effective heavy quark potential in which the quarks propagate with a final dynamical recombination to white clusters. These clusters are then mapped to known hadrons and resonances that are later allowed to decay. Note that qMD is a recombination model that does not violate energy and momentum conservation and does not reduce the entropy in the hadronization process. The reader is referred to [19, 20] for more details about the qMD model.

$C_{BS}$ is evaluated from the event-by-event fluctuation analyses following [11]:

$$C_{BS} = -3 \frac{\frac{1}{N} \sum_n B^{(n)} S^{(n)} - (\frac{1}{N} \sum_n B^{(n)})(\frac{1}{N} \sum_n S^{(n)})}{\frac{1}{N} \sum_n (S^{(n)})^2 - (\frac{1}{N} \sum_n S^{(n)})^2} \tag{2}$$

$B^{(n)}$ and $S^{(n)}$ stand for the baryon number and strangeness in a given event $n$.

If a QGP is created, the signal given by the $C_{BS}$ coefficient should survive the hadronic phase only if the flow is strong enough. I.e. strangeness and baryon number within a given rapidity range should be frozen in. The rapidity window used must not be too wide in order to avoid global baryon number and strangeness conservation which will lead to a vanishing correlation. Nevertheless, the acceptance window must be wide enough to avoid smearing due to hadronization. A suggested reasonable width is of the order of $y_{\text{max}} = 0.25 - 0.5$.

The energy scan of $C_{BS}$ for central Au+Au/Pb+Pb collisions as calculated with UrQMD is shown as full circles in Figure 1. As discussed in [11], $C_{BS}$ increases with an increase of the baryon chemical potential $\mu_B$, i.e. when going to lower beam energies. With increasing collision energy, and therefore decreasing $\mu_B$, $C_{BS}$ goes down to

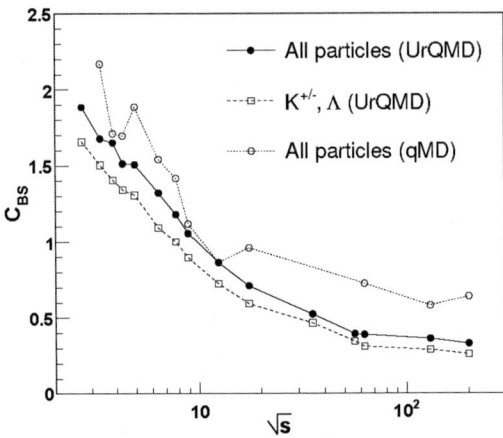

**FIGURE 1.** Correlation coefficient $C_{BS}$ for central Au+Au/Pb+Pb as a function of $\sqrt{s}$ calculated with UrQMD with all particles taken into account (full circles) and only $\Lambda$'s and charged kaons (open squares). Open circles are calculated with the qMD model using all particles. The rapidity window is $y_{\max} = 0.25$.

$C_{BS} \approx 0.4$ at the highest RHIC energy available and is slightly lower than the value for a fully thermalized hadron gas. Unfortunately it is difficult to explore $C_{BS}$ directly in experiment, because it includes contributions from neutrons and other difficult to measure hadrons. It is therefore desirable to test, if also a better accessible subset of particles can be used to explore this correlation. Therefore, we study next, how $C_{BS}$ is modified if only charged kaons and $\Lambda$'s are taken into account. As shown in Fig. 1 (open squares) one observes that this subset of particles leads in good approximation to the same results for $C_{BS}$ at high energies as for the full set of hadrons. Therefore, we conclude that a measurement of the energy dependence of the $C_{BS}$ correlation extracted out of charged kaons and $\Lambda$'s only might be sufficient to measure the correlation between baryon number and strangeness in heavy-ions collisions.

Let us finally discuss how a model with quark degrees of freedom compares to the hadron/string dynamics results. As shown in Fig. 1 (open circles) the result from the quark molecular dynamics model follows roughly the shape of the UrQMD values. Especially towards the highest RHIC energy $C_{BS}$ decreases below the QGP expectation of $C_{BS} = 1$. This surprising result strongly contrasts with the expected value for an ideal quark-gluon-plasma and might be a first indication that smearing due to hadronization process itself might have drastic effects on fluctuation observables.

Let us explore this important question further by studying the time evolution of $C_{BS}$ within the qMD for Au+Au collisions at $\sqrt{s_{NN}} = 200$ GeV as shown in Fig. 2. At early times one observes that $C_{BS} \approx 1$, in agreement with the expectations for a quark-gluon-plasma. However, around 6 fm/c (when the hadronization starts) $C_{BS}$ decreases strongly and reaches its final value $C_{BS} \approx 0.6 - 0.7$. One should note that there is no hadronic rescattering stage in qMD, thus, the decrease of the correlation is solely related to the

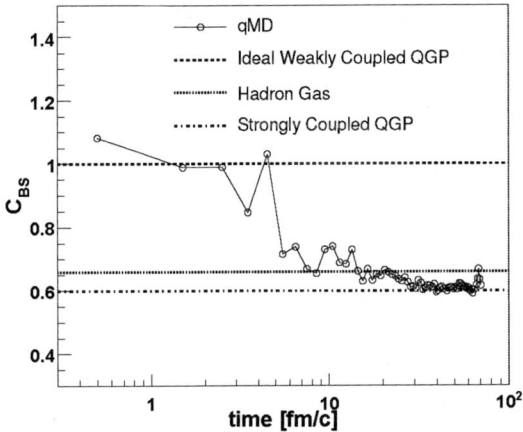

**FIGURE 2.** Correlation coefficient from the quark molecular dynamics calculation for Au+Au collisions at $\sqrt{s_{NN}} = 200$ GeV as a function of time at midrapidity. The maximum rapidity accepted is $y_{max} = 0.25$. Also shown here are the values for an ideal weakly coupled QGP, a strongly coupled QGP and for a hadron gas.

recombination-like hadronization process in the model. Thus, the present investigation might explain why no signal of the phase transition has been observed in the data up to now. Because even if the initial state consists of a quark-gluon-plasma with the expected fluctuations, these fluctuations might be completely blurred in the hadronization process.

To summarize, we have studied the dependence of the baryon-strangeness correlation coefficient as a function of energy from $E_{lab} = 4A$ GeV to $\sqrt{s_{NN}} = 200$ GeV for central Au+Au/Pb+Pb reactions with two different models. The UrQMD model is based on string-hadronic degrees of freedom, whereas the qMD model contains an explicit quark phase and a transition from quark to hadronic matter. $C_{BS}$ is found to decrease from the lower energies towards the top RHIC energy in both approaches. At the highest RHIC energy the $C_{BS}$ value from the hadron/string transport model is roughly half the one expected in the case of a QGP. However, the calculation including a phase transition gives similar results as without phase transition, in clear contradiction with what has been expected in case a plasma of quarks and gluons as the initial matter. This finding is traced back to the hadronization process itself that destroys the initially present correlations.

# ACKNOWLEDGMENTS

This work has been supported by GSI and BMBF. The computational resources were provided by the Center for Scientific Computing (CSC) in Frankfurt.

# REFERENCES

1. L. Stodolsky, Phys. Rev. Lett. **75** (1995) 1044.
2. E. V. Shuryak, Phys. Lett. B **423** (1998) 9 [arXiv:hep-ph/9704456].
3. M. Bleicher *et al.*, Nucl. Phys. A **638** (1998) 391.
4. M. A. Stephanov, K. Rajagopal and E. V. Shuryak, Phys. Rev. Lett. **81** (1998) 4816 [arXiv:hep-ph/9806219].
5. S. Mrowczynski, Phys. Lett. B **459** (1999) 13 [arXiv:nucl-th/9901078].
6. A. Capella, E. G. Ferreiro and A. B. Kaidalov, Eur. Phys. J. C **11** (1999) 163 [arXiv:hep-ph/9903338].
7. S. Mrowczynski, Phys. Lett. B **465** (1999) 8 [arXiv:nucl-th/9905021].
8. M. Asakawa, U. W. Heinz and B. Muller, Phys. Rev. Lett. **85** (2000) 2072 [arXiv:hep-ph/0003169].
9. B. Muller, Nucl. Phys. A **702** (2002) 281 [arXiv:nucl-th/0111008].
10. L. Cunqueiro, E. G. Ferreiro, F. del Moral and C. Pajares, Phys. Rev. C **72**, 024907 (2005) [arXiv:hep-ph/0505197].
11. V. Koch, A. Majumder and J. Randrup, Phys. Rev. Lett. **95**, 182301 (2005) [arXiv:nucl-th/0505052].
12. M. Bleicher *et al.*, Phys. Lett. B **435** (1998) 9 [arXiv:hep-ph/9803345].
13. M. Bleicher, S. Jeon and V. Koch, Phys. Rev. C **62** (2000) 061902 [arXiv:hep-ph/0006201].
14. M. Bleicher, J. Randrup, R. Snellings and X. N. Wang, Phys. Rev. C **62** (2000) 041901 [arXiv:nucl-th/0006047].
15. S. Jeon, L. Shi and M. Bleicher, Phys. Rev. C **73**, 014905 (2006) [arXiv:nucl-th/0506025].
16. S. Haussler, H. Stoecker and M. Bleicher, Phys. Rev. C **73**, 021901 (2006) [arXiv:hep-ph/0507189].
17. S. A. Bass *et al.*, Prog. Part. Nucl. Phys. **41** (1998) 225 [arXiv:nucl-th/9803035].
18. M. Bleicher *et al.*, J. Phys. G **25** (1999) 1859 [arXiv:hep-ph/9909407].
19. M. Hofmann, M. Bleicher, S. Scherer, L. Neise, H. Stoecker and W. Greiner, Phys. Lett. B **478**, 161 (2000) [arXiv:nucl-th/9908030].
20. S. Scherer, M. Hofmann, M. Bleicher, L. Neise, H. Stoecker and W. Greiner, New J. Phys. **3**, 8 (2001) [arXiv:nucl-th/0106036].

# Using the Balance Function to study the charge correlations of hadrons

P. Christakoglou, A. Petridis, M. Vassiliou for the NA49 and ALICE collaborations

*Physics Department - University of Athens - 15771 - Athens, Greece*

**Abstract.**

We present the recent Balance Function (BF) results obtained by the NA49 collaboration for the pseudo-rapidity dependence of non-identified charged particle correlations for two SPS energies. Experimental results indicate a clear centrality dependence only in the mid-rapidity region. The results of an energy dependence study of the BF throughout the whole SPS energy range will also be discussed. In addition, the correlation of identified hadrons is studied and presented for the first time. The study of hadron correlation has also been extended in order to cope with the high multiplicity environment that is expected to be seen at LHC. We will present the latest results from simulations concerning the extension of these studies to the ALICE experiment.

**Keywords:** Balance Function,correlations
**PACS:** <25.75.Gz>

The study of correlations and fluctuations on an event by event basis is expected to provide additional information on the reaction mechanism of high energy nuclear collisions [1]. The Balance Function (BF), introduced by Bass, Danielewicz and Pratt [2], provides the means to explore the space-time evolution of the emitted hadrons after such a collision. The method was initially proposed [2] to be related to the time of hadronization: if a pair of oppositely charged particles was produced late in the reaction, then the particles are tightly correlated in space and their relative momenta can be determined by the breakup temperature. On the other hand, if such a pair is produced at an early stage of the reaction, then the particles may diffuse apart from one another. The previous two opposite mechanisms are reflected in the BF's width and result in a narrow distribution for the late stage hadronization and a wide one for the early stage hadron production [3, 4].

The BF was studied by the NA49 collaboration in order to investigate the system size and the centrality dependence of the width at two SPS energies ($\sqrt{s} = 17.3$ GeV and $\sqrt{s} = 8.8$ GeV) and in two different rapidity intervals (mid-rapidity and forward rapidity region). Fig. 1 shows the width of the BF distributions for real, UrQMD [5] and shuffled data [3, 4] as a function of the mean number of wounded nucleons, for the two rapidity regions analyzed for both energies. There is an apparent centrality dependence for experimental data only in the mid-rapidity regions (plots a and c).

In addition, an attempt was made by NA49 to study the energy dependence of the Balance Function, by analyzing the most central Pb+Pb events throughout the whole available SPS energy range. The upper right plot of fig. 1 shows the dependence of the normalized W parameter, defined as $W = \frac{100 \cdot (\langle \Delta \eta \rangle_{shuffled} - \langle \Delta \eta \rangle_{data})}{\langle \Delta \eta \rangle_{shuffled}}$ [4], on $\sqrt{s_{NN}}$. There is a first indication of an energy dependence for experimental points which is

CP892, *Quark Confinement and the Hadron Spectrum VII*
edited by J. E. F. T. Ribeiro

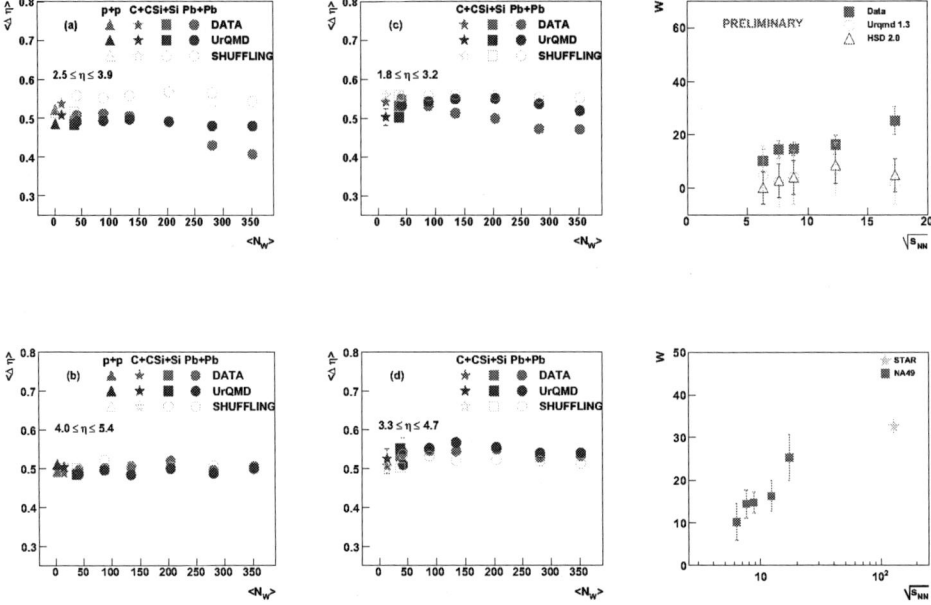

**FIGURE 1.** The system size and centrality dependence of the measured width of the BF for charged particles at $\sqrt{s_{NN}} = 17.3$ GeV (plots a and b) and $\sqrt{s_{NN}} = 8.8$ GeV (plots c and d) as a function of $\langle N_W \rangle$ for the two different rapidity intervals. The two right plots show the dependence of the normalized parameter W, which indicates the relative decrease of the width of the BF between experimental and shuffled data, on $\sqrt{s_{NN}}$ for central Pb+Pb collisions in the SPS energy range (upper plot) and its evolution towards higher RHIC energies (lower plot).

not reproduced by the microscopic models studied [5, 6]. The lower right plot of fig. 1 shows the evolution of the W parameter from SPS to RHIC energies.

Finally, we studied the rapidity correlations of identified pion and kaon pairs, by selecting them using the dE/dx information from the TPCs, for the highest SPS energy and the corresponding results are summarized in fig. 2. The main conclusion is that there is a narrowing of the BF's width with centrality for the experimental data of pion pairs but not for kaon pairs. A similar behavior has also been reported by STAR [3].

The method was also extended to LHC energies by studying p+p PYTHIA events at $\sqrt{s} = 14$ TeV, generated and analyzed with the software framework of the ALICE experiment [7]. Fig. 3 shows the width of the BF for non identified particles as a function of the analyzed pseudo-rapidity interval (left plot) and of the mean number of tracks (right plot). A linear increase of the width on the analyzed interval is observed while no dependence on the mean multiplicity is found.

In summary, we presented the latest experimental results of the BF obtained by the NA49 collaboration. The rapidity dependence study revealed that the narrowing of the BF with centrality is restricted to the mid-rapidity region. In addition, the energy scan

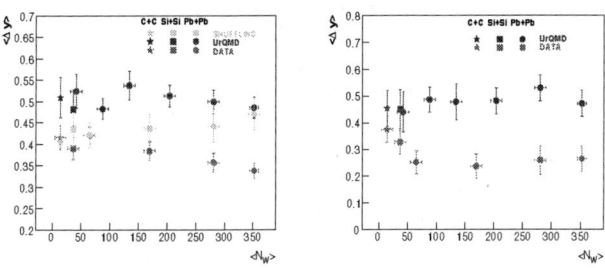

**FIGURE 2.** The dependence of the width of the BF for identified pion (left plot) and kaon (right plot) pairs on $\langle N_W \rangle$ for A+A collisions at the highest SPS energy.

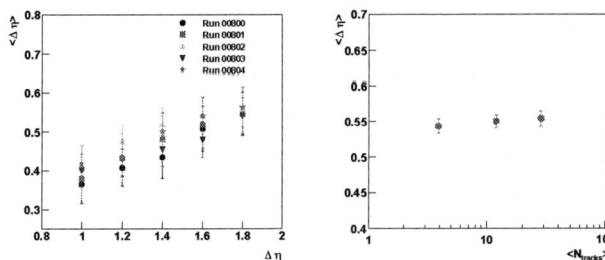

**FIGURE 3.** The dependence of the BF's width on the analyzed interval (left plot) and on the mean number of tracks (right plot) for PYTHIA p+p events at $\sqrt{s} = 14$ TeV, using the framework of the ALICE experiment.

showed a first indication of a dependence of the normalized parameter W on $\sqrt{s_{NN}}$. The investigation of the rapidity correlations for identified pion and kaon pairs showed that we observe a narrowing of the BF's width with centrality for pions but not for kaons. Finally, we showed that the method has been extended to LHC energies by studying the hadron correlations for events simulated within the ALICE environment.

# REFERENCES

1. M. Gazdzicki (*NA49 Collaboration*) Proc. of Quark Matter 2004, *J. Phys. G* **30**, S701 (2004).
2. S. A. Bass, P. Danielewicz and S. Pratt, *Phys. Rev. Lett.* **85**, 2689 (2000).
3. J. Adams et al., (STAR Collaboration) *Phys. Rev. Lett.* **90**, 172301 (2003); G. Westfall et al., (STAR Collaboration) *J. Phys. G* **30**, S345-S349 (2004).
4. C. Alt et al., (NA49 Collaboration), *Phys. Rev. C* **71**, 034903 (2005); P. Christakoglou et al., (NA49 Collaboration), *Nucl. Phys. A* **749**, 279-282 (2005); P. Christakoglou et al., (NA49 Collaboration), *AIP Conf.Proc.* **828**, 107-112 (2006).
5. M. Bleicher et al., *Nucl. Part. Phys.* **25**, 1859-189 (1999).
6. W. Ehehalt, W. Cassing, *Nucl. Phys. A* **602**, 449-486 (1996).
7. B. Alessandro et al., (ALICE Colalboration) *J. Phys. G* **30**, 1295-2040 (2006).

# Electromagnetic Probes at RHIC:
# The Present and the Future

## G. David

*Brookhaven National Laboratory, Upton, NY 11973, USA*

**Abstract.** In this paper we briefly review the importance of electromagnetic probes in understanding the evolution of the system and the new form of matter created in relativistic heavy ion collisions at RHIC. We highlight two very important recent results. While progress has been impressive both on theoretical and experimental side, many questions remain unanswered and new ones were raised. They can be grouped in two major categories: where and how does the phase transition occur and what physical processes give the new matter its observed properties? In parallel with completing the upgrades of the two major RHIC detectors the accelerator is planning to increase its luminosity by a factor of 10 over current values (which is already a significantly above design). This project is called RHIC-II, and it will open the possibility of a detailed energy and species scan going as low as AGS energies if needed to map out the QCD phase transition as well as to access rare probes that so far eluded observation due to limited statistics.

**Keywords:** Direct photon, dielectron, quark-gluon plasma, relativistic heavy ion collisions
**PACS:** 25.75.-q, 25.75.Nq, 12.38.Mh, 13.85.Qk

## INTRODUCTION

There can be little doubt that the hot, dense, strongly interacting matter produced in heavy ion collisions at the Relativistic Heavy Ion Collider (RHIC) is something qualitatively new[1, 2, 3, 4]. Not only is it very opaque, suppressing jets by as much as a factor of 5, but it also exhibits such collectivity (apparently already at the partonic level) that it behaves as an almost perfect fluid. In this new medium (often called strongly interacting Quark-Gluon Plasma or sQGP) even heavy quarks lose significant amount of their initial energy and exhibit flow. With some simplification the first three years of RHIC operations established the discovery *qualitatively* in Au+Au collisions and made the conclusions solid by measuring the p+p and d+Au "baselines" in the very same detectors (*i.e. with essentially the same systematic errors*), whereas the next years have produced an order of magnitude higher statistics for more precise measurements as well as the first steps of a system size (Cu+Cu) and energy (200, 130, 62, 22GeV) scan.

The plethora of results, some of them quite unexpected, triggered a rapid evolution of theory but it is fair to say that we are still far from a *coherent* and *quantitative* description of the sQGP. The first major group of open questions can be summarized as "when and how does deconfinement occur"? If the interpretations are correct, so far we have seen the two extremes only, either no sQGP or a fully developed sQGP formed in the collision. However, we don't know yet where exactly the transition occurs, at what system size and energy (we didn't "map out" a phase transition). Also, sQGP appears to thermalize on a timescale (0.1-0.5fm/c) that is very challenging to explain, and we don't know yet if chiral symmetry is indeed restored in it (at least partially), and so on.

CP892, *Quark Confinement and the Hadron Spectrum VII*
edited by J. E. F. T. Ribeiro

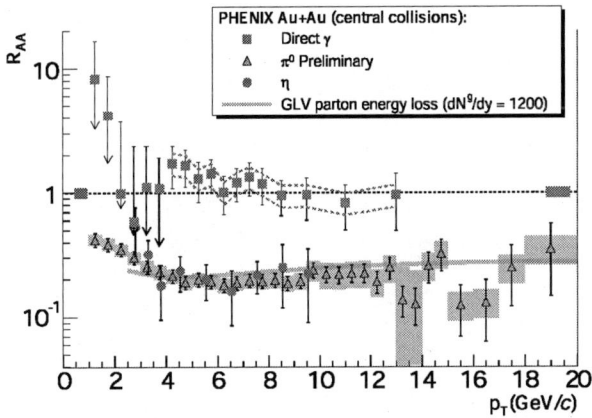

**FIGURE 1.** Nuclear modification factor $R_{AA}$ in central Au+Au collisions ($\sqrt{sNN} = 200$GeV) for $\pi^0$, $\eta$ and direct photons, compared to a GLV calculation with $dN^g/dy = 1200$ gluon density. The mesons - independent of their mass - are suppressed by a factor of 5 with respect to their yield in p+p scaled with the nuclear thickness $T_{AB}$, whereas the direct photons are not suppressed.

The other major group of questions revolve around the properties of (physics mechanisms in) the new matter. We have established its strong collectivity and high opacity, but it is still unclear what exactly causes jet quenching? Does it harbor bound states, resonances? Do vector meson masses drop or broaden, and what happens to their yields?

Electromagnetic probes will be crucial in finding many of the answers. They are penetrating probes, *emitted at all stages* of the collision and once created, emerging mostly unaltered from the collision since $\alpha_e \ll \alpha_s$. Encoding all subprocesses at all times they are in principle the best "historians" - unfortunately, for the very same reason their message is hard to decipher: contributions from different subprocesses often are of comparable magnitude and the background from final state hadrons is large. Still, they already led to important discoveries at RHIC and in the future their role will be crucial in understanding both the phase transition to and the properties of the new matter.

## IN-MEDIUM ENERGY LOSS

One cannot do justice in a short review to all results obtained so far at RHIC involving electromagnetic probes, but arguably the two most important, almost emblematic ones so far are the direct photon measurements and the energy loss and flow of heavy quarks *via* single electrons in Au+Au collisions. Early observation of the suppression[1] at $y = 0$ of $\pi^0$ in Au+Au [5], later extended to $p_T = 20$GeV/c and confirmed for $\eta$ as well (Fig. 1), along with the non-suppression in d+Au collisions [6] made it virtually certain that there

---

[1] Suppression with respect to hard scattering as measured in p+p scaled by the probability of a hard scattering occuring in a nucleus-nucleus collision

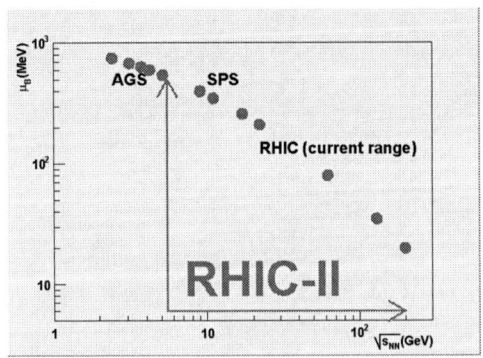

**FIGURE 2.** Baryon chemical potential *vs.* $\sqrt{s_{NN}}$ (based upon a plot by H. Stocker).

is a severe parton energy loss and it is a final state effect, but the ultimate confirmation came only when a *penetrating probe* (direct photons), created in the same process as jets but unaffected by the medium showed indeed no suppression. This is summarized on Fig. 1 where the nuclear modification factor $R_{AA}$ - the ratio of the observed yield in Au+Au and the expectation from p+p multiplied by the nuclear thickness function $T_{AB}$ - is plotted for $\pi^0$, $\eta$ and direct photon production in Au+Au collisions. Whereas there is a strong (factor of 5) suppression for mesons in the entire range where hard scattering dominates, no such suppression is observed for direct photons. It is important that the experiment uses its own p+p reference rather than some theoretical calculation - in fact recent preliminary results from PHENIX suggest that NLO pQCD underpredicts the direct photon cross-section at moderate to high $p_T$.

While the in-medium energy loss of light quarks has been firmly established, until recently it had been expected that heavy quarks will lose considerably less energy or none at all. However, recent measurements of single electrons from heavy flavor decay [7] show at high $p_T$ a suppression comparable to that of light quarks along with an (equally unexpected) strong flow suggesting some degree of thermalization and strong interaction with the medium.

## RHIC-II: FROM DISCOVERY TO EXPLORATION

Despite impressive progress several major questions remain open at RHIC and electro-magnetic probes will play an important role answering them. As already pointed out in the Introduction, the phase transition itself has not been mapped out yet, and it requires the continuous "dialing" of the baryon chemical potential $\mu_B$ by changing $\sqrt{s_{NN}}$, as shown on Fig. 2. The physics case for an energy scan in the very same experiments (same systematics!) ranging from the AGS to top RHIC energies has been laid out at [8, 9] and summarized in the statement that "a growing body of theoretical and experimental evidence shows that the critical point in the QCD phase diagram, if it exists, should appear on the QGP transition boundary at $\mu_B = 100 - 500\text{MeV}$, corresponding to heavy ion collisions in the energy range $\sqrt{s_{NN}}$=5-50GeV/u". RHIC being a very flexible machine

382

this c.m. energy range is accessible (incidentally, the lower limit by *deceleration* of the beam injected from the AGS). However, there is an approximately quadratic decrease in luminosity with decreasing $\sqrt{S_{NN}}$. Therefore, in order to do the measurements on a reasonable timescale a 10-fold increase of the current luminosities is needed (which are already considerably above original design). This is the primary goal of the accelerator upgrade project called RHIC-II and the primary tool to achieve it is electron cooling of ion beams [8].

Equally important is the need for high luminosities in the quest for rare signals. Whereas the need is obvious in the heavy flavor sector due to the low cross-sections, many electromagnetic probes require high luminosity as well. High $p_T$ photon HBT will give access to the size of the system at the earliest times. Azimuthal asymmetries of direct photon production up to very high $p_T$ have to be measured to investigate contributions from parton fragmentation, jet-thermal interactions and Bremsstrahlung off quarks. Dielectron pairs from internal conversions (near-zero mass virtual photons), a novel technique to measure low and medium $p_T$ direct photons requires very large statistics. The same is true for finding possible resonances above $T_c$, predicted by lattice QCD calculations. (For more details, presentations and write-ups see [10].)

Significant detector upgrades have already been installed in the two major RHIC experiments (PHENIX and STAR), and further new subsystems are under construction or development. They will broaden our physics reach considerably: examples include (but are not limited to!) a precision measurement of in-medium modification of vector mesons or access photon-jet correlations in more than six units of rapidity. The new capabilities along with the luminosities of RHIC-II will ensure a quantum leap in our understanding of strong interactions.

## ACKNOWLEDGMENTS

This work has been supported by the DOE.

## REFERENCES

1. I. Arsene *et al.* [BRAHMS Collaboration], Nucl. Phys. A **757**, 1 (2005)
2. B.B. Back *et al.* [PHOBOS Collaboration], Nucl. Phys. A **757**, 28 (2005)
3. J. Adams *et al.* [STAR Collaboration], Nucl. Phys. A **757**, 102 (2005)
4. K. Adcox *et al.* [PHENIX Collaboration], Nucl. Phys. A **757**, 184 (2005)
5. K. Adcox *et al.*, Phys. Rev. Lett. **88**, 022301 (2002)
6. S.S. Adler *et al.*, Phys. Rev. Lett. **91**, 072303 (2003)
7. S.S. Adler *et al.*, Phys. Rev. Lett. **94**, 082301 (2005)
8. The homepage of the workshop "Can We Discover the QCD Critical Point at RHIC?" held at BNL, March 2006 can be found at https://www.bnl.gov/riken/QCDRhic/
9. The homepage of the Joint EIC2006 and hot QCD meeting (July 2006, BNL) can be found at http://www.phenix.bnl.gov/WWW/publish/abhay/qcdfp2006/.
10. Information on the RHIC-II science workshops can be found at http://www.bnl.gov/physics/rhicIIscience/default.asp and work done in the RHIC-II Electromagnetic Probes Working Group is documented at http://www.phenix.bnl.gov/WWW/publish/david/rhicii_em/.

# Correlations at RHIC and the Clustering of Color Sources

E. G. Ferreiro

*Departamento de Física de Partículas, Universidad de Santiago de Compostela, 15782 Santiago de Compostela, Spain*

**Abstract.** We present our results on transverse momentum fluctuations and multiplicity fluctuations in the framework of the clustering of color sources. In this approach, elementary color sources - strings- overlap forming clusters, so the number of effective sources is modified. These clusters decay into particles with mean transverse momentum that depends on the number of elementary sources that conform each cluster and the area occupied by the cluster. We find a non-monotonic dependence of the $p_T$ and multiplicity fluctuations with the number of participants. In our approach, the physical mechanism responsible of these fluctuations is the same: the formation of clusters of strings that introduces correlations between the produced particles.

Non-statistical event-by-event fluctuations in relativistic heavy ion collisions have been proposed as a probe of phase instabilities near de QCD phase transition. In a thermodynamical picture of the strongly interacting system formed in heavy-ion collisions, the fluctuations of the mean transverse momentum or mean multiplicity are related to the fundamental properties of the system, such the specific heat, so they may reveal information about the QCD phase boundary. In particular, a phase transition in the evolution of the system created in relativistic heavy ion collisions may lead to a divergence of the specific heat which could be observed as event-by-event fluctuations. Here I am going to present our results, in the framework of clustering of color sources, concerning event-by-event $p_T$ and multiplicity fluctuations.

Event-by-event fluctuations of the transverse momentum have been measured at SPS and RHIC energies. The non-statistical fluctuations show a particular behaviour as a function of the centrality of the collision: they grow as the centrality increases, achieving a maximum at mid centralities, followed by a decrease at larger centralities. Different mechanisms have been proposed in order to explain those data: complete or partial equilibration, critical phenomena, as string clustering or string percolation, and jets production.

Let us concentrate on the results obtained in the framework of clustering of color sources [1]. In this framework, we consider that in each collision color strings are stretched between the projectile and the target. Those strings act as the sources of particle production: particles are created via sea $q - \bar{q}$ production in the field of the string. Moreover, in the transverse space, the color strings correspond to small areas filled with the color field created by the colliding partons.

With growing energy and/or atomic number of the colliding nuclei, the number of sources grows, so the elementary color sources start to overlap, forming clusters, very much like disk in the 2-dimensional percolation theory. The density of strings is

CP892, *Quark Confinement and the Hadron Spectrum VII*
edited by J. E. F. T. Ribeiro

expressed by $\eta = N_{st}\frac{S_1}{S_A}$, where $N_{st}$ corresponds to the total number of strings, $S_1 = \pi r_0^2$ with $r_0 = 0.2$ fm is the area of each individual string and $S_A$ is the nuclear overlap area. In particular, at a certain critical density, $\eta_c = 1.1 \div 1.2$, a macroscopic cluster appears, which marks the percolation phase transition. Percolation means that a cluster is formed through the whole collision area.

Taking into account that the color charge of a cluster is the vectorial sum of the string charges that come into the cluster, one can calculate, for a cluster of $n$ overlapping strings covering an area $S_n$, the multiplicity and $p_T$ of the produced particles :

$$Q_n = \sqrt{\frac{nS_n}{S_1}}Q_1, \quad \mu_n = \sqrt{\frac{nS_n}{S_1}}\mu_1, \quad \langle p_T^2 \rangle_n = \sqrt{\frac{nS_1}{S_n}}\langle p_T^2 \rangle_1. \tag{1}$$

In the clustering approach, the behaviour of the transverse momentum fluctuations can be understood as follows: at low density, most of the particles are produced by individual strings with the same $< p_T >_1$, so fluctuations are small. At large density, above the critical point, we have only one cluster, so fluctuations are not expected either -equilibration-. The fluctuations will be maximal just below the percolation critical density, where there are a large number of clusters formed by different number of strings with different size and different $< p_T >_n$.

In orther to measure the event-by-event $p_T$ fluctuations, the proposed variables are $F_{p_T}$ and $\phi$, which quantify the deviation of the observed fluctuations from statistically independent particle emission:

$$F_{p_T} = \frac{\omega_{data} - \omega_{random}}{\omega_{random}}, \quad \omega = \frac{\sqrt{< p_T^2 > - < p_T >^2}}{< p_T >}, \quad \phi = \sqrt{\frac{< Z^2 >}{< \mu >}} - \sqrt{< z^2 >}. \tag{2}$$

$z_i = p_{Ti} - < p_T >$ is defined for each particle, and $Z_i = \sum_{j=1}^{N_i} z_j$ is defined for each event.

Both variables are related: $F_{p_T} = \frac{\phi}{\sqrt{<z^2>}} = \frac{1}{\sqrt{<z^2>}}\sqrt{\frac{<Z^2>}{<\mu>}} - 1$. We have computed $F_{p_T}$ [2] using a Monte Carlo code to evaluate the cluster formation and the analytical expressions (1) for the transverse momentum and the multiplicities of the clusters. The behaviour of the transverse momentum fluctuations with the centrality of the collision shown by the RHIC data is naturally explained by the clustering of color sources. In this framework, elementary color sources -strings- overlap forming clusters, so the number of effective sources is modified. These clusters decay into particles with mean transverse momentum that depends on the number of elementary sources that conform each cluster, and the area occupied by the cluster. The transverse momentum fluctuations in this approach correspond to the fluctuations of the transverse momentum of these clusters, and they behave essentially as the number of effective sources. In a jet production scenario, the mean $p_T$ fluctuations are attributed to jet production in peripheral events, combined with jet suppression at larger centralities.

A way to discriminate between the two approaches is to study the fluctuations at SPS energies [3], where jet production cannot play a fundamental role. Recently, the NA49 Collaboration have presented their data on multiplicity fluctuations as a function of centrality at SPS energies. In order to measure these fluctuations, the variance of the multiplicity distribution scaled to the mean value of the multiplicity, $Var(N) = \frac{<N^2>-<N>^2}{<N>}$,

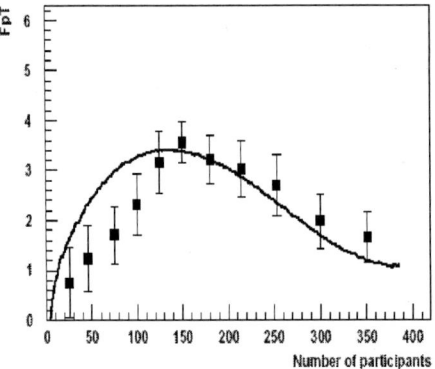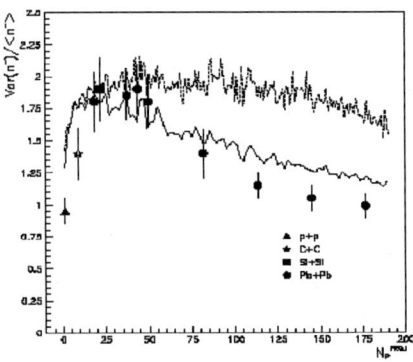

**FIGURE 1.** Left: $F_{p_T}(\%)$ versus the number of participants. Experimental data from PHENIX at $\sqrt{s} = 200$ GeV are compared with our results (solid line). Right: Our results for the scaled variance of negatively charged particles in Pb+Pb collisions at $P_{lab} = 158$ AGeV/c compared to NA49 experimental data. The dashed line corresponds to our result when clustering formation is not included, the continuous line takes into account clustering.

has been used. A non-monotonic centrality -system size- dependence was found. In fact, its behaviour is similar to the one obtained for $\Phi(p_T)$ -used by the NA49 Collaboration to quantify the $p_T$-fluctuations-, suggesting that they are related to each other. We find a non-monotonic dependence of the multiplicity fluctuations with the number of participants. The centrality behaviour of these fluctuations is very similar to the one found for the mean $p_T$ fluctuations. In our approach, the mechanism responsible for multiplicity and mean $p_T$ fluctuations is the formation of clusters of strings that introduces correlations between the produced particles. On the other hand, the mean $p_T$ fluctuations have been also attributed to jet production in peripheral events, combined with jet suppression in central events. However, this hard-scattering interpretation, based on jet production and jet suppression, can not be applied to SPS energies, so it does not explain the non-monotonic behaviour of the mean $p_T$ fluctuations neither the relation between mean $p_T$ and multiplicity fluctuations at SPS energy. Other possible mechanisms are: combination of strong and electromagnetic interaction, dipole-dipole interaction and non-extensive thermodynamics. Still, it is not clear if these fluctuations have a kinematic or dynamic origin, but clustering of colour sources remains a good possibility.

## REFERENCES

1. N. Armesto, M. A. Braun, E. G. Ferreiro and C. Pajares, Phys. Rev. Lett. 77 (1996) 3736.
2. E. G. Ferreiro, F. del Moral and C. Pajares, Phys. Rev. C69 (2004) 034901.
3. L. Cunqueiro, E. G. Ferreiro, F. del Moral and C. Pajares, Phys. Rev. C72 (2005) 024907.

# Quasi-Particle Perspective on Equation of State[1]

M. Bluhm*, R. Schulze*, D. Seipt* and B. Kämpfer*,[†]

*Forschungszentrum Dresden-Rossendorf, PF 510119, 01314 Dresden, Germany
[†]Institut für Theoretische Physik, TU Dresden, 01062 Dresden, Germany

**Abstract.** We propose a procedure for determining the equation of state of strongly interacting matter needed in a hydrodynamical description of relativistic heavy-ion collisions.

**Keywords:** QCD equation of state, elliptic flow, quasi-particle model
**PACS:** 12.38.Mh;25.75-q;25.75.Ld

## INTRODUCTION

Relativistic heavy-ion collisions are aimed at investigating that part of the phase diagram of strongly interacting matter which is otherwise not accessible. Thus, the theoretical interpretation of experimental data contributes to the verification of fundamental issues of QCD, such as deconfinement, phase structure of matter, criticality etc. One important notion for describing strongly interacting matter in equilibrium is the equation of state (EOS). In the EOS, all information of a system is condensed, for instance, in the pressure $p$ at a given temperature $T$ and net baryon density $n_B$ or equivalently chemical potential $\mu_B$, $p(T, \mu_B)$. Once knowing this function, one can, within the framework of relativistic hydrodynamics, calculate the expansion dynamics of the early universe (here $\mu_B \ll T$ applies and the contribution of electro-weakly interacting particles must be accounted for), gross properties of neutron or quark or hybrid stars (here $\mu_B \gg T$ applies) and a series of observables in heavy-ion collisions.

In this contribution we resume our knowledge about the EOS in a region relevant for heavy-ion collisions at RHIC and LHC energies. We briefly discuss possible modifications when including the critical point conjectured in the region of modest baryon density. Then we consider transverse momentum and elliptic flow observables.

## FROM LATTICE QCD TO A PHENOMENOLOGICAL EOS

A few recent lattice QCD results for the scaled pressure as a function of the scaled temperature are exhibited in left panel of Fig. 1 for 2+1 flavors at $\mu_B = 0$. Due to different discretization schemes, cut-offs, employed actions (mimicking the proper QCD action) etc. a concise and unique function $p(T/T_c)$ is not yet at our disposal. Even the value of $T_c$ is under debate. The various lattice results can be fitted individually by our quasi-particle model [1, 2]. This model allows to translate the EOS $p(T)$ into the

---

[1] Supported by BMBF, GSI, EU-I3HP.

CP892, *Quark Confinement and the Hadron Spectrum VII*
edited by J. E. F. T. Ribeiro
© 2007 American Institute of Physics 978-0-7354-0396-3/07/$23.00

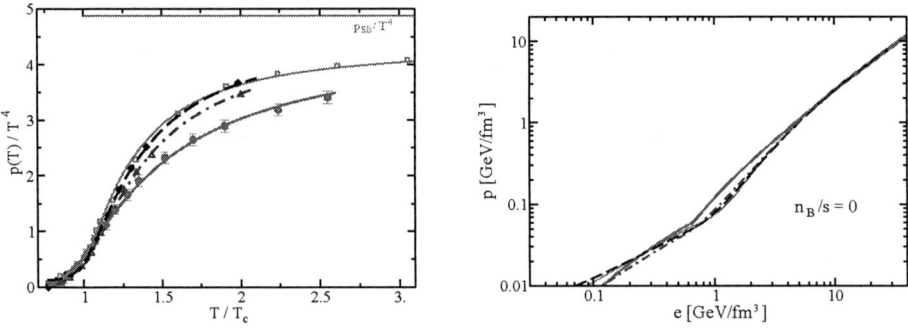

**FIGURE 1.** Lattice QCD calculations (squares: [3], rhombi and triangles: [4], circles: [5]) of the scaled pressure as a function of the scaled temperature for 2+1 flavors and for $\mu_B = 0$ (left panel). Using the quasi-particle model [1, 2] (curves), the lattice data can be translated into the EOS of the form $p(e)$ at $n_B = 0$ (right panel).

form $p(e)$ in a thermodynamically consistent manner, where $e$ is the energy density. Furthermore, an extrapolation to larger values of $T/T_c$ or $e$ is straightforward. The result is displayed in right panel of Fig. 1 The astonishing observation is that above $e = 4$ GeV/fm$^3$ all EOSs coincide. This statement is robust, as long as $T_c = 170 \pm 15$ MeV. Below $e = 4$ GeV/fm$^3$, the various lattice results differ. In the region around 1 GeV/fm$^3$ the confinement transition is expected. Below $e = 1$ GeV/fm$^3$, the lattice results are hampered by noisy signals and/or unphysically heavy quark masses.

Our strategy to arrive at an EOS useful for hydrodynamics is as follows: (i) use below $e_1 = 0.45$ GeV/fm$^3$ a resonance gas EOS motivated by the arguments that a lot of strong interaction is encoded in resonances according to the Dashen-Ma-Bernstein theorem [6] and that the resonance gas model and lattice OCD results agree once the same quark masses are employed [3]; (ii) use a linear interpolation above $e_1$ and below $e_2$ from $p(e_1)$ to $p(e_2)$ with the matching point $e_2$ below 4 GeV/fm$^3$; (iii) use the unique lattice EOS above $e_2$. In such a way we arrive at a family of EOSs which we label as $QPM(e_2)$. Examples are displayed in Fig. 2. The EOS $p(e)$ can be extended to $p(e, n_B)$ with our quasi-particle model which was proven [2] to describe the $n_B$ dependence in agreement with lattice QCD data for 2 flavors.

## THE CRITICAL POINT ?

With respect to discussions on the location of the critical point [7] we refer to a phenomenological procedure of accounting for critical point effects in [8]. Since we focus here on RHIC and LHC top energies, we can assume, in line with [7], that for these conditions neither the conjectured critical point nor the first-order deconfinement region are relevant as they are important at larger values of $\mu_B$. The EOS relevant for RHIC and LHC is hardly modified when including critical point features [8].

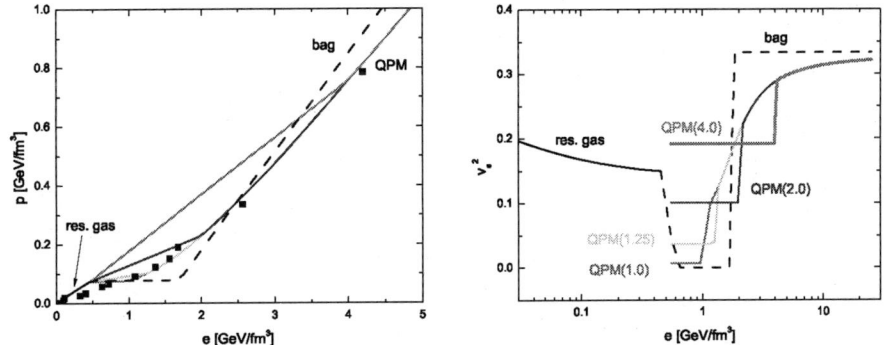

**FIGURE 2.** A family of EOSs interpolating between the resonance gas model and lattice QCD based high-energy density part (left panel, symbols: lattice data from [3]). The corresponding sound velocity squared as a function of the energy density is exhibited in the right panel (labelled by the matching point $e_2$). For comparison, a bag model EOS with constructed phase transition of first order is displayed too.

## EMPLOYING THE LATTICE QCD BASED EOS

We solve the hydrodynamical evolution equations with the code employed in [9] for fixed initial conditions (i) RHIC: $e_0 = 29.8$ GeV/fm$^3$, $p_0 = 9.4$ GeV/fm$^3$, $T_0 = 357$ MeV, $\tau_0 = 0.6$fm/c, (ii) LHC: $e_0 = 127$ GeV/fm$^3$, $p_0 = 42$ GeV/fm$^3$, $T_0 = 515$ MeV, $\tau_0 = 0.6$fm/c, and unique freeze-out conditions $e_{f.o.} = 0.075$ GeV/fm$^3$.

Examples for transverse momentum spectra and elliptic flow of strange baryons are exhibited in Fig. 3. One observes both for RHIC and for LHC a sensitivity to details of the EOS in the confinement region. At LHC no new uncertainty concerning the EOS enters the calculations, as the EOS at high energy density is uniquely determined by lattice QCD (highlighted in Fig. 1). At LHC, baryon density effects are very tiny.

## SUMMARY

In summary we propose a family of equations of state suitable for a hydrodynamical description of the expansion stage of strongly interacting matter in relativistic heavy-ion collisions. A remaining uncertainty in the confinement region can be constrained by comparison with data on transverse momentum spectra and elliptic flow along the previous investigations in [10]. The equation of state in the form $p(e)$ is rather robust w.r.t. the numerical value of the pseudo-critical temperature $T_c$ and a chiral extrapolation of lattice QCD data. Note that hydrodynamics needs only the dependence $p(e)$; baryon density effects entering $p(e,n_B)$ turn out to be negligible for RHIC and LHC. Only if initial conditions need to be formulated via the entropy density (determining the particle multiplicities), an explicit $T_c$ dependence would show up.

An extension of the equation of state to higher baryon densities, including a possible critical point and the related region of first-order phase transitions, needs separate studies in view of applying this framework to SPS energies and the envisaged low-energy runs at RHIC as well as the planned SIS/100/300 accelerators at FAIR (CBM experiment).

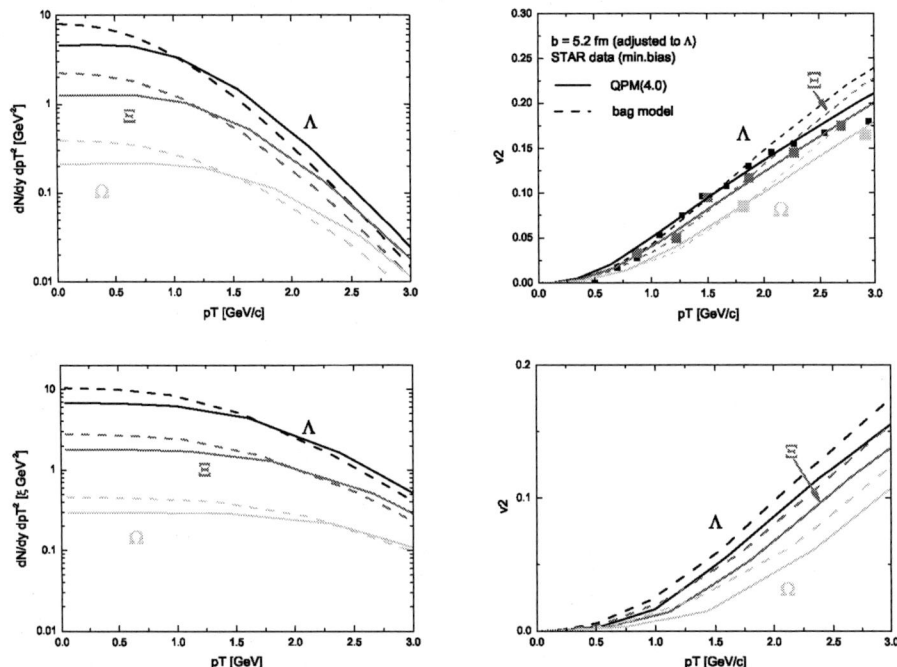

**FIGURE 3.** Transverse momentum spectra (left column, w/o proper normalization) and differential elliptic flow (right column) for RHIC top energy (top row) and LHC (bottom row). Solid curves are for a matching point $e_2 = 4.0$ GeV/fm$^3$, while the dashed curves depict results for the bag model (cf. Fig. 2). Au + Au collisions at impact parameter 5.2 fm. Initial conditions as described in text. The resonance gas EOS includes chemical off-equilibrium effects [11].

# REFERENCES

1. A. Peshier et al., *Phys. Lett. B* **337**, 235 (1994); *Phys. Rev. D* **54**, 2399 (1996); *Phys. Rev. C* **61**, 045203 (2000); *Phys. Rev. D* **66**, 094003 (2002).
2. M. Bluhm, B. Kämpfer, and G. Soff, *Phys. Lett. B* **620**, 131 (2005).
3. F. Karsch, K. Redlich, and A. Tawfik, *Eur. Phys. J. C* **29**, 549 (2003); *Phys. Lett. B* **571**, 67 (2003).
4. C. Bernard et al., *Phys. Rev. D* **55**, 6861 (1997); *PoS* **LAT2005**, 156 (2005).
5. Y. Aoki, Z. Fodor, S. D. Katz, and K. K. Szabo, *JHEP* **0601**, 089 (2006).
6. R. Dashen, S. Ma, and H. J. Bernstein, *Phys. Rev.* **187**, 345 (1969).
7. M. A. Stephanov, *Prog. Theor. Phys. Suppl.* **153**, 139 (2004); *Int. J. Mod. Phys. A* **20**, 4387 (2005); Z. Fodor, and S. D. Katz, *JHEP* **0203**, 014 (2002); *JHEP* **0404**, 050 (2004).
8. M. Bluhm, and B. Kämpfer, to appear in *PoS* **CPOD2006**.
9. P. F. Kolb, J. Sollfrank, and U. Heinz, *Phys. Rev. C* **62**, 054909 (2000); U. W. Heinz, *J. Phys. G* **31**, 17 (2005).
10. D. Teaney, J. Lauret, and E. V. Shuryak, *Phys. Rev. Lett.* **86**, 4783 (2001); P. Huovinen, *Nucl. Phys. A* **761**, 296 (2005).
11. P. F. Kolb, and R. Rapp, *Phys. Rev. C* **67**, 044903 (2003).

# The flux-tube phase transition

## G. A. Kozlov

*Joint Institute for Nuclear Research, Dubna, 141980 Moscow Region, Russia*

**Abstract.** We consider the phase transition in the "hot" dual long-distances Yang-Mills theory at finite temperature $T$. This phase transition is associated with a change of symmetry.

**PACS:** 12.38.Aw, 11.15.Kc, 12.38Aw, 12.38Lg, 12.39Mk, 12.39Pn

1. Most expositions of dual model focus on its possible use as a framework for quark confinement in nature. Rather than the confinement of color charges, I will here describe what one might call the phase transition in the dual Yang-Mills (Y-M) theory at finite temperature $T$.

There is a general statement that the color confinement is supported by the idea that the vacuum of quantum Y-M theory is realized by a condensate of monopole-antimonopole pairs [1]. In such a vacuum the interacting field between two colored sources located in $\vec{x}_1$ and $\vec{x}_2$ is squeezed into a tube whose energy $E_{tube} \sim |\vec{x}_1 - \vec{x}_2|$. This is a complete dual analogy to the magnetic monopole confinement in the Type II superconductor. Since there is no monopoles as classical solutions with finite energy in a pure Y-M theory, it has been suggested by 't Hooft [2] to go into the Abelian projection where the gauge group SU(2) is broken by a suitable gauge condition to its maximal Abelian subgroup U(1).

The aim of this talk is to consider the phase transition in the "hot" four-dimensional model based on the dual description of a long-distance Y-M theory which shows some kind of confinement. We study the model of Lagrangian where the fundamental variables are an octet of dual potentials coupled minimally to three octets of monopole (Higgs-like) fields [3]. The flux distribution in the tubes formed between two heavy color charges is understood via the following statement: the Abelian Higgs-like monopoles are excluded from the string region while the Abelian electric flux is squeezed into the string region.

In the model there are the dual gauge field $\hat{C}_\mu^a(x)$ and the scalar field $\hat{B}_i^a(x)$ ($i = 1,...,N_c(N_c - 1)/2$; $a=1,...,8$ is a color index) which are relevant modes for infrared behaviour. The scope of commutation relations, two-point Wightman functions and Green's functions as well-defined distributions in the space $S(\Re^d)$ of complex Schwartz test functions on $\Re^d$, the monopole- and dual gauge-field propagations, the asymptotic transverse behaviour of both the dual gauge field and the color-electric field, the analytic expression for the static potential can be found in [3].

2. Phase transitions in dual models are associated with a change in symmetry or

CP892, *Quark Confinement and the Hadron Spectrum VII*
edited by J. E. F. T. Ribeiro
© 2007 American Institute of Physics 978-0-7354-0396-3/07/$23.00

more correctly these transitions are related with the breaking of symmetry. The dual description of the Y-M theory is simply understood by switching on the dual gauge field $\hat{C}_\mu(x)$ and the three scalar octets $\hat{B}_i(x)$ (necessary to give mass to all $C_\mu^a$ and carrying color magnetic charge) in the Lagrangian density (LD) $L$ [3]

$$L = 2Tr\left[-\frac{1}{4}\hat{F}^{\mu\nu}\hat{F}_{\mu\nu} + \frac{1}{2}(D_\mu\hat{B}_i)^2\right] - W(\hat{B}_i),$$ (1)

where $\hat{F}_{\mu\nu} = \partial_\mu\hat{C}_\nu - \partial_\nu\hat{C}_\mu - ig[\hat{C}_\mu, \hat{C}_\nu]$, $D_\mu\hat{B}_i = \partial_\mu\hat{B}_i - ig[\hat{C}_\mu, \hat{B}_i]$. The Higgs-like fields develop their vacuum expectation values (v.e.v.) $\hat{B}_{0_i}$ and the Higgs potential $W(\hat{B}_i)$ has a minimum at $\hat{B}_{0_i}$ of the order O (100 MeV) defined by the string tension and the Higgs-like potential $W(\hat{B}_i)$ has a minimum at $\hat{B}_{0_i}$. The interaction of the dual field with other fields (scalar nonobservable fields) is due to monopole current $J_\mu^{mon}(x)$ in the Higgs-like condensate $(\chi + B_0)$ in terms of the dual gauge coupling $g$ up to divergence of the local fase of the Higgs-like field, $\partial_\mu f(x)$:

$$gC_\mu(x) = \frac{J_\mu^{mon}(x)}{4g(\chi + B_0)^2} + \partial_\mu f(x).$$

We introduce the canonical partition function

$$Z_c = \sum_{flux\ tube\ configurations} \exp[-\beta E(R)] = \sum_R N(R)\exp[-\beta E(R)],$$

$$E(R) \simeq \frac{\xi \vec{Q}_\alpha^2}{16\pi}m^2 R(12.4 - 6\ln\tilde{\mu}R)$$ (2)

for ensembles of systems with a single static flux tube, where $N(R)$ is the number of configurations of the flux tube of length $R$. In the rest of physics, for $l \sim \mu^{-1} = (\sqrt{2\lambda}\,\hat{B}_0)^{-1} << R$ the number of configurations $N(R)$ is interpreted in terms of the entropy density $s$ of the flux tube by a fundamental formula $\tilde{N}(R) = exp(\tilde{s})$, where $\tilde{N}(R) = N(R)cl^3/V$, $\tilde{s} = sR/l$, $c \sim O(1)$. Can one by some sort of calculation count the number of configurations of flux tube and reproduce the formula for the entropy density? For this, we need a quantum theory of confinement, so, at present at least, dual Y-M theory is the only candidate. Even in this theory, the question was out of reach for last three decades.

Finally, one gets $Z_c = \frac{V}{l^3}\sum_R \exp[-\beta\,\sigma_{eff}(\beta)R]$, where $\sigma_{eff}(\beta) = \tilde{\sigma}_0 - s/(l\beta)$ is the order parameter of the phase transition and $\tilde{\sigma}_0 = \sigma_0\left(1 - \frac{1}{4}\ln\frac{\tilde{\mu}^2}{m_R^2}\right)$, $\sigma_0 = \frac{3}{4}\alpha(Q)m^2 = \frac{3}{4}\frac{\pi}{g^2}m^2$ with $\alpha(Q)$ being the running coupling constant. We got $\sigma_0 \simeq 0.18\ GeV^2$ [3] for the mass of the dual $C_\mu$-field $m = 0.85\ GeV$ and $\alpha = e^2/(4\pi){=}0.37$.

It is evident that the flux tube picture cannot work at $T > T_0$, where

$$T_0 = \frac{3}{4}\frac{1}{s}\alpha(Q)\frac{m^2}{\mu}\left(1 - \frac{1}{4}\ln\frac{\tilde{\mu}^2}{m_R^2}\right),\ \sigma_{eff}(T_0) = 0.$$ (3)

The T-dependent mass $m(\beta)$ of the field $\hat{C}_\mu$ looks like $m^2(\beta) = 4\sigma_{eff}(\beta)/[3\alpha(Q,\beta)]$.

3. The T-dependent flux-tube solution for the dual gauge field along the z-axis has the following asymptotic transverse behaviour

$$\tilde{C}(r,\beta) \simeq \frac{4n}{7g(\beta)} - \sqrt{\frac{\pi m(\beta)r}{2\kappa}}\, e^{-\kappa m(\beta)r}\left[1 + \frac{3}{8\kappa m(\beta)r}\right], \qquad (4)$$

where $r$ is the radial coordinate (the distance from the center of the flux-tube), $n$ is the integer number associated with the topological charge, $\kappa = \sqrt{21}$.

The color-electric field $E$ inside the quark-antiquark bound state is given by the rotation of the dual gauge field $\vec{E} = \vec{\nabla} \times \vec{C} = E_z(r) \cdot \vec{e}_z$, where $\vec{e}_z$ is a unit vector along the z-axis, and the $T$-dependent $E_z(r,\beta)$ looks like

$$E_z(r,\beta) = \sqrt{\frac{\pi m(\beta)}{2\kappa r}}\, e^{-\kappa m(\beta)r}\left[\kappa m(\beta) - \frac{1}{2r}\right]. \qquad (5)$$

The lower bound on $r = r_0$ can be estimated from the relation $r_0 > [2\kappa m(\beta)]^{-1}$ which leads to $r_0 > 0.03$ fm at $T = 0$. Obviously, $r_0 \to \infty$ as $m(\beta) \to 0$ at $T \to T_0$ (deconfinement).

In Fig. 1, we show the dependence of $m$ as a function of $T$ at different scale parameters $M$. No dependence found on quark current masses. No essential dependence found for different numbers of flavors $N_f$ and fermions $N_F$.

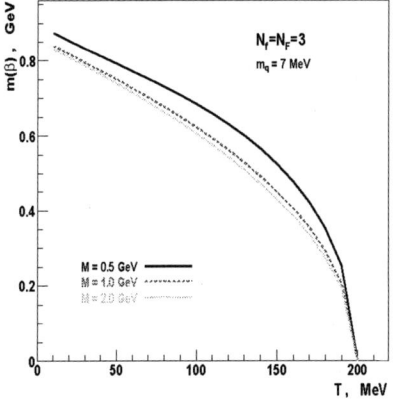
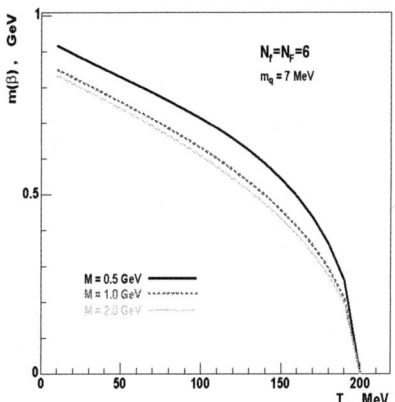

FIGURE 1.

In Fig. 2 and Fig. 3, we show numerical solutions of the flux tube, namely, the profiles of the transverse behaviour of $\tilde{C}(r,\beta)$ and the color electric field $E_z(r,\beta)$, respectively, as functions of radial variable $r$ at different temperatures. We found rather sharp increasing of $\tilde{C}(r,\beta)$ at small values of $r$. No essential dependence of $r$ emerges in the region $r > 0.1$ fm. The field $E(r,\beta)$ disappears when the temperature close to $T_0$.

4. In conclusion, we observed that the flux tube can be produced abundantly when the phase transition emerges at the temperature $T = T_0$, obeying the condition $\sigma_{eff}(T =$

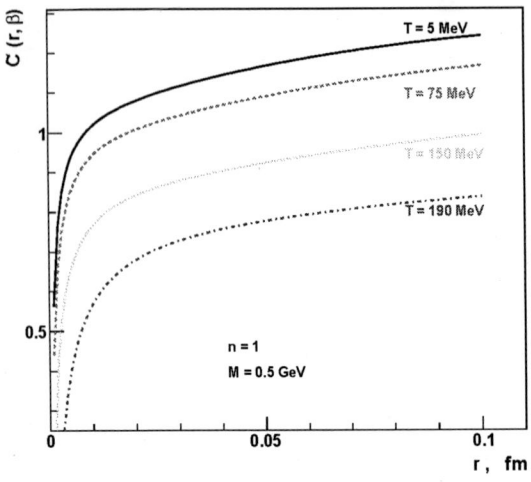

**FIGURE 2.**

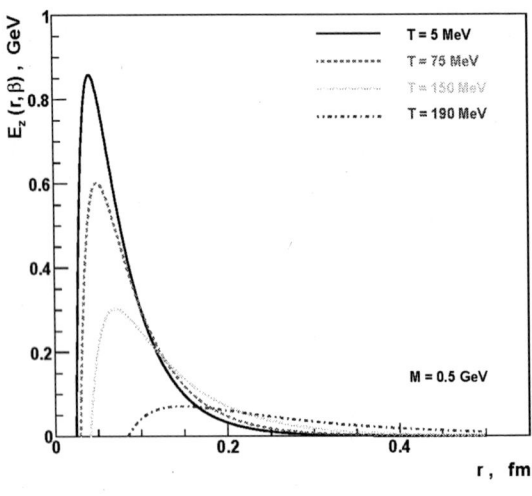

**FIGURE 3.**

$T_0) = 0$. We found that the phase transition temperature essentially depends on $\alpha(Q)$ and the mass of the dual gauge field $m$.

# REFERENCES

1. Y. Nambu, Phys. Rev. D **10**, 4262 (1974) ; S. Mandelstam, Phys. Rep. C **23**, 245 (1976); Phys. Rev. D **19**, 2391 (1979).
2. G.'t Hooft, Nucl. Phys. B [FS3] **190**, 455 (1981).
3. G.A. Kozlov and M. Baldicchi, New J. Phys. **4**, 16.1 (2002).

# Dynamics of the deconfinement transition

G. Ananos*, E.S. Fraga†, G. Krein* and A.J. Mizher†

*Instituto de Física Teórica, Universidade Estadual Paulista
Rua Pamplona 145, 01405-900 São Paulo, SP, Brazil
†Instituto de Física, Universidade Federal do Rio de Janeiro
Caixa Postal 68528, 21941-972 Rio de Janeiro, RJ Brazil

**Abstract.** We estimate the dissipation coefficient $\Gamma$ that appears in Ginzburg-Landau-Langevin equations that describe phenomenologically the deconfinement transition in QCD. This is done through the implementation of Glauber dynamics of pure SU(3) lattice gauge theory. The coefficient $\Gamma$ is extracted from the short-time exponential growth of the equal time correlation function of the order parameter. Although the absolute determination of $\Gamma$ is ambiguous due to the difficulties in relating real time and Monte Carlo time, its relative temperature dependence can be obtained with much less arbitrariness.

**Keywords:** Dynamics of color deconfinement, lattice QCD, Ginzburg-Landau-Langevin equations
**PACS:** 11.15.Ha, 12.38.Gc, 25.75.Nq, 25.75-q

## INTRODUCTION

The Polyakov loop $L(x)$ provides an adequate order parameter for the deconfinement transition at high temperatures $T$ in pure gauge QCD [1]. It is well established [2] that at a temperature $T = T_d \simeq 267$ MeV pure gauge QCD undergoes a deconfining phase transition. $T_d$ can be calculated by looking at the value of $T$ where the thermal ensemble average of the trace of $L(x)$ in the fundamental representation of the gauge group, $\langle \mathrm{Tr}_F L(x) \rangle$, changes very rapidly from zero to a nonzero value. Less understood, however, is the dynamics of this phase transition. The theoretical description of the dynamics of such a deconfinement transition directly from QCD is not yet possible and the use of effective models is one viable alternative for making progress in the field.

One promising class of models considers the real time evolution of the eigenvalues $\psi(x, \tau)$ of the Polyakov loop $L(x)$ through stochastic Ginzburg-Landau-Langevin (GLL) equations [3, 4, 5]. The use of GLL equations to describe the relaxation of an order parameter is common practice in studies of nonequilibrium properties of condensed matter systems, and might be equally well suited for QCD applications.

For SU(3) $L(x)$ in the fundamental representation can be written as

$$L(x) = P \exp\left[ ig \int_0^{1/T} d\tau A_0(x, \tau) \right] = \begin{pmatrix} e^{i\phi(x)} & 0 & 0 \\ 0 & 1 & 0 \\ 0 & 0 & e^{-i\phi(x)} \end{pmatrix}, \qquad (1)$$

where $P$ means path-ordering in $\tau$, $g$ is the coupling constant, and $A_0$ is the $\mu = 0$ component of the gluon field $A_\mu$. In the confined phase, $L_F \equiv \mathrm{Tr}_F L(x) = 1 + 2\cos\phi = 0$ and therefore $\phi = 2\pi/3$. Here the subscript $F$ means fundamental representation. For convenience we change variables such that $\phi(x) = 2\pi/3 - \psi(x)$, and the confined phase

CP892, *Quark Confinement and the Hadron Spectrum VII*
edited by J. E. F. T. Ribeiro
© 2007 American Institute of Physics 978-0-7354-0396-3/07/$23.00

is then for $\psi = 0$. For the description of the dynamics of $\psi$, when $\psi$ becomes also a function of time $\tau$, we postulate Langevin dynamics as

$$\frac{\partial \psi}{\partial \tau} = -\Gamma \frac{\delta S_{eff}[\psi]}{\delta \psi} + \eta, \tag{2}$$

where $S_{eff}[\psi]$ is the coarse-grained effective action, $\Gamma$ is a dissipation coefficient and $\eta$ is a noise term that mimics the thermal fluctuations of the system – for GLL equations that incorporate multiplicative noise and memory see [6] and [7]. $\Gamma$ is an important quantity, since its value determines the rate at which the system reaches equilibrium. Once a model for the effective action $S_{eff}$ is given and $\Gamma$ is known, one can calculate the space-time evolution of $\psi$ by solving Eq. (2). Although models for $S_{eff}$ can be found in the literature [8, 9], the value of $\Gamma$ is unknown. For example, the model of Ref. [8] gives for $S_{eff}$

$$S_{eff} = \frac{4T^2}{g^2} \int d^3x \left[ \frac{1}{2} (\nabla \psi)^2 + \left( \frac{3g^2}{8\pi^2} M^2 - \frac{g^2}{6} T^2 \right) \psi^2 \; \frac{g^2 T^2}{6\pi} \psi^3 + \frac{3g^2 T^2}{8\pi^2} \psi^4 \right], \tag{3}$$

where $M$ is as phenomenological constant related to the deconfinement temperature by $T_d = 9\sqrt{10}M/20\pi$. The large-time solution of Eq. (2) gives the equilibrium probability distribution $P[\psi] \sim \exp[-S_{eff}/T]$.

In the present communication we report on a lattice estimate of $\Gamma$ and its temperature dependence by fitting the early time exponential growth of the structure function $S(k,t)$ – to be defined below. We do so by employing Glauber dynamics of pure SU(3) lattice gauge theory, in the lines of the earlier studies of Refs. [3, 10]. The value of $\Gamma$ extracted in this way will be expressed in terms of $T_d$. Since the relation between the real time variable in the GLL equation and the Monte Carlo time of the Glauber dynamics is unknown, one needs independent phenomenological input to fix the physical value of $\Gamma$ to be used in the GLL equation, but this issue will not be addressed here.

## LATTICE SIMULATIONS

We have performed lattice simulations for pure SU(3) gauge QCD. Glauber dynamics is implemented by starting from thermalized gauge field configurations at a temperature $T < T_d$. Then the temperature of the entire lattice is changed (quenched) to $T > T_d$ and the gauge fields are updated using the heath-bath algorithm of Ref. [11] (without over-relaxation). A "time" unit is defined as one update of the entire lattice by visiting systematically one site a time. The structure function is defined as

$$S(k,\tau) = \langle \widetilde{L}_F(k,\tau)\widetilde{L}_F(-k,\tau) \rangle, \tag{4}$$

where $\widetilde{L}_F(k,\tau)$ is the Fourier transform of $L_F(x,\tau)$. Extraction of the value of $\Gamma$ that appears in Eq. (2) is done as follows. At early times, right after the quench, $\psi \simeq 0$ and one can neglect the terms proportional to $\psi^3$ and $\psi^4$ in Eq. (3). It is not difficult to show that at early times when $\psi$ is small the structure function can be written as

$$S(k,\tau) = S(k,0) \exp[2\omega(k)\,\tau] \quad \text{with} \quad \omega(k) = \frac{4T^2}{g^2}\Gamma \left( k_c^2 - k^2 \right), \tag{5}$$

where, for the model action of Eq. (3), $k_c^2$ is given by

$$k_c^2 = \frac{g^2}{3}\left(T^2 - \frac{9M^2}{4\pi^2}\right).\tag{6}$$

One sees that for momenta smaller than the critical momentum $k_c$, one has the familiar exponential growth, signalling spinodal decomposition. Plotting $\ln S(k,\tau)/\tau$ for different values of $k$ allows one to extract $2\omega(k)$, and in particular the value of $k_c^2$. Once one has extracted these values, then $\Gamma$ can be obtained from the formula

$$\Gamma = 2\omega(k)\frac{g^2}{8T^2(k^2 - k_c^2)}.\tag{7}$$

Now, in Monte Carlo simulations one does not have a time variable in physical units and so in plotting $\ln S$ from the lattice, one obtains a $2\omega(k)$ that does not include the (unknown) scale that relates real time $\tau$ and Monte Carlo time. But, if one assumes that the relation between the Langevin time variable $\tau$ and the Monte Carlo time is linear, one can parameterize this relation in terms of the lattice spacing $a$ as $\tau = a\lambda_{MC}$, where $\lambda_{MC}$ is a dimensionless parameter that gives this relation in units of the lattice spacing. In Fig. 1, we present a typical set of results for $S(k,\tau)$ and the corresponding $2\omega(k)$. The results are for a lattice with volume $N_\tau \times N_s^3 = 4 \times 64^3$ for a quench from $\beta = 6/g^2 = 5.5$ to $\beta = 5.92$. In physical scales [12], the final temperature corresponds to $T_f = 1.57 T_d$. For this particular lattice one obtains (errors are not included)

$$\lambda_{MC}\,\Gamma = 10^{-3}\,T_d^{-3}.\tag{8}$$

We have also estimated the temperature dependence of $\Gamma$. We simulated on a smaller spatial lattices, $N_\tau \times N_s^3 = 4 \times 32^3$, starting from the same initial temperature and extracted $\Gamma$ for four final temperatures $T_f$. The results are shown in Table 1. One clearly sees that $\Gamma$ decreases as the final temperature increases.

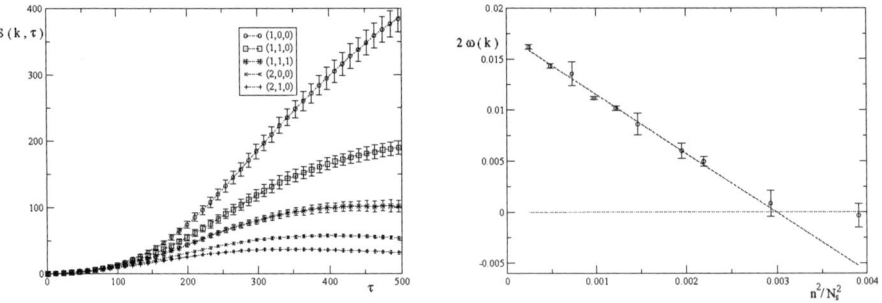

**FIGURE 1.** Results of simulations corresponding to a quench from $\beta = 6/g^2 = 5.5$ to $\beta = 5.92$. Left: $S(k,\tau)$ as function of Monte Carlo time for several momenta $\vec{k} = (n_x, n_y, n_z)$. Right: $2\omega(k)$ as extracted from $\ln S(k,\tau)$ for 10 different values of $n^2/N_s^2$.

**TABLE 1.** $\Gamma$ for different final quenching temperatures $T_f$ for a $4 \times 32^3$ lattice.

| $\beta = 6/g^2$ | $\beta_f$ | $T_f/T_d$ | $\lambda_{MC}\Gamma T_d^3$ |
|---|---|---|---|
| 5.5 | 5.80 | 1.25 | $1.50 \times 10^{-3}$ |
| 5.5 | 5.85 | 1.38 | $1.17 \times 10^{-3}$ |
| 5.5 | 5.92 | 1.56 | $8.27 \times 10^{-4}$ |
| 5.5 | 5.95 | 1.66 | $6.83 \times 10^{-4}$ |

# CONCLUSIONS AND PERSPECTIVES

We have estimated the dissipation coefficient of the phenomenological GLL equation in Eq. (2) that describes the dynamics of the deconfinement transition. These are admittedly rough estimates. First, it is clear that the absolute determination of $\Gamma$ is ambiguous due to the lack of direct information on $\lambda_{MC}$, although the ratio of two $\Gamma$'s is less ambiguous. Also, the continuum limit should be eventually taken for the estimate of $\Gamma$ and its temperature dependence. The same procedure presented here can be applied to the dynamics of the chiral transition [13], and to the study of the coupled problem of the deconfinement and chiral transition, as well as the dynamics of conserved charges [14].

# ACKNOWLEDGMENTS

Work partially supported by CNPq, FAPERJ and FAPESP (Brazilian Agencies).

# REFERENCES

1.  A. M. Polyakov, *Phys. Lett. B* **72**, 477–480 (1978).
2.  E. Laermann and O. Philipsen, *Annu. Rev. Nucl. Part. Sci* **53**, 163–198 (2003).
3.  T.R. Miller and M.C. Ogilvie, *Pys. Lett. B* **488**, 313–318 (2000); *Nucl. Phys. B (Proc. Suppl.)* **106**, 537–539 (2002).
4.  G. Krein, *AIP Conf. Proc.* **756**, 419–421 (2005).
5.  A.J.Mizher, E.S. Fraga, and G. Krein, hep-ph/0608099, to be published in the *Braz. J. Phys.* as proceedings of 1st Latin American Workshop on High Energy Phenomenology (LAWHEP), Porto Alegre, Brazil, 1-3 Dec 2005.
6.  E.S. Fraga, G. Krein, and R.O. Ramos, *AIP Conf. Proc.* **814**, 621–628 (2006).
7.  E.S. Fraga, T. Kodama, G. Krein, A.J. Mizher, and L.F. Palhares, hep-ph/0608132, to appear in the proceedings of International Conference on Strong and Electroweak Matter (SEWM 2006), Upton, New York, 10-13 May 2006.
8.  P.N. Meisinger, T.N. Miller, and M.C. Ogilvie, *Phys. Rev. D* **65**, 034009 (2002).
9.  R.D. Pisarski, *Phys. Rev. D* **62**, 111501 (2000).
10. E. Tomboulis and A. Velytsky, *Phys. Rev. D* **72**, 074509 (2005); A. Bazanov, B.A. Berg, and A. Velytsky, *Phys. Rev. D* **74**, 014501 (2006).
11. N. Cabibbo and E. Marinari, *Phys. Lett. B* **119**, 387–390 (1982); A.D. Kennedy and B.J. Pendleton. *Phys. Lett. B* **156**, 393–399 (1985).
12. G. Boyd, J. Engels, F. Karsch, E. Laermann, C. Legeland, M. Lutgemeier, and B. Petersson, *Nucl. Phys. B* **469**, 419–444 (1996).
13. E.S. Fraga and G. Krein, *Phys. Lett. B* **614**, 181–186 (2005).
14. T. Koide, G. Krein, and R.O. Ramos, *Phys. Lett. B* **636**, 96–100 (2006).

# Multiplicity Fluctuations in Heavy Ion Collisions at CERN SPS

Benjamin Lungwitz
for the NA49 Collaboration

*Fachbereich Physik der Universität, Frankfurt, Germany.*

**Abstract.** The system size and centrality dependence of multiplicity fluctuations in nuclear collisions at 158$A$ GeV as well as the energy dependence for the most central $Pb + Pb$ collisions were studied by the NA49 experiment at CERN SPS. A strong increase of fluctuations was observed with decreasing centrality in $C + C$, $Si + Si$ and $Pb + Pb$ collisions. The string hadronic models (UrQMD, Venus, HIJING, HSD) can not reproduce the observed increase. This may indicate a strong mixing of target and projectile contribution in a broad rapidity range. For the most central collisions at all SPS energies multiplicity distributions are significantly narrower than a corresponding Poisson one both for negatively and positively charged hadrons. The UrQMD model seems to reproduce the measured values on scaled variance. Statistical model calculations overpredict results when conservation laws are not taken into account.

**Keywords:** Multiplicity Fluctuations, NA49, CERN SPS
**PACS:** 24.60.Ky

## INTRODUCTION

At high energy densities ($\approx 1 \, GeV / fm^3$) a phase transition from hadron gas to quark-gluon-plasma (QGP) is expected to occur. There are indications that at top SPS energies quark-gluon-plasma is created at the early stage of heavy ion collisions [1]. Lattice QCD calculations suggest furthermore the existence of a critical point in the phase diagram of strongly interacting matter which separates the line of the first order phase transition from a crossover. Models predict an increase of multiplicity fluctuations near the onset of deconfinement [2] or the critical point [3]. Statistical model calculations [4] showed a non-trivial decrease of fluctuations due to conservation laws. These reduced fluctuations would then serve as a "background" in the search for the critical point.

## THE NA49 EXPERIMENT

The NA49 detector is a large acceptance fixed target hadron spectrometer described in [5]. The centrality of a collision is determined using a downstream Veto calorimeter which measures the energy in the projectile spectator domain [6]. This allows a determination of the number of projectile participant nucleons $N_P^{proj}$. The number of target participants is not fixed in the experiment, model calculations show large fluctuations of their number in peripheral collisions [7].

CP892, *Quark Confinement and the Hadron Spectrum VII*
edited by J. E. F. T. Ribeiro
© 2007 American Institute of Physics 978-0-7354-0396-3/07/$23.00

# Analysis Procedure

Since detector effects like track reconstruction efficiency might have a large influence on multiplicity fluctuations, it is important to select a very clean track sample for the analysis. Therefore the acceptance for this analysis is limited to a part of the forward hemisphere, where the NA49 detector has the highest tracking efficiency. This was done by restricting the analysis to the rapidity interval $1 < y(\pi) < y_{beam}$ [1] for $20A$ to $80A$ GeV and $1.08 < y(\pi) < 2.57$ for $158A$ GeV. In addition a cut on transverse momentum according to [8] was applied, which is dependent both on rapidity and azimuthal angle. For such tracks the reconstruction efficiency is larger than 98%.

The basic measure of multiplicity fluctuations used in this analysis is the scaled variance: $\omega = \frac{Var(n)}{<n>}$, where $Var(n)$ and $<n>$ are variance and mean of the multiplicity distribution, respectively. In this paper only the results for negatively charged hadrons are shown.

The data is corrected for the finite width of centrality bins. Data on centrality and system size dependence at $158A$ GeV is corrected for the resolution of the Veto calorimeter, for most central $Pb + Pb$ data this correction is small (it would decrease the results by less than 5%) and was neglected.

The total systematic error is estimated to be $+2\%$ respectively $-5\%$ [6] for the most central $Pb + Pb$ collisions, the systematic error for $p + p$, $C + C$, $Si + Si$ and non-central $Pb + Pb$ collisions is shown in figure 1.

# CENTRALITY AND SYSTEM SIZE DEPENDENCE

The centrality dependence of the scaled variance for negatively charged hadrons ($h^-$) at $158A$ GeV is shown in figure 1. Scaled variance increases with decreasing centrality of a collision. For very peripheral collisions there is a hint that it decreases again, but the systematic errors are large in this region. Data is compared to predictions of various string hadronic models for $Pb + Pb$ collisions; they all predict, in contradiction to data, a flat centrality dependence. Scaled variance behaves similar in $p + p$, $C + C$, $Si + Si$ and $Pb + Pb$ collisions if plotted against centrality defined as $N_P^{proj}/A$, where $A$ is the number of nucleons in nuclei. This is a hint that not the size of the collision system is the correct scaling parameter for the observed increase of $\omega$ but the fraction of the colliding nucleons in the projectile nuclei. In [9] it is suggested that target participants contribute to particle production in the projectile hemisphere and their fluctuations therefore result in an increase of multiplicity fluctuations. The predictions of this effect called mixing is in approximate agreement with data as shown in figure 1 (right). Interparticle correlations [10] and percolation models [11] provide alternative explanations of the results.

---

[1] Rapidity is calculated in the center of mass system assuming pion mass.

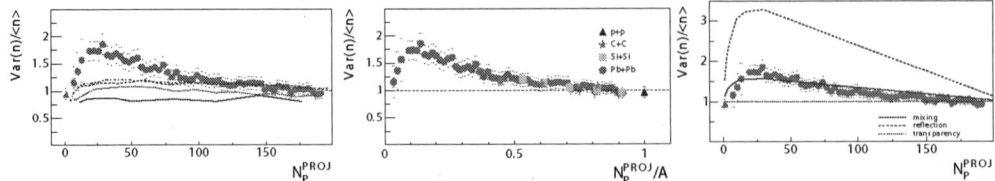

**FIGURE 1.** Left: Comparison of scaled variance of $h^-$ in $Pb + Pb$ collisions (circles) with string hadronic models: HIJING [12], UrQMD, HSD [7] and Venus [13]. Middle: Centrality dependence of scaled variance of $h^-$ in $p + p$, $C + C$, $Si + Si$ and $Pb + Pb$ collisions at 158$A$ GeV. Right: Scaled variance of $h^-$ in comparison to model calculations for transparency, mixing and reflection of matter in the early stage [9]. The outer errors correspond to a sum of statistical and systematical uncertainties.

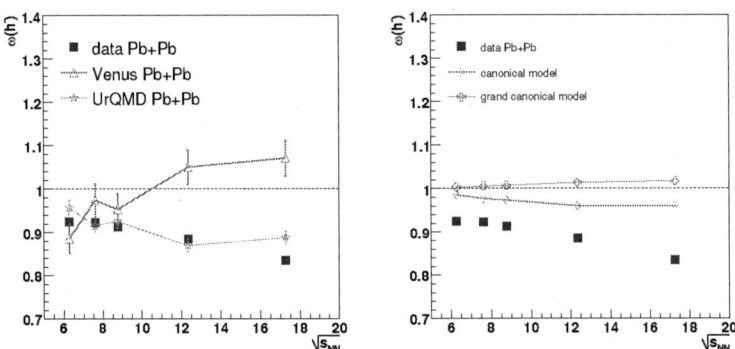

**FIGURE 2.** Energy dependence of multiplicity fluctuations of $h^-$ in $Pb + Pb$ collisions in comparison to string-hadronic models Venus [13] and UrQMD [14] (left) and in comparison to canonical and grand-canonical statistical hadron-resonance gas models [15] (right). Only statistical errors are shown, the systematical errors (not shown) are $\sigma_{sys} \approx {}^{+2}_{-5}$ %.

## ENERGY DEPENDENCE IN CENTRAL COLLISIONS

At all energies the scaled variance of $h^-$ is significantly smaller than one, the value for a corresponding Poisson distribution (see figure 2) [6]. A direct quantitative comparison of $\omega$ at different energies is not possible due to different experimental acceptance. The UrQMD model is in approximate agreement with data. The Venus model overpredicts $\omega$ at higher energies. No significant increase of multiplicity fluctuations due to critical point or onset of deconfinement is observed.

Predictions of a statistical model [15] for canonical and grand-canonical ensembles are compared to data. The measured scaled variance is much lower than predicted by the grand-canonical ensemble. The canonical model predicts $\omega$ smaller than 1, in qualitative agreement with data. Energy momentum conservation and the finite volume of hadrons are expected to cause an additional suppression of fluctuations. Thus the observed small fluctuations seem to be a non-trivial effect of conservation laws in a relativistic hadron gas [4].

# REFERENCES

1. U. W. Heinz, and M. Jacob (2000), nucl-th/0002042.
2. M. Gazdzicki, M. I. Gorenstein, and S. Mrowczynski, *Phys. Lett.* **B585**, 115–121 (2004), hep-ph/0304052.
3. M. A. Stephanov, K. Rajagopal, and E. V. Shuryak, *Phys. Rev.* **D60**, 114028 (1999), hep-ph/9903292.
4. V. V. Begun, M. Gazdzicki, M. I. Gorenstein, and O. S. Zozulya, *Phys. Rev.* **C70**, 034901 (2004), nucl-th/0404056.
5. S. Afanasev, et al., *Nucl. Instrum. Meth.* **A430**, 210–244 (1999).
6. B. Lungwitz (2006), nucl-ex/0610046.
7. V. P. Konchakovski, et al., *Phys. Rev.* **C73**, 034902 (2006), nucl-th/0511083.
8. C. Alt, et al., *Phys. Rev.* **C70**, 064903 (2004), nucl-ex/0406013.
9. M. Gazdzicki, and M. Gorenstein, *Phys. Lett.* **B640**, 155–161 (2006), hep-ph/0511058.
10. M. Rybczynski, and Z. Wlodarczyk, *J. Phys. Conf. Ser.* **5**, 238–245 (2005), nucl-th/0408023.
11. L. Cunqueiro, E. G. Ferreiro, F. del Moral, and C. Pajares, *Phys. Rev.* **C72**, 024907 (2005), hep-ph/0505197.
12. M. Gyulassy, and X.-N. Wang, *Comput. Phys. Commun.* **83**, 307 (1994), nucl-th/9502021.
13. K. Werner, *Phys. Rept.* **232**, 87–299 (1993).
14. S. A. Bass, et al., *Prog. Part. Nucl. Phys.* **41**, 225–370 (1998), nucl-th/9803035.
15. V. V. Begun, M. I. Gorenstein, M. Hauer, V. P. Konchakovski, and O. S. Zozulya, *Phys. Rev.* **C74**, 044903 (2006), nucl-th/0606036.

# Overview of Charm Physics at RHIC

## M.J. Leitch

*Los Alamos National Laboratory, Los Alamos NM 87545 USA, leitch@lanl.gov*

**Abstract.** Heavy-quark production provides a sensitive probe of the gluon structure of nucleons and its modication in nuclei. It is also a key probe of the hot-dense matter created in heavy-ion collisions. We will discuss the physics issues involved, as seen in quarkonia and open heavy-quark production, starting with those observed in proton-proton collisions. Then cold nuclear matter effects on heavy-quark production including shadowing, gluon saturation, energy loss and absorption will be reviewed in the context of recent proton-nucleus and deuteron-nucleus measurements. Next we survey the most recent measurements of open-charm and $J/\psi$s in heavy-ion collisions at RHIC and their interpretation. We discuss the high-$p_T$ suppression and flow of open charm in terms of energy loss and thermalization and, for $J/\psi$, contrast explanations in terms of screening in a deconfined medium vs. recombination models.

**Keywords:** quarkonia, gluon saturation, quark gluon plasma
**PACS:** 24.85.+p, 25.75.-q, 25.75.Nq

## CHARM PRODUCTION IN P+P COLLISIONS AT RHIC

Gluon fusion dominates the production of quarkonia, but the configuration of the produced state and how it hadronizes remain uncertain. Absolute cross sections can be reproduced by NRQCD models that involve a color octet state[1], but these models predict transverse polarization of the $J/\psi$ at large $p_T$ that is not seen in the data[2]. A general complication in understanding $J/\psi$ results is the fact that $\sim$40% of the $J/\psi$s come from decays of higher mass resonances ($\psi'$ and $\chi_C$)[3] - a feature that may contribute to the lack of polarization seen. One exception to this feature is the maximal transverse polarization observed for the $\Upsilon_{2S+3S}$ states[4]; where the lack of feed-down for these states may allow the polarization to persist.

$J/\psi$ cross section measurements for p+p collisions at $\sqrt{s} = 200$ GeV from PHENIX[5] are shown in Fig. 1. These results, based on approximately 500 $J/\psi$s from the 2003 run, provide the baseline for both CNM studies in d+Au collisions and QGP studies in A+A collisions at RHIC, and are presently one of the limiting factors in obtaining precise nuclear modifications. However p+p data from the 2005 and 2006 runs will soon improve this baseline significantly with over 40,000 $J/\psi$s.

Open charm measurements at RHIC suffer from large systematics and statistical uncertainties due to the statistical subtraction methods that are used. Measurements by PHENIX and by STAR differ substantially[6] on the size of the charm cross section. Also the measured cross sections lie substantially higher than current theoretical predictions as shown in Fig. 2.

CP892, *Quark Confinement and the Hadron Spectrum VII*
edited by J. E. F. T. Ribeiro
2007 American Institute of Physics 978-0-7354-0396-3/07/$23.00

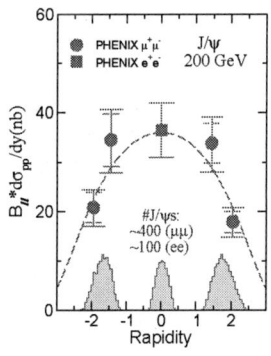

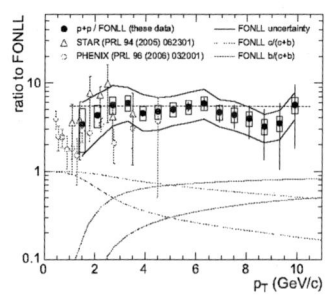

**FIGURE 1.** $J/\psi$ cross section vs rapidity for 200 GeV p+p collisions at RHIC[5].

**FIGURE 2.** Open charm plus beauty cross section from prompt electrons divided by FONLL theory vs $p_T$ for 200 GeV p+p collisions at RHIC[9].

## NUCLEAR EFFECTS ON CHARM

When quarkonia are produced in nuclei their yields per nucleon-nucleon collision are known to be significantly modified. This modification, shown vs. $x_F$ in Fig. 3 for 800 GeV p+A fixed target measurements and in Fig. 4 at RHIC energy, is thought to be due to several CNM effects including gluon shadowing, initial-state gluon energy loss and multiple scattering, and absorption (or dissociation) of the $c\bar{c}$ in the final-state before it can form a $J/\psi$.

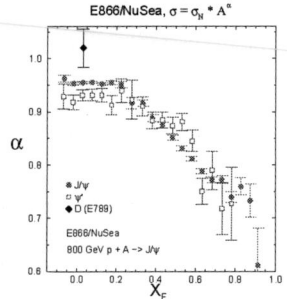

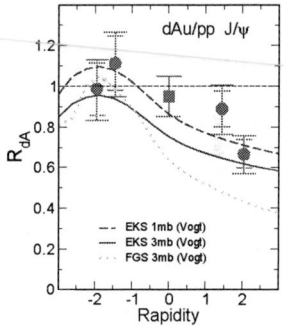

**FIGURE 3.** Nuclear moification factor $\alpha$ vs $x_F$ for $J/\psi$ and $\psi'$ production in $\sqrt{s} = 38$ GeV collisions in E866/NuSea[7], and for $D^0$ from E789[8].

**FIGURE 4.** Rapidity dependence of the $J/\psi$ nuclear modification factor, $R_{dAu}$ for 200 GeV d+Au collisions at RHIC[5].

Shadowing is the depletion of low-momentum partons (gluons in this case) in a nucleon embedded in a nucleus compared to their population in a free nucleon. The strength of the depletion differs between numerous models by up to a factor of three. Some models are based on phenomenological fits to deep-inelastic scattering and Drell-Yan data[10], while others obtain shadowing from coherence effects in the nuclear medium[11, 12]. In addition, models such as the Color Glass Condensate (CGC)[13] yield shadowing through gluon saturation pictures where the large gluon populations at very small x in a nucleus generate a deficit of gluons at small x.

In the final state, the produced $c\bar{c}$ can be disassociated or absorbed on either the nucleus itself, or on light co-moving partons produced when the projectile proton or deuteron enters the nucleus. The latter is probably only important in nucleus-nucleus collisions as the number of co-movers created in a p+A or d+A collisions is small.

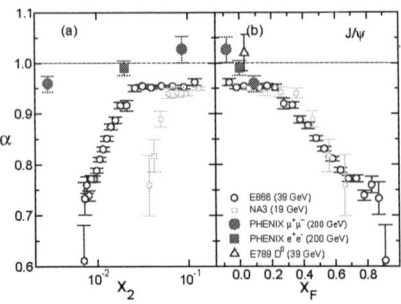

**FIGURE 5.** Test of scaling vs $x_2$ and $x_F$ for $J/\psi$ suppression data for three different collision energies. Data is from Refs.[7, 15, 5]

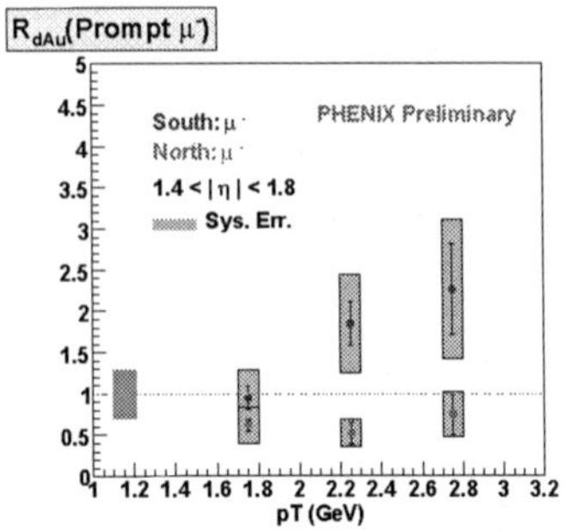

**FIGURE 6.** Nuclear dependence of heavy quark suppression vs $p_T$ from single muons in PHENIX.

However, $J/\psi$ suppression in p(d)+A collisions remains a puzzle given that one does not find a universal suppression vs $x_2$ as would be expected from shadowing, Fig. 5a; while vs. $x_F$ the dependence is similar for all energies, Fig. 5b. This apparent $x_F$ scaling supports explanations that involve initial-state energy loss or Sudakov suppression[14].

On the open-charm front, there are no substantial modifications seen at central rapidity in d+Au collisions, but for forward rapidity (shadowing region) - as shown in Fig. 6 - substantial suppression is seen, while some enhancement is see at backward rapdity (Au-going direction).

## $J/\psi$ IN HEAVY-ION COLLISIONS - A QUARK GLUON PLASMA SIGNATURE?

One of the leading predictions for the hot-dense matter created in high-energy heavy-ion collisions was that if a deconfined state of quarks and gluons is created, i.e. a quark-gluon plasma (QGP), the heavy-quark bound states would be screened by the deconfined colored medium and destroyed before they could be formed[16]. This screening would depend on the particular heavy-quark state, with the $\psi'$ and $\chi_C$ being dissolved first; next the $J/\psi$ and then the $\Upsilon$'s only at the highest QGP temperatures. The CERN SPS measurements[18] showed a suppression for the $J/\psi$ and $\psi'$ beyond what was expected from CNM effects - as represented by a simple absorption model constrained to p+A data. In addition to explanations involving creation of a QGP, a few theoretical models[17] were also able to explain the data without including a QGP, so the evidence that a QGP was formed was controversial.

The first measurements from PHENIX at RHIC in 2004 are beginning to yield results - see Fig. 8 for preliminary results for Au+Au and Cu+Cu collisions[19]. First it is important to understand what the normal CNM $J/\psi$ suppression should look like in these A+A collisions. This is illustrated by the blue error bands for A+A collisions in Fig. 8 which represent identical theoretical calculations to the analogous blue error band in Fig. 7 for d+Au collisions. As can be seen the present d+Au data lacks enough precision to provide a good constraint on the CNM effects. As a result it is difficult to be very quantitative about the amount of "anomalous" suppression observed in A+A collisions, although there does seem to be a clear suppression beyond CNM for the most central collisions.

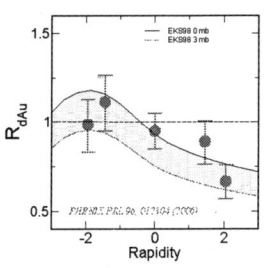

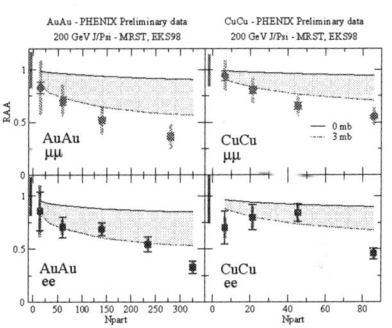

**FIGURE 7.** Results for $J/\psi$ suppression in d+Au collisions[5] compared to a theoretical calculation that includes absorption and EKS shadowing[20].

**FIGURE 8.** $J/\psi$ suppression in Au+Au and Cu+Cu collisions for forward rapidity and central rapdity[19] compared to predictions for CNM from the same calculations as shown in Fig. 7[20].

On the other hand, all of the models[17, 21, 22] that were successful in describing the lower energy SPS data over-predict the suppression compared to the preliminary data at RHIC - unless a "regeneration" mechanism is added as was done by Rapp[22] and by Thews[23]. The regeneration models assert that if the total production of charm is high enough then densities in the final state will be sufficient to have substantial formation of $J/\psi$s from the large number of independent charm quarks created in the collision. This production mechanism was almost insignificant at SPS energies but at RHIC may be substantial. This leads to a scenario in which strong screening or dissociation by a very high-density gluon density occurs to a level of suppression stronger than the RHIC data shows, but the regeneration mechanism compensates for this and brings the net suppression back up to where the data lies. This is shown in Fig. 9.

An alternative interpretation of the preliminary results, sequential screening, is given by Karsch, Kharzeev and Satz[24]. In this picture, they assume that the $J/\psi$ is never screened, as supported by recent Lattice QCD calculations for the $J/\psi$ - not at SPS nor at RHIC. Then the observed suppression comes from screening of the higher-mass states alone ($\psi\prime$ and $\chi_C$) that, by their decay, normally provide $\sim40\%$ of the observed $J/\psi$s. This scenario is consistent with the apparently identical suppression patterns seen at the SPS and RHIC shown in Fig. 10.

As a result we are left for the moment with two different scenarios that provide explanations for the RHIC A+A data. Both include the QGP in their picture, either through color screening in the QGP or through severe suppression of the $J/\psi$ by a very high gluon density. Further tests from the data will be necessary to clarify the picture. Regeneration models predict narrowing of both the rapidity and $p_T$ distributions, but so far the preliminary data shows little or no change in the rapidity shape from ordinary p+p and only a hint of narrowing of the $p_T$. We are also trying to extract a measurement of flow for the $J/\psi$, since emerging results for single charm are beginning to show flow and the $J/\psi$'s, if they were from regeneration, would inherit this flow. These tests await the more precise final analysis of the 2004,5 Au+Au and Cu+Cu data; and higher statistics runs for Au+Au and d+Au in the near future.

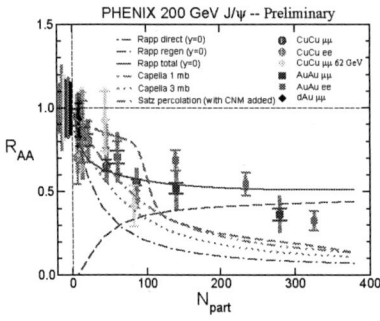
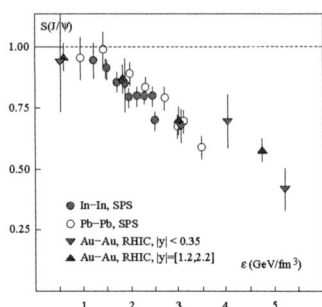

**FIGURE 9.** Theories that agree with SPS data do not agree with RHIC data, unless regeneration is added as in the Rapp (solid blue) curve.

**FIGURE 10.** Universal dependence on energy density of $J/\psi$ suppression measurements at RHIC and at the SPS[24].

## OPEN CHARM IN AU+AU COLLISIONS

In Au+Au collisions, open charm (and beauty) together, are suppressed due to energy loss in the dense medium, with gluon densities per unit rapidity of up to 1000 infered in some theoretical analysis. However, as shown in Fig. 11, calculations that include both radiative and collisional energy loss[25] predict too small a suppression when both charm and beauty are included. Flow has also been oberved for heavy quark production. As shown in Fig. 12, the flow is similar to that of light quarks at small $p_T$, but at higher $p_T$ the data with large uncertainties hints at vanishing flow, consistent with simple expectations that higher $p_T$ charm simply punches out of the medium and never thermalizes.

## SUMMARY

Substantial uncertainties remain in the understanding of charm production cross sections, and the polarization of charmonia. There are also a number of cold nuclear matter effects that influence their production in nuclei and cloud our understanding of the suppression seen in nucleus-nucleus collisions. Two competing pictures are able to explain the $J/\psi$ suppression seen in nucleus-nucleus collisions at RHIC - one involving sequential screening in the plasma of the various charmonia states; the other with strong dissociation of all charmonia states by a dense gluon field but recombination of independently produced charm quarks. For open charm, the the energy loss observed in the dense medium from nucleus-nucleus collisions is larger than that expected from theoretical models that include radiative and collisional energy loss of both charm and beauty. Higher statistics data with higher luminosity runs as well as RHIC vertex detector upgrades will enable more precise data in the future that will give a clearer understanding of the rich physics in charm production.

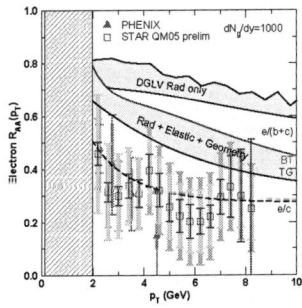

**FIGURE 11.** Energy loss calculations compared to open heavy (charm + beauty) data vs $p_T$[25].

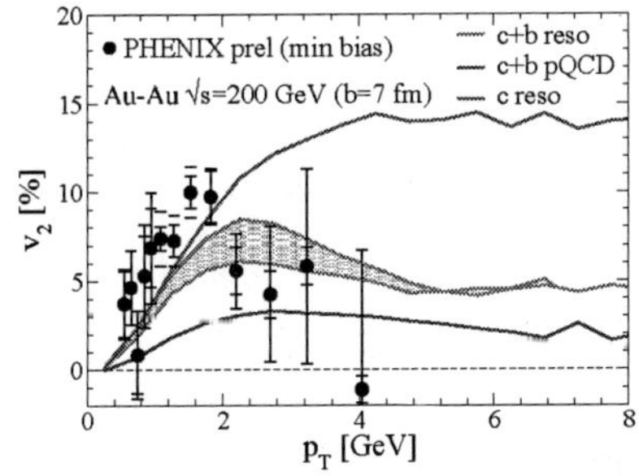

**FIGURE 12.** Elliptic flow of open heavy (charm + beauty) compared to Rapp calculations[26].

# REFERENCES

1. M. Beneke, M. Kraemer, *Phys. Rev.* **D55**, 5269 (1997).
2. M. Beneke, I Z. Rothstein, *Phys. Rev.* **D54**, 2005 (1996).
3. I. Abt et al. (HERA-B), *Phys. Lett.* **B561**, 61 (2003).
4. C. N. Brown, et al (E866/NuSea), *Phys. Rev. Lett.* **86**, 2529 (2001).
5. S.S. Adler, et al. (PHENIX), *Phys. Rev. Lett.* **96**, 012304 (2006).
6. A. Adare, et al. (PHENIX), hep-ex/0609010.
7. M.J. Leitch et al. (E866/NuSea), *Phys. Rev. Lett.* **84**, 3256-3260 (2000)
8. M.J. Leitch et al. (E789), *Phys. Rev. Lett.* **72**, 2542 (1994).
9. B.I. Abelev, et al. (STAR), nucl-ex/0607012.
10. K.J. Eskola, V.J. Kolhinen, R. Vogt, *Nucl. Phys.* **A696**, 729 (2001).
11. L. Frankfurt, M. Strikman, *Eur. Phys. J* **A5**, 293 (1999).
12. B. Kopeliovich, A. Tarasov, and J. Hufner, *Nucl. Phys.* **A696**, 669 (2001).
13. L. McLerran and R. Venugopalan, *Phys. Rev.* **D49**, 2233 (1994); *Phys. Rev.* **D49**, 3352 (1994).
14. B.Z. Kopeliovich et al , *Phys. Rev.* **C72**, 054606 (2005); hep-ph/0501260 (2005).
15. J. Badiër et al., *Z. Phys.* **C20**, 101 (1983).
16. T. Matsui, H. Satz, *Phys. Lett.* **B178**, 416 (1986).
17. A. Capella, D. Sousa, *Eur. Phys. J* **C30**, 117 (2003).
18. M.C. Abreu et al. (NA50) *Phys. Lett.* **B477**, 28 (2000) ; *Phys. Lett.* **B521**, 195 (2001).
19. H. Pereira da Costa, et al. (PHENIX) Quark Matter 2005, nucl-ex/0510051.
20. R. Vogt, M.J. Leitch, private communication.
21. S. Digal, S. Fortunator, H. Satz, *Eur. Phys. J* **C32**, 547 (2004); hep-ph/0310354
22. L. Grandchamp, R. Rapp, G.E. Brown, *Phys. Rev. Lett.* **92**, 212301 (2004); hep-ph/0306077.
23. R.L. Thews, *Eur. Phys. J* **C43**, 97 (2005).
24. F. Karsch, D. Kharzeev, H. Satz, *Phys. Lett.* **B637**, 75 (2006); hep-ph/0512239.
25. S. Wicks, W. Horowitz, M. Djordjevic, M. Gyulassy, nucl-th/0512076.
26. H. van Hees, V. Greco, R. Rapp, *Phys. Rev.* **C73** 034913 (2006); and nucl-th/0608033.

# J/ψ production and suppression in heavy-ion collisions at the CERN SPS

Pedro Martins for the NA60 Collaboration[1]

*IST-CFTP and LIP, Lisbon, Portugal, and CERN, Geneva, Switzerland*

**Abstract.**
The NA60 experiment has studied J/ψ production in Indium-Indium collisions at 158 A GeV. This paper presents results obtained with the complete statistics and the final alignment of the vertex tracker. The centrality dependence of the J/ψ suppression, obtained by comparing the measured distribution to the "normal nuclear absorption" curve, shows that the J/ψ is suppressed in In-In collisions beyond the absorption induced by cold nuclear matter.

**Keywords:** Ultra-relativistic heavy-ion collisions, J/ψ suppression
**PACS:** 13.20.Gd,25.75.Dw, 25.75.Nq

Some of the most interesting observables explored so far in search for the quark-gluon plasma have been studied through measurements of dilepton production in heavy-ion collisions at the CERN SPS. In particular, J/ψ suppression has been proposed as a signature of the formation of a deconfined QCD phase [2]. The NA38 and NA50 experiments studied J/ψ production in various collision systems, including p-A, S-U and Pb-Pb [3]. The proton-nucleus data allow us to study the normal nuclear absorption affecting the charmonium states crossing cold nuclear matter. The NA60 experiment, with an improved apparatus, studied J/ψ production in In-In collisions. The comparison between different colliding systems should indicate the physics mechanism at the origin of the J/ψ suppression. NA60 complemented the muon spectrometer previously used by NA38/NA50 with a high-granularity and radiation-tolerant *silicon pixel telescope* in the vertex region, inside a 2.5 T dipole magnet, to measure the muons *before* they suffer multiple scattering and energy loss. A Zero Degree Calorimeter (ZDC) estimates the centrality of the collisions, by measuring the energy ($E_{ZDC}$) released by the beam spectators. A more detailed description of the apparatus can be found in [4]. NA60 collected $\sim 230$ million dimuon triggers in the In-In run of 2003, running with beam intensities around $5 \times 10^7$ per 5-second spill. This paper shows results from the full statistics available and using a final alignment of the vertex tracker. The event selection criteria ensures the selection of a clean sample of J/ψ events produced in In-In collisions [1].

The centrality dependence of the J/ψ production in Indium-Indium collisions was first studied with the analysis procedure already used by the NA38 and NA50 experiments, where the J/ψ yield is normalised to the yield of high-mass dimuons (Drell-Yan). This is a good procedure since DY production is proportional to the number of nucleon-nucleon collisions and is not affected by final state effects. The J/ψ/DY cross-section

---

[1] See Ref. [1] for the full list of authors

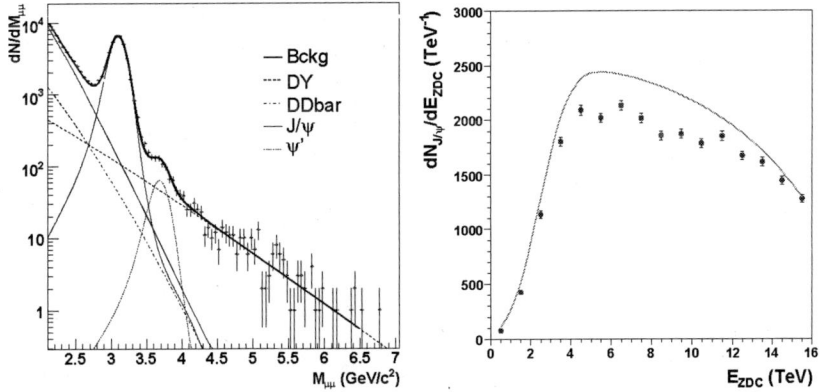

**FIGURE 1.** Left: Opposite-sign dimuon mass distribution measured in In-In collisions. Right: Measured centrality distribution of the $J/\psi$ yield and normal nuclear absorption curve calculated with the Glauber model [5] using an absorption cross section of $4.18 \pm 0.35$ mb [3].

ratio is obtained by fitting the opposite-sign dimuon mass spectrum (see Fig. 1-left) to a superposition of several contributions: the $J/\psi$ and the $\psi'$ resonances, the Drell-Yan continuum, correlated muon pairs from semimuonic decays of D and $\overline{\text{D}}$ mesons, and uncorrelated muon pairs from $\pi$ and K decays. The expected mass spectra were evaluated through Monte Carlo simulation. The combinatorial background was obtained through an event mixing technique, using single muons from the measured sample of like-sign muon pairs. The study of the ratio between the $J/\psi$ and DY cross sections has the advantage of being free from systematic errors related to experimental inefficiencies and to the integrated luminosity, but is affected by a large statistical error, due to the small number of high mass DY events.

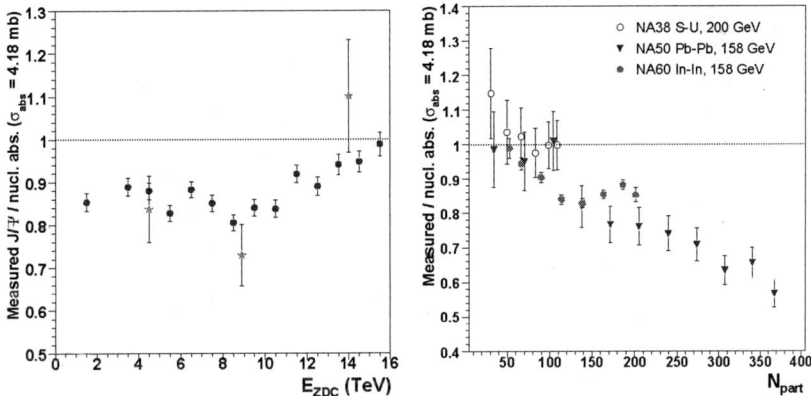

**FIGURE 2.** Left: Ratio between the measured $J/\psi$/DY ratio and the normal nuclear absorption curve, versus $E_{ZDC}$ (stars: "standard analysis"). Right: Ratio between the measured $J/\psi$ yield and the normal nuclear absorption curve, as a function of $N_{part}$, for S-U, Pb-Pb and In-In collisions.

We have also studied the $J/\psi$ suppression pattern without the limitation of the Drell-Yan statistics, by directly comparing the measured $J/\psi$ yield, as a function of centrality, with the distribution expected from normal nuclear absorption (see Fig. 1-right). In this study we only used matched dimuons. We checked that the efficiency of the matching between a muon track and a vertex track, for $J/\psi$ events, increases by less than 2 % from central to peripheral collisions, inducing a negligible bias on the $J/\psi$ centrality distribution. The matching improves the $J/\psi$ mass resolution from 105 to 70 MeV and reduces the combinatorial background from 3 to 1 % in the $J/\psi$ mass region. The vertexing efficiency for $J/\psi$ events decreases by less than 1 % from central to peripheral collisions. After event selection, the $J/\psi$ distribution as a function of centrality has been obtained, in 1 TeV $E_{ZDC}$ bins, by means of a simple fitting procedure that allows to subtract the small amount of Drell-Yan and combinatorial background under the resonance peak. The $J/\psi$ centrality distribution is then divided by the nuclear absorption curve, and normalised such that the obtained ratio, integrated over centrality, is the same as in the "standard" $J/\psi$ / DY analysis ($0.87 \pm 0.05$). The result is shown in Fig. 2, as a function of $E_{ZDC}$ (left) and of the number of participant nucleons, $N_{part}$ (right), together with the patterns measured in S-U and Pb-Pb collisions. The stars in Fig. 2-left show the measured $J/\psi$/DY values, divided by the normal nuclear absorption curve. The measured suppression pattern shows that the "anomalous" $J/\psi$ suppression in In-In collisions sets in at around $N_{part} \sim 80$ and remains rather flat for more central events.

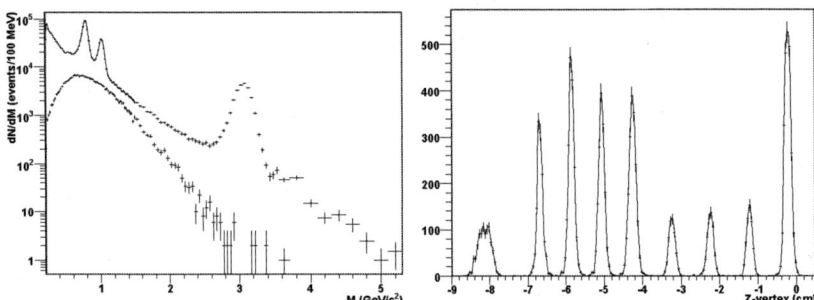

**FIGURE 3.** Opposite-sign and like-sign matched dimuon invariant mass distributions (left) and distribution of dimuon vertices along the beam line (right) for the 158 GeV p-A data collected in 2004.

In year 2004, NA60 collected p-nucleus data with 7 nuclear targets (Be, Al, Cu, In, W, Pb and U) and two beam energies (400 and 158 GeV). Figure 3 shows preliminary distributions from the 158 GeV data. The right panel shows that we can easily separate the 11 000 $J/\psi$ events (after dimuon matching) among the different targets.

## REFERENCES

1. R. Arnaldi *et al.* (NA60 Coll.), Proceedings of the Hard Probes 2006 Conf., Asilomar, June 2006.
2. T. Matsui and H. Satz, Phys. Lett. B178 (1986) 416.
3. B. Alessandro *et al.* (NA50 Coll.), Eur. Phys. J. C39 (2005) 335.
4. G. Usai *et al.* (NA60 Coll.), Eur. Phys. J. C43 (2005) 415; M. Keil *et al.*, Nucl. Instrum. Meth. A539 (2005) 137; A546 (2005) 448.
5. D. Kharzeev *et al.*, Z. Phys. C74 (1997) 307.

# Direct Photons in Heavy-Ion Collisions

## Klaus Reygers

*University of Münster, Institut für Kernphysik,*
*Wilhelm-Klemm-Straße 9, 48149 Münster, Germany*

**Abstract.** A brief overview of direct-photon measurements in ultra-relativistic nucleus-nucleus collisions is given. The results for Pb+Pb collisions at $\sqrt{s_{NN}} = 17.3$ GeV and for Au+Au collisions at $\sqrt{s_{NN}} = 200$ GeV are compared to estimates of the direct-photon yield from hard scattering. Both results leave room for a significant thermal photon component. A description purely based on hard scattering processes, however, is not ruled out so far.

**Keywords:** Direct Photons, Heavy-Ion Collisions, CERN SPS, RHIC
**PACS:** 13.85.Qk, 25.75.-q

## INTRODUCTION

In ultra-relativistic heavy-ion collisions it is expected that for a brief period of several fm/$c$ a thermalized medium is created whose relevant degrees of freedom are quarks and gluons. It has long been suggested that the initial temperature of this quark-gluon plasma (QGP) can be determined via the measurement of direct photons, *i.e.*, photons not coming from late hadron decays like $\pi^0 \to \gamma\gamma$ [1]. The virtue of direct photons is that they escape the hot and dense medium unscathed. The experimental challenge is to extract a direct-photon signal above the large decay-photon background and to identify other sources of direct photons which are not of thermal origin.

A brief summary of known and presumed photon sources in nucleus-nucleus collisions is given in Fig. 1 [6]. Photons from hard scattering of quarks and gluons, analogous to the production mechanisms in p+p-collisions, dominate the direct-photon spectrum at high transverse momenta ($p_T$). The main motivation for the measurement of high-$p_T$ photons in heavy-ion collisions is to test perturbative QCD models and to measure the rate of initial hard scatterings.

The QGP expands and cools and at a temperature of $T_c \approx 190$ MeV a phase transition to a hadron gas takes place [5]. During the entire evolution of the QGP and the hadron gas thermal direct photon are produced. The shape of their $p_T$ spectra reflects the temperature of the medium. Thermal photon are expected to contribute to the direct photon spectrum significantly at low $p_T$ ($\lesssim 3$ GeV/$c$). For model comparisons and the extraction of the initial temperature model calculations need to convolve photon rates for the QGP and the hadron gas with realistic scenarios of the space-time evolution of the fireball. Initial temperatures $T_i > T_c$ would provide evidence for the creation of a QGP.

Direct photons might furthermore be produced in interactions of quarks or gluons from early hard scattering processes with soft quarks and gluons from the QGP. One suggested mechanism is jet-photon conversion in processes like $q_{hard} + g_{QGP} \to \gamma + q$ and $q_{hard} + \bar{q}_{QGP} \to \gamma + g$ in which the photon obtains a large fraction of the momentum

CP892, *Quark Confinement and the Hadron Spectrum VII*
edited by J. E. F. T. Ribeiro
© 2007 American Institute of Physics 978-0-7354-0396-3/07/$23.00

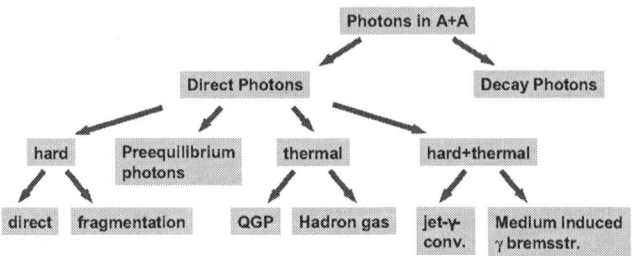

**FIGURE 1.** Known and presumed photon sources in nucleus-nucleus collisions.

of $q_{hard}$ [7]. In Au+Au collisions at $\sqrt{s_{NN}} = 200$ GeV jet-photon conversion might be a significant direct-photon source for $p_T \lesssim 6$ GeV/$c$. Direct photons might furthermore be produced due to multiple scattering of quarks in the medium. These interesting ideas, however, still require a experimental verification.

## MEASUREMENTS: WA98 AND PHENIX

Direct photons were measured by the fixed-target experiment WA98 at the CERN SPS in central Pb+Pb collisions at $\sqrt{s_{NN}} = 17.3$ GeV [4] and by the PHENIX experiment at the Relativistic Heavy-Ion Collider (RHIC) in Au+Au collisions at $\sqrt{s_{NN}} = 200$ GeV [2, 3] (see Fig. 2). One of the basic questions in both cases is whether thermal photons or photons from jet-plasma interactions are needed on top of the hard direct-photon component in order to explain the data.

In both experiments the direct-photon spectra are determined by a statistical subtraction of the calculated yield of photons from hadron decays from the total photon yield. The WA98 measurement was made with a highly segmented lead-glass calorimeter. PHENIX measured high-$p_T$ direct photons ($p_T \gtrsim 4$ GeV/$c$) in a similar way with its electromagnetic calorimeters (see Fig. 3). The preliminary low-$p_T$ direct-photon spectrum shown in Fig. 2b was obtained by measuring virtual photons via their decay into of $e^+e^-$ pairs with the aid of a Ring Imaging Cherenkov Detector. The spectrum of real direct photons can then be obtained under the assumption that the fraction $\gamma_{direct}/\gamma_{all}$ of real direct photons is identical to the fraction $\gamma^*_{direct}/\gamma^*_{all}$ of virtual direct photons with small mass ($\leq 30$ MeV) [3].

The spectrum of direct photons in central Pb+Pb collisions at $\sqrt{s_{NN}} = 17.3$ GeV in Fig. 2a is compared to p+p and p+A direct-photon data measured at slightly higher $\sqrt{s_{NN}}$. These data sets have been scaled to $\sqrt{s_{NN}} = 17.3$ GeV and furthermore scaled by the respective number of nucleon-nucleon collisions in p+A and central Pb+Pb [4]. The underlying assumption in both cases is that direct photons in p+p and p+A are produced in hard scattering processes. This comparison shows that for $p_T \gtrsim 2.5$ GeV/$c$ the direct-photon yield in central Pb+Pb collisions is consistent with the expected yield from hard scattering. Another possibility to pin down the hard scattering contribution is a comparison to a perturbative QCD (pQCD) calculation and to a parameterization of p+p direct-photon data. Fig. 2a indicates that a thermal photon signal might be present

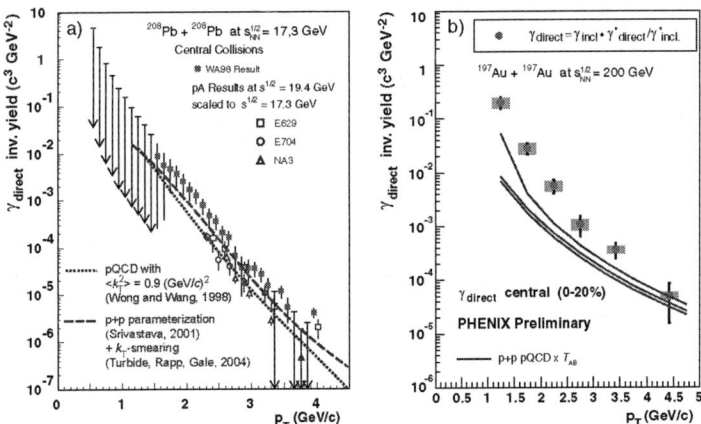

**FIGURE 2.** Direct-photon spectra measured in Pb+Pb collisions at the CERN SPS (WA98 experiment) and in Au+Au collisions at RHIC (PHENIX experiment). Both spectra are compared to estimates of the contribution of hard scattering processes in order determine whether an additional thermal photon contribution is needed.

below $p_T \approx 2.5$ GeV/$c$. However, a solid estimate of the hard scattering contribution at CERN SPS energies remains difficult. Firm conclusions can only be drawn if, *e.g.*, a better understanding of the modification of the hard scattering yield in Pb+Pb due to multiple soft scattering of the incoming partons prior to the hard process ("Cronin" or "nuclear $k_T$" effect) can be achieved.

The PHENIX low-$p_T$ direct-photon spectrum in Fig. 2b is compared to a next-to-leading-order p+p pQCD calculation scaled by the number of nucleon-nucleon collisions. The three different pQCD curves correspond to different scales used in the calculation and reflect theoretical uncertainties. An advantage at RHIC energies is that the modification of the hard scattering yield due to the Cronin effect is expected to be small [9]. The difference between the data and the hard scattering yield as estimated by the pQCD calculation hints at the presence of significant thermal photon signal. This will be confirmed or disproved with forthcoming low-$p_T$ direct-photon measurements in p+p and d+Au at the same energy.

Despite these difficulties several attempts have been made to describe the WA98 and PHENIX direct-photon spectra with a combination of a hard and a thermal component and to extract the initial temperature of the thermalized fireball. Both measurements are consistent with a QGP scenario. Initial temperatures for central Pb+Pb collisions at $\sqrt{s_{NN}} = 17.3$ GeV roughly range from $200 \lesssim T_i \lesssim 370$ MeV. For central Au+Au collisions at $\sqrt{s_{NN}} = 200$ GeV the extracted initial temperatures tend to be higher and cover the range $370 \lesssim T_i \lesssim 570$ MeV [8].

The high-$p_T$ direct-photon measurement in central Au+Au collisions at $\sqrt{s_{NN}} = 200$ GeV is presented in Fig. 3 in terms of the nuclear modification factor

$$R_{AA}(p_T) = \frac{dN/dp_T|_{A+A}}{\langle T_{AA} \rangle \times d\sigma/dp_T|_{p+p}} \, . \tag{1}$$

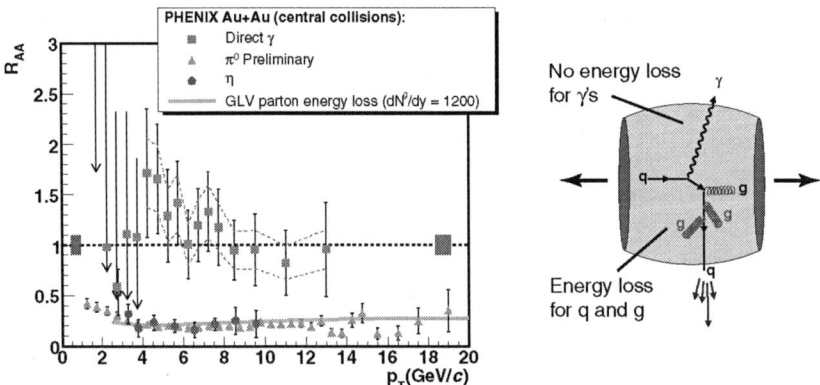

**FIGURE 3.** Nuclear modification factor $R_{AA}$ for direct-photons, neutral pions, and $\eta$ mesons in central Au+Au collisions at $\sqrt{s_{NN}} = 200$ GeV. Pions and $\eta$-mesons are suppressed whereas direct photons are not. The cartoon illustrates the most popular explanation: energetic quarks and gluons which fragment into hadrons suffer energy loss in the medium, direct photons don't.

The nuclear overlap function $T_{AA}$ is related to the number of inelastic nucleon-nucleon collisions according to $\langle T_{AA} \rangle = \langle N_{coll} \rangle / \sigma_{inel}^{NN}$. Fig. 3 shows that unlike pions and $\eta$-mesons high-$p_T$ direct photons are not suppressed, $i.e.$, they follow $N_{coll}$ scaling as expected for hard processes. This is in line with jet-quenching models which attribute the hadron suppression to energy loss of highly-energetic quarks and gluons from initial hard scattering processes in the QGP.

## CONCLUSIONS

Direct-photon measurements in nucleus-nucleus collisions from WA98 (Pb+Pb at $\sqrt{s_{NN}} = 17.3$ GeV) and PHENIX (Au+Au at $\sqrt{s_{NN}} = 200$ GeV) have been discussed. Both measurements are consistent with a thermal photon signal and initial temperatures $T_i > T_c$. A description purely based on hard scattering processes, however, is not ruled out so far. The observation that hadrons at high-$p_T$ are suppressed whereas direct photons are not supports jet-quenching models.

## REFERENCES

1. P. Stankus, Ann. Rev. Nucl. Part. Sci. **55** (2005) 517.
2. S. S. Adler *et al.* [PHENIX Collaboration], Phys. Rev. Lett. **94**, 232301 (2005)
3. S. Bathe [PHENIX Collaboration], Nucl. Phys. A **774** (2006) 731
4. M. M. Aggarwal *et al.* [WA98 Collaboration], Phys. Rev. Lett. **85** (2000) 3595
5. M. Cheng *et al.*, Phys. Rev. D **74** (2006) 054507
6. S. Turbide, C. Gale, S. Jeon and G. D. Moore, Phys. Rev. C **72** (2005) 014906
7. R. J. Fries, B. Muller and D. K. Srivastava, Phys. Rev. Lett. **90** (2003) 132301
8. K. Reygers [PHENIX Collaboration], arXiv:nucl-ex/0608043.
9. S. S. Adler [PHENIX Collaboration], arXiv:nucl-ex/0610036.

# Supersonic Jets in Relativistic Heavy-Ion Collisions

Fuqiang Wang

*Department of Physics, Purdue University, West Lafayette, Indiana 47907, USA*

**Abstract.** Mach-cone shock waves were proposed to explain the broad and perhaps double-peaked away-side 2-particle jet-correlations at RHIC; however, other mechanisms cannot be ruled out. Three-particle jet-correlation is needed in order to distinguish various physics mechanisms. In this talk the 3-particle jet-correlation measurements are presented and their implications are discussed.

**Keywords:** Heavy-ion, Azimuthal correlation, Three-particle, Mach-cone
**PACS:** 25.75.-q, 25.75.Gz

## INTRODUCTION

Jets and jet-correlations are good probes to study the medium created in relativistic heavy-ion collisions because their properties in vacuum can be calculated by perturbative quantum chromodynamics. Modifications to their properties in nuclear medium can be used to study the nature of the medium [1]. While exclusive jet reconstruction is difficult in central heavy-ion collisions at RHIC, two-particle azimuthal correlations with a high transverse momentum ($p_\perp$) trigger particle have proved to be a powerful alternative [2].

The first study of azimuthal correlations between low $p_\perp$ hadrons and modest high $p_\perp$ particles has revealed rich information [3]. The correlated hadrons on the away side of the trigger particle are found to be broadly distributed; their energy distribution is similar to that of the bulk medium particles indicating partial equilibration. The away-side broadening becomes more prominent with lower trigger $p_\perp^{trig}$ and higher associated $p_\perp$; the azimuthal correlations with $1 < p_\perp < 2.5 < p_\perp^{trig} < 4$ GeV/$c$ are shown in Fig. 1 for central Au+Au collisions from PHENIX [4] (left panel) and STAR [5, 6] (middle panel). Moreover, the average $\langle p_\perp \rangle$ of the away-side correlated hadrons shows a novel behavior as depicted in Fig. 1 (right panel) – those more collimated with the trigger possess a lower $\langle p_\perp \rangle$ [5], contrary to what is expected from jet fragmentation in vacuum.

The broad distribution stimulated many theoretical investigations. In particular, Mach-cone shock waves were suggested as a possible physics mechanism [7] – particles are emitted on a cone due to collective excitations, and their projection onto the azimuth results in a double-peak structure. The generation of Mach-cone shock waves seems inevitable given that the medium is hydrodynamic [2], the jets are supersonic, and there are strong interactions between the jets and the medium [2]. The Mach-cone angle is determined by the speed of sound of the medium and is independent of the associated particle $p_\perp$. Recently, Čerenkov gluon radiation was suggested as an alternative mechanism for conical emission [8]; the cone angle in this case is dependent of $p_\perp$.

The double-peak structure of the away-side correlation is consistent not only with

CP892, *Quark Confinement and the Hadron Spectrum VII*
edited by J. E. F. T. Ribeiro
© 2007 American Institute of Physics 978-0-7354-0396-3/07/$23.00

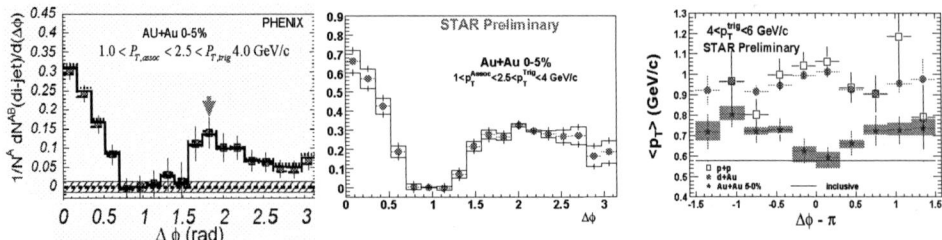

**FIGURE 1.** (color online) Left panel: $\Delta\phi$ correlations in central 5% Au+Au collisions with $1 < p_\perp <$ 2.5 $< p_\perp^{trig}| < 4$ GeV/$c$ from (a) PHENIX with $|\eta^{trig}| < 0.35$ and $|\eta| < 0.35$ [4] and (b) STAR with $|\eta^{trig} < 0.7$ and $|\eta| < 1.0$ [6]. The histograms indicate systematic uncertainties. Right panel: Away-side correlated hadrons $\langle p_\perp \rangle$ versus $\Delta\phi$ for $4 < p_\perp^{trig} < 6$ GeV/$c$ from STAR [6]. Result for $3 < p_\perp^{trig} < 4$ GeV/$c$ is similar. Shaded areas indicate systematic uncertainties for the central Au+Au data.

conical emission, but also with other scenarios including large angle gluon radiation [9] and jet "deflection" due to radial flow or the preferential selection of particles by the pathlength dependent energy loss mechanism [10]. In order to distinguish between conical emission and other mechanisms, 3-particle azimuthal correlations are needed.

## THREE-PARTICLE RESULTS AND DISCUSSIONS

In 2- and 3-particle correlation analyses, a trigger particle at large $p_\perp$ is selected. Given a trigger particle, the event is composed of two parts: one directly correlated with the trigger (the so called "dijet"), and the other not directly correlated (background). The background is *indirectly* correlated with the trigger via the reaction plane (flow correlation). The background is normalized to the 2-particle correlation signal by the common practice of ZYA1 or ZYAM (zero yield at 1 radian or minimum). Two combinatorial backgrounds are present in 3-particle correlation to a trigger particle: pairs of background particles and pairs of a correlated particle and a background particle. In the STAR results shown below, these backgrounds have been subtracted. The details of the 3-particle jet-correlation analysis is described in [11].

Since the backgrounds are large in 3-particle correlations, it is critical to carefully construct the backgrounds. Analysis without careful background construction, for instance the 3-particle cumulant analysis [12] where the background is blindly taken as the constant, average multiplicity density, can result in complex structures that are practically impossible to interpret [13].

Figure 2 shows the 3-particle correlations between a trigger charged particle with $3 < p_\perp^{trig} < 4$ GeV/$c$ and two associated charged particles of $1 < p_\perp < 2$ GeV/$c$ measured by the STAR TPC [14]. The $pp$, d+Au and peripheral 50-80% Au+Au results are similar. Peaks are clearly visible for the near-side, the away-side and the two cases of one particle on the near-side and the other on the away-side. The peak at $(\pi,\pi)$ displays a diagonal elongation, consistent with $k_T$ broadening. The additional broadening in Au+Au may be due to deflected jets. The more central Au+Au collisions display off-diagonal structure, at about $\pi \pm 1.3$ radian, that is consistent with conical emission. The structure increases

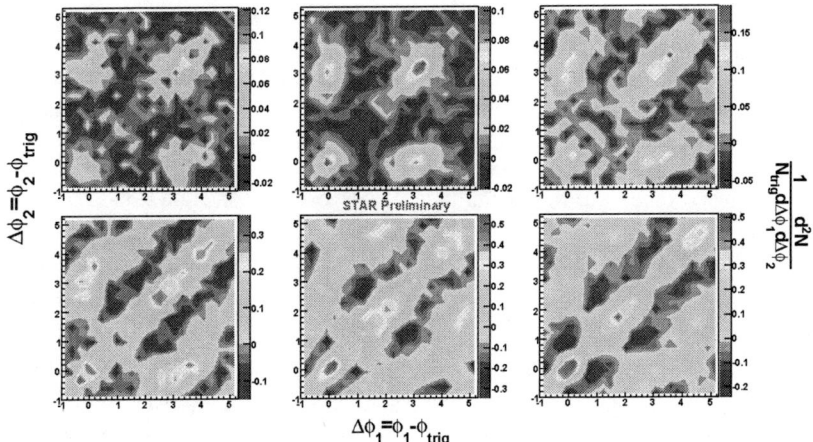

**FIGURE 2.** (color online) Background subtracted 3-particle jet-like azimuthal correlations from STAR for $pp$ (top left), d+Au (top middle), and Au+Au 50-80% (top right), 30-50% (bottom left), 10-30% (bottom center), and ZDC triggered 0-12% (bottom right). The figure is taken from [14].

in magnitude with centrality and is quite clear in the high statistics 12% central data [14].

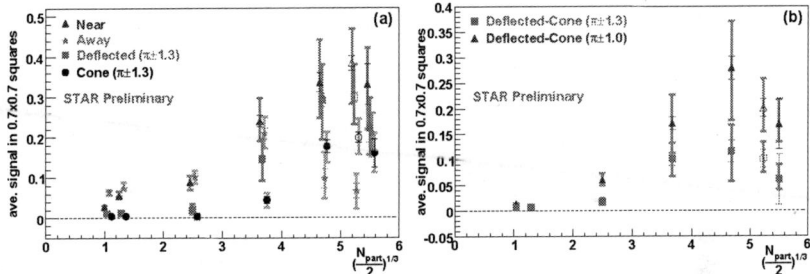

**FIGURE 3.** (color online) (a) Average signals in $0.7 \times 0.7$ boxes at $(0,0)$ (triangle), $(\pi,\pi)$ (star), $(\pi\pm1.3,\pi\pm1.3)$ (square), and $(\pi\pm1.3,\pi\mp1.3)$ (circle). (b) Differences between average signals, between $(\pi\pm1.3,\pi\pm1.3)$ and $(\pi\pm1.3,\pi\mp1.3)$ (square), and between $(\pi\pm1.0,\pi\pm1.0)$ and $(\pi\pm1.0,\pi\mp1.0)$ (triangle). Solid error bars are statistical and shaded are systematic. $N_{part}$ is the number of participants. The ZDC 0-12% points (open symbols) are shifted to the left for clarity. The figure is taken from [14].

Figure 3a shows the centrality dependence of the average signal strengths in different regions [14]. The off-diagonal signals (circle) increase with centrality and significantly deviate from zero in central Au+Au collisions. Figure 3b shows the differences between on-diagonal signals, where both conical emission and deflected jets may contribute, and off-diagonal signals, where only conical emission contributes. Since conical emission signals are of equal magnitude on-diagonal as off-diagonal, the difference may indicate the contribution from deflected jets. The difference decreases with distance from $(\pi,\pi)$.

The measured 3-particle jet-correlation structure in central Au+Au collisions is consistent with conical emission. The discrimination between Mach-cone shock waves and Čerenkov radiation needs future $p_\perp$-dependent studies. If the measured structure is indeed from Mach-cone shock waves, then it is possible to extract the conical emission angle, thereby the speed of sound of the medium created in these collisions. It is highly likely that the system evolves through different stages which the initially produced dijet probes: the partonic stage quark-gluon plasma, the mixed phase, and the hadronic stage. The measured conical emission is likely a net effect of all these stages; the extracted speed of sound is, therefore, an average over the evolution of the medium. The nature (or the equation of state) of the medium, however, needs careful investigations.

## CONCLUSION

Broad, and for some kinematic regions even double-peaked, structures were observed on the away side of the 2-particle jet-correlations. The average transverse momentum of the away-side correlated hadrons is the lowest in the most collimated region. Mach-cone shock waves were proposed to explain the observations, however, other physics mechanisms cannot be ruled out without the knowledge of 3-particle azimuthal correlations. The 3-particle jet-correlation results from STAR are discussed. The central collision data show clear evidence of conical emission; they also indicate the presence of deflected jets. If Mach-cone shock waves are confirmed, further studies should be possible to extract the speed of sound (and the equation of state) of the medium, thereby providing crucial evidence for the creation of the quark-gluon plasma at RHIC.

## ACKNOWLEDGMENTS

The author would like to thank Dr. Yiota Foka for the kind invitation. This work is supported by U.S. DOE under Grants DE-FG02-02ER41219 and DE-FG02-88ER40412.

## REFERENCES

1.  R. Baier, D. Schiff, B.G. Zakharov, Annu. Rev. Nucl. Part. Sci. **50** (2000) 37; X.-N. Wang and M. Gyulassy, Phys. Rev. Lett. **68** (1992) 1480.
2.  J. Adams *et al.* (STAR Collaboration), Nucl. Phys. **A757** (2005) 102; K. Adcox *et al.* (PHENIX Collaboration), Nucl. Phys. **A757** (2005) 184.
3.  J. Adams *et al.* (STAR Collaboration), Phys. Rev. Lett. **95** (2005) 152301.
4.  S.S. Adler *et al.* (PHENIX Collaboration), Phys. Rev. Lett. **97** (2006) 052301.
5.  F. Wang (STAR Collaboration), J. Phys. Conf. Ser. **27** (2005) 32 [nucl-ex/0508021]; F. Wang (STAR Collaboration), Nucl. Phys. **A774** (2006) 129 [nucl-ex/0510068].
6.  J.G. Ulery (STAR Collaboration), Nucl. Phys. **A774** (2006) 581 [nucl-ex/0510055].
7.  H. Stoecker, Nucl. Phys. **A750** (2005) 121; J. Casalderrey-Solana, E. Shuryak and D. Teaney, J. Phys. Conf. Ser. **27** (2005) 23.
8.  I.M. Dremin, Nucl. Phys. **A767** (2006) 233; V. Koch, A. Majumder and X.-N. Wang, Phys. Rev. Lett. **96** (2006) 172302.
9.  I. Vitev, Phys. Lett. B **630** (2005) 78; A.D. Polosa and C.A. Salgado, hep-ph/0607295.
10. R. Hwa, nucl-th/0609017.
11. J.G. Ulery and F. Wang, nucl-ex/0609016.
12. C. Pruneau, nucl-ex/0608002.
13. J.G. Ulery and F. Wang, nucl-ex/0609017.
14. J.G. Ulery (STAR Collaboration), nucl-ex/0609047.

# Multiplicity distributions inside parton cascades developing in a medium

Nicolas Borghini

*Physics Department, Theory Division, CERN, CH-1211 Geneva 23, Switzerland*

**Abstract.** The jet-quenching explanation of the suppressed high-$p_T$ hadron yields at RHIC implies that the multiplicity distributions of particles inside a jet and jet-like particle correlations differ strongly in heavy-ion collisions at RHIC or at the LHC from those observed at $e^+e^-$ or hadron colliders. We present a framework for describing the medium-induced modification, which has a direct interpretation in terms of a probabilistic medium-modified parton cascade, and which treats leading and subleading partons on an equal footing. We show that our approach implies a characteristic distortion of the single inclusive distribution of soft partons inside the jet. We determine, as a function of the jet energy, to what extent the soft fragments within a jet can be measured above some momentum cut.

**Keywords:** Relativistic heavy-ion collisions, jet quenching
**PACS:** 12.38.Mh, 25.75.-q

**Introduction.** Among the most notable results from the first years of running at RHIC stand the deficit in high transverse-momentum hadrons and the suppression of leading back-to-back hadron correlations observed in central Au–Au collisions with respect to expectations from scaling the yields measured in *pp* collisions [1]. These observations are consistent with the "jet-quenching" picture: before they hadronize in the vacuum, partons produced in the dense matter created in head-on Au–Au collisions lose a significant fraction of their energy through an enhanced radiation of soft gluons [2, 3, 4].

Irrespective of the details of the implementation of the medium-enhanced radiation of gluons — either through coherent multiple soft-momentum transfers [5, 6], or through single hard scattering [3] — jet-quenching models of inelastic (radiative) energy loss are quite successful in explaining present light-hadron data from RHIC [7, 8, 9]. However, there remains much room for technical improvement over the existing formulations of inelastic energy loss. Thus, a generic feature of these approaches is that they only consider the medium-induced enhancement in gluon radiation for the leading parton, discarding the medium influence on subleading partons. Such an approximation may remain under control when dealing with leading-hadron production; yet, predictions involving subleading particles become questionable, be it for jet shapes, which may become experimentally accessible at the LHC, or for intrajet two-particle correlations. Similarly, in existing models energy-momentum conservation is not explicitly conserved at each parton splitting, but only globally, through various *ad hoc* corrections.

A novel formulation of medium-induced parton energy loss was recently introduced in Ref. [10], which aims at correcting some of the shortcomings of standard approaches. Thus, it is the first one that deals equally with the various splittings of both leading and subleading partons inside a shower. Furthermore, it automatically conserves energy-momentum at each parton splitting.

CP892, *Quark Confinement and the Hadron Spectrum VII*
edited by J. E. F. T. Ribeiro
© 2007 American Institute of Physics 978-0-7354-0396-3/07/$23.00

**Formalism.** One of the most testing ground of the color structure of QCD is provided by the jets that are created in $e^+e^-$ or in $pp/p\bar{p}$ collisions. The asymptotic shape of the distribution of hadron momenta inside a jet can be computed exactly, especially at small momentum fractions $x = p/E_{\text{jet}}$, by resumming infrared-singular terms to all orders, within the so-called Modified Leading Logarithmic Approximation (MLLA) of QCD [11, 12, 13]. Color coherence thus results in destructive interference between partons, leading to a suppression of small-$x$ hadrons. This amounts, to double and single logarithmic accuracy in $\ln(1/x)$ and $\ln(Q/\Lambda_{\text{eff}})$ — where $Q \sim E_{\text{jet}}$ is the jet virtuality and $\Lambda_{\text{eff}}$ an infrared cutoff which is eventually fitted to experimental data — to an angular ordering of the sequential parton decays within the shower, with leading-order splitting functions. An important prediction of this angular-ordered probabilistic parton cascade is, to next-to-leading order $\sqrt{\alpha_S}$, the characteristic "hump-backed plateau" shape of the distribution of parton momenta inside a jet, represented as a function of $\ln(1/x)$. The parton shower, evolved down to an infrared cutoff $\sim \Lambda_{\text{eff}}$, is eventually identified to a hadron jet, by mapping locally each parton onto a hadron ("Local Parton–Hadron Duality", LPHD): for each hadron type, the hadron distribution equals $K^h$ times the parton distribution, where $K^h$ is a proportionality factor of order unity. This resummation and the LPHD prescription give a good description of the measured longitudinal distributions of hadrons $D^h(x, Q^2)$ over a wide energy range, both in $e^+e^-$ [14, 15] and in $p\bar{p}$ [16] collisions. For instance, Fig. 1 shows $D^h(x, Q^2)$ for inclusive hadrons inside 17.5 GeV jets in $e^+e^-$ annihilations [14], together with the MLLA prediction with $K^h = 1.35$.

The formalism developed in Ref. [10] to describe the medium-induced distortion of jets reduces to the MLLA baseline in the absence of a medium. This new approach involves different approximations from the standard models of parton energy loss that are currently used in the phenomenology of RHIC data. Thus, present model comparisons to RHIC data start with a medium-modified energy spectrum of radiated gluons, $\mathrm{d}I^{\text{tot}} = \mathrm{d}I^{\text{vac}} + \mathrm{d}I^{\text{med}}$ [2, 3, 4]. The part corresponding to the "normal" vacuum radiation shows a double logarithmic dependence $\mathrm{d}I^{\text{vac}} = \frac{\alpha_s}{\pi^2} \frac{\mathrm{d}\omega}{\omega} \frac{\mathrm{d}k}{k^2}$; its integral over $\mathbf{k}$ gives rise to the leading $\ln Q^2$ term in the DGLAP evolution equation. This contrasts to the $\mathbf{k}$-integration of $\mathrm{d}I^{\text{med}}$, which is infrared- and ultraviolet-safe [6] and leads to a nuclear-enhanced "higher-twist" contribution, $\propto \hat{q}L/Q^2$, where $\hat{q}$ is the transport coefficient that characterizes the medium, subleading in an expansion in $1/Q^2$, but enhanced with respect to other such terms by a factor proportional to the geometrical extension $\sim L$ of the target. In practice, however, the parton virtuality does not enter the existing comparisons to experimental data, where one rather considers the $\mathbf{k}$-integrated gluon distribution $\omega \frac{\mathrm{d}I^{\text{med}}}{\mathrm{d}\omega}$, neglecting the $Q^2$-dependence. In addition, existing approximations only include the extra source of gluon radiation $\mathrm{d}I^{\text{med}}$ for the leading parton, dropping it for the further medium-induced splittings of subleading partons in the shower.

The obvious way to improve over this state of the art is to replace the double differential gluon spectrum $\mathrm{d}I^{\text{vac}}$ by $\mathrm{d}I^{\text{tot}}$ in *all* leading and subleading splitting processes of a medium-modified parton cascade. This can only be done within a Monte-Carlo approach, which we intend to develop in future studies. The first, still fully analytical step in that direction consists in using an extra approximation: instead of using the computed $\mathbf{k}$-integrated medium-induced distribution, we replaced it by a constant $f_{\text{med}}$. In the kinematic regime tested at RHIC, this assumption amounts to a similar uncertainty as that

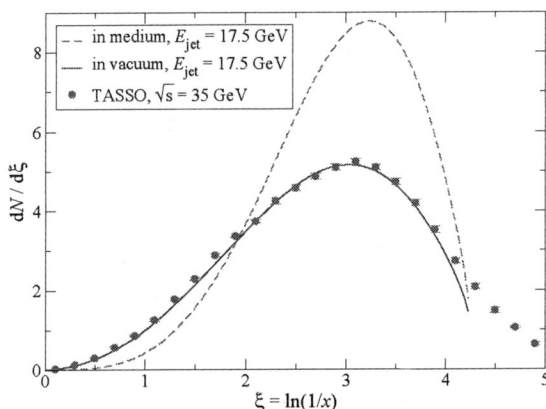

**FIGURE 1.** Longitudinal distribution $dN/d\ln(1/x)$ of inclusive hadrons inside a jet of energy $E_{jet} = 17.5$ GeV, as a function of $\ln(1/x) = \ln(E_{jet}/p)$, as measured by TASSO [14] and within MLLA (solid curve: $f_{med} = 0$; dashed curve: $f_{med} = 0.8$).

arising from whether one should use the multiple-soft scattering approach or the single hard scattering picture. We have then used the medium-induced spectrum $\omega \frac{d I^{med}}{d\omega}$ on the same level as $\omega \frac{d I^{vac}}{d\omega}$, i.e., as a leading logarithmic correction [10]. With this ansatz, our formalism ensures energy-momentum conservation at each parton splitting, and treats all leading and subleading parton splittings on the same footing.

**Phenomenological predictions.** The replacement of the medium-induced contribution to the gluon spectrum $\omega \frac{d I^{med}}{d\omega}$ by a constant $f_{med}$ in the kinematically relevant range of $\omega$ amounts to considering "medium-modified parton splitting functions" that differ from the standard ones by enhancing their singular parts by a factor $(1 + f_{med})$.[1] This formulation allows us to follow the same line of technical arguments as that used for the calculation of jet multiplicity distributions in the absence of a medium [13], and to compute the momentum distribution of partons within a parton cascade. To exemplify the effect of the medium-enhanced gluon radiation on the hump-backed plateau of particle production, we compare in Fig. 1 the longitudinal distribution inside a jet with energy $E_{jet} = 17.5$ GeV in the cases $f_{med} = 0$ (no medium) and $f_{med} = 0.8$ (which allows us to reproduce the light-hadron suppression measured at RHIC [10]). One clearly sees that the effect of the medium is a strong distortion of the distribution, with a depletion of the number of particles at large $x$, and correspondingly a largely enhanced emission of particles at small $x$: due to energy-momentum conservation in the parton cascade, the energy which in the vacuum is taken by a single large-$x$ parton is redistributed over many small-$x$ partons in the presence of a medium.

Once the longitudinal multiplicity distribution inside a jet is known, a straightforward integration yields the number of hadrons inside the jet with transverse momenta larger

---

[1] Such a modification of parton splitting functions was discussed in Ref. [17], where it results from considering nuclear-enhanced twist-four parton matrix elements in studies of deeply inelastic $eA$ scattering.

than a given cut. One can calculate this multiplicity for jets with the same energy both in the presence of medium effects (in which case, the lower cut gives some control on the high-multiplicity soft background over which the jet develops) and in vacuum, and compute their ratio. For jets with $E_{jet} = 17.5$ GeV and medium effects modeled by a constant coefficient $f_{med} = 0.8$, one finds [10] that the ratio is smaller than 1 for $p_T^{cut} \gtrsim 1.5$ GeV/$c$, while the medium-induced enhancement in soft-particle production becomes dominant for smaller values of the transverse-momentum cut. The crossover value is close to that reported by the STAR Collaboration in attempts at measuring the excess of particles inside the back jet over the soft background [18]. Although we did not consider several effects (varying $E_{jet}$ and in-medium path lengths, geometry...) that should be included in a more thorough comparison between our calculation and the STAR data, the reasonable agreement we find is a further hint that energy is indeed redistributed from high- to low-$x$ partons through the influence of the medium. For jets of energy $E_{jet} = 100 - 200$ GeV, which should be accessible at the LHC, the crossover between enhancement and depletion should take place at transverse momenta $p_T^{cut} \sim 4 - 7$ GeV/$c$ [10]. This should leave a window above the upper kinematic boundary of the soft background, in which there is an enhancement of the jet multiplicity, thereby allowing a more detailed characterization of the medium-enhanced radiation.

**Conclusion.** We have reported a first step towards a description of parton cascades developing in a medium, which conserves energy-momentum at each successive parton splitting, and treats all partons in the shower on the same footing [10]. The simplified analytical formalism we have presented, which will serve as a reference for future more realistic Monte-Carlo implementations, is able to reproduce semi-quantitatively several characteristic features of RHIC data, such as the suppression of high-momentum particle yields and the enhanced soft-particle distribution associated to high-$p_T$ trigger particles.

# REFERENCES

1. See P. Jacobs and X.-N. Wang, Prog. Part. Nucl. Phys. **54**, 443 (2005) and references therein.
2. R. Baier, D. Schiff and B. G. Zakharov, Ann. Rev. Nucl. Part. Sci. **50**, 37 (2000).
3. M. Gyulassy, I. Vitev, X.-N. Wang and B. W. Zhang, in *Quark Gluon Plasma III* edited by R. C. Hwa and X.-N. Wang, World Scientific, Singapore, 2004, p. 123.
4. A. Kovner and U. A. Wiedemann, in *Quark Gluon Plasma III* edited by R. C. Hwa and X.-N. Wang, World Scientific, Singapore, 2004, p. 192.
5. R. Baier, Yu. L. Dokshitzer, A. H. Mueller, S. Peigné and D. Schiff, Nucl. Phys. B **484**, 265 (1997).
6. C. A. Salgado and U. A. Wiedemann, Phys. Rev. D **68**, 014008 (2003).
7. S. Turbide, C. Gale, S. Jeon and G. D. Moore, Phys. Rev. C **72**, 014906 (2005).
8. A. Dainese, C. Loizides and G. Paic, Eur. Phys. J. C **38**, 461 (2005).
9. K. J. Eskola, H. Honkanen, C. A. Salgado and U. A. Wiedemann, Nucl. Phys. A **747**, 511 (2005).
10. N. Borghini and U. A. Wiedemann, hep-ph/0506218.
11. A. H. Mueller, Nucl. Phys. B **213** (1983) 85.
12. A. Bassetto, M. Ciafaloni and G. Marchesini, Phys. Rep. **100**, 201 (1983).
13. Yu. L. Dokshitzer, V. A. Khoze and S. I. Troian, Adv. Ser. Direct. High Energy Phys. **5**, 241 (1988).
14. W. Braunschweig *et al.* [TASSO Collaboration], Z. Phys. C **47**, 187 (1990).
15. G. Abbiendi *et al.* [OPAL Collaboration], Eur. Phys. J. C **27**, 467 (2003).
16. D. Acosta *et al.* [CDF Collaboration], Phys. Rev. D **68**, 012003 (2003).
17. X. F. Guo and X. N. Wang, Phys. Rev. Lett. **85**, 3591 (2000).
18. J. Adams *et al.* [STAR Collaboration], Phys. Rev. Lett. **95**, 152301 (2005).

# Probing QCD with photons at the Tevatron

Mikołaj Ćwiok[1]

*University College Dublin, School of Physics, Belfield, Dublin 4, Ireland*

**Abstract.** Prompt photons with high transverse momenta in p$\bar{\text{p}}$ collisions at $\sqrt{s} = 1.96\,\text{TeV}$ at the Tevatron collider have been studied by CDF and DØ collaborations. In this paper I present results on production cross-sections for: isolated photons (DØ), photons with heavy flavor jets (CDF) and di-photon final states (CDF). The experimental data are compared against predictions of several QCD models.

**Keywords:** Perturbative QCD, Direct photons, Jets, Hadronic Collisions, CDF, D0
**PACS:** 12.38.Qk , 13.85.Qk

## MOTIVATION

Studying prompt photon production at high transverse momenta in hadronic collisions allows for precise testing of the next-to-leading order (NLO) and resummed QCD calculations as well as phenomenological models of: gluon radiation, photon isolation and fragmentation processes. Events with energetic isolated photons accompanied by an identified $b$ jet are interesting not only for testing QCD models of $b$ jet production at the Tevatron, but also for new physics studies such as searches for light stop or techniomega particles. The QCD di-photon final states are interesting as a dominant source of background for many discovery channels, including: Higgs boson, large extra dimensions and cascade decays of heavy supersymmetric particles.

## ISOLATED PHOTON PRODUCTION

At the Tevatron production of direct photons via quark-gluon Compton scattering $(qg \to q\gamma)$ dominates over quark-antiquark annihilation $(q\bar{q} \to g\gamma)$ up to $p_T^\gamma$ of $150\,\text{GeV}$. These direct photon rates are overwhelmed by rates due to photons originating from fragmentation of energetic $\pi^0$ and $\eta$ mesons produced inside jets, especially at low $p_T^\gamma$. Such a background can be significantly suppressed by requiring photons in the event to be isolated from other particles. At hight $p_T^\gamma$ isolated electrons from the electroweak production of $W$ and $Z$ bosons also contribute to the background.

DØ has measured the differential cross-section for isolated photons in the range of $23 < p_T^\gamma < 300\,\text{GeV}$ with pseudorapidity[2] $|\eta_\gamma| < 0.9$ using data sample corresponding to integrated luminosity of $326\,\text{pb}^{-1}$ [1]. In DØ detector [2] photon candidates

---

[1] For CDF and DØ collaborations.
[2] Pseudorapidity is defined as $\eta = -\ln\tan\frac{\theta}{2}$, where $\theta$ is the polar angle w.r.t. the proton beam direction.

CP892, *Quark Confinement and the Hadron Spectrum VII*
edited by J. E. F. T. Ribeiro
© 2007 American Institute of Physics 978-0-7354-0396-3/07/\$23.00

are formed from clusters of electromagnetic calorimeter cells inside a cone of radius $R = \sqrt{(\Delta\eta)^2 + (\Delta\phi)^2} = 0.4$ provided that they have no spatially-matched tracks and isolation requirements are satisfied. An artificial neural network is used to further suppress backgrounds as well as to estimate the purity of the resulting photon sample.

The measured differential cross-section as a function of $p_T^\gamma$ corrected for finite transverse momentum resolution of the DØ detector is shown in Fig. 1 (left) along with predictions of the NLO perturbative QCD (pQCD) calculations using JETPHOX [3] with CTEQ6.1M parton distribution functions (PDFs) and BFG [4] fragmentation functions. The ratio of data and theory is depicted in Fig. 1 (right). The NLO pQCD calculations agree with the measurements within uncertainties. The scale dependency is estimated by varying renormalization, factorization and fragmentation scales by a factor of two. The experimental errors are comparable with the scale sensitivity in the whole $p_T^\gamma$ range studied, however, they exceed uncertainties due to proton PDFs, especially at low $p_T^\gamma$ region.

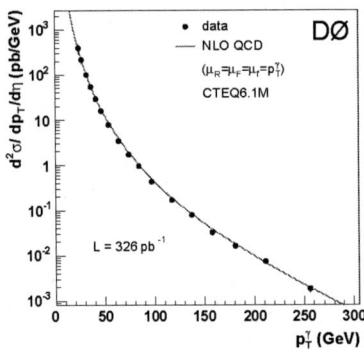 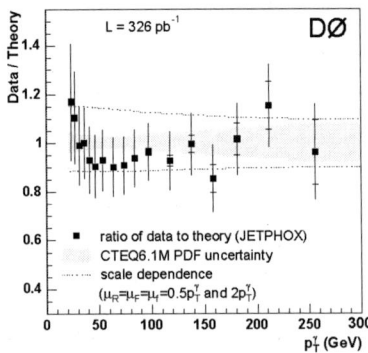

**Figure 1.** Left: the inclusive production cross-section for isolated photons as a function of $p_T^\gamma$ compared with JETPHOX calculations. Right: the ratio of data to theory (full points) along with changes in the cross-section due to: scale variatons (dashed) and CTEQ6.1M PDFs uncertainties (shaded area).

## PHOTON WITH HEAVY FLAVOR JET PRODUCTION

The differential cross-section for events with isolated photon and $b$-tagged jet has been measured by CDF in the range of $26 < E_T^\gamma < 70\,\text{GeV}$ using $340\,\text{pb}^{-1}$ of data [6]. In CDF detector [7] photon candidates are selected based on: the ratio of hadronic to electromagnetic energy deposited, comparison of the shower profile with the profile of electrons measured in a test beam, absence of a spatially-matched track and isolation criteria from adjacent calorimeter cells. Heavy flavor jets are identified using displaced secondary vertex technique. The following kinematical cuts were imposed on $\gamma$ + jet events in this analysis : $E_T^\gamma > 26\,\text{GeV}$, $E_T^{jet} > 20\,\text{GeV}$, $|\eta_\gamma| < 1.1$, $|\eta_{jet}| < 1.5$ and photon candidates being isolated from tagged jets by $\Delta R > 0.7$.

The invariant mass distribution of the tagged secondary vertex for the full event sample is shown in Fig. 2 (left) along with fitted Monte Carlo templates for $b$, $c$ and light quark contributions. The event sample composition resulting from this fit is: $39 \pm 2\%$,

$39 \pm 3\%$ and $22 \pm 3\%$ for $b$, $c$ and light jets, respectively. The cross-section for $\gamma + b$ jet production is shown in Fig. 2 (right) together with the LO predictions of PYTHIA (Tune A) [8] and HERWIG [9] generators using CTEQ5L PDFs. Both models agree with data, although statistical and systematical errors are large at low $E_T^\gamma$.

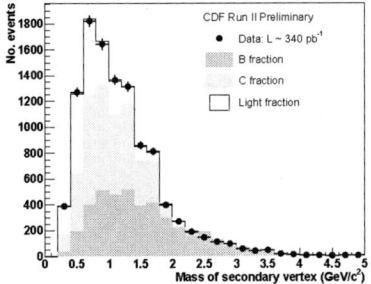
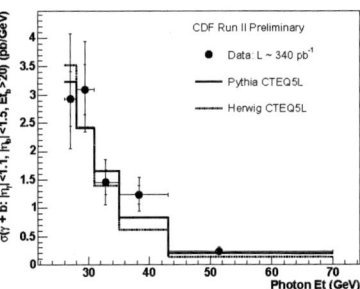

**Figure 2.** Left: the secondary vertex mass distribution for the full event sample for data (full dots) and MC template fits to the data for: $b$, $c$ and light quark contributions. Right: the $\gamma + b$ production cross-section as a function of $E_T^\gamma$ for: data (full dots) and the LO predictions of PYTHIA (solid) and HERWIG (dashed).

## DI-PHOTON PRODUCTION

The leading pQCD contributions to di-photon production at the Tevatron are from: quark-antiquark annihilation ($q\bar{q} \to \gamma\gamma$) and gluon-gluon scattering ($gg \to \gamma\gamma$). Although the latter subprocess is suppressed by a factor of $\alpha_s^2$ since final photons couple to initial gluons via a quark box, it becomes important at kinematic regions where gluon densities are high, such as at low invariant masses of the $\gamma\gamma$ system ($M_{\gamma\gamma}$). The inclusive di-photon cross section also receives contributions from events where one or two photons originated from fragmentation of neutral mesons inside jets. In addition, the total transverse momentum of the $\gamma\gamma$ system ($q_T$) is sensitive to initial state soft gluon radiation.

CDF has measured the di-photon production cross-section with respect to three kinematic variables: $q_T$, $M_{\gamma\gamma}$ and azimuthal angle between the two photons ($\Delta\phi_{\gamma\gamma}$) using $207\,\mathrm{pb}^{-1}$ data sample [10]. Events selected for this analysis were required to have: $E_T^{\gamma_1} > 13\,\mathrm{GeV}$, $E_T^{\gamma_2} > 14\,\mathrm{GeV}$ and both photons in the central pseudorapidity region of $|\eta| < 0.9$.

In Fig. 3 the resulting cross-sections as a function of $\Delta\phi_{\gamma\gamma}$ (left) and $q_T$ (right) are compared against predictions of: DIPHOX [3, 11] (direct photons at NLO, gluon-gluon contribution included, fragmentation at NLO), RESBOS [12] (direct photons at NLO, fragmentation at LO, initial state soft gluon resummation) and PYTHIA [8] (all diagrams at LO). The latter calculation has to be increased by a factor of 2 to agree with the total measured cross section. At low $q_T$ and at $\Delta\phi_{\gamma\gamma} > \frac{\pi}{2}$ regions RESBOS describes data better than other models thanks to soft gluon resummations taken into account. By contrast, only DIPHOX reproduces shoulders observed in the data at large $q_T$ and at $\Delta\phi_{\gamma\gamma} < \frac{\pi}{2}$ regions where the NLO fragmentation processes contribute significantly.

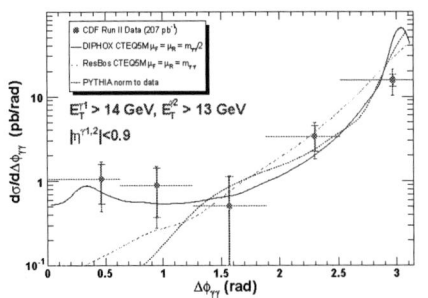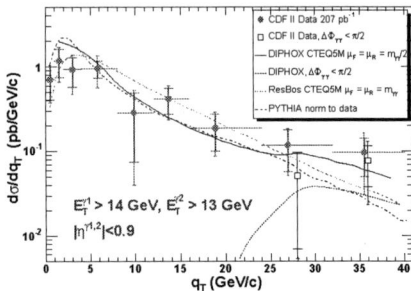

**Figure 3.** The differential cross-sections with respect to $\Delta\phi_{\gamma\gamma}$ (left) and $q_T$ (right) from data (full dots) and calculated by: DIPHOX (solid), RESBOS (dashed) and PYTHIA (dotted, scaled by a factor of 2). At large $q_T$ are also shown: data (open squares) and DIPHOX prediction (dot-dashed) for $\gamma\gamma$ configurations having $\Delta\phi_{\gamma\gamma} < \frac{\pi}{2}$.

# CONCLUSIONS

The DØ result on prompt isolated photon cross-section is well described by the NLO prediction of JETPHOX model within experimental errors and present theoretical uncertainties. The measurement of $\gamma + b$ jet cross-section by CDF is consistent with the LO predictions of PYTHIA and HERWIG generators. CDF results on inclusive di-photon production are well described by resummed and NLO pQCD predictions in different regions of the phase space. However, neither DIPHOX nor RESBOS model alone can reproduce data in every critical region of the phase space.

In the near future updated CDF and DØ analyzes based on full Run IIa statistics ($\sim 1\,\text{fb}^{-1}$ per experiment) should provide better quality data for more precise testing of QCD photon production models.

# ACKNOWLEDGMENTS

I would like to thank CDF and DØ collaborators for providing recent results of the analyzes. I thank the staffs at Fermilab and collaborating institutions and acknowledge support from SFI (Ireland).

# REFERENCES

1. V. M. Abazov et al. (DØ Collaboration), *Phys. Letters* **B639**, 151 (2006).
2. V. M. Abazov et al. (DØ Collaboration), *Nucl. Instr. Methods* **A565**, 463 (2006).
3. T. Binoth et al., *Eur. Phys. Journal* **C16**, 311 (2000).
4. L. Bourhis, M. Fontannaz and J. P. Guillet, *Eur. Phys. Journal* **C2**, 529 (1998).
5. S. Catani et al., *JHEP* **05**, 028 (2002).
6. D. Acosta et al. (CDF Collaboration), CDF NOTE 8377 (July 17, 2006).
7. D. Acosta et al. (CDF Collaboration), *Phys. Review* **D71**, 032001 (2005).
8. T. Sjöstrand et al., *Comput. Phys. Commun.* **135**, 238 (2001).
9. G. Corcella et al., *JHEP* **01**, 010 (2001).
10. D. Acosta et al. (CDF Collaboration), *Phys. Rev. Letters* **95**, 022003 (2005).
11. Z. Bern, L. J. Dixon and C. Schmidt, *Nucl. Phys. B Proc. Suppl.* **116**, 178 (2003).
12. C. Balazs et al., *Phys. Review* **D57**, 6934 (1998).

# Probing QCD with jets at the Tevatron

## Mikołaj Ćwiok[1]

*University College Dublin, School of Physics, Belfield, Dublin 4, Ireland*

**Abstract.** The selected results on QCD jets measured in $p\bar{p}$ collisions at $\sqrt{s} = 1.96\,$TeV at the Tevatron Collider at Fermilab are presented in this conference note. The experimental data from CDF and DØ detectors are compared against perturbative QCD calculations.

**Keywords:** QCD, Jets, Hadronic Collisions, PDF, CDF, D0
**PACS:** 12.38.Qk , 13.85.Ni , 13.87.Ce

## INTRODUCTION

In Run II of the Fermilab $p\bar{p}$ Tevatron Collider the center of mass energy was increased to 1.96 TeV together with the instantaneous luminosity. Nowadays, the production of particle jets at high transverse momenta ($p_T^{jet}$) with respect to the beam can be probed beyond 600 GeV/c. This allows one to verify perturbative QCD (pQCD) calculations over 8 orders of magnitude in the cross section. The inclusive jet rates are also sensitive to the non-perturbative structure of the proton as parameterized in the parton distribution functions (PDFs). In addition, the angular jet distributions are affected by QCD radiative effects. Measurements of the departure of the two leading jets in the event from the back to back topology provide a test of predicted soft- and hard radiation components without the need to reconstruct additional jets. Furthermore, the measurement of inclusive heavy flavor jet cross sections provide an important quantitative test of the next-to-leading order (NLO) pQCD calculations.

## INCLUSIVE JET PRODUCTION

Recently the CDF and DØ collaborations have measured the inclusive jet cross sections using data samples corresponding to the integrated luminosity of $\sim 1\,\mathrm{fb}^{-1}$ [1]. Both CDF [2] and DØ [3] experiments use their calorimeters as a primary tool to identify jets. Thus, the systematic uncertainty of measured cross sections is dominated by precision of the jet energy scale calibration. The analyzes employ infrared safe jet finding algorithms, such as $k_T$ and *midpoint cone*, which are described in details elsewhere [4].

CDF measured the inclusive cross sections for jets with rapidity[2] $|y_{jet}| < 2.1$ and transverse momentum in the range $54 < p_T^{jet} < 700\,$GeV/c based on 0.98 fb$^{-1}$ of data.

---

[1] For CDF and DØ collaborations.
[2] The rapidity $y$ is defined as $y = -\frac{1}{2}\ln\frac{E+p_z}{E-p_z}$ where $E$ and $p_z$ denote the energy and the momentum component along the proton beam direction, respectively.

CP892, *Quark Confinement and the Hadron Spectrum VII*
edited by J. E. F. T. Ribeiro
© 2007 American Institute of Physics 978-0-7354-0396-3/07/$23.00

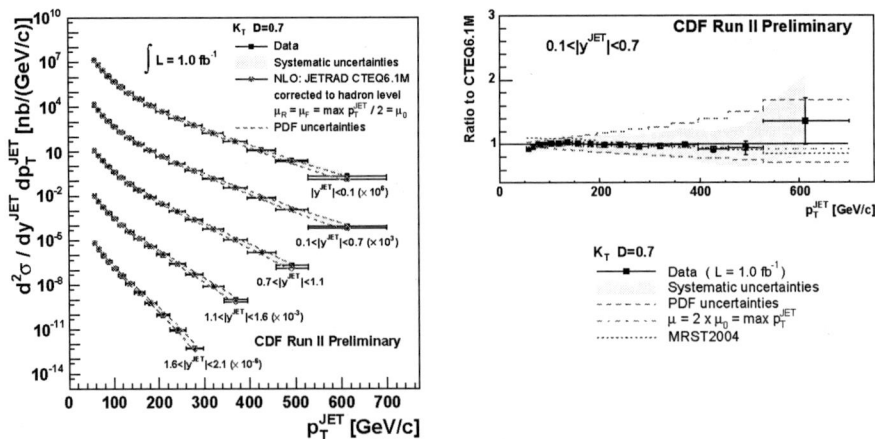

**Figure 1.** Left: Measured inclusive jet cross sections using the $k_T$ algorithm (filled squares) compared with the NLO pQCD predictions (open circles). Right: Ratio of data and theory for rapidity region $0.1 < |y_{jet}| < 0.7$.

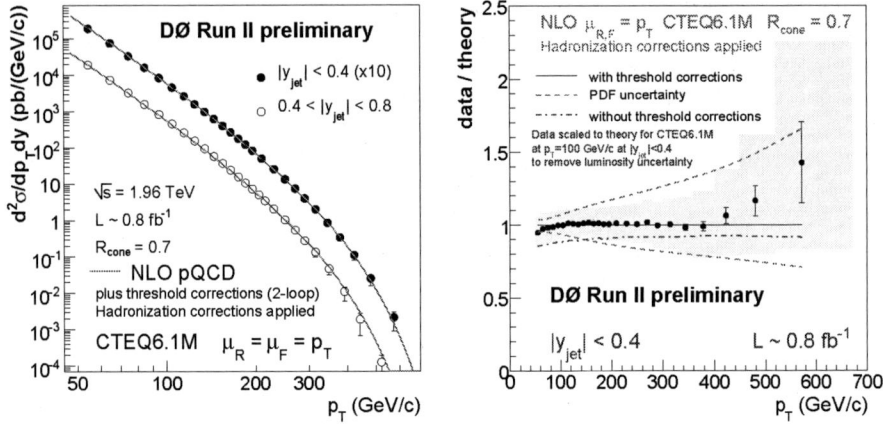

**Figure 2.** Left: Measured inclusive jet cross sections using the midpoint cone algorithm (filled and open circles) compared with the NLO pQCD predictions (solid lines). Right: Ratio of data and theory for central rapidity region $|y_{jet}| < 0.4$.

In Fig. 1 (left) the results for jets reconstructed using $k_T$ algorithm and corrected to the hadron level are shown for 5 rapidity bins (filled squares). The data is in good agreement with the NLO pQCD predictions (open circles) computed using JETRAD [5]. The results from CDF based on the midpoint cone algorithm agree with the rates predicted by the NLO calculations using EKS [6] program, as well.

DØ measured inclusive jet cross sections using midpoint cone algorithm and $0.8\,\mathrm{fb}^{-1}$

430

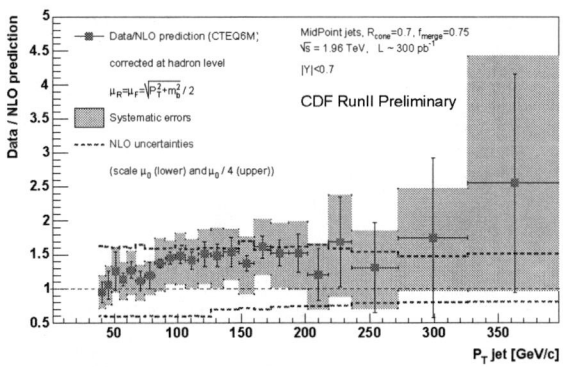

**Figure 3.** Ratio of the measured and theoretical inclusive $b$-jet cross sections.

of data for 2 central rapidity bins up to $|y_{jet}| = 0.8$ and for $p_T^{jet}$ range similar to that of CDF. The resulting cross sections corrected to the hadron level and scaled to theory at $p_T^{jet} = 100\,\text{GeV/c}$ for $y_{jet} < 0.4$ are shown in Fig. 2 (left). Theory predictions (solid lines) combine the NLO calculations obtained with NLOJET++ [7] with 2-loop threshold corrections [8] what reduced uncertainties due to the choice of renormalization and factorization scales. The observed $p_T^{jet}$ spectra are well described by theory.

The ratios of measured and theoretical cross sections are depicted in Figs. 1 (right) and 2 (right) for CDF and DØ analyzes, respectively. It can be seen that experimental systematic errors (shaded areas) are comparable with uncertainties due to CTEQ6.1M [9] PDFs (dashed bands). Therefore, the latest results from the Tevatron can help to further constrain the gluon content in the proton since most of the PDF uncertainty at high $p_T^{jet}$ region comes from the gluon density function at large fractional momentum.

## INCLUSIVE B-JET PRODUCTION

The inclusive $b$-jet cross section has been measured by CDF for jets with transverse momentum in the range $38 < p_T^{jet} < 400\,\text{GeV/c}$ and rapidity $|y_{jet}| < 0.7$ using data sample corresponding to $300\,\text{pb}^{-1}$ [10]. Jets were reconstructed using the midpoint cone algorithm and their momenta were corrected for detector effects back to the hadron level. Heavy flavor jets were tagged via a displaced secondary vertex technique. The measured secondary vertex mass distribution was fitted to a linear combination of $b$ and non-$b$ templates generated by PYTHIA [11] (ver. 6.203, Tune A) in order to extract flavor composition of the final event sample.

The ratio of data to the NLO pQCD calculations [12] is shown in Fig. 3. Systematic errors (shaded area) include contributions from: $p_T^{jet}$ resolution, energy scale, flavor composition, $b$-tagging efficiency, luminosity, jet algorithm and CTEQ6M PDFs. Theoretical uncertainties (dashed lines) are estimated by varying renormalization ($\mu_R$) and factorization ($\mu_F$) scales by a factor of 2. The NLO calculations agree with data in the whole $p_T^{jet}$ range, although their dependence on the choice of $\mu_R$ and $\mu_F$ is strong.

# AZIMUTHAL DECORRELATIONS

In $p\bar{p}$ collisions the distribution of the azimuthal angle[3] between the two leading $p_T$ jets ($\Delta\phi_{dijet}$) is a single convenient observable for testing QCD radiation effects. In the vicinity of $\Delta\phi_{dijet} \approx \pi$ such a distribution is strongly affected by soft radiation effects. On the contrary, hard parton emissions result in significantly larger deviations from the back to back topology. Namely, exclusive 3-jet events populate the region of $\frac{2}{3}\pi < \Delta\phi_{dijet} < \pi$, while events with at least 4 jets populate the region of $\Delta\phi_{dijet} < \frac{2}{3}\pi$.

DØ has measured inclusive dijet production rates as a functions of $\Delta\phi_{dijet}$ and the largest transverse momentum of the two leading jets in the event ($p_T^{max}$) using $150\,\mathrm{pb}^{-1}$ of data [13]. Events were required to have: $p_T^{max} > 75\,\mathrm{GeV/c}$, the second leading $p_T$ jet with $p_T^{jet} > 40\,\mathrm{GeV/c}$ and rapidity of each jet $|y_{jet}| < 0.5$. Jets were defined using midpoint cone algorithm.

The resulting $\Delta\phi_{dijet}$ distributions are shown in Fig. 4 (left) in 4 different $p_T^{max}$ ranges. It can be seen that decorrelations increase towards smaller $p_T$ values. Predictions of HERWIG [14] (ver. 6.505) and PYTHIA (ver. 6.225) generators which use $2 \rightarrow 2$ leading order pQCD matrix elements and CTEQ6L [15] PDFs are also superimposed

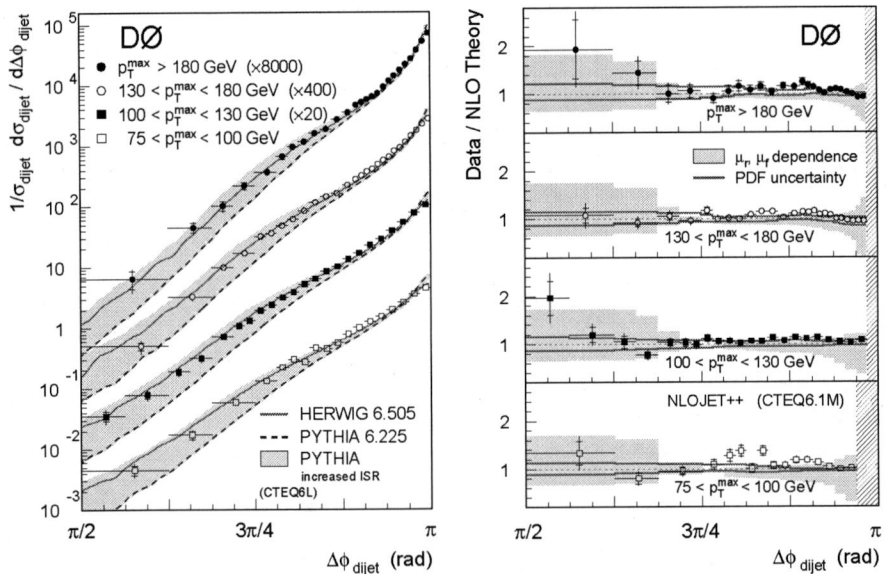

**Figure 4.** Left: Measured distributions of $\Delta\phi_{dijet}$ in different $p_T^{max}$ ranges along with predictions of: HERWIG (solid lines), default PYTHIA (dashed lines) and tuned PYTHIA (shaded areas). Right: Ratios of data to the NLO pQCD calculations in different $p_T^{max}$ ranges.

---

[3] The azimuthal angle $\phi$ is defined on a plane transverse to the beam axis.

on the figure. While HERWIG (solid lines) correctly describes data over the whole phase space studied, the PYTHIA predictions (dashed lines) are too narrowly peaked at $\Delta\phi_{dijet} \approx \pi$. However, a fourfold increase of the maximum allowed virtuality in the initial-state parton shower in PYTHIA resulted in a reasonable agreement with the data (shaded bands).

The ratios of data to the NLO pQCD calculations using the parton-level generator NLOJET++ with CTEQ6.1M PDFs are shown in Fig. 4 (right) in different $p_T^{max}$ regions. Such ratios are insensitive to hadronization and underlaying event corrections. The NLO pQCD predictions describe data with accuracy of 5-10 %, however at large $\Delta\phi_{dijet}$ angles calculations diverge and those regions were excluded from the plots (dashed areas).

## SUMMARY

Results from the Tevatron Collider at Fermilab on: inclusive jet and inclusive $b$-jet cross sections as well as on dijet azimuthal decorrelations have been presented in this note. The NLO pQCD calculations are in good agreement with the experimental data over wide region of jet transverse momenta and rapidities. Recent measurements of the inclusive jet production can contribute to a better understanding of the PDFs of the proton.

## ACKNOWLEDGMENTS

I would like to thank CDF and DØ collaborators for providing recent results of the analyzes. I thank the staffs at Fermilab and collaborating institutions and acknowledge support from SFI (Ireland).

## REFERENCES

1. A. Kupčo, *Inclusive jet production from the Tevatron*, to appear in the Proceedings of the XXXIIIrd International Conference on High Energy Physics, Moscov, Russia (2006).
2. F. Abe et al. (CDF Collaboration), *Nucl. Instr. Methods* **A271**, 387 (1988).
   R. Blair et al. (CDF Collaboration), FERMILAB-PUB-96/390-E.
   D. Acosta et al. (CDF Collaboration), *Phys. Review* **D71**, 032001 (2005).
3. S. Abachi et al. (DØ Collaboration), *Nucl. Instr. Methods* **A338**, 185 (1994).
   V. M. Abazov et al. (DØ Collaboration), *Nucl. Instr. Methods* **A565**, 463 (2006).
4. G. C. Blazey et al., hep-ex/0005012.
5. W. T. Giele, E. W. N. Glover and D. A. Kosower, *Nucl. Physics* **B403**, 633 (1993).
6. S. D. Ellis, Z. Kunszt and D. E. Soper, *Phys. Rev. Letters* **64**, 2121 (1990).
7. Z. Nagy, *Phys. Rev. Letters* **88**, 122003 (2002).
   Z. Nagy, *Phys. Review* **D68**, 094002 (2003).
8. N. Kidonakis and J. F. Owens, *Phys. Review* **D63**, 054019 (2001).
9. D. Stump et al., *JHEP* **10**, 046 (2003).
10. A. Abulencia et al. (CDF Collaboration), CDF NOTE 8418 (July 25, 2006).
11. T. Sjöstrand et al., *Comput. Phys. Commun.* **135**, 238 (2001).
12. S. Frixione and M. Mangano, *Nucl. Physics* **B483**, 321 (1997).
13. V. M. Abazov et al. (DØ Collaboration), *Phys. Rev. Letters* **94**, 221801 (2005).
14. G. Corcella et al., *JHEP* **01**, 010 (2001).
15. J. Pumplin et al., *JHEP* **07**, 012 (2002).

# Standard Model Physics at the Tevatron

## Tommaso Dorigo

*Dipartimento di Fisica "G.Galilei", Via Marzolo 8, 35131 Padova, Italy*

**Abstract.** The CDF and DØ collaborations at the Tevatron have been producing exquisite precision measurements on high-$P_T$ physics with their large datasets of $p\bar{p}$ collisions. The Higgs boson is being sought in all available channels, and the Tevatron experiments will have a chance to discover it before LHC starts operating. The top quark is being studied in great detail, and a precision of 1.2% in the measurement of its mass has been achieved. In this brief report I will provide an overview of the most interesting recent results and hot topics.

**Keywords:** Standard Model, Higgs boson, Top quark, hadron collisions
**PACS:** 12.15.-y, 14.80.Bn, 14.65.Ha

## THE FACILITIES

The Tevatron accelerator has been subjected at the turn of the millennium to a massive upgrade, with the construction of an entirely new ring, the Main Injector, and several improvements in the facility producing and storing antiprotons – the most challenging part of the whole project. The collider has recently surpassed the peak luminosity of $2.3 \times 10^{32} cm^{-2} s^{-1}$. An integrated luminosity of $2.0 fb^{-1}$ has been delivered to CDF and DØ at the time of writing, and 5 to 8 inverse femtobarns are expected by end of 2009. Up-to-date information on the performance of the machine can be found in [1].

An overview of the CDF and DØ detectors for Run II at the Tevatron can be found in [2]. In what follows their most important features for high-$P_T$ physics are briefly mentioned.

Both detectors are all-purpose, near-hermetic devices consisting of a tracker immersed in a solenoidal field and an outer shell of calorimeters and muon chambers. In DØ an excellent system of silicon microstrip detectors has been installed in Run II. Six barrels of silicon sensors organized in four concentrical layers provide coverage for central tracks, while a total of sixteen silicon disks allow reconstruction of large rapidity tracks. A similar set of seven barrels of silicon strips is organized in the core of CDF.

Outside of the silicon barrels CDF features a large gas tracking chamber, and DØ has a compact scintillating fiber tracker. Associated hits in the silicon strips allow the determination of track impact parameter with accuracy sufficient to reconstruct $B$-hadron decay and enable $\sim 45\%$ efficient tagging of $b$-quark originated jets with fake rates well below 1%. Calorimeters are finely segmented in projective towers and are divided in a inner electromagnetic and an outer hadronic section. Electrons within the pseudorapidity interval $|\eta| < 2.0$ are identified with high purity and efficiency, with a resolution of about $15\%/\sqrt{E}$ in both detectors, and hadronic jets are reconstructed with resolutions better than $100\%/\sqrt{E}$. Muon chambers cover the rapidity region $|\eta| < 1.5$ in CDF and $|\eta| < 2.0$ in DØ.

CP892, *Quark Confinement and the Hadron Spectrum VII*
edited by J. E. F. T. Ribeiro
© 2007 American Institute of Physics 978-0-7354-0396-3/07/$23.00

Both detectors have a sophisticated trigger system that reduces the 2.5MHz collision rate to about 100 Hz of events written to tape. Of particular relevance for the ongoing Higgs boson search in CDF is the Silicon Vertex Tracker (SVT), a device that provides precise online tracking. The SVT identifies track candidates by comparing hit patterns to a predetermined array of possible roads stored in associative memory banks. A linearized $R - \phi$ fit of track hits in the silicon layers yields track momentum and impact parameter with precision close to that attainable offline in less than $10\mu s$, thanks to a highly parallelized architecture. This allows the collection of datasets based on the presence of $b$-quarks in the final state, enhancing the $B$-physics program of CDF but also providing higher efficiency for several Higgs boson signatures.

## SEARCHES FOR THE STANDARD MODEL HIGGS BOSON

The search for the Higgs boson at the Tevatron is carried out by looking for its two main decay signatures, depending on the particle mass: if $M_H < 135$ GeV the dominant decay is $H \rightarrow bb$, while at higher masses the $H \rightarrow WW$ decay provides the most promising signature.

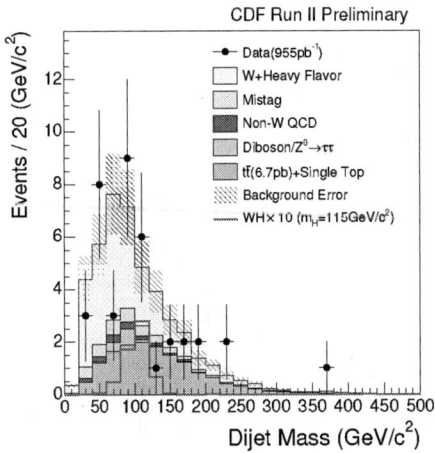 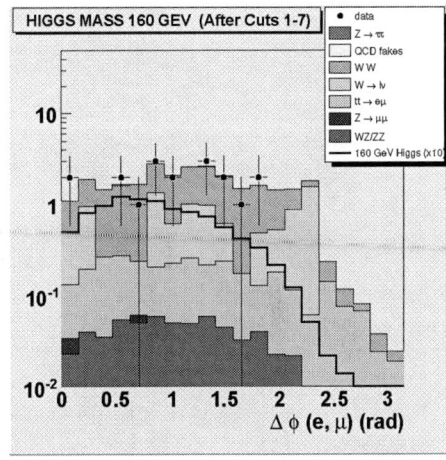

**FIGURE 1.** *Left: dijet mass distribution of b-tagged jets in WH candidate events collected by CDF in $955pb^{-1}$ of run 2 data (black points) compared to background expectations. The red line shows the expected Higgs signal multiplied ten times for clarity. Right: azimuthal angle between the two charged lepton candidates in the dielectron final state of the DØ $H \rightarrow WW$ analysis.*

Both CDF and DØ search their datasets for $WH$ and $ZH$ associated production with a low mass Higgs boson decaying to a pair of $b$-quark jets, while the vector boson is tagged by the reconstruction of two charged leptons (for $Z \rightarrow ee$ or $Z \rightarrow \mu\mu$ decays), a lepton and missing transverse energy (to select $W \rightarrow ev$ or $W \rightarrow \mu\nu$), or missing $E_T$ alone (for $Z \rightarrow \nu\nu$). The two critical parameters affecting signal significance are the efficiency of $b$-quark jet identification -performed through the reconstruction of a secondary vertex in

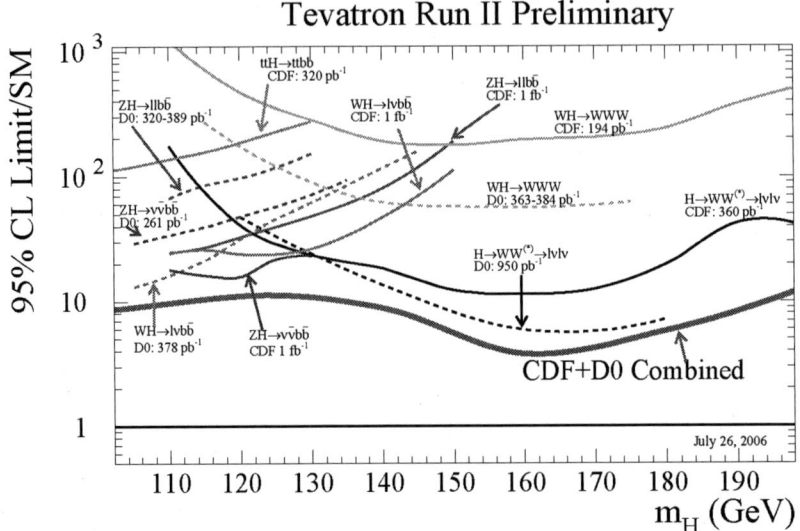

**FIGURE 2.** *Summary of 95% C.L. limits on the ratio of Higgs boson cross section divided by standard model prediction, for all studied search channels. Individual limit curves obtained by the different analyses are shown together with the combined limit of all displayed CDF and DØ searches.*

the jet- and the resolution of a reconstructed dijet mass peak. The dijet mass distribution is studied in search for an excess over backgrounds, which are mainly due to real vector boson production associated with jets from QCD radiation and top quark production (see Fig. 1, left).

If the Higgs mass is higher than 135 GeV, its $WW$ decay becomes dominant. In that case, both direct production and associated production of a Higgs and an electroweak boson -yielding three vector bosons in the final state- are promising search channels.

In direct $H \rightarrow WW$ searches, when both $W$ bosons decay to an electron-neutrino or muon-neutrino pair the final state is quite clean, with reducible backgrounds mostly due to Drell-Yan production of lepton pairs. To discriminate direct production of a Higgs boson from non-resonant $WW$ production - which in the standard model has a sizeable cross section [3] – it is useful to study the azimuthal angle $\Delta\Phi_{ll}$ between the two charged leptons (Fig. 1, right), since the zero spin of the Higgs boson and helicity conservation conspire to produce leptons in the same direction in the transverse plane.

The sum of all known processes accounts nicely for the number of events found in each search channel, and 95% limits are set to the production cross section of the Higgs boson as a function of its mass. Figure 2 summarizes the present status of Higgs boson searches at the Tevatron. It is necessary to note that most searches are based on relatively small amounts of data, with larger datasets still awaiting to be analyzed. The standard model prediction for Higgs production appears still far away: nevertheless, results are roughly in line with what was predicted by the 2003 Higgs sensitivity study [4]. That

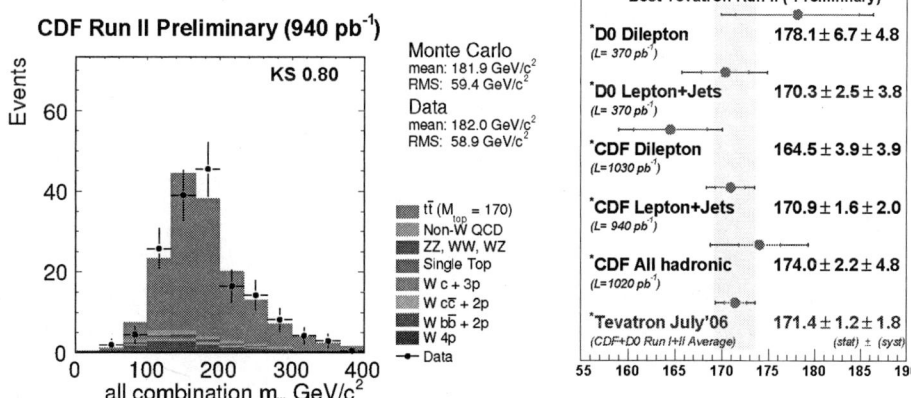

**FIGURE 3.** Left: Reconstructed top quark mass in the CDF lepton+jets matrix element analysis. Right: summary of the best results on the top mass obtained by the CDF and DØ collaborations. The combined result is in blue.

study foresees that by the end of 2009 a light Higgs boson is likely to be discovered at the Tevatron, or excluded for all masses below 135 GeV. In order to reach that goal, a combination of all results by CDF and DØ is mandatory. The two experiments are collaborating fruitfully in these searches, and new limits will be produced periodically as more data are analyzed.

## TOP QUARK MEASUREMENTS

The large datasets of $p\bar{p}$ collisions collected by CDF and DØ allow for many precision measurements of top quark properties. Top quarks are particularly interesting as a laboratory of perturbative QCD, because of their large mass and short lifetime. The most interesting measurement is the one of the top mass, which is a fundamental parameter of the standard model. The top quark mass $M_{top}$ has a large impact through radiative corrections in the global fits to electroweak observables attempting to verify the internal consistency of the model and predict the unknown mass of the Higgs.

There are by now tens of determinations of $M_{top}$, using different techniques and final states of top quark pair production. The single most precise measurement has been obtained by CDF by reconstructing the top mass in the single lepton final state of top pair production from $940 pb^{-1}$ of collisions, resulting in 166 events containing a high-$P_T$ lepton, missing transverse energy, and four jets, at least one of which originated by $b$-quark hadronization. A likelihood is calculated for each event using the matrix element for leading order top pair production and a parametrization of parton showering yielding the observed jets. The final measured top mass is then extracted from a joint likelihood of the product of the individual event likelihoods, where the jet energy scale uncertainty

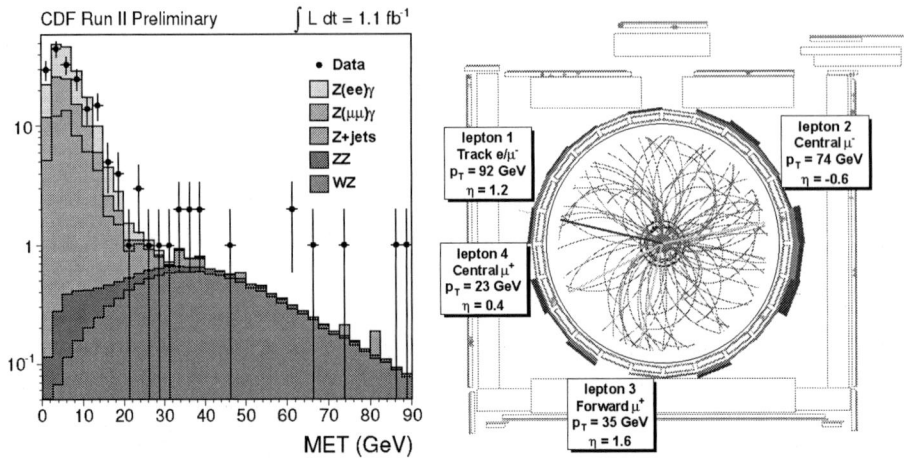

**FIGURE 4.** Left: The signal of $WZ$ production discovered by CDF in the tail of the missing transverse energy distribution. Right: a candidate for $ZZ \to \mu\mu\mu\mu$ production found by CDF.

is convoluted with the statistical error using an *in-situ* measurement of the hadronic $W$ boson mass. The use of the $W$ mass as a calibration point allows to reduce the dominant source of systematics, and the final measurement is $M_{top} = 170.9 \pm 2.2 \pm 1.4 GeV$, where the first error is the statistical plus jet energy scale uncertainty, and the second accounts for all other systematics. Fig. 3 (left) shows the reconstructed top quark mass in the selected sample.

The measurement just discussed, along with selected others, has been combined into a world average of $M_{top} = 171.4 \pm 2.1 GeV$ (see Fig. 3, right). Before the end of Run 2 the two experiments are likely to obtain a 1 GeV accuracy on the top quark mass.

## PRECISION ELECTROWEAK PHYSICS AT THE TEVATRON

In the remainder of this paper it is only possible to mention one electroweak physics result among the dozens of new measurements: namely, the recent discovery of production of pairs of $WZ$ bosons, a rare process of high relevance for Higgs searches. CDF obtained 16 $WZ$ candidates by an optimized search for triplets of charged leptons and missing transverse energy in $1.1 fb^{-1}$ of data, where only 2.7 events were expected from background processes. Figure 4 displays the distribution of the missing transverse energy in the events before a cut on that variable selects the final candidates: the signal of $WZ$ production emerges clearly at large missing $E_T$. Now one more process of that kind is still missing, associated $ZZ$ production. CDF did observe one event with the required characteristics, but it will take some more data to claim definitive observation of that process as well. An event display of the $ZZ$ candidate is shown in Fig. 4 (right).

# CONCLUSIONS

The CDF and DØ experiments are producing remarkable high-$P_T$ physics results with the large datasets of $p\bar{p}$ collisions they collected so far during Run 2. If the facilities continue to perform as expected, there is a chance that the Tevatron beat the LHC in the quest for the Higgs boson. One less ambitious and more certain target is reaching a precision in the top quark mass which will remain the most precise measurement for many years to come. CDF and DØ look forwards to the last few years of running with a lot of enthusiasm for the forthcoming challenges.

# ACKNOWLEDGMENTS

I wish to thank Helge Krueger for his editorial advice.

# REFERENCES

1. See http://www.fnal.gov/pub/now/index.html
2. T. LeCompte and H.T. Diehl, *The CDF and DØ upgrades for Run II*, Ann. Rev. Nucl. Part. Sci. 50 (2000), 71.
3. The latest measurement by CDF is $\sigma_{WW} = 13.6 \pm 2.8_{stat.} \pm 1.6_{syst.} \pm 1.2_{lum.} pb$, see hep-ex/0605066.
4. CDF and DØ collaborations, *Results of the Tevatron Higgs Sensitivity Study*, FERMILAB-PUB-03/320-E.

# Colored SUSY and R-hadron Physics in Atlas

Rasmus Mackeprang

*Niels Bohr Institute, Blegdamsvej 17, 2100 Copenhagen E, Denmark*

**Abstract.** We summarize the strategy for searching and understanding for the R- parity conserving SUSY particles at the ATLAS detector at LHC. Results are focused on the mass reconstruction of SUSY particles, especially for the colored SUSY particles, which is critical for understanding SUSY-QCD physics. Also the ATLAS detector has good potential to discover R-hadrons, and the search strategy for this will be summarized.

**Keywords:** Supersymmetry,R-hadrons,Atlas
**PACS:** 12.39.-x,12.60.Jv,13.75.-n,13.85.-t

## THE PHENOMENOLOGY OF A SUSY FOCUS POINT

The search for SUSY in Atlas is focused on a specific set of points and regions in the SUSY parameter space. These points reflect different types of SUSY phenomenologies. One of these points, SPS 1a [1] is an mSUGRA point that has been studied in Atlas [2]. It is characterized by a mass hierarchy that satisfies $m_{\tilde{l}} > m_{\tilde{\chi}_2^0} > m_{\tilde{\chi}_1^0}$ and in terms of SUSY parameters it is defined as:

$$m_0 = -A_0 = 0.4m_{1/2}, \qquad \tan\beta = 10, \qquad \mu > 0. \tag{1}$$

As squarks are expected to be produced abundantly at the LHC in this particular scenario, the study of decay chain of squarks at the LHC becomes of particular interest. Specifically the decay chain, where the decay is via a slepton as in the process:

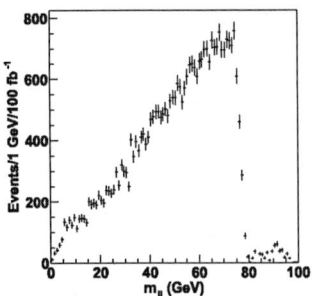

$$\tilde{q}_L \rightarrow \tilde{\chi}_2^0 q \rightarrow \tilde{l}_R^{\mp} l_{near}^{\pm} q \rightarrow \tilde{\chi}_1^0 l_{far}^{\mp} l_{near}^{\pm} q, \tag{2}$$

which has been studied in Atlas.

In this scenario LSP escapes the detector. Hence, one may reconstruct the following observables: $m_{ll}$, $m_{l_1q}$, $m_{l_2q}$ and $m_{llq}$. These are connected to the SUSY mass spectrum through the endpoints of their kinematical distributions. For example, the endpoint of the $m_{ll}$ distribution is connected to the SUSY masses through the relation:

1: Invariant mass distribution of lepton pairs from squark decays after background subtraction [2].

$$(m_{ll}^2)^{edge} = \frac{(m_{\tilde{\chi}_2^0}^2 - m_{\tilde{l}_R}^2)(m_{\tilde{l}_R}^2 - m_{\tilde{\chi}_1^0}^2)}{m_{\tilde{l}_R}^2} \tag{3}$$

CP892, *Quark Confinement and the Hadron Spectrum VII*
edited by J. E. F. T. Ribeiro

In conclusion, Atlas is able to measure the SUSY mass spectra for this class of mass SUSY hierarchies with a precision at the percent level for an integrated luminosity of 300 fb$^{-1}$.

## LIGHT $\tilde{t}$ SCENARIO

Supersymmetric models have recently been proposed fitting the MSSM parameters to cosmological observables. The requirement is that the matter / anti-matter asymmetry is generated at the electroweak scale ([3], [4], [5]). These models favor a light stop squark. One such model has been studied in Atlas [6].

The study utilizes an event topology with 2 jets + 2 $b$-jets as well as one isolated lepton and missing transverse energy. As the mass splittings between the $\tilde{t}_1$, $\chi_1^+$ and $\tilde{\chi}_1^0$ do not allow for resonant W production, kinematic constraints exist and can be used for background rejection. The signature is good for triggering due to the isolated lepton but much Standard Model background survives the cuts. This is mainly W production.

Imposing hard cuts to reject resonant W production, the signal to background ratio can be brought up to $\sim$1/10. Figure 2 shows the minimum invariant mass distributions of one $b$-jet combined with either the to light jets or the lepton. The SM background has been subtracted in the plot.

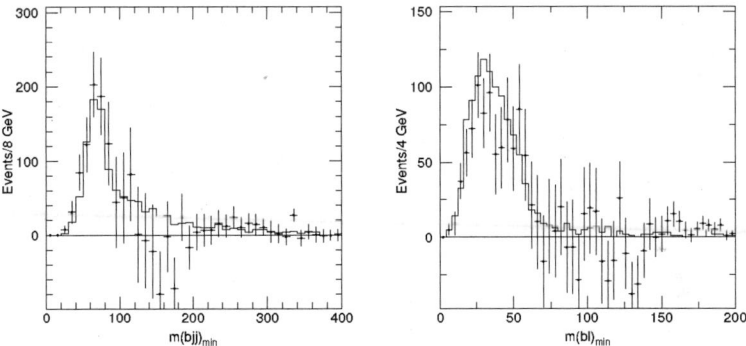

**FIGURE 2.** Invariant mass distributions for 1.8 fb$^{-1}$ [6].

This study has demonstrated that it is possible to determine the kinematical structure of the $\tilde{t}$ decays. It can then be used to extract the underlying SUSY parameters.

## R-HADRON PHENOMENOLOGY

Not all SUSY models are characterized by the missing transverse energy signal of the $\tilde{\chi}_1^0$. Models exist in which the gluino is the LSP [7] and thus is stable by R-parity. Other models such as Split SUSY [8] have the $\tilde{\chi}_1^0$ as LSP and a light gluino. In this scenario the light gluino decays to the LSP through a squark which is very heavy. Hence the gluino is allowed to long-lived, and it may thus be considered stable from the point of view of a HEP detector. Other models again predict long-lived $\tilde{t}$-squarks.

These scenarios all produce a heavy colored object interacting in the detector. Assuming color confinement, they will form hadrons that are quasi-stable by R-parity, hence the name "R-hadrons". In an R-hadron the heavy sparticle will carry most of the momentum, and interactions will take place via the surrounding light quark system [9].

The hadronic energy loss per interaction is only a few GeV as is seen in figure 3. A geometric cross section is assumed and yields a hadronic interaction length on the order of 10 cm in iron [9] allowing for R-hadrons to change charge in the detector. The phenomenology of R-hadrons is thus a combination of R-hadrons charge flipping and their punch-through in a HEP detector due to a limited energy loss per interaction. The charge flipping can be quantified conveniently using the signed $p_t$ in the inner detector vs. that found in the muon chambers.

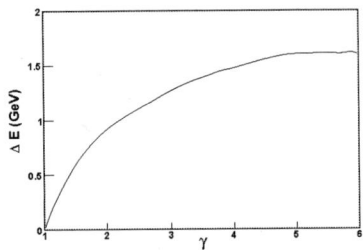

3: Energy loss per hadronic interaction for a gluino R-hadron in iron as a function of the $\gamma$-factor.

Figure 4 shows a comparison of the signed $p_t$ in the inner detector vs. that found in the muon chambers for different event samples. As is evident, the R-hadrons in the muons chambers have no memory of their initial charge in contrast to all the SM samples.

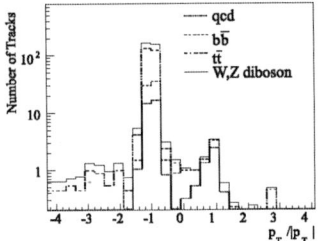
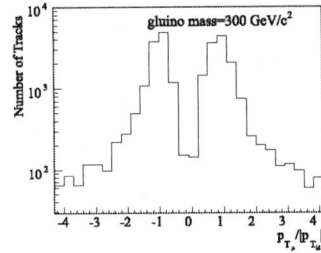

**FIGURE 4.**    Signed $p_T$ in the muon system compared to that found in the inner detector. All tracks were required to have negative charge in the inner detector [10].

## Triggering

The anticipated high mass of either the $\tilde{t}$ or the $\tilde{g}$ will manifest itself as R-hadrons with very high momentum. As R-hadrons will punch through the detector, they will specifically satisfy both low and high energy muon triggers in Atlas. This of course requires them to be in a charged state when they reach the muon system of the detector.

Considering the $\beta$ distributions of gluino R-hadrons in a Split SUSY scenario, a non-negligible fraction of the R-hadrons are produced with values of $\beta$ significantly smaller than 1. For the bunch-crossing rate of the LHC (40 MHz) an R-hadron in Atlas with $\beta < 0.7$ will arrive too late to the muon system to be assigned to the correct event. The corresponding losses of signal are 25% for $m_{\tilde{g}} = 300$ GeV/$c^2$ rising to 60% for $m_{\tilde{g}} = 1$ TeV/$c^2$. Hadronic interactions in the muons system leads to a further loss in signal which is estimated conservatively to be 50% under the assumption that any hadronic interaction in the muon system causes the muon trigger to fail.

## Discovery Potential

A variety of event topologies can be studied and subsequently combined into a coherent search for R-hadrons. As is studied in [10], the R-hadron signature is very distinct. In a Split SUSY scenario the gluino is expected to be discovered already with an integrated luminosity of 2 fb$^{-1}$ for gluino masses up to 1 TeV.

## R-HADRON MEASUREMENTS

Once an R-hadron signature has been detected, it is important to measure its mass, as this constrains the SUSY parameter space. As the R-hadron does not decay in the detector, we have to measure its mass by other means than measuring the decay kinematics. One method that can be used, is to use the Time of Flight (ToF) for the R-hadron to reach the muon system. One can then use the relation $p = mc\beta\gamma$ to fit the mass to the measured momenta and velocities. The mass of the heavy sparticle can be determined through this method to an accuracy of order a few percent.

## CONCLUSIONS

The Atlas detector has a wide spectrum of capabilities for SUSY studies. It is sensitive to an wide spectrum of phenomenologies, and one year of nominal running provides enough statistics to discover SUSY at the TeV scale in many models.

## REFERENCES

1.  B. C. Allanach *et al.*, in *Proc. of the APS/DPF/DPB Summer Study on the Future of Particle Physics (Snowmass 2001)* ed. N. Graf, Eur. Phys. J. C **25** (2002) 113 [eConf **C010630** (2001) P125] [arXiv:hep-ph/0202233].
2.  B. K. Gjelsten, E. Lytken, D. J. Miller, P. Osland and G. Polesello, ATL-PHYS-2004-007.
3.  M. Carena, M. Quiros and C. E. M. Wagner, Nucl. Phys. B **524** (1998) 3 [arXiv:hep-ph/9710401].
4.  C. Balazs, M. Carena and C. E. M. Wagner, Phys. Rev. D **70** (2004) 015007 [arXiv:hep-ph/0403224].
5.  C. Balazs, M. Carena, A. Menon, D. E. Morrissey and C. E. M. Wagner, Phys. Rev. D **71** (2005) 075002 [arXiv:hep-ph/0412264].
6.  T. Lari and G. Polesello, ATL-PHYS-CONF-2006-001.
7.  H. Baer, K. m. Cheung and J. F. Gunion, Phys. Rev. D **59**, 075002 (1999) [arXiv:hep-ph/9806361].
8.  G. F. Giudice and A. Romanino, "Split supersymmetry," Nucl. Phys. B **699** (2004) 65 [Erratum-ibid. B **706** (2005) 65] [arXiv:hep-ph/0406088].
9.  A. C. Kraan, Eur. Phys. J. C **37** (2004) 91 [arXiv:hep-ex/0404001].
10. S. Hellman, D. Milstead and M. Ramstedt, ATL-PHYS-PUB-2006-005; A. C. Kraan, J. B. Hansen and P. Nevski, arXiv:hep-ex/0511014.

# The Quantum and Local Polyakov loop in Chiral Quark Models at Finite Temperature [1]

E. Megías [2], E. Ruiz Arriola and L.L. Salcedo

*Departamento de Física Atómica, Molecular y Nuclear, Universidad de Granada.*
*18071-Granada (Spain)*

**Abstract.** We describe results for the confinement-deconfinement phase transition as predicted by the Nambu–Jona-Lasinio model where the local and quantum Polyakov loop is coupled to the constituent quarks in a minimal way (PNJL). We observe that the leading correlation of two Polyakov loops describes the chiral transition accurately. The effects of the current quark mass on the transition are also analysed.

**Keywords:** Chiral Quark Model, Polyakov loop, Finite Temperature, Phase Transition
**PACS:** 12.39.Fe,11.10.Wx,12.38.Lg

The simultaneous ocurrence of two phase transitions where chiral symmetry is restored and hadrons become deconfined seems a unique and misterious feature of QCD matter at finite temperature (for an early review see e.g. [1]). Besides direct QCD lattice simulations, there exist theoretical constraints below and above the deconfinement phase transition. At low temperatures the leading thermal excitations correspond to a gas of weakly interacting pions [2]. Moreover, in the large $N_c$ limit with the temperature $T$ kept fixed, if a chiral phase transition takes place it should be first order [3]. The use of resonance hadron Lagrangians implies that thermal corrections are $1/N_c$ suppressed [4]. At high temperatures one has a weakly interacting quark-gluon plasma (for a review see e.g. [5]). However, the previous powerful constraints assume from the start a given phase and do not provide a clue on how chiral symmetry restoration and deconfinement are intertwined.

The coupling of relevant order parameters such as the quark condensate for chiral symmetry breaking and the Polyakov loop for deconfinement at finite temperature can be made explicit in Polyakov chiral quark models, an amalgamate of colour and flavour degrees of freedom where the simultaneous chiral-deconfinement crossover can be quantitatively studied with an acceptable phenomenological success [6, 7, 8, 9, 10, 11, 12, 13, 14, 15]. Actually, in our recent work [10] we have shown why and how Polyakov loops must be coupled to Chiral Quark Models (CQM) to comply with large gauge invariance at finite temperature and how the quantum and local nature of the Polyakov loop generates a rather sharp crossover at about the observed critical temperature, although uncertainties are expected. More specifically, ChPT and large $N_c$ constraints are

[1] Supported by funds provided by the Spanish DGI and FEDER founds with Grant No. FIS2005-00810, Junta de Andalucía Grant No. FM-225, and EURIDICE Grant No. HPRN-CT-2003-00311.
[2] Speaker at Quark Confinement and the Hadron Spectrum VII, Ponta Delgada, Portugal, 2-7 IX, 2006.

CP892, *Quark Confinement and the Hadron Spectrum VII*
edited by J. E. F. T. Ribeiro
© 2007 American Institute of Physics 978-0-7354-0396-3/07/$23.00

444

naturally accomodated [13] within those models, hence solving a long standing puzzle which was ignored for a long time; traditional CQM did produce a chiral phase transition while violating those restrictions. An immediate consequence of this new coupling is an upward shift of the critical temperature referred to as *Polyakov cooling* in Ref. [10].

In this work we will deal with the PNJL model for definiteness. After bosonization the NJL Lagrangian reads

$$\Gamma_Q[S,P] = \operatorname{Tr}\log\left(i\slashed{\partial} + \hat{M}_0 + S + i\gamma^5 P\right) + \frac{1}{4G_S}\int d^4x\, \operatorname{tr}_f\left(S^2 + P^2\right). \tag{1}$$

We use Tr for the full functional trace, $\operatorname{tr}_f$ for the trace in flavour space, and $\operatorname{tr}_c$ for the trace in colour space. The UV divergencies in Eq. (1) from the Dirac determinant only affect in practice the zero temperature contributions [16]. In the Polyakov gauge $\Omega = e^{iA_4/T}$, where $A_4$ is time independent and diagonal, and the minimal coupling is made $\partial_4 \to \partial_4 - iA_4$. Integrating further over the $A_4$ gluon field in a gauge invariant manner [17] yields a generic partition function of the form

$$Z = \int DS\,DP\,D\Omega\, e^{-\Gamma_G[\Omega]}e^{-\Gamma_Q[S,P;\Omega]}, \tag{2}$$

where $D\Omega$ is the Haar measure of the SU($N_c$) colour group, $\Gamma_G$ is the effective gluon action and $\Gamma_Q$ stands for the quark effective action. In general $\Omega(x)$ is a local and quantum variable since the gluon field itself depends on the point. As argued in [10] mean field approximations [7, 8, 11, 12] generate a spurious gauge orbit dependence and a possibly complex $\Omega$ (violating colour charge conjugation) and they necessarily imply a non-vanishing value of the Polyakov loop in the adjoint representation, in contradiction with lattice results [18].

For a constant value of the Polyakov loop, $\Omega$, and the scalar field $S = M$ (which we identify with the constituent quark mass) the quark effective action is given by

$$\frac{T}{V}\Gamma_Q(M,\Omega,T) = \frac{1}{4G_S}\operatorname{Tr}_f(M - \hat{M}_0)^2 - 2N_f\int\frac{d^3k}{(2\pi)^3}\left[N_c\varepsilon_k\right.$$
$$\left. + T\operatorname{Tr}_c\left\{\log\left(1 + e^{-\varepsilon_k/T}\Omega\right) + \log\left(1 + e^{-\varepsilon_k/T}\Omega^\dagger\right)\right\}\right], \tag{3}$$

where we have only retained the vacuum contribution, so there is no contribution of meson fields $(S,P)$ (see [13] for a chiral expansion up to $\mathcal{O}(p^4)$), $V$ is three dimensional volume and $\varepsilon_k = +\sqrt{k^2 + M^2}$ is the energy of a constituent quark with mass $M$. We define the Polyakov-loop averaged action

$$e^{-\Gamma_Q(M,T)} = \int d\Omega\, e^{-\Gamma_G[\Omega]}e^{-\Gamma_Q(M,\Omega,T)}. \tag{4}$$

The value of $M$ is determined by minimization of $\Gamma_Q(M,T)$ with respect to $M$, $\partial\Gamma_Q(M,T)/\partial M = 0$, which corresponds to computing the integration in $DSDP$ at the mean field level and determines $M$ at a given temperature $T$, denoted as $M^* = M(T)$. In addition, the relation between the (single flavour) chiral quark condensate, $\langle \bar{q}q \rangle$, and the constituent quark mass, reads

$$2G_S N_f \langle \bar{q}q \rangle^* = (M^* \quad m_q).\tag{5}$$

Any observable is obtained by using $M^*$ and averaging over $\Omega$. The integral in $d\Omega$ in the case $N_c = 3$ and in the Polyakov gauge was computed numerically in Ref. [10]. Here we show a much simpler method which is based on evaluating the integral analytically for any $N_c$ in the low temperature limit and corresponds to take $\Omega$ small. (To see this use the formula $\int d^3 k\, e^{-\varepsilon_k/T} = 4\pi M^2 T K_2\left(\frac{M}{T}\right)$, where $K_2(x)$ is the modified Bessel function and $K_2(x) \quad \sqrt{\pi/2x}\,e^{-x}$ for $x \to \infty$). From Eq. (3) we get

$$\Gamma_Q(M,\Omega,T) = \Gamma_Q(M,0) + 2N_f \sum_{n=1}^{\infty} \frac{(\ 1)^n}{n} \int \frac{d^3x\, d^3k}{(2\pi)^3} e^{-n\varepsilon_k/T} \mathrm{Tr}_c[\Omega^n + \Omega^{\dagger n}].\tag{6}$$

Expanding the exponent in Eq. (2) one obtains a power series in terms of $\Omega$ and $\Omega^\dagger$. The simplest correlation of two Polyakov loops is taken to be [10]

$$\int d\Omega\, \mathrm{Tr}_c\Omega(\vec{x})\mathrm{Tr}_c\Omega^\dagger(\vec{y}) = e^{-\sigma|\vec{x}-\vec{y}|/T},\tag{7}$$

with $\sigma$ the string tension, and yields the leading thermal contribution to the effective action

$$\frac{T}{V}\Gamma_Q(M,T) = \frac{T}{V}\Gamma_Q(M,0) \quad TV_\sigma \left(2N_f \int \frac{d^3k}{(2\pi)^3} e^{-\varepsilon_k/T}\right)^2 + \cdots,\tag{8}$$

where $V_\sigma = 8\pi T^3/\sigma^3$ is the equivalent confinement correlation volume. As we see, the effect of quantum corrections on the Polyakov loop lowers the vacuum energy as it should. Moreover, they are $1/N_c$ suppressed, as one would expect in Chiral Perturbation Theory or a resonance gas model but unlike traditional chiral quark models without Polyakov loop.

Minimizing with respect to the mass we get the effective temperature dependent mass $M^*$ and from Eq. (5) the corresponding $\langle \bar{q}q \rangle^*$ condensate can be evaluated. The approximate result is presented and compared to the full result [10] in Fig. 1 and, as we can see the approximation is quite efficient and very easy to implement in standard chiral quark models. For the Polyakov loop expectation value similar manipulations hold, yielding the leading order contribution

$$L = \left\langle \frac{1}{N_c}\mathrm{tr}_c\Omega \right\rangle = \frac{N_f}{N_c}V_\sigma \frac{M^2 T}{\pi^2}K_2(M/T) + \cdots \quad \frac{N_f}{N_c}\frac{V_\sigma}{T}\sqrt{\frac{M^3 T^5}{2\pi^3}}e^{-M/T}.\tag{9}$$

The full result and the approximated formula are compared in Fig. 1. In this case the agreement is only up to temperatures about $0.75 T_D$.

The analysis above is done with physical current quark masses, $m_q = 5.5\,\mathrm{MeV}$ where one obtains $T_\chi = T_D = 256(1)\,\mathrm{MeV}$. This remarkable coincidence between transitions is not accidental nor depends on the particular choice of $m_q$ as can be seen in Fig. 1, where we show the temperature dependence of $\langle \bar{q}q \rangle^*$ and $L$ for $m_q = 0$, 5.5, 40, 80, 120 and 300 MeV. The corresponding susceptibilities are displayed in Fig. 2.

446

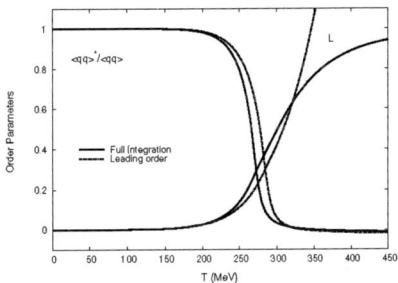
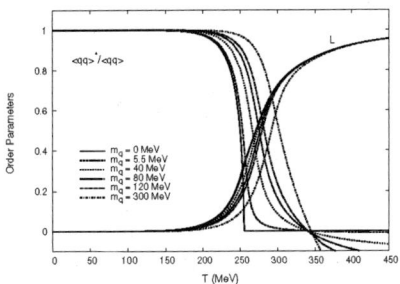

**FIGURE 1.** Chiral condensate $\langle\bar{q}q\rangle^*$ and Polyakov loop $L = \langle\mathrm{tr}_c\Omega\rangle/N_c$. Left: Leading Polyakov loop correlation approximation (see Eqs. (8) and (9)). We take the 2-flavor PNJL model, and $\sqrt{\sigma} = 425\,\mathrm{MeV}$, $f_\pi = 93\,\mathrm{MeV}$, $M = 300\,\mathrm{MeV}$, $m_u = m_d \equiv m_q = 5.5\,\mathrm{MeV}$. Right: Current quark mass dependence.

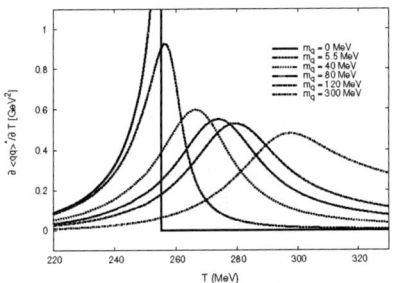
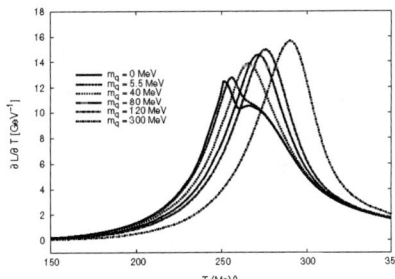

**FIGURE 2.** Temperature dependence of $\partial L/\partial T$ (left) and $\partial\langle\bar{q}q\rangle^*/\partial T$ (right) for several values of the current quark mass. The transition temperatures are $T_\chi = T_D = 255(1)$, $256(1)$, $266(1)$ MeV for $m_q = 0$, $5.5$, $40$ MeV respectively, and $T_\chi = 297(1)$ MeV, $T_D = 290(1)$ MeV for $m_q = 300$ MeV.

# REFERENCES

1. F. Karsch, *PoS* **Corfu98**, 008 (1999).
2. P. Gerber, and H. Leutwyler, *Nucl. Phys.* **B321**, 387 (1989).
3. A. Gocksch, and F. Neri, *Phys. Rev. Lett.* **50**, 1099 (1983).
4. D. Toublan, and J. B. Kogut, *Phys. Lett.* **B605**, 129–136 (2005), hep-ph/0409310.
5. U. Kraemmer, and A. Rebhan, *Rept. Prog. Phys.* **67**, 351 (2004), hep-ph/0310337.
6. P. N. Meisinger, and M. C. Ogilvie, *Phys. Lett.* **B379**, 163–168 (1996).
7. P. N. Meisinger, T. R. Miller, and M. C. Ogilvie, *Nucl. Phys. Proc. Suppl.* **129**, 563–565 (2004).
8. K. Fukushima, *Phys. Lett.* **B591**, 277–284 (2004), hep-ph/0310121.
9. E. Megías, E. Ruiz Arriola, and L. L. Salcedo (2004), hep-ph/0410053.
10. E. Megías, E. Ruiz Arriola, and L. L. Salcedo, *Phys. Rev.* **D74**, 065005 (2006), hep-ph/0412308.
11. C. Ratti, M. A. Thaler, and W. Weise, *Phys. Rev.* **D73**, 014019 (2006), hep-ph/0506234.
12. S. K. Ghosh, T. K. Mukherjee, M. G. Mustafa, and R. Ray, *Phys. Rev.* **D73**, 114007 (2006).
13. E. Megías, E. Ruiz Arriola, and L. L. Salcedo (2006), hep-ph/0607338.
14. H. Hansen, et al. (2006), hep-ph/0609116.
15. C. Ratti, S. Roessner, M. A. Thaler, and W. Weise (2006), hep-ph/0609218.
16. C. V. Christov, E. Ruiz Arriola, and K. Goeke, *Acta Phys. Polon.* **B22**, 187–202 (1991).
17. H. Reinhardt, *Mod. Phys. Lett.* **A11**, 2451–2462 (1996), hep-th/9602047.
18. A. Dumitru, Y. Hatta, J. Lenaghan, K. Orginos, and R. D. Pisarski, *Phys. Rev.* **D70**, 034511 (2004).

# Neutron structure function moments at leading twist

M. Osipenko*, S. Simula†, S. Kulagin**, G. Ricco‡ and CLAS
Collaboration

*Istituto Nazionale di Fisica Nucleare, Sezione di Genova, Genoa, Italy 16146
†Istituto Nazionale di Fisica Nucleare, Sezione Roma III, Roma, Italy 00146
**Institute for Nuclear Research of Russian Academy of Science, Moscow, Russia 117312
‡Università di Genova, Genoa, Italy 16146

**Abstract.** The experimental data on $F_2$ structure functions of the proton and deuteron were used to construct their moments. In particular, recent measurements performed with CLAS detector at Jefferson Lab allowed to extend our knowledge of structure functions in the large-$x$ region. The phenomenological analysis of these experimental moments in terms of the Operator Product Expansion permitted to separate the leading and higher twist contributions. Applying nuclear corrections to extracted deuteron moments we obtained the contribution of the neutron. Combining leading twist moments of the neutron and proton we found $d/u$ ratio at $x \to 1$ approaching 0, although $1/5$ value could not be excluded. The twist expansion analysis suggests that the contamination of higher twists influences the extraction of the $d/u$ ratio at $x \to 1$ even at $Q^2$-scale as large as 12 (GeV/c)$^2$.

**Keywords:** nucleon structure functions, moments, twist expansion
**PACS:** 13.60.Hb, 12.38.Lg, 24.85.+p

## DATA ANALYSIS

QCD through the Operator Product Expansion (OPE) allows to relate measurable moments of nucleon structure functions to the series of local operators, so called twists. The Leading Twist (LT) (first term in the series) represents the asymptotic freedom domain. This term is completely determined by perturbative calculations and Lattice simulations[1]. Higher Twists (HT) (all further terms) describe the virtual photon scattering off interacting partons. The complexity of this interaction and therefore of corresponding QCD operators increases with twist order. Calculations of these terms have been performed only in a few cases.

The experimental data on the structure function moments were obtained recently for the proton and deuteron in Refs. [1] and [2], respectively. Moments were extracted from a combined analysis of the structure functions $F_2$ measured at CLAS and other world data on the inclusive lepton-nucleon scattering. This was performed by integrating experimental data points independently at each fixed $Q^2$ value. Hence, the obtained $Q^2$-evolution of the moments is not affected by any model dependence and can be directly compared to pQCD predictions. Example of measured moments is shown in Fig. 1. As one can see the proton and deuteron moments have similar $Q^2$-behavior, but different

---

[1] Lattice simulations up to now are limited to a few lower moments

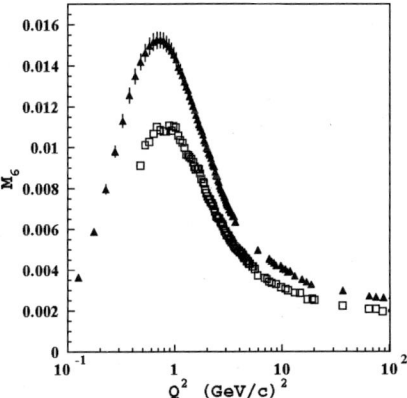

**FIGURE 1.** Total experimental $n = 6$ moments of the proton (full triangles) and deuteron per nucleon (open squares) structure functions $F_2$.

absolute value. Indeed, in the QCD at the LT two moments should have the same $Q^2$-evolution. This observation suggests that the duality is valid for both targets and the contribution of nuclear HTs[2] is small in the covered $Q^2$-range.

We analyzed these experimentally extracted moments of the proton and deuteron structure functions $F_2$ [1, 2] to separate LT and HT terms. This was performed by fitting the data with the following expression:

$$M_n(Q^2) = \int_0^1 dx x^{n-2} F_2(x, Q^2) = LT_n(\alpha_S) + \sum_{\tau=4}^{k} a_n^\tau \left( \frac{\alpha_S(Q^2)}{\alpha_S(\mu^2)} \right)^{\gamma_n^\tau} \left( \frac{\mu^2}{Q^2} \right)^{\frac{\tau-2}{2}}, \quad (1)$$

where $\alpha_S$ is the running coupling constant, $\mu^2$ is an arbitrary scale (taken to be 10 (GeV/c)$^2$), $a_n^\tau$ is the matrix element of corresponding QCD operators, $\gamma_n^\tau$ is the anomalous dimension, $\tau$ is the order of the twist and $k$ is the maximum HT order considered. The number of HT terms ($k$) in the expansion is of course arbitrary because we don't know at which $1/Q^2$ power the series converges. This prevents an evaluation of each separate HT term from the data. Instead, *the total contribution of HTs can be extracted with good precision*. In Fig. 2 one can see that taking two, three or four HT terms in Eq. 1 does not change the total HT contribution. Moreover, this result was expected because the total HT contribution represents simply the difference between the data and calculated LT.

Extracted LT components of the proton and deuteron moments can be combined now to obtain moments of the neutron structure function $F_2$. In the Euclidean space of moments the convolution of the nuclear Impulse Approximation (IA) transforms into a product of moments. This allows for a simple extraction of neutron moments from the

---

[2] nuclear higher twists are mostly related to Final State Interactions of the nucleon in the nucleus

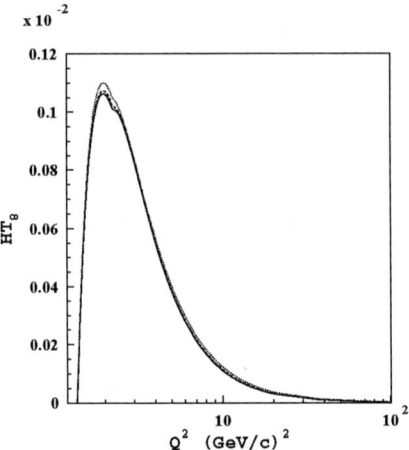

**FIGURE 2.** Total higher twist contribution in $n = 8$ moment obtained with the present procedure for different number of terms in the OPE series (see Eq. 1): solid line - two HT terms, dashed line - three HT terms, dotted line - four HT terms.

following algebraic relation:

$$M_n^n(Q^2) = \frac{2M_n^D(Q^2)}{N_n^D} - M_n^p(Q^2) \, , \qquad (2)$$

where $M_n^p$, $M_n^n$ and $M_n^D$ are moments of the proton, neutron and deuteron, respectively. $N_n^D$ is the moment of the nuclear momentum distribution $f^D$ i.e. the structure function of the deuteron composed of point-like nucleons (see Ref. [3] for details). This nuclear structure function $f^D$ was obtained from the data on the deuteron wave function.

The data on proton and neutron moments can be used to study the contribution of $u$ and $d$ quarks in the proton at $x \to 1$. Assuming that $u$ and $d$ quark distributions in the proton and neutron are the same[3] the $d/u$ ratio *at the leading twist accuracy* can be related to the ratio of neutron to proton structure functions $F_2^n/F_2^p(x \to 1)$. This ratio, in turn, is equal to the ratio of moments of these structure functions $M_n^n/M_n^p(n \to \infty)$. The latter equality requires only that $F_2^{p,n}(x \to 1) \to a^{p,n}(1-x)^{b^{p,n}}$, which follows from the analyticity of the forward Compton amplitude in OPE. Instead, the existence of a finite $F_2^n/F_2^p(x \to 1)$ limit guarantees that exponents $b^p$ and $b^n$ are exactly the same. We constructed these ratios of moments and results are shown in Fig. 3. As one can see our data tend to the "standard" $1/4$ value used in most of parton distribution fits. This value corresponds to the vanishing $d/u$ ratio. However, the precision of our data does not allow to exclude $3/7$ value corresponding to $d/u$ ratio $1/5$. In the Fig. 3 one also can see the impact of the HT contribution on the ratio at $Q^2 = 12$ (GeV/c)$^2$. The previous analysis [4] performed in the $x$-space showed that applying similar nuclear corrections the ratio of structure functions $F_2^n/F_2^p(x \to 1)$ goes to $3/7$ at $Q^2 = 12$ (GeV/c)$^2$. No corrections on

---

[3] though the number of $u$ and $d$ quarks is of course different

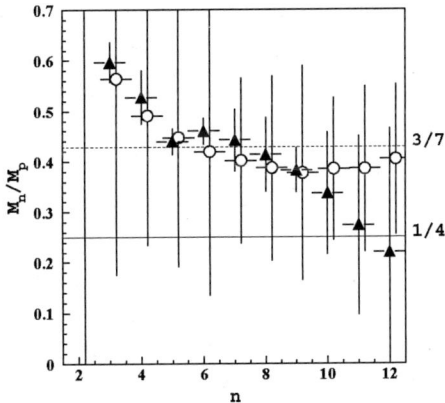

**FIGURE 3.** Ratio of the neutron to proton moments as a function of $n$ at $Q^2 = 12$ (GeV/c)$^2$: full triangles show the ratio at the leading twist obtained in Ref. [3], open circles represent the ratio of measured inelastic moments including the higher twist contribution.

the possible HT contamination has been applied in this article, assuming that at such large $Q^2$ they are negligible. Therefore, if instead of LT part we take measured inelastic moments, containing also HT terms, and construct the same ratios we should confirm the result from Ref. [4]. Indeed, in the Fig. 3 one can see that the ratio of moments including HTs tends to $3/7$ at largest $n$ ($n = 12$ corresponds to $x$ values about 0.75). This result could also be deduced from the isospin independence of HTs observed in the Ref. [3].

## CONCLUSIONS

We obtained the experimental data on moments of proton and deuteron structure function $F_2$. In these data contributions of the leading and higher twists were separated via OPE analysis. By combining the proton and deuteron moments and applying nuclear corrections we extracted moments of the neutron structure function $F_2$. The ratio of the neutron and proton moments at large $n$ is related to the ratio of $d$ and $u$ quark contributions in the proton at large-$x$. The obtained ratio is consistent with the asymptotic limit $d/u \to 0$ at $x \to 1$, which originates from the dominance of soft, non-perturbative physics at large-$x$. Nevertheless the alternative value of $1/5$, derived from helicity conservation arguments, is not excluded. The HT contamination to the ratio, if not subtracted as in the present analysis, influences significantly the extracted $d/u$ ratio also at large $Q^2$-values.

## REFERENCES

1. M. Osipenko, et al., *Phys. Rev.* **D67**, 092001 (2003).
2. M. Osipenko, et al., *Phys. Rev.* **C73**, 045205 (2006).
3. M. Osipenko, et al., *Nucl. Phys.* **A766**, 142 (2006).
4. W. Melnitchouk, and A. Thomas, *Acta Phys. Polon.* **B27**, 1407 (1996).

# ISR study at Belle

G. Pakhlova

(*For the Belle Collaboration*)

*Institute for Theoretical and Experimental Physics, Moscow, Russia*

**Abstract.** We present a study of $Y(4260)$ properties and measurements of the exclusive $e^+e^- \to D^{(*)}\overline{D}^*$ cross section as a function of center-of-mass energy near the $D^{(*)}\overline{D}^*$ threshold using the initial state radiation.

**Keywords:** Charmonium, hadronic cross section
**PACS:** 13.66.Bc, 13.87.Fh, 14.40.Gx

Exclusive $e^+e^-$ hadronic cross sections for specific charmed final states are of particular interest because they provide information on the spectrum of $J^{PC} = 1^{--}$ charmonium states above the open-charm threshold, which are not well understood. Furthermore, the observation by BaBar of a structure in the $J/\psi\pi^+\pi^-$ invariant mass distribution near $4.26\,\mathrm{GeV}/c^2$ in initial state radiation data, denoted as $Y(4260)$ [1] has stimulated a further interest in this field. The existence of $Y(4260)$ has been confirmed with CLEO III data [2]. Surprisingly the $Y(4260)$ peak position corresponds to a minimum of $e^+e^- \to hadrons$ cross section [3]. A large $\mathscr{B}(Y(4260) \to J/\psi\pi^+\pi^-)$ calculated using inclusive cross section is unexpected for a conventional charmonium state of this mass.

In this paper we report on a search for $Y(4260) \to J/\psi\pi^+\pi^-$ production and a study of exclusive cross sections for $e^+e^- \to D^{*+}D^{*-}$ and $e^+e^- \to D^+D^{*-}$ using initial state radiation (ISR). ISR offers the unique possibility of measuring cross sections over a continuum of energies. The high luminosity of the $B$-factories compensates for the suppression associated with the emission of a hard photon. The data used for this analysis correspond to an integrated luminosity of $\sim 550\,\mathrm{fb}^{-1}$ collected by the Belle detector at the $\Upsilon(4S)$ resonance and nearby continuum at the KEKB asymmetric-energy $e^+e^-$ collider.

To search for $Y(4260)$ in $e^+e^- \to J/\psi\pi^+\pi^-\gamma_{isr}$ production, we select events with four charged tracks only. Two identified leptons are combined to form $J/\psi$ candidate; the signal window is defined by $|M(\ell^+\ell^-) - M_{J/\psi}| < 0.03\,\mathrm{GeV}/c^2$. The $J/\psi$ candidates are then constrained to a common vertex and the nominal $J/\psi$ mass. The dipion combination is required to have mass larger than $0.4\,\mathrm{GeV}/c^2$ to suppress the contribution of misidentified $\gamma \to e^+e^-$. The recoil mass squared of the $J/\psi\pi^+\pi^-$ combination,

$$M_{\mathrm{rec}}^2 = (\sqrt{s} - E_{J/\psi\pi^+\pi^-})^2 - p_{J/\psi\pi^+\pi^-} \tag{1}$$

is required to satisfy $|M_{\mathrm{rec}}^2| < 1\,\mathrm{GeV}^2/c^4$ corresponding to the ISR production. A prominent structure is evident in $J/\psi\pi^+\pi^-$ invariant mass distribution near $4.3\,\mathrm{GeV}/c^2$ presented in Fig. 1 a). A fit to this distribution with the signal function as a Breit-Wigner, corrected for the phase space and the detector efficiency, yields $165 \pm 23$ events. The

CP892, *Quark Confinement and the Hadron Spectrum VII*
edited by J. E. F. T. Ribeiro
© 2007 American Institute of Physics 978-0-7354-0396-3/07/$23.00

$Y(4260)$ mass is found to be $M = 4295 \pm 10^{+10}_{-3}$ and width is $\Gamma = 133 \pm 26^{+13}_{-6}$; both are marginally consistent with the BaBar measurements. The squared recoil mass distribution is shown in Fig. 1 b) for the sideband subtracted $Y(4260)$ region and compared with the signal Monte Carlo simulation. The electronic width is estimated to be $\Gamma_{ee} \times \mathscr{B}(Y(4260) \to J/\psi \pi^+ \pi^-) = (8.7 \pm 1.1^{+0.3}_{-0.9})eV$.

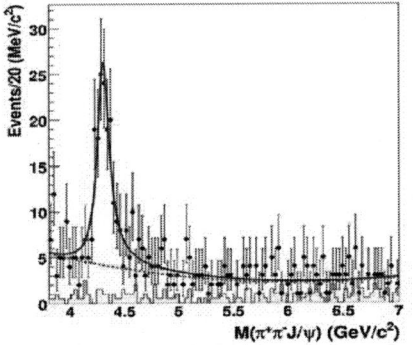 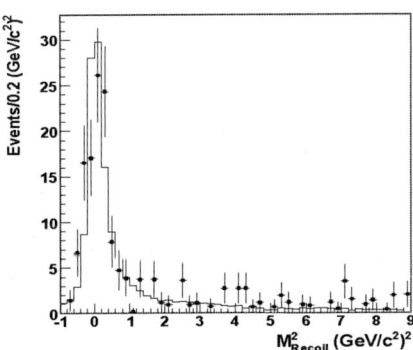

**FIGURE 1.** a) The $M(J/\psi\pi^+\pi^-)$ distribution in $J/\psi$ signal window (points) and scaled sidebands (shaded histogram); b) The $M^2_{\text{rec}}(J/\psi\pi^+\pi^-)$ distribution for the data (points) and MC (histogram).

The study of exclusive open charm cross sections presented here is limited by the $D^{*+}D^{*-}$ and $D^+D^{*-}$ final states where the partial reconstruction of the final state can be used. The method of partial reconstruction achieves higher efficiency by requiring full reconstruction of only one of $D^{(*)+}$ mesons, the $\gamma_{isr}$, and the slow $\pi^-_{slow}$ from the other $D^{*-}$. In this case the spectrum of masses recoiling against the $D^{(*)+}\gamma_{isr}$ system:

$$M_{\text{rec}}(D^{(*)+}\gamma_{isr}) = \sqrt{(E_{CM} - E_{D^{(*)+}\gamma_{isr}})^2 - p^2_{D^{(*)+}\gamma_{isr}}} \qquad (2)$$

peaks at the $D^{*-}$ mass. Resolution of this peak estimated from the Monte Carlo (MC) simulation to be $\sim 300\,\text{MeV}/c^2$ is not sufficient to separate $D\overline{D}^*$, $D^*\overline{D}^*$ or $D^{(*)}\overline{D}^*\pi$ final states. To disentangle the contributions from these final states and to suppress combinatorial backgrounds, we use the slow pion from the unreconstructed $D^{*-}$. The difference between the mass recoiling against $D^{(*)+}\gamma_{isr}$ and $D^{(*)+}\pi^-_{slow}\gamma_{isr}$:

$$\Delta M_{\text{rec}} = M_{\text{rec}}(D^{(*)+}\gamma_{isr}) - M_{\text{rec}}(D^{(*)+}\pi^-_{slow}\gamma_{isr}), \qquad (3)$$

has a narrow distribution ($\sigma \sim 1.4\,\text{MeV}/c^2$) around the nominal $M(D^{*-}) - M(\overline{D}^0)$ mass difference, since the uncertainty in $\gamma_{isr}$ momentum partially cancels out. For the measurement of the exclusive cross section, one needs to determine the $D^{(*)+}D^{*-}$ mass when one of the $D^*$'s is not reconstructed. In the absence of higher-order QED processes, $M(D^{(*)+}D^{*-})$ is the mass recoiling against the $\gamma_{isr}$. However, the photon energy resolution results in a typical $M_{\text{rec}}(\gamma_{isr})$ resolution of $\sim 100\,\text{MeV}$, which is too wide for the study of relatively narrow $D^{(*)+}D^{*-}$ mass states. We significantly improve the $M_{\text{rec}}(\gamma_{isr})$ resolution by applying a refit that constrains $M_{\text{rec}}(D^{(*)+}\gamma_{isr})$ to the nominal

$D^{*-}$ mass. In this way we use the well measured properties of the fully reconstructed $D^{(*)+}$ to correct the energy of the $\gamma_{isr}$. As a result, the $M_{D^{(*)}\overline{D}^*}$ ($\equiv M_{\text{rec}}(\gamma_{isr})$) resolution is improved by a factor of $\sim 10$. Higher-order ISR processes are suppressed by a tight cut on $M_{\text{rec}}(D^{(*)+}\gamma_{isr})$. The final $M_{D^{(*)+}+D^{*-}}$ resolution, varies from $\sim 6\,\text{MeV}/c^2$ around threshold to $\sim 12\,\text{MeV}/c^2$ at $M_{D^{(*)+}+D^{*-}} = 5.0\,\text{GeV}/c^2$. The recoil mass difference after the refit procedure ($\Delta M_{\text{rec}}^{\text{fit}}$) has a resolution improved by a factor of $\sim 2$. The detailed description of the method and the analysis procedure can be found elsewhere [4].

A $\gamma_{isr}$ candidate is required to have energy greater than $2.5\,\text{GeV}$. $D^0$ candidates are reconstructed using five decay modes: $K^-\pi^+$, $K^-K^+$, $K^-\pi^-\pi^+\pi^+$, $K^0_S\pi^+\pi^-$ and $K^-\pi^+\pi^0$. A $\pm 15\,\text{MeV}/c^2$ mass window is used for all modes except for $K^-\pi^-\pi^+\pi^+$ where a $\pm 10\,\text{MeV}/c^2$ requirement is applied. $D^+$ candidates are reconstructed using decay modes: $K^0_S\pi^+$, $K^-\pi^+\pi^+$ and $K^-K^+\pi^+$ with $\pm 10\,\text{MeV}/c^2$ mass window. To improve $D$ meson candidates momentum resolution, final tracks are fitted to a common vertex applying the nominal $D^0$ or $D^+$ mass as a constraint. $D^*$ candidates are selected via $D^{*+} \to D^0\pi^+$ decay modes with a $\pm 2\,\text{MeV}/c^2$ mass-difference window. The mass and vertex constrained fit is also applied to $D^*$ candidates.

A clean peak corresponding to $e^+e^- \to D^{*+}D^{*-}$ is evident in the distribution of $M_{\text{rec}}(D^{*+}\gamma_{isr})$ shown in Fig. 2a) after requirement on the slow pion from the unreconstructed $D^{*-}$ and the tight requirement on $\Delta M_{\text{rec}}^{\text{fit}}$: *i.e.* within $\pm 2\,\text{MeV}/c^2$ of the nominal $M(D^{*-}) - M(\overline{D}^0)$ mass difference. The background from the other processes is substantially suppressed. We define the signal region by the requirement that $M_{\text{rec}}(D^{*+}\gamma_{isr})$

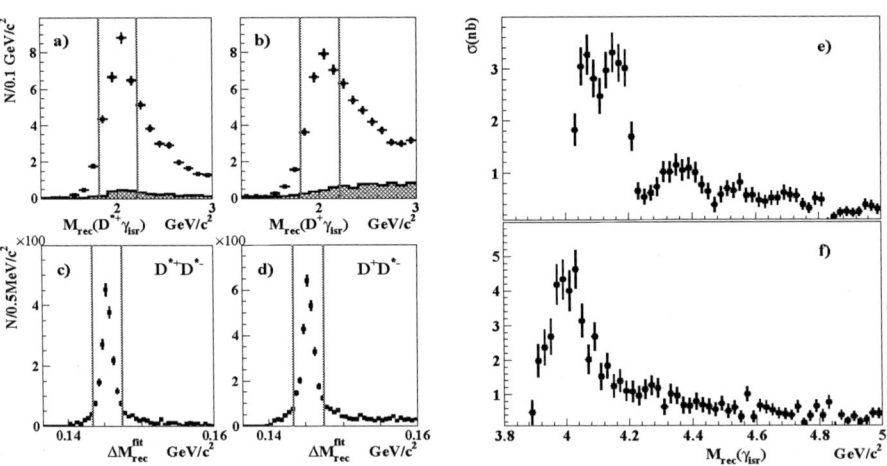

**FIGURE 2.** a), b): The $M_{\text{rec}}(D^{(*)+}\gamma_{isr})$ distribution for the data with $\Delta M_{\text{rec}}^{\text{fit}}$ requirement. Histograms show the normalized $M_{D^{(*)+}}$ sidebands contributions. c), d) The distribution of $\Delta M_{\text{rec}}^{\text{fit}}$ in the data after the refit procedure. The selected signal windows are indicated by the vertical lines. e), f): The exclusive cross-sections $e^+e^- \to D^{*+}D^{*-}$ and $e^+e^- \to D^+D^{*-}$, respectively.

be within $\pm 0.2\,\text{GeV}/c^2$ of the nominal $D^{*-}$ mass to suppress $e^+e^- \to D^{*+}D^{*-}\pi^0\gamma_{isr}$ events. The spectrum of recoil-mass differences for the signal $M_{\text{rec}}(D^{*+}\gamma_{isr})$ window after the $\Delta M_{\text{rec}}^{\text{fit}}$ refit procedure in data is shown in Fig. 2 c).

The analysis of the $e^+e^- \to D^+D^{*-}$ exclusive cross section is identical to that described above with the fully reconstructed $D^{*+}$ meson replaced by a fully reconstructed $D^+$ meson. The requirement of a detected slow pion from the unreconstructed $D^{*-}$ and a tight requirement on $\Delta M_{\text{rec}}^{\text{fit}}$ provides the clean $e^+e^- \to D^+D^{*-}$ signal peak that is shown in Fig. 2 b). The recoil mass difference distribution for the signal $M_{\text{rec}}(D^+\gamma_{isr})$ window, after the $\Delta M_{\text{rec}}^{\text{fit}}$ refit procedure, is shown in Fig. 2 d).

The following sources of background for $e^+e^- \to D^{(*)+}D^{*-}$ are considered and subtracted using the data: combinatorial backgrounds (either under the reconstructed $D^{(*)+}$ or random slow pion); reflection from the process $e^+e^- \to D^{(*)+}D^{*-}\pi^0_{miss}\gamma_{isr}$ with an extra $\pi^0$ in the final state; contribution of $e^+e^- \to D^{(*)+}D^{*-}\pi^0$ when an energetic $\pi^0$ is misidentified as a single $\gamma_{isr}$. The first background source is reliably estimated using the sidebands of the reconstructed $D^{(*)+}$ and $\Delta M_{\text{rec}}^{\text{fit}}$ sidebands. The second background is extracted from the study of the isospin conjugated final state $e^+e^- \to D^{(*)0}D^{*-}\pi^+_{miss}$. This background is found to be negligible in case of $D^{*+}D^{*-}$ final state, while in case of $D^+D^{*-}$ final state this background is found to be as large as $\sim 1/5$ of the $D^+D^{*-}$ signal. The last background contribution is found to be negligibly small in both cases.

The resulting exclusive $e^+e^- \to D^{(*)+}D^{*-}$ cross-sections extracted from the $D^{(*)+}D^{*-}$ mass distributions after background subtraction and taking into account the differential ISR luminosity and the total efficiency are shown in Fig. 2 e),f). Since the bin width is much larger than the resolution, the distributions are not needed to be corrected for resolution. The systematic errors for the $\sigma(e^+e^- \to D^{(*)+}D^{*-})$ measurements are found to be 9%(11%).

In summary, we report on the study of the properties of $Y(4260)$ and the first measurements of exclusive $e^+e^- \to D^{*+}D^{*-}$ and $e^+e^- \to D^+D^{*-}$ cross sections at $\sqrt{s}$ around the $D^{*+}D^{*-}$ and $D^+D^{*-}$ thresholds using initial state radiation. The preliminary $Y(4260)$ mass and width measured with our study are marginally consistent with BaBar measurement. The shape of the $e^+e^- \to D^+D^{*-}$ cross section is complicated with several local maximum and minimum, one of the minimum is close both to the $Y(4260)$ mass and to the $D_s^{*+}D_s^{*-}$ threshold. The $e^+e^- \to D^{*+}D^{*-}$ cross section is more smooth, with a prominent excess near the $\psi(4040)$. The measured cross sections are compatible within errors with the $D^{(*)}\overline{D}^*$ exclusive cross section in the energy region up to 4.260 GeV measured by CLEO-c [5].

## REFERENCES

1. B. Aubert *et al.* (BABAR Collab.), *Phys.Rev.Lett.* **95**, 142001 (2005).
2. T. E. Coan *et al.* (CLEO Collab.), *Phys.Rev.Lett.* **96**, 162003 (2006).
3. X. H. Mo *et al.* , *Phys.Lett.* **B640**, 182-187 (2006).
4. K. Abe *et al.* (Belle Collab.), arXiv:hep-ex/0608018.
5. R. Poling (for CLEO collab.),arXiv:hep-ex/0606016.

# New charmonium and charm states at B*A*B*AR*

S. Ricciardi

(for the B*A*B*AR* Collaboration)

*Rutherford Appleton Laboratory, Chilton, Didcot, Oxon, OX11 0QX, United Kingdom*

**Abstract.** We review the latest B*A*B*AR* results on charm spectroscopy focussing on recently observed states with both hidden and open charm content.

**Keywords:** hadron spectroscopy, charm quark
**PACS:** 14.20.Lq, 14.40.Lb

## INTRODUCTION

Charmonium and charm studies at $B$ factories have led to a series of new states that are challenging conventional heavy-flavour spectroscopy. The B*A*B*AR*/PeP-II $B$ Factory at SLAC has undertaken a number of studies to look for additional states and understand the nature of the recently discovered ones.

Here, we review the latest experimental results on meson and baryon spectroscopy focussing on recently observed states with both hidden and open charm content.

## CHARMONIUM-LIKE STATES

Five new states have been discovered recently at $B$ factories above open-charm threshold with properties not well-fitting quark model and Lattice QCD calculations for $c\bar{c}$ bound states. Among these charmonium-like states, particularly puzzling are two resonances decaying to $J/\psi\ \pi^+\ \pi^-$: $X(3872)$, first observed by Belle [1] in $B^+ \to K^+ J/\psi \pi^+ \pi^-$, and $Y(4260)$, discovered by B*A*B*AR* [2] in $e^+e^-$ events with initial state radiation (ISR), $e^+e^- \to (\gamma)J/\psi \pi^+ \pi^-$.

## $X(3872)$

The existence of $X(3872)$ is well established. Soon after the discovery by Belle [1] in 2003, this state was confirmed by CDF [3], D0 [4], and B*A*B*AR* [5]. Other decay modes of the $X(3872)$ have also been observed. Quantum numbers are not firmly established yet, though angular distribution studies [6] in the $J/\psi\ \pi^+\ \pi^-$ decay channel favour $J^{PC} = 1^{++}$.

Recently, both Belle and B*A*B*AR*, have measured the radiative decay $X(3872) \to J/\psi\gamma$ [7, 8] with an average branching ratio of $(19\pm7)\%$ relative to the $X(3872) \to J/\psi\pi^+\pi^-$ decay mode. The observation of this mode firmly establishes that the $X(3872)$ has even C-parity. The identification of $X(3872)$ with $\chi'_{c1}$, the only remaining possibility

CP892, *Quark Confinement and the Hadron Spectrum VII*
edited by J. E. F. T. Ribeiro

for a conventional state with $J^{PC} = 1^{++}$, is not well consistent with the measured mass $(3871.2 \pm 0.6) \text{MeV}/c^2$. Hence, mainly non-conventional interpretations remain, in particular that of a loose-bound $D^0 \bar{D}^{*0}$ molecule or a tetra-quark, which are currently viewed as the most likely explanations [9].

## $Y(4260)$

The production mechanism, through single photon annihilation in ISR events, requires that $Y(4260)$ is vector state with $J^{PC} = 1^{--}$. The most exceptional feature for this state is its relatively large coupling to $J/\psi\pi\pi$ compared to open charm decays. BABAR has searched for the $Y(4260)$ decay to a neutral or charged pair of $D$ mesons [10]. No signal has been found and an upper limit has been set: $\frac{\mathscr{B}(Y(4260) \to D\bar{D})}{Y(4260) \to J/\psi\pi^+\pi^-} < 7.6$ at 95% C.L.. This is another indication that $Y(4260)$ is not a conventional vector charmonium state. Currently, the most favoured explanation, consistent with all observations, is that of a $c\bar{c}g$ hybrid [9].

BABAR has also searched for $Y(4260) \to \psi(2S)\pi^+\pi^-$ [11] in ISR events and observed a structure in the $\psi(2S)$ $\pi^+$ $\pi^-$ invariant mass spectrum at slightly higher mass, near $4.32 \text{GeV}/c^2$. A fit with a single resonance well describes the spectrum in the region below $5.7 \text{GeV}/c^2$. The fitted values of the Breit-Wigner parameters are a mass of $(4324 \pm 24) \text{MeV}/c^2$ and a width of $(172 \pm 33) \text{MeV}$, where the errors are statistical only. The goodness-of-fit, measured by a $\chi^2$, deteriorates significantly if mass and width are fixed to the BABAR values for $Y(4260)$, hence the hypothesis has been made that the observed structure is not due to the $Y(4260)$ but indicates the presence of another new state. However, no definitive conclusions can be drawn from the present data and an independent confirmation by other experiments is eagerly awaited.

## A NEW CHARMED-STRANGE MESON

Spectroscopy of $c\bar{s}$ states has also generated a lot of interest recently thanks to the discovery of $D^*_{sJ}(2317)^+$ and $D_{sJ}(2460)^+$ by the BABAR [12] and CLEO [13] collaborations. In particular, the identification of $D^*_{sJ}(2317)^+$ and $D_{sJ}(2460)^+$ with two $c\bar{s}$ P-wave states, $J^P = 0^+$ and $1^+$, has been very much debated, since these states were predicted at higher mass and to decay broadly to a charm meson and a kaon.

News in this sector comes from the studies of the inclusive $DK$ production in $e^+e^-$ events by BABAR [14], which show the presence of a new broad resonance in the $D^0K^+$ and in the $D^+K^0_s$ invariant mass spectra. Several checks have been made to rule out fake signals due to possible reflections. No signal is seen in simulated events and studies of the $D$ mass sidebands do not show any extra structure. Therefore, this is a genuine signal of a new state with a mass of $(2856.6 \pm 1.5 \pm 5.0) \text{MeV}/c^2$ and a width of $(48 \pm 7 \pm 10) \text{MeV}$, as determined by a combined fit to the $D^0K^+$ and $D^+K^0_s$ mass spectra. The spin-parity of this state has not yet been established, but theoretical speculations indicate that it could be a $J^P = 3^-$ [15] D-wave state or a radial excitation of the $D^*_{sJ}(2317)^+$ [16].

A broad peak is also observed just below $2.7\,\mathrm{GeV}/c^2$. This bump can be well described by an additional Breit-Wigner in the fit model with a mass of $(2688\pm4\pm3)\,\mathrm{MeV}/c^2$ and a width of $(112\pm7\pm36)\,\mathrm{MeV}/c^2$. However, in this case, a hint of a structure could be seen in the same mass region when using candidates from a $D^0$ mass sideband. Therefore BABAR is not able to establish whether this is a new state. We note that this bump could coincide with a $D^0K^+$ resonance reported by Belle [17] in $B^+ \to D^0\bar{D}^0K^+$ at a mass of about $2715\,\mathrm{MeV}/c^2$.

## CHARMED BARYONS

The particle spectrum of baryons with a single charm quark is also expected to have a rich structure of radial and orbitally-excited states. All nine $J^P = \frac{1}{2}^+$ with $L = 0$ are well-known. Here we report the observation of the last of the six $J^P = \frac{3}{2}^+$ states without internal orbital angular momentum, the $\Omega_c^*$, and of three new $L = 1$ charmed baryons, recently observed by BABAR.

### $\Omega_c^*$

The $\Omega_c^*$ has been observed by BABAR [18] in the radiative decay $\Omega_c^* \to \Omega_c^0\gamma$. The mass difference between $\Omega_c^*$ and $\Omega_c^0$ is determined to be $(70.8\pm1.0\pm1.1)\,\mathrm{MeV}/c^2$, in agreement with theoretical calculations [19]. The ratio of the production cross-section of $e^+e^- \to \Omega_c^*X$ relative to $e^+e^- \to \Omega_c^0X$, where $X$ denotes the rest of the event, is measured to be $1.01\pm0.23\pm0.11$.

### $\Lambda_c(2940)^+$

The study of the invariant mass spectrum of $D^0$-proton pairs in $e^+e^- \to c\bar{c}$ events has led to the observation of a new charmed baryon with mass $(2939.8\pm1.3\pm1.0)\,\mathrm{MeV}/c^2$ and width $(17.5\pm5.2\pm5.9)\,\mathrm{MeV}/c^2$ [20]. A second larger peak in the $D^0p$ mass spectrum, at a mass of $(2881.9\pm0.1\pm0.5)\,\mathrm{MeV}/c^2$ and a width of $(5.8\pm1.5\pm1.1)\,\mathrm{MeV}$ is identified as the $\Lambda_c(2880)^+$, previously discovered in decays to $\Lambda_c^+\pi^+\pi^-$. These are the first charmed baryons observed to decay to a charmed meson and a charmless baryon. Because of the absence of a charge partner in the $D^+p$ mass spectrum, the new state is identified as the isospin scalar $\Lambda_c(2940)^+$. Its spin-parity has yet to be determined.

### $\Xi_c(2980)^+$ and $\Xi_c(3077)^+$

BABAR [21] has also confirmed the discovery of two new $\Xi_c$ states by Belle [22]: $\Xi_c(2980)^+$ and $\Xi_c(3077)^+$, decaying into $\Lambda_c^+K^-\pi^+$. The masses and widths are determined in a 2D fit to the $\Lambda_c^+K^-\pi^+$ and $\Lambda_c^+\pi^+$ mass spectra. Results are in general consistent with Belle's measurements, except for the mass and width of $\Xi_c(2980)^+$,

$(2967 \pm 1.9 \pm 1.0)\,\mathrm{MeV}/c^2$ and $(23.6 \pm 2.8 \pm 1.3)\,\mathrm{MeV}/c^2$ respectively, which are both significantly lower than Belle's values [22]. The discrepancy is probably due to the different signal model, in particular to the inclusion of a mass-dependent phase-space factor in the BABAR fit, which affects mainly the shape near threshold. The second dimension in the fit allows BABAR to determine that quasi-two body decays via a $\Sigma_c^{++}$ are dominant for $\Xi_c(3077)^+$ and represent about 50% of $\Xi_c(2980)^+$ decays.

## SUMMARY

A surge of new states with charm content have been recently observed at BABAR. In the hidden charm sector, the new available results on $X(3872)$ and $Y(4260)$ support the hypothesis that these are not conventional charmonium states. In the open charm sector new insights on the understanding of the hadron spectra come from the discovery of a new charmed-strange meson, $D_{sJ}^*(2860)$, and of several charmed baryons. Other sightings, in particular the structure at $\sim 4.32\,\mathrm{GeV}/c^2$ in the $\psi(2S)\pi^+\pi^-$ ISR spectrum, are also very interesting and call for confirmation.

## ACKNOWLEDGMENTS

I am grateful to Bill Dunwoodie and Shuwei Ye for the illuminating discussions in the preparation of the talk. I thank the organisers for *confining* theorists, experimentalists, and myself in a *charming* environment.

## REFERENCES

1. S.-K. Choi et al., *Phys. Rev. Lett.* **91**, 262001(2003).
2. B. Aubert et al., *Phys. Rev. Lett.* **95**, 142001 (2005).
3. D. Acosta et al., *Phys. Rev. Lett.* **93**, 072001 (2004).
4. V. M. Abazov et al., *Phys. Rev. Lett.* **93**, 162002 (2004).
5. B. Aubert et al., *Phys. Rev. D* **71**, 071103 (2005).
6. K. Abe et al., hep-ex/0505038; I. Kravchenko et al., eCOnf C060409, 016 (2006).
7. K. Abe et al., hep-ex/0505037.
8. B. Aubert et al., *Phys. Rev. D* **74**, 071101 (2006).
9. E. Swanson, *Phys. Rept.* **429**, 243-305 (2006).
10. B, Aubert et al., hep-ex/0607083.
11. B. Aubert et al., hep-ex/0610057.
12. B. Aubert et al., *Phys. Rev. Lett.* **92**, 242001 (2003).
13. D. Besson et al., *Phys. Rev. D* **68**, 032002 (2003).
14. B. Aubert et al., *sub. to Phys. Rev. Lett.*, hep-ex/0607082.
15. P. Colangelo, F. De Fazio and S. Nicotri, hep-ph/0607245.
16. E. Beveren and G.Rupp, hep-ph/0606110.
17. K.Abe et al, hep-ex/0608031.
18. B. Aubert et al., *sub. to Phys. Rev. Lett.*, hep-ex/0608055.
19. N. Mathur et al., *Phys. Rev.D*, **66**, 014502 (2002).
20. B. Aubert et al., *sub. to Phys. Rev. Lett.*, hep-ex/0603052.
21. B. Aubert et al., hep-ex/0607042.
22. R. Chistov et al., *sub. to Phys. Rev. Lett.*, hep-ex/0606051.

# Recent Results from BES
# on
# Charmonium Decays

Shuxian Du for the BES Collaboration

*Institute of High Energy Physics, Chinese Academy of Science,
Beijing 100049, People's Republic of China*

**Abstract.** Based on 14 M $\psi(2S)$ events at BES-II, the decays of $\psi(2S)$ into $\tau^+\tau^-$, baryon pairs, multibody final states and the hadronic decays of $\chi_{cJ}$ are studied. The branching ratios of these decays are reported. These results contribute to understand QCD.

**Keywords:** Charmonium decay; $\psi(2S)$; $\chi_{cJ}$
**PACS:** 13.25.Gv

## 1. INTRODUCTION

The BES is a conventional solenoidal magnet detector that is described in detail elsewhere [1] [2]. In this paper, we report the leptonic decay, the baryon pairs decays and multibody decays of $\psi(2S)$ and the hadronic decays of $\chi_{cJ}$ which is based on 14 M $\psi(2S)$ events collected by BES-II detector.

## 2. $\psi(2S)$ DECAYS

### 2.1. $\psi(2S) \rightarrow \tau^+\tau^-$

$\psi(2S)$ decays provides a opportunity to compare the three lepton generations by studying the leptonic decays $\psi(2S) \rightarrow e^+e^-, \mu^+\mu^-, \tau^+\tau^-$. The sequential lepton hypothesis leads to a relationship between the branching fractions of these decays: $\mathscr{B}_{e^+e^-} = \mathscr{B}_{\mu^+\mu^-} = \mathscr{B}_{\tau^+\tau^-}/0.3885$, which is in good agreement with BES-I result [3]. BES-II remeasured $\mathscr{B}(\psi(2S) \rightarrow \tau^+\tau^-)$ with 14 M $\psi(2S)$ events [4]. The $\tau^+\tau^-$ pair are reconstructed with $\tau^+ \rightarrow \mu^+\nu_\mu\bar{\nu}_\tau$ and $\tau^- \rightarrow e^-\bar{\nu}_e\nu_\tau$, respectively. After event selection, 1015 signal events are observed at $\psi(2S)$ peak. Subtracting the contribution of QED including interference and backgrounds, the branching fraction is calculated to be $(0.310 \pm 0.021 \pm 0.038)\%$, where the first error is statistical and the second is systematic. Compared with BES-I result, this measurement is improved in the QED contribution measurement, efficiency and background estimation and reasonable interference subtraction.

*CP892, Quark Confinement and the Hadron Spectrum VII*
edited by J. E. F. T. Ribeiro

**TABLE 1.** Summary of $\psi(2S)$ hadronic decays and radiative decays and comparision with "12% rule", where $Q_h = \frac{\mathcal{B}(\psi(2S) \to h)}{\mathcal{B}(J/\psi \to h)}$.

| mode | $\mathcal{B}(\psi(2S) \to X)(\times 10^{-4})$ | $\mathcal{B}(J/\psi \to X)(\times 10^{-4})^*$ | $Q_h(\%)$ |
|---|---|---|---|
| $p\bar{p}$ | $3.36 \pm 0.09 \pm 0.24$ | $2.12 \pm 0.10$ | $14.9 \pm 1.4$ |
| $\Lambda\bar{\Lambda}$ | $3.39 \pm 0.20 \pm 0.32$ | $1.30 \pm 0.12$ | $16.7 \pm 2.1$ |
| $\Sigma^0\bar{\Sigma}^0$ | $2.35 \pm 0.36 \pm 0.32$ | $1.27 \pm 0.17$ | $16.8 \pm 3.6$ |
| $\Xi^-\bar{\Xi}^+$ | $3.03 \pm 0.40 \pm 0.32$ | $1.8 \pm 0.4$ | $16.8 \pm 4.7$ |
| $p\bar{n}\pi^-$ | $2.45 \pm 0.11 \pm 0.21$ | $2.02 \pm 0.17$ | $12.1 \pm 1.6$ |
| $\bar{p}n\pi^+$ | $2.52 \pm 0.12 \pm 0.22$ | $1.93 \pm 0.17$ | $13.1 \pm 1.8$ |
| $p\bar{p}\pi^0$ | $1.32 \pm 0.10 \pm 0.15$ | $1.09 \pm 0.09$ | $12.1 \pm 1.9$ |
| $p\bar{p}\eta$ | $0.58 \pm 0.11 \pm 0.07$ | $2.09 \pm 0.18$ | $2.8 \pm 0.7$ |
| $K^+K^-\pi^+\pi^-\pi^0$ | $11.7 \pm 1.0 \pm 1.5$ | $120 \pm 28$ | $9.8 \pm 2.8$ |
| $\omega K^+K^-$ | $0.59 \pm 0.20 \pm 0.09$ | $16.8 \pm 2.1$ | $14.5 \pm 3.4$ |
| $\omega f_0(1710) \to \omega K^+K^-$ | $0.59 \pm 0.20 \pm 0.09$ | $6.6 \pm 1.3$ | $8.9 \pm 3.8$ |

* PDG 2004

## 2.2. $\psi(2S)$ Baryon Pairs and Multibody Decays

In perturbative QCD picture, hadronic decays of $\psi(2S)$ proceed via an annihilations of $c\bar{c}$ quarks into three gluons or a virtual photon. This model leads to the prediction that the ratio of the branching fraction into a specific final state to the branching fraction of the $J/\psi$ into the same final state should be a constant value of approximately 12%, the correspond ratio to the dilepton final state [5]. This rule, which was previously referred to as the "12% rule", is roughly obeyed for several channels, but fails for others [6]. Table 1 summarizes recent measurements on $\psi(2S)$ decays. These $\psi(2S)$ decays obey the "12% rule" except the $p\bar{p}\eta$ channel [7, 8, 9, 10].

## 3. $\chi_{cJ}$ HADRONIC DECAYS

### 3.1. $\chi_{cJ} \to K^+K^-K^+K^-$

Decays of $\chi_{cJ} \to K^+K^-K^+K^-$ are measured recently [11]. The branching fractions including intermediate states are given in Table 2. The decays of $\chi_{cJ} \to \phi K^+K^-$ is observed for the first time, and the precision of the branching fractions of $\chi_{cJ} \to \phi\phi$ and $\chi_{cJ} \to K^+K^-K^+K^-$ are improved compared with PDG values.

The branching fractions of $\chi_{cJ} \to \phi\phi$ together with BES previous measurements on $\chi_{cJ} \to \omega\omega$ [12] and $\chi_{cJ} \to K^*(892)\bar{K}^*(892)$ [13] are used to predict the decay branching fractions of $\chi_{cJ}$ to other vector meson pairs, such as $\rho\rho$ and $\omega\phi$ [14], large double OZI suppressed amplitude is expected.

## 3.2. $\chi_{cJ} \to K_S^0 K\pi, \eta\pi^+\pi^-$

Decays of $\chi_{c0}$ and $\chi_{c2}$ into three pseudoscalars are highly suppressed by the spin-parity selection rule. BES recently measures the branching fraction of $\chi_{c1}$ decays into $K_S^0 K^+\pi^- + c.c.$ and $\eta\pi^+\pi^-$ including intermediate states [15]. The branching fraction or upper limits at 90% CL are summarized in Table 2.

The $K_S^0 K^+\pi^- + c.c.$ events are mainly produced via $K^*(892)$ intermediate state, and the $\eta\pi^+\pi^-$ events via $f_2(1270)\eta$ or $a_0(980)\pi$. The branching fractions with these resonances are $\mathscr{B}(\chi_{c1} \to K^*(892)\bar{K}^0 + c.c.) = (1.1 \pm 0.4 \pm 0.1) \times 10^{-3}$, $\mathscr{B}(\chi_{c1} \to K^*(892)^+K^- + c.c.) = (1.6 \pm 0.7 \pm 0.2) \times 10^{-3}$, $\mathscr{B}(\chi_{c1} \to a_0(980)^+\pi^- + c.c.) = (2.0 \pm 0.5 \pm 0.5) \times 10^{-3}$, $\mathscr{B}(\chi_{c1} \to f_2(1270)\eta) = (3.0 \pm 0.7 \pm 0.5) \times 10^{-3}$. Except for $\chi_{c1} \to K_S^0 K^+\pi^- + c.c.$, all other models are first observation.

## 3.3. $\chi_{cJ} \to p\bar{n}\pi^-\pi^0, p\bar{n}\pi^-$

For $\psi(2S) \to p\bar{n}\pi^-, \bar{p}n\pi^+$ and $p\bar{n}\pi^-\pi^0$, the $\chi_{cJ}$ decays into the same finale states are backgrounds which be considered [8]. Compared with the decays of $\chi_{cJ}$ into $\pi^+\pi^- p\bar{p}$ and $\pi^+\pi^-\Lambda\bar{\Lambda}$, the $\chi_{cJ}$ decays into $p\bar{n}\pi^-\pi^0$ are highly suppressed such that no significant signals are observed. The upper limit at 90% CL is put on the decay $\Sigma_{J=0}^2 \mathscr{B}(\psi(2S) \to \gamma\chi_{cJ} \to \gamma p\bar{n}\pi^-\pi^0) < 1.2 \times 10^{-4}$. The joint branching fractions are $\mathscr{B}(\psi(2S) \to \gamma\chi_{c0} \to \gamma p\bar{n}\pi^-) = (1.10 \pm 0.24 \pm 0.18) \times 10^{-4}$ and $\mathscr{B}(\psi(2S) \to \gamma\chi_{c2} \to \gamma p\bar{n}\pi^-) = (0.97 \pm 0.20 \pm 0.26) \times 10^{-4}$.

## 3.4. $\chi_{cJ} \to \Xi^-\bar{\Xi}^+$

The importance of the Color Octet Mechanism (COM) for $\chi_{cJ}$ decays has been pointed out for many years [16], and the theoretical predictions on the two-body exclusive decays have been made based on it. Recently, some experimental results on $\chi_{cJ}$ exclusive decays have been reported[17, 18]. COM predictions on some decays of $\chi_{cJ}$ into meson pairs are in agreement with the experimental values, while predictions from some decays into baryon pairs are not. For further testing of the COM of P-wave charmonia, the measurement of other baryon pair decays of $\chi_{cJ}$, such as $\chi_{cJ} \to \Xi^-\bar{\Xi}^+$ and $\Sigma^0\bar{\Sigma}^0$ are desired.

The measurement of $\chi_{cJ} \to \Xi^-\bar{\Xi}^+$ is helpful for understanding the Helicity Selection Rule (HSR) [19]. The measured branching fractions for $\chi_{c0}$ decays into $p\bar{p}$ and $\Lambda\bar{\Lambda}$ do not vanish, demonstrating a strong violation of HSR in charmonium decays. Measurements of $\chi_{c0}$ decays into baryon antibaryon pairs provide additional test of the HSR.

The measured results are summarized in Table 2. Theoretically, the quark creation model [20] predicts $\mathscr{B}(\chi_{c0} \to \Xi^-\bar{\Xi}^+) = (2.3 \pm 0.7) \times 10^{-4}$, which is consistent with the experimental value within $1\sigma$. For $\chi_{c1}$ and $\chi_{c2}$ decays into $\Xi^-\bar{\Xi}^+$, the measured upper limits cover both the COM and the quark creation model predictions. Within $2\sigma$, the

**TABLE 2.** Summary of $\chi_{cJ}$ hadronic decays. The upper limit are given at 90% CL

| mode | $\mathscr{B}(\chi_{c0} \to X)(\times 10^{-3})$ | $\mathscr{B}(\chi_{c1} \to X)(\times 10^{-3})$ | $\mathscr{B}(\chi_{c2} \to X)(\times 10^{-3})$ |
|---|---|---|---|
| $2(K^+K^-)$ | $3.47 \pm 0.22 \pm 0.48$ | $0.68 \pm 0.13 \pm 0.10$ | $1.88 \pm 0.18 \pm 0.25$ |
| $\phi K^+K^-$ | $1.02 \pm 0.22 \pm 0.15$ | $0.44 \pm 0.14 \pm 0.07$ | $1.46 \pm 0.21 \pm 0.22$ |
| $\phi\phi$ | $0.94 \pm 0.21 \pm 0.14$ | $-$ | $1.48 \pm 0.26 \pm 0.23$ |
| $K_S^0 K^+ \pi^- + c.c.$ | $< 0.35$ | $4.0 \pm 0.3 \pm 0.5$ | $2.46 \pm 0.44 \pm 0.65$ |
| $\eta\pi^+\pi^-$ | $< 1.1$ | $5.9 \pm 0.7 \pm 0.8$ | $< 1.7$ |
| $\Xi^-\bar{\Xi}^+$ | $0.53 \pm 0.27 \pm 0.09$ | $< 0.34$ | $< 0.37$ |

branching fraction of $\chi_{c0} \to \Xi^-\bar{\Xi}^+$ doesn't vanish. For further testing of the violation of the HSR in this decay, higher accuracy measurement are required.

## ACKNOWLEDGMENTS

I would like to thank the organizer rot the invitation to give the talk. I thank my colleagues in BES Collaboration for the helpful discussions. This work was supported in part by National Science Foundation of China (10491303).

## REFERENCES

1. BES Collaboration, J. Z. Bai *et al.*, *Nuc. Inst. Meth.* **A344**, 319–334 (1994).
2. BES Collaboration, J. Z. Bai *et al.*, *Nuc. Inst. Meth.* **A458**, 627–637 (2001).
3. BES Collaboration, J. Z. Bai *et al.*, *Phys. Rev.* **D65**, 052004 (2002).
4. BES Collaboration, M. Ablikim *et al.*, hep-ex/0609023.
5. W. S. Hou and A. Soni, *Phys. Rev. Lett.* **50**, 569 (1983); Mark II Collaboration, M. E. B. Franklin *et al.*, *Phys. Rev. Lett.* 51, 963 (1983); W. S. Hou, *Phys. Rev.* **D55**, 6952 (1997); Y. F. Gu and X. H. Li, *Phys. Rev.* **D63**, 114019 (2001).
6. BES Collaboration, J. Z. Bai *et al.*, *Phys. Rev.* **D67**, 052002 (2003); 69, 072001 (2004); BES Collaboration, M. Ablikim *et al.*, *Phys. Rev.* **D70**, 112003 (2004); 70, 112007 (2004); CLEO Collaboration, N. E. Adam *et al.*, *Phys. Rev. Lett.* **94**, 012005 (2005).
7. BES Collaboration, M. Ablikim *et al.*, hep-ex/0610079.
8. BES Collaboration, M. Ablikim *et al.*, *Phys. Rev.* **D74**, 12004 (2006).
9. BES Collaboration, M. Ablikim *et al.*, *Phys. Rev.* **D71**, 072006 (2005).
10. BES Collaboration, M. Ablikim *et al.*, *Phys. Rev.* **D73**, 052004 (2006).
11. BES Collaboration, M. Ablikim *et al.*, hep-ex/0607025.
12. BES Collaboration, M. Ablikim *et al.*, *Phys. Lett.* **B630**, 7 (2005).
13. BES Collaboration, M. Ablikim *et al.*, *Phys. Rev.* textbfD70, 092003 (2004).
14. Q. Zhao, *Phys. Rev.* **D72**, 12004 (2006).
15. BES Collaboration, M. Ablikim *et al.*, *Phys. Rev.* textbfD74, 072001 (2006).
16. G. D. Bodwin, E. Braaten and G. P. Lepage, *Phys. Rev.* **D51**, 1129 (1995).
17. BES Collaboration, M. Ablikim *et al.*, *Phys. Lett.*, **B630**, 21 (2005); *Phys. Rev.* **D72**, 092002 (2005).
18. BES Collaboration, M. Ablikim *et al.*, *Phys. Rev.* **D67**, 032004 (2003); *Phys. Rev.* **D67**, 112001 (2003); *Phys. Rev.* **D60**, 072001 (1999); *Phys. Rev. Lett.* **81**, 3091 (1998).
19. S. J. Brodsky, G. P. Lepage, *Phys. Rev.* **D24**, 2848 (1981).
20. R. G. Ping, B. S. Zou and H. C. Chiang, *Eur. Phys. J.* **A23**, 129 (2005).

# R measurements with ISR in BaBar - hadronic part of muon magnetic dipole moment

## Paul Taras

*Université de Montréal, Physique des Particules, Montréal, Québec, Canada H3C 3J7*
*representing the BaBar Collaboration*

**Abstract.** Recent measurements of the quantity $R$, the ratio of annihilation $\sigma$, including those following Initial State Radiation, are discussed in the context of the hadronic part of $\mu$, the muon magnetic dipole moment. The data indicate that more precise theoretical and experimental values of $\mu$ are needed to establish whether new physics has been observed in the measurement of $\mu$.

**Keywords:** annihilation cross sections, ISR, muon magnetic moment
**PACS:** 13.66.Bc, 13.25.Gv, 13.25.Jx, 14.40.Cs, 14.60.Ef

## INTRODUCTION

The recent [1] very precise measurement of the muon g-factor hinted at a possibility of new physics because of a small discrepancy with its value expected in the framework of the Standard Model. This presentation will dwell on the extraction of the leading order part of the hadronic component of the muon magnetic dipole moment from the various available values of $R$, the ratio of annihilation cross sections.

## MEASUREMENTS OF R AND THEIR IMPACT ON THE MUON MAGNETIC DIPOLE MOMENT PROBLEM

### The magnetic anomaly in the framework of the Standard Model

It is well known that the magnetic dipole moment of a spin $\frac{1}{2}$ particle is given by: $\vec{\mu} = g\frac{e}{2m}\vec{s}$ where in the case of the muon, $e$, $m$ and $\vec{s}$ are the charge, the mass and the spin of the muon while, in quantum mechanics, the gyromagnetic factor $g \equiv 2$ for all spin $\frac{1}{2}$ particles. However, soon after quantum electrodynamics was developped, it was realized that the value of $g$ needed to be slightly modified. This was written as $g = 2 + 2a$, where by definition, $a \equiv (g-2)/2$ is known as the magnetic anomaly. In the framework of the Standard model, the value of $a$ has three components :

$$a_\mu^{SM} = a_\mu^{QED} + a_\mu^{EW} + a_\mu^{had} \tag{1}$$

The values of $a_\mu^{QED}$, the QED component [2] and $a_\mu^{EW}$, the electroweak processes'component [3] are calculated with high precision, to 5 and 2 loops, respectively, and need not concern us here. The hadronic component, $a_\mu^{had}$ can be broken down into

CP892, *Quark Confinement and the Hadron Spectrum VII*
edited by J. E. F. T. Ribeiro
© 2007 American Institute of Physics 978-0-7354-0396-3/07/$23.00

three parts :

$$a_\mu^{had} = a_\mu^{had,Lo} + a_\mu^{had,Ho} + a_\mu^{had,LBL} \tag{2}$$

$a_\mu^{had,Lo}$, the leading order part of the hadronic interaction arising from the contribution of the hadronic vacuum polarization [4], $a_\mu^{had,Ho}$, the three-loop part of the hadronic interaction involving one hadronic vacuum polarization insertion [4], and $a_\mu^{had,LBL}$, the so-called hadronic "light-by-light" scaterring part [5, 6, 4].

## The relation between $R$ and the magnetic anomaly

The contribution of the hadronic vacuum polarization is calculated via the dispersion integral [4, 7] :

$$a_\mu^{had,Lo} = \left(\frac{\alpha m_\mu}{3\pi}\right)^2 \int_{4m_\pi^2}^{\infty} \frac{K(s)}{s^2} R(s) ds \tag{3}$$

where :

$$R(s) = \frac{\sigma_f(e^+e^- \to hadrons)}{\sigma_f(e^+e^- \to \mu^+\mu^-)} \tag{4}$$

is the ratio of the direct annihilation cross section, $\sigma_f$, to a given hadronic final state $f$ to that of the muon pair production. This ratio $R(s)$ is measured as a function of $\sqrt{s}$, the center of mass energy of the final system $f$. The QED kernel, $K(s)$ varies very little over the whole range [7], from 0.63 at $s = 4m_\pi^2$ to 1 at $s = \infty$. There is, however, a very strong dependence on energy. Because the main theoretical error in $a_\mu^{SM}$ arises from the uncertainties in $a_\mu^{had}$, it is of the utmost importance to measure the values of $R(s)$ as precisely as possible, specially at lower energies since about 91% of the total contribution to $a_\mu^{had,Lo}$ is provided by the cross sections measured at energies $\sqrt{s} \le 1.8$ $GeV$. Monte Carlo calculations show that many channels open up as the energy available for the final state increases : $\pi^+\pi^-\pi^0$, $\pi^+\pi^-\pi^0\pi^0$, $K^+K^-\pi^+\pi^-$, etc ... but the most important final state is $\pi^+\pi^-$ which contributes up to 73% of the total value of $a_\mu^{had,Lo}$.

## Annihilation cross sections

The best existing $e^+e^- \to \pi^+\pi^-$ data are provided by the CMD-2, SND and KLOE collaborations. Although they can be combined to provide a precise contribution of $(376.5 \pm 0.8 \pm 2.4) \times 10^{-10}$ to the value of $a_\mu^{had,Lo}$, there is some discrepancy between the KLOE data set and the others which needs to be resolved. In addition, the pion form factor extracted from the annihilation cross sections using the conserved vector current hypothesis is not in agreement [4] with that obtained from the branching fraction $BF(\tau^- \to \pi^-\pi^0\nu_\tau)$, once all sources of isospin symmetry breaking have been taken into account. This may be due to the use of incorrect spectral functions and merits further investigation. In any case, it is clear that an independent measurement of this very important cross section is required.

**TABLE 1.** Samples of contributions to $a_\mu^{had,Lo}$ including *BaBar* data following ISR

| Integrated luminosity $fb^{-1}$ | Final state | $a_\mu^{had,Lo} \times 10^{10}$ without *BaBar* data | $a_\mu^{had,Lo} \times 10^{10}$ with *BaBar* data |
|---|---|---|---|
| 89.3 | $\pi^+\pi^-\pi^0$ | $2.45 \pm 0.26 \pm 0.03$ | $3.25 \pm 0.09 \pm 0.01$ |
| 89.3 | $2\pi^+2\pi^-$ | $14.20 \pm 0.87 \pm 0.24$ | $13.09 \pm 0.44 \pm 0.00$ |
| 232 | $3\pi^+3\pi^-$ | $0.10 \pm 0.10$ | $0.108 \pm 0.016$ |
| 232 | $2\pi^+2\pi^-2\pi^0$ | $1.42 \pm 0.30 \pm 0.03$ | $0.890 \pm 0.093$ |

In *BaBar*, the annihilation cross sections for the final states of interest are measured following Initial State Radiation. Depending on the energy of the radiated photon, the cross section for the reconstructed final state can be obtained from threshold to nominally $\sqrt{s}$, the c.m. energy of the PEP-II collider in a single experiment at $\sqrt{s}$. This program is made possible by the high luminosity of our collider and the excellent *BaBar* detector. The statistics obtained are comparable to those of CMD-2 and SND at $E_{cm} \leq 1.4$ *GeV*, and much better than those of DM1 and DM2 at higher energies. The measurements following ISR contain little background and should thus lead to high accuracy even for most of the exclusive decays which have a rather low cross section.

Only a portion of the available data is presented here to illustrate the precision achieved with ISR in the *BaBar* experiment. These are given in Table 1 for the $e^+e^- \rightarrow \pi^+\pi^-\pi^0$ [9], $e^+e^- \rightarrow 2\pi^+2\pi^-$ [10], $e^+e^- \rightarrow 3\pi^+3\pi^-$ [11] and $e^+e^- \rightarrow 2\pi^+2\pi^-2\pi^0$ [11] channels. In the table, we list the contributions to the leading order hadronic part, $a_\mu^{had,Lo}$, first without taking into account the *BaBar* data, and then doing so. The considerable gain in precision in doing so (factor of 2 to 5) results from the fact that the present *BaBar* data is more accurate and cover a larger energy range than previous data for these same final states. We should note that the large improvement in accuracy in the $\pi^+\pi^-\pi^0$ channel is due in part to the removal of the DM2 data shown to be incorrect by our data.

## DISCUSSION AND CONCLUSIONS

The precisely measured [1] value of the magnetic anomaly : $a_\mu^{exp} = 11659208.0 \pm 6.3) \times 10^{-10}$ can be compared to the value computed in the framework of the standard model, $a_\mu^{SM} = 11659180.5 \pm 5.6_{had} \pm 0.2_{QED+EW}) \times 10^{-10}$ using all known annihilation cross section data, as reported by Eidelman at ICHEP06 [7]. This value includes the three hadronic contributions : $a_\mu^{had,Lo} = (690.9 \pm 3.9 \pm 2.0) \times 10^{-10}$ [7], $a_\mu^{had,Ho} = -(9.8 \pm 0.1) \times 10^{-10}$ [4] and $a_\mu^{had,LBL} = +(12.0 \pm 3.5) \times 10^{-10}$ [4, 5, 6]. The last two terms are rather small but still significant in light of the high precision achieved experimentally. Comparing the values, it appears that there is a difference of $3.3\sigma$ between the measured and the phenomenological values. Is this a sign of New Physics?

Let us consider the results shown in Fig. 1 where the band represents the experimental value while the full lines show the values obtained from the annihilation cross sections and the dashed line that from the $\tau$ decay branching fraction [4]. It is clear that although

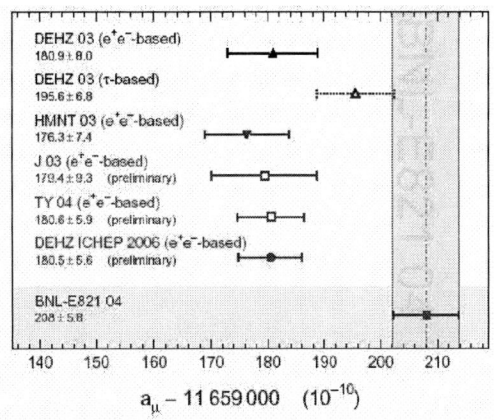

$$a_\mu - 11\,659\,000 \quad (10^{-10})$$

**FIGURE 1.** Experimental and SM values of the magnetic anomaly

there is a significant difference between the experimental value and the one based on the cross sections, the value based on the branching fraction is compatible with the measurement. In addition, as reported [12] at ICHEP06, recent lattice QCD calculations could also be compatible with the measurement. Furthermore, my own logic dictates that the theoretical errors generated in the computation of $a_\mu^{had}$ must be increased. The same point was made by the summary conference speaker [13] at ICHEP06. Thus, more precise theoretical and experimental values of $\mu$ are needed to establish whether new physics has indeed been observed in the measurement of the muon magnetic moment.

## ACKNOWLEDGMENTS

M. Davier, S. Eidelman and E. Solodov are thanked for many fruitful discussions. This work is partially supported by NSERC (Canada).

## REFERENCES

1. G.W. Bennett *et al.*, *Phys. Rev.* **D73**, 072003 (2006).
2. G. Gabrielse *et al.*, *Phys. Rev. Lett.* **97**, 030802 (2006).
3. A. Czarnecki, W. Marciano, A. Vainshtein, *Phys. Rev.* **D67**, 073006 (2003)
4. M. Davier, W. Marciano, *Annu. Rev. Nucl. Part. Sci.* **54**, 115 (2004).
5. M. Knecht, A. Nyffeler, *Phys. Rev. Lett.* **88**, 071802 (2002).
6. K. Melnikov, A. Vainshtein, *hep-ph/0312226*.
7. S. Eidelman, *International Conference on High Energy Physics*, Moscow, 2006, contributed paper.
8. R. Akhmetsin *et al.*, CMD-2 collaboration, hep-ex/0308008
9. B. Aubert *et al.* (*BaBar* collaboration), *Phys. Rev.* **D70**, 072004 (2004).
10. B. Aubert *et al.* (*BaBar* collaboration), *Phys. Rev.* **D71**, 052001 (2005).
11. B. Aubert *et al.* (*BaBar* collaboration), *Phys. Rev.* **D73**, 052003 (2006).
12. G. Schierholz , *International Conference on High Energy Physics*, Moscow, 2006, invited paper.
13. V. A. Rubakov, *International Conference on High Energy Physics*, Moscow, 2006, summary paper.

# Test of non point-like behavior of Fermions

Urs Burch[*], Chih-Hsun Lin[†], André Rubbia [*], Alexander S. Sakharov[**,*],
Jürgen Ulbricht[*], Jian Wu[‡] and Jiawei Zhao[‡]

[*]Swiss Institute of Technology, ETH-Zürich, 8093 Zürich, Switzerland
[†]National Central University, Jhungli 320, Taiwan
[**]TH Division, Department of Physics, CERN, CH-1211 Geneva 23, Switzerland
[‡]University of Science and Technology of China, Hefei Anhui 230029, China

**Abstract.** We search in measurements from LEP, TRISTAN, CDF, D0 and UA1 for those phenomena beyond the Standard Model where the fundamental particles can exhibit non point-like behavior. The total and differential cross sections for the reaction $e^+e^- \to \gamma\gamma(\gamma)$ measured at center-of-mass energies from 55 GeV to 207 GeV are used for a global fit to test the hypothesis of an excited electron and contact interaction induced terms. The best value for the fit exhibits an approximately 5 $\sigma$ effect for the scale factor of an excited electron of $(1/\Lambda_\pm^4)_{best} = -(1.11 \pm 0.20) \times 10^{-10}$ $GeV^{-4}$ and the scale of contact interaction of $(1/\Lambda^4)_{best} = -(4.05 \pm 0.73) \times 10^{-13}$ $GeV^{-4}$ but with the opposite to the expected in the modified cross section sign. An empirical scheme is discussed to encompass all fermions and bosons in one framework, if these particles would be geometrically extended objects.

**Keywords:** Experimental tests, Composite models, Leptons, Quarks
**PACS:** PACS numbers: 11.10.Ef, 12.15.-y, 12.15.Ff, 12.20.-m, 12.20.Fv, 12.60.Rc, 13.66.Jn.

**1.** To test the finite size of fundamental particles (FP), experiments are performed to search for compositeness or to investigate a non-point like behavior in strong, electromagnetic and weak interaction. Each test is assumed to be characterized by its own energy scale, which can be related to the size of the interaction region.

**2.** In the case of the strong interaction the CDF $p\bar{p}$ data analysis [1] exclude excited quarks $q^*$ with a mass between 200 and 760 GeV at 95% CL. The UA2 data [2] exclude $u^*$ and $d^*$ quark masses smaller than 288 GeV at 90% CL. In this case characteristic energy scale is given by the mass of the excited quark and can be converted into characteristic size $r_q \sim \hbar/(m_q^* c) < 3.5 \times 10^{-17}$ cm.

**3.** In the case of electromagnetic interaction the process $e^+e^- \to \gamma\gamma(\gamma)$ is ideal to test the QED because it is not interfered by the $Z^0$ decay. We test the model of an exited electron [3] and model of contact interaction of the electron [4].

In the model of the exited electron the heavy electron with mass $m_{e^*}$ couple to an electron and photon via magnetic interaction with an effective Lagrangian of

$$\mathcal{L}_{\text{excited}} = ((e\lambda)/(2m_{e^*}))\overline{\psi_{e^*}}\sigma_{\mu\nu}\psi_e F^{\mu\nu} \tag{1}$$

In this equation $\lambda$ is the coupling constant, $F^{\mu\nu}$ the electromagnetic field tensor, $\psi_{e^*}$ and $\psi_e$ are the wave function of the heavy electron and the electron respectively. The model has $\lambda$ and $m_{e^*}$ as parameters. The corresponding differential cross-section for the reaction $e^+e^- \to \gamma\gamma(\gamma)$ used for the result reported here is given by the QED differential cross-section calculated up to $O(\alpha^3)$ radiative effects modified by adding a deviation

CP892, *Quark Confinement and the Hadron Spectrum VII*
edited by J. E. F. T. Ribeiro
© 2007 American Institute of Physics 978-0-7354-0396-3/07/$23.00

term $\delta_{new}$

$$(d\sigma/d\Omega)_{theo} = (d\sigma/d\Omega)_{O(\alpha^3)}(1+\delta_{new}). \tag{2}$$

If the center-of-mass energy $\sqrt{s}$ satisfies the condition $s/m_{e^*}^2 << 1$, then $\delta_{new}$ would read as

$$\delta_{new} = \pm s^2/2(1/\Lambda_{\pm}^4)(1-\cos^2\Theta) \tag{3}$$

In this approximation, the parameters $\Lambda_{\pm}$ are the QED cut-off parameters with $\Lambda_+^2 = m_{e^*}^2/\lambda$. The angle $\Theta$ is the scattering angle of the two most energetic photons emitted.

In the case of effective contact interaction with non-standard coupling a cut-off parameter $\Lambda$ is introduced to describe the scale of a dimension 6 operator [4]

$$\mathscr{L}_{contact} = i\overline{\psi_e}\gamma_\mu(D_v\psi_e)(\sqrt{4\pi})((1/\tilde{\Lambda}_6^2)F^{\mu\nu} + (1/\tilde{\Lambda}_6^2)\tilde{F}^{\mu\nu}), \tag{4}$$

where $\psi_e$ stands again for the wave function of the electron and $D_v$ is the QED covariant derivative. As in the case of excited electron the $e^+e^- \to \gamma\gamma(\gamma)$ differential cross-section is given by Eq.(2) where $\delta_{new}$ reads as

$$\delta_{new} = s^2/(2\alpha)(1/\Lambda_6^4 + 1/\tilde{\Lambda}_6^4)(1-\cos^2\Theta). \tag{5}$$

We performed a global analysis of all published measured differential cross sections for the reaction $e^+e^- \to \gamma\gamma(\gamma)$. The data has been taken in the energy range from $\sqrt{s} = 55$ GeV to $\sqrt{s} = 207$ GeV with VENUS [5], TOPAZ [6], ALEPH [7], DELPHI [8], L3 [9] and OPAL [10] detector during the period from 1989 to 2003. We applied a $\chi^2$ test to the whole available data set using $1/\Lambda_{\pm}^4$ for the hypothesis of an exited electron and $1/\Lambda^4$ for the contact interaction term Eq.(4) as fit parameters (the condition $\Lambda_6 = \tilde{\Lambda}_6 = \Lambda$ is imposed). Both hypothesis Eq.(3) and Eq.(5) assume an increase of the differential and total QED-$\alpha^3$ cross section Eq.(2). On the contrary we found that the best fit values $(1/\Lambda_{\pm}^4)_{best} = -(1.11 \pm 0.20) \times 10^{-10}$ $GeV^{-4}$ [11] and $(1/\Lambda^4)_{best} = -(4.05 \pm 0.73) \times 10^{-13}$ $GeV^{-4}$ are negative. The fit does not allow to distinguishing between both hypothesis. The results indicate that the used data set prefers to decrease the cross section of $e^+e^- \to \gamma\gamma(\gamma)$ with respect to that predicted by pure QED calculations. Using the best values of $(1/\Lambda_{\pm})^4$ and $(1/\Lambda)^4$ we can calculate the mass of an exited electron to $m_{e^*} = 308$ GeV and the scale factor $\Lambda = 1253.2$ GeV which translates in a size of the interaction area of $r \approx 15.7 \times 10^{-18}$ cm.

**3.** The $ep$ accelerator HERA and the LEP tested the exited and non-point like couplings of quarks and leptons. In the initial state the reaction involved in these tests proceeds via magnetic and weak interaction while in the final state the strong interaction contributes as well. The LEP groups [12] uses the reaction $e^+e^- \to$ ff to search for the non-point like coupling. Nine models of contact interaction in the reaction $e^+e^- \to$ ff with four possible helicity amplitudes have been tested and 95% CL limits have been established for quarks ranging from $R_-^q < 2.5 \times 10^{-18}$ cm to $R_+^q < 2.2 \times 10^{-18}$ cm and for leptons from $R_-^l < 1.3 \times 10^{-18}$ cm to $R_+^l < 0.9 \times 10^{-18}$ cm. A 2.6 $\sigma$ effect for axial-vector contact interaction in the data on $e^+e^- \to e^+e^-(\gamma)$ at center- of-mass energies $192 - 208$ GeV is reported [13].

**4.** The construction of a geometrically extended FP is still illusive goal. Nevertheless, the experimental expectations described above in particular the electromagnetic tests inspire us to try to understand what could be the simplest geometrical structure to model an extended core of a FP if there exists one if any. Here we present some empirical toy ansatz of that how such an object could be constructed to encompass some basic properties of known FPs. We assume that the core of a FP can appear at very early cosmic time when the gravity was dominating over the other interactions. The credible structure of the object is presented in Fig.1 (left panel). In the context of unification of all interactions [14] including gravity the core mass density $\rho$ scales from a very high, probably the Plank one, down to zero at the surface of the object. Thus, the object under consideration should contain every energy state a cross of its radius which the Universe passed through during its evolution [15].

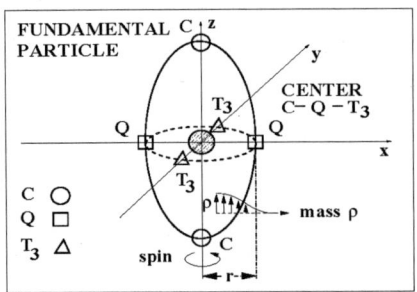 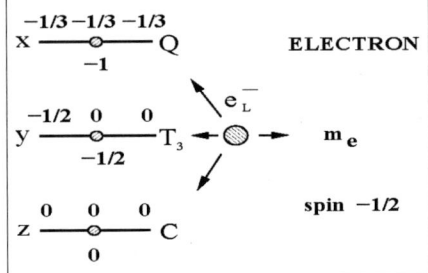

**FIGURE 1.** Extension of a FP ( left geometry; right scheme )

Further, to assign the interaction properties of a FP modeled with the object described above we represent it as an aggregate of electric, weak and color charges distributed along the different space-like axes. For simplicity, we consider only pointlike charges while a continues charge distributions are also possible. It would be possible to place on the x-coordinate the electric charges Q in three subunits ($Q = \pm 1/3$ left, $Q = \pm 1/3$ center, $Q = \pm 1/3$ right; the spin we used to tag the z-axis), similar on the y-coordinate three places for the weak charge $T_3 = \pm 1/2$ are open and at the z-axis it is possible in the same manner to place three color charges C (Red $R, \bar{R}$; Green $G, \bar{G}$ and Blue $B, \bar{B}$). In Fig.1 (right panel) we simplify the three dimensional picture to a scheme for sorting all FP with respective to their properties and demonstrate this for the case of a left handed electron $e_L^-$. Above the x-axis are the three electric sub-charges $Q = -1/3$ (left, center and right) below is the integral of the charge $Q = -1$ seen from an observer far outside, similar above the y-axis one open position is taken from the weak charge $T_3 = -1/2$ the integral below is the same as above and no color exist on the z-axis. We sort all fermions, eight gluons, the $\gamma$ and the $Z^0$ with the $W^\pm$ about the rules of this scheme, as example Fig.2 (left) displays the left handed lightest fermion family. The two more heavy families differing in the rest mass of the FP from the first family, all features of Fig.2 (left) stay the same. To complete the scheme it is sufficient to sort the masses of all fermions according there rest mass. We use the three possible ground states of the x- ,y-and z-axis of a harmonic oscillator to calculate the fermion and anti-fermion masses of all three families

$$E(k_i; Q) = (A + B|Q| + CQ^2 + D|Q|^3)(k_i)^{f(Q,k_i)} \tag{6}$$
$$f(Q,k_i) = (R + |Q|V(k_i - 1) + |Q|(|Q| - 1) + (S(|Q| - 1/3) + W(|Q| - 1/3)(k_i - 1)$$
$$+ T(|Q| - 2/3) + Z(|Q| - 2/3)(k_i - 1))).$$

470

The parameter $k_i$, which stands for the number of the fermion family and the charge $Q$ allow to calculate the mass of the three families of fermions. The constant factors $A < 3 \times 10^{-6} MeV$, $B = 42.1358\,MeV$, $C = -87.7995\,MeV$, $D = 46.1747\,MeV$, $R = 7.9617$, $V = -0.2698$, $S = 5.2528$, $W = -19.3806$, $T = -77.3429$ and $Z = 26.8191$ are adjusted to the measured masses of the fermions [16]. The comparison between these masses and the calculation of Eq. 6 is displayed in Fig. 2 (right). For the neutrinos $v_e$, $v_\mu$ and $v_\tau$ exist mass limits only. We adjust the calculation (solid line) to the limit of $v_e$. We like to stress that Eq. 6 has only the intention to sort the masses of all fermions.

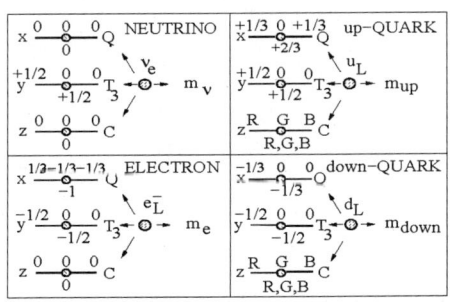 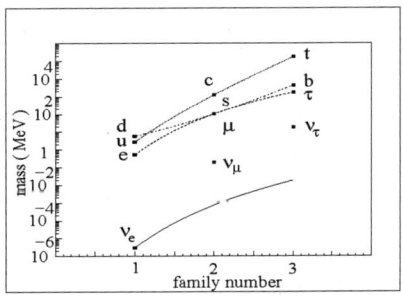

**FIGURE 2.** Scheme of the family of lightest fermions ( left ) and mass of all three families ( right ). Squares are the experimental masses of FPs and the lines are the calculation of Eq. 6

We conclude the scheme would be a heritage of the time development of the Big Bang and Standard Theory frozen in the finite size of the FP. The three space coordinates are linked with the three charges strong, electromagnetic and weak. The scheme interpretes the three particle families as different modes of a three dimensional oscillator.

## REFERENCES

1. F. Abe *et al.* (CDF), Phys. Rev. D **55** (1997) R5263; M. P. Giordani (CDF and D0), Eur. Phys. J. C **33** (2004) 785.
2. J. Allitti *et al.* (UA2), Nucl. Phys. B **400** (1993) 3.
3. A. M. Litke, PhD Thesis (Harvard University, 1970); S. D. Drell, Ann. Phys. **4** (1958) 75 ; F. E. Low, Phys. Rev. Lett. **14** (1965) 238.
4. O. J. P. Eboli *et al.* , Phys. Lett. B **271** (1991) 274; P. Mery, M. Perrottet and F.M. Renard, Z. Phys. C **38** (1980) 579; S.J. Brodsky and S.D. Drell, Phys. Rev. D **22** (1980) 2236.
5. K. Abe *et al.* (VENUS), Z. Phys. C **45** (1989) 175.
6. K. Shimozawa *et al.* (TOPAS) Phys. Lett. B **284** (1992) 144.
7. D. Decamp *et al.* (ALEPH) Phys. Rep. **216** (1992) 253.
8. P. Abreu *et al.* (DELPHI) Phys. Lett. B **327** (1994) 386; P. Abreu *et al.* (DELPHI) Phys. Lett. B **433** (1998) 429; P. Abreu *et al.* (DELPHI) Phys. Lett. B **491** (2000) 67.
9. P. Achard *et al.* (L3) Phys. Lett. B **531** (2002) 28.
10. M.Z. Akrawy *et al.* (OPAL) Phys. Lett. B **275** (1991) 531; G. Abbiendi *et al.* (OPAL) Eur. Phys. J. C **26** (2003) 331.
11. U. Burch *et al.* , Proc. 9th Conf. of Astroparticle, Particles and Space Physics, Detectors and Medical Physics Application (Como-2005) (2006) 643.
12. The LEP Collaborations *et al.* , hep-ex/0511027.
13. D. Bourilkov, Phys. Rev. D **64** (2001) R071701.
14. W. de Boer and J. H. Kühn, Phys. Bl. **47** ( 1991 ) No.11 995.
15. O. Lahav and A.R Liddle, astro-ph/0601168.
16. Particle Data Group, Phys. Lett. B **592** (2004) 1.

# New charmonium-like resonances at Belle

A. Zupanc

(for the Belle collaboration)

*Jozef Stefan Institute, Jamova 21, 1000 Ljubljana, Slovenia.*

**Abstract.** We give an overview on $X(3872)$, $Z(3930)$, $Y(3940)$ and $X(3940)$ resonances observed by the Belle detector at asymmetric energy $e^+e^-$ KEKB collider and consider their possible interpretations.

**Keywords:** charmonium, new resonances, Belle
**PACS:** 14.40.Gx, 13.20.He, 13.66.Bc

## INTRODUCTION

The Belle detector [1] at the KEKB factory [2] has accumulated around 630 fb$^{-1}$ by Summer 2006. While the main goal of $B$ factories are measurements of $CP$ violation in the $B$ meson system, the excellent detector performance also makes possible searches for new hadronic (bound) states as well as studies of their properties. There are several possible mechanisms of the particle production at $B$ factories: production in the $B$ meson decays, fragmentation of quarks in $e^+e^-$ annihilation or creation of $C$ even states in two photon processes. In this review we report on several new states observed recently, namely $X(3872)$, $Z(3930)$, $Y(3940)$ and $X(3940)$.

## OBSERVATION AND PROPERTIES OF X(3872)

In 2003 Belle published the first observation of a narrow charmonium-like state, named $X(3872)$, decaying to $\pi^+\pi^-J/\psi$ in the exclusive decay of $B^\pm \to K^\pm X(3872)$ [3]. The observation was later confirmed by CDF [4], D0 [5] and BaBar [6] experiments. The world average on the mass is currently $M(X(3872)) = 3871.2 \pm 0.5$ MeV/c$^2$ [7] and the upper limit on its width, as measured by Belle, is $\Gamma < 2.3$ MeV [3]. The most recent update from Belle is shown in Fig. 1(a) [8].

Several interpretations of $X(3872)$ resonance have been suggested, including charmonium hypothesis [9, 10], $D^0\bar{D}^{*0}$ molecule [11] and tetraquarks [12]. Several dedicated studies to determine possible quantum numbers of $X(3872)$ and its nature were performed at Belle.

In 2005 Belle reported a strong evidence for the radiative decay of $X(3872) \to \gamma J/\psi$ [13]. The fitted yield of reconstructed B mesons, as obtained from the simultaneous fit to the $\Delta E$ and $M_{bc}$ distributions, in the $B \to K\gamma J/\psi$ is shown in Fig. 1(b) as a function of the $M(\gamma J/\psi)$. The signal has a significance above $4\sigma$ and is used to determine $\mathcal{B}(X(3872) \to \gamma J/\psi)/\mathcal{B}(X(3872) \to \pi^+\pi^-J/\psi) = 0.14 \pm 0.05$, which is not in

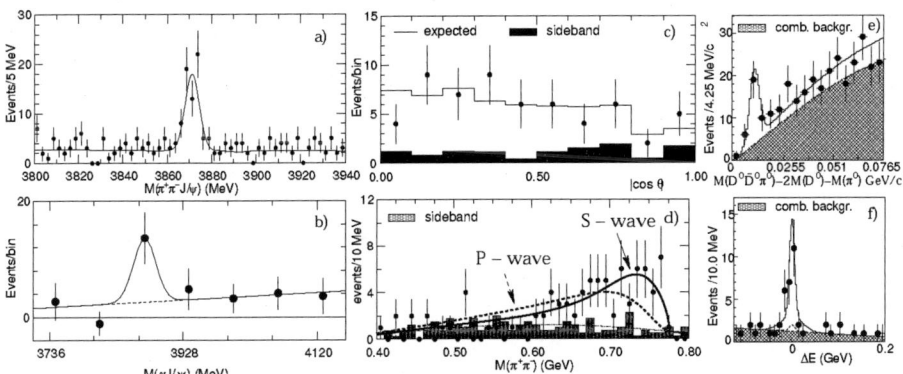

**FIGURE 1.** (a) $\pi^+\pi^- J/\psi$ invariant mass obtained by Belle using a sample of 275 million $B\bar{B}$ pairs. (b) Yield of $B$ mesons in $B \to K\gamma J/\psi$ as a function of $M(\gamma J/\psi)$. (c) Distribution of angle in $X(3872)$ decays described in the text. The full histogram represents the expectation for $J^{PC} = 1^{++}$ assignment and the hatched histogram is the contribution of background as obtained from the scaled sidebands of $M(\pi^+\pi^- J/\psi)$. (d) $M(\pi^+\pi^-)$ distribution for events in $X(3872)$ signal region. The solid (dashed) shows the results of the fits that used $\rho$ Breit-Wigner function that assumes $J/\psi$ and $\rho$ to be in a relative S- or in P-wave. The histogram indicates the sideband determined background. (e) and (f) Near-threshold $D^0\bar{D}^0\pi^0$ enhancement in $B \to KD^0\bar{D}^0\pi^0$ decay. The solid line is the fitting function and the dashed line indicates the total background. Plots a, c and d are reprinted from [8], plot b from [13], and plots e and f from [14].

agreement with the expectations for charmonium interpretation of $X(3872)$. The existence of this radiative decay establishes positive charge conjugation parity of $X(3872)$.

Belle examined possible $J^{PC} = 0, 1, 2$ quantum number assignments of $X(3872)$ by studying angular correlations between final state particles in the $X(3872) \to \pi^+\pi^- J/\psi$ decays [8]. One example is presented in Fig. 1(c): the measured distribution of the angle between the negative $B$ meson direction and $\pi$ from $X(3872)$ in the $X(3872)$ frame, is in agreement with the expectation for the $1^{++}$ state. The $\pi^+\pi^-$ invariant mass distribution (for the events in the $X(3872) \to \pi^+\pi^- J/\psi$ signal region [8]), shown in Fig 1(d), peaks at the upper kinematical limit indicating the $\rho J/\psi$ intermediate state and favors S-wave over P-wave as the relative orbital angular momentum between the final-state dipion and $J/\psi$. As a consequence, $J^{PC} = 1^{++}$ is strongly favored for the $X(3872)$, but the $2^{++}$ was not ruled out by these tests.

The latter possibility could be ruled out by the recent observation of a near-threshold $D^0\bar{D}^0\pi^0$ invariant mass enhancement at $(3875.4 \pm 0.7 \pm 1.1)$ MeV/c$^2$ in $B \to KD^0\bar{D}^0\pi^0$ decays (see Fig. 1(e, f)) [14]. The observed mass is around $2\sigma$ higher than the world average value for $X(3872)$. If the observed near-threshold enhancement is due to the $X(3872)$, the $J^{PC} = 1^{++}$ quantum number assignment for the $X(3872)$ is favored, since near-threshold decays $X(3872) \to D^0\bar{D}^{*0}/D^0\bar{D}^0\pi^0$ are expected to be strongly suppressed for $J = 2$.

While currently available data agree with the hypothesis that $X(3872)$ is a $D^0\bar{D}^{*0}$ molecule [11], some spin assignments corresponding to more conventional interpretations cannot be ruled out (see for example [15]). The molecular interpretation, however, successfully explains the mass, $1^{++}$ quantum numbers and the decay modes.

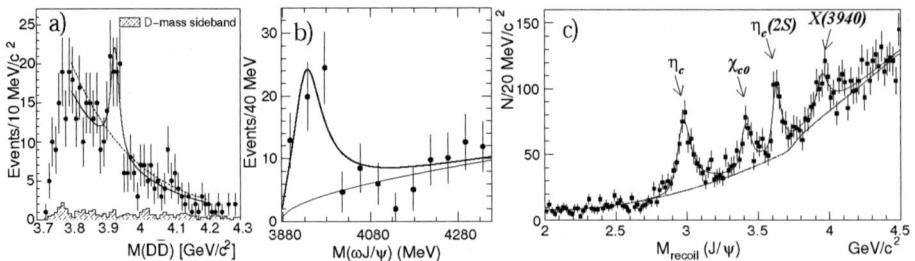

**FIGURE 2.** (a) Invariant mass of $D\bar{D}$ produced in two photon reactions. (b) Yield of $B$ mesons in $B \to K\omega J/\psi$ as a function of $M(\omega J/\psi)$. (c) Spectrum of mass recoiling against the $J/\psi$. Plots a, b and c are reprinted from [16], [17] and [19], respectively.

## OBSERVATION OF Z(3930) IN TWO PHOTON PRODUCTION

A search for the $\chi'_{cJ}$ ($J = 0$ or 2) states and other C-even charmonium states in the mass range of 3.73 GeV/c$^2$ – 4.3 GeV/c$^2$ was carried out in the $\gamma\gamma \to D\bar{D}$ transitions [16]. The two-photon process $e^+e^- \to e^+e^- D\bar{D}$ was studied in the un-tagged mode, where final state positron and electron are not detected, and the $D\bar{D}$ system has a very small transverse momentum, to enhance the exclusive production from quasi-real two-photon collisions. The $D$ mesons were reconstructed in decays of $D^0 \to K^-\pi^+$, $K^-\pi^+\pi^0$, $K^-\pi^+\pi^-\pi^+$ and $D^+ \to K^-\pi^+\pi^+$ (and their charge conjugations). The obtained $D\bar{D}$ invariant mass distribution is shown in Fig. 2(a). The measured mass and width of the observed resonance are $(3929 \pm 5 \pm 2)$ MeV/c$^2$ and $29 \pm 10 \pm 2$ MeV, respectively. A product of the two-photon decay width and branching fraction of the $Z(3930)$ is found to be $\Gamma(Z(3930))\mathscr{B}(Z(3930) \to D\bar{D}) = 0.18 \pm 0.05 \pm 0.03$ keV. An angular analysis, performed by Belle [16], showed that spin-2 assignment is strongly favored over spin-0 assignment, making all measurements consistent with the expectations for the $\chi'_{c2}$, a radial excitation of $^3P_2$ charmonium.

## OBSERVATION OF TWO NEW STATES AT M ≈ 3940 MeV/c$^2$

Belle performed an analysis of the $\omega J/\psi$ system produced in exclusive $B \to K\omega J/\psi$ decays [17]. Events with $M(K\omega) < 1.6$ GeV/c$^2$ are rejected in order to exclude $K^* \to K\omega$ contribution. The plot of signal yield of $B$ decays, as obtained from the fit to the $M_{bc}$ distribution, in bins of $M(\omega J/\psi)$ shows a strong enhancement above the phase space expectation (see Fig. 2(b)). The fit with an S-wave Breit-Wigner function yields the mass of new resonance named Y of $(3943 \pm 11 \pm 13)$ MeV/c$^2$ and the width of $(87 \pm 22 \pm 26)$ MeV, with a significance above $8\sigma$.

The charmonium state with mass around 3940 MeV, that is above $D\bar{D}^*$ threshold, is expected to dominantly decay to $D\bar{D}^{(*)}$ which for $Y(3940)$ were not observed yet. Adding that the hadronic charmonium transition is expected to be very small, one can conclude that the $Y(3940)$ is probably not a charmonium state. This two observed properties of $Y(3940)$ and its measured width are consistent with expectations for $c\bar{c}g$

hybrid state, but the measured mass is several hundreds of MeV too low [18].

Another resonance above $D\bar{D}^{(*)}$ threshold, denoted as $X(3940)$, was observed in the $e^+e^- \to J/\psi X$ process as a system recoiling against the $J/\psi$ [19]. The mass recoilinig against the $J/\psi \to l^+l^-$ is determined as $M_{\text{recoil}} = \sqrt{(E_{cms} - E^*_{J/\psi})^2 - (cp^*_{J/\psi})^2}/c^2$, where $E^*$ is the $J/\psi$ CM energy and $E_{cms}$ is the CM energy of the event. A previously unobserved peak can be seen in a recoil mass spectrum at $(3943 \pm 6 \pm 6)$ MeV/c$^2$ and with width $\Gamma < 52$ MeV, together with three known peaks corresponding to $\eta_c$, $\chi_{c0}$ and $\eta_c(2S)$ (see Fig. 2(c)). At Belle a search for $X(3940) \to D\bar{D}^{(*)}$ decays was performed and a significant signal was found only for $X(3940) \to DD^*$ decays. No significant signal was found $X(3940) \to \omega J/\psi$ decays. Since $X(3940)$ state does not share decay modes with the $Y(3940)$, these two states are probably not the same. A possible interpretation of $X(3940)$ is radially excited charmonium state $\eta_c(3S)$.

## CONCLUSION

The large data sample collected by the Belle experiment at KEKB provides an excellent opportunity for the search of new particles. During the Belle operation more than ten new states have been discovered. In this paper we reported only on four of them, namely $X(3872)$, $Y(3940)$, $X(3940)$ and $Z(3930)$. The latter two resonances can be interpreted as charmonium states, $\eta_c(3S)$ and $\chi'_{c2}$, respectively. None of the existing measurements contradicts the $X(3872)$ interpretation as a $D^0\bar{D}^{*0}$ molecule. The nature of $Y(3940)$ remains to be addressed.

## REFERENCES

1.   A. Abashian *et al.* [Belle Collaboration], Nucl. Instrum. Meth. A **479**, 117 (2002).
2.   S. Kurokawa and E. Kitunani, Nucl. Instrum. Meth. A **499**, 1 (2003) and other papers in the volume.
3.   S. K. Choi *et al.* [Belle Collaboration], Phys. Rev. Lett. **91**, 262001 (2003).
4.   D. Acosta *et al.* [CDF II Collaboration], Phys. Rev. Lett. **93**, 072001 (2004).
5.   V. M. Abazov *et al.* [D0 Collaboration], Phys. Rev. Lett. **93**, 162002 (2004).
6.   B. Aubert *et al.* [BABAR Collaboration], Phys. Rev. D **71**, 071103 (2005).
7.   W. M. Yao *et al.* [Particle Data Group], J. Phys. G **33**, 1 (2006).
8.   K. Abe *et al.* [Belle collaboration], arXiv:hep-ex/0505038.
9.   E. J. Eichten, K. Lane and C. Quigg, Phys. Rev. D **69**, 094019 (2004).
10.   T. Barnes and S. Godfrey, Phys. Rev. D **69**, 054008 (2004).
11.   E. S. Swanson, Phys. Lett. B **588**, 189 (2004).
12.   L. Maiani, F. Piccinini, A. D. Polosa and V. Riquer, Phys. Rev. D **71**, 014028 (2005).
13.   K. Abe *et al.* [Belle colaboration], arXiv:hep-ex/0505037.
14.   G. Gokhroo *et al.* [Belle collaboration], Phys. Rev. Lett. **97**, 162002 (2006).
15.   A. Abulencia *et al.* [CDF Collaboration], Phys. Rev. Lett. **96**, 102002 (2006).
16.   S. Uehara *et al.* [Belle Collaboration], Phys. Rev. Lett. **96**, 082003 (2006).
17.   S.K. Choi *et al.* [Belle Collaboration], Phys. Rev. Lett. **94**, 182002 (2005).
18.   Z. H. Mei and X. Q. Luo, Int. J. Mod. Phys. A **18**, 5713 (2003).
19.   K. Abe *et al.* [Belle Collaboration], submitted to Phys. Rev. Lett., arXiv:hep-ex/0507019.

# Phase diagram of quark matter under compact star conditions

Michael Buballa

*Institut für Kernphysik, TU Darmstadt, Schlossgartenstr. 9, D-64289 Darmstadt, Germany*

**Abstract.** The present picture of the phase diagram of neutral quark matter at large, but non-asymptotic densities is briefly reviewed. Model studies reveal a rich phase structure, but details are very sensitive to uncontrolled parameters. These problems may be overcome by Dyson-Schwinger approaches.

**Keywords:** color superconductivity
**PACS:** 12.39.-x,12.38.Aw,26.60.+c

## INTRODUCTION

Theoretical studies suggest that strongly interacting matter at low temperatures and extremely high densities is a color superconductor in the color-flavor locked (CFL) phase [1]. This is a rather safe statement because at large densities the momentum scale, set by the quark chemical potential $\mu$, is large and therefore the QCD running coupling $\alpha_s(\mu)$ becomes small. The problem can thus be treated from first principles, employing a weak-coupling expansion of QCD.

In the CFL phase up, down, and strange quarks form Cooper pairs of spin-0 in a color antitriplet channel, corresponding to the most attractive channel for single gluon exchange. As a consequence of the Pauli principle, the pairing pattern must then be antisymmetric in flavor, i.e., the diquark condensates consist of quark pairs with unequal flavors.

Of course, it would be interesting to know whether color superconducting phases are not only a theoretical solution of QCD in a certain limit, but also exist in Nature. In this context the most promising candidates are the centers of compact stars, where densities of several times nuclear matter density are reached. Unfortunately, these densities are not large enough to allow for a weak-coupling expansion. Therefore, and since this regime is not accessible to lattice calculations, our picture of the phase structure of strongly interacting matter under compact star conditions is much less clear.

This is also related to the role of strange quarks: At asymptotic densities, the strange quark mass $M_s$ can be neglected as compared to the chemical potential, and up, down, and strange quarks can pair in the CFL phase. However, when $\mu$ gets of the order of $M_s$, the strange quarks are suppressed and their pairing with non-strange quarks eventually becomes unfavorable. On the other hand compact star matter must be electrically neutral (at least globally). Therefore, if the fraction of strange quarks is decreased, this must be compensated by a larger fraction of down quarks in order to neutralize the up quarks. This means that all quark flavors have different Fermi momenta, and other phases with

CP892, *Quark Confinement and the Hadron Spectrum VII*
edited by J. E. F. T. Ribeiro
© 2007 American Institute of Physics 978-0-7354-0396-3/07/$23.00

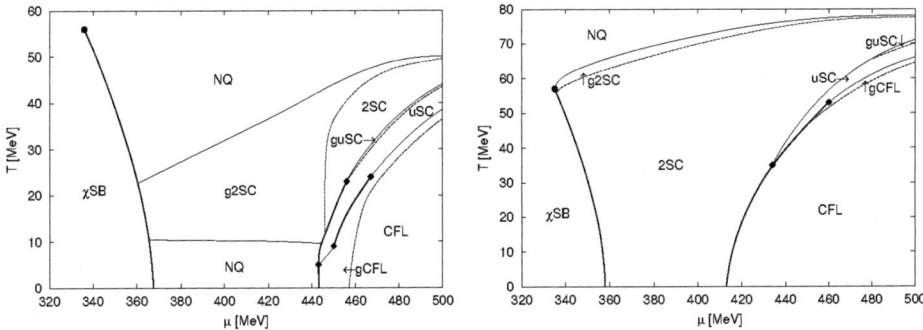

**FIGURE 1.** Phase diagram for homogeneous electrically and color neutral quark matter in an NJL-type model with "intermediate" (left) and "strong" diquark coupling (right). From Ref. [3].

cross-flavor pairing could be unfavorable as well. It can be shown that this problem is unavoidable for spin-0 color antitriplet pairing [2]. There is thus no obvious candidate for the most favored quark matter phase under compact star conditions, but the answer depends on the details of the interaction.

## NJL-MODEL CALCULATION

Since weak-coupling schemes are not valid at "moderate" densities, most studies in this regime have been performed within models. In Ref. [3] the phase diagram of homogeneous neutral quark matter has been investigated within an NJL-type model. In this model, despite being much simpler than QCD, the diquark gaps as well as the effective quark masses are generated dynamically. It is therefore well suited for studying competing phases [4].

The main results of Ref. [3] are displayed in Fig. 1, revealing a rich phase structure: Besides the CFL phase, there is 2SC phase where only up and down quarks are paired, a uSC phase (with $ud$ and $us$ pairing) and the corresponding "gapless" phases (gCFL, g2SC, guSC). The latter are probably unstable and may indicate the existence of non-homogeneous phases. Finally, there are also non-superconducting phases with chiral symmetry being spontaneously broken ($\chi$SB) or restored (NQ = normal quark matter). Recently, the analysis has been extended to include the possibility of kaon condensation in the CFL phase [5].

Unfortunately, the detailed structure of the phase diagram depends strongly on the model parameters, which are not really under control. This can be seen by comparing the left panel of Fig. 1 with the right one. The latter corresponds to the same model, but with a 30% larger diquark coupling. Whereas the $\chi$SB phase at low $\mu$ and the CFL phase at high $\mu$ are relatively robust, there are obviously big differences in between, just in the regime which is presumably most relevant for compact stars. In particular, there is a large non-superconducting region in the left panel which becomes replaced by a 2SC phase at larger coupling (right panel).

# SPIN-1 PAIRING

As pointed out above, spin-0 pairing is subject to stress which results from the finite strange quark mass and the requirement of electric neutrality. As a result all spin-0 phases could be disfavored, as for instance in the left panel of Fig. 1 at $\mu \approx 400$ MeV and low temperature, where a normal quark matter phase was found. In this case, the pairing of quarks with *equal* flavors in a spin-1 channel is a promising alternative [6].

A particularly interesting pattern is the so-called CSL phase, where color and spin degrees of freedom are locked to each other. In Ref. [7], this phase has been studied in an NJL-type model. It was found that all quasiparticle modes are gapped with the smallest gap of the order of 100 keV, in rather good agreement with constraints previously extracted from cooling data of neutron stars. Again, however, the model results depend strongly on the choice of the parameters.

# DISCUSSION

The NJL-model calculations discussed above should be viewed as explorative studies, which may give hints about possibly interesting scenarios. In this sense, they should be considered as "state of the art". On the other hand, the large uncertainties related to the model parameters are of course unsatisfactory, and more QCD based approaches are desirable. In this context, there has recently been considerable progress in applying the Dyson-Schwinger formalism to QCD at finite chemical potential [8, 9]. In that approach, a truncation scheme is used which has been tested against lattice data at $\mu = 0$, while at very large $\mu$ the weak-coupling results are recovered. In this way being constrained from two ends, it has then been applied to the "interesting" regime of moderate chemical potentials. Here the authors find considerable deviations from both, the naively extrapolated weak-coupling results of the pairing gaps [8] and the NJL-model behavior of the dynamical quark masses [9].

As the Dyson-Schwinger method is rather involved, it has not yet been applied to study the phase diagram under compact star conditions. However, the first steps in this direction are certainly promising. Whether quark matter really exists in compact stars remains a different story.

# REFERENCES

1. M. G. Alford, K. Rajagopal, and F. Wilczek, *Nucl. Phys.* **B537**, 443–458 (1999), hep-ph/9804403.
2. K. Rajagopal, and A. Schmitt, *Phys. Rev.* **D73**, 045003 (2006), hep-ph/0512043.
3. S. B. Rüster, V. Werth, M. Buballa, I. A. Shovkovy, and D. H. Rischke, *Phys. Rev.* **D72**, 034004 (2005), hep-ph/0503184.
4. M. Buballa, *Phys. Rept.* **407**, 205–376 (2005), hep-ph/0402234.
5. H. J. Warringa (2006), hep-ph/0606063.
6. A. Schmitt, *Phys. Rev.* **D71**, 054016 (2005), nucl-th/0412033.
7. D. N. Aguilera, D. Blaschke, M. Buballa, and V. L. Yudichev, *Phys. Rev.* **D72**, 034008 (2005), hep-ph/0503288.
8. D. Nickel, J. Wambach, and R. Alkofer, *Phys. Rev.* **D73**, 114028 (2006), hep-ph/0603163.
9. D. Nickel, R. Alkofer, and J. Wambach (2006), hep-ph/0609198.

# Fermion mass and the pressure of dense matter

Eduardo S. Fraga and Letícia F. Palhares

*Instituto de Física, Universidade Federal do Rio de Janeiro*
*C.P. 68528, Rio de Janeiro, RJ 21941-972, Brazil*

**Abstract.** We consider a simple toy model to study the effects of finite fermion masses on the pressure of cold and dense matter, with possible applications in the physics of condensates in the core of neutron stars and color superconductivity.

**Keywords:** Finite-temperature field theory; Yukawa theory; Equation of state
**PACS:** 11.15.Bt, 11.10.Wx

The role of finite quark masses in QCD thermodynamics has received increasing attention in the last few years. In the case of cold and dense QCD, it was generally believed that effects of nonzero quark masses on the equation of state were of the order of 5%, thereby yielding only minor corrections to the mass-radius diagram of compact stars [1]. In fact, mass, as well as color superconductivity gap, contributions to the pressure are supressed by two powers of the chemical potential as compared to zero-mass interacting quark gas terms. Therefore, assuming a critical chemical potential for the chiral transition of the order of a few hundred MeV, naively those terms should not matter. However, recent results for the thermodynamic potential to one loop have shown that corrections are sizable, and may dramatically affect the structure of compact stars [2]. Moreover, the situation in which mass (as well as gap) effects are significant corresponds to the critical region for chiral symmetry breakdown in the phase diagram of QCD. Hence, not only the value of the critical chemical potential will be affected, but also the nature of the chiral transition. In particular, if the latter is strongly first-order there might be a new class of compact stars, smaller and denser, with a deconfined quark matter core [3]. Of course, contributions due to color superconductivity [4] as well as chiral condensation [5] will also affect this picture.

In what follows, we study a simple toy model – cold and dense Yukawa theory – to investigate the influence of fermion masses on the pressure. Here, we present a one-loop calculation of the pressure with massive fermions in the modified minimal subtraction ($\overline{MS}$) renormalization scheme [6], and briefly comment on possible implications to the physics of condensates in the core of neutron stars and effective models for color superconductivity. Higher-order corrections and a thorough analysis of renormalization group effects will be presented elsewhere [7].

We consider a gas of massive fermions whose interaction is mediated by a real scalar field, $\phi$, with an interaction Lagrangian of the Yukawa form, $\mathscr{L}_I = g\,\overline{\psi}\psi\phi$, where $g$ is the coupling constant. In the zero-temperature limit, the perturbative pressure results in

CP892, *Quark Confinement and the Hadron Spectrum VII*
edited by J. E. F. T. Ribeiro
© 2007 American Institute of Physics 978-0-7354-0396-3/07/$23.00

a power series of $\alpha_Y \equiv g^2/4\pi$. [1] Up to $O(\alpha_Y)$, the first non-trivial contributions to the pressure are given by the free massive gas term, $P_0$, and the "exchange diagram", $P_1$. Using standard methods of field theory at finite temperature and density [8], one can derive the free gas pressure for fermions of mass $m$, obtaining in the zero-temperature limit the following form:

$$\lim_{T \to 0} P_0 = \frac{1}{12\pi^2} \left[ \mu p_f \left( \mu^2 - \frac{5}{2}m^2 \right) + \frac{3}{2}m^4 \ln \left( \frac{\mu + p_f}{m} \right) \right], \tag{1}$$

where $\mu$ is the chemical potential and $p_f = \sqrt{\mu^2 - m^2}$ denotes the Fermi momentum. The $O(\alpha_Y)$ renormalized correction reads [6]:

$$\lim_{T \to 0} P_1 = -\frac{\alpha_Y}{4\pi^3} \left[ \frac{3}{4}u^2 - p_f^4 + m^2 \left( 3 + 2\ln \frac{\Lambda^2}{m^2} \right) u \right], \tag{2}$$

where $u = \mu p_f - m^2 \ln[(\mu + p_f)/m]$ and $\Lambda$ is the renormalization scale in the $\overline{MS}$ scheme.

Fig. 1 illustrates the effect of modifying the mass on the total pressure to $O(\alpha_Y)$, $P = P_0 + P_1$. The choice of range for $\mu$, and accordingly for the masses, are inspired by the scales found in the case of QCD [2]. In the same vein, the coupling is fixed to $\alpha_Y = 0.3$. It is clear from the figure that mass corrections bring significant changes to the pressure, even in the absence of renormalization group (RG) running for the coupling and the mass. The figure also shows the dependence on the renormalization scale $\Lambda$. The values chosen are motivated by the ones which appear in QCD, as before. Although the effects of varying $\Lambda$ appear to be relatively small, it would be premature to conclude that this feature will remain after implementing the RG flow. In fact, the results presented in Fig. 2 most probably underestimate the scale dependence of the full correction, since not only the coupling but also the mass will run with $\Lambda$. In the Yukawa theory, in contrast to QCD, the effect will become larger as we increase the chemical potential. For fixed coupling, larger values of $\Lambda$ yield larger modifications in the pressure. However, after the inclusion of RG running, this behavior can be mantained, as should be the case here, or become the opposite, as is the case in QCD, depending on the sign of the beta function. Since the $\Lambda$-dependence comes from the term $\sim m^2 \alpha_Y \ln(\Lambda/m)$ in (2), there will be a competition between the behavior of the renormalization scale $\Lambda$ and that of $m$ and $\alpha_Y$ as functions of $\mu$.

Even at one loop order mass effects bring into play logarithmic corrections originated in the $\overline{MS}$ subtraction scheme. As usual, they bring about a non-physical dependence on the renormalization scale $\Lambda$, since one has to cut the perturbative series at some order. Higher-order computations in this framework are in progress [7], and will give a better handle on the choice of this scale, which in our case should be a function of $\mu$ and $m$. On the other hand, one can also choose the scale in a phenomenological way in a given model, imposing physical constraints to the equation of state, as was done in Ref. [3] to model the non-ideality of QCD at finite density with massless quarks.

---

[1] Since we are concerned only with the zero-temperature limit, there are no odd powers of $g$ coming from resummed contributions of the zero Matsubara mode for bosons in the perturbative series.

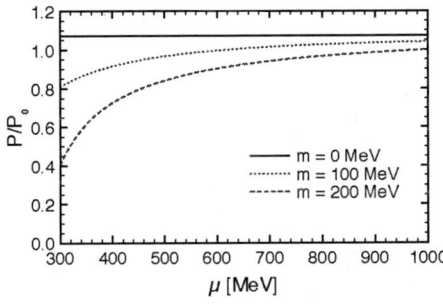
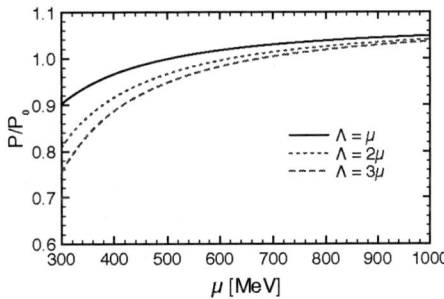

**FIGURE 1.** Pressure normalized by the free fermion gas pressure as a function of the fermion chemical potential. Left: $\Lambda = 2\mu$ and different values of the fermion mass. Right: $m = 100$ MeV and different values of the renormalization scale $\Lambda$.

The points discussed above might be relevant in the study of effective models for the cold and dense matter found in the interior of compact stars, especially because the effects seem to be significant near the critical region. In the context of the NJL model, e.g., it was shown that a self-consistent treatment of quark masses strongly affects the competition between different phases [5]. And the mechanism of pairing in color superconductivity will certainly be influenced [9] by the running of nonzero quark masses. The investigation of these issues, as well as the effect of nonzero fermion masses in the formation of other condensates in neutron star matter, is under way [7].

## ACKNOWLEDGMENTS

We thank R. D. Pisarski, J. Schaffner-Bielich and C. Villavicencio for fruitful discussions. This work was partially supported by CAPES, CNPq, FAPERJ and FUJB/UFRJ.

## REFERENCES

1. E. Witten, Phys. Rev. D **30**, 272 (1984); E. Farhi and R. L. Jaffe, Phys. Rev. D **30**, 2379 (1984); C. Alcock, E. Farhi and A. Olinto, Astrophys. J. **310**, 261 (1986); P. Haensel, J. L. Zdunik, and R. Schaeffer, Astron. Astrophys. **160**, 121 (1986).
2. E. S. Fraga and P. Romatschke, Phys. Rev. D **71**, 105014 (2005).
3. E. S. Fraga, R. D. Pisarski and J. Schaffner-Bielich, Phys. Rev. D **63**, 121702 (2001); Nucl. Phys. A **702**, 217 (2002).
4. M. Alford and S. Reddy, Phys. Rev. D **67**, 074024 (2003); M. Alford *et al.*, Astrophys. J. **629**, 969 (2005).
5. M. Buballa and M. Oertel, Nucl. Phys. A **703**, 770 (2002); S. B. Ruster *et al.*, Phys. Rev. D **72**, 034004 (2005); D. Blaschke *et al.*, Phys. Rev. D **72**, 065020 (2005).
6. L. F. Palhares and E. S. Fraga, to appear in Braz. J. Phys. (2006).
7. L. F. Palhares and E. S. Fraga, work in progress.
8. J. I. Kapusta, *Finite-temperature field theory* (Cambridge University Press, 1989).
9. K. Rajagopal and A. Schmitt, Phys. Rev. D **73**, 045003 (2006).

# Saturation properties of nuclear matter in a relativistic mean field model constrained by quark dynamics

R. Huguet, J.C. Caillon and J. Labarsouque

*Centre d'Etudes Nucléaires de Bordeaux-Gradignan, CNRS-IN2P3*
*Université Bordeaux 1, Le Haut-Vigneau, 33170 Gradignan Cedex, France*

**Abstract.** We have built an effective Walecka-type hadronic Lagrangian in which the hadron masses and the density dependence of the coupling constants are deduced from the quark dynamics using a Nambu-Jona-Lasinio model. The parameters of this Nambu-Jona-Lasinio model have been determined using the meson properties in the vacuum but also in the medium through the omega meson mass in nuclei measured by the TAPS collaboration. Realistic properties of nuclear matter have been obtained.

**Keywords:** Nuclear matter; effective hadronic models; Nambu-Jona-Lasinio model
**PACS:** 21.65.+f; 24.10.Jv; 24.85.+p; 12.39.Fe; 12.39.Ki

## Introduction

One of the most fascinating challenges of nuclear physics is the description of nuclear matter and nuclei starting from Quantum ChromoDynamics (QCD). Even if important progress have been made in QCD calculations on the lattice, such a description is not yet available. Models incorporating the most prominent features of QCD have to be used.

A possibility is to apply the strategy of effective field theories where low energy effective hadronic Lagrangians are obtained by integrating out the degrees of freedom lying above the energy scale considered. This decimation leads to density-dependent masses and couplings in the hadronic Lagrangian [1]. Such a Lagrangian with density dependent masses and coupling constants determined according to Brown and Rho scaling[2] has been proposed by Brown, Song, Min and Rho[3]. The calculation reported in [3], assuming a scaling law leading to a decreasing of the vector meson mass of approximately 20% at saturation, enables a realistic description of bulk properties of nuclear matter.

Recently, new experimental results from TAPS collaboration [4] and KEK/PS [5] for the in-medium $\omega$ meson mass would suggest a small decreasing of approximately 10-15% at saturation density. We have explored the possibility of obtaining a realistic description of bulk properties of nuclear matter in the model of Song et al. [3] with a density dependence of the in-medium $\omega$ mass in accordance with recent experimental indications. In addition, the density dependencies of the lagrangian have been deduced directly from the quark dynamics using an in-medium NJL model, since, despite the lack of confinement, it allows to take into account an important part of quark dynamics with the in-medium chiral symmetry restoration.

CP892, *Quark Confinement and the Hadron Spectrum VII*
edited by J. E. F. T. Ribeiro
© 2007 American Institute of Physics 978-0-7354-0396-3/07/$23.00

# Model

We use an effective hadronic Lagrangian similar to the Walecka one but with density dependent masses $M_N^*$, $m_\omega^*$, $m_\sigma^*$ and couplings $g_{\sigma NN}^*$, $g_{\omega NN}^*$. The vacuum values $g_{\omega NN}$, $g_{\sigma NN}$ are the only free-parameters, which will be adjusted at the end of the calculation to reproduce the position of the saturation point. This Lagrangian is treated at the mean-field level for infinite symmetric nuclear matter.

The density-dependent couplings and mass parameters are determined directly from the quark dynamics using a two flavor NJL model, which includes two chirally invariant four-quark terms in scalar/pseudoscalar and vector channel, and a scalar/vector eight quark term. At the quark level, the free parameters are the bare quark mass $m_0$, the couplings $g_1$, $g_2$ and $g_3$ and the three momentum cut-off $\Lambda$. The meson masses and couplings to quarks $m_{\sigma,\omega}^*$, $g_{\sigma,\omega qq}$ are determined by solving the Bethe-Salpeter equation in quark-antiquark appropriate channels. We assume that $M_N^*$ is directly related to the quark condensate, with the same relation as found in finite-density QCD sum-rule calculations[6], and that the quark-meson and nucleon-meson couplings in medium are proportional :

$$\frac{M_N^*}{M_N} = \frac{\langle \bar{q}q \rangle}{\langle \bar{q}q \rangle_0}, \quad \frac{g_{\sigma NN}^*}{g_{\sigma NN}} = \frac{g_{\sigma qq}^*}{g_{\sigma qq}}, \quad \frac{g_{\omega NN}^*}{g_{\omega NN}} = \frac{g_{\omega qq}^*}{g_{\omega qq}}, \tag{1}$$

where $\langle \bar{q}q \rangle_0$ represents the quark condensate in vacuum, and $M_N = 939$ MeV is the free nucleon mass. More details of the model are available in [7].

# Results

At the quark level, for a given value of the cutoff $\Lambda$, we impose to reproduce the pion mass $m_\pi = 135$ MeV, the pion decay constant $f_\pi = 92.4$ MeV and the $\omega$ meson mass $m_\omega = 782$ MeV in vacuum. We have chosen to take into account the result obtained by the TAPS collaboration[4] for the $\omega$ meson mass in nuclei, $m_\omega^*(\rho_B = 0.6\rho_0) = 722^{+4}_{-4}$ (stat)$^{+35}_{-5}$(syst) MeV in order to constrain the eight-quark term $g_3$ parameter.

At the hadronic level, the free parameters $g_{\sigma NN}$, $g_{\omega NN}$ are fixed to reproduce the saturation point. For each value of $\Lambda$ considered, in order to probe the description of nuclear matter obtained, we have calculated the effective nucleon mass $m_N^*/M_N = (M_N^* - (g_{\sigma NN}^*/m_\sigma^*)^2 \rho_S)/M_N$ (with $\rho_S$ the nucleon scalar density), the incompressibility parameter $K$ and the slope of the real part of the energy dependence of the nucleon-nucleus optical potential $U_0/M_N$ at saturation, which are expected empirically to be respectively of order $0.58 - 0.6$, $250 \pm 50$ MeV, $0.25 - 0.40$. We have plotted on Fig.1 these three quantities as function of the constituent quark mass in vacuum $m$, where the shaded areas correspond to the bound on the empirical value.

As we can see, for $m \approx 465 - 470$ MeV ($\Lambda = 572 \pm 1$ MeV), the three physical quantities $m_N^*/M_N$, $K$ and $U_0/M_N$ are all in good agreement with the empirical values, and keep reasonable values at least up to $m \approx 500$ MeV. The saturation curve is realistic for density around the saturation point. At higher densities, it is somewhat harder that what is expected, but this is not surprising since the relationship between $M_N^*$ and $\langle \bar{q}q \rangle$ should be valid only at low densities.

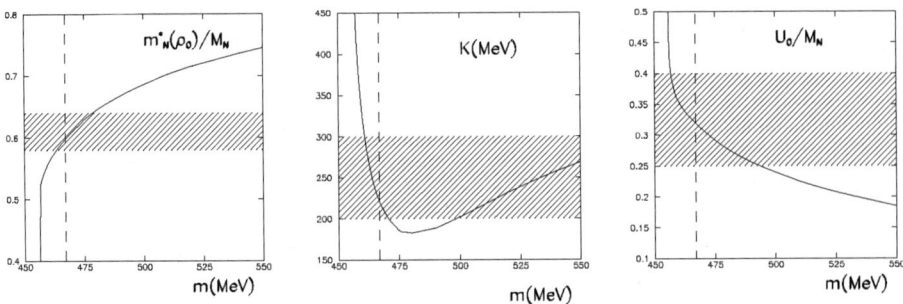

**FIGURE 1.** Saturation properties as function of the constituent quark mass $m$.

## Conclusion

We have investigated the properties of nuclear matter in a relativistic mean field model with density-dependent masses and couplings, similar to that used in [3], in which we have replaced the Brown and Rho scaling by a direct calculation of meson masses and couplings in a NJL quark model. The NJL model including four and eight quark interaction terms has been constrained to reproduce the recent TAPS result for the in medium $\omega$ meson mass in addition to the vacuum pion and $\omega$ meson properties. At the hadronic level, the two free parameters have been fixed to reproduce the empirical saturation point.

At the quark level, the eight quark term is essential for obtaining realistic saturation properties. At saturation, the nucleon effective mass, the incompressibility parameter and the slope of the energy dependence of the nucleon-nucleus optical potential obtained are all in good agreement with the empirical data.

Even if more work is needed in this direction and in a more fundamental one, this result is a very encouraging one since with only a few free-parameters, a realistic description of saturation properties of nuclear matter has been obtained from a hadronic Lagrangian constrained by a quark model which reproduces vacuum and in-medium meson properties.

## REFERENCES

1. G.E. Brown and M. Rho *Phys. Rep.* **396** (2004) 1
2. G. E. Brown and M. Rho, *Phys. Rev. Lett.* **66** (1991) 2720
3. C. Song, *Phys. Rep.* **347** (2001) 289 and references therein.
4. D. Trnka et al., *Phys. Rev. Lett.* **94** (2005) 192303
5. M. Naruki and al. *Phys. Rev. Lett.* **96** (2006) 092301
6. T. D. Cohen, R. J. Furnstahl and D. K. Griegel, *Phys. Rev.* **C45** (1992) 1881
7. R. Huguet, J.C. Caillon and J. Labarsouque, *nucl-th/0610099*

# Dressed Dibaryon as Carrier of Short-Range NN and 3N Interactions[1]

## V.I. Kukulin

*Institute of Nuclear Physics, Moscow State University, 119899 Moscow, Russia*

**Abstract.** Six-quark dibaryon model for fundamental $NN$- and $3N$-forces at intermediate and short ranges are discussed briefly. The structure of the intermediate dibaryon dressed with $\sigma$-meson field is similar to those of Roper resonance where the $\sigma$-meson field plays a very important role. The $t$-channel $\sigma$-exchange between dibaryon and third nucleon is shown to result in new strong scalar attractive $3N$ force which contributes at least a half of nuclear binding in $3N$ system. This dibaryon induced $\sigma$-exchange leads also to a new spin-orbit attractive $3N$-force.

**Keywords:** Dibaryon, nuclear force, $3N$-interactions
**PACS:** 13.75.Cs, 21.30-x, 21.45.+v

## MOTIVATION FOR THE DIBARYON MODEL OF BASIC NUCLEAR FORCE

(i) There were found very clear and numerous discrepancies between modern high-quality experimental data and predictions of conventional nuclear force models: typical examples are evident disagreements for many electromagnetic observables for few-nucleon systems like the $d(\gamma, \vec{n})p$, $pp \rightarrow pp\gamma$, ${}^3\text{He}(e, e'pp)$, ${}^3\text{He}(e, e'pn)$ etc. processes [1-3]. The real situation is even much worse: the attempts to remove the discrepancies in one kinematical area leads often to an enlargement of the gap between the theory and experiment in other kinematical areas. (ii) It was established in recent years that some basic assumptions adopted in traditional force models being treated more carefully and consistently failed completely. One of the most evident examples of this sort is a traditional believing that the basic attractive $NN$ force is induced by two-pion $t$-channel meson exchange in scalar-isoscalar channel (see Fig. 1). Contrary to the believing, three

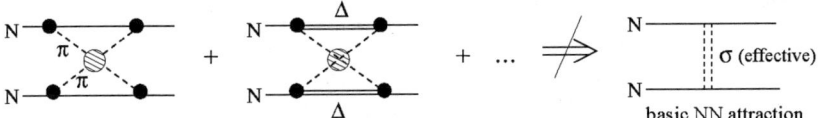

**FIGURE 1.** Puzzle with scalar meson exchange

independent groups have found recently that $2\pi$-exchange with intermediate $\pi - \pi$ s-wave interaction leads to strong short- and intermediate-range *repulsion* and only very moderate peripheral attraction (see Fig. 2).

---

[1] This work was partially supported by Russian Foundation for Basic Research (grants 05-02-04000 and 05-02-17407) and Deutsche Forschungsgemeinschaft (grant 436 RUS 113/790/).

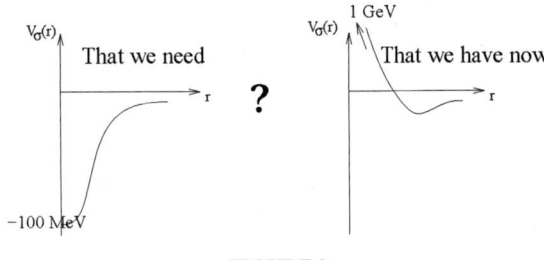

**FIGURE 2.**

As a result, we have at the moment *no consistent mechanism for basic internucleon attraction* within the traditional meson-exchange picture. In this point the intermediate dressed dibaryons appear on the scene!

## THE CONCEPT OF DIBARYON INDUCED $NN$- AND $3N$-INTERACTIONS

The basic mechanism of $NN$-interaction induced by intermediate dressed dibaryon can be illustrated by the graphs:

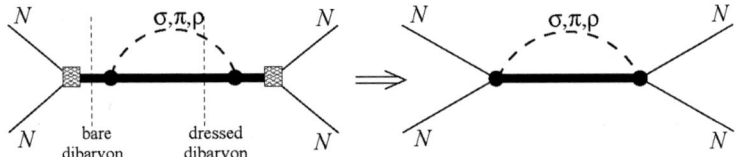

**FIGURE 3.** The $\sigma$- and other meson dressing in $6q$-bag (dibaryon).

We found [4-6] that the $\sigma$-dressing of intermediate dibaryon shifts its mass downward noticeably ($\Delta \sim 0.5 - 0.7$ GeV). The similar $\sigma$-dressing of the Roper resonance: $|s^2(2s)[3]\rangle \Rightarrow |s^3[3] + \sigma\rangle$ (see Fig. 4) reduces its mass about 0.5 GeV! Moreover, just

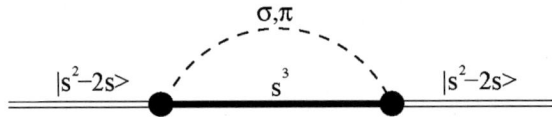

**FIGURE 4.** The $\sigma$-meson dressing in $3q$-bag (Roper resonance).

this scalar-meson mechanism have been found recently [7] to be responsible for the level inversion between the Roper resonance and negative parity $S$- and $D$-isobar states in nucleon spectrum. So, the dressed dibaryon in our case may be viewed as a some generalization of the $3q$ Roper-resonance structure over $6q$ system. We were able [4-6] with the above dibaryon model, using only a few free parameters with natural sense, to describe quantitatively the $NN$ phase shifts until energy 1 GeV and higher.

An effective Hamiltonian for $3N$ system is $H^{\text{eff}} = T + \sum_i \{v_i^{\text{ex}} + W_i(E)\}$, where $v_i^{\text{ex}}$ are Yukawa meson-exchange components and each of three effective potentials $W_i(E)$ takes the form: $W_i(E) = \delta(\mathbf{q}_i - \mathbf{q}_i')w_i(E - q_i^2/2\overline{m})$. In the pole approximation, this three-body effective interaction reduces to a sum of two-body separable potentials with the coupling constants depending on the three-body energy $E$ and the third-particle momentum $\mathbf{q}_i$:

$$W_i(\mathbf{p}_i, \mathbf{p}_i', \mathbf{q}_i, \mathbf{q}_i'; E) = \delta(\mathbf{q}_i - \mathbf{q}_i') \sum_{J_i M_i, L_i, L_i'} \varphi_{L_i}^{J_i M_i}(\mathbf{p}_i) \lambda_{L_i L_i'}^{J_i} \left( E - \frac{q_i^2}{2\overline{m}} \right) \varphi_{L_i'}^{J_i M_i}(\mathbf{p}_i')$$

– "two-body" interactions in three-body system. Due to the fact the two-body coupling constant $\lambda_{L_i L_i'}^{J_i}(\varepsilon)$ is decreasing function at $\varepsilon \to -\infty$ the two-body force in three-( or many-)body system (when the average magnitude of $q_i^2/2m$ is rising with the increase of the total nucleon density) is effectively weaker than in two-body $NN$ system. And the higher average nucleon density in the system, the weaker the effective two-body interaction. So this specific mechanism leads to a natural saturation properties in nuclear matter at higher density.

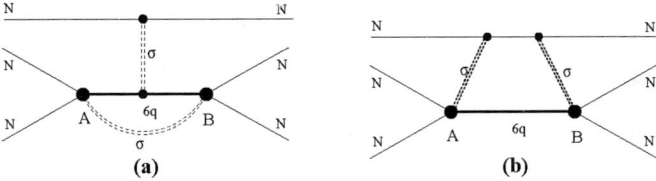

**FIGURE 5.**   The new three-body forces in $3N$ system.

However in our case there appears a new genuine three-body attractive force in $3N$ or many-nucleon systems between dibaryon and surrounding nucleons [8-9] (see Fig. 5). This scalar $3N$ force results also in additional spin-orbit interaction between dibaryon and third nucleon. The detailed three-body calculations [8] have revealed that the new attractive $3N$ central force contributes at least a half of total nuclear binding in $3N$ (and likely in other few-nucleon) systems and is able to give an excellent description of bound state energies and other characteristics of $^3$H and $^3$He nuclei including the Coulomb energy displacement [8-9].

**To summarize:** we developed a new dibaryon-induced model for $NN$- and $3N$ nuclear forces which occurred to be very fruitful for explanation of many long-standing puzzles in both few-nucleon and hadronic physics as well.

# REFERENCES

1.   R. Schiavilla, Phys. Rev. **C72**, 034001 (2005).
2.   M Cozma, O. Scholten, R.G.E. Timmermans and J.E. Tjon, Phys. Rev. **C68**, 044003 (2003).
3.   D.L. Groep et al., Phys. Rev. C **63**, 014005 (2001).
4.   V.I. Kukulin, I.T. Obukhovsky, V.N. Pomerantsev, and A. Faessler, J.Phys. G: Nucl. Part. Phys. **27**, 1851 (2001).
5.   V.I. Kukulin, I.T. Obukhovsky, V.N. Pomerantsev, and A. Faessler, Int. J. Mod. Phys. E **11**, 1 (2002).
6.   A. Faessler, V.I. Kukulin, and M.A. Shikhalev, Ann. Phys. (NY) **320**, 71 (2005).
7.   P. Stassart, Fl. Stancu, and J.-M. Richard, nucl-th/9905015.
8.   V.I. Kukulin, V.N. Pomerantsev, M. Kaskulov, and A. Faessler, J. Phys. G **30**, 287 (2004); V.I. Kukulin, V.N. Pomerantsev, and A. Faessler, J. Phys. G **30**, 309 (2004).

# Light plasmon mode in the CFL phase

H. Malekzadeh* and Dirk H. Rischke†

*Frankfurt International Graduate School for Science.
†Institut für Theoretische Physik and Frankfurt Institute for Advanced Studies.
J.W. Goethe-Universität, D-60438 Frankfurt am Main, Germany

**Abstract.** The self-energies and the spectral densities of longitudinal and transverse gluons at zero temperature in the color-flavor-locked (CFL) phase are calculated. There appears a collective excitation, a light plasmon, at energies smaller than two times the gap parameter and momenta smaller than about eight times the gap. The minimum in the dispersion relation of this mode at some nonzero value of momentum corresponds to the van Hove singularity.

In cold and dense quark matter, due to asymptotic freedom, at quark chemical potentials $\mu \ll \lambda_{QCD}$ single-gluon exchange is the dominant interaction between quarks. Since this interaction is attractive in the color-antitriplet channel therefore quark matter is a color superconductor [1]. While there are, in principle, many different color-superconducting phases, corresponding to the different possibilities to form quark Cooper pairs, the ground state of color-superconducting quark matter is the so-called color-flavor-locked (CFL) phase [2].

At asymptotically large $\mu$, the QCD coupling constant $g \ll 1$, thus the gluon self-energy is dominated by the contributions from one quark and one gluon loop. The quark loop is $\sim g^2\mu^2$, while the gluon loops are $\sim g^2 T^2$. Since the color-superconducting gap parameter is $\phi \sim \mu \exp(-1/g) \ll \mu$ [3], and since the transition temperature to the normal conducting phase is $T_c \sim \phi$, for temperatures where quark matter is in the color-superconducting phase, $T$ less than $T_c \ll \mu$, the gluon loop contribution can be neglected. The full description for the 2SC phase is given in Sec. II of Ref. [8]. The full energy-momentum dependence of the one-loop gluon self-energy has also been computed, but so far only for the 2SC phase [7, 8]. Here we want to do the same calculations for the CFL phase. The detailed computation of the individual components and projections can be found in the appendix of Ref.[9].

Fig. 1 shows the imaginary part of several components of the gluon self-energy for a gluon momentum $p = 4\phi$ as a function of the gluon energy $p_0$. The corresponding results for the gluon self-energy in the "hard-dense loop" (HDL) limit, $\Pi_0^{\mu\nu}$, are also shown with the dotted lines. The imaginary parts are quite similar to those of the 2SC case, cf. Fig. 1 of Ref. [8]. Nevertheless, there are subtle differences due to appearance of two kinds of gapped quark excitations, one so-called singlet excitation with a gap $\phi_1$, and eight so-called octet excitations with a gap $\phi_8 \equiv \phi$ [2]. In weak coupling, the singlet gap is approximately twice as large as the octet gap, $\phi_1 \simeq 2\,\phi_8 \equiv 2\,\phi$ [11, 10]. Therefore, the one-loop gluon self-energy in the CFL phase has two types of contributions, depending on whether the quarks in the loop correspond to singlet or octet excitations, cf. Eq. (23b) of Ref. [6]. For the first type, both quarks in the loop are octet excitations, and for the

CP892, *Quark Confinement and the Hadron Spectrum VII*
edited by J. E. F. T. Ribeiro
© 2007 American Institute of Physics 978-0-7354-0396-3/07/$23.00

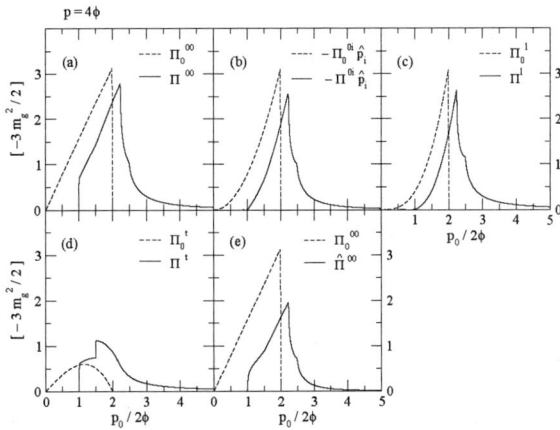

**FIGURE 1.** The imaginary parts of (a) $\Pi^{00}$, (b) $-\Pi^{0i}\hat{p}_i$, (c) $\Pi^\ell$, (d) $\Pi^t$, and (e) $\hat{\Pi}^{00}$ as a function of energy $p_0$ for a gluon momentum $p = 4\phi$. The solid lines are for the CFL phase, the dotted lines correspond to the HDL self-energy.

second, one is an octet and the other a singlet excitation. There is no contribution from singlet-singlet excitations.

Nonvanishing octet-octet excitations require gluon energies to be larger than $2\phi_8 \equiv 2\phi$, while octet-singlet excitations require a larger gluon energy, $p_0 \geq \phi_1 + \phi_8 \equiv 3\phi$. This introduces some additional structure in the imaginary parts at $p_0 = 3\phi$, which can be seen particularly well in Figs. 1 (d) and (e). In the normal phase, the imaginary parts of the gluon self-energies vanish above $p_0 = p$. In color-superconducting phases, the imaginary parts do not vanish but fall off rapidly. This has already been noted for the 2SC phase [8], and is confirmed here by the results for the CFL phase.

The real parts of the gluon self-energy are shown in Fig. 2. When computing the real part from a dispersion integral over the imaginary part, a change of gradient in the imaginary part leads to a cusp-like structure in the real part. As one expects, for large energies $p_0 \gg \phi$ the real parts of the self-energies approach the corresponding HDL limit. Deviations from the HDL limit occur only for gluon energies $p_0 \sim \phi$.

The spectral densities are obtained from the real and imaginary parts of the gluon self-energies [8, 9]. Note that, in Fig. 3 at an energy $p_0 \simeq 0.21 m_g$, there is a delta function-like peak in the transverse spectral density. This peak corresponds to a collective excitation, the so-called "light plasmon" predicted in Ref. [13, 14]. We show the dispersion relation of this collective mode in Fig. 4 (b). The mass $m_{coll} \simeq 1.35\phi$ is roughly in agreement with the value $m_{coll} \simeq 1.362\phi$ of Ref. [13]. As the momentum increases, the energy of the light plasmon excitation approaches $2\phi$ from below. For momenta larger than $\sim 8\phi$, the location of this excitation branch becomes numerically indistinguishable from the continuum in the spectral density above $p_0 = 2\phi$, cf. Fig. 3. Close inspection reveals that the dispersion relation of the light plasmon has a minimum at a nonzero value of $p \simeq 1.33\phi$, indicating a van Hove singularity.

In Fig. 4 we also show the dispersion relations for the "regular" longitudinal and

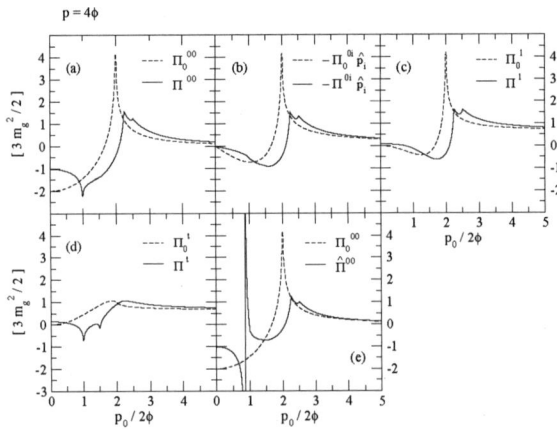

**FIGURE 2.** The real parts of (a) $\Pi^{00}$, (b) $-\Pi^{0i}\hat{p}_i$, (c) $\Pi^\ell$, (d) $\Pi^t$, and (e) $\hat{\Pi}^{00}$ as a function of energy $p_0$ for a gluon momentum $p = 4\phi$. The solid lines are for the CFL phase, the dotted lines correspond to the HDL self-energy.

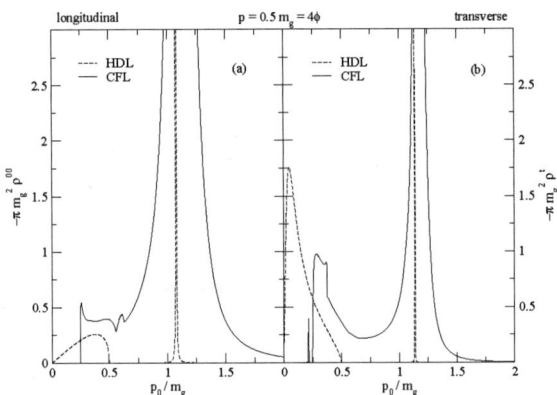

**FIGURE 3.** The spectral densities for (a) longitudinal and (b) transverse gluons for a gluon momentum $p = m_g/2$, with $m_g = 8\phi$. The dashed lines correspond to the HDL limit.

transverse excitations, as well as for the Nambu-Goldstone excitation defined by the root of $P^\mu \Pi_{\mu\nu}(P)P^\nu = 0$ [8, 12]. For our choice of gauge, the gluon propagator is 4-transverse and this mode does not mix with the longitudinal component of the gauge field [8]. Therefore, the Nambu-Goldstone mode does not appear as a peak in the longitudinal spectral density, cf. Fig. 3. We finally note that other collective excitations have been investigated in Ref. [15].

In conclusion, we have computed the gluon self-energy in the CFL phase as a function of energy and momentum. While the imaginary parts of the gluon self-energy could be expressed analytically in terms of elliptic functions (see appendix of [9]), the real parts had to be computed numerically with the help of dispersion integrals. From the real and

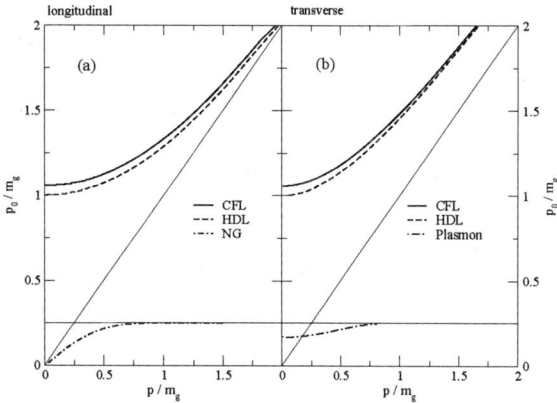

**FIGURE 4.** The dispersion relations for (a) longitudinal and (b) transverse excitations in the CFL phase. The full lines correspond to the regular longitudinal and transverse excitations. The dashed lines are for the HDL limit. The dash-dotted line in part (a) shows the dispersion relation for the Nambu-Goldstone excitation. The light plasmon dispersion relation is shown by the dash-dotted line in part (b). As in Fig. 3, the value of the gap is chosen such that $m_g = 8\phi$.

imaginary parts we constructed the spectral densities. We confirmed the existence of a low-energy collective excitation, the so-called "light plasmon" predicted in Ref. [13].

# ACKNOWLEDGMENTS

H. M. thanks the Frankfurt International Graduate School of Science for support.

# REFERENCES

1.  D. Bailin and A. Love, Phys. Rept. **107**, 325 (1984).
2.  M.G. Alford, K. Rajagopal, and F. Wilczek, Nucl. Phys. **B537**, 443 (1999); H. Malekzadeh, Phys. Rev. D **74**, 065011 (2006).
3.  D.T. Son, Phys. Rev. D **59**, 094019 (1999).
4.  G.W. Carter and D. Diakonov, Nucl. Phys. **B582**, 571 (2000); D.H. Rischke, Phys. Rev. D **62**, 034007 (2000).
5.  D.T. Son, M.A. Stephanov, Phys. Rev. D **61**, 074012 (2000).
6.  D.H. Rischke, Phys. Rev. D **62**, 054017 (2000).
7.  D.H. Rischke, Phys. Rev. D **64**, 094003 (2001).
8.  D.H. Rischke and I.A. Shovkovy, Phys. Rev. D **66**, 054019 (2002).
9.  H. Malekzadeh and D. H. Rischke, Phys. Rev. D **73**, 114006 (2006).
10. I.A. Shovkovy and L.C.R. Wijewardhana, Phys. Lett. B **470**, 189 (1999).
11. T. Schäfer, Nucl. Phys. **B575**, 269 (2000).
12. K. Zarembo, Phys. Rev. D **62**, 054003 (2000).
13. V.P. Gusynin, I.A. Shovkovy, Nucl. Phys. **A700**, 577 (2002).
14. R. Casalbuoni, R. Gatto, and G. Nardulli, Phys. Lett. **B498**, 179 (2001), Erratum ibid. **B517**, 483 (2001).
15. K. Fukushima and K. Iida, Phys. Rev. D **71**, 074011 (2005).

# The Nucleon Parton Distribution for Finite Density

## J. Rożynek

*J.Rożynek Nuclear Theory Department, Sołtan Institute for Nuclear Studies, Hoża 69*
*PL-00-681 Warsaw, Poland*
*rozynek@fuw.edu.pl*

**Abstract.** We present the evolution of the nucleon structure function with barion density in the phenomenological model for the parton distribution in nuclei. The sea parton distributions are described by additional virtual pions in hadron in such a way as to reproduce the nuclear lepton pair production data and saturate the energy-momentum sum rule. The influence of Fermi motion changes the nucleon rest energy and consequently the transverse momentum square of partons inside bound nucleons. Finally we estimate in the Sigma model the critical density where the fluctuations of pion field lead to the chiral symmetry restoration.

**Keywords:** DIS, Nuclear Matter, EOS, Nucleon Structure
**PACS:** 13.60.Hb

Convolution in the nuclear Deep Inelastic Scattering (DIS) tries to describe the deep inelastic scattering as a two step process. Electrons interact with partons (quarks) which constitute nucleon - constituent of a nucleus. Other components of a nuclear wave function like pions will be also consider in this work. Thus we have two extended objects: nucleon and nucleus. We construct a nuclear quark distribution function from quark distribution function $F_2^N(x)$ in the nucleon and nucleon distribution function $F_2^A(x_A)$ inside nucleus. Additional degrees of freedom present only in nuclear matter, usually attributed to additional mesonic [1] could be present. It can be described by changing accordingly the nuclear distribution $\rho^A$. However, mesons alone cannot account for data on lepton pair production on nuclei [2].

Let us introduce the partonic mean free path $z = 1/M_N x$ which measure the uncertainty of the time when the hit quark is propagating through nuclear matter. The $x_B$ is the Bjorken x corresponding to z equal to mucleon radius. There are 2 different regions.

1. In the $x > x_B$ regime, where the partonic mean free paths $z$ are much shorter then the average distances between nucleons. With such a resolution we can "see" the nuclear pions which form the nuclear interaction. Let $M_B$, identified with the $p^+$ component in the nucleon rest frame, denote the rest energy of the nucleon in this case. To calculate it, we assume that nuclear longitudinal momentum $P_A^+$ component is given as a sum of all partonic momenta $k_{Ai}^+$. Thus we have energy of $A$ nucleons and additional nuclear pions with the average energy $E_m$ plus nuclear binding potential $V_N$. In the large x limit where parton are on shell:

$$\sqrt{p^2} \equiv M_B \xrightarrow[rest]{} p^+. \tag{1}$$

CP892, *Quark Confinement and the Hadron Spectrum VII*
edited by J. E. F. T. Ribeiro
© 2007 American Institute of Physics 978-0-7354-0396-3/07/$23.00

the $E_m$ and $V_N$ should cancel each other, and we have the nucleon rest energy given by[3]:

$$M_B \cong M_N + -E_{Fermi} \tag{2}$$

with the average nucleon Fermi energy $E_{Fermi}$ and mass defect $\varepsilon$.

2. Let us switch to the small x. Notice that for sufficiently small values of $x$, in the region of $x < x_N \simeq 0.3$, the uncertainties in the lifetime of an intermediate parton state are so big that one should include exchanges of nuclear mesons (like two $\pi$, but also $\sigma$, $\omega$ and $\rho$) between nucleons, in the nucleon mass rather then separate objects. In standard low energy nuclear physics this is usually done by incorporating now the meson energy $E_m$ to the nucleon mass in (2) which reset the effective nucleon mass to $M_N$.

In summary for $x > x_B = 0.6$ we have $z(x) = 1/(xM_N) < r_C = 0.35$ fm. In this region, the nearby second nucleon, which is separated by the average distance $r_h \simeq 1.7$ fm (obtained from independent pair approximation), will not affect collisions proceeding on the active nucleon. The situation changes when $x$ is getting smaller and the uncertainty $z$ exceeds the nucleon radius $r_N = 0.85$ fm $\simeq r_h/2$. For such an $x$, the single nucleon approximation is no more applicable. Now we can take into account the effect of two body FSI in $F_2^{2N}(x)$ by neglecting meson energy $E_M$ in the equation (2) because for small x this meson energy is already included (contribute) in (to) the nucleon mass $M_B$. Assuming the dominant role of two-body short range $NN$ correlation, we have a gas of quasi deutrons and binding effects reset nucleon the rest energy to the standard low energy value $M_N$. It gives the dominant change of the nucleon Structure Function (SF) for smaller $x < x_B$ coming from FSI. We are not including other effects of the $NN$ interaction. Without free parameters we were able to describe the EMC effect . The results of pure convolution model is given by dashed line which is out of date. The solid line present the best fit obtained in our dual picture of bound nucleons for small x and non interactiong nucleons with nuclear pions for large x which is model by x dependent nucleon mass in the medium[3].

The Equation of State (EOS) for nuclear matter has to match the saturation point but then the behavior for higher densities is different for different RMF models. In[4] there is presented comparison between stiff (big compressibility around 400MeV) EOS for Walecka Model[5] and soft EOS for Zimanyi-Moszkowski models [6] (ZM) with compressibility slightly below 200 MeV. It was shown[7] that in the linear Walecka model , the ratio of the quark condensate of the nucleon in the medium, to that of the vacuum,

$$R_{\bar{q}q} = \frac{\langle N|\bar{q}q|N\rangle_\rho}{\langle 0|\bar{q}q|0\rangle_0} \tag{3}$$

is not going to zero, contrary to expectations based on ideas of chiral symmetry restoration. But in Zimanyi and Moszkowski models, where the scalar and the scalar-vector mesons interact non-linearly and couplings are $\rho$ - dependent, the effective nucleon remains massive but $\langle \bar{q}q \rangle$ goes to zero at high $\rho$. The central result of the work[7] tells that the non linear coupling to meson field depends effectively on $\rho$ to give a soft EOS.

The extension of calculation[3] of the nuclear to nucleon SF ratios R(x) to density $\rho$ three times bigger then the saturation density $\rho_0$ are presented in Fig. 1. We see that for Walecka model the nucleon SF is changed in saturation point but for higher density

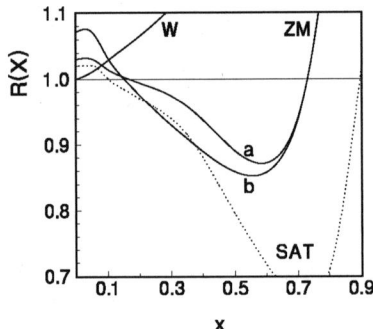

**FIGURE 1.** Results for the ratio $R(x) = F_2^{NM}(x)/F_2^N(x)$ which shows the evolution of the nucleon SF for Walecka (W) and ZM models of RMF for density $\rho = .51\,fm^{-3}$. Results for equilibrium density (SAT) $\rho_0 = .17\,fm^{-3}$ same for both models are shown for reference.

became very similar to the free nucleon SF except strong Fermi motion which rise kinematically (W) the ratio R(x) for higher x. For ZM3 model (ZM) we have two curves because for non-equilibrium nucleon dynamic we are not sure if nucleon rest mass is reset to its free value $M_N$ like is saturation (SAT) case. The "a" curve display the SF when $M_B$ is going to $M_N$ for small $x$ at $\rho = 0.51\,fm^{-3}$. The "b" curve is for the same density when nucleon mass is equal $M_B$ for all x. We see response of the pion contributions for the small x which increases with density in comparison to dotted curve. For higher densities, when the fluctuation of the pion field will be bigger we can expect the chiral phase transition to the Wigner phase and it is consistent with $\langle \bar{q}q \rangle$ going to zero in ZM RMF models at $\rho$ around four times $\rho_0$. In Walecka model, which was chosen rather for reference we have also consistency between non zero value of quark condensates and unaffected nucleon SF for high density. The main reason is the dominance of the strong repulsive vector field in this region. In contrary the nonlinear meson-nucleon coupling in the Relativistic Mean Field models looks much more realistic with respect to possible chiral phase transition for (3-4) $\rho_0$, connected as was shown, with the change of the nucleon SF inside dense nuclear matter.

# REFERENCES

1. D. Geesaman, K. Saito, A. W. Thomas, *Ann. Rev. Nucl. Part. Sci.*, **45** 337 (1995).
2. A. Alde at al. *Phys. Rev. Lett.* **64**, 2479 (1990).
3. J. Rozynek, G. Wilk, *Phys. Rev.* **C71**, 068202 (2005).
4. A. Delfino, J. Dey, M. Dey and M. Malheiro, *Phys. Rev.* **C51**, 2188 (1995).
5. B. D. Serot and J. D. Walecka, *Adv. Nucl. Phys.* Vol.16 (Plenum, N. Y. 1986).
6. J. Zimanyi and S. A. Moszkowski, *Phys. Rev.* **C44**, 178 (1990).
7. A. Delfino, J. Dey, M. Dey and M. Malheiro, *Phys. Lett.* **B363**, 17 (1995); see also T. D. Cohen, R. J. Furnstahl, D. K. Griegel, *Phys. Rev. Lett.* **67**, 961 (1991).

# Inhomogeneous color superconductivity and the cooling of compact stars

## M. Ruggieri

*Dipartimento di Fisica, Università degli Studi di Bari, Italy*
*and*
*Istituto Nazionale di Fisica Nucleare, Sezione di Bari, Italy*

**Abstract.** In this talk I discuss the inhomogeneous (LOFF) color superconductive phases of Quantum Chromodynamics (QCD). In particular, I show the effect of a core of LOFF phase on the cooling of a compact star.

**Keywords:** KEYWORDS
**PACS:** PACS

Study of the phase diagram of Quantum Chromodynamics (QCD) in extreme condition of density and/or temperature has attracted a lot of interest in recent years. In particular, high density and low temperature conditions make room for a new state of deconfined quark matter known as Color Superconductor [1] (see [2] for reviews). Understanding this phase is an important challenge both for the purely theoretical aspects and for the phenomenological implications. As a matter of fact, the study of Color Superconductivity (CSC) allows for a deeper knowledge of the phase diagram of QCD; moreover, one is expected to find high baryon densities and low temperatures in the core of the compact stellar objects: as a consequence, it is interesting to understand the way CSC modifies the properties of such stars (equations of state, transport coefficient, cooling properties), in order to get a more accurate knowledge of these intriguing stellar objects.

In three flavor QCD and at asymptotically high density the ground state of CSC is known to be the Color-Flavor-Locked (CFL) state [3]. In this state of matter the color and the flavor degrees of freedom are linked together and the ground state is invariant under transformations in the diagonal group $SU(3)_{c+V}$. At moderate densities, as can be found in the core of compact stars, one has to keep into account electrical and color neutrality conditions and finite mass effects of the quarks [4]. As a consequence, the Fermi spheres of the pairing quarks are likely to be mismatched and the CFL state can be disfavored. In this case more exotic patterns of condensation can occur, and the ground state of QCD in these conditions is still a matter of debate (see for example [5] and references therein).

Among the various candidates I discuss here the crystalline color superconductor, known in literature as the LOFF phase [6]; the LOFF state is characterized by a non vanishing total momentum of the pair. In particular for the three flavor case I consider

CP892, *Quark Confinement and the Hadron Spectrum VII*
edited by J. E. F. T. Ribeiro
© 2007 American Institute of Physics 978-0-7354-0396-3/07/$23.00

here the simplest one-plane wave structure defined by

$$\langle \psi_{\alpha i}(x) C \gamma_5 \psi_{\beta j}(x) \rangle \propto \sum_{I=1}^{3} \Delta_I e^{2i\mathbf{q}_I \cdot \mathbf{r}} \varepsilon_{\alpha\beta I} \varepsilon_{ijI} \tag{1}$$

($i, j = 1, 2, 3$ flavor indices, $\alpha, \beta = 1, 2, 3$ color indices); it has been considered for the first time in the three flavor QCD contest in Ref. [7] and it was found energetically favored with respect to other phases of QCD in a certain range of values of the strange quark mass $M_s$. In Eq. (1), $2 \mathbf{q}_I$ represents the momentum of the Cooper pair and the gap parameters $\Delta_1, \Delta_2, \Delta_3$ describe respectively $d-s$, $u-s$ and $u-d$ pairing. For sufficiently large $\mu$ the energetically favored phase is characterized by $\Delta_1 = 0$, $\Delta_2 = \Delta_3$ and $\mathbf{q}_2 = \mathbf{q}_3$. This phase turns out to be also chromomagnetically stable [8]. In [9] more sophisticated ansatz have been considered, and the window of $M_s$ where the LOFF phase exists has been enlarged.

If LOFF matter is present in the core of a compact star then it affects the neutrino emissivity, and consequently the cooling process of the star itself. In the following I discuss the role of the LOFF phase on the cooling of neutron stars.

Neutrino emissivity is defined as the energy loss by $\beta$-decay per volume unit per time unit [10]. In [11] a simplified approach based on the study of three different toy models of stars has been used. The first model (denoted as I) is a star consisting of noninteracting nuclear matter (neutrons, protons and electrons) with mass $M = 1.4 M_\odot$, radius $R = 12$ km and uniform density $n = 1.5 n_0$, where $n_0 = 0.16$ fm$^{-3}$ is the nuclear equilibrium density. The nuclear matter is assumed to be electrically neutral and in beta equilibrium. The second model (II) is a star containing a core of radius $R_1 = 5$ km of neutral unpaired quark matter at $\mu = 500$ MeV, with a mantle of noninteracting nuclear matter with uniform density $n$. Solution of the Tolman-Oppenheimer-Volkov equations gives a mass-radius relation so that a mass $M = 1.4 M_\odot$ corresponds to a star radius $R_2 = 10$ km. The model III is represented by a compact star containing a core of electric and color neutral three flavors quark matter in the LOFF phase, with $\mu = 500$ MeV and $M_s^2/\mu = 140$ MeV.

The main processes of cooling are dominated by neutrino emission in the early stage of the lifetime of the pulsar and by photon emission at later ages. The cooling rate is governed by the following differential equation:

$$\frac{dT}{dt} = -\frac{L_\nu + L_\gamma}{V_{nm} c_V^{nm} + V_{qm} c_V^{qm}} = -\frac{V_{nm} \varepsilon_\nu^{nm} + V_{qm} \varepsilon_\nu^{qm} + L_\gamma}{V_{nm} c_V^{nm} + V_{qm} c_V^{qm}}. \tag{2}$$

Here $T$ is the inner temperature at time $t$; $L_\nu$ and $L_\gamma$ are neutrino and photon luminosities, i.e. emissivity by the corresponding volume. The superscripts $nm$ and $qm$ refer, respectively, to nuclear matter and quark matter including the superconductive phase; $c_V^{nm}$ and $c_V^{qm}$ denote specific heats of the two forms of hadronic matter. Eq. (2) is solved imposing a given temperature $T_0$ at a fixed early time $t_0$ (we use $T_0 \to \infty$ for $t_0 \to 0$). To compute the neutrino emissivity of nuclear and unpaired quark matter the standard textbook results are used [10, 12]; for the LOFF phase I refer to [11].

In Fig. 1 the star surface temperature as a function of time is shown (see [13] for similar results obtained in other models). Solid line (black online) is for model I; dashed

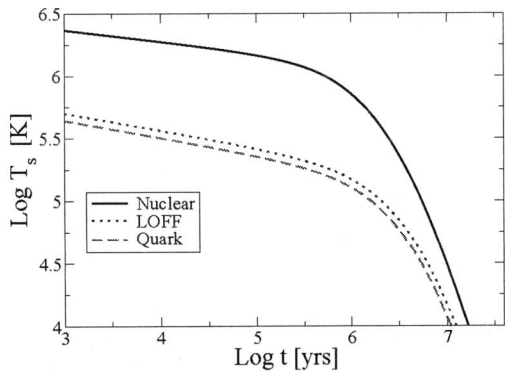

**FIGURE 1.** (Color online) Surface temperature $T_s$, in Kelvin, as a function of time, in years, for the three toy models of pulsars described. Solid black curve refers to a neutron star formed by nuclear matter with uniform density $n = 0.24$ fm$^{-3}$ and radius $R = 12$ Km (model I); dashed line (red online) refers to a star with $R_2 = 10$ km, having a mantle of nuclear matter and a core of radius $R_1 = 5$ Km of unpaired quark matter, interacting *via* gluon exchange (model II); dotted curve (blue online) refers to a star like model II, but in the core there is quark matter in the LOFF state; see [11] for more details. All stars have $M = 1.4 M_\odot$.

curve (red online) refers to model II; the dotted line (blue online) is for model III and it is obtained for the following values of the parameters: $\mu = 500$ MeV, $M_s^2/\mu = 140$ MeV, $\Delta_1 = 0, \Delta_2 = \Delta_3 \simeq 6$ MeV. For unpaired quark matter $\alpha_s \simeq 1$, accordingly to the one loop beta function of QCD, corresponding to $\mu = 500$ MeV and $\Lambda_{QCD} = 250$ MeV. The use of perturbative QCD at such small momentum scales is however questionable. Therefore the results for model II should be considered with some caution and the curve is plotted only to allow a comparison with the other models. In any case it is important to remark that the apparent similarity between the LOFF curve and the unpaired quark curve depends on the fact that the LOFF phase is gapless. This yields a parametric dependence on temperature analogous to that of the unpaired quark matter: $c_V \sim T$ and $\varepsilon_v \sim T^6$. However the similarity between the curves of models II and III should be considered accidental because emissivity of unpaired quark matter depends on the value we assumed for the strong coupling constant.

A final remark: the improvement of the pairing condensation ansatz and, more important, of the model of the stars would allow a direct comparison with the observational data. Nevertheless we expect our results capture the essential physics: indeed from our knowledge of the *two* flavor LOFF phase [14] we may argue that fermion gapless excitations are peculiar of the crystalline color superconductivity; since these gapless excitations are responsible for the rapid cooling, a neutron star with a LOFF core should cool faster than the cooling of a star made only of nuclear matter. If a careful comparison with the observational data (see for example [15]) could allow to rule out slow cooling for star masses in the range we have considered, this would favor either the presence of condensed mesons [16] or quark matter in a gapless state in the core (since gapped quarks emit neutrinos very slowly).

# ACKNOWLEDGMENTS

I would like to thank R. Anglani, R. Casalbuoni, N. Ippolito, R. Gatto, G. Nardulli and M. Mannarelli for the fruitful collaboration. Moreover I thank M. Alford, H. Malekzadeh, S. Reddy and A. Schmitt for enlightening discussions during the conference.

# REFERENCES

1. M. G. Alford, K. Rajagopal and F. Wilczek, Phys. Lett. B **422**, 247 (1998) [arXiv:hep-ph/9711395]; R. Rapp, T. Schäfer, E. V. Shuryak and M. Velkovsky, Phys. Rev. Lett. **81**, 53 (1998) [arXiv:hep-ph/9711396]; D. T. Son, Phys. Rev. D **59**, 094019 (1999) [arXiv:hep-ph/9812287]; R. D. Pisarski and D. H. Rischke, Phys. Rev. D **61**, 074017 (2000) [arXiv:nucl-th/9910056].
2. K. Rajagopal and F. Wilczek, arXiv:hep-ph/0011333; M. G. Alford, Ann. Rev. Nucl. Part. Sci. **51**, 131 (2001) [arXiv:hep-ph/0102047]; G. Nardulli, Riv. Nuovo Cim. **25N3**, 1 (2002) [arXiv:hep-ph/0202037]; S. Reddy, Acta Phys. Polon. B **33**, 4101 (2002) [arXiv:nucl-th/0211045]; T. Schäfer, arXiv:hep-ph/0304281; D. H. Rischke, Prog. Part. Nucl. Phys. **52**, 197 (2004) [arXiv:nucl-th/0305030]; M. Alford, Prog. Theor. Phys. Suppl. **153**, 1 (2004) [arXiv:nucl-th/0312007]; M. Buballa, Phys. Rept. **407**, 205 (2005) [arXiv:hep-ph/0402234]; H. c. Ren, arXiv:hep-ph/0404074; I. Shovkovy, arXiv:nucl-th/0410091; T. Schäfer, arXiv:hep-ph/0509068.
3. M. G. Alford, K. Rajagopal and F. Wilczek, Nucl. Phys. B **537**, 443 (1999) [arXiv:hep-ph/9804403].
4. A. W. Steiner, S. Reddy and M. Prakash, Phys. Rev. D **66**, 094007 (2002) [arXiv:hep-ph/0205201].
5. I. Shovkovy and M. Huang, Phys. Lett. B **564**, 205 (2003) [arXiv:hep-ph/0302142]; M. Alford, C. Kouvaris and K. Rajagopal, Phys. Rev. Lett. **92**, 222001 (2004) [arXiv:hep-ph/0311286]; A. Schmitt, Phys. Rev. D **71**, 054016 (2005) [arXiv:nucl-th/0412033]; S. B. Ruster, V. Werth, M. Buballa, I. A. Shovkovy and D. H. Rischke, Phys. Rev. D **72**, 034004 (2005) [arXiv:hep-ph/0503184].
6. A. I. Larkin and Yu. N. Ovchinnikov, Zh. Eksp. Teor. Fiz. **47**, 1136 (1964); P. Fulde and R. A. Ferrell, Phys. Rev. **135**, A550 (1964); M. G. Alford, J. A. Bowers and K. Rajagopal, Phys. Rev. D **63**, 074016 (2001); R. Casalbuoni and G. Nardulli, Rev. Mod. Phys. **76**, 263 (2004).
7. R. Casalbuoni, R. Gatto, N. Ippolito, G. Nardulli and M. Ruggieri, Phys. Lett. B **627**, 89 (2005) [Erratum-ibid. B **634**, 565 (2006)]; M. Mannarelli, K. Rajagopal and R. Sharma, Phys. Rev. D **73**, 114012 (2006) [arXiv:hep-ph/0603076].
8. M. Ciminale, G. Nardulli, M. Ruggieri and R. Gatto, Phys. Lett. B **636**, 317 (2006) [arXiv:hep-ph/0602180].
9. K. Rajagopal and R. Sharma, hep-ph/0605316.
10. S. L. Shapiro and S. A. Teukolsky, **Black Holes, White Dwarfs and Neutron Stars**, (New York: Wiley, 1983).
11. R. Anglani, G. Nardulli, M. Ruggieri and M. Mannarelli, Phys. Rev. D **74**, 074005 (2006) [arXiv:hep-ph/0607341].
12. N. Iwamoto, Phys. Rev. Lett. **44**, 1637 (1980); Ann. Phys. **141**, 1 (1982).
13. M. Alford, P. Jotwani, C. Kouvaris, J. Kundu and K. Rajagopal, Phys. Rev. D **71**, 114011 (2005) [arXiv:astro-ph/0411560]; A. Schmitt, I. A. Shovkovy and Q. Wang, Phys. Rev. D **73**, 034012 (2006) [arXiv:hep-ph/0510347]; P. Jaikumar, C. D. Roberts and A. Sedrakian, Phys. Rev. C **73**, 042801 (2006) [arXiv:nucl-th/0509093].
14. R. Casalbuoni, R. Gatto, M. Mannarelli, G. Nardulli, M. Ruggieri and S. Stramaglia, "Quasi-particle specific heats for the crystalline color superconducting Phys. Lett. B **575**, 181 (2003) [Erratum-ibid. B **582**, 279 (2004)] [arXiv:hep-ph/0307335].
15. D. Blaschke, H. Grigorian and D. N. Voskresensky, astro-ph/0009120. D. Page, M. Prakash, J. M. Lattimer and A. Steiner, Phys. Rev. Lett. **85**, 2048 (2000); T. Klahn et al., nucl-th/0609067.
16. A. Kryjevski, arXiv:hep-ph/0508180; T. Schafer, Phys. Rev. Lett. **96**, 012305 (2006) [arXiv:hep-ph/0508190].

# Stress-free BCS pairing in color superconductors is impossible

Krishna Rajagopal[*] and Andreas Schmitt[†]

[*]Center for Theoretical Physics, Massachusetts Institute of Technology, Cambridge, MA 02139
[†]Department of Physics, Washington University St Louis, MO 63130

**Abstract.** Cold, asymptotically dense three-flavor quark matter is in the color-flavor locked (CFL) phase, in which all quarks pair in a particularly symmetric fashion. At smaller densities, taking into account a nonzero strange quark mass and electric and color neutrality, the CFL phase requires pairing of quarks with mismatched Fermi momenta. We present a classification of all other possible, less symmetric, pairing patterns and prove that none of them can avoid this mismatch. This result suggests unconventional, e.g., spatially inhomogeneous, superconducting phases for moderate densities.

**Keywords:** deconfined quark matter, color superconductivity
**PACS:** 12.38.Mh,24.85.+p

At asymptotically large densities and sufficiently cold temperatures, three-flavor quark matter is in the color-flavor locked (CFL) phase [1]. This phase is a color superconductor because the color gauge group $SU(3)_c$ is spontaneously broken, due to the formation of quark Cooper pairs. The underlying mechanism is an attractive QCD interaction between the quarks that are antisymmetric in color. Consequently, the order parameter is a color antitriplet and, due to the overall antisymmetry of the two-fermion wave function, also an antitriplet in flavor space (assuming pairing in the spin-0 channel). Hence we can write the order parameter as

$$\mathcal{M} = \Delta_{ij}J_i \otimes I_j, \qquad (1)$$

with the antisymmetric color and flavor matrices $(J_i)_{jk} = (I_i)_{jk} = -i\varepsilon_{ijk}$. The complex $3 \times 3$ matrix $\Delta$ determines the pairing pattern. In the CFL phase, $\Delta_{ij} \propto \delta_{ij}$, and all nine quasiquarks acquire a gap in their excitation spectrum. The reason is that $\mathcal{M}\mathcal{M}^\dagger$ has nine nonzero eigenvalues.

At large densities the quark chemical potential $\mu$ is much larger than all three quark masses and we may approximate $m_u \simeq m_d \simeq m_s \simeq 0$. In this case, the Lagrangian of the system is invariant under color and flavor transformations $SU(3)_c \times SU(3)_f$. Therefore, any order parameter $\Delta$ is equivalent to rotated order parameters $U^T \Delta V$, with $U \in SU(3)_c$, $V \in SU(3)_f$. It is thus sufficient to consider diagonal matrices $\Delta$, and it is easy to see that the preferred phase (= lowest in free energy) is the CFL phase.

Things get more complicated for moderate densities. For instance, in the interior of compact stars we expect the quark chemical potential to be of the order of 500 MeV at most. In this case, we may not neglect the strange mass anymore. For our purpose this has two important consequences. First, color and electric neutrality become important nontrivial constraints for the system (for $m_s = 0$ these constraints are trivially fulfilled

CP892, *Quark Confinement and the Hadron Spectrum VII*
edited by J. E. F. T. Ribeiro
© 2007 American Institute of Physics 978-0-7354-0396-3/07/$23.00

in the CFL phase). In the CFL phase, the neutrality conditions in the case of a small but nonzero strange mass lead to pairing of quarks with mismatched Fermi surfaces. The difference in chemical potentials is of the order $m_s^2/\mu$. This imposes a stress on the pairing. A cost in free energy of the order of $m_s^4$ has to be paid in order to fill both participating quark states up to a common Fermi momentum $\nu$. Only then, conventional BCS pairing with zero-momentum Cooper pairs can be achieved, gaining condensation energy of the order of $\Delta^2\mu^2$, $\Delta$ being the energy gap in the quasiparticle spectrum. At some particular value of the parameter $m_s^2/\mu$, the cost exceeds the gain and conventional pairing is no longer favored. For the CFL phase, this value is $m_s^2/\mu = 2\Delta$.

The second important consequence of a nonzero strange mass is the explicit breaking of the flavor symmetry down to $SU(2)_f$. There exist not necessarily transformations $U \in SU(3)_c$ and $V \in SU(2)_f$, which diagonalize a given complex matrix $\Delta$. Hence, there are non-diagonal order parameters which describe physically distinct phases. Such order parameters are known from systems with similar mathematical structure, cf. the $A$-phase in a spin-1 color superconductor [2] or in $^3$He. An interesting question arises regarding the fate of these phases at moderate densities. While it is clear that the CFL phase is favored for large $\mu$, for smaller $\mu$ there might be a phase which is more comfortable with the neutrality constraints. More precisely, we may ask whether there is some less symmetric, but still conventional pairing pattern in which, once electric and color neutrality are imposed, pairing only occurs among those quarks whose Fermi momenta would be equal in the absence of pairing.

In order to answer this question, we have to set up formal conditions for ($i$) electric and color neutrality and ($ii$) stress-free pairing, and have to find a way to exhaust all possible pairing patterns systematically. Conditions ($i$) and ($ii$) are derived upon introducing an electric chemical potential $\mu_e$ and color chemical potentials $\mu_3$ and $\mu_8$. In general, a color-superconducting phase may require additional or other nonzero color chemical potentials. For our purpose, $\mu_3$ and $\mu_8$ are sufficient since any pattern can be transformed into one where only $\mu_3$ and $\mu_8$ are nonzero [3]. Then, neglecting higher order terms in $m_s, \Delta \ll \mu$, conditions ($i$) and ($ii$) are [3]

$$0 = \sum_{i=1}^{9} \nu_i q_{ik}, \qquad \nu_i = \mu_i^{\mathrm{eff}}. \tag{2}$$

Here, $\nu_i$ are the common Fermi momenta, assigned to each of the nine quarks according to their pairing partners. Moreover, $k = e, 3, 8$, and $\mu_i^{\mathrm{eff}} = \mu_i$ for up and down quarks and $\mu_i^{\mathrm{eff}} = \mu_i - m_s^2/(2\mu)$ for strange quarks, where $\mu_i = \mu + q_{ie}\mu_e + q_{i3}\mu_3 + q_{i8}\mu_8$. The charges of the quarks are denoted by $q_{ik}$.

We observe that the only information we need to know about a particular pairing pattern are the common Fermi momenta $\nu_i$ as a function of $\mu, m_s, \mu_e, \mu_3, \mu_8$. A common Fermi momentum is the arithmetic mean of the effective chemical potentials of the quarks that pair with each other [3]. Consequently, all we need to know is which quark pairs with which other quarks. This information is given by setting the 9 entries $\Delta_{ij}$ of the order parameter to zero or nonzero (say 1) in all possible combinations. Hence, respecting the symmetries given by the structure of $\mathcal{M}$, we have to investigate $2^9 = 512$ patterns. One of these, $\Delta_{ij} = 0$ for all $i, j$, corresponds to unpaired quark matter. Note that many of the 512 patterns are related by color rotations, meaning that not all of them are physically distinct.

In order to automatize the treatment of all patterns, we have translated each pattern into a graph. The vertices of this graph are the nine quarks while there are at most $9 \cdot 2 = 18$ edges, depending on the matrix $\Delta$. Every nonzero entry $\Delta_{ij}$ yields two edges, meaning that the corresponding quarks form Cooper pairs with each other. Then, a pattern is, for our purposes, given by the connected components of the graph. Any pattern uniquely defines its set of components while any given set of components may corresponds to several patterns; it turns out that the 512 patterns yield 149 distinct sets of components. Once the patterns are classified in this way, we may use a computer program to test each set of components on the conditions (2): we first determine the common Fermi momenta for each set and then search for a simultaneous solution $(\mu_e, \mu_3, \mu_8)$ of Eqs. (2). We find that none of the patterns allows for such a solution (except for unpaired quark matter). Consequently, *neutral stress-free BCS pairing is impossible.*

We emphasize that this result does not exclude any less symmetric phase in the phase diagram. We rather conclude that any of these phases must break down at a point in density that is parametrically the same as for the CFL phase and given by a quark chemical potential $\mu$ of the order of $m_s^2/\Delta$. If this value of $\mu$ is larger than the one at which the phase transition from quark matter to nuclear matter occurs, one can expect some form of unconventional pairing which succeeds CFL down in density. Assuming the CFL pairing pattern and a spatially homogeneous system, the CFL phase is succeeded by a phase in which some quasiparticle excitations become gapless despite a nonzero order parameter. This "gapless CFL" phase, however, is unstable with respect to the formation of counter-propagating currents. In the simplest situation, there are two opposite currents, provided by the Cooper pair condensate (or possibly a kaon condensate) and ungapped fermionic modes. In a more complicated version, crystalline structures arise [4]. Other possibilities have been proposed, for instance single-flavor pairing [2]. Most of the proposed phases have ungapped modes. Since transport properties (e.g., neutrino emissivity) are very sensitive to these ungapped modes [5], related astrophysical observables (e.g., cooling curves) might help in the search for the ground state.

## ACKNOWLEDGMENTS

The authors acknowledge support by the U.S. Department of Energy under contracts DE-FG02-91ER50628, DE-FG01-04ER0225 (OJI), and DF-FC02-94ER40818.

## REFERENCES

1. M. G. Alford, K. Rajagopal and F. Wilczek, Nucl. Phys. B **537**, 443 (1999) [arXiv:hep-ph/9804403].
2. A. Schmitt, Phys. Rev. D **71**, 054016 (2005) [arXiv:nucl-th/0412033].
3. K. Rajagopal and A. Schmitt, Phys. Rev. D **73**, 045003 (2006) [arXiv:hep-ph/0512043].
4. K. Rajagopal and R. Sharma, Phys. Rev. D, to appear [arXiv:hep-ph/0605316].
5. A. Schmitt, I. A. Shovkovy and Q. Wang, Phys. Rev. D **73**, 034012 (2006) [arXiv:hep-ph/0510347].

# Superfluidity in neutron stars and cold atoms

Achim Schwenk

*TRIUMF, 4004 Wesbrook Mall, Vancouver, BC, Canada, V6T 2A3*
*Department of Physics, University of Washington, Seattle, WA 98195-1560*

**Abstract.** We discuss superfluidity in neutron matter, with particular attention to induced interactions and to universal properties accessible with cold atoms.

**Keywords:** Superfluidity, induced interactions, neutron matter, resonant Fermi gases
**PACS:** 26.60.+c, 03.75.Ss

Superfluidity plays a central role in strongly-interacting many-body systems. Nuclear pairing shows striking trends in neutron-proton asymmetric systems [1]. The $\beta$ decay of the two-neutron halo in $^{11}$Li is suppressed due to pairing [2] similar to neutrino emission in neutron star cooling [3]. Ultracold atoms exhibit vortices and superfluid characteristics in thermodynamic and spectroscopic properties [4].

The physics of dilute Fermi gases with large scattering lengths is universal, independent of atomic or nuclear details. For neutrons the scattering length is also large, $a_{nn} = -18.5 \pm 0.3$ fm, and therefore cold atom experiments constrain low-density neutron matter. For instance, for two spin states with equal populations, the S-wave superfluid pairing gap of resonant gases of $^6$Li atoms, $^{40}$K atoms or neutrons is given by $\Delta/\varepsilon_F = \zeta$, where $\varepsilon_F = k_F^2/(2m)$ is the Fermi energy and $\zeta$ is a universal number.

For relative momenta $k \lesssim 2$ fm$^{-1}$, nucleon-nucleon (NN) interactions are well constrained by the existing scattering data [5]. In Fig. 1, we show superfluid pairing gaps in neutron matter obtained by solving the BCS gap equation with a free spectrum. At low densities (in the crust of neutron stars), neutrons form a $^1S_0$ superfluid. At higher densities, the S-wave interaction is repulsive and neutrons pair in the $^3P_2$ channel (with a small coupling to $^3F_2$ due to the tensor force). Fig. 1 demonstrates that the $^1S_0$ BCS gap is practically independent of nuclear interactions, and therefore strongly constrained by the NN phase shifts [6]. This includes a very weak cutoff dependence for the class of low-momentum interactions $V_{\text{low}\,k}$ [5] with sharp or sufficiently narrow smooth regulators with $\Lambda > 1.6$ fm$^{-1}$. The model dependence for larger momenta shows up prominently in Fig. 1 for the $^3P_2$–$^3F_2$ gaps at Fermi momenta $k_F > 2$ fm$^{-1}$ [7].

Polarization effects ("induced interactions") due to particle-hole screening and vertex corrections are crucial for superfluidity. They lead to a reduction of the S-wave gap, which is significant $[(4e)^{-1/3} \approx 0.45]$ even in the perturbative $k_F a$ limit [8]:

$$\frac{\Delta}{\varepsilon_F} = \frac{8}{e^2} \exp\left\{ \left( \diagram_1 + \diagram_2 + \diagram_3 + ... \right)^{-1} \right\} = (4e)^{-1/3} \frac{8}{e^2} \exp\left\{ \frac{\pi}{2 k_F a} + \mathcal{O}(k_F a) \right\}.$$

This reduction is due to spin fluctuations, which are repulsive for spin singlet pairing and overwhelm attractive density fluctuations. In finite systems, the spin and density

CP892, *Quark Confinement and the Hadron Spectrum VII*
edited by J. E. F. T. Ribeiro
© 2007 American Institute of Physics 978-0-7354-0396-3/07/$23.00

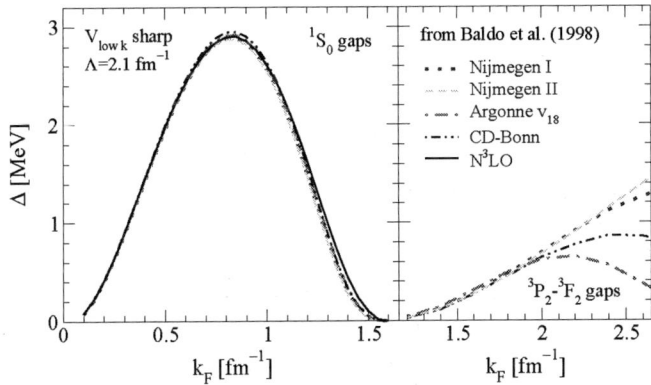

**FIGURE 1.** The $^1S_0$ (left) and $^3P_2-^3F_2$ (right) superfluid pairing gaps $\Delta \equiv \Delta(k_F)$ versus Fermi momentum $k_F$, based on various charge-dependent NN interactions at the BCS level. The results are for low-momentum interactions $V_{low k}$ with $\Lambda = 2.1\,\text{fm}^{-1}$ [6] (left) or taken from Baldo *et al.* [7] (right).

response differs. In nuclei with cores, the low-lying response is due to surface vibrations. Consequently, induced interactions may be attractive, since the spin response is weaker.

The renormalization group (RG) provides a systematic tool to reduce a physical system to a simpler, equivalent problem focusing on relevant degrees of freedom. Following Shankar [9], we have applied the RG to neutron matter, restricting the effective interaction to low-lying states in the vicinity of the Fermi surface [10]. Starting from the low-momentum interaction $V_{low k}$ [5], we solve a one-loop RG equation in the particle-hole channels ("phRG") that includes contributions from successive ph momentum shells. The RG builds up many-body effects similar to the two-body parquet equations, and efficiently includes induced interactions on superfluidity beyond the perturbative result.

The phRG results for the $^1S_0$ gap are shown in Fig. 2. We find a factor $3-4$ reduction to a maximal gap $\Delta \approx 0.8\,\text{MeV}$. At the larger densities, the dotted band indicates the uncertainty due to an approximate self-energy treatment in [10]. For the lowest densities, the phRG is consistent with the dilute result $\Delta/\Delta_0 = (4e)^{-1/3}$. This is similar to the GFMC calculations of Carlson *et al.* [12] for cold atoms in the unitary regime, which are also consistent with the extrapolated dilute result to a good approximation. On the lower side of Fig. 2, there are differences between neutron matter and unitary gases: For $k_F \approx 0.4\,\text{fm}^{-1}$, one has $k_F r_e \approx 1$ (with effective range $r_e$), and pairing is weaker, $\Delta/\varepsilon_F \approx 0.1$. For these densities, neutron matter is close to the unitary regime, but theoretically simpler due to an appreciable effective range [13]. Note that the (low order) CBF results of [14] do not include long-range polarization effects, and therefore are close to the BCS gap at low densities.

The RG approach is widely used in condensed matter physics to study the interference of different instabilities, especially in the context of the 2d Hubbard model. Similar competing instabilities are present in color superconductivity at intermediate densities. Here, the RG method seems ideal to resolve the zoo of possible phases.

Non-central spin-orbit and tensor interactions are crucial for $^3P_2-^3F_2$ superfluidity. Without a spin-orbit interaction, neutrons would form a $^3P_0$ superfluid instead. The

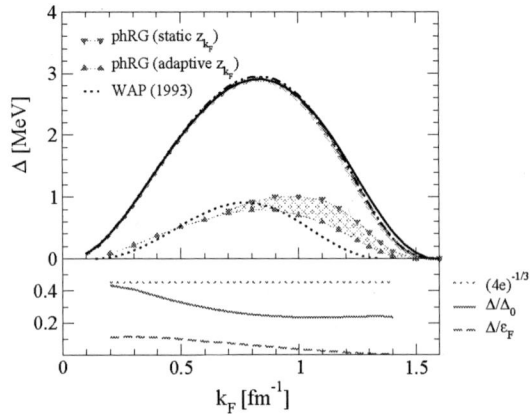

**FIGURE 2.** Top panel: Comparison of the $^1S_0$ BCS gap to the results including polarization effects through the phRG, for details see [10], and to the results of Wambach *et al.* [11]. Lower panel: Comparison of the full superfluid gap $\Delta$ to the BCS gap $\Delta_0$ and to the Fermi energy $\varepsilon_F$.

first perturbative calculation of non-central induced interactions shows that $^3P_2$ gaps below $10\,\mathrm{keV}$ are possible (while $\langle V_{\mathrm{ind}}\rangle/\langle V_{\mathrm{low}\,k}\rangle < 0.5$) [15]. This arises from a repulsive induced spin-orbit interaction due to the mixing with the (large) spin-spin interaction. Our result impacts the cooling of neutron stars [3] and would imply that core neutrons are only superfluid at late times ($t \sim 10^5\,\mathrm{yrs}$).

I would like to thank B. Friman, R. Furnstahl, K. Hebeler, C. Horowitz, C. Pethick and P. Reuter for discussions and the organizers of QCHS VII for the stimulating meeting. This work is supported in part by NSERC and US DOE Grant DE–FG02–97ER41014. TRIUMF receives federal funding via a contribution agreement through the NRC.

# REFERENCES

1. Yu.A. Litvinov et al., *Phys. Rev. Lett.* **95**, 042501 (2005).
2. F. Sarazin et al., *Phys. Rev.* **C70**, 031302(R) (2004).
3. D.G. Yakovlev and C.J. Pethick, *Ann. Rev. Astron. Astrophys.* **42**, 169 (2004).
4. C. Chin et al., *Science* **305**, 1128 (2004); J. Kinast et al., *Science* **307**, 1296 (2005);
   M.W. Zwierlein, A. Schirotzek, C.H. Schunck and W. Ketterle, *Nature* **435**, 1047 (2005).
5. S.K. Bogner, T.T.S. Kuo and A. Schwenk, *Phys. Rept.* **386**, 1 (2003).
6. K. Hebeler, A. Schwenk and B. Friman, nucl-th/0611024.
7. M. Baldo et al., *Phys. Rev.* **C58**, 1921 (1998); M. Hjorth-Jensen, private communication (2005).
8. L.P. Gorkov and T.K. Melik-Barkhudarov, *Sov. Phys. JETP* **13**, 1018 (1961);
   H. Heiselberg, C.J. Pethick, H. Smith and L. Viverit, *Phys. Rev. Lett.* **85**, 2418 (2000).
9. R. Shankar, *Rev. Mod. Phys.* **66**, 129 (1994).
10. A. Schwenk, G.E. Brown and B. Friman, *Nucl. Phys.* **A703**, 745 (2002).
11. J. Wambach, T.L. Ainsworth and D. Pines, *Nucl. Phys.* **A555**, 128 (1993).
12. S.Y. Chang, J. Carlson, V.R. Pandharipande and K.E. Schmidt, *Phys. Rev.* **A70**, 043602 (2004).
13. A. Schwenk and C.J. Pethick, *Phys. Rev. Lett.* **95**, 160401 (2005).
14. A. Fabrocini, S. Fantoni, A.Yu. Illarionov and K.E. Schmidt, *Phys. Rev. Lett.* **95**, 192501 (2005).
15. A. Schwenk and B. Friman, *Phys. Rev. Lett.* **92**, 082501 (2004).

# Strange Quark Star Crusts

Andrew W. Steiner

*Theoretical Division, Los Alamos National Laboratory, Los Alamos, NM 87545*
*Joint Institute for Nuclear Astrophysics, Michigan State University, East Lansing, MI 48824*

**Abstract.** If strange quark matter is absolutely stable, some neutron stars may be strange quark stars. Strange quark stars are usually assumed to have a simple liquid surface. We show that if the surface tension of droplets of quark matter in the vacuum is sufficiently small, droplets of quark matter on the surface of a strange quark star may form a solid crust on top of the strange quark star. This solid crust can significantly modify the predictions for the photon emission for the surface in an observable way.

**Keywords:** strange quark stars, quark matter, strange crusts
**PACS:** 25.75.Nq, 26.60.+c, 97.60.Jd

## Heterogeneous Phases in Dense Matter

The putative solid crust on top of a strange quark star is part of the generic phenomenon of heterogeneous phases which may be present in dense matter. Heterogeneous phases can be present in dense matter because (1) the characteristic length scale for the attraction for the strong interaction operates is much smaller than the characteristic length scale for the repulsive Coulomb interaction, and (2) the "clumps" that are preferred by the strong interaction are often charged, requiring an oppositely-charged "background" of electrons for charge neutrality. Matter near terrestrial densities is naturally heterogeneous because the electron screening length is larger than the size of strongly-bound nuclei, which are positively charged.

Heterogeneous phases are stable only if the associated energy cost of creating a surface is sufficiently small. Nuclei are stable, in part, because the energy contribution from surface tension $\sim 1.1$ MeV/fm$^2$ is small compared to the binding energy. The heterogeneous mixed phase between quark matter and hadronic matter in normal neutron stars is also subject to the energy cost of creating a surface between hadronic matter and quark matter at high density [1]. This surface tension is not well known, but could be significantly larger, $\sim 50$ MeV/fm$^2$, than the surface tension in nuclei [2]. On the other hand, if this surface tension is sufficiently small, then the heterogeneous phase (if present) likely extends to the highest densities in neutron star interiors [3].

## Quarks in Dense Matter

One of the critical questions regarding neutron star matter is whether the central density of neutron stars is large enough so that hadronic matter is deconfined and quarks become the relevant degrees of freedom [4, 5]. This question remains unanswered be-

CP892, *Quark Confinement and the Hadron Spectrum VII*
edited by J. E. F. T. Ribeiro
© 2007 American Institute of Physics 978-0-7354-0396-3/07/$23.00

cause QCD is still not sufficiently well understood at neutron star densities. Recently, significant progress has been made in understanding QCD at asymptotically high densities. In that regime, where pertubative studies are reliable, quark matter is believed to be in a superconducting color-flavor-locked (CFL) phase, characterized by quark pairing and a completely gapped spectrum. Such a phase is an electromagnetic insulator in bulk and admits no electrons, even when stressed by small quark masses [6, 7]. If dense quark matter indeed exists inside neutron stars, where densities are well above nuclear matter density but below the density where perturbative QCD is expected to be valid, the ground state of quark matter is uncertain [8, 9]. Nevertheless, in such a "hybrid star", attractive interactions between quarks will lead to the formation of a color superconducting state [10, 11, 12], characterized by quark pairing and superfluidity with singlet pairing gaps as large as 100 MeV.

## Strange Quark Stars

Over thirty years ago, it was conjectured that at sufficiently high density, macroscopic quark matter composed only of up and down $(u, d)$ quarks might be stabilized by the introduction of strange $(s)$ quarks, and constitute the true ground state of matter, as it would be more bound than nuclear matter [13, 14, 15]. The traditional paradigm is that the positively charged quark matter is compensated by an electron fluid which extends to a slightly larger radius than the homogenous quark matter underneath. A large electric field is formed at the surface because of the electron layer, leading to novel spectral features [16]. This picture of the quark star surface, involving homogenous quark matter and electrons, has been recently challenged [17]. Matter may satisfy charge neutrality globally rather than locally, provided surface and Coulomb costs are not prohibitively large in a heterogenous mixed phase. This mixed phase would then be qualitatively similar to the mixed phase of nuclei and electrons in the crust of normal neutron stars and would share several features with the mixed phase of quark drops and nuclear matter in hybrid stars [18].

To understand this mixed phase, we note that since the electron chemical potential, $\mu_e$, is significantly smaller than the quark chemical potential $\mu$ for all known models of quark matter, a general parameterization of the EoS can be obtained by expanding in powers of $\mu_e/\mu$ [17],

$$p_{QM} = p_0(\mu, m_s) - n_Q(\mu, m_s)\mu_e + \frac{1}{2}\chi_Q(\mu, m_s)\mu_e^2 + \dots \tag{1}$$

where $p_0$, $n_Q$, and $\chi_Q$ are well-defined and calculable functions of $\mu$ and the strange quark mass, $m_s$. This second-order expansion, which neglects the electron pressure $p_e \sim \mu_e^4$, can be used for any model EoS or for that predicted by QCD.

The structure of droplets in the crust of a strange quark star can be obtained from the Poisson equation. At zero temperature and pressure, the Gibbs free energy per quark for droplets can be compared with the Gibbs free energy per quark for homogeneous matter. If the surface tension, i.e. the energy cost of creating a droplet surface, is small enough, then the crustal phase is preferred over homogeneous quark matter. The critical surface tension is [19]

$$\sigma_{\text{crit}} = \frac{0.8n_Q^2}{12\sqrt{\pi}\alpha\chi_Q^{3/2}} \, . \tag{2}$$

In the context of the Bag model for dense quark matter, the condition for forming a mixed phase becomes

$$\sigma \lesssim 12 \left(\frac{m_s}{150 \, \text{MeV}}\right)^3 \frac{m_s}{\mu} \, \text{MeV/fm}^2 \, . \tag{3}$$

Using two estimates of the surface energy of strangelets: (i) $\sigma \simeq 8 \, \text{MeV/fm}^2$ for $m_s = 150 \, \text{MeV}$ and $\mu \simeq 300 \, \text{MeV}$; and (ii) $\sigma \simeq 5 \, \text{MeV/fm}^2$ for $m_s = 200 \, \text{MeV}$ at $\mu \simeq 300$ MeV, the condition in Eq. 3 implies that a homogeneous phase is marginally favored for $m_s = 150 \, \text{MeV}$ while the structured mixed phase is favored for $m_s = 200 \, \text{MeV}$. The sensitivity to $m_s$ in Eq. 3 and uncertainty in other finite size effects can alter these quantitative estimates. If the structured phase is favored, it will be composed of quark nuggets immersed in a sea of electrons. The size of the quark nuggets in this phase is determined by minimizing the surface, Coulomb and other finite size contributions to the energy. At low temperature, this mixed phase will be a solid with electrons contributing to the pressure while quarks contribute to the energy density - much like the mixed phase with electrons and nuclei in crust of a conventional neutron star. This modified picture of the strange star surface has a much reduced density gradient and negligible electric field unlike the old paradigm. The observed photon spectrum from such a surface will be very different than from the traditional one with a large electric field.

## REFERENCES

1. S. Reddy, G. Bertsch, and M. Prakash, *Phys. Lett. B* **475**, 1 (2000).
2. M. G. Alford, K. Rajagopal, S. Reddy, and F. Wilczek, *Phys. Rev. D* **64**, 074017 (2001), hep-ph/0105009.
3. A. W. Steiner, M. Prakash, and J. M. Lattimer, *Phys. Lett. B* **486**, 239–248 (2000), nucl-th/0003066.
4. D. Ivanenko, and D. G. Kurdgelaidze, *Nuovo Cim. Lett.* **2**, 13 (1969).
5. J. C. Collins, and M. J. Perry, *Phys. Rev. Lett.* **30**, 1353 (1975).
6. K. Rajagopal, and F. Wilczek, *Phys. Rev. Lett.* **86**, 3492 (2000).
7. A. W. Steiner, S. Reddy, and M. Prakash, *Phys. Rev. D* **66**, 094007 (2002).
8. M. Alford, M. Braby, M. W. Paris, and S. Reddy, *Astrophys. J.* **629**, 969–978 (2005), nucl-th/0411016.
9. A. W. Steiner, *Phys. Rev. D* **72**, 054024 (2005).
10. D. Bailin, and A. Love, *Phys. Rep.* **107**, 325 (1984).
11. M. G. Alford, K. Rajagopal, and F. Wilczek, *Nucl. Phys. B* **537**, 443 (1999).
12. R. Rapp, T. Schaefer, E. V. Shuryak, and M. Velkovsky, *Phys. Rev. Lett.* **81**, 53 (1998).
13. A. R. Bodmer, *Phys. Rev. D* **4**, 1601 (1971).
14. E. Witten, *Phys. Rev. D* **30**, 272 (1984).
15. E. Farhi, and R. L. Jaffe, *Phys. Rev. D* **30**, 2379 (1984).
16. D. Page, and V. V. Usov, *Phys. Rev. Lett.* **89**, 131101 (2002).
17. P. Jaikumar, S. Reddy, and A. W. Steiner, *Phys. Rev. Lett.* **96**, 041101 (2006), nucl-th/0507055.
18. N. Glendenning, *Phys. Rev. D* **48**, 1274 (1992).
19. M. Alford, K. Rajagopal, S. Reddy, and A. W. Steiner, *Phys. Rev. D* **73**, 114016 (2006).

# The neutron matter equation of state from low-momentum interactions

L. Tolós[*], B. Friman[*] and A. Schwenk[†,**]

[*]*Gesellschaft für Schwerionenforschung, Planckstrasse 1, D-64291 Darmstadt, Germany*
[†]*TRIUMF, 4004 Wesbrook Mall, Vancouver, BC, Canada, V6T 2A3*
[**]*Department of Physics, University of Washington, Seattle, WA 98195-1560*

**Abstract.** We calculate the neutron matter equation of state at finite temperature based on low-momentum nucleon-nucleon and three-nucleon interactions. Our results are compared to the model-independent virial equation of state and to variational calculations. We provide a simple estimate for the theoretical error, important for extrapolations to astrophysical conditions.

**Keywords:** Neutron matter, finite temperature
**PACS:** 21.65.+f, 26.60.+c

The nuclear equation of state plays a central role in the physics of neutron stars [1] and core-collapse supernovae [2, 3]. Renormalization group methods coupled with effective field theory offer the possibility of a new and systematic approach to nuclear matter: For low-momentum interactions $V_{low\,k}$ [4] with cutoffs around $2\,fm^{-1}$, the strong short-range repulsion in conventional nucleon-nucleon (NN) interactions and the tensor force are tamed [5, 6]. At sufficient density, Pauli blocking eliminates the shallow bound states, and thus the particle-particle channel becomes perturbative in nuclear matter [5]. In addition, the corresponding leading-order chiral three-nucleon (3N) interaction becomes perturbative in light nuclei for $\Lambda \lesssim 2\,fm^{-1}$ [7]. Consequently, the Hartree-Fock (HF) approximation is a good starting point, and perturbation theory (in the sense of a loop expansion) around the HF energy becomes tractable [5]. The perturbative character is due to a combination of Pauli blocking and an appreciable effective range (see also [8]).

At finite temperature, the loop expansion around the HF free energy can be realized, based on the work of Kohn, Luttinger and Ward [9, 10], by the perturbative expansion of the free energy, where the momentum dependence of the self-energy is treated perturbatively. In this work, we include the first-order NN and 3N contributions, as well as anomalous and normal second-order diagrams with NN interactions. The pressure, entropy and energy are calculated using standard thermodynamic relations. We use the cutoff dependence to provide simple error estimates, and find that the cutoff dependence is reduced significantly, when second-order contributions are included.

We start from the perturbative expansion of the grand-canonical potential $\Omega(\mu, T, V)$, where $\mu$ is the chemical potential, $T$ the temperature and $V$ the volume:

$$\Omega = \Omega_0 + \Omega_{1,NN} + \Omega_{1,3N} + \Omega_{2,a} + \Omega_{2,n} + \dots . \tag{1}$$

The non-interacting system is given by $\Omega_0$, $\Omega_1 = \Omega_{1,NN} + \Omega_{1,3N}$ denotes the first-order NN and 3N, and $\Omega_{2,a} + \Omega_{2,n}$ are the second-order anomalous and normal contributions.

CP892, *Quark Confinement and the Hadron Spectrum VII*
edited by J. E. F. T. Ribeiro

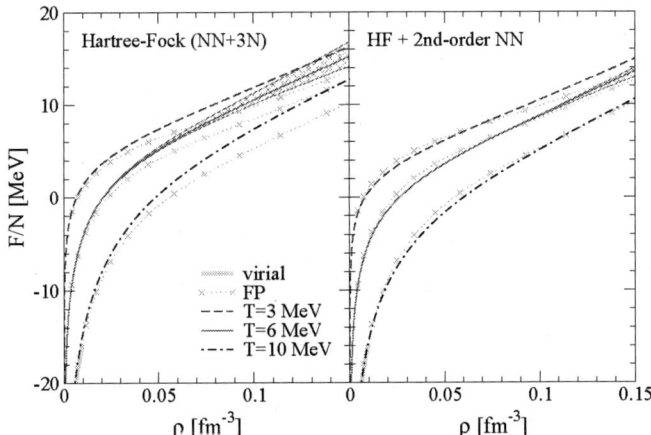

**FIGURE 1.** The free energy per particle $F/N$ as a function of density $\rho$. The left figure gives the first-order NN and 3N contributions with a free single-particle spectrum. Second-order anomalous and normal NN contributions are included in the right figure. Our results are compared to the virial equation of state (virial) [12] and to the variational calculations of Friedman and Pandharipande (FP) [13]. The virial curve ends where the fugacity $z = e^{\mu/T}$ is 0.5.

The free energy $F(N,T,V)$ is obtained by a Legendre transformation of the grand-canonical potential with respect to the chemical potential, $F(N,T,V) = \Omega(\mu,T,V) + \mu N$, with mean particle number $N$. Following Kohn and Luttinger [9], we have

$$F(N) = F_0(N) + \Omega_1(\mu_0) + \Omega_{2,\mathrm{n}}(\mu_0) + \left[\Omega_{2,\mathrm{a}}(\mu_0) - \frac{1}{2}\frac{(\partial\Omega_{1,\mathrm{NN}}/\partial\mu)^2}{\partial^2\Omega_0/\partial\mu^2}\bigg|_{\mu_0}\right] + \ldots, \quad (2)$$

where $\mu_0$ is the chemical potential of a non-interacting system with the same density $\rho = N/V$ as the interacting system, $N = -[\partial\Omega_0/\partial\mu]_{\mu_0}$, and $F_0(N) = \Omega_0(\mu_0) + \mu_0 N$ is the free energy of the non-interacting system. The above expansion ensures that the $T \to 0$ limit is correctly reproduced [9, 10]. The anomalous second-order diagram accounts for perturbative corrections to the free single-particle spectrum.

Our results [11] for the free energy per particle are shown in Fig. 1 for temperatures $T = 3\,\mathrm{MeV}, 6\,\mathrm{MeV}$ and $10\,\mathrm{MeV}$, where the low-momentum interaction $V_{\mathrm{low}\,k}$ is obtained from the Argonne $v_{18}$ potential for a cutoff $\Lambda = 2.1\,\mathrm{fm}^{-1}$. For the 3N contribution at the HF level, we find that only the $c_1$ and $c_3$ terms of the long-range $2\pi$-exchange part survive (for details on the 3N interaction, see [5, 7]). For the $T = 6\,\mathrm{MeV}$ results, we provide error estimates by varying the cutoff over the range $\Lambda = 1.9\,\mathrm{fm}^{-1}$ (lower curve) to $\Lambda = 2.5\,\mathrm{fm}^{-1}$ (upper curve). As expected the error grows with increasing density. From Fig. 1, we observe that the equation of state becomes significantly less cutoff dependent with the inclusion of the second-order NN contributions.

In Fig. 1, we also compare our results for the free energy to the model-independent virial equation of state [12] and to the variational calculations of Friedman and Pand-haripande [13] (FP, based on the Argonne $v_{14}$ and a 3N potential). We find a very good agreement with the virial free energy, and for the densities in Fig. 1 similar results as FP.

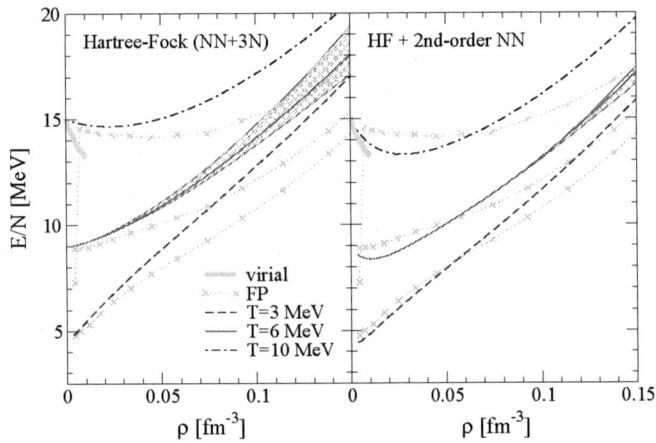

**FIGURE 2.** The energy per particle $E/N$ as a function of density $\rho$ to first and second order as in Fig. 1.

Our results [11] for the energy per particle are presented in Fig. 2. As for the free energy, we observe additional binding and a significantly reduced cutoff dependence at second order. In contrast to the variational calculation of FP [13], the low-density behavior at second order is in good agreement with the virial equation of state [12]. This highlights the importance of a correct finite-temperature treatment of second and higher-order contributions. This work is part of a program to improve the nuclear equation of state input for astrophysics, and to provide error estimates, for example, for the neutron star mass and radius predictions.

## ACKNOWLEDGMENTS

This work was supported in part by the Virtual Institue VH-VI-041 of the Helmholtz Association, NSERC and the US DOE Grant DE–FG02–97ER41014. TRIUMF receives federal funding via a contribution agreement through the NRC.

## REFERENCES

1. J.M. Lattimer and M. Prakash, *Astrophys. J.* **550**, 426 (2001).
2. A. Mezzacappa, *Annu. Rev. Nucl. Part. Sci.* **55**, 467 (2005).
3. H.T. Janka, R. Buras, F.S. Kitaura Joyanes, A. Marek and M. Rampp, astro-ph/0405289.
4. S.K. Bogner, T.T.S. Kuo and A. Schwenk, *Phys. Rept.* **386**, 1 (2003).
5. S.K. Bogner, A. Schwenk, R.J. Furnstahl and A. Nogga, *Nucl. Phys.* **A763**, 59 (2005).
6. S.K. Bogner, R.J. Furnstahl, S. Ramanan and A. Schwenk, *Nucl. Phys.* **A773**, 203 (2006).
7. A. Nogga, S.K. Bogner and A. Schwenk, *Phys. Rev.* **C70**, 061002(R) (2004).
8. A. Schwenk and C.J. Pethick, *Phys. Rev. Lett.* **95**, 160401 (2005).
9. W. Kohn and J.M. Luttinger, *Phys. Rev.* **118**, 41 (1960).
10. J.M. Luttinger and J.C. Ward, *Phys. Rev.* **118**, 1417 (1960).
11. L. Tolós, B. Friman and A. Schwenk, in preparation.
12. C.J. Horowitz and A. Schwenk, *Phys. Lett.* **B638**, 153 (2006).
13. B. Friedman, and V.R. Pandharipande, *Nucl. Phys.* **A361**, 502 (1981).

# Hadron-Hadron Bound States From First Principles

Paulo A. Faria da Veiga[1] and Michael O'Carroll

*Departamento de Matemática Aplicada e Estatística, ICMC-USP*
*C.P. 668, 13560-970 São Carlos SP, Brazil*

**Abstract.** We present a summary of our results on the hadron spectrum and hadron-hadron bound state spectrum which were obtained as part of a program which tries to bridge the gap between QCD and Nuclear Physics. We analyzed SU(3) lattice QCD models in the strong coupling regime (small hopping parameter $\kappa > 0$ and large glueball mass). We considered an imaginary time formulation for $2 + 1$ and $3 + 1$ dimensions models with one and two flavors. For the more algebraically complex and more realistic model in $3 + 1$ dimensions, $4 \times 4$ spin matrices, and two flavors, we show the existence of three-quark isospin $1/2$ particles (proton and neutron) and isospin $3/2$ baryons (delta particles), with asymptotic masses $-3 \ln \kappa$ and isolated dispersion curves. Baryon-baryon bound states of isospin zero are found with binding energy of order $\kappa^2$. The two-particle analysis is performed using a ladder approximation to a lattice Bethe-Salpeter equation written with the help of relative coordinates adapted to the lattice. The dominant baryon-baryon interaction is an energy-independent spatial range-one attractive potential with an $\mathscr{O}(\kappa^2)$ strength. There is also attraction arising from gauge field correlations associated with six overlapping bonds, but it is counterbalanced by Pauli repulsion to give a vanishing zero-range potential. The overall range-one potential results from a quark, antiquark exchange with *no* meson exchange interpretation since the spin indices are not of a meson particle; we call it a quasimeson exchange. The repulsive or attractive nature of the interaction depends on the isospin and spin of the two-baryon state.

**Keywords:** Spectral Analysis, Lattice QCD, Bethe-Salpeter Equation, Excitation Spectrum
**PACS:** 11.15.Ha, 02.30.Tb, 11.10.St, 24.85.+p

It is a fundamental problem in particle physics to determine the low-lying energy-momentum (EM) spectrum of Quantum Chromodynamics (QCD). A convenient ultra-violet cutoff version is given by the QCD models defined on a lattice [1, 2, 3, 4]. In the strong coupling regime, the infinite volume limit can be reached, hadrons are seen as bound states of quarks, and confinement is manifested.

As part of a program which tries to bridge the gap between QCD and Nuclear Physics, here we briefly summarize our results - from first principles - on the hadron spectrum and hadron-hadron bound state spectrum in strong coupling SU(3) lattice QCD with one and two flavors, using the Wilson action and an imaginary-time functional integral formulation. Whenever two-quark flavors are present, our model has a global SU(2) isospin symmetry. Our analysis of the low-lying spectrum in lattice QCD with strong coupling (small hopping parameter $0 < \kappa \ll 1$, and large glueball mass, such that $\beta \equiv 1/(2g_0^2) \ll \kappa$) appears in a recent series of papers of Refs. [5, 6, 7, 8, 9, 10, 11, 12, 13] and [14]. Because of space limitations, all details of the quite complicate machinery used

---

[1] Email: veiga@icmc.usp.br

CP892, *Quark Confinement and the Hadron Spectrum VII*
edited by J. E. F. T. Ribeiro
© 2007 American Institute of Physics 978-0-7354-0396-3/07/$23.00

in this analysis are omitted here. Our goal is to understand when and how bound states occur and how their binding related to the effective Yukawa meson-exchange theory.

We determined the low-lying EM spectrum for increasingly complex SU(3) QCD lattice models. As a necessary step to obtain the two-hadron spectrum, we must first analyze the one-hadron sector. Mesons correspond, as expected, to tightly bound bound states of a quark and an antiquark. Baryons are given by tightly bound bound states of three quarks. These particles are manifested by isolated dispersion curves in the EM spectrum. Mesons have asymptotic masses of order $-2\ln\kappa$ and baryon asymptotic masses are of order $-3\ln\kappa$. Their dispersion curves are convex and increasing functions of each momentum component, for small momenta. We also made an analysis of the hadron mass splitting for models with one flavor in spatial dimension $d = 2, 3$, using $4 \times 4$ spin matrices. For mesons, mass splitting between the $1/2$ and $3/2$ total spin states occur at order $\kappa^4$. For baryons, it is of order $\kappa^6$ for $d = 2$ and, if any, is at least of $\mathscr{O}(\kappa^7)$, for $d = 3$.

Concerning the two-baryon spectrum, for the case of $d = 2$, $2 \times 2$ Pauli spin matrices and only *one* flavor, *no* baryon-baryon, meson-meson or meson-baryon bound state was detected. This is so because Pauli repulsion is too strong if only one quark flavor is present. The simplest case that we analyzed and in which a two-hadron bound state is detected is for some particular total isospin sectors for the two-flavor model with $2 \times 2$ spin matrices and in $2 + 1$ dimensions: two-baryon and two-meson bound states appear, and there should be no restriction for a meson-baryon bound state to be present, as well. However, this model is not complex enough to accommodate protons and neutrons in the one-particle spectrum.

More recently, we analyzed the baryon sector for the two-flavor model with *up* and *down* quarks in $3 + 1$ dimensions and $4 \times 4$ spin matrices. We first showed the existence of twenty, three-quark, one-particle states with isolated dispersion curves (upper gap property), and also their associated antiparticles, which includes the proton ($p$), the neutron ($n$) and the delta ($\Delta$) particles. These one-baryon spectral results are exact and, making the lower order explicit, their asymptotic masses are $-3\ln\kappa - 3\kappa^3/4 + \mathscr{O}(\kappa^7)$; if there is mass splitting it is due to contributions of $\mathscr{O}(\kappa^7)$ or higher. The upper gap property in the Hamiltonian formulation is unknown.

Next, we determined the two-baryon bound states in the $I = 0$ sector and below the two-baryon threshold, which is given by twice the smallest of the baryon masses. We find several bound states with binding energies of order $\kappa^2$. The most strongly bound, bound states are given by $\Delta - \Delta$, total spin $S = 3$ states, and also by a *superposition* of $p - n$ and $\Delta - \Delta$ total spin $S = 1$ states. The later corresponds to the deuteron, which turns out to be given by a superposition of two states. The more weakly bound, bound states are associated with a *superposition* of $p - n$ and $\Delta - \Delta$ total spin $S = 0$ states, and also with $\Delta - \Delta$, $S = 2$ bound states. In contrast to the $I = 0$ states treated here, we have found that for the maximum isospin $I = 3$ sector there are bound states in the lowest total spin sectors $S = 0, 1$ and *no* bound states if $S = 2, 3$. These results are in agreement with our previous results that the attraction between the two particles decreases with increasing $I$ (more alignment). Moreover, as before, there are two sources of attraction, namely, *i)* the exchange of a quark and an antiquark, which is *not* a meson particle exchange, and *ii)* gauge field correlation effects associated with six overlapping bonds. We note that the exchange of a *quasimeson* particle alone is not enough for a baryon-baryon bound state to be formed, and that the effects of gauge field correlations are essential.

We note that although our baryon-baryon bound state results are obtained using a complicate machinery, in the end a simple picture emerges for the formation of a baryon-baryon bound state. The two-baryon dynamics in relative coordinates behaves approximately like that of a non-relativistic one-particle lattice hamiltonian $T + V$ with lattice kinetic energy $-\kappa^3 \Delta/8$, where $\Delta$ is the spatial lattice Laplacian and the potential energy $V$ is $\kappa^2 V'$, the quasi-meson exchange space range-one potential which dominates the kinetic energy for small $\kappa$. The attractive or repulsive nature of the interaction depends on the isospin, spin spectral structure of $V$ at a single site of space-range one. Because of the $\kappa^3$ dependence of the kinetic energy $T$ and the $\kappa^2$ dependence of the potential energy, there is no minimal critical value of the interaction strength needed for the presence of a bound state.

We also remark that the detection of particle (and their bound states) masses from the exponential decay rate of suitable correlations, *without* a spectral representation, is meaningless, and the resulting values may be far from the correct ones, especially in cases where degeneracies are broken with small separations.

Three ingredients are basic in our method, that makes it differs significantly from other approaches to the particle spectrum, both theoretical or numerical (see Refs. [1, 15, 16, 17, 18, 19, 20, 21, 22, 23, 24, 25, 26, 27]):

- It incorporates a hyperplane decoupling technique, which has the nice feature of revealing the form of the fields that create the particles of the model (and their multiplicities!), and which is used to obtain temporal decays for suitable two- and four-hadron correlations. No a priori guesswork is needed!
- We derive spectral representations for suitable two- and four-hadron correlations, via Feynman-Kac formulas. It is these new spectral representations that allow us to relate complex momentum singularities of the two- and four-hadron functions with the one- and two-hadron EM spectrum. The one-particle spectrum is manifested as isolated dispersion curves, and we obtain convergent expansions for masses. Furthermore, we show that the particle spectrum is only spectrum up to close the two-particle thresholds;
- We use a lattice version of a Bethe-Salpeter (B-S) equation for the four-hadron function expressed in special lattice relative coordinates and in the leading (ladder) approximation in $\kappa$. We emphasize, however, that our method is tuned to the full control of the model beyond the ladder approximation, as we did for other models (see Refs. [28, 29]).

To conclude, we observe that from our method we obtain naturally an exponentially decreasing interacting potential for the two hadrons in the bound state with decay rate $\approx -2\ln\kappa$, as for the Yukawa theory. It would be interesting to look for the expected *distance*$^{-1}$ Ornstein-Zernicke like correction to this potential and to determine the spin and isospin dependence of the binding energy, and the effect of the number of flavors.

This work was supported by CNPq and FAPESP. PAFdV thanks the organizers and Prof. T.D. Cohen for the invitation to deliver a talk at the QCHS7.

# REFERENCES

1.  K. Wilson, in *New Phenomena in Subnuclear Physics*, Part A, A. Zichichi ed. (Plenum Press, NY, 1977).
2.  M. Creutz, *Quarks, Gluons and Lattices* (Cambridge University Press, Cambridge, 1983). **109**, 279 (2002).
3.  E. Seiler, Lect. Notes in Phys. **159**, *Gauge Theories as a Problem of Constructive Quantum Field Theory and Statistical Mechanics* (Springer, New York, 1982).
4.  I. Montvay and G. Münster, *Quantum Fields on a Lattice* (Cambridge University Press, Cambridge, 1997).
5.  P.A. Faria da Veiga, M. O'Carroll, and R. Schor, Phys. Rev. **D 67**, 017501 (2003).
6.  P.A. Faria da Veiga, M. O'Carroll, and R. Schor, Commun. Math. Phys. 245, 383 (2004).
7.  A. Francisco Neto, P.A. Faria da Veiga, and M. O'Carroll, J. Math. Phys. **45**, 628 (2004).
8.  P.A. Faria da Veiga, M. O'Carroll, and R. Schor, Phys. Rev. **D 68**, 037501 (2003).
9.  P.A. Faria da Veiga, M. O'Carroll, and A. Francisco Neto, Phys. Rev. **D 69**, 097501 (2004).
10. P.A. Faria da Veiga and M. O'Carroll, Phys. Rev. **D 71**, 017503 (2005).
11. P.A. Faria da Veiga, M. O'Carroll, and A. Francisco Neto, Phys. Rev. **D 72**, 034507 (2005).
12. P.A. Faria da Veiga and M. O'Carroll, Phys. Lett. B, to appear.
13. P.A. Faria da Veiga and M. O'Carroll, *Baryon-Baryon Bound States From First Principles* (2006), arXiv:hep-lat/0610070.
14. A.F. Neto, Phys. Rev. **D 70**, 037502(2004).
15. T. Banks et al., Phys. Rev. **D 15**, 1111 (1977).
16. J. Fröhlich and C. King, Nucl. Phys **B 290**, 157 (1987).
17. I. Montvay, Rev. Mod. Phys. **59**, 263 (1987).
18. F. Myhrer and J. Wroldsen, Rev. Mod. Phys. **60**, 629 (1988).
19. D. Schreiber, Phys. Rev. **D 48**, 5393 (1993).
20. M. Creutz, Nucl. Phys. **B** (Proc. Suppl.) **94** 219 (2001).
21. R. Machleidt, Nucl. Phys. **A 689**, 11c (2001).
22. R. Machleidt, K. Holinde, and Ch. Elster, Phys. Rep. **149**, 1 (1987).
23. M.J. Savage, *Nuclei from QCD : Strategy, challenges and status*, PANIC 05, arXiv:nucl-th/0601001.
24. H.R. Fiebig, H. Markum, A. Mihály, and K. Rabitsch, Nucl. Phys. Proc. Suppl. **53**, 804 (1997).
25. C. Stewart and R. Koniuk, Phys. Rev. **D 57**, 5581 (1998).
26. H.R. Fiebig and H. Markum, in *International Review of Nuclear Physics, Hadronic Physics from Lattice QCD*, A.M. Green ed. (World Scientific, Singapoure, 2003).
27. Ph. de Forcrand and S. Kim, *The Spectrum of lattice QCD with staggered fermions at strong coupling* (2006), arXiv:hep-lat/0608012.
28. P.A. Faria da Veiga, M. O'Carroll, E. Pereira, and R. Schor, Commun. Math. Phys. **220**, 377 (2001).
29. R.S. Schor and M. O'Carroll, Phys. Rev. **E 62**, 1521 (2000). J. Stat. Phys. **99**, 1207 (2000); **99**, 1265 (2000); and **109**, 279 (2002).

# Neutron star interiors and the equation of state of ultra-dense matter

F. Weber*, R. Negreiros*, P. Rosenfield* and Andreu Torres i Cuadrat[†]

*Dept. of Physics, San Diego State University, 5500 Campanile Drive, San Diego, CA 92182, USA
[†]Physics Department, Universitat Autonoma de Barcelona, Spain

**Abstract.** There has been much recent progress in our understanding of quark matter, culminating in the discovery that if such matter exists in the cores of neutron stars it ought to be in a color superconducting state. This paper explores the impact of superconducting quark matter on the properties (e.g., masses, radii, surface gravity, photon emission) of compact stars.

**Keywords:** stars, equation of state, quarks, color superconductivity, phase transition
**PACS:** 12.38.A, 12.38.M, 26.60, 82.60.F, 97.60.G, 97.60.J

Exploring the composition of matter inside compact stars has become a forefront area of modern physics [1, 2]. Despite the progress that was made over the years, the physical properties of the matter in the ultra-dense core of compact stars is only poorly known. Recently it has been theorized that, if quark matter exists in the core, it ought to be a color superconductor [3, 4]. This paper reviews the consequences of color superconducting quark matter cores on the properties of neutron stars. The study is based on three sample

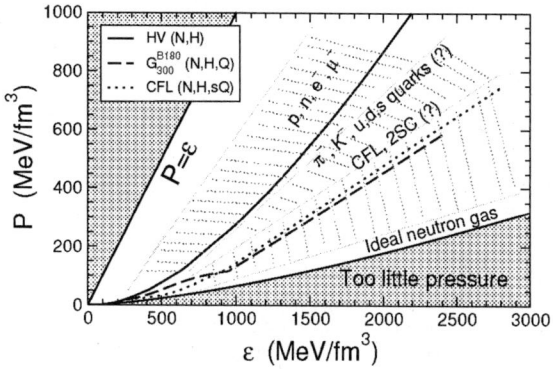

**FIGURE 1.** Eos considered in this study. The shaded areas reflect the uncertainties in the eos originating from different many-body treatments and competing assumption about the particle composition.

models for the nuclear equation of state (eos). The first model, HV[5], treats the core matter as made of conventional hadronic particles (nucleons and hyperons) in chemical equilibrium with leptons (electrons and muons). The second eos, $G_{300}^{B180}$ [1], additionally accounts for non-superconducting quark matter. Finally, the third model accounts for quark matter in the superconducting color-flavor locked (CFL) phase [6]. Figure 1 shows these eos graphically.

CP892, *Quark Confinement and the Hadron Spectrum VII*
edited by J. E. F. T. Ribeiro
© 2007 American Institute of Physics 978-0-7354-0396-3/07/$23.00

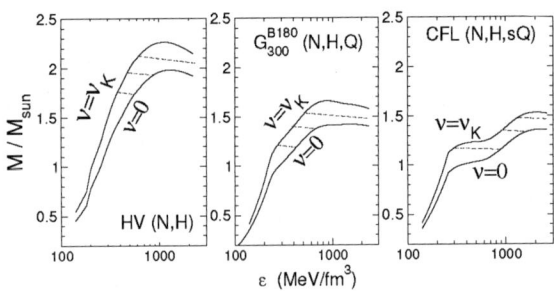

**FIGURE 2.** Mass–central energy density relations for the three model stars of this study.

Figure 2 shows the evolutionary (constant stellar baryon number, $A$) paths that isolated rotating neutron stars would follow during their stellar spin-down caused by the emission of magnetic dipole radiation and a wind on $e^+$–$e^-$ pairs. Figure 2 reveals that CFL stars may spend considerably more time in the spin-down phase than their competitors of the same mass. Figure 3 shows the general relativistic effect of frame dragging [1, 2, 7],

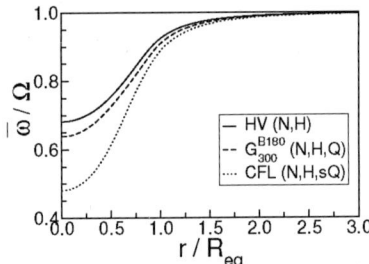

**FIGURE 3.** Lense–Thirring effect caused by $\sim 1.4\,M_\odot$ stars rotating at 2 ms.

which is considerably more pronounced for the CFL stars because of their much greater densities. This may be of great importance for binary millisecond neutron stars in their final accretion stages, when the accretion disk is closest to the neutron star. Table 1 summarizes the impact of strangeness on several intriguing properties of non-rotating as well as rotating neutron stars. The latter spin at their respective Kepler frequencies. One sees that the central energy density, $\varepsilon_c$, spans a very wide range, depending on particle composition. The surface redshift is of importance since it is connected to observed neutron star temperatures through the relation $T^\infty/T_{\text{eff}} = 1/(1+Z)$. CFL quark stars may have redshifts that are up to 50% higher than those of conventional stars. Finally, we also show in Table 1 the surface gravity of stars, $g_{s,14}$ [8], which again may be up to 50% higher for CFL stars. The other quantities listed are the rotational kinetic energy in units of the total energy of the star, $T/W$, the stellar binding energy, $BE$, and the rotational velocity of a particle at the star's equator [2].

**TABLE 1.** Properties of neutron stars composed of nucleons and hyperons (HV), nucleons, hyperons, normal quarks ($G^{B180}_{300}$), and nucleons, hyperons, superconducting quarks (CFL) [2].

| | HV $\nu = 0$ | $G^{B180}_{300}$ $\nu = 0$ | CFL $\nu_K = 0$ | HV $\nu_K = 850$ Hz | $G^{B180}_{300}$ $\nu_K = 940$ Hz | CFL $\nu_K = 1400$ Hz |
|---|---|---|---|---|---|---|
| $\varepsilon_c$ (MeV/fm³) | 361.0 | 814.3 | 2300.0 | 280.0 | 400.0 | 1100.0 |
| $I$ (km³) | 0 | 0 | 0 | 223.6 | 217.1 | 131.8 |
| $M$ ($M_\odot$) | 1.39 | 1.40 | 1.36 | 1.39 | 1.40 | 1.41 |
| $R$ (km) | 14.1 | 12.2 | 9.0 | 17.1 | 16.0 | 12.6 |
| $Z_p$ | 0.1889 | 0.2322 | 0.3356 | 0.2374 | 0.2646 | 0.3618 |
| $Z_F$ | 0.1889 | 0.2322 | 0.3356 | −0.1788 | −0.1817 | −0.2184 |
| $Z_B$ | 0.1889 | 0.2322 | 0.3356 | 0.6046 | 0.6502 | 0.9190 |
| $g_{s,14}$ (cm/s²) | 1.1086 | 1.5447 | 3.0146 | 0.7278 | 0.8487 | 1.4493 |
| $T/W$ | 0 | 0 | 0 | 0.0894 | 0.0941 | 0.0787 |
| $BE$ ($M_\odot$) | 0.0937 | 0.1470 | 0.1534 | 0.0524 | 0.1097 | 0.1203 |
| $V_{eq}/c$ | 0 | 0 | 0 | 0.336 | 0.353 | 0.424 |

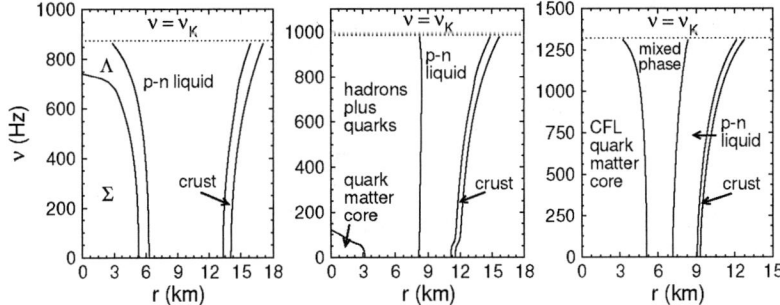

**FIGURE 4.** Particle profiles of the neutron stars of our collection.

# ACKNOWLEDGMENTS

This work is supported by the National Science Foundation under Grant PHY-0457329, and by the Research Corporation.

# REFERENCES

1. N. K. Glendenning, *Compact Stars, Nuclear Physics, Particle Physics, and General Relativity*, 2nd ed. (Springer-Verlag, New York, 2000).
2. F. Weber, *Pulsars as Astrophysical Laboratories for Nuclear and Particle Physics*, High Energy Physics, Cosmology and Gravitation Series (IOP Publishing, Bristol, Great Britain, 1999).
3. K. Rajagopal and F. Wilczek, *The Condensed Matter Physics of QCD*, At the Frontier of Particle Physics / Handbook of QCD, ed. M. Shifman, (World Scientific) (2001).
4. M. Alford, Ann. Rev. Nucl. Part. Sci. **51** (2001) 131.
5. N. K. Glendenning, Astrophys. J. **293** (1985) 470.
6. M. Alford and S. Reddy, Phys. Rev. D **67** (2003) 074024.
7. F. Weber, Prog. Part. Nucl. Phys. **54** (2005) 193, (astro-ph/0407155).
8. M. Bejger and P. Haensel, Astron. & Astrophys. **420** (2004) 987.

# Vacuum Energy, EoS, and the Gluon Condensate at Finite Baryon Density in QCD

Ariel R. Zhitnitsky

*Department of Physics and Astronomy, University of British Columbia, Vancouver, BC, Canada, V6T 1Z1*

**Abstract.** The Equation of States (EoS) plays the crucial role in all studies of neutron star properties. Still, a microscopical understanding of EoS remains largely an unresolved problem. We use 2-color QCD as a model to study the dependence of vacuum energy (gluon condensate in QCD) as function of chemical potential $\mu \ll \Lambda_{QCD}$ where we find very strong and unexpected dependence on $\mu$. We present the arguments suggesting that similar behavior may occur in 3-color QCD in the color superconducting phases. Such a study may be of importance for analysis of EoS when phenomenologically relevant parameters (within such models as MIT Bag model or NJL model) are fixed at zero density while the region of study lies at much higher densities not available for terrestrial tests.

**PACS:**

## INTRODUCTION

This talk is based on few recent publications with Max Metlitski [1]. Neutron stars represent one of the densest concentrations of matter in our universe. The properties of super dense matter are fundamental to our understanding of nature of nuclear forces as well as the underlying theory of strong interactions, QCD. Unfortunately, at present time, we are not in a position to answer many important questions starting from fundamental QCD lagrangian. Instead, this problem is usually attacked by using some phenomenological models such as MIT Bag model or NJL model. Dimensional parameters (e.g. the vacuum energy) for these models are typically fixed by using available experimental data at zero baryon density. Once the parameters are fixed, the analysis of EoS or other quantities is typically performed by assuming that the parameters of the models (e.g. bag constant) at nonzero $\mu$ are the same as at $\mu = 0$.

The main lesson to be learned from the calculations presented below can be formulated as follows: the standard assumption (fixing the parameters of a model at $\mu = 0$ while calculating the observables at nonzero $\mu$) may be badly violated in QCD.

The problem of density dependence of the chiral and gluon condensates in QCD has been addressed long ago in[2]. The main motivation of ref.[2] was the application of the QCD sum rules technique to study some hadronic properties in the nuclear matter environment. The main result of that studies is– the effect is small. More precisely, at nuclear matter saturation density the change of the gluon condensate is only about 5%. Indeed, in the chiral limit the variation of the gluon condensate with density can be expressed as follows[2],

$$\langle \frac{bg^2}{32\pi^2} G^a_{\mu\nu} G^{\mu\nu a} \rangle_{\rho_B} - \langle \frac{bg^2}{32\pi^2} G^a_{\mu\nu} G^{\mu\nu a} \rangle_0 = -m_N \rho_B + 0(\rho_B^2), \quad b = \frac{11N_c - 2N_f}{3}, \quad (1)$$

CP892, *Quark Confinement and the Hadron Spectrum VII*
edited by J. E. F. T. Ribeiro

where the standard expression for the conformal anomaly is used, $\Theta^\mu_\mu = -\frac{bg^2}{32\pi^2}G^a_{\mu\nu}G^{a\mu\nu}$. We should note here that the variation of the gluon condensate is well defined observable (in contrast with the gluon condensate itself) because the perturbative (divergent) contribution cancels in eq.(1). The most important consequences of this formula: a) the variation of the gluon condensate is small numerically, and b) the absolute value of the condensate decreases when the baryon density increases. Such a behavior can be interpreted as due to the suppression of the non-perturbative QCD fluctuations with increase of the baryon density.

Our ultimate goal here is to understand the behavior of the vacuum energy (gluon condensate ) as a function of $\mu$ for color superconducting (CS) phases[3], [4]. It is clear that the problem in this case is drastically different from nuclear matter analysis [2] because the system becomes relativistic and binding energy ($\sim \Delta$) per baryon charge is order of $\Lambda_{QCD}$ in contrast with $\leq 2\%$ of the nucleon mass at nuclear saturation density. The quark-quark interaction also becomes essential in CS phases such that the small density expansion (valid for dilute noninteracting nuclear matter) used to derive (1) can not be justified any more.

Unfortunately, we can not answer the questions on $\mu$ dependence of the vacuum energy in real $QCD(N_c = 3)$. However, these questions can be formulated and can be answered in more simple model $QCD(N_c = 2)$ due to the extended symmetry of this model. Some lessons for the real life with $N_c = 3$ can be learned from our analysis, see below.

## GLUON CONDENSATE FOR $QCD(N_c = 2)$

We start from the equation for the conformal anomaly,

$$\Theta^\mu_\mu = -\frac{bg^2}{32\pi^2}G^a_{\mu\nu}G^{a\mu\nu} + \bar\psi M \psi, \qquad b = \frac{11}{3}N_c - \frac{2}{3}N_f = 6 \tag{2}$$

For massless quarks and in the absence of chemical potential, eq. (2) implies that the QCD vacuum carries a negative non-perturbative vacuum energy due to the gluon condensate.

Now, we can use the effective Lagrangian [5]

$$\mathscr{L} = \frac{F^2}{2}Tr\nabla_\nu\Sigma\nabla_\nu\Sigma^\dagger, \quad \nabla_0\Sigma = \partial_0\Sigma - \mu\left[B\Sigma + \Sigma B^T\right] \tag{3}$$

to calculate the change in the trace of the energy-momentum tensor $\langle\theta^\mu_\mu\rangle$ due to a finite chemical potential $\mu \ll \Lambda_{QCD}$. The energy density $\varepsilon$ and pressure $p$ are obtained from the free energy density $\mathscr{F}$,

$$\varepsilon = \mathscr{F} + \mu n_B, \quad p = -\mathscr{F}. \tag{4}$$

Therefore, the conformal anomaly implies,

$$\langle\frac{bg^2}{32\pi^2}G^a_{\mu\nu}G^{\mu\nu a}\rangle_{\mu,m} - \langle\frac{bg^2}{32\pi^2}G^a_{\mu\nu}G^{\mu\nu a}\rangle_0 =$$
$$-4\left(\mathscr{F}(\mu,m) - \mathscr{F}_0\right) - \mu n_B(\mu,m) + \langle\bar\psi M \psi\rangle_{\mu,m}, \tag{5}$$

where the subscript 0 on an expectation value means that it is evaluated at $\mu = m = 0$. Now we notice that all quantities on the right hand side are known from the previous calculations [5], therefore the variation of $\langle G_{\mu\nu}^2 \rangle$ with $\mu$ can be explicitly calculated. As expected, $\langle G_{\mu\nu}^2 \rangle$ does not depend on $\mu$ in the normal phase $\mu < m_\pi$ while in the superfluid phase $\mu > m_\pi$ this dependence can be represented as follows [1],

$$\langle \frac{bg^2}{32\pi^2} G_{\mu\nu}^a G^{\mu\nu a} \rangle_{\mu,m} - \langle \frac{bg^2}{32\pi^2} G_{\mu\nu}^a G^{\mu\nu a} \rangle_{\mu=0,m} = 4F^2(\mu^2 - m_\pi^2)\left(1 - 2\frac{m_\pi^2}{\mu^2}\right). \quad (6)$$

The behavior of the condensate is quite interesting: it decreases with $\mu$ for $m_\pi < \mu < 2^{1/4}m_\pi$ and increases afterwards. The qualitative difference in the behaviour of the gluon condensate for $\mu \approx m_\pi$ and for $m_\pi \ll \mu \ll \Lambda_{QCD}$ can be explained as follows. Right after the normal to superfluid phase transition occurs, the baryon density $n_B$ is small and our system can be understood as a weakly interacting gas of diquarks. The pressure of such a gas is negligible compared to the energy density, which comes mostly from diquark rest mass. Thus, $\langle \Theta_\mu^\mu \rangle$ increases with $n_B$ in precise correspondence with the "dilute" nuclear matter case (1). On the other hand, for $\mu \gg m_\pi$, energy density is approximately equal to pressure, and both are mostly due to self-interactions of the diquark condensate. Luckily, the effective Chiral Lagrangian (3) gives us control over these self-interactions as long as $\mu \ll \Lambda_{QCD}$.

The main lesson to be learned for real $QCD(N_c = 3)$ from exact results discussed above is as follows. The transition to the CS phases is expected to occur[1] at $\mu_c \simeq 2.3 \cdot \Lambda_{QCD}$[6],[7] in contrast with $\mu_c = m_\pi$ for transition to superfluid phase for $N_c = 2$ case. The binding energy, the gap, the quasi -particle masses are also expected to be the same order of magnitude $\sim \mu_c$. This is in drastic contrast with nuclear matter case when binding energy is very small. At the same time, $QCD(N_c = 2)$ represents a nice model where the binding energy, the gap, the masses of quasi -particles carrying the baryon charge are the same order of magnitude. This model explicitly shows that the gluon condensate can experience extremely nontrivial behavior as function of $\mu$. We expect a similar behavior for $QCD(N_c = 3)$ in CS phases when function of $m_\pi^2/\mu^2$ in (6) is replaced by some function of $\mu_c/\mu^2$ for $N_c = 3$. We should note in conclusion that the recent lattice calculations [8],[9] are consistent with our prediction (6).

## REFERENCES

1. M. A. Metlitski and A. R. Zhitnitsky, Nucl. Phys. B **731**, 309 (2005); Phys. Lett. B **633**, 721 (2006).
2. T. D. Cohen, R. J. Furnstahl and D. K. Griegel, Phys. Rev. C **45**, 1881 (1992).
3. M. Alford, K. Rajagopal, and F. Wilczek, Phys. Lett. **B 422** (1998) 247.
4. R. Rapp, T. Schäfer, E. V. Shuryak, and M. Velkovsky, Phys. Rev. Lett. **81** (1998) 53.
5. J.B. Kogut et al Nucl. Phys. **B 582** (2000) 477-513.
6. D. Toublan and A. R. Zhitnitsky, Phys. Rev. D **73**, 034009 (2006)
7. A. R. Zhitnitsky, arXiv:hep-ph/0601057, Proceedings, Light-Cone QCD, Australia, 7-15 Jul 2005.
8. S. Hands, S. Kim and J. I. Skullerud, arXiv:hep-lat/0604004.
9. B. Alles, M. D'Elia and M. P. Lombardo, Nucl. Phys. B **752**, 124 (2006)

---

[1] $\mu$ here for $QCD(N_c = 3)$ is normalized as the quark (rather than the baryon) chemical potential

# Ultrasoft Quark Damping in High-$T$ QCD

Abdessamad Abada[1], Nacéra Daira-Aifa[2] and Karima Bouakaz[3]

[1] *Physics Department, United Arab Emirates University, POB 17551 Al Ain, United Arab Emirates*
[2] *Facutlé de Physique, USTHB, BP32 Al Alia, Bab Ezzouar 16111 Alger, Algeria*
[3] *Département de Physique, Ecole Normale Supérieure, BP92 Vieux-Kouba, 16052 Alger, Algeria*

**Abstract.** We determine the ultrasoft quark damping rates in the context of next-to-leading order hard-thermal-loop summed perturbation of high-temperature QCD. Three types of divergences are encountered: infrared, light-cone and at specific points determined by the gluon energies. The infrared divergence persists and is logarithmic whereas the two others are circumvented.

**Keywords:** ultrasoft quark damping. hard thermal loops. logarithmic infrared sensitivity.
**PACS:** 11.10.Wx 12.38.Bx 12.38.Cy 12.38.Mh.

RHIC results indicate that hadronic matter changes at $T_c \approx 170\text{MeV}$ into the so-called strongly coupled quark-gluon plasma (sQGP) [1]. As the temperature increases, the sQGP becomes a weakly coupled quark-gluon plasma (wQGP) of quasi quarks and gluons with chromo-neutralizing isotropic Debye clouds. Perturbative QCD is to apply in the wQGP regime, but it is known for some time that the standard loop-expansion would break at some order, depending on the quantity under consideration. It has also proved inadequate when describing slow-moving particles since it does not reflect an expansion in powers of the coupling. An important improvement has been the dressing of the lowest-order propagators and vertices with the so-called hard thermal loops (HTL) [2].

The present work finishes the calculation of the ultrasoft quark damping rates in the context of HTL-dressed perturbation started in [3]. For three colors and three flavors, the result is:

$$\gamma_{\pm}(p) = \frac{g^2 T}{12\pi}\left[5.7057 \pm 1.0568\bar{p} - (5.968\ln\bar{\eta} - 8.5536)\bar{p}^2 + O(\bar{p}^3)\right]$$

In this relation, $\bar{p}$ and $\bar{\eta}$ are the quark momentum and the infrared cutoff respectively, normalized to the soft quark mass, $g$ the coupling constant and $T$ the temperature. The details of the calculation and a discussion of the result are in [4].

# REFERENCES

1. E. Shuryak, `Toward the theory of strongly coupled Quark-Gluon Plasma', plenary talk at Quark Matter 05, Budapest, August 2005, hep-ph/0510123.
2. E. Braaten and R.D. Pisarski, Nucl. Phys. B337 (1990) 569; M. Le Bellac, `Thermal Field Theory', Cambridge University Press, 1996.
3. A. Abada, K. Bouakaz and N. Daira-Aifa, Eur. Phys. J. C18 (2001) 765.
4. A. Abada, N. Daira-Aifa and K. Bouakaz, hep-ph/0511258, Int. Jour. Mod. Phys. A, in press.

CP892, *Quark Confinement and the Hadron Spectrum VII*
edited by J. E. F. T. Ribeiro
© 2007 American Institute of Physics 978-0-7354-0396-3/07/$23.00

# Center vortices and the Atiyah-Singer index theorem

Gerald Jordan*, Rainer Pullirsch*, Urs Heller[†] and Manfried Faber*

*Atomic Institute, Vienna University of Technology, Wiedner Hauptstr. 8-10, 1040 Vienna, Austria
[†]American Physical Society, One Research Road, Box 9000, Ridge, NY 11961-9000, USA

**Abstract.** The lattice index theorem is checked for classical center vortices. For non-orientable spherical vortices, lattice results differ from continuum-based expectations, possibly because of coarse discretization.

**Keywords:** center vortices, Atiyah-Singer index theorem, overlap operator, lattice gauge theory
**PACS:** 11.15.Ha; 12.38.Gc

The center vortex model, which has been proposed to explain confinement in QCD, could also account for the topological charge, fermionic zero-modes and spontaneous chiral symmetry breaking.

As a first step, we looked into lattice topology for center vortices. Classical vortex configurations of an SU(2) gauge field were created on a $12^4$ lattice. We compare different definitions of the lattice topological charge: First, the gluonic charge in the plaquette and hypercube definition, before and after cooling. Second, the index of a Ginsparg-Wilson operator, in particular the overlap operator.

At each intersection of center vortices, topological charge of modulus $|Q| = \frac{1}{2}$ arises, which is correctly reproduced by all definitions of the lattice topological charge.

After abelian projection, an orientation may be assigned to the surface patches of the vortex. We investigated sphere-shaped vortices with one patch (orientable) and two patches (non-orientable). Geometrically, both types are spatial spherical shells, the only non-trivial links being $t$-links in one time slice. Lacking intersection and writhing points, these vortices are expected to yield $Q = 0$. This is indeed the case for the orientable vortex, both for gluonic and fermionic definition.

However, this is not so for the non-orientable spherical vortex. While the gluonic topological charge found before cooling is indeed zero, during cooling, it rises near to one, in agreement with the index of the overlap operator, which is equal to one. More generally, the following empirical rule can be formulated: a non-orientable spherical vortex contributes to topological charge and index with an integer given by the winding number of the links, seen as a map (not from $\mathbb{T}^4$) but from the compactified time-slice (which is homeomorphic to $S^3$) to SU(2).

This seeming inapplicability of the lattice index theorem could be due to the "inadmissiblity" of our gauge field configuration. The topological charge of a continuum gauge field can only faithfully be reconstructed from its lattice counterpart if the plaquettes are not too "large". A similar condition applies to the index of the overlap operator. Further investigations of this issue are under way.

CP892, *Quark Confinement and the Hadron Spectrum VII*
edited by J. E. F. T. Ribeiro

# Improved staggered eigenvalues in SU(2).

E. Follana*, A. Hart† and C.T.H. Davies*

*SUPA, Department of Physics and Astronomy, University of Glasgow, Glasgow G12 8QQ, U.K.
†SUPA, School of Physics, University of Edinburgh, King's Buildings, Edinburgh EH9 3JZ, U.K.

**Abstract.** We study the low-lying modes of staggered Dirac operators for quenched SU(2) and show that improvement changes the distribution from lattice-like to continuum-like at lattice spacings $a = 0.07 - 0.10$ fm, which are representative of current dynamical SU(3) simulations.

**Keywords:** Lattice QCD, improvement, chiral symmetry, epsilon regime
**PACS:** 11.15.Ha, 12.38.Gc

If lattice QCD is to explain low energy phenomena such as the $U_A(1)$ axial anomaly and spontaneous chiral symmetry breaking, discretisation effects in the low-lying Dirac operator spectrum must be small and respect the gluonic topological charge. We have shown this is so for improved staggered fermions on a quenched SU(3) background [1].

We are now studying SU(2) [2] as it is qualitatively and quantitatively similar to SU(3). Again, we use $\varepsilon$-regime eigenvalue distributions to measure the restoration of the continuum chiral symmetry group. Fig. 1 shows this is more obvious for SU(2) and occurs for $a = 0.07 - 0.10$ fm as we improve the theory, much more suddenly than for SU(3). We also see a good Index Theorem and eigenvalue scaling with $a^2$.

A.H. and E.F. thank the UK Royal Society and PPARC for financial support.

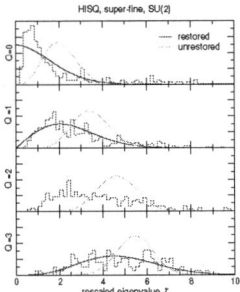
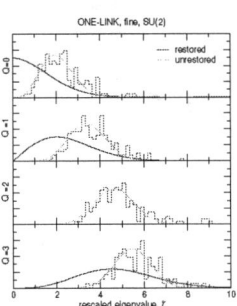

**FIGURE 1.** Comparison of the distribution of the smallest non-zero eigenvalue with $\varepsilon$-regime predictions: "restored" denotes the continuum–like chOE and "unrestored" the chSE for the discrete case.

## REFERENCES

1. E. Follana, A. Hart, C. T. H. Davies, *Phys. Rev. Lett.* **93**, 241601 (2004), hep-lat/0406010; *Nucl. Phys. Proc. Suppl.* **140**, 141–147 (2005), hep-lat/0409062; *Phys. Rev.* **D72**, 054501 (2005), hep-lat/0507011; *PoS* **LAT2005**, 298 (2006), hep-lat/0509177; *Nucl. Phys. Proc. Suppl.* **153**, 106–113 (2006).
2. E. Follana, A. Hart, C. T. H. Davies, *PoS* **LAT2006**, 051 (2006).

CP892, *Quark Confinement and the Hadron Spectrum VII*
edited by J. E. F. T. Ribeiro
© 2007 American Institute of Physics 978-0-7354-0396-3/07/$23.00

# Generalized Instantaneous Bethe–Salpeter Equation and Exact Quark Propagators

Wolfgang Lucha* and Franz F. Schöberl†

*Institute for High Energy Physics, Austrian Academy of Sciences,
Nikolsdorfergasse 18, A-1050 Vienna, Austria
E-mail: wolfgang.lucha@oeaw.ac.at
†Department of Theoretical Physics, University of Vienna,
Boltzmanngasse 5, A-1090 Vienna, Austria
E-mail: franz.schoeberl@univie.ac.at

**Keywords:** Bethe–Salpeter equation, relativistic bound state, instantaneous limit, exact propagator
**PACS:** 11.10.St, 03.65.Pm, 03.65.Ge, 12.38.Lg, 12.39.Ki, 14.40.Aq, 14.65.Bt

Assuming all interactions to be instantaneous *and* the exact fermion propagator $S$ to be approximately given by $iS^{-1}(p) = A(\mathbf{p}^2)\,\slashed{p} - B(\mathbf{p}^2)$, i.e., with its scalar functions $A$ and $B$ depending only on $\mathbf{p}$, a three-dimensional reduction of the homogeneous Bethe–Salpeter equation retaining, in contrast to the Salpeter equation, the exact propagators (crucial for, e.g., a proper incorporation of dynamical chiral symmetry breakdown) is constructed [1]. For spherically symmetric interactions the resulting bound-state equation reduces to a set of (coupled) integral equations for the independent radial components of the bound-state amplitude [2], which may be solved by conversion into an equivalent matrix problem [3]. Adopting as light-quark propagator the solution of the quark Dyson–Schwinger equation found within a model [4] consistent with the QCD axial-vector Ward–Takahashi identity, a tentative application [5] to "pion-like" pseudoscalar ($J^{PC} = 0^{-+}$) mesons considered as quark–antiquark bound states formed by linear confining interactions of time-component Lorentz-vector Dirac structure $\gamma^0 \otimes \gamma^0$ indicates that our "exact-propagator instantaneous formalism" yields significantly smaller spacings of its bound-state mass eigenvalues than those obtained from Salpeter's equation for "reasonable" constituent light-quark masses.

## REFERENCES

1. W. Lucha and F. F. Schöberl, *J. Phys. G: Nucl. Part. Phys.* **31**, 1133 (2005) [hep-th/0507281].
2. J.-F. Lagaë, *Phys. Rev. D* **45**, 305 (1992); M. G. Olsson, S. Veseli, and K. Williams, *Phys. Rev. D* **52**, 5141 (1995) [hep-ph/9503477].
3. W. Lucha, K. Maung Maung, and F. F. Schöberl, *Phys. Rev. D* **63**, 056002 (2001) [hep-ph/0009185]; "Instantaneous Bethe–Salpeter Equation: (Semi-)Analytical Solution," in *Proceedings of the Int. Conf. on "Quark Confinement and the Hadron Spectrum IV,"* edited by W. Lucha and K. Maung Maung, World Scientific, Singapore, 2002, pp. 340 [hep-ph/0010078]; *Phys. Rev. D* **64**, 036007 (2001) [hep-ph/0011235]; W. Lucha and F. F. Schöberl, *Int. J. Mod. Phys. A* **17**, 2233 (2002) [hep-ph/0109165].
4. P. Maris and C. D. Roberts, *Phys. Rev. C* **56**, 3369 (1997) [nucl-th/9708029]; P. Maris and P. C. Tandy, *Phys. Rev. C* **60**, 055214 (1999) [nucl-th/9905056]; P. Maris, "Continuum QCD and Light Mesons," in *Proceedings of the Int. Conference on "Quark Confinement and the Hadron Spectrum IV,"* edited by W. Lucha and K. Maung Maung, World Scientific, Singapore, 2002, pp. 163 [nucl-th/0009064].
5. Li Z.-F., W. Lucha, and F. F. Schöberl, *Mod. Phys. Lett. A* **21**, 1657 (2006) [hep-ph/0510372].

CP892, *Quark Confinement and the Hadron Spectrum VII*
edited by J. E. F. T. Ribeiro
© 2007 American Institute of Physics 978-0-7354-0396-3/07/$23.00

# Quark–Hadron Duality and Properties of Pseudoscalar Mesons from QCD Sum Rules

Wolfgang Lucha and Dmitri Melikhov

*Institute for High Energy Physics, Austrian Academy of Sciences,
Nikolsdorfergasse 18, A-1050 Vienna, Austria*

**Keywords:** operator product expansion, OPE, chiral symmetry, QCD sum rules
**PACS:** 12.38.Aw, 11.30.Rd, 12.38.Lg

We discussed the operator product expansion (OPE) and the quark–hadron duality for 2- and 3-point correlators of the axial ($A$) and pseudoscalar ($P$) currents of the light ($u$, $d$) quarks [1]. Our principal goals and main results may be briefly summarized as follows.

1. A detailed study of the OPE and the duality for the correlators $\langle PP \rangle$, $\langle AA \rangle$, and $\langle AP \rangle$ reveals that the quark–hadron duality for $\langle AP \rangle$, similar to the longitudinal structure of $\langle AA \rangle$, has an interesting feature: in the chiral limit the pion contribution proves to be dual to a single $\bar{\psi}\psi$ term in the OPE. We discussed the sum rules for $\langle PP \rangle$, $\langle AA \rangle$, and $\langle AP \rangle$. The sum rule for the correlator $\langle PP \rangle$ contains contributions of the ground-state as well as excited light pseudoscalars. According to our estimates, this sum rule receives a sizeable contribution from the excited $\pi'(1300)$ and therefore provides a promising possibility to extract the corresponding (weak) decay constant $f_{\pi'}$ [1].

2. We studied properties of the 3-point correlators $\langle PVP \rangle$ and $\langle AVP \rangle$, where $V$ denotes the vector current, derived the Ward identities for these correlators, and demonstrated the way the normalization of the pion form factor at zero momentum transfer arises due to these Ward identities. We analyzed the region of large momentum transfers and the way quark–hadron duality works in this case. We showed that the OPE for $\langle PVP \rangle$ and $\langle AVP \rangle$ at large momentum transfers squared, $q^2$, is dominated by nonperturbative corrections, and identified the operators responsible for providing the correct large-$q^2$ asymptotics of the pion form factor in accordance with pQCD: the four-quark condensate $\langle \bar{\psi}\psi\bar{\psi}\psi \rangle$ in the case of $\langle PVP \rangle$ and the mixed condensate $\langle \bar{\psi}\sigma G\psi \rangle$ in the case of $\langle AVP \rangle$ [1]. We have thus disproved the recent statement in the literature that the pion form factor as extracted from the correlator $\langle PVP \rangle$ does not have the right asymptotics required by pQCD.

3. For the correlator $\langle AVA \rangle$, we find that the local duality representation for the pion form factor with the double spectral density, which includes the radiative corrections, is applicable for all spacelike momentum transfers and has several interesting features [1]: at $q^2 = 0$ the form factor is properly normalized due to the vector Ward identity, and at large $q^2 < 0$ it reproduces the pQCD asymptotic behavior. Therefore, this parameter-free representation should yield reliable predictions for all spacelike momentum transfers [1].

## REFERENCES

1. W. Lucha and D. Melikhov, *Phys. Rev. D* **73**, 054009 (2006) [hep-ph/0602217].

CP892, *Quark Confinement and the Hadron Spectrum VII*
edited by J. E. F. T. Ribeiro
© 2007 American Institute of Physics 978-0-7354-0396-3/07/$23.00

# Are light hadronic coherent-like states possible?

A. V. Nefediev[*] and J. E. F. T. Ribeiro[†]

[*]Institute of Theoretical and Experimental Physics, 117218, B.Cheremushkinskaya 25, Moscow,
Russia
[†]Centro de Física das Interacções Fundamentais (CFIF), Departamento de Física, Instituto
Superior Técnico, Av. Rovisco Pais, P-1049-001 Lisboa, Portugal

**Abstract.** We investigate a possibility of building light quasistable coherent-like states of pions.

We argue any relativistic quark model with confinement which is able to describe spontaneous breaking of chiral symmetry to possess, as a rule rather than as an exception, multiple solutions (replicas) of the nonlinear equation defining the vacuum with broken chiral symmetry (we consider only one replica for simplicity) [1, 2]. While the solution with the lowest energy defines the physical vacuum, we interpret the replica encapsulated in a finite volume as the scalar excitation of a certain type which is simply a strongly correlated coherent-like pionic cloud with the quantum numbers of the vacuum. If a formal procedure of building the Fock space of quark–antiquark states on top of the "replicated" vacuum is performed [3], then every "replicated" hadron is a normal hadron plus the coherent-like cloud of correlated pions.

Consider the radially excited pion $\pi'$ placed into the replica bubble of a sufficient size. From the point of view of the replica Fock space, the pion $\pi'$ is the Goldstone mode $\tilde{\pi}$ in the "excited" vacuum since it possesses the same quantum numbers and the same (one–node) radial w.f. On the other hand, the state $\tilde{\pi}$ is massless due to chiral symmetry. Thus the energy balance is (we use the results [2] obtained in the Generalised Nambu–Jona-Lasinio quark model [4, 5]): $\Delta E = \Delta E_{\text{vac}} + \Delta E_{\text{had}}$ with $\Delta E_{\text{vac}} \approx V_{\tilde{\pi}} \Delta \varepsilon \approx 400 MeV$ and $\Delta E_{\text{had}} = M_{\tilde{\pi}} - M_{\pi'} = -1300 MeV$. Then $\Delta E \approx -900 MeV$ (to reduce to about $-500 MeV$ beyond the chiral limit). Therefore, it is energetically favourable for the radially excited pion $\pi'$ to be dressed with the coherent pionic cloud with the energy released in the form of correlated pions. We call this process the chiral fusion. Basically, the main message of this work is that it may be energetically more favourable to store energy through the formation of coherent-like hadronic states than through the radial excitation of hadrons.

## REFERENCES

1. A. Le Yaouanc, L. Oliver, O. Pene, and J.-C. Raynal, *Phys.Rev.* **D29**, 1233 (1984).
2. P.J.A. Bicudo, A.V. Nefediev, and J.E.F.T. Ribeiro, Phys.Rev. D **65**, 085026 (2002); A.V. Nefediev and J.E.F.T. Ribeiro, *Phys.Rev.* **D67**, 034028 (2003); P.J.A. Bicudo and A.V. Nefediev, *Phys.Rev.* **D68**, 065021 (2003).
3. A.V. Nefediev and J.E.F.T. Ribeiro, *Phys.Rev.* **D70**, 094020 (2004).
4. A. Amer, A. Le Yaouanc, L. Oliver, O. Pene, and J.-C. Raynal, *Phys.Rev.Lett.* **50**, 87 (1983); A. Le Yaouanc, L. Oliver, S. Ono, O. Pene, and J.-C. Raynal, *Phys.Rev.* **D31**, 137 (1985).
5. P. Bicudo and J.E.Ribeiro, *Phys.Rev.* **D42**, 1611 (1990); *ibid.*, 1625 (1990); *ibid.*, 1635 (1990).

CP892, *Quark Confinement and the Hadron Spectrum VII*
edited by J. E. F. T. Ribeiro
© 2007 American Institute of Physics 978-0-7354-0396-3/07/$23.00

# Coulombic contribution and fat center vortex model

Shahnoosh Rafibakhsh and Sedigheh Deldar

Department of Physics, University of Tehran, P.O. Box 14395/547, Tehran 1439955961, Iran

**Abstract.** The fat (thick) center vortex model is one of the phenomenological models which is fairly successful to interpret the linear potential between static sources. However, the Coulombic part of the potential has not been investigated by the model yet. In an attempt to get the Coulombic contribution and to remove the concavity of the potentials, we are studying different vortex profiles and vortex sizes.

Using the thick center vortices model, it is possible to get the linear potentials at intermediate distances which are qualitatively proportional to the Casimir scaling. However, the model has some shortcomings like the concavity of the potential in some region. In addition, because the vortex profiles have been quantized, the Coulombic part of the potential is not observed. We have started calculating the potentials with various fluxes or vortex profiles. Even though, we have not seen the Coulombic part yet, the concavity of the potentials has been removed. In fact, by changing some parameters of the model, we have produced the fluxes which are fluctuating continuously (the left plot of figure 1). The potential obtained between the sources in the adjoint representation is shown in the right plot of figure 1. The concavity of the potential has been removed compared to the potential shown in the same plot. The wiggles in the modified potential would be removed by higher statistics.

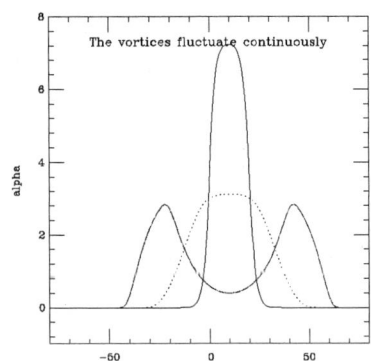
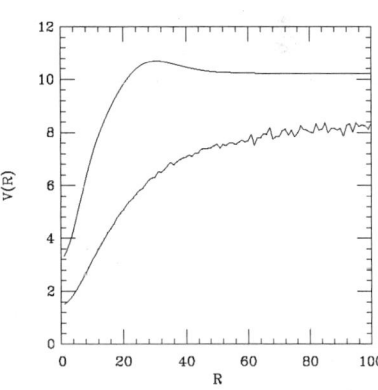

**FIGURE 1.** Vortex profiles (left plot) and the potential between adjoint sources (right plot). The concavity of the potential has been removed by using various fluxes.

CP892, *Quark Confinement and the Hadron Spectrum VII*
edited by J. E. F. T. Ribeiro
© 2007 American Institute of Physics 978-0-7354-0396-3/07/$23.00

# Nucleon and N-Delta Electromagnetic Transition Form Factors

Franz Gross[*], G. Ramalho[†] and M. T. Peña[†,**]

[*]College of William and Mary, Williamsburg, Virginia 23185, USA and
Thomas Jefferson National Accelerator Facility, Newport News, VA 23606, USA
[†]Centro de Física Teórica e de Partículas, Av. Rovisco Pais, 1049-001 Lisboa, Portugal
[**]Department of Physics, Instituto Superior Técnico, Av. Rovisco Pais, 1049-001 Lisboa, Portugal

Keywords: Baryon form factors, relativistic dynamics, electromagnetic transitions
PACS: 12.39.Ki, 13.40.Em, 13.40.Gp, 14.20.Dh, 21.10.Ft

We consider the baryons as three constituent valence quark systems. Their dynamics is described by the covariant Spectator formalism [1,2] for a quark-diquark system, where the diquark is always on its mass-shell. The electromagnetic interaction is considered in the relativistic impulse interaction (RIA) where the photon couples with the quark through the current $j^\mu = j_1 \gamma^\mu + j_2 \frac{i\sigma^{\mu\nu}q_\nu}{2m}$ ($m$ is the nucleon mass). The two form factors $j_1$ and $j_2$ account for all QCD mechanisms ($q\bar{q}$ pairs, pion cloud and gluon sea effects). Only the $j_1$ form factor includes pion cloud effects in its isovector part.

The nucleon wave function consists of spin-0 (isospin-0) and spin-1 (isospin-1) components written in terms of the diquark polarization vectors and the nucleon Dirac spinor [1]. Furthermore, it verifies the Dirac equation and generates the correct structure for its non-relativistic limit. Current conservation is also satisfied. The Jlab polarization data of the electromagnetic nucleon elastic form factors are described [3,4] when one assumes an S-state for the quark-diquark system [1], which means that the data does not signal any angular dependence in the wave function. The results show that spherical charge and matter distributions are compatible with the data, even when we consider Spin Direction Dependent density definitions [5]. The explicit consideration of the pion cloud effects definitively improves the description of the nucleon form factors [1].

We also calculated the N-Delta transition form factors. Preliminary results considering the Delta wave function as a mixture of a S and a D state explain the magnetic dipole $G_M^*$ and the electric quadrupole $G_E^*$ data [6]. Improvements are underway in order to describe also the Coulomb quadrupole form factor $G_C^*$.

[1] F. Gross, G. Ramalho and M. T. Peña, arXiv:nucl-th/0606029.
[2] F. Gross, Phys. Rev. 186, 1448 (1969); F. Gross and D. O. Riska, Phys. Rev. C36, 1928 (1987)
[3] V. Punjabi et al., Phys. Rev. C 71, 055202 (2005)
[4] New proton data analysis were supplied by J. Arrington, private communication.
[5] A. Kvinikhidze and G. A. Miller, Phys. Rev. C 73, 065203 (2006)
[6] V. V. Frolov et al., Phys. Rev. Lett. 82, 45 (1999); K. Joo et al. [CLAS Collaboration], Phys. Rev. Lett. 88, 122001 (2002); M. Ungaro, P. Stoler, I. Aznauryan, V. D. Burkert, K. Joo and L. C. Smith [CLAS Collaboration], arXiv:hep-ex/0606042.

CP892, Quark Confinement and the Hadron Spectrum VII
edited by J. E. F. T. Ribeiro
© 2007 American Institute of Physics 978-0-7354-0396-3/07/$23.00

# Fitting the lattice gluon propagator and the question of positivity violation

P. J. Silva and O. Oliveira

*Centro de Física Computacional, Departamento de Física, Universidade de Coimbra, P-3004-516 Coimbra, Portugal*

**Abstract.** The solution of Alkofer *et al* for the gluon propagator in Landau gauge of the Dyson-Schwinger equations was fitted with several functional forms [1, 2]. We study if the lattice gluon propagator, computed using large asymmetric lattices, is well reproduced by such functional forms. Furthermore, we also check for the violation of positivity of the lattice gluon propagator.

**Keywords:** lattice QCD; Landau gauge; confinement; gluon propagator
**PACS:** 12.38.-t; 11.15.Ha; 12.38.Gc; 12.38.Aw; 14.70.Dj

In [7] large asymmetric lattices were used to study the lattice Landau gauge gluon propagator. In particular, the value of the infrared exponent $\kappa$ was estimated by fitting the lattice data, in the deep infrared (IR) region, to a pure power law. Here we reported on the results of fitting the gluon propagator with functional forms which were able to reproduce a particular solution of the DSE [1, 2]. These results can be seen in [5]. We observed that the same functional forms describe the DSE gluon propagator and the lattice gluon propagator. However, the parameters obtained in both studies are not compatible. The measured infrared exponents are larger than those obtained assuming a pure power law for the IR region [7], supporting an infrared vanishing gluon propagator.

We were also able to verify positivity violation using asymmetric lattices. These results can be seen in [6]. Positivity violation of the gluon propagator is by now well established, being observed both at DSE [1] and lattice propagators [3, 4].

## ACKNOWLEDGMENTS

This work was supported by FCT via grant SFRH/BD/10740/2002, and project POCI/FP/63436/2005.

## REFERENCES

1. R. Alkofer, W. Detmold, C.S. Fischer, P. Maris, *Phys. Rev.* **D70** (2004) 014014 [hep-ph/0309077].
2. C. S. Fischer, R. Alkofer, *Phys. Rev.* **D67** (2003) 094020 [hep-ph/0301094].
3. A. Cucchieri, T. Mendes, A. Taurines , *Phys. Rev.* **D71** (2005) 051902(R) [hep-lat/0406020].
4. A. Sternbeck, E.-M. Ingelfritz, M. Müller-Preussker, A. Schiller, I. L. Bogolubsky, PoS(LAT2006)076 [hep-lat/0610053].
5. O. Oliveira, P. J. Silva, hep-lat/0609036, to appear in Brazilian Journal of Physics.
6. P. J. Silva, O. Oliveira, PoS(LAT2006)075 [hep-lat/0609069].
7. P. J. Silva, O. Oliveira, "Studying the infrared behaviour of gluon and ghost propagators using large asymmetric lattices", these proceedings.

# Measurement of the Photon Structure Function $F_2^\gamma$ with the L3 Detector at LEP

Igor Vorobiev

(on behalf of the L3 Collaboration)

*Carnegie Mellon University, Pittsburgh, PA 15213*

**Keywords:** LEP, two-photon physics, hadrons.
**PACS:** 13.60.Le

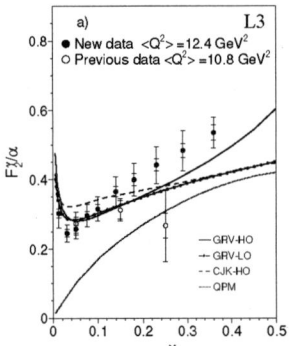 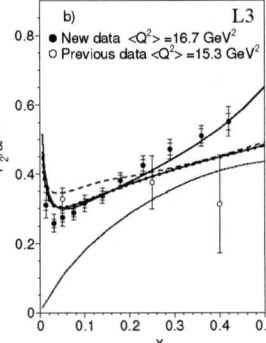 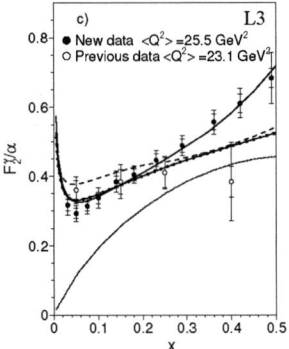

The $e^+e^- \to e^+e^-$ *hadrons* reaction, where one of the two electrons is detected in a low polar-angle calorimeter, is analysed in order to measure the hadronic photon structure function $F_2^\gamma$. The full high-energy and high-luminosity data set, collected with the L3 detector at centre-of-mass energies 189 GeV $\leq \sqrt{s} \leq$ 209 GeV, corresponding to an integrated luminosity of 608 pb$^{-1}$ is used. The $Q^2$ range 11 GeV$^2 \leq Q^2 \leq$ 34 GeV$^2$ and the $x$ range $0.006 \leq x \leq 0.556$ are considered.

The data are better reproduced by the higher-order parton density function of GRV than by other parton distribution functions determined from the low energy data.

Combining the present results with previous L3 measurements, the $Q^2$ evolution is studied from 1.5 GeV$^2$ to 120 GeV$^2$ in the low-$x$ region, $0.01 \leq x \leq 0.1$, and from 12.4 GeV$^2$ to 225 GeV$^2$ in the higher-$x$ region, $0.1 < x \leq 0.5$. The measurements at different centre-of-mass energies are consistent and the $\ln Q^2$ evolution of $F_2^\gamma$ is clearly confirmed.

# REFERENCES

1. L3 Coll., P. Achard *et al*, *Phys. Lett.* **B 622**, 249 (2005).

CP892, *Quark Confinement and the Hadron Spectrum VII*
edited by J. E. F. T. Ribeiro
© 2007 American Institute of Physics 978-0-7354-0396-3/07/$23.00

# Study of Resonance Formation in the Mass Region $1400 - 1500$ MeV through the Reaction $\gamma\gamma \rightarrow K_S^0 K^{\pm} \pi^{\mp}$

Igor Vorobiev
(on behalf of the L3 Collaboration)

*Carnegie Mellon University, Pittsburgh, PA 15213*

**Keywords:** LEP, two-photon physics, exclusive meson production.
**PACS:** 13.60.Le

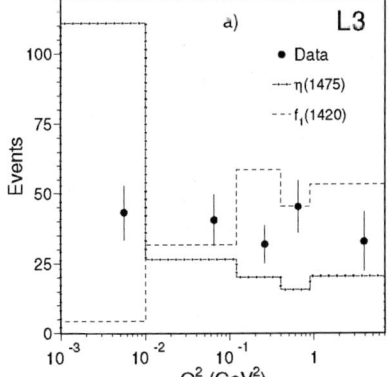 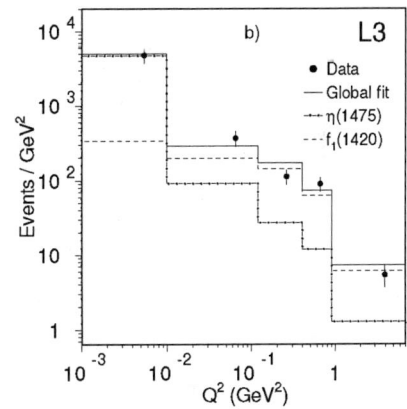

The exclusive $K_S^0 K^{\pm} \pi^{\mp}$ final state, studied in two-photon interactions with the full high energy statistics collected by L3 at LEP, shows a significant enhancement in the mass spectra in the region $1.35 - 1.55$ GeV for the $Q^2$ range $0 - 7$ GeV$^2$. The $Q^2$ - dependent mass spectra cannot be described by formation of only a pseudoscalar or an axial-vector meson. Contributions of both the $\eta(1475)$ and $f_1(1420)$ resonances are required. The $\eta(1475)$ signal dominantes at $Q^2 < 0.01$ GeV$^2$ and has a statistical significance of 4.6 standard deviations. The $f_1(1420)$ dominantes for $Q^2 > 0.1$ GeV$^2$ and decays entirely through $K^*(892)K$. The two-photon coupling and the form factor of these resonances are well described by the formalism of G. A. Schuler, F. A. Berends, R. van Gulik, *Nucl. Phys.* **B 523**, 423 (1998) with the following parameters:

$\Gamma_{\gamma\gamma}(\eta(1475))\mathrm{BR}(\overline{KK}\pi) = 230 \pm 50(stat) \pm 50(sys)$ eV,
$\Gamma_{\gamma\gamma}(f_1(1420))\mathrm{BR}(\overline{KK}\pi) = 3.2 \pm 0.6(stat) \pm 0.7(sys)$ keV,
$\Lambda_1 = 926 \pm 72(stat) \pm 32(sys)$ MeV.
The production of $f_1(1285)$ is also observed, but not that of $\eta(1405)$ or $f_1(1510)$.

CP892, *Quark Confinement and the Hadron Spectrum VII*
edited by J. E. F. T. Ribeiro
© 2007 American Institute of Physics 978-0-7354-0396-3/07/$23.00

# LIST OF PARTICIPANTS

| | | |
|---|---|---|
| Abdessamad | Abada | United Arab Emirates University |
| Conrado | Albertus Torres | University de Granada |
| Mark | Alford | Washington Univ, Saint Louis |
| Dmitry | Antonov | University of Heidelberg |
| Sinya | Aoki | University of Tsukuba |
| Yuting | Bai | NIKHEF |
| Marshall | Baker | University of Washington |
| Gordon | Baym | University of Illinois |
| Wolfgang | Bentz | Tokai University |
| Joao Pacheco | Bicudo Cabral de Melo | UNICSUL - CETEC and IFT- UNESP |
| Cesare | Bini | Univ. ' "La Sapienza" and INFN Roma |
| Deirdre | Black | University of Cambridge |
| Marcus | Bleicher | Frankfurt University |
| Geoffrey | Bodwin | Argonne National Lab |
| Nicolas | Borghini | CERN |
| Nora | Brambilla | Univesity of Milano |
| Falk | Bruckmann | Leiden University |
| Michael | Buballa | TU Darmstadt |
| Panos | Christakoglou | University of Athens - CERN |
| Thomas | Cohen | University of Maryland |
| Pedro | Costa | Universit of Coimbra |
| Michael | Creutz | BNL |
| Attilio | Cucchieri | University of Sao Paulo |
| Mikolaj | Cwiok | University College Dublin |
| Alessio | D'Alessandro | Universita' di Genova & INFN |
| Karin | Daum | Wuppertal University/DESY |
| Gabor | David | BNL |
| Philippe | de Forcrand | ETH Zurich |
| Sedigheh | Deldar | University of Tehran |
| Dmitri | Diakonov | St Petersburg Nuclear Physics Institute |
| Chaden | Djalali | University of South Carolina |
| Tommmaso | Dorigo | INFN Padova |
| Shuxian | Du | IHEP |
| Gerhard | Ecker | Universiy Wien |
| Estia | Eichten | Fermilab |
| Michael | Engelhardt | NMSU |
| Paul | Eugenio | Florida State University |
| Manfried | Faber | TU Wien |
| Paulo Afonso | Faria da Veiga | ICMC-USP |
| Elena | Ferreiro | University of Santiago de Compostela |
| Yiota | Foka | GSI |
| Harald | Fox | University of Freiburg |
| Eduardo | Fraga | University Federal do Rio deJaneiro |

| | | |
|---|---|---|
| Marco | Ghiotti | University of Adelaide |
| Ferdinando | Gliozzi | Torino University |
| Leonid | Glozman | University of Graz |
| Ralf | Gothe | University of South Carolina |
| Jeff | Greensite | San Francisco State University |
| Georges | Grunberg | Ecole Polytechnique |
| Alistair | Hart | University of Edinburgh |
| Brigitte | Hiller | University of Coimbra |
| Boris | Hippolyte | IPHC |
| Charles | Horowitz | Indiana University |
| Rémi | Huguet | CENBG |
| Edmond | Iancu | Saclay |
| Ernst-Michael | Ilgenfritz | Humboldt-Universitaet Berlin |
| Yulia | Kalashnikova | ITEP |
| Burkhard | Kampfer | Research Center Rossendorf/Dresden |
| David | Kaplan | Institute for Nuclear Theory |
| Dmitri | Kharzeev | Brookhaven National Laboratory |
| Valentin | Khoze | University of Durham |
| Ayse | Kizilersu | CSSM (University of Adelaide) |
| Gennady | Kozlov | Joint Institute for Nuclear Research |
| Gastao | Krein | Instituto de Fisica Teorica |
| Vladimir | Kukulin | Moscow State University |
| Julius | Kuti | University of California |
| Youngjoon | Kwon | Yonsei University |
| Jean-Philippe | Lansberg | CPhT, Ecole Polytechnique |
| Emmanuel | Latour | LLR-Ecole Polytechnique |
| Rob | Leigh | University of Illinois |
| Michael | Leitch | LANL |
| Tadeusz | Lesiak | Institute of Nuclear Physics PAN |
| Heinrich | Leutwyler | University of Bern |
| Michael | Lisa | Ohio State University |
| Kenneth | Livingston | University of Glasgow |
| Felipe Jose | Llanes Estrada | University Complutense de Madrid |
| Wolfgang | Lucha | Austrian Academy of Sciences |
| Benjamin | Lungwitz | IKF Universitaet Frankfurt |
| Axel | Maas | University of de São Paulo |
| Rasmus | Mackeprang | Niels Bohr Institute |
| Hossein | Malekzadeh | Frankfurt Int. Grad School for Science |
| Spiridon | Margetis | Kent University |
| Pieter | Maris | University of Pittsburg |
| Guido | Martinelli | University of Rome "La Sapienza" |
| Pedro | Martins | CFTP |
| Nicolas | Matagne | University of Liege |
| Khin | Maung | University of Southern MS |
| Eugenio | Megias | University of Granada |
| Dmitry | Melikhov | HEPHY |
| Tereza | Mendes | University of Sao Paulo |
| Tito | Mendonca | IST |

| | | |
|---|---|---|
| J. Guilherme | Milhano | CENTRA-IST/UALG |
| Rita | Monteiro | University of Coimbra |
| Michael | Muller-Preussker | Humboldt-University Berlin |
| Giuseppe | Nardulli | University of Bari |
| Alexey | Nefediev | ITEP |
| Antti | Niemi | Uppsala UNiversity |
| Roberto | Onofrio | Dartmouth College |
| Peter | Orland | Baruch College, CUNY |
| Mikhail | Osipenko | INFN, Genoa |
| Carlos | Pajares | University of Santiago de Compostela |
| Galina | Pakhlova | ITEP |
| Stephane | Peigne | Subatech |
| Jose | Pelaez | University Complutense de Madrid |
| Timofei | Piatenko | Caltech |
| Antonio | Pineda | University of Barcelona |
| Mikhail | Polikarpov | ITEP |
| Giovanni | Prosperi | University of Milano |
| Jianwei | Qiu | Iowa State University |
| Shahnoosh | Rafibakhsh | University of Teheran |
| Antonio | Rago | University of Milano |
| Gilberto | Ramalho | CFTP/IST |
| Sanjay | Reddy | Los Alamos National Laboratory |
| Hugo | Reinhardt | University of Tuebingen |
| Klaus | Reygers | University of Muenster |
| Jose | Ribeiro | IST, Technical University of Lisbon |
| Giulia | Ricciardi | University degli Studi di Napoli Federico II |
| Stefania | Ricciardi | RAL |
| Jacek | Rozynek | Institute for Nuclear Studies |
| Marco | Ruggieri | University degli Studi di Bari |
| Chris | Sachrajda | University of Southampton |
| Hagop | Sazdjian | University Paris XI |
| Andreas | Schmitt | Washington University, St Louis |
| Achim | Schwenk | University of Washington |
| Kamal | Seth | Northwestern University |
| Mikhail | Shifman | University of Minnesota |
| Leonid | Shifrin | SUNY Stony Brook US and Brunel U. London UK |
| Ian | Shipsey | University of Purdue |
| Yakov | Shnir | University of Oldenburg |
| Paulo | Silva | Centro de Física Computacional |
| Emanuele | Sorace | INFN |
| Joan | Soto | University of Barcelona |
| Andrew | Steiner | LANL |
| Joerg | Stelzer | CERN |
| Eric | Swanson | University of Pittsburgh |
| Kalman | Szabo | University of Wuppertal |
| Adam | Szczepaniak | Indiana University |
| Paul | Taras | Universite de Montreal |
| Harry | Thacker | University of Virginia |

| Charles | Thorn | University of Florida |
| Laura | Tolos | GSI |
| Hisayuki | Torii | Hiroshima University |
| Nils A. | Tornqvist | University of Helsinki |
| Juergen | Ulbricht | ETH-Hoenggerberg |
| Antonio | Vairo | University of Milano |
| Pierre | van Baal | Institute Lorentz for Theoretical Physics |
| Jose Maria | Verde Velasco | University of Salamanca |
| Igor | Vorobiev | CMU, Pittsburg USA |
| Marc | Wagner | University of Erlangen-Nürnberg |
| Michael | Walker | Chiba University |
| Fuqiang | Wang | Purdue University |
| Peter | Watson | University Tuebingen |
| Fridolin | Weber | San Diego State University |
| Edmond | Yancu | SACLAY |
| Francisco | Yndurain | University Autonoma de Madrid |
| Hanqing | Zheng | Peking University |
| Ariel | Zhitnitsky | University of British Columbia |
| Anze | Zupanc | Jozef Stefan Institute |
| Daniel | Zwanziger | New York University |

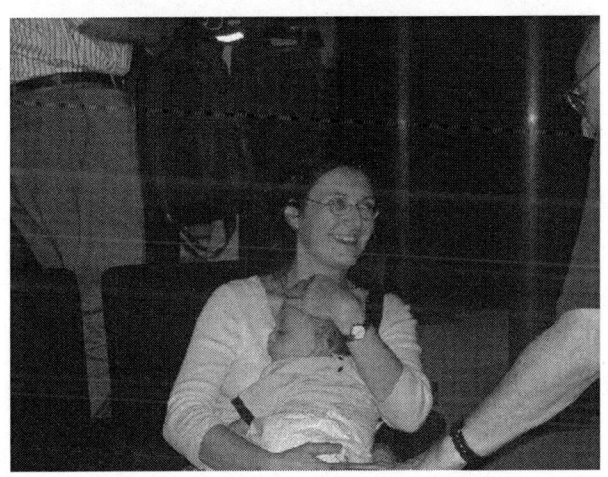

# W

Wagner, M., 231
Walker, M. L., 235
Wang, F., 417
Watson, P., 238
Weber, F., 515
Weinberg, V., 187
Weygand, D. P., 262
Williams, A. G., 180
Wood, M. H., 262
Wu, J., 468

# X

Xiao, L. Y., 308

# Y

Yelnykov, A., 196
Yndurain, F. J., 305
Yu, C., 315

# Z

Zhao, J., 468
Zheng, H. Q., 308
Zhitnitsky, A. R., 518
Zschocke, S., 274
Zupanc, A., 472
Zwanziger, D., 121